DATE DUE

MY 2X '05			
DE 2 0 08			
AP 2 1 '08			

DEMCO 38-296

Memoirs

of

The Second World War

The Six Volumes in One

The Gathering Storm
Their Finest Hour
The Grand Alliance
The Hinge of Fate
Closing the Ring
Triumph and Tragedy

ABRIDGEMENT BY DENIS KELLY

Winston S. Churchill

Memoirs

of

The Second World War

An ABRIDGEMENT of the six volumes of
The Second World War with an EPILOGUE
by the author on the postwar years written
for this volume

Illustrated with maps and diagrams

HOUGHTON MIFFLIN COMPANY · BOSTON

For information about permission to reproduce selections from this book,
write to Permissions, Houghton Mifflin Company, 2 Park Street, Boston,
Massachusetts 02108.

This abridgement has been prepared by Denis Kelly under arrangement
with Cassell & Co., Ltd., London, England, publishers of the British edition,
and Houghton Mifflin Company.

Library of Congress Cataloging-in-Publication Data

Churchill, Winston, Sir, 1874–1965.
    Memoirs of the Second World War : an abridgement of the six volumes
of The Second World War with an epilogue by the author on the postwar
years written for this volume / Winston S. Churchill.
        p.   cm.
    ISBN 0–395–52262–5   ISBN 0–395–59968–7 (pbk.)
    1. World War, 1939–1945. 2. World War, 1939–1945 — Great Britain.
3. Churchill, Winston, Sir, 1874–1965. 4. World War, 1939–1945 — Personal narratives, British. I. Churchill, Winston, Sir, 1874–1965. Second World War. II. Title.
D743.C484   1990   90–38253
940.53 — dc20   CIP

Printed in the United States of America

BP 10 9 8 7 6 5 4

☆

Publisher's Note

THIS VOLUME is an abridgement of Sir Winston Church-ill's great work *The Second World War,* which consists of the six volumes: *The Gathering Storm, Their Finest Hour, The Grand Alliance, The Hinge of Fate, Closing the Ring,* and *Triumph and Tragedy.*

The text has been reduced to approximately one quarter of its original length, yet it graphically presents the war events and planning in the words of the magnificent leader who for his writing was awarded the Nobel Prize in Literature. The result is not so much a condensation of history as the quintessence of the war as it was seen by its greatest protagonist.

☆

Extract from the Preface
to *The Gathering Storm*

I MUST regard these volumes as a continuation of the story of the First World War which I set out in *The World Crisis, The Eastern Front,* and *The Aftermath.* Together they cover an account of another Thirty Years' War.

I have followed, as in previous volumes, as far as I am able, the method of Defoe's *Memoirs of a Cavalier,* in which the author hangs the chronicle and discussion of great military and political events upon the thread of the personal experiences of an individual. I am perhaps the only man who has passed through both the two supreme cataclysms of recorded history in high cabinet office. Whereas, however, in the First World War I filled responsible but subordinate posts, I was for more than five years in this second struggle with Germany the head of His Majesty's Government. I write, therefore, from a different standpoint and with more authority than was possible in my earlier books. I do not describe it as history, for that belongs to another generation. But I claim with confidence that it is a contribution to history which will be of service to the future.

These thirty years of action and advocacy comprise and express my life-effort, and I am content to be judged upon them. I have adhered to my rule of never criticising any measure of war or policy after the event unless I had before expressed publicly or formally my opinion or warning about it. Indeed in the after-light I have softened many of the severities of contemporary controversy. It has given me pain to record these disagreements with so many men whom I liked or respected; but it would be wrong not to lay the lessons of the past before the future. Let no one look down on those honourable, well-meaning men whose actions are chronicled in these pages, without searching his own heart, reviewing his own discharge of public duty, and applying the lessons of the past to his future conduct.

It must not be supposed that I expect everybody to agree with what I say, still less that I only write what will be popular. I give my testimony according to the lights I follow. Every possible care has been taken to verify the facts; but much is constantly coming to light from the disclosure of captured documents or other revelations which may present a new aspect to the conclusions which I have drawn.

One day President Roosevelt told me that he was asking publicly for suggestions about what the war should be called. I said at once "the Unnecessary War." There never was a war more easy to stop than that which has just wrecked what was left of the world from the previous struggle. The human tragedy reaches its climax in the fact that after all the exertions and sacrifices of hundreds of millions of people and the victories of the Righteous Cause, we have still not found Peace or Security, and that we lie in the grip of even worse perils than those we have surmounted. It is my earnest hope that pondering upon the past may give guidance in days to come, enable a new generation to repair some of the errors of former years, and thus govern, in accordance with the needs and glory of man, the awful unfolding scene of the future.

WINSTON SPENCER CHURCHILL

Chartwell,
Westerham,
Kent
March 1948

MORAL OF THE WORK

In War: Resolution

In Defeat: Defiance

In Victory: Magnanimity

In Peace: Goodwill

☆

BOOK I

Milestones to Disaster

1919–May 10, 1940

BOOK II

Alone

May 10, 1940–June 22, 1941

BOOK III

The Grand Alliance

Sunday, December 7, 1941, and Onward

BOOK IV

Triumph and Tragedy

1943–1945

CONTENTS

☆

Maps and Diagrams

BOOK I

Milestones
to Disaster

1919–May 10, 1940

One day President Roosevelt told me that he was asking publicly for suggestions about what the war should be called. I said at once "the Unnecessary War." There never was a war more easy to stop than that which has just wrecked what was left of the world from the previous struggle.

☆

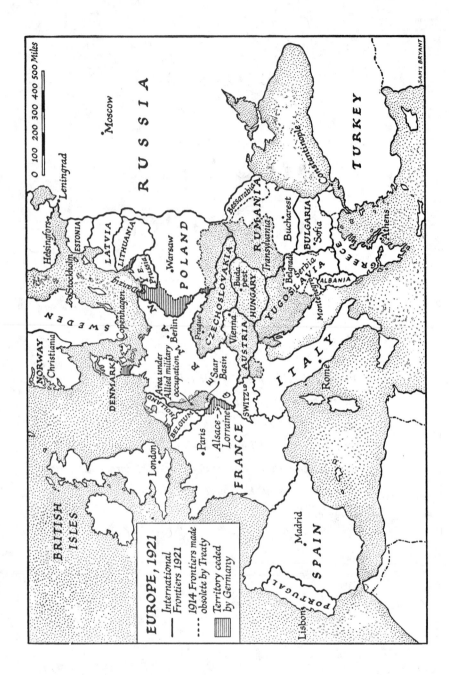

EUROPE, 1921

—— International
      Frontiers 1921
----- 1914 Frontiers made
      obsolete by Treaty
▦ Territory ceded
      by Germany

0 100 200 300 400 500 Miles

RUSSIA

Moscow

Leningrad

Helsingfors

Stockholm

ESTONIA
LATVIA
LITHUANIA

Copenhagen

Danzig

E. Prussia

Warsaw

POLAND

Bessarabia

Bucharest

RUMANIA

Constantinople

TURKEY

Sofia

BULGARIA

Transylvania

Athens

GREECE

ALBANIA

Montenegro

Belgrade

YUGOSLAVIA

Serbia

Budapest

HUNGARY

Vienna

AUSTRIA

CZECHOSLOVAKIA

Prague

Berlin

Area under
Allied military
occupation

Saar Basin

Alsace-
Lorraine

SWITZ'D

ITALY

Rome

SWEDEN

NORWAY

Christiania

DENMARK

NORTH GERMANY

HOLLAND

BELGIUM

Paris

FRANCE

London

BRITISH
ISLES

SPAIN

Madrid

PORTUGAL

Lisbon

SAM'L BRYANT

I

The Follies of the Victors,

1919–1929

AFTER THE END of the World War of 1914 there was a deep conviction and almost universal hope that peace would reign in the world. This heart's desire of all the peoples could easily have been gained by steadfastness in righteous convictions, and by reasonable common sense and prudence. The phrase "the war to end war" was on every lip, and measures had been taken to turn it into reality. President Wilson, wielding, as was thought, the authority of the United States, had made the conception of a League of Nations dominant in all minds. The Allied Armies stood along the Rhine, and their bridgeheads bulged deeply into defeated, disarmed, and hungry Germany. The chiefs of the victor Powers debated and disputed the future in Paris. Before them lay the map of Europe to be redrawn almost as they might resolve. After fifty-two months of agony and hazards the Teutonic coalition lay at their mercy, and not one of its four members could offer the slightest resistance to their will. Germany, the head and front of the offence, regarded by all as the prime cause of the catastrophe which had fallen upon the world, was at the mercy or discretion of conquerors, themselves reeling from the torment they had endured. Moreover, this had been a war not of Governments but of peoples. The whole life energy of the greatest nations had been poured out in wrath and slaughter. The war leaders assembled in Paris in the summer of 1919 had been borne thither upon the strongest and most furious tides that have ever flowed in human history. Gone were the days of the treaties of Utrecht and Vienna, when aristocratic statesmen and diplomats, victor and vanquished alike, met in polite and courtly disputation, and, free from the clatter and babel of democracy, could reshape systems upon the fundamentals of which they were all agreed. The peoples, transported by their sufferings and by the mass teachings with which they had been inspired, stood around in scores of millions to demand that retribution should be exacted to the full. Woe betide the leaders now perched on their dizzy pinnacles of triumph if they cast away at the conference table what the soldiers had won on a hundred bloodsoaked battlefields.

France, by right alike of her efforts and her losses, held the leading place. Nearly a million and a half Frenchmen had perished defending the soil of France on which they stood against the invader. Five times in a hundred years, in 1814, 1815, 1870, 1914, and 1918, had the towers of Notre Dame seen the flash of Prussian guns and heard the thunder of their cannonade. Now for four horrible years thirteen provinces of France had lain in the rigorous grip of Prussian military rule. Wide regions had been systematically devastated by the enemy or pulverised in the encounter of the armies. There was hardly a cottage or a family from Verdun to Toulon that did not mourn its dead or shelter its cripples. To those Frenchmen — and there were many in high authority — who had fought and suffered in 1870 it seemed almost a miracle that France should have emerged victorious from the incomparably more terrible struggle which had just ended. All their lives they had dwelt in fear of the German Empire. They remembered the preventive war which Bismarck had sought to wage in 1875; they remembered the brutal threat which had driven Delcassé from office in 1905; they had quaked at the Moroccan menace in 1906, at the Bosnian dispute of 1908, and at the Agadir crisis of 1911. The Kaiser's "mailed fist" and "shining armour" speeches might be received with ridicule in England and America: they sounded a knell of horrible reality in the hearts of the French. For fifty years almost they had lived under the terror of the German arms. Now, at the price of their lifeblood, the long oppression had been rolled away. Surely here at last was peace and safety. With one passionate spasm the French people cried "Never again!"

But the future was heavy with foreboding. The population of France was less than two-thirds that of Germany. The French population was stationary, while the German grew. In a decade or less the annual flood of German youth reaching the military age must be double that of France. Germany had fought nearly the whole world, almost single-handed, and she had almost conquered. Those who knew the most knew best the several occasions when the result of the Great War had trembled in the balance, and the accidents and chances which had turned the fateful scale. What prospect was there in the future that the Great Allies would once again appear in their millions upon the battlefields of France or in the East? Russia was in ruin and convulsion, transformed beyond all semblance of the past. Italy might be upon the opposite side. Great Britain and the United States were separated by the seas or oceans from Europe. The British Empire itself seemed knit together by ties which none but its citizens could understand. What combination of events could ever bring back again to France and Flanders the formidable Canadians of the Vimy Ridge; the glorious Australians of Villers-Bretonneux; the dauntless New Zealanders of the crater fields of Passchendaele; the steadfast Indian Corps which in the cruel winter of 1914 had held the line by Armentières? When again would peaceful, careless, anti-militarist Britain tramp the plains of Artois and Picardy with armies of two or three million men? When again would the ocean bear two millions of the spendid manhood of America to Champagne and the Argonne? Worn down, doubly decimated, but undisputed masters of the hour, the French nation peered into the future in thankful wonder and haunting dread. Where then was that SECURITY without which all that had been gained seemed valueless, and life itself, even amid the rejoicings of victory, was almost unendurable? The mortal need was Security at all costs and by all methods, however stern or even harsh.

On Armistice Day the German armies had marched homeward in good order. "They fought well," said Marshal Foch, Generalissimo of the Allies, with the laurels bright upon his brow, speaking in soldierly mood: "let them keep their weapons." But he demanded that the French frontier should henceforth be the Rhine. Germany might be disarmed; her military system shivered in fragments; her fortresses dismantled: Germany might be impoverished; she might be loaded with measureless indemnities; she might become a prey to internal feuds: but all this would pass in ten years or in twenty. The indestructible might "of all the German tribes" would rise once more and the unquenched fires of warrior Prussia glow and burn again. But the Rhine, the broad, deep, swift-flowing Rhine, once held and fortified by the French Army, would be a barrier and a shield behind which France could dwell and breathe for generations. Very different were the sentiments and views of the English-speaking world, without whose aid France must have succumbed. The territorial provisions of the Treaty of Versailles left Germany practically intact. She still remained the largest homogeneous racial block in Europe. When Marshal Foch heard of the signing of the Peace Treaty of Versailles he observed with singular accuracy: "This is not Peace. It is an Armistice for twenty years."

*   *   *

The economic clauses of the Treaty were malignant and silly to an extent that made them obviously futile. Germany was condemned to pay reparations on a fabulous scale. These dictates gave expression to the anger of the victors, and to the failure of their peoples to understand that no defeated nation or community can ever pay tribute on a scale which would meet the cost of modern war.

The multitudes remained plunged in ignorance of the simplest economic facts, and their leaders, seeking their votes, did not dare to undeceive them. The newspapers, after their fashion, reflected and emphasised the prevailing opinions. Few voices were raised to explain that payment of reparations can only be made by services or by the physical transportation of goods in wagons across land frontiers or in ships across salt water; or that when these goods arrive in the demanding countries they dislocate the local industry except in very primitive or rigorously controlled societies. In practice, as even the Russians have now learned, the only way of pillaging a defeated nation is to cart away any movables which are wanted, and to drive off a portion of its manhood as permanent or temporary slaves. But the profit gained from such processes bears no relation to the cost of the war. No one in great authority had the wit, ascendancy, or detachment from public folly to declare these fundamental, brutal facts to the electorates; nor would anyone have been believed if he had. The triumphant Allies continued to assert that they would squeeze Germany "till the pips squeaked." All this had a potent bearing on the prosperity of the world and the mood of the German race.

In fact, however, these clauses were never enforced. On the contrary, whereas about £1000 millions of German assets were appropriated by the victorious Powers, more than £1500 millions were lent a few years later to Germany, principally by the United States and Great Britain, thus enabling the ruin of the war to be rapidly repaired in Germany. As this apparently magnanimous process was still accompanied by the machine-made howlings of the unhappy and embittered populations in the victorious countries, and the assurances of their statesmen that Germany should be made to pay "to the uttermost farthing," no gratitude or goodwill was to be expected or reaped.

History will characterise all these transactions as insane. They helped to breed both the martial curse and "economic blizzard," of which more later. All this is a sad story of complicated idiocy in the making of which much toil and virtue was consumed.

*   *   *

The second cardinal tragedy was the complete breakup of the Austro-Hungarian Empire by the Treaties of St. Germain and Trianon. For centuries this surviving embodiment of the Holy Roman Empire had afforded a common life, with advantages in trade and security, to a large number of peoples, none of whom in our own time has had the strength or vitality to stand by themselves in the face of pressure from a revivified Germany or Russia. All these races wished to break away from the federal or imperial structure, and to encourage their desires was deemed a liberal policy. The Balkanisation of Southeastern Europe proceeded apace, with the consequent relative aggrandisement of Prussia and the German Reich, which, though tired and war-scarred, was intact and locally overwhelming. There is not one of the peoples or provinces that constituted the Empire of the Habsburgs to whom gaining their independence has not brought the tortures which ancient poets and theologians had reserved for the damned. The noble capital of Vienna, the home of so much long-defended culture and tradition, the centre of so many roads, rivers, and railways, was left stark and starving, like a great emporium in an impoverished district whose inhabitants have mostly departed.

The victors imposed upon the Germans all the long-sought ideals of the liberal nations of the West. They were relieved from the burden of compulsory military service and from the need of keeping up heavy armaments. The enormous American loans were presently pressed upon them, though they had no credit. A democratic constitution, in accordance with all the latest improvements, was established at Weimar. Emperors having been driven out, nonentities were elected. Beneath this flimsy fabric raged the passions of the mighty, defeated, but substantially uninjured German nation. The prejudice of the Americans against monarchy had made it clear to the beaten Empire that it would have better treatment from the Allies as a republic than as a monarchy. Wise policy would have crowned and fortified the Weimar Republic with a constitutional sovereign in the person of an infant grandson of the Kaiser, under a council of regency. Instead, a gaping void was opened in the national life of the German people. All the strong elements, military and feudal, which might have rallied to a constitutional monarchy and for its sake respected and sustained the new democratic and Parliamentary processes were for the time being unhinged. The Weimar Republic, with all its liberal trappings and blessings, was regarded as an imposition of the enemy. It could not hold the loyalties or the imagination of the German people. For a spell they sought to cling as in desperation to the aged Marshal Hindenburg. Thereafter mighty forces were adrift, the void was open, and into that void after a pause there strode a maniac of ferocious genius, the repository and expression of the most virulent hatreds that have ever corroded the human breast — Corporal Hitler.

*   *   *

France had been bled white by the war. The generation that had dreamed since 1870 of a war of revenge had triumphed, but at a deadly cost in national life-strength. It was a haggard France that greeted the dawn of victory. Deep fear of Germany pervaded the French nation on the morrow of their dazzling success. It was this fear that had prompted Marshal Foch to demand the Rhine frontier for the safety of France against her far larger neighbour. But the British and American statesmen held that the absorption of German-populated districts in French territory was contrary to the Fourteen Points and to the principles of nationalism and self-determination upon which the Peace Treaty was to be based. They therefore withstood Foch and France. They gained Clemenceau by promising, first, a joint Anglo-American guarantee for the defence of France; secondly, a demilitarised zone; and, thirdly, the total, lasting disarmament of Germany. Clemenceau accepted this in spite of Foch's protests and his own instincts. The Treaty of Guarantee was signed accordingly by Wilson and Lloyd George and Clemenceau. The United States Senate refused to ratify the treaty. They repudiated President Wilson's signature. And we, who had deferred so much to his opinions and wishes in all this business of peacemaking, were told without much ceremony that we ought to be better informed about the American Constitution.

In the fear, anger, and disarray of the French people the rugged, dominanting figure of Clemenceau, with his world-famed authority, and his special British and American contacts, was incontinently discarded. "Ingratitude towards their great men," says Plutarch, "is the mark of strong peoples." It was imprudent for France to indulge this trait when she was so grievously weakened. There was little compensating strength to be found in the revival of the group intrigues and ceaseless changes of governments and ministers which were the characteristic of the Third Republic, however profitable or diverting they were to those engaged in them.

Poincaré, the strongest figure who succeeded Clemenceau, attempted to make an independent Rhineland under the patronage and control of France. This had no chance of success. He did not hesitate to try to enforce reparations on Germany by the invasion of the Ruhr. This certainly imposed compliance with the Treaties on Germany; but it was severely condemned by British and American opinion. As a result of the general financial and political disorganisation of Germany, together with reparation payments during the years 1919 to 1923, the mark rapidly collapsed. The rage aroused in Germany by the French occupation of the Ruhr led to a vast, reckless printing of paper notes with the deliberate object of destroying the whole basis of the currency. In the final stages of the inflation the mark stood at forty-three million millions to the pound sterling. The social and economic consequences of this inflation were deadly and far-reaching. The savings of the middle classes were wiped out, and a natural following was thus provided for the banners of National Socialism. The whole structure of German industry was distorted by the growth of mushroom trusts. The entire working capital of the country disappeared. The internal national debt and the debt of industry in the form of fixed capital charges and mortgages were of course simultaneously liquidated or repudiated. But this was no compensation for the loss of working capital. All led directly to the large-scale borrowings of a bankrupt nation abroad which were the feature of ensuing years. German sufferings and bitterness marched forward together — as they do today.

The British temper towards Germany, which at first had been so fierce, very soon went as far astray in the opposite direction. A rift opened between Lloyd George and Poincaré, whose bristling personality hampered his firm and farsighted policies. The two nations fell apart in thought and action, and British sympathy or even admiration for Germany found powerful expression.

*   *   *

The League of Nations had no sooner been created than it received an almost mortal blow. The United States abandoned President Wilson's offspring. The President himself, ready to do battle for his ideals, suffered a paralytic stroke just as he was setting forth on his campaign, and lingered henceforward a futile wreck for a great part of two long and vital years, at the end of which his party and his policy were swept away by the Republican Presidential victory of 1920. Across the Atlantic on the morrow of the Republican success isolationist conceptions prevailed. Europe must be left to stew in its own juice, and must pay its lawful debts. At the same time tariffs were raised to prevent the entry of the goods by which alone these debts could be discharged. At the Washington Conference of 1921 far-reaching proposals for naval disarmament were made by the United States, and the British and American Governments proceeded to sink their battleships and break up their military establishments with gusto. It was argued in odd logic that it would be immoral to disarm the vanquished unless the victors also stripped themselves of their weapons. The finger of Anglo-American reprobation was presently to be pointed at France, deprived alike of the Rhine frontier and of her treaty guarantee, for maintaining, even on a greatly reduced scale, a French Army based upon universal service.

The United States made it clear to Britain that the continuance of her alliance with Japan, to which the Japanese had punctiliously conformed, would constitute a barrier in Anglo-American relations. Accordingly this alliance was brought to an end. The annulment caused a profound impression in Japan, and was viewed as the spurning of an Asiatic Power by the Western world. Many links were sundered which might afterwards have proved of decisive value to peace. At the same time, Japan could console herself with the fact that the downfall of Germany and Russia had, for a time, raised her to the third place among the world's naval Powers, and certainly to the highest rank. Although the Washington Naval Agreement prescribed a lower ratio of strength in capital ships for Japan than for Britain and the United States (5:5:3), the quota assigned to her was well up to her building and financial capacity for a good many years, and she watched with an attentive eye the two leading naval Powers cutting each other down far below what their resources would have permitted and what their responsibilities enjoined. Thus, both in Europe and in Asia, conditions were swiftly created by the victorious Allies which, in the name of peace, cleared the way for the renewal of war.

While all these untoward events were taking place, amid a ceaseless chatter of well-meant platitudes on both sides of the Atlantic, a new and more terrible cause of quarrel than the imperialism of czars and kaisers became apparent in Europe. The Civil War in Russia ended in the absolute victory of the Bolshevik Revolution. The Soviet armies which advanced to subjugate Poland were indeed repulsed in the Battle of Warsaw, but Germany and Italy nearly succumbed to Communist propaganda and designs, and Hungary actually fell for a while under the control of the Communist dictator Bela Kun. Although Marshal Foch wisely observed that "Bolshevism had never crossed the frontiers of victory," the foundations of European civilisation trembled in the early postwar years. Fascism was the shadow or ugly child of Communism. While Corporal Hitler was making himself useful to the German officer class in Munich by arousing soldiers and workers to fierce hatred of Jews and Communists, on whom he laid the blame for Germany's defeat, another adventurer, Benito Mussolini, provided Italy with a new theme of government which, while it claimed to save the Italian people from Communism, raised himself to dictatorial power. As Fascism sprang from Communism, so Nazism developed from Fascism. Thus were set on foot those kindred movements which were destined soon to plunge the world into even more hideous strife, which none can say has ended with their destruction.

*   *   *

Nevertheless one solid security for peace remained. Germany was disarmed. All her artillery and weapons were destroyed. Her fleet had already sunk itself in Scapa Flow. Her vast army was disbanded.

By the Treaty of Versailles only a professional long-service army not exceeding one hundred thousand men, and unable on this basis to accumulate reserves, was permitted to Germany for purposes of internal order. The annual quotas of recruits no longer received their training; the cadres were dissolved. Every effort was made to reduce to a tithe the officer corps. No military air force of any kind was allowed. Submarines were forbidden, and the German Navy was limited to a handful of vessels under ten thousand tons. Soviet Russia was barred off from Western Europe by a cordon of violently anti-Bolshevik states, who had broken away from the former Empire of the Czars in its new and more terrible form. Poland and Czechoslovakia raised independent heads, and seemed to stand erect in Central Europe. Hungary had recovered from her dose of Bela Kun. The French Army, resting upon its laurels, was incomparably the strongest military force in Europe, and it was for some years believed that the French Air Force was also of a high order.

Up until the year 1934 the power of the conquerors remained unchallenged in Europe, and indeed throughout the world. There was no moment in these sixteen years when the three former allies, or even Britain and France with their associates in Europe, could not in the name of the League of Nations and under its moral and international shield have controlled by a mere effort of the will the armed strength of Germany. Instead, until 1931 the victors, and particularly the United States, concentrated their efforts upon extorting by vexatious foreign controls their annual reparations from Germany. The fact that these payments were made only from far larger American loans reduced the whole process to the absurd. Nothing was reaped except ill-will. On the other hand, the strict enforcement at any time till 1934 of the disarmament clauses of the Peace Treaty would have guarded indefinitely, without violence or bloodshed, the peace and safety of mankind. But this was neglected while the infringements remained petty, and shunned as they assumed serious proportions. Thus the final safeguard of a long peace was cast away. The crimes of the vanquished find their background and their explanation, though not, of course, their pardon, in the follies of the victors. Without these follies crime would have neither temptation nor opportunity.

*   *   *

In these pages I attempt to recount some of the incidents and impressions which form in my mind the story of the coming upon mankind of the worst tragedy in its tumultuous history. This presented itself not only in the destruction of life and property inseparable from war. There had been fearful slaughter of soldiers in the First World

War, and much of the accumulated treasure of the nations was consumed. Still, apart from the excesses of the Russian Revolution, the main fabric of European civilisation remained erect at the close of the struggle. When the storm and dust of the cannonade passed suddenly away, the nations, despite their enmities, could still recognise each other as historic racial personalities. The laws of war had on the whole been respected. There was a common professional meeting ground between military men who had fought one another. Vanquished and victors alike still preserved the semblance of civilised states. A solemn peace was made which, apart from unenforceable financial aspects, conformed to the principles which in the nineteenth century had increasingly regulated the relations of enlightened peoples. The reign of law was proclaimed, and a World Instrument was formed to guard us all, and especially Europe, against a renewed convulsion.

In the Second World War every bond between man and man was to perish. Crimes were committed by the Germans under the Hitlerite domination to which they allowed themselves to be subjected which find no equal in scale and wickedness with any that have darkened the human record. The wholesale massacre by systematised processes of six or seven millions of men, women, and children in the German execution camps exceeds in horror the rough-and-ready butcheries of Genghis Khan, and in scale reduces them to pigmy proportions. Deliberate extermination of whole populations was contemplated and pursued by both Germany and Russia in the Eastern war. The hideous process of bombarding open cities from the air, once started by the Germans, was repaid twentyfold by the ever-mounting power of the Allies, and found its culmination in the use of the atomic bombs which obliterated Hiroshima and Nagasaki.

We have at length emerged from a scene of material ruin and moral havoc the like of which had never darkened the imagination of former centuries. After all that we suffered and achieved we find ourselves still confronted with problems and perils not less but far more formidable than those through which we have so narrowly made our way.

It is my purpose, as one who lived and acted in these days, to show how easily the tragedy of the Second World War could have been prevented; how the malice of the wicked was reinforced by the weakness of the virtuous; how the structure and habits of democratic states, unless they are welded into larger organisms, lack those elements of persistence and conviction which can alone give security to humble masses; how, even in matters of self-preservation, no policy is pursued for even ten or fifteen years at a time. We shall see how the counsels of prudence and restraint may become the prime agents of mortal danger; how the middle course adopted from desires for safety and a quiet life may be found to lead direct to the bull's-eye of disaster. We shall see how absolute is the need of a broad path of international action pursued by many states in common across the years, irrespective of the ebb and flow of national politics.

It was a simple policy to keep Germany disarmed and the victors adequately armed for thirty years, and in the meanwhile, even if a reconciliation could not be made with Germany, to build ever more strongly a true League of Nations capable of making sure that treaties were kept, or changed only by discussion and agreement. When three or four powerful Governments acting together have demanded the most fearful sacrifices from their peoples, when these have been given freely for the common cause, and when the longed-for result has been attained, it would seem reasonable that concerted action should be preserved so that at least the essentials would not be cast away. But this modest requirement the might, civilisation, learning, knowledge, science, of the victors were unable to supply. They lived from hand to mouth and from day to day, and from one election to another, until, when scarcely twenty years were out, the dread signal of the Second World War was given, and we must write of the sons of those who had fought and died so faithfully and well:

> *Shoulder to aching shoulder, side by side,*
> *They trudged away from life's broad wealds of light.*[1]

[1] Siegfried Sassoon.

2

Peace at Its Zenith, 1922–1931

DURING THE YEAR 1922 a new leader arose in Britain. Mr. Stanley Baldwin had been unknown or unnoticed in the world drama and played a modest part in domestic affairs. He had been Financial Secretary to the Treasury during the war, and was at this time President of the Board of Trade. He became the ruling force in British politics from October 1922, when he ousted Mr. Lloyd George, until May 1937, when, loaded with honours and enshrined in public esteem, he laid down his heavy task and retired in dignity and silence to his Worcestershire home. My relations with this statesman are a definite part of the tale I have to tell. Our differences at times were serious, but in all these years and later I never had an unpleasant personal interview or contact with him, and at no time did I feel we could not talk together in good faith and understanding as man to man.

Early in 1923 he became the Conservative Prime Minister, and thus began that period of fourteen years which may well be called "the Baldwin-MacDonald Régime." Mr. Ramsay MacDonald was the leader of the Socialist Party, and at first in alternation but eventually in political brotherhood, these two statesmen governed the country. Nominally the representatives of opposing parties, of contrary doctrines, of antagonistic interests, they proved in fact to be more nearly akin in outlook, temperament, and method than any other two men who had been Prime Ministers since that office was known to the Constitution. Curiously enough, the sympathies of each extended far into the territory of the other. Ramsay MacDonald nursed many of the sentiments of the old Tory. Stanley Baldwin, apart from a manufacturer's ingrained approval of Protection, was by disposition a truer representative of mild Socialism than many to be found in the Labour ranks.

In 1924 there was a general election. The Conservatives were returned by a majority of 222 over all other parties combined. I myself became the member for Epping by a ten thousand majority, but as a "Constitutionalist." I would not at that time adopt the name "Conservative." I had had some friendly contacts with Mr. Baldwin in the interval; but I did not think he would survive to be Prime Minister. Now on the morrow of his victory I had no idea how he felt towards me. I was surprised, and the Conservative Party dumbfounded, when he invited me to become Chancellor of the Exchequer, the office which my father had once held. A year later, with the approval of my constituents, not having been pressed personally in any way, I formally rejoined the Conservative Party and the Carlton Club, which I had left twenty years before.

For almost five years I lived next door to Mr. Baldwin at No. 11 Downing Street, and nearly every morning on my way through his house to the Treasury I looked in upon him for a few minutes' chat in the Cabinet Room. As I was one of his leading colleagues, I take my share of responsibility for all that happened. These five years were marked by very considerable recovery at home. This was a capable sedate Government during a period in which marked improvement and recovery were gradually effected year by year. There was nothing sensational or controversial to boast about on the platforms, but measured by every test, economic and financial, the mass of the people were definitely better off, and the state of the nation and of the world was easier and more fertile by the end of our term than at its beginning. Here is a modest but a solid claim.

It was in Europe that the distinction of the Administration was achieved.

*   *   *

Hindenburg now rose to power in Germany. At the end of February 1925 Friedrich Ebert, leader of the prewar German Social-Democrat Party, and first President of the German Republic after the defeat, died. A new President had to be chosen. All Germans had long been brought up under paternal despotism, tempered by far-reaching customs of free speech and Parliamentary opposition. Defeat had brought them on its scaly wings democratic forms and liberties in an extreme degree. But the nation was rent and bewildered by all it had gone through, and many parties and groups contended for precedence and office. Out of the turmoil emerged a strong desire to turn to old Field-Marshal von Hindenburg, who was dwelling in dignified retirement. Hindenburg was faithful to the exiled Emperor, and favoured a restoration of the imperial monarchy "on the English model." This of course was much the most sensible though least fashionable thing to do. When he was besought to stand as a candidate for the Presidency under the Weimar Constitution he was profoundly disturbed. "Leave me in peace," he said again and again.

However, the pressure was continuous, and only Grand-Admiral von Tirpitz at last was found capable of persuading him to abandon both his scruples and his inclinations at the call of duty, which he had always obeyed. Hindenburg's opponents were Marx of the Catholic Centre and Thaelmann the Communist. On Sunday, April 26, all Germany voted. The result was unexpectedly close: Hindenburg, 14,655,766; Marx, 13,751,615; Thaelmann, 1,931,151. Hindenburg, who towered above his opponents by being illustrious, reluctant, and disinterested, was elected by less than a million majority, and with no absolute majority on the total poll. He rebuked his son Oskar for waking him at seven to tell him the news: "Why did you want to wake me up an hour earlier? It would still have been true at eight." And with this he went to sleep again till his usual calling time.

In France the election of Hindenburg was at first viewed as a renewal of the German challenge. In England there was an easier reaction. Always wishing as I did to see Germany recover her honour and self-respect and to let war bitterness die, I was not at all distressed by the news. "He is a very sensible old man," said Lloyd George to me when we next met; and so indeed he proved as long as his faculties remained. Even some of his most bitter opponents were forced to admit "Better a Zero than a Nero." [1] However, he was seventy-seven, and his term of office was to be seven years. Few expected him to be returned again. He did his best to be impartial between the various parties, and certainly his tenure of the Presidency gave a sober strength and comfort to Germany without menace to her neighbors.

*   *   *

Meanwhile in February 1925 the German Government suggested a pact by which the Powers interested in the Rhine, above all England, France, Italy, and Germany, should enter into a solemn obligation for a lengthy period towards the Government of the United States, as trustees, not to wage war against each other. They also proposed a pact expressly guaranteeing the existing territorial status on the Rhine. This was a remarkable event. The British Dominions were not enthusiastic. General Smuts was anxious to avoid regional arrangements. The Canadians were lukewarm, and only New Zealand was unconditionally prepared to accept the view of the British Government. Nevertheless we persevered. To me the aim of ending the thousand-year

[1] Theodore Lessing (murdered by the Nazis, September 1933).

strife between France and Germany seemed a supreme object. If we could only weave Gaul and Teuton so closely together economically, socially, and morally as to prevent the occasion of new quarrels, and make old antagonisms die in the realisation of mutual prosperity and interdependence, Europe would rise again. It seemed to me that the supreme interest of the British people in Europe lay in the assuagement of the Franco-German feud, and that they had no other interests comparable or contrary to that. This is still my view today.

In August the French, with the full agreement of Great Britain, replied officially to Germany. Germany must enter the League without reservations as the first and indispensable step. The German Government accepted this stipulation. This meant that the conditions of the Treaties were to continue in force unless or until modified by mutual arrangement, and that no specific pledge for a reduction of Allied armaments had been obtained. Further demands by the Germans, put forward under intense nationalistic pressure and excitement, for the eradication from the Peace Treaty of the "war guilt" clause, for keeping open the issue of Alsace-Lorraine, and for the immediate evacuation of Cologne by Allied troops, were not pressed by the German Government, and would not have been conceded by the Allies.

On this basis the Conference at Locarno was formally opened on October 4. By the waters of this calm lake the delegates of Britain, France, Germany, Belgium, and Italy assembled. The Conference achieved: first, a Treaty of Mutual Guarantee between the five Powers; secondly, arbitration treaties between Germany and France, Germany and Belgium, Germany and Poland, Germany and Czechoslovakia; thirdly, special agreements between France and Poland, and France and Czechoslovakia, by which France undertook to afford them assistance if a breakdown of the Western Pact were followed by an unprovoked resort to arms. Thus did the Western European democracies agree to keep the peace among themselves in all circumstances, and to stand united against any one of their number who broke the contract and marched in aggression upon a brother land. As between France and Germany, Great Britain became solemnly pledged to come to the aid of whichever of these two states was the object of unprovoked aggression. This far-reaching military commitment was accepted by Parliament and endorsed warmly by the nation. The histories may be searched in vain for a parallel to such an undertaking.

The question whether there was any obligation on the part of France or Britain to disarm, or to disarm to any particular level, was not affected. I had been brought into these matters as Chancellor of the Exchequer at an early stage. My own view about this two-way guarantee was that while France remained armed and Germany disarmed, Germany could not attack her; and that on the other hand France would never attack Germany if that automatically involved Britain becoming Germany's ally. Thus, although the proposal seemed dangerous in theory — pledging us in fact to take part on one side or the other in any Franco-German war that might arise — there was little likelihood of such a disaster ever coming to pass; and this was the best means of preventing it. I was therefore always equally opposed to the disarmament of France and to the rearmament of Germany, because of the much greater danger this immediately brought on Great Britain. On the other hand, Britain and the League of Nations, which Germany joined as part of the agreement, offered a real protection to the German people. Thus there was a balance created in which Britain, whose major interest was the cessation of the quarrel between Germany and France, was to a large extent umpire and arbiter. One hoped that this equilibrium might have lasted twenty years, during which the Allied armaments would gradually and naturally have dwindled under the influence of a long peace, growing confidence, and financial burdens. It was evident that danger would arise if ever Germany became more or less equal with France, still more if she became stronger than France. But all this seemed excluded by solemn treaty obligations.

The Pact of Locarno was concerned only with peace in the West, and it was hoped that what was called an "Eastern Locarno" might be its successor. We should have been very glad if the danger of some future war between Germany and Russia could have been controlled in the same spirit and by similar measures as the possibility of war between Germany and France. Even the Germany of Stresemann was however disinclined to close the door on German claims in the East, or to accept the territorial treaty position about Poland, Danzig, the Corridor, and Upper Silesia. Soviet Russia brooded in her isolation behind the *cordon sanitaire* of anti-Bolshevik states. Although our efforts were continued, no progress was made in the East. I did not at any time close my mind to an attempt to give Germany greater satisfaction on her eastern frontier. But no opportunity arose during those brief years of hope.

*   *   *

There were great rejoicings about the treaty which emerged at the end of 1925 from the Conference at Locarno. Mr. Baldwin was the first to sign it at the Foreign Office. The Foreign Secretary, Mr. Austen Chamberlain, having no official residence, asked me to lend my dining room at No. 11 Downing Street for his intimate friendly luncheon with Herr Stresemann.[2] We all met together in great amity,

[2] The German Foreign Minister.

and thought what a wonderful future would await Europe if its greatest nations became truly united and felt themselves secure. After this memorable instrument had received the cordial assent of Parliament, Mr. Austen Chamberlain was given the Garter and the Nobel Peace Prize. His achievement was the high-water mark of Europe's restoration, and it inaugurated three years of peace and recovery. Although old antagonisms were but sleeping, and the drumbeat of new levies was already heard, we were justified in hoping that the ground thus solidly gained would open the road to a further forward march.

By 1929 the state of Europe was tranquil, as it had not been for twenty years, and was not to be for at least another twenty. A friendly feeling existed towards Germany following upon our Treaty of Locarno, and the evacuation of the Rhineland by the French Army and Allied contingents at a much earlier date than had been prescribed at Versailles. The new Germany took her place in the truncated League of Nations. Under the genial influence of American and British loans Germany was reviving rapidly. Her new ocean liners gained the Blue Riband of the Atlantic. Her trade advanced by leaps and bounds, and internal prosperity ripened. France and her system of alliances also seemed secure in Europe. The disarmament clauses of the Treaty of Versailles were not openly violated. The German Navy was non-existent. The German Air Force was prohibited and still unborn. There were many influences in Germany strongly opposed, if only on grounds of prudence, to the idea of war, and the German High Command could not believe that the Allies would allow them to rearm. On the other hand, there lay before us what I later called the "economic blizzard." Knowledge of this was confined to rare financial circles, and these were cowed into silence by what they foresaw.

*   *   *

The general election of May 1929 showed that the "swing of the pendulum" and the normal desire for change were powerful factors with the British electorate. The Socialists had a small majority over the Conservatives in the new House of Commons. Mr. Baldwin tendered his resignation to the King. We all went down to Windsor in a special train to give up our seals and offices; and on June 7 Mr. Ramsay MacDonald became Prime Minister at the head of a minority Government depending upon Liberal votes.

The Socialist Prime Minister wished his new Labour Government to distinguish itself by large concessions to Egypt, by a far-reaching constitutional change in India, and by a renewed effort for world, or at any rate British, disarmament. These were aims in which he could count upon Liberal aid, and for which he therefore commanded a

Parliamentary majority. Here began my differences with Mr. Baldwin, and thereafter the relationship in which we had worked since he chose me for Chancellor of the Exchequer five years before became sensibly altered. We still of course remained in easy personal contact, but we knew we did not mean the same thing. My idea was that the Conservative Opposition should strongly confront the Labour Government on all great imperial and national issues, should identify itself with the majesty of Britain as under Lord Beaconsfield and Lord Salisbury, and should not hesitate to face controversy, even though that might not immediately evoke a response from the nation. So far as I could see, Mr. Baldwin felt that the times were too far gone for any robust assertion of British imperial greatness, and that the hope of the Conservative Party lay in accommodation with Liberal and Labour forces and in adroit, well-timed manoeuvres to detach powerful moods of public opinion and large blocks of voters from them. He certainly was very successful. He was the greatest party manager the Conservatives had ever had. He fought, as their leader, five general elections, of which he won three.

It was on India that our definite breach occurred. The Prime Minister, strongly supported and even spurred by the Conservative Viceroy, Lord Irwin, afterwards Lord Halifax, pressed forward with his plan of Indian self-government. A portentous conference was held in London, of which Mr. Gandhi, lately released from commodious internment, was the central figure. There is no need to follow in these pages the details of the controversy which occupied the sessions of 1929 and 1930. On the release of Mr. Gandhi in order that he might become the envoy of Nationalist India to the London conference I reached the breaking-point in my relations with Mr. Baldwin. He seemed quite content with these developments, was in general accord with the Prime Minister and the Viceroy, and led the Conservative Opposition decidedly along this path. I felt sure we should lose India in the final result and that measureless disasters would come upon the Indian peoples. I therefore after a while resigned from the Shadow Cabinet upon this issue, but assured Mr. Baldwin that I would give him whatever aid was in my power in opposing the Socialist Government in the House of Commons, and do my utmost to secure their defeat at any General Election.

*   *   *

The year 1929 reached almost the end of its third quarter under the promise and appearance of increasing prosperity, particularly in the United States. Extraordinary optimism sustained an orgy of speculation. Books were written to prove that economic crisis was a phase which expanding business organisation and science had at last mastered. "We are apparently finished and done with economic cycles as we have known them," said the President of the New York Stock Exchange in September. But in October a sudden and violent tempest swept over Wall Street. The intervention of the most powerful agencies failed to stem the tide of panic sales. A group of leading banks constituted a milliard-dollar pool to maintain and stabilise the market. All was vain.

The whole wealth so swiftly gathered in the paper values of previous years vanished. The prosperity of millions of American homes had grown upon a gigantic structure of inflated credit now suddenly proved phantom. Apart from the nation-wide speculation in shares which even the most famous banks had encouraged by easy loans, a vast system of purchase by instalment of houses, furniture, cars, and numberless kinds of household conveniences and indulgences had grown up. All now fell together. The mighty production plants were thrown into confusion and paralysis. But yesterday there had been the urgent question of parking the motorcars in which thousands of artisans and craftsmen were beginning to travel to their daily work. Today the grievous pangs of falling wages and rising unemployment afflicted the whole community, engaged till this moment in the most active creation of all kinds of desirable articles for the enjoyment of millions. The American banking system was far less concentrated and solidly based than the British. Twenty thousand local banks suspended payment. The means of exchange of goods and services between man and man was smitten to the ground, and the crash on Wall Street reverberated in modest and rich households alike.

It should not however be supposed that the fair vision of far greater wealth and comfort ever more widely shared which had entranced the people of the United States had nothing behind it but delusion and market frenzy. Never before had such immense quantities of goods of all kinds been produced, shared, and exchanged in any society. There is in fact no limit to the benefits which human beings may bestow upon one another by the highest exertion of their diligence and skill. This splendid manifestation had been shattered and cast down by vain imaginative processes and greed of gain which far outstripped the great achievement itself. In the wake of the collapse of the stock market came during the years between 1929 and 1932 an unrelenting fall in prices and consequent cuts in production causing widespread unemployment.

The consequences of this dislocation of economic life became worldwide. A general contraction of trade in the face of unemployment and declining production followed. Tariff restrictions were imposed to protect the home markets. The general crisis brought with it acute monetary difficulties, and paralysed internal credit. This spread ruin and unemployment far and wide throughout the globe. Mr. Mac-Donald's Labour-Socialist Government, with all their promises behind them, saw unemployment during 1930 and 1931 bound up in their faces from one million to nearly three millions. It was said that in the United States ten million persons were without work. The entire banking system of the great Republic was thrown into confusion and temporary collapse. Consequential disasters fell upon Germany and other European countries. However, nobody starved in the English-speaking world.

It is always difficult for an administration or party which is founded upon attacking capital to preserve the confidence and credit so important to the highly artificial economy of an island like Britain. Mr. MacDonald's Government were utterly unable to cope with the problems which confronted them. They could not command the party discipline or produce the vigour necessary even to balance the budget. In such conditions a Government already in a minority and deprived of all financial confidence could not survive.

The failure of the Labour Party to face this tempest, the sudden collapse of British financial credit, and the break-up of the Liberal Party, with its unwholesome balancing power, led to a national coalition. It seemed that only a Government of all parties was capable of coping with the crisis. Mr. MacDonald and his Chancellor of the Exchequer, on a strong patriotic emotion, attempted to carry the mass of the Labour Party into this combination. Mr. Baldwin, always content that others should have the function so long as he retained the power, was willing to serve under Mr. MacDonald. It was an attitude which, though deserving respect, did not correspond to the facts. Mr. Lloyd George was still recovering from an operation — serious at his age — and Sir Herbert Samuel led the bulk of the Liberals into the all-party combination.

I was not invited to take part in the Coalition Government. I was politically severed from Mr. Baldwin about India. I was an opponent of the policy of Mr. MacDonald's Labour Government. Like many others, I had felt the need of a national concentration. But I was neither surprised nor unhappy when I was left out of it. Indeed, I remained painting at Cannes while the political crisis lasted. What I should have done if I had been asked to join I cannot tell. It is superfluous to discuss doubtful temptations that have never existed. But I was awkwardly placed in the political scene. I had had fifteen years of Cabinet office, and was now busy with my *Life of Marlborough*. Political dramas are very exciting at the time to those engaged in the clatter and whirlpool of politics, but I can truthfully affirm that I never felt resentment, still less pain, at being so decisively discarded in a moment of national stress. There was however an inconvenience. For all these years since 1905 I had sat on one or the other of the Front Benches, and always had the advantage of speaking from the box, on which you can put your notes and pretend with more or less success to be making it up as you go along. Now I had to find with some difficulty a seat below the gangway on the Government side, where I had to hold my notes in my hand whenever I spoke, and take my chance in debate with other well-known ex-Cabinet Ministers. However, from time to time I got called.

The formation of the new Government did not end the financial crisis, and I returned from abroad to find everything unsettled in the advent of an inevitable general election. The verdict of the electorate was worthy of the British nation. A National Government had been formed under Mr. Ramsay MacDonald, the founder of the Labour-Socialist Party. They proposed to the people a programme of severe austerity and sacrifice. It was an earlier version of "blood, toil, tears, and sweat," without the stimulus or the requirements of war and mortal peril. The sternest economy must be practised. Everyone would have his wages, salary, or income reduced. The mass of the people were asked to vote for a régime of self-denial. They responded as they always do when caught in the heroic temper. Although, contrary to their declarations, the Government abandoned the gold standard, and although Mr. Baldwin was obliged to suspend, as it proved for ever, those very payments on the American debt which he had forced on the Cabinet of 1923, confidence and credit were restored. There was an overwhelming majority for the new Administration. Mr. MacDonald as Prime Minister was only followed by seven or eight members of his own party; but barely fifty of his Labour opponents and former followers were returned to Parliament. His health and powers were failing fast, and he reigned in increasing decrepitude at the summit of the British system for nearly four fateful years. And very soon in these four years came Hitler.

3

Adolf Hitler

IN OCTOBER 1918 a German corporal had been temporarily blinded by mustard gas in a British attack near Comines. While he lay in hospital in Pomerania defeat and revolution swept over Germany. The son of an obscure Austrian customs official, he had nursed youthful dreams of becoming a great artist. Having failed to gain entry to the Academy of Art in Vienna, he had lived in poverty in that capital and later in Munich. Sometimes as a house-painter, often as a casual labourer, he suffered physical privations and bred a harsh though concealed resentment that the world had denied him success. These misfortunes did not lead him into Communist ranks. By an honourable inversion he cherished all the the more an abnormal sense of racial loyalty and a fervent and mystic admiration for Germany and the German people. He sprang eagerly to arms at the outbreak of the war, and served for four years with a Bavarian regiment on the Western Front. Such were the early fortunes of Adolf Hitler.

As he lay sightless and helpless in hospital during the winter of 1918 his own personal failure seemed merged in the disaster of the whole German people. The shock of defeat, the collapse of law and order, the triumph of the French, caused this convalescent regimental orderly an agony which consumed his being, and generated those portentous and measureless forces of the spirit which may spell the rescue or the doom of mankind. The downfall of Germany seemed to him inexplicable by ordinary processes. Somewhere there had been a gigantic and monstrous betrayal. Lonely and pent within himself, the little soldier pondered and speculated upon the possible causes of the catastrophe, guided only by his narrow personal experiences. He had mingled in Vienna with extreme German Nationalist groups, and here he had heard stories of sinister, undermining activities of another race, foes and exploiters of the Nordic world — the Jews. His patriotic anger fused with his envy of the rich and successful into one overpowering hate.

When at length, as an unnoted patient, he was released from hospital, still wearing the uniform in which he had an almost schoolboyish pride, what scenes met his newly unscaled eyes! Fearful are the convulsions of defeat. Around him in the atmosphere of despair and frenzy glared the lineaments of Red Revolution. Armoured cars dashed through the streets of Munich scattering leaflets or bullets upon the fugitive wayfarers. His own comrades, with defiant red armbands on their uniform, were shouting slogans of fury against all that he cared for on earth. As in a dream everything suddenly became clear. Germany had been stabbed in the back and clawed down by the Jews, by the profiteers and intriguers behind the front, by the accursed Bolsheviks in their international conspiracy of Jewish intellectuals. Shining before him he saw his duty, to save Germany from these plagues, to avenge her wrongs, and lead the master race to its long-decreed destiny.

The officers of his regiment, deeply alarmed by the seditious and revolutionary temper of their men, were very glad to find one, at any rate, who seemed to have the root of the matter in him. Corporal Hitler desired to remain mobilised, and found employment as a "political education officer" or agent. In this guise he gathered information about mutinous and subversive designs. Presently he was told by the security officer for whom he worked to attend meetings of the local political parties of all complexions. One evening in September 1919 the Corporal went to a rally of the German Workers' Party in a Munich brewery, and here he heard for the first time people talking in the style of his secret convictions against the Jews, the speculators, the "November criminals" who had brought Germany into the abyss. On September 16 he joined this party, and shortly afterwards, in harmony with his military work, undertook its propaganda. In February 1920 the first mass meeting of the German Workers' Party was held in Munich, and here Adolf Hitler himself dominated the proceedings and in twenty-five points outlined the party programme. He had now become a politician. His campaign of national salvation had been opened. In April he was demobilised, and the expansion of the party absorbed his whole life. By the middle of the following year he had ousted the original leaders, and by his passion and genius forced upon the hypnotised company the acceptance of his personal control. Already he was "the Fuehrer." An unsuccessful newspaper, the *Voelkischer Beobachter*, was bought as the party organ.

The Communists were not long in recognising their foe. They tried to break up Hitler's meetings, and in the closing days of 1921 he

organised his first units of storm troopers. Up to this point all had
moved in local circles in Bavaria. But in the tribulation of German life
during these first postwar years many began here and there throughout the Reich to listen to the new gospel. The fierce anger of all Germany at the French occupation of the Ruhr in 1923 brought what was
now called the National Socialist Party a broad wave of adherents.
The collapse of the mark destroyed the basis of the German middle
class, of whom many in their despair became recruits of the new party
and found relief from their misery in hatred, vengeance, and patriotic
fervour.

At the beginning Hitler had made it clear that the path to power lay
through aggression and violence against a Weimar Republic born
from the shame of defeat. By November 1923 "the Fuehrer" had a determined group around him, among whom Goering, Hess, Rosenberg,
and Roehm were prominent. These men of action decided that the
moment had come to attempt the seizure of authority in the State of
Bavaria. General von Ludendorff, Chief of Staff of the German army
for most of the First World War, lent the military prestige of his name
to the venture, and marched forward in the *Putsch*. It used to be said
before the war: "In Germany there will be no revolution, because in
Germany all revolutions are strictly forbidden." This precept was revived on this occasion by the local authorities in Munich. The police
troops fired, carefully avoiding the General, who marched straight forward into their ranks and was received with respect. About twenty of
the demonstrators were killed. Hitler threw himself upon the ground,
and presently escaped with other leaders from the scene. In April 1924
he was sentenced to four years' imprisonment.

Although the German authorities had maintained order, and the
German court had inflicted punishment, the feeling was widespread
throughout the land that they were striking at their own flesh and
blood, and were playing the foreigners' game at the expense of Germany's most faithful sons. Hitler's sentence was reduced from four
years to thirteen months. These months in the Landsberg fortress were
however sufficient to enable him to complete in outline *Mein Kampf*,
a treatise on his political philosophy inscribed to the dead of the recent
Putsch. When eventually he came to power there was no book which
deserved more careful study from the rulers, political and military, of
the Allied Powers. All was there — the programme of German resurrection, the technique of party propaganda; the plan for combating
Marxism; the concept of a National-Socialist State; the rightful position of Germany at the summit of the world. Here was the new Koran
of faith and war: turgid, verbose, shapeless, but pregnant with its
message.

The main thesis of *Mein Kampf* was simple. Man is a fighting animal; therefore the nation, being a community of fighters, is a fighting unit. Any living organism which ceases to fight for its existence is doomed to extinction. A country or race which ceases to fight is equally doomed. The fighting capacity of a race depends on its purity. Hence the need for ridding it of foreign defilements. The Jewish race, owing to its universality, is of necessity, pacifist and internationalist. Pacifism is the deadliest sin, for it means the surrender of the race in the fight for existence. The first duty of every country is therefore to nationalise the masses. The ultimate aim of education is to produce a German who can be converted with the minimum of training into a soldier. The greatest upheavals in history would have been unthinkable had it not been for the driving force of fanatical and hysterical passions. Nothing could have been effected by the bourgeois virtues of peace and order. The world is now moving towards such an upheaval, and the new German State must see to it that the race is ready for the last and greatest decisions on this earth.

Foreign policy may be unscrupulous. It is not the task of diplomacy to allow a nation to founder heroically, but rather to see that it can prosper and survive. England and Italy are the only two possible allies for Germany. So long as Germany does not fend for herself, nobody will fend for her. Her lost provinces cannot be regained by solemn appeals to Heaven or by pious hopes in the League of Nations, but only by force of arms. Germany must not repeat the mistake of fighting all her enemies at once. To attack France for purely sentimental reasons would be foolish. What Germany needs is increase of territory in Europe. Germany's prewar colonial policy was a mistake and should be abandoned. Germany must look for expansion to Russia, and especially to the Baltic States. No alliance with Russia can be tolerated. To wage war together with Russia against the West would be criminal, for the aim of the Soviets is the triumph of international Judaism. Such were the "granite pillars" of his policy.

The ceaseless struggles and gradual emergence of Adolf Hitler as a national figure were little noticed by the victors, oppressed and harrassed as they were by their own troubles and party strife. A long interval passed before National Socialism, or the "Nazi Party," as it came to be called, gained so strong a hold of the masses of the German people, of the armed forces, of the machinery of the State, and among industrialists not unreasonably terrified of Communism, as to become a power in German life of which world-wide notice had to be taken. When Hitler was released from prison at the end of 1924 he said that it would take him five years to reorganise his movement.

*   *   *

One of the democratic provisions of the Weimar Constitution prescribed elections to the Reichstag every four years. It was hoped by
this provision to make sure that the masses of the German people
should enjoy a complete and continuous control over their Parliament.
In practice of course it only meant that they lived in a continual atmosphere of febrile political excitement and ceaseless electioneering. The
progress of Hitler and his doctrines is thus registered with precision.
In 1928 he had but twelve seats in the Reichstag. In 1930 this became
107; in 1932, 230. By that time the whole structure of Germany had
been permeated by the agencies and discipline of the National Socialist
Party, and intimidation of all kinds and insults and brutalities towards
the Jews were rampant.

It is not necessary in this account to follow year by year this complex
and formidable development, with all its passions and villainies and all
its ups and downs. The pale sunlight of Locarno shone for a while
upon the scene. The spending of the profuse American loans induced a
sense of returning prosperity. Marshal Hindenburg presided over the
German State, and Stresemann was his Foreign Minister. The stable,
decent majority of the German people, responding to their ingrained
love of massive and majestic authority, clung to him till his dying
gasp. But other powerful factors were also active in the distracted nation to which the Weimar Republic could offer no sense of security, and
no satisfaction of national glory or revenge.

Behind the veneer of republican governments and democratic institutions, imposed by the victors and tainted with defeat, the real political
power in Germany and the enduring structure of the nation in the postwar years had been the General Staff of the German army or Reichswehr. They it was who secretly laid the basis for German rearmament
and who made and unmade presidents and cabinets. They had found
in Marshal Hindenburg a symbol of their power and an agent of their
will. But Hindenburg in 1930 was eighty-three years of age. From
this time his character and mental grasp steadily declined. He became
increasingly prejudiced, arbitrary, and senile. An enormous image had
been made of him in the war, and patriots could show their admiration
by paying for a nail to drive into it. This illustrates effectively what
he had now become — "The Wooden Titan." It had for some time
been clear to the generals that a satisfactory successor to the aged Marshal would have to be found. The search for the new man was however
overtaken by the vehement growth and force of the National-Socialist
movement. After the failure of the 1923 *Putsch* in Munich Hitler had
professed a programme of strict legality within the framework of the
Weimar Republic. Yet at the same time he had encouraged and planned
the expansion of the military and para-military formations of the Nazi

Party. From very small beginnings the S.A., the Storm Troops or Brownshirts, with their small disciplinary core, the S.S., grew in numbers and vigour to the point where the Reichswehr viewed their activities and potential strength with grave alarm.

At the head of the Storm Troop formations stood a German soldier of fortune, Ernst Roehm, the comrade and hitherto the close friend of Hitler through all the years of struggle. Roehm, Chief of Staff of the S.A., was a man of proved ability and courage, but dominated by personal ambition and sexually perverted. His vices were no barrier to Hitler's collaboration with him along the hard and dangerous path to power. Pondering most carefully upon the tides that were flowing in the nation, the Reichswehr convinced themselves with much reluctance that, as a military caste and organisation in opposition to the Nazi movement, they could no longer maintain control of Germany. Both factions had in common the resolve to raise Germany from the abyss and avenge her defeat; but while the Reichswehr represented the ordered structure of the Kaiser's Empire, and gave shelter to the feudal, aristocratic, land-owning, and well-to-do classes in German society, the S.A. had become to a large extent a revolutionary movement fanned by the discontents of temperamental or embittered subversives and the desperation of ruined men. They differed from the Bolsheviks whom they denounced no more than the North Pole does from the South.

For the Reichswehr to quarrel with the Nazi Party was to tear the defeated nation asunder. The Army chiefs in 1931 and 1932 felt they must, for their own sake and for that of the country, join forces with those to whom in domestic matters they were opposed with all the rigidity and severeness of the German mind. Hitler, for his part, although prepared to use any battering-ram to break into the citadels of power, had always before his eyes the leadership of the great and glittering Germany which had commanded the admiration and loyalty of his youthful years. The conditions for a compact between him and the Reichswehr were therefore present and natural on both sides. The Army chiefs gradually realised that the strength of the Nazi Party was such that Hitler was the only possible successor to Hindenburg as head of the German nation. Hitler on his side knew that to carry out his programme of German resurrection an alliance with the governing élite of the Reichswehr was indispensable. A bargain was struck, and the German Army leaders began to persuade Hindenburg to look upon Hitler as eventual Chancellor of the Reich. Thus by agreeing to curtail the activities of the Brownshirts, to subordinate them to the General Staff, and ultimately, if unavoidable, to liquidate them, Hitler gained the allegiance of the controlling forces in Germany, official exe-

cutive dominance, and the apparent reversion of the headship of the
German State. The Corporal had travelled far.

There was however an inner and separate complication. If the key
to any master-combination of German internal forces was the General
Staff of the Army, several hands were grasping for that key. General
Kurt von Schleicher at this time exercised a subtle and on occasions a
decisive influence. He was the political mentor of the reserved and
potentially dominating military circle. He was viewed with a measure
of distrust by all sections and factions, and regarded as an adroit and
useful political agent possessed of much knowledge outside the General
Staff manuals and not usually accessible to soldiers. Schleicher had long
been convinced of the significance of the Nazi movement and of the
need to stem and control it. On the other hand, he saw that in this
terrific mob-thrust, with its ever-growing private army of S.A., there
was a weapon which, if properly handled by his comrades of the
General Staff, might reassert the greatness of Germany, and perhaps
even establish his own. In this intention during the course of 1931
Schleicher began to plot secretly with Roehm. There was thus a
major double process at work, the General Staff making their arrangements with Hitler, and Schleicher in their midst pursuing his personal
conspiracy with Hitler's principal lieutenant and would-be rival, Roehm.
Schleicher's contacts with the revolutionary element of the Nazi Party,
and particularly with Roehm, lasted until both he and Roehm were
shot by Hitler's orders three years later. This certainly simplified the
political situation, and also that of the survivors.

*   *   *

Meanwhile the economic blizzard smote Germany in her turn. The
United States banks, faced with increasing commitments at home, refused to increase their improvident loans to Germany. This reaction
led to the widespread closing of factories, and the sudden ruin of many
enterprises on which the peaceful revival of Germany was based. Unemployment in Germany rose to 2,300,000 in the winter of 1930. The
Allies offered a far-reaching and benevolent easement of Reparations.
Stresemann, the Foreign Minister, who was now a dying man, gained
his last success in the agreement for the complete evacuation of the
Rhineland by the Allied armies, long before the Treaty required.

But the German masses were largely indifferent to the remarkable
concessions of the victors. Earlier, or in happier circumstances, these
would have been acclaimed as long steps upon the path of reconciliation and a return to true peace. But now the ever-present overshadowing fear of the German masses was unemployment. The middle classes
had already been ruined and driven into violent courses by the flight

from the mark. Stresemann's internal political position was undermined by the international economic stresses, and the vehement assaults of Hitler's Nazis and certain capitalist magnates led to his overthrow. On March 28, 1930, Bruening, the leader of the Catholic Centre Party, became Chancellor. Bruening was a Catholic from Westphalia and a patriot, seeking to re-create the former Germany in modern democratic guise. He pursued continuously the scheme of factory preparation for war. He had also to struggle towards financial stability amid mounting chaos. His programme of economy and reduction of Civil Service numbers and salaries was not popular. The tides of hatred flowed ever more turbulently. Supported by President Hindenburg, Bruening dissolved a hostile Reichstag, and the election of 1930 left him with a majority. He now made the last recognisable effort to rally what remained of the old Germany against the resurgent, violent, and debased nationalist agitation. For this purpose he had first to secure the re-election of Hindenburg as President. Chancellor Bruening looked to a new but obvious solution. He saw the peace, safety, and glory of Germany only in the restoration of an Emperor. Could he then induce the aged Marshal Hindenburg, if and when re-elected, to act for his last term of office as Regent for a restored monarchy to come into effect upon his death? This policy, if achieved would have filled the void at the summit of the German nation towards which Hitler was now evidently making his way. In all the circumstances this was the right course. But how could Bruening lead Germany to it? The Conservative element, which was drifting to Hitler, might have been recalled by the return of Kaiser Wilhelm; but neither the Social Democrats nor the trade union forces would tolerate the return of the old Kaiser or the Crown Prince. Bruening's plan was not to recreate a Second Reich. He desired a constitutional monarchy on English lines. He hoped that one of the sons of the Crown Prince might be a suitable candidate.

In November 1931 he confided his plans to Hindenburg, on whom all depended. The aged Marshal's reaction was at once vehement and peculiar. He was astonished and hostile. He said that he regarded himself solely a trustee of the Kaiser. Any other solution was an insult to his military honour. The monarchical conception, to which he was devoted, could not be reconciled with picking and choosing among royal princes. Legitimacy must not be violated. Meanwhile, as Germany would not accept the return of the Kaiser, there was nothing left but he himself, Hindenburg. On this he rested. No compromise for him! *"J'y suis, j'y reste."* Bruening argued vehemently and perhaps over-long with the old veteran. The Chancellor had a strong case. Unless Hindenburg would accept this monarchical solution, albeit unorthodox, there must be a revolutionary Nazi dictatorship. No agreement was reached. But whether or not Bruening could convert Hindenburg, it was imperative to get him re-elected as President, in order at least to stave off an immediate political collapse of the German State. In its first stage Bruening's plan was successful. At the Presidential elections held in March 1932 Hindenburg was returned, after a second ballot, by a majority over his rivals, Hitler and the Communist Thaelmann. Both the economic position in Germany and her relations with Europe had now to be faced. The Disarmament Conference was sitting at Geneva, and Hitler throve upon a roaring campaign against the humiliation of Germany under Versailles.

In careful meditation Bruening drafted a far-reaching plan of Treaty revision; and in April he went to Geneva and found an unexpectedly favourable reception. In conversations between him and MacDonald, and Mr. Stimson and Mr. Norman Davis from America, it seemed that agreement could be reached. The extraordinary basis of this was the principle, subject to various reserved interpretations, of "equality of armaments" between Germany and France. It is indeed surprising, as future chapters will explain, that anyone in his senses should have imagined that peace could be built on such foundations. If this vital point were conceded by the victors, it might well pull Bruening out of his plight, and then the next step — and this one wise — would be the cancelling of reparations for the sake of European revival. Such a settlement would of course have raised Bruening's personal position to one of triumph.

Norman Davis, the American Ambassador-at-Large, telephoned to the French Premier, Tardieu, to come immediately from Paris to Geneva. But, unfortunately for Bruening, Tardieu had other news. Schleicher had been busy in Berlin, and had just warned the French Ambassador not to negotiate with Bruening because his fall was imminent. It may well be also that Tardieu was concerned with the military position of France on the formula of "equality of armaments." At any rate, Tardieu did not come to Geneva, and on May 1 Bruening returned to Berlin. To arrive there empty-handed at such a moment was fatal to him. Drastic and even desperate measures were required to cope with the threatened economic collapse inside Germany. For these measures Bruening's unpopular Government had not the necessary strength. He struggled on through May, and meanwhile Tardieu, in the kaleidoscope of French Parliamentary politics, was replaced by M. Herriot.

The new French Premier declared himself ready to discuss the formula reached in the Geneva conversations. The American Ambassador in Berlin was instructed to urge the German Chancellor to go to Geneva without a moment's delay. This message was received by Bruening early on May 30. But meanwhile Schleicher's influence had prevailed. Hindenburg had already been persuaded to dismiss the Chancellor. In the course of that very morning, after the American invitation, with all its hope and imprudence, had reached Bruening, he learned that his fate was settled, and by midday he resigned to avoid actual dismissal. So ended the last Government in postwar Germany which might have led the German people into the enjoyment of a stable and civilised constitution and opened peaceful channels of intercourse with their neighbours. The offers which the Allies had made to Bruening would, but for Schleicher's intrigue and Tardieu's delay, certainly have saved him. These offers had presently to be discussed with a different system and a different man.

4

The Locust Years,[1] 1931–1933

THE BRITISH GOVERNMENT which resulted from the General Election of 1931 was in appearance one of the strongest and in fact one of the weakest in British records. Mr. Ramsay MacDonald, the Prime Minister, had severed himself, with the utmost bitterness on both sides, from the Socialist Party which it had been his life's work to create. Henceforward he brooded supinely at the head of an Administration which, though nominally National, was in fact overwhelmingly Conservative. Mr. Baldwin preferred the substance to the form of power, and reigned placidly in the background. The Foreign Office was filled by Sir John Simon, one of the leaders of the Liberal contingent. The main work of the Administration at home was done by Mr. Neville Chamberlain, who soon became Chancellor of the Exchequer. The Labour Party, blamed for its failure in the financial crisis and sorely stricken at the polls, was led by the extreme pacifist, Mr. George Lansbury. During the period of four and a quarter years of this Administration, from August 1931 to November 1935, the entire situation on the Continent of Europe was reversed.

*   *   *

All Germany was astir and great events marched forward. Papen, who succeeded Bruening as Chancellor, and the political general, Schleicher, had hitherto attempted to govern Germany by cleverness and intrigue. The time for these had now passed. Papen hoped to rule with the support of the entourage of President Hindenburg and of the extreme Nationalist group in the Reichstag. On July 20 a decisive

---

[1] Four years later Sir Thomas Inskip, Minister for Co-ordination of Defence, who was well versed in the Bible, used the expressive phrase about this dismal period, of which he was the heir: "The years that the locust hath eaten" (Joel, ii, 25).

34

step was taken. The Socialist Government in Prussia was forcibly ousted from office. But Papen's rival was eager for power. In Schleicher's calculations the instrument lay in the dark, hidden forces storming into German politics behind the rising power and name of Adolf Hitler. He hoped to make the Hitler Movement a docile servant of the Reichswehr, and in so doing to gain the control of both himself. The contacts between Schleicher and Roehm, the leader of the Nazi Storm Troopers, which had begun in 1931, were extended in the following year to more precise relations between Schleicher and Hitler himself. The road to power for both men seemed to be obstructed only by Papen and by the confidence displayed by Hindenburg in him.

In August 1932 Hitler came to Berlin on a private summons from the President. The moment for a forward step seemed at hand. Thirteen million German voters stood behind the Fuehrer. A vital share of office must be his for the asking. He was now in somewhat the position of Mussolini on the eve of the march on Rome. But Papen did not care about recent Italian history. He had the support of Hindenburg and had no intention of resigning. The old Marshal saw Hitler. He was not impressed, *"That* man for Chancellor? I'll make him a postmaster and he can lick stamps with my head on them." In palace circles Hitler had not the influence of his competitors.

In the country the vast electorate was restless and adrift. In November 1932, for the fifth time in a year, elections were held throughout Germany. The Nazis lost ground and their 230 seats were reduced to 196, the Communists gaining the balance. The bargaining power of the Fuehrer was thus weakened. Perhaps General Schleicher would be able to do without him after all. The General gained favour in the circle of Hindenburg's advisers. On November 17 Papen resigned and Schleicher became Chancellor in his stead. But the new Chancellor was found to have been more apt at pulling wires behind the scenes than at the open summit of power. He had quarrelled with too many people. Hitler, together with Papen and the Nationalists, now ranged themselves against him; and the Communists, fighting the Nazis in the streets and the Government by their strikes, helped to make his rule impossible. Papen brought his personal influence to bear on President Hindenburg. Would not after all the best solution be to placate Hitler by thrusting upon him the responsibilities and burdens of office? Hindenburg at last reluctantly consented. On January 30, 1933, Adolf Hitler took office as Chancellor of Germany.

The hand of the Master was soon felt upon all who would or might oppose the New Order. On February 2 all meetings or demonstrations of the German Communist Party were forbidden, and throughout Germany a roundup of secret arms belonging to the Communists began.

The climax came on the evening of February 27, 1933. The building of the Reichstag broke into flames. Brownshirts, Blackshirts, and their auxiliary formations were called out. Four thousand arrests, including the Central Committee of the Communist Party, were made overnight. These measures were entrusted to Goering, now Minister of the Interior of Prussia. They formed the preliminary to the forthcoming elections and secured the defeat of the Communists, the most formidable opponents of the régime. The organising of the electoral campaign was the task of Goebbels, and he lacked neither skill nor zeal.

But there were still many forces in Germany reluctant, obstinate, or actively hostile to Hitlerism. The Communists, and many who in their perplexity and distress voted with them, obtained 81 seats, the Socialists 118, the Centre party 73, and the Nationalist allies of Hitler under Papen and Hugenberg 52. Thirty-three seats were allotted to minor Right Centre groups. The Nazis obtained a vote of 17,300,000 with 288 seats. These results gave Hitler and his Nationalist allies control of the Reichstag. Thus, and thus only, did Hitler obtain by hook and crook a majority vote from the German people. Under the ordinary processes of civilised Parliamentary government, so large a minority would have had great influence and due consideration in the State. But in the new Nazi Germany minorities were now to learn that they had no rights.

On March 21, 1933, Hitler opened, in the garrison church at Potsdam, hard by the tomb of Frederick the Great, the first Reichstag of the Third Reich. In the body of the church sat the representatives of the Reichswehr, the symbol of the continuity of German might, and the senior officers of the S.A. and S.S., the new figures of resurgent Germany. On March 24 the majority of the Reichstag, overbearing or overawing all opponents, confirmed by 441 votes to 94 complete emergency powers to Chancellor Hitler for four years. As the result was announced Hitler turned to the benches of the Socialists and cried, "And now I have no further need of you."

Amid the excitement of the election the exultant columns of the Nationalist Socialist Party filed past their leader in the pagan homage of a torchlight procession through the streets of Berlin. It had been a long struggle, difficult for foreigners, especially those who had not known the pangs of defeat, to comprehend. Adolf Hitler had at last arrived. But he was not alone. He had called from the depths of defeat the dark and savage furies latent in the most numerous, most serviceable, ruthless, contradictory, and ill-starred race in Europe. He had conjured up the fearful idol of an all-devouring Moloch of which he was the priest and incarnation. It is not within my scope to describe the inconceivable brutality and villainy by which this apparatus of hatred and tyranny had been fashioned and was now to be perfected. It is necessary, for the purpose of this account, only to present to the reader the new and fearful fact which had broken upon the still unwitting world: GERMANY UNDER HITLER, AND GERMANY ARMING.

*   *   *

While these deadly changes were taking place in Germany the MacDonald-Baldwin Government felt bound to enforce for some time the severe reductions and restrictions which the financial crisis had imposed upon our already modest armaments, and steadfastly closed their eyes and ears to the disquieting symptoms in Europe. In vehement efforts to procure a disarmament of the victors equal to that which had been enforced upon the vanquished by the Treaty of Versailles, Mr. MacDonald and his Conservative and Liberal colleagues pressed a series of proposals forward in the League of Nations and through every other channel that was open. The French, although their political affairs still remained in constant flux and in motion without particular significance, clung tenaciously to the French Army as the centre and prop of the life of France and of all her alliances. This attitude earned them rebukes both in Britain and in the United States. The opinions of the press and public were in no way founded upon reality; but the adverse tide was strong.

The German Government were emboldened by the British demeanour. They ascribed it to the fundamental weakness and inherent decadence imposed even upon a Nordic race by the democratic and parliamentary form of society. With all Hitler's national drive behind them, they took a haughty line. In July 1932, their delegation had gathered up its papers and quitted the Disarmament Conference. To coax them back became the prime political objective of the victorious Allies. In November the French, under severe and constant British pressure, proposed what was somewhat unfairly called "the Herriot Plan." The essence of this was the reconstitution of all European defence forces as short-service armies with limited numbers, admitting equality of status but not necessarily accepting equality of strength. In fact and in principle, the admission of equality of status made it impossible ultimately not to accept equality of strength. This enabled the Allied Governments to offer to Germany "Equality of rights in a system which would provide security for all nations." Under certain safeguards of an illusory character the French were reduced to accepting this meaningless formula. On this the Germans consented to return to the Disarmament Conference. This was hailed as a notable victory for peace.

Fanned by the breeze of popularity, His Majesty's Government now produced on March 16, 1933, what was called, after its author and inspirer, "the MacDonald Plan." It accepted as its starting-point the adoption of the French conception of short-service armies — in this case of eight months' service — and proceeded to prescribe exact figures for the troops of each country. The French Army should be reduced from its peacetime establishment of 500,000 men to 200,000 and the Germans should increase to parity at that figure. By this time the German military forces, though not yet provided with the mass of trained reserves which only a succession of annual conscripted quotas could supply, may well have amounted to the equivalent of over a million ardent volunteers, partially equipped, and with many forms of the latest weapons coming along through the convertible and partially converted factories to arm them. The result was unexpected. Hitler, now Chancellor and Master of all Germany, having already given orders on assuming power to drive ahead boldly on a nation-wide scale, both in the training-camps and the factories, felt himself in a strong position. He did not even trouble to accept the quixotic offers pressed upon him. With a gesture of disdain he directed the German Government to withdraw both from the Conference and from the League of Nations.

It is difficult to find a parallel to the unwisdom of the British and weakness of the French Governments, who none the less reflected the opinion of their Parliaments in this disastrous period. Nor can the United States escape the censure of history. Absorbed in their own affairs and all the abounding interests, activities, and accidents of a free community, they simply gaped at the vast changes which were taking place in Europe, and imagined they were no concern of theirs. The considerable corps of highly competent, widely trained professional American officers formed their own opinions, but these produced no noticeable effect upon the improvident aloofness of American foreign policy. If the influence of the United States had been exerted, it might have galvanised the French and British politicians into action. The League of Nations, battered though it had been, was still an august instrument which would have invested any challenge to the new Hitler war-menace with the sanctions of International Law. Under the strain the Americans merely shrugged their shoulders, so that in a few years they had to pour out the blood and treasures of the New World to save themselves from mortal danger.

Seven years later when at Tours I witnessed the French agony all this was in my mind, and that is why, even when proposals for a separate peace were mentioned, I spoke only words of comfort and reassurance, which I rejoice to feel have been made good.

*   *   *

I had arranged at the beginning of 1931 to undertake a considerable lecture tour in the United States, and travelled to New York. Here I suffered a serious accident, which nearly cost me my life. On December 13, when on my way to visit Mr. Bernard Baruch, I got out of my car on the wrong side, and walked across Fifth Avenue without bearing in mind the opposite rule of the road which prevails in America, or the red lights, then unused in Britain. There was a shattering collision. For two months I was a wreck. I gradually regained at Nassau in the Bahamas enough strength to crawl around. In this condition I undertook a tour of forty lectures throughout the United States, living all day on my back in a railway compartment, and addressing in the evening large audiences. On the whole I consider this was the hardest time I have had in my life. I lay pretty low all through this year, but in time my strength returned.

The years from 1931 to 1935, apart from my anxiety on public affairs, were personally very pleasant to me. I earned my livelihood by dictating articles which had a wide circulation not only in Great Britain and the United States, but also, before Hitler's shadow fell upon them, in the most famous newspapers of sixteen European countries. I lived in fact from mouth to hand. I produced in succession the various volumes of the *Life of Marlborough*. I meditated constantly upon the European situation and the rearming of Germany. I lived mainly at Chartwell, where I had much to amuse me. I built with my own hands a large part of two cottages and extensive kitchen-garden walls, and made all kinds of rockeries and waterworks and a large swimming-pool which was filtered to limpidity and could be heated to supplement our fickle sunshine. Thus I never had a dull or idle moment from morning till midnight, and with my happy family around me dwelt at peace within my habitation.

During these years I saw a great deal of Frederick Lindemann, Professor of Experimental Philosophy at Oxford University. Lindemann was already an old friend of mine. I had met him first at the close of the previous war, in which he had distinguished himself by conducting in the air a number of experiments, hitherto reserved for daring pilots, to overcome the then almost mortal dangers of a "spin." We came much closer together from 1932 onward, and he frequently motored over from Oxford to stay with me at Chartwell. Here we had many talks into the small hours of the morning about the dangers which seemed to be gathering upon us. Lindemann, "the Prof," as he was called among his friends, became my chief adviser on the scientific aspects of modern war and particularly air defence, and also on questions involving statistics of all kinds. This pleasant and fertile association continued throughout the war.

Another of my close friends was Desmond Morton.[2] When in 1917 Field-Marshal Haig filled his personal staff with young officers fresh from the firing line, Desmond was recommended to him as the pick of the artillery. To his Military Cross he added the unique distinction of having been shot through the heart, and living happily ever afterwards with the bullet in him. I formed a great regard and friendship for this brilliant and gallant officer, and in 1919, when I became Secretary of State for War and Air, I appointed him to a key position in the Intelligence, which he held for many years. He was a neighbour of mine, dwelling only a mile away from Chartwell. He obtained from the Prime Minister, Mr. MacDonald, permission to talk freely to me and keep me well informed. He became, and continued during the war to be, one of my most intimate advisers till our final victory was won.

I had also formed a friendship with Ralph Wigram, then the rising star of the Foreign Office and in the centre of all its affairs. He had reached a level in that department which entitled him to express responsible opinions upon policy, and to use a wide discretion in his contacts, official and unofficial. He was a charming and fearless man, and his convictions, based upon profound knowledge and study, dominated his being. He saw as clearly as I did, but with more certain information, the awful peril which was closing in upon us. This drew us together. Often we met at his little house in North Street, and he and Mrs. Wigram came to stay with us at Chartwell. Like other officials of high rank, he spoke to me with complete confidence. All this helped me to form and fortify my opinion about the Hitler Movement.

It was of great value to me, and it may be thought also to the country, that I should have the means of conducting searching and precise discussions for so many years in this very small circle. On my side however I gathered and contributed a great deal of information from foreign sources. I had confidential contacts with several of the French Ministers and with the successive chiefs of the French Government. Mr. Ian Colvin was the *News Chronicle* correspondent in Berlin. He plunged very deeply into German politics, and established contacts of a most secret character with some of the important German generals, and also with independent men of character and quality in Germany who saw in the Hitler Movement the approaching ruin of their native land. Several visitors of consequence came to me from Germany and poured their hearts out in their bitter distress. Most of these were executed by Hitler during the war. From other directions I was able to check and furnish information on the whole field of our air defence.

[2] Now Major Sir Desmond Morton, K.C.B., M.C.

In this way I became as well instructed as many Ministers of the Crown. All the facts I gathered from every source, including especially foreign connections, I reported to the Government from time to time. My personal relations with Ministers and also with many of their high officials were close and easy, and although I was often their critic we maintained a spirit of comradeship. Later on I was made officially party to much of their most secret technical knowledge. From my own long experience in high office I was also possessed of the most precious secrets of the State. All this enabled me to form and maintain opinions which did not depend on what was published in the newspapers, though these brought many items to the discriminating eye.

The reader will pardon a personal digression in a lighter vein.

In the summer of 1932, for the purposes of my *Life of Marlborough* I visited his old battlefields in the Low Countries and Germany. Our family expedition, which included "the Prof," journeyed agreeably along the line of Marlborough's celebrated march in 1705 from the Netherlands to the Danube, passing the Rhine at Coblenz. As we wended our way through these beautiful regions from one ancient, famous city to another, I naturally asked questions about the Hitler Movement, and found it the prime topic in every German mind. I sensed a Hitler atmosphere. After passing a day on the field of Blenheim, I drove into Munich, and spent the best part of a week there.

At the Regina Hotel a gentleman introduced himself to some of my party. He was Herr Hanfstaengl, and spoke a great deal about "the Fuehrer," with whom he appeared to be intimate. As he seemed to be a lively and talkative fellow, speaking excellent English, I asked him to dine. He gave a most interesting account of Hitler's activities and outlook. He spoke as one under the spell. He had probably been told to get in touch with me. He was evidently most anxious to please. After dinner he went to the piano and played and sang many tunes and songs in such remarkable style that we all enjoyed ourselves immensely. He seemed to know all the English tunes that I liked. He was a great entertainer, and at that time, as is known, a favourite of the Fuehrer. He said I ought to meet him, and that nothing would be easier to arrange. Herr Hitler came every day to the hotel about five o'clock, and would be very glad indeed to see me.

I had no national prejudices against Hitler at this time. I knew little of his doctrine or record and nothing of his character. I admire men who stand up for their country in defeat, even though I am on the other side. He had a perfect right to be a patriotic German if he chose. I always wanted England, Germany, and France to be friends. However, in the course of conversation with Hanfstaengl I happened to say, "Why is your chief so violent about the Jews? I can quite understand being angry with Jews who have done wrong or are against the country, and I understand resisting them if they try to monopolise power in any walk of life; but what is the sense of being against a man simply because of his birth? How can any man help how he is born?" He must have repeated this to Hitler, because about noon the next day he came round with rather a serious air and said that the appointment he had made with me to meet Hitler could not take place as the Fuehrer would not be coming to the hotel that afternoon. This was the last I saw of "Putzi" — for such was his pet name — although we stayed several more days at the hotel. Thus Hitler lost his only chance of meeting me. Later on, when he was all-powerful, I was to receive several invitations from him. But by that time a lot had happened, and I excused myself.

*   *   *

All this while the United States remained intensely preoccupied with its own vehement internal affairs and economic problems. Europe and far-off Japan watched with steady gaze the rise of German warlike power. Disquietude was increasingly expressed in Scandinavian countries and the states of the Little Entente, Czechoslovakia, Yugoslavia and Rumania, and in some Balkan countries. Deep anxiety ruled in France, where a large amount of knowledge of Hitler's activities and of German preparations had come to hand. There was, I was told, a catalogue of breaches of the Treaties of immense and formidable gravity, but when I asked my French friends why this matter was not raised in the League of Nations, and Germany invited, or even ultimately summoned, to explain her action and state precisely what she was doing, I was answered that the British Government would deprecate such an alarming step. Thus, while Mr. MacDonald, with Mr. Baldwin's full authority, preached disarmament to the French and practised it upon the British, the German might grew by leaps and bounds, and the time for overt action approached.

In justice to the Conservative Party it must be mentioned that at each of the Conferences of the National Union of Conservative Associations from 1932 onwards resolutions in favour of an immediate strengthening of our armaments to meet the growing danger from abroad were carried almost unanimously. But the parliamentary control by the Government Whips in the House of Commons was at this time so effective, and the three parties in the Government, as well as the Labour Opposition, so sunk in lethargy and blindness, that the warnings of their followers in the country were as ineffective as were the signs of the times and the evidence of the Secret Service. This was one of those awful periods which recur in our history, when the noble

British nation seems to fall from its high estate, loses all trace of sense or purpose, and appears to cower from the menace of foreign peril, frothing pious platitudes while foemen forge their arms.

In this dark time the basest sentiments received acceptance or passed unchallenged by the responsible leaders of the political parties. In 1933 the students of the Oxford Union, under the inspiration of a Mr. Joad, passed their ever-shameful resolution, "That this House will in no circumstances fight for its King and Country." It was easy to laugh off such an episode in England, but in Germany, in Russia, in Italy, in Japan, the idea of a decadent, degenerate Britain took deep root and swayed many calculations. Little did the foolish boys who passed the resolution dream that they were destined quite soon to conquer or fall gloriously in the ensuing war, and prove themselves the finest generation ever bred in Britain. Less excuse can be found for their elders, who had no chance of self-redemption in action.

*   *   *

While this fearful transformation in the relative war-power of victors and vanquished was taking place in Europe, a complete lack of concert between the non-aggressive and peace-loving states had also developed in the Far East. This story forms a counterpart to the disastrous turn of events in Europe, and arose from the same paralysis of thought and action among the leaders of the former and future Allies.

The economic blizzard of 1929 to 1931 had affected Japan not less than the rest of the world. Since 1914 her population had grown from fifty to seventy millions. Her metallurgical factories had increased from fifty to one hundred and forty-eight. The cost of living had risen steadily. The production of rice was stationary, and its importation expensive. The need for raw material and for external markets was clamant. In the violent depression Britain and forty other countries felt increasingly compelled, as the years passed, to apply restrictions or tariffs against Japanese goods produced under labour conditions unrelated to European or American standards. China was more than ever Japan's principal export market for cotton and other manufactures, and almost her sole source of coal and iron. A new assertion of control over China became therefore the main theme of Japanese policy.

In September 1931, on a pretext of local disorders, the Japanese occupied Mukden and the zone of the Manchurian Railway. In January 1932 they demanded the dissolution of all Chinese associations of an anti-Japanese character. The Chinese Government refused, and on the 28th the Japanese landed to the north of the International Concession at Shanghai. The Chinese resisted with spirit, and, although without aeroplanes or anti-tank guns or any of the modern weapons, maintained their resistance for more than a month. At the end of February, after suffering very heavy losses, they were obliged to retire from their forts in the Bay of Wu-Sung, and took up positions about twelve miles inland. Early in 1932 the Japanese created the puppet State of Manchukuo. A year later the Chinese province of Jehol was annexed to it, and Japanese troops, penetrating deeply into defenceless regions, had reached the Great Wall of China. This aggressive action corresponded to the growth of Japanese power in the Far East and her new naval position on the oceans.

From the first shot the outrage committed upon China aroused the strongest hostility in the United States. But the policy of isolation cut both ways. Had the United States been a member of the League of Nations, she could undoubtedly have led that assembly into collective action against Japan, of which the United States would herself have been the principal mandatory. The British Government on their part showed no desire to act with the United States alone; nor did they wish to be drawn into antagonism with Japan further than their obligations under the League of Nations Charter required. There was a rueful feeling in some British circles at the loss of the Japanese Alliance and the consequential weakening of the British position with all its long-established interests in the Far East. His Majesty's Government could hardly be blamed if in their grave financial and growing European embarrassments they did not seek a prominent role at the side of the United States in the Far East without any hope of corresponding American support in Europe.

China however was a member of the League, and although she had not paid her subscription to that body she appealed to it for what was no more than justice. On September 30, 1931, the League called on Japan to remove her troops from Manchuria. In December a Commission was appointed to conduct an inquiry on the spot. The League of Nations entrusted the chairmanship of the Commission to the Earl of Lytton, the worthy descendant of a gifted line. He had had many years' experience in the East as Governor of Bengal and as acting Viceroy of India. The Report, which was unanimous, was a remarkable document, and forms the basis of any serious study of the conflict between China and Japan. The whole background of the Manchurian affair was carefully presented. The conclusions drawn were plain: Manchukuo was the artificial creation of the Japanese General Staff, and the wishes of the population had played no part in the formation of this puppet state. Lord Lytton and his colleagues in their report not only analysed the situation but put forward concrete pro-

posals for an international solution. These were for the declaration
of an autonomous Manchuria. It would still remain part of China,
under the aegis of the League, and there would be a comprehensive
treaty between China and Japan regulating their interests in Manchuria.
The fact that the League could not follow up these proposals in no way
detracts from the value of the Lytton Report. In February 1933 the
League of Nations declared that the State of Manchukuo could not
be recognised. Although no sanctions were imposed upon Japan, nor
any other action taken, she thereupon withdrew from the League of
Nations. Germany and Japan had been on opposite sides in the war;
they now looked towards each other in a different mood. The moral
authority of the League was shown to be devoid of any physical support at a time when its activity and strength were most needed.

*   *   *

We must regard as deeply blameworthy before history the conduct
not only of the British National and mainly Conservative Government, but of the Labour-Socialist and Liberal Parties, both in and out
of office, during this fatal period. Delight in smooth-sounding platitudes, refusal to face unpleasant facts, desire for popularity and electoral success irrespective of the vital interests of the State, genuine
love of peace and pathetic belief that love can be its sole foundation,
obvious lack of intellectual vigour in both leaders of the British Coalition Government, marked ignorance of Europe and aversion from its
problems in Mr. Baldwin, the strong and violent pacifism which at
this time dominated the Labour-Socialist Party, the utter devotion of
the Liberals to sentiment apart from reality, the failure and worse
than failure of Mr. Lloyd George, the erstwhile great wartime leader,
to address himself to the continuity of his work, the whole supported
by overwhelming majorities in both Houses of Parliament: all these
constituted a picture of British fatuity and fecklessness which, though
devoid of guile, was not devoid of guilt, and, though free from wickedness or evil design, played a definite part in the unleashing upon the
world of horrors and miseries which, even so far as they have unfolded,
are already beyond comparison in human experience.

5

The Darkening Scene, 1934

HITLER'S ACCESSION to the Chancellorship in 1933 had not been regarded with enthusiasm in Rome. Nazism was viewed as a crude and brutalised version of the Fascist theme. The ambitions of a Greater Germany towards Austria and in Southeastern Europe were well known. Mussolini foresaw that in neither of these regions would Italian interests coincide with those of the new Germany. Nor had he long to wait for confirmation.

The acquisition of Austria by Germany was one of Hitler's most cherished ambitions. The first page of *Mein Kampf* contains the sentence, "German Austria must return to the great German Motherland." From the moment, therefore of the acquisition of power in January 1933, the Nazi German Government cast its eyes upon Vienna. Hitler could not afford as yet to clash with Mussolini, whose interests in Austria had been loudly proclaimed. Even infiltration and underground activities had to be applied with caution by a Germany as yet militarily weak. Pressure on Austria however began in the first few months. Unceasing demands were made on the Austrian Government to force members of the satellite Austrian Nazi Party both into the Cabinet and into key posts in the Administration. Austrian Nazis were trained in an Austrian legion organised in Bavaria. Bomb outrages on the railways and at tourist centres, German aeroplanes showering leaflets over Salzburg and Innsbruck, disturbed the daily life of the Republic. The Austrian Chancellor Dollfuss was equally opposed both by Socialist pressure within and external German designs against Austrian independence. Nor was this the only menace to the Austrian State. Following the evil example of their German neighbours, the Austrian Socialists had built up a private army with which to override the decision of the ballot box. Both dangers loomed upon Dollfuss

46

during 1933. The only quarter to which he could turn for protection and whence he had already received assurances of support was Fascist Italy. In August he met Mussolini at Riccione. A close personal and political understanding was reached between them. Dollfuss, who believed that Italy would hold the ring, felt strong enough to move against one set of his opponents — the Austrian Socialists.

In January 1934 Suvich, Mussolini's principal adviser on foreign affairs, visited Vienna as a gesture of warning to Germany; and declared that Italy publicly favoured the independence of Austria. Three weeks later the Dollfuss Government took action against the Socialist organisations of Vienna. The Heimwehr, under Major Fey, belonging to Dollfuss's own party, received orders to disarm the equivalent and equally illegal body controlled by the Austrian Socialists. The latter resisted forcibly, and on February 12 street fighting broke out in the capital. Within a few hours the Socialist forces were broken. This event not only brought Dollfuss closer to Italy but strengthened him in the next stage of his task against the Nazi penetration and conspiracy. On the other hand, many of the defeated Socialists or Communists swung over to the Nazi camp in their bitterness. In Austria as in Germany the Catholic-Socialist feud helped the Nazis.

*   *   *

Until the middle of 1934 the control of events was still largely in the hands of His Majesty's Government without the risk of war. They could at any time, in concert with France and through the agency of the League of Nations, have brought an overwhelming power to bear upon the Hitler Movement, about which Germany was profoundly divided. This would have involved no bloodshed. But this phase was passing. An armed Germany under Nazi control was approaching the threshold. And yet, incredible though it may seem, far into this cardinal year Mr. MacDonald, armed with Mr. Baldwin's political power, continued to work for the disarmament of France. There was indeed a flicker of European unity against the German menace. On February 17, 1934, the British, French, and Italian Governments made a common declaration upon the maintenance of Austrian independence and a month later Italy, Hungary, and Austria signed the so-called Rome Protocols, providing for mutual consultation in the event of a threat to any of the three parties. But Hitler was growing steadily stronger, and in May and June subversive activities increased throughout Austria. Dollfuss immediately sent reports on these terrorist acts to Suvich, with a note deploring their depressive effect upon Austrian trade and tourists.

It was with this dossier in his hand that Mussolini went to Venice on

June 14 to meet Hitler for the first time. The German Chancellor stepped from his aeroplane in a brown mackintosh and Homburg hat into an array of sparkling Fascist uniforms, with a resplendent and portly Duce at their head. As Mussolini caught sight of his guest, he murmured to his aide, *"Non mi piace."* ("I don't like the look of him.") At this strange meeting only a general exchange of ideas took place, with mutual lectures upon the virtues of dictatorship on the German and Italian models. Mussolini was clearly perplexed both by the personality and language of his guest. He summed up his final impression in these words, "A garrulous monk." He did however extract some assurances of relaxation of German pressure upon Dollfuss. Ciano, Mussolini's son-in-law, told the journalists after the meeting, "You'll see. Nothing more will happen."

But the pause in German activities which followed was due not to Mussolini's appeal but to Hitler's own internal preoccupations.

*   *   *

The acquisition of power had opened a deep divergence between the Fuehrer and many of those who had borne him forward. Under Roehm's leadership the S.A. increasingly represented the more revolutionary elements of the party. There were senior members of the party, such as Gregor Strasser, ardent for social revolution, who feared that Hitler in arriving at the first place would simply be taken over by the existing hierarchy, the Reichswehr, the bankers, and the industrialists. He would not have been the first revolutionary leader to kick down the ladder by which he had risen to exalted heights. To the rank and file of the S.A. ("Brownshirts") the triumph of January 1933 was meant to carry with it the freedom to pillage not only the Jews and profiteers but also the well-to-do, established classes of society. Rumours of a great betrayal by their leader soon began to spread in certain circles of the party. Chief-of-Staff Roehm acted on this impulse with energy. In January 1933 the S.A. had been four hundred thousand strong. By the spring of 1934 he had recruited and organised nearly three million men. Hitler in his new situation was uneasy at the growth of this mammoth machine, which, while professing fervent loyalty to his name, and being for the most part deeply attached to him, was beginning to slip from his own personal control. Hitherto he had possessed a private army. Now he had the national army. He did not intend to exchange the one for the other. He wanted both, and to use each, as events required, to control the other. He had now therefore to deal with Roehm. "I am resolved," he declared to the leaders of the S.A. in these days, "to repress severely any attempt to overturn the existing order. I will oppose with the sternest energy a second revolutionary wave, for it would bring with it inevitable chaos. Whoever raises his head against the established authority of the State will be severely treated, whatever his position."

In spite of his misgivings Hitler was not easily convinced of the disloyalty of his comrade of the Munich *Putsch,* who for the last seven years had been the Chief of Staff of his Brownshirt army. When, in December 1933, the unity of the party with the State had been proclaimed Roehm became a member of the German Cabinet. One of the consequences of such a union was to be the merging of the Brownshirts with the Reichswehr. The rapid progress of national rearmament forced the issue of the status and control of all the German armed forces into the forefront of politics. In February 1934 Mr. Eden arrived in Berlin, and in the course of conversation Hitler agreed provisionally to give certain assurances about the non-military character of the S.A. Roehm was already in constant friction with General von Blomberg, the Chief of the General Staff. He now feared the sacrifice of the party army he had taken so many years to build, and, in spite of warnings of the gravity of his conduct, he published on April 18 an unmistakable challenge:

> The Revolution we have made is not a national revolution, but a National *Socialist* Revolution. We would even underline this last word, "Socialist." The only rampart which exists against reaction is represented by our Assault Groups, for they are the absolute incarnation of the revolutionary idea. The militant in the Brown Shirt from the first day pledged himself to the path of revolution, and he will not deviate by a hairbreadth until our ultimate goal has been achieved.

He omitted on this occasion the "Heil Hitler!" which had been the invariable conclusion of Brownshirt harangues.

During the course of April and May, Blomberg continually complained to Hitler about the insolence and activities of the S.A. The Fuehrer had to choose between the generals who hated him and the Brownshirt thugs to whom he owed so much. He chose the generals. At the beginning of June, Hitler, in a five-hour conversation, made a last effort to conciliate and come to terms with Roehm. But with this abnormal fanatic, devoured by ambition, no compromise was possible. The mystic hierarchic Greater Germany of which Hitler dreamed and the Proletarian Republic of the People's Army desired by Roehm were separated by an impassable gulf.

Within the framework of the Brownshirts there had been formed a small and highly trained élite, wearing black uniforms and known as the S.S., or later as Blackshirts. These units were intended for the personal protection of the Fuehrer and for special and confidential tasks. They were commanded by an unsuccessful ex-poultry-farmer, Heinrich Himmler. Foreseeing the impending clash between Hitler and the Army on the one hand and Roehm and the Brownshirts on the other, Himmler took care to carry the S.S. into Hitler's camp. On the other hand, Roehm had supporters of great influence within the party, who, like Gregor Strasser, saw their ferocious plans for Social Revolution being cast aside. The Reichswehr also had its rebels. Ex-Chancellor von Schleicher had never forgiven his disgrace in January 1933 and the failure of the Army chiefs to choose him as successor to Hindenburg. In a clash between Roehm and Hitler, Schleicher saw an opportunity. He was imprudent enough to drop hints to the French Ambassador in Berlin that the fall of Hitler was not far off. This repeated the action he had taken in the case of Bruening. But the times had become more dangerous.

It will long be disputed in Germany whether Hitler was forced to strike by the imminence of a plot by Roehm, or whether he and the generals, fearing what might be coming, resolved on a clean-cut liquidation while they had the power. Hitler's interest and that of the victorious faction was plainly to establish the case for a plot. It is improbable that Roehm and the Brownshirts had actually got as far as this. They were a menacing movement rather than a plot, but at any moment this line might have been crossed. It is certain they were drawing up their forces. It is also certain they were forestalled.

Events now moved rapidly. On June 25 the Reichswehr was confined to barracks, and ammunition was issued to the Blackshirts. On the opposite side the Brownshirts were ordered to stand in readiness, and Roehm with Hitler's consent called a meeting for June 30 of all their senior leaders to meet at Wiessee, in the Bavarian lakes. Hitler received warning of grave danger on the 29th. He flew to Godesberg, where he was joined by Goebbels, who brought alarming news of impending mutiny in Berlin. According to Goebbels, Roehm's adjutant, Karl Ernst, had been given orders to attempt a rising. This seems unlikely. Ernst was actually at Bremen, about to embark from that port on his honeymoon.

On this information, true or false, Hitler took instant decisions. He ordered Goering to take control in Berlin. He boarded his aeroplane for Munich, resolved to arrest his main opponents personally. In this life-or-death climax, as it had now become, he showed himself a terrible personality. Plunged in dark thought, he sat in the co-pilot's seat throughout the journey. The plane landed at an airfield near Munich at four o'clock in the morning of June 30. Hitler had with him, besides Goebbels, about a dozen of his personal bodyguard. He drove to the Brown House in Munich, summoned the leaders of the local S.A. to his presence, and placed them under arrest. At six o'clock, with Goebbels and his small escort only, he motored to Weissee.

Roehm was ill in the summer of 1934 and had gone to Wiessee to take a cure. At seven o'clock the Fuehrer's procession of cars arrived in front of Roehm's chalet. Alone and unarmed, Hitler mounted the stairs and entered Roehm's bedroom. What passed between the two men will never be known. Roehm was taken completely by surprise, and he and his personal staff were arrested without incident. The small party, with its prisoners, now left by road for Munich. It happened that they soon met a column of lorries of armed Brownshirts on their way to acclaim Roehm at the conference convened at Weissee for noon. Hitler stepped out of his car, called for the commanding officer, and, with confident authority, ordered him to take his men home. He was instantly obeyed. If he had been an hour later, or they had been an hour earlier, great events would have taken a different course.

On arrival at Munich, Roehm and his entourage were imprisoned in the same gaol where he and Hitler had been confined together ten years before. That afternoon the executions began. A revolver was placed in Roehm's cell, but as he disdained the invitation the cell door was opened within a few minutes and he was riddled with bullets. All the afternoon the executions proceeded in Munich at brief intervals. The firing parties of eight had to be relieved from time to time on account of the mental stress of the soldiers. But for several hours the recurrent volleys were heard every ten minutes or so.

Meanwhile in Berlin, Goering, having heard from Hitler, followed a similar procedure. But here, in the capital, the killing spread beyond the hierarchy of the S.A. Schleicher and his wife, who threw herself in front of him, were shot in their house. Gregor Strasser was arrested and put to death. Papen's private secretary and immediate circle were also shot; but for some unknown reason he himself was spared. In the Lichterfelde barracks in Berlin Karl Ernst, clawed back from Bremen, met his fate; and here, as in Munich, the volleys of the executioners were heard all day. Throughout Germany, during these twenty-four hours, many men unconnected with the Roehm plot disappeared as the victims of private vengeance, sometimes for very old scores. The total number of persons "liquidated" is variously estimated as between five and seven thousand.

Late in the afternoon of this bloody day Hitler returned by air to Berlin. It was time to put an end to the slaughter, which was spreading every moment. That evening a certain number of the S.S., who through excess of zeal had gone a little far in shooting prisoners, were themselves led out to execution. About one o'clock in the morning of July 1 the sounds of firing ceased. Later in the day the Fuehrer appeared on the balcony of the Chancellery to receive the acclamations of the Berlin crowds, many of whom thought that he himself had been a victim. Some say he looked haggard, others triumphant. He may well have been both. His promptitude and ruthlessness had saved his purpose and no doubt his life. In that "Night of the Long Knives," as it was called, the unity of National-Socialist Germany had been preserved to carry its curse throughout the world.

This massacre, however explicable by the hideous forces at work, showed that the new Master of Germany would stop at nothing, and that conditions in Germany bore no resemblance to those of a civilised state. A dictatorship based upon terror and reeking with blood had confronted the world. Anti-Semitism was ferocious and brazen, and the concentration-camp system was already in full operation for all obnoxious or politically dissident classes. I was deeply affected by the episode, and the whole process of German rearmament, of which there was now overwhelming evidence, seemed to me invested with a ruthless, lurid tinge. It glittered and it glared.

*   *   *

During the early part of July 1934 there was much coming and going over the mountain paths leading from Bavaria into Austrian territory. At the end of the month a German courier fell into the hands of the Austrian frontier police. He carried documents, including cipher keys, which showed that a complete plan of revolt was reaching fruition. The organiser of the *coup d'état* was to be Anton von Rintelen, at that time Austrian Minister to Italy. Dollfuss and his Ministers were slow to respond to the warnings of an impending crisis, and to the signs of imminent revolt which became apparent in the early hours of July 25. The Nazi adherents in Vienna mobilised during the morning. Just before one o'clock in the afternoon a party of armed rebels entered the Chancellery, and Dollfuss, hit by two revolver bullets, was left to bleed slowly to death. Another detachment of Nazis seized the broadcasting station and announced the resignation of the Dollfuss Government and the assumption of office by Rintelen.

But the other members of the Dollfuss Cabinet reacted with firmness and energy. President Dr. Miklas issued a formal command to restore order at all costs. Dr. Schuschnigg assumed the administration. The majority of the Austrian Army and police rallied to his Government, and beseiged the Chancellery building, where, surrounded by a small party of rebels, Dollfuss was dying. The revolt had also broken out in the provinces, and parties from the Austrian legion in Bavaria crossed the frontier. Mussolini had by now heard the news. He telegraphed at once promising Italian support for Austrian independence. Flying specially to Venice, the Duce received the widow of Dr. Dollfuss with every circumstance of sympathy. At the same time three Italian divisions were dispatched to the Brenner Pass. On this Hitler, who knew the limits of his strength, recoiled. The German Minister in Vienna and other German officials implicated in the rising, were recalled or dismissed. The attempt had failed. A longer process was needed. Papen, newly spared from the blood bath, was appointed as German Minister to Vienna, with instructions to work by more subtle means.

Amid these tragedies and alarms the aged Marshal Hindenburg, who had for some months been almost completely senile, and so more than ever a tool of the Reichswehr, expired. Hitler became the head of the German State while retaining the office of Chancellor. He was now the Sovereign of Germany. His bargain with the Reichswehr had been sealed and kept by the blood purge. The Brownshirts had been reduced to obedience and reaffirmed their loyalty to the Fuehrer. All foes and potential rivals had been extirpated from their ranks. Henceforward they lost their influence and became a kind of special constabulary for ceremonial occasions. The Blackshirts, on the other hand, increased in numbers, and strengthened by privileges and discipline, became under Himmler a Praetorian Guard for the person of the Fuehrer, a counterpoise to the Army leaders and military caste, and also political troops to arm with considerable military force the activities of the expanding Secret Police or Gestapo. It was only necessary to invest these powers with the formal sanction of a managed plebiscite to make Hitler's dictatorship absolute and perfect.

*   *   *

Events in Austria drew France and Italy together, and the shock of the Dollfuss assassination led to General Staff contacts. The menace to Austrian independence promoted a revision of Franco-Italian relations, and this had to comprise not only the balance of power in the Mediterranean and North Africa, but the relative positions of France and Italy in Southeastern Europe. But Mussolini was anxious not only to safeguard Italy's position in Europe against the potential German threat but also to secure her imperial future in Africa. Against Germany, close relations with France and Great Britain would be useful; but in the Mediterranean and Africa disagreements with both these Powers might be inevitable. The Duce wondered whether the common need for security felt by Italy, France, and Great Britain might not induce the two former allies of Italy to accept the Italian imperialist programme in Africa. At any rate this seemed a hopeful course for Italian policy.

France, now presided over by M. Doumergue as Premier and M. Barthou as Foreign Minister, had long been anxious to reach formal agreement on security measures in the East. British reluctance to undertake commitments beyond the Rhine, the German refusal to make binding agreements with Poland and Czechoslovakia, the fears of the Little Entente as to Russian intentions, Russian suspicion of the capitalist West, all united to thwart such a programme. In September 1934, however, Louis Barthou determined to go forward. His original plan was to propose an Eastern Pact, grouping together Germany, Russia, Poland, Czechoslovakia, and the Baltic States on the basis of a guarantee by France of the European frontiers of Russia, and by Russia of the eastern borders of Germany. Both Germany and Poland were opposed to an Eastern Pact; but Barthou succeeded in obtaining the entry of Russia into the League of Nations on September 18, 1934. This was an important step. Litvinov, who represented the Soviet Government, was versed in every aspect of foreign affairs. He adapted himself to the atmosphere of the League of Nations and spoke its moral language with so much success that he soon became an outstanding figure.

In her search for allies against the new Germany that had been allowed to grow up, it was natural that France should turn her eyes to Russia and try to re-create the balance of power which had existed before the war. But in October a tragedy occurred. King Alexander of Yugoslavia had been invited to pay an official visit to Paris. He landed at Marseilles, was met by M. Barthou, and drove with him and General Georges through the welcoming crowds who thronged the streets, gay with flags and flowers. Once again from the dark recesses of the Serbian and Croat underworld a hideous murder plot sprang upon the European stage, and, as at Sarajevo in 1914, a band of assassins, ready to give their lives, were at hand. The French police arrangements were loose and casual. A figure darted from the cheering crowds, mounted the running-board of the car, and discharged his automatic pistol into the King and its other occupants, all of whom were stricken. The murderer was immediately cut down and killed by the mounted Republican guardsman behind whom he had slipped. A scene of wild confusion occurred. King Alexander expired almost immediately. General Georges and M. Barthou stepped out of the carriage streaming with blood. The General was too weak to move, but soon received medical aid. The Minister wandered off into the crowd. It was twenty minutes before he was attended to. He had already lost much blood; he was seventy-two, and he died in a few hours. This was a heavy blow to French foreign policy, which under him was beginning to take a coherent form. He was succeeded as Foreign Secretary by Pierre Laval.

Laval's later shameful record and fate must not obscure the fact of his personal force and capacity. He had a clear and intense view. He believed that France must at all costs avoid war, and he hoped to secure this by arrangements with the dictators of Italy and Germany, against whose systems he entertained no prejudice. He distrusted Soviet Russia. Despite his occasional protestations of friendship, he disliked England and thought her a worthless ally. At that time indeed British repute did not stand very high in France. Laval's first object was to reach a definite understanding with Italy, and he deemed the moment ripe. The French Government was obsessed by the German danger, and was prepared to make solid concessions to gain Italy. In January 1935 Laval went to Rome and signed a series of agreements with the object of removing the main obstacles between the two countries. Both Governments were united upon the illegality of German rearmament. They agreed to consult each other in the event of future threats to the independence of Austria. In the colonial sphere France undertook to make administrative concessions about the status of Italians in Tunisia, and handed over to Italy certain tracts of territory on the borders both of Libya and of Somaliland, together with a twenty per cent share in the Jibuti–Addis-Ababa Railway. These conversations were designed to lay the foundations for more formal discussions between France, Italy, and Great Britain about a common front against the growing German menace. Across them all there cut in the ensuing months the fact of Italian aggression in Abyssinia.

In December 1934, a clash took place between Italian and Abyssinian soldiers on the borders of Abyssinia and Italian Somaliland. This was to be the pretext for the ultimate presentation before the world of Italian claims upon the Ethiopian kingdom. Thus the problem of containing Germany in Europe was henceforth confused and distorted by the fate of Abyssinia.

6

Air Parity Lost, 1934–1935

THE GERMAN GENERAL STAFF did not believe that the
German Army could be formed and matured on a scale greater
than that of France, and suitably provided with arsenals and equipment,
before 1943. The German Navy, except for U-boats, could not be rebuilt in its old state under twelve or fifteen years, and in the process
would compete heavily with all other plans. But owing to the unlucky
discovery by an immature civilisation of the internal combustion engine
and the art of flying, a new weapon of national rivalry had leapt upon
the scene capable of altering much more rapidly the relative war power
of States. Granted a share in the ever-accumulating knowledge of mankind and in the march of Science, only four or five years might be required by a nation of the first magnitude, devoting itself to the task, to
create a powerful, and perhaps a supreme, air force. This period would
of course be shortened by any preliminary work and thought.

As in the case of the German Army, the re-creation of the German
air power was long and carefully prepared in secret. As early as 1923
it had been decided that the future German Air Force must be a part of
the German war machine. For the time being the General Staff were
content to build inside the "airforceless army" a well-articulated air-
force skeleton which could not be discerned, or at any rate was not
discerned in its early years, from without. Air power is the most difficult of all forms of military force to measure, or even to express in
precise terms. The extent to which the factories and training-grounds
of civil aviation have acquired a military value and significance at
any given moment cannot easily be judged and still less exactly defined.
The opportunities for concealment, camouflage, and treaty evasion are
numerous and varied. The air, and the air alone, offered Hitler the
chance of a short cut, first to equality and next to predominance, in a

vital military arm over France and Britain. But what would France and Britain do?

By the autumn of 1933 it was plain that neither by precept nor still less by example would the British effort for disarmament succeed. The pacifism of the Labour and Liberal Parties was not affected even by the grave event of the German withdrawal from the League of Nations. Both continued in the name of peace to urge British disarmament, and anyone who differed was called "warmonger" and "scaremonger." It appeared that their feeling was endorsed by the people, who of course did not understand what was unfolding. At a by-election which occurred in East Fulham on October 25 a wave of pacifist emotion increased the Socialist vote by nearly 9000, and the Conservative vote fell by over 10,000. The successful candidate said after the poll that "British people demand . . . that the British Government shall give a lead to the whole world by initiating immediately a policy of general disarmament." And Mr. Lansbury, then Leader of the Labour Party, said that all nations must "disarm to the level of Germany as a preliminary to total disarmament." This election left a deep impression upon Mr. Baldwin, and he referred to it in a remarkable speech three years later. In November came the Reichstag election, at which no candidates except those endorsed by Hitler were tolerated, and the Nazis obtained ninety-five per cent of the votes polled.

It would be wrong in judging the policy of the British Government not to remember the passionate desire for peace which animated the uninformed, misinformed majority of the British people, and seemed to threaten with political extinction any party or politician who dared to take any other line. This, of course, is no excuse for political leaders who fall short of their duty. It is much better for parties or politicians to be turned out of office than to imperil the life of the nation. Moreover, there is no record in our history of any Government asking Parliament and the people for the necessary measures for defence and being refused. Nevertheless, those who scared the timid MacDonald-Baldwin Government from their path should at least keep silent.

The air estimates of March 1934 totalled only twenty millions, and contained provision for four new squadrons, or an increase in our first-line air strength from 850 to 890. The financial cost involved in the first year was £130,000.

On this I said in the House of Commons:

> We are, it is admitted, the fifth air Power only — if that. We are but half the strength of France, our nearest neighbour. Germany is arming fast and no one is going to stop her. That seems quite clear. No one proposes a preventive war to stop Germany

breaking the Treaty of Versailles. She is going to arm; she is doing it; she has been doing it. . . . There is time for us to take the necessary measures, but it is the measures we want. We want the measures to achieve parity. No nation playing the part we play and aspire to play in the world has a right to be in a position where it can be blackmailed. . . .

I called upon Mr. Baldwin, as the man who possessed the power, for action. His was the power, and his the responsibilty.

In the course of his reply Mr. Baldwin said:

If all our efforts for an agreement fail, and if it is not possible to obtain this equality in such matters as I have indicated, then any Government of this country — a National Government more than any, and *this* Government — will see to it that in air strength and air power this country shall no longer be in a position inferior to any country within striking distance of its shores.

Here was a most solemn and definite pledge, given at a time when it could almost certainly have been made good by vigorous action on a large scale. Nevertheless, when on July 20 the Government brought forward some belated and inadequate proposals for strengthening the Royal Air Force by 41 squadrons, or about 820 machines, *only to be completed in five years,* the Labour Party, supported by the Liberals, moved a vote of censure upon them in the House of Commons. Mr. Attlee, as he then was, speaking in their name, said: "We deny the need for increased air armaments . . . We deny the proposition that an increased British Air Force will make for the peace of the world, and we reject altogether the claim to parity." The Liberal Party, supported this censure motion, and their leader, Sir Herbert Samuel, said: "What is the case in regard to Germany? Nothing we have so far seen or heard would suggest that our present air force is not adequate to meet any peril at the present time from this quarter."

When we remember that this was language used after careful deliberation by the responsible heads of parties, the danger of our country becomes apparent. This was the formative time when by extreme exertions we could have preserved the air strength on which our independence of action was founded. If Great Britain and France had each maintained quantitative parity with Germany they would together have been double as strong, and Hitler's career of violence might have been nipped in the bud without the loss of a single life. Thereafter it was too late. We cannot doubt the sincerity of the leaders of the Socialist and Liberal Parties. They were completely wrong and mistaken, and they bear their share of the burden before history. It is indeed astonishing that the Socialist Party should have endeavoured in after years to claim superior foresight and should have reproached their opponents with failing to provide for national safety.

I now enjoyed for once the advantage of being able to urge rearmament in the guise of a defender of the Government. I therefore received an unusually friendly hearing from the Conservative Party.

> I do not suppose there has ever been such a pacifist-minded Government. There is the Prime Minister [Mr. Ramsay Mac-Donald] who in the war proved in the most extreme manner and with very great courage his convictions and the sacrifices he would make for what he believed was the cause of pacifism. The Lord President of the Council [Mr. Baldwin] is chiefly associated in the public mind with the repetition of the prayer, "Give peace in our time." One would have supposed that when Ministers like these come forward and say that they feel it their duty to ask for some small increase in the means they have of guaranteeing the public safety, it would weigh with the Opposition and would be considered as a proof of the reality of the danger from which they seek to protect us. . . . We are a rich and easy prey. No country is so vulnerable, and no country would better repay pillage than our own. . . . *With our enormous metropolis here, the greatest target in the world, a kind of tremendous, fat, valuable cow tied up to attract the beast of prey,* we are in a position in which we have never been before, and in which no other country is at the present time.
>
> *Let us remember this: our weakness does not only involve ourselves; our weakness involves also the stability of Europe.*

I then proceeded to argue that Germany was already approaching air parity with Britain.

> I first assert that Germany has already, in violation of the Treaty, created *a military air force which is now nearly two-thirds as strong as our present home defence air force.* That is the first statement which I put before the Government for their consideration. The second is that Germany is rapidly increasing this air force, not only by large sums of money which figure in her estimates, but also by public subscriptions — very often almost forced subscriptions — which are in progress and have been in progress for some time all over Germany. *By the end of* 1935 *the German air force will be nearly equal in numbers and efficiency to our home defence air force at that date even if the Government's present proposals are carried out.*

The third statement is that if Germany continues this expansion and if we continue to carry out our scheme, then some time in 1936 Germany will be definitely and substantially stronger in the air than Great Britain. Fourthly, and this is the point which is causing anxiety, once they have got that lead we may never be able to overtake them. . . . *If the Government have to admit at any time in the next few years that the German air forces are stronger than our own, then they will be held, and I think rightly held, to have failed in their prime duty to the country. . . .*

The Labour Party's Vote of Censure was of course defeated by a large majority, and I have no doubt that the nation, had it been appealed to with proper preparation on these issues, would equally have sustained the measures necessary for national safety.

*   *   *

It is not possible to tell this story without recording the milestones which we passed on our long journey from security to the jaws of Death. Looking back, I am astonished at the length of time that was granted to us. It would have been possible in 1933, or even in 1934, for Britain to have created an air power which would have imposed the necessary restraints upon Hitler's ambition, or would perhaps have enabled the military leaders of Germany to control his violent acts. More than five whole years had yet to run before we were to be confronted with the supreme ordeal. Had we acted even now with reasonable prudence and healthy energy, it might never have come to pass. Based upon superior air power, Britain and France could safely have invoked the aid of the League of Nations, and all the states of Europe would have gathered behind them. For the first time the League would have had an instrument of authority.

When the winter session opened on November 28, 1934, I moved in the name of some of my friends [1] an amendment to the Address, declaring that "the strength of our national defences and especially of our air defences is no longer adequate to secure the peace, safety, and freedom of Your Majesty's faithful subjects." The House was packed and very ready to listen. After using all the arguments which emphasised the heavy danger to us and to the world, I came to precise facts:

"I assert, first, that Germany already, at this moment, has a military air force . . . and that this . . . is rapidly approaching equality with our own. Secondly . . . the German military air force will this time next year be in fact at least as strong as our own, and it may be even

---

[1] The amendment stood in the names of Mr. Churchill, Sir Robert Horne, Mr. Amery, Captain F. E. Guest, Lord Winterton, and Mr. Boothby.

stronger. Thirdly . . . by the end of 1936, that is, one year further on, and two years from now, the German military air force will be nearly fifty per cent stronger, and in 1937 nearly double."

Mr. Baldwin, who followed me at once, faced this issue squarely, and, on the case made out by his Air Ministry advisers, met me with direct contradiction:

"It is not the case that Germany is rapidly approaching equality with us. . . . Germany is actively engaged in the production of service aircraft, but her real strength is not fifty per cent of our strength in Europe today. As for the position this time next year . . . *so far from the German military air force being at least as strong as, and probably stronger than, our own, we estimate that we shall have a margin in Europe alone of nearly fifty per cent.* I cannot look farther forward than the next two years. Mr. Churchill speaks of what may happen in 1937. Such investigations as I have been able to make lead me to believe that his figures are considerably exaggerated."

This sweeping assurance from the virtual Prime Minister soothed most of the alarmed, and silenced many of the critics. Everyone was glad to learn that my precise statements had been denied upon unimpeachable authority. I was not at all convinced. I believed that Mr. Baldwin was not being told the truth by his advisers, and anyhow that he did not know the facts.

Thus the winter months slipped away, and it was not till the spring that I again had the opportunity of raising the issue. Before doing so I gave full and precise notice to Mr. Baldwin, and when, on March 19, 1935, the air estimates were presented to the House, I reiterated my statement of November, and again directly challenged the assurances which he had then given. A very confident reply was made by the Under-Secretary for Air. However, at the end of March the Foreign Secretary and Mr. Eden paid a visit to Hitler in Germany, and in the course of an important conversation, the text of which is on record, they were told personally by him that the German Air Force had already reached parity with Great Britain. This fact was made public by the Government on April 3. At the beginning of May the Prime Minister wrote an article in his own organ, *The Newsletter,* in which he emphasised the dangers of German rearmament in terms akin to those which I had so often expressed since 1932. He used the revealing word "ambush," which must have sprung from the anxiety of his heart. We had indeed fallen into an ambush. Mr. MacDonald himself opened the debate. After referring to the declared German intention to build a navy beyond the Treaty and submarines in breach of it, he admitted that Hitler claimed to have reached parity with Great Britain in the air. "Whatever may be the exact interpretation of this phrase in terms of air strength, it undoubtedly indicated that the German force has been expanded to a point considerably in excess of the estimates which we were able to place before the House last year. That is a grave fact, with regard to which both the Government and the Air Ministry have taken immediate notice."

When in due course I was called I said:

"Even now we are not taking the measures which would be in true proportion to our needs. The Government have proposed these increases. They must face the storm. They will have to encounter every form of unfair attack. Their motives will be misrepresented. They will be calumniated and called warmongers. Every kind of attack will be made upon them by many powerful, numerous, and extremely vocal forces in this country. They are going to get it anyway. Why, then, not fight for something that will give us safety? Why, then, not insist that the provision for the Air Force should be adequate, and then, however severe may be the censure and however strident the abuse which they have to face, at any rate there will be this satisfactory result — that His Majesty's Government will be able to feel that in this, of all matters the prime responsibility of a Government, they have done their duty."

Although the House listened to me with close attention, I felt a sensation of despair. To be so entirely convinced and vindicated in a matter of life and death to one's country, and not to be able to make Parliament and the nation heed the warning, or bow to the proof by taking action, was an experience most painful.

It was not until May 22, 1935, that Mr. Baldwin made his celebrated confession. I am forced to cite it:

> First of all, with regard to the figure I gave in November of German aeroplanes, nothing has come to my knowledge since that makes me think that figure was wrong. I believed at that time it was right. *Where I was wrong was in my estimate of the future. There I was completely wrong. We were completely misled on that subject.* . . .
>
> I would repeat here that there is no occasion, in my view, in what we are doing, for panic. But I will say this deliberately, with all the knowledge I have of the situation, that I would not remain for one moment in any Government which took less determined steps than we are taking today. I think it is only due to say that there has been a great deal of criticism, both in the press and verbally, about the Air Ministry, as though they were responsible for possibly an inadequate programme, for not having gone ahead faster, and for many other things. I only want to repeat that what-

ever responsibility there may be — and we are perfectly ready to meet criticism — *that responsibility is not that of any single Minister, it is the responsibility of the Government as a whole, and we are all responsible and we are all to blame.*

I hoped that this shocking confession would be a decisive event, and that at the least a parliamentary committee of all parties would be set up to report upon the facts and upon our safety. The House of Commons had a different reaction. The Labour and Liberal Oppositions, having nine months earlier moved or supported a vote of censure even upon the modest steps the Government had taken, were ineffectual and undecided. They were looking forward to an election against "Tory armaments." Neither the Labour nor the Liberal spokesmen had prepared themselves for Mr. Baldwin's disclosures and admission, and they did not attempt to adapt their speeches to this outstanding episode. Nothing they said was in the slightest degree related to the emergency in which they admitted we stood, or to the far graver facts which we now know lay behind it.

The Government majority for their part appeared captivated by Mr. Baldwin's candour. His admission of having been utterly wrong, with all his sources of knowledge, upon a vital matter for which he was responsible was held to be redeemed by the frankness with which he declared his error and shouldered the blame. There was even a strange wave of enthusiasm for a Minister who did not hesitate to say that he was wrong. Indeed, many Conservative Members semed angry with me for having brought their trusted leader to a plight from which only his native manliness and honesty had extricated him; but not, alas, his country.

A disaster of the first magnitude had fallen upon us. Hitler had already obtained parity with Great Britain. Henceforward he had merely to drive his factories and training schools at full speed not only to keep his lead in the air but steadily to improve it. Henceforward all the unknown, immeasurable threats which overhung London from air attack would be a definite and compelling factor in all our decisions. Moreover, we could never catch up; or at any rate the Government never did catch up. Credit is due to them and to the Air Ministry for the high efficiency of the Royal Air Force. But the pledge that air parity would be maintained was irretrievably broken. It is true that the immediate further expansion of the German Air Force did not proceed at the same rate as in the period when they gained parity. No doubt a supreme effort had been made by them to achieve at a bound this commanding position and to assist and expoit it in their diplomacy. It gave Hitler the foundation for the successive acts of aggression which he had planned and which were now soon to take place. Very considerable efforts were made by the British Government in the next four years. The first prototypes of the ever-famous Hurricane and Spitfire fighters flew in November 1935 and March 1936 respectively. Immediate large-scale production was ordered, and they were ready in some numbers none too soon. There is no doubt that we excelled in air quality; but quantity was henceforth beyond us. The outbreak of the war found us with barely half the German numbers.

7

Challenge and Response, 1935

THE YEARS of underground burrowings, of secret or disguised preparations, were now over, and Hitler at length felt himself strong enough to make his first open challenge. On March 9, 1935, the official constitution of the German Air Force was announced, and on the 16th it was declared that the German Army would henceforth be based on national compulsory service. The laws to implement these decisions were soon promulgated, and action had already begun in anticipation. The French Government, who were well informed of what was coming, had actually declared the consequential extension of their own military service to two years a few hours earlier on the same momentous day. The German action was an open formal affront to the treaties of peace upon which the League of Nations was founded. As long as the breaches had taken the form of evasions or calling things by other names, it was easy for the responsible victorious Powers, obsessed by pacifism and preoccupied with domestic politics, to avoid the responsibility of declaring that the Peace Treaty was being broken or repudiated. Now the issue came with blunt and brutal force. Almost on the same day the Ethiopian Government appealed to the League of Nations against the threatening demands of Italy. When, on March 24, against this background, Sir John Simon with the Lord Privy Seal, Mr. Eden, visited Berlin at Hitler's invitation, the French Government thought the occasion ill-chosen. They had now themselves at once to face, not the reduction of their Army, so eagerly pressed upon them by Mr. MacDonald the year before, but the extension of compulsory military service from one year to two. In the prevailing state of public opinion this was a heavy task. Not only the Communists but the Socialists had voted against the measure. When M. Léon Blum said, "The workers of France will rise to resist Hitlerite aggression,"

Thorez replied, amid the applause of his Soviet-bound faction, "We will not tolerate the working classes being drawn into a so-called war in defence of democracy against fascism."

The United States had washed their hands of all concern with Europe, apart from wishing well to everybody, and were sure they would never have to be bothered with it again. But France, Great Britain, and also — decidedly — Italy, in spite of their discordances, felt bound to challenge this definite act of treaty-violation by Hitler. A conference of the former principal Allies was summoned under the League of Nations at Stresa, and all these matters were brought to debate.

There was general agreement that open violation of solemn treaties, for the making of which millions of men had died, could not be borne. But the British representatives made it clear at the outset that they would not consider the possibility of sanctions in the event of treaty-violation. This naturally confined the Conference to the region of words. A resolution was passed unanimously to the effect that "unilateral" — by which they meant one-sided — breaches of treaties could not be accepted, and the Executive Council of the League of Nations was invited to pronounce upon the situation disclosed. On the second afternoon of the Conference Mussolini strongly supported this action, and was outspoken against aggression by one Power upon another. The final declaration was as follows:

> The three Powers, the object of whose policy is the collective maintenance of peace within the framework of the League of Nations, find themselves in complete agreement in opposing, by all practicable means, any unilateral repudiation of treaties which may endanger the peace of Europe, and will act in close and cordial collaboration for this purpose.

The Italian Dictator in his speech had stressed the words *"peace of Europe,"* and paused after "Europe" in a noticeable manner. This emphasis on Europe at once struck the attention of the British Foreign Office representatives. They pricked up their ears, and well understood that while Mussolini would work with France and Britain to prevent Germany from rearming he reserved for himself any excursion in Africa against Abyssinia on which he might later resolve. Should this point be raised or not? Discussions were held that night among the Foreign Office officials. Everyone was so anxious for Mussolini's support in dealing with Germany that it was felt undesirable at that moment to warn him off Abyssinia, which would obviously have very much annoyed him. Therefore the question was not raised; it passed by default, and Mussolini felt, and in a sense had reason to feel, that the Allies had acquiesced in his statement and would give him a free hand against Abyssinia. The French remained mute on the point, and the Conference separated.

In due course, on April 15–17, the Council of the League of Nations examined the alleged breach of the Treaty of Versailles committed by Germany in decreeing universal compulsory military service. The following Powers were represented on the Council: the Argentine Republic, Australia, Great Britain, Chile, Czechoslovakia, Denmark, France, Italy, Mexico, Poland, Portugal, Spain, Turkey, and the U.S.S.R. All voted for the principle that treaties should not be broken by "unilateral" action, and referred the issue to the Plenary Assembly of the League. At the same time the Foreign Ministers of the three Scandinavian countries, Sweden, Norway, and Denmark, being deeply concerned about the naval balance in the Baltic, also met together in general support. In all, nineteen countries formally protested. But how vain was all their voting without the readiness of any single Power or any group of Powers to contemplate the use of *force* even in the last resort!

*   *   *

Laval was not disposed to approach Russia in the firm spirit of Barthou. But in France there was now an urgent need. It seemed, above all, necessary to those concerned with the life of France to obtain national unity on the two years' military service which had been approved by a narrow majority in March. Only the Soviet Government could give permission to the important section of Frenchmen whose allegiance they commanded. Besides this, there was a general desire in France for a revival of the old alliance of 1895, or something like it. On May 2, 1935, the French Government put their signature to a Franco-Soviet pact. This was a nebulous document guaranteeing mutual assistance in the face of aggression over a period of five years.

To obtain tangible results in the French political field Laval now went on a three days' visit to Moscow, where he was welcomed by Stalin. There were lengthy discussions, of which a fragment not hitherto published may be recorded. Stalin and Molotov were of course anxious to know above all else what was to be the strength of the French Army on the Western Front: how many divisions? what period of service? After this field had been explored Laval said: "Can't you do something to encourage religion and the Catholics in Russia? It would help me so much with the Pope." "Oho!" said Stalin. "The Pope! How many divisions has *he* got?" Laval's answer was not reported to me; but he might certainly have mentioned a number of legions not always visible on parade. Laval had never intended to commit France to any of the specific obligations which it is the habit of the Soviet to demand. Nevertheless he obtained a public declaration from Stalin on May 15 approving the policy of national defence carried out by France in order to maintain her armed forces at the level of security. On these instructions the French Communists immediately turned about and gave vociferous support to the defence programme and the two years' service. As a factor in European security the Franco-Soviet Pact, which contained no engagements binding on either party in the event of German aggression, had only limited advantages. No real confederacy was achieved with Russia. Moreover, on his return journey the French Foreign Minister stopped at Cracow to attend the funeral of Marshal Pilsudski. Here he met Goering, with whom he talked with much cordiality. His expressions of distrust and dislike of the Soviets were duly reported through German channels to Moscow.

*   *   *

Mr. MacDonald's health and capacity had now declined to a point which made his continuance as Prime Minister impossible. He had never been popular with the Conservative Party, who regarded him, on account of his political and war records and Socialist faith, with long-bred prejudice, softened in later years by pity. No man was more hated, or with better reason, by the Labour-Socialist Party, which he had so largely created and then laid low by what they viewed as his treacherous desertion in 1931. In the massive majority of the Government he had but seven party followers. The disarmament policy to which he had given his utmost personal efforts had now proved a disastrous failure. A general election could not be far distant, in which he could play no helpful part. In these circumstances there was no surprise when, on June 7, it was announced that he and Mr. Baldwin had changed places and offices and that Mr. Baldwin had become Prime Minister for the third time. The Foreign Office also passed to another hand. Sir Samuel Hoare's labours at the India Office had been crowned by the passing of the Government of India Bill, and he was now free to turn to a more immediately important sphere. For some time past Sir John Simon had been bitterly attacked for his foreign policy by influential Conservatives closely associated with the Government. He now moved to the Home Office, with which he was well acquainted, and Sir Samuel Hoare became Secretary of State for Foreign Affairs.

At the same time Mr. Baldwin adopted a novel expedient. He appointed Mr. Anthony Eden to be Minister for League of Nations Affairs. Eden had for nearly ten years devoted himself almost entirely to the study of foreign affairs. Taken from Eton at eighteen to the World War, he had served for four years with distinction in the 60th Rifles through many of the bloodiest battles, and risen to the position of Brigade-Major, with the Military Cross. He was to work in the Foreign Office with equal status to the Foreign Secretary and with full access to the dispatches and the departmental staff. Mr. Baldwin's object was no doubt to conciliate the strong tide of public opinion associated with the League of Nations Union by showing the importance which he attached to the League and to the conduct of our affairs at Geneva. When about a month later I had the opportunity of commenting on what I described as "the new plan of having two equal Foreign Secretaries," I drew attention to its obvious defects.

While men and matters were in this posture a most surprising act was committed by the British Government. Some at least of its impulse came from the Admiralty. It is always dangerous for soldiers, sailors, or airmen to play at politics. They enter a sphere in which the values are quite different from those to which they have hitherto been accustomed. Of course they were following the inclination or even the direction of the First Lord and the Cabinet, who alone bore the responsibility. But there was a strong favourable Admiralty breeze. There had been for some time conversations between the British and German Admiralties about the proportions of the two navies. By the Treaty of Versailles the Germans were not entitled to build more than six armoured ships of 10,000 tons, in addition to six light cruisers not exceeding 6000 tons. The British Admiralty had recently found out that the last two pocket battleships being constructed, the *Scharnhorst* and the *Gneisenau*, were of a far larger size than the Treaty allowed, and of a quite different type. In fact, they turned out to be 26,000-ton light battle cruisers, or commerce destroyers of the highest class, and were to play a significant part in the Second World War.

In the face of this brazen and fraudulent violation of the Peace Treaty, carefully planned and begun at least two years earlier (1933), the Admiralty actually thought it was worth while making an Anglo-German naval agreement. His Majesty's Government did this without consulting their French ally or informing the League of Nations. At the very time when they themselves were appealing to the League and enlisting the support of its members to protest against Hitler's violation of the military clauses of the Treaty they proceeded by a private agreement to sweep away the naval clauses of the same treaty.

The main feature of the agreement was that the German Navy should not exceed one-third of the British. This greatly attracted the Admiralty, who looked back to the days before the First World War when we had been content with a ratio of sixteen to ten. For the sake of that prospect, taking German assurances at their face value, they proceeded to concede to Germany the right to build U-boats, explicitly denied to her in the Peace Treaty. Germany might build sixty per cent of the British submarine strength, and if she decided that the circumstances were exceptional she might build to one hundred per cent. The Germans, of course, gave assurances that their U-boats would never be used against merchant ships. Why, then, were they needed? For, clearly, if the rest of the agreement was kept, they could not influence the naval decision, so far as warships were concerned.

The limitation of the German Fleet to a third of the British allowed Germany a programme of new construction which would set her yards to work at maximum activity for at least ten years. There was therefore no practical limitation or restraint of any kind imposed upon German naval expansion. They could build as fast as was physically possible. The quota of ships assigned to Germany by the British project was, in fact, far more lavish than Germany found it expedient to use, having regard partly no doubt to the competition for armour plate arising between warship and tank construction. Hitler, as we now know, informed Admiral Raeder that war with England would not be likely till 1944–45. The development of the German Navy was therefore planned on a long-term basis. In U-boats alone did they build to the full paper limits allowed. As soon as they were able to pass the sixty per cent limit they invoked the provision allowing them to build to one hundred per cent, and fifty-seven were actually constructed when war began.

In the design of new battleships the Germans had the further advantage of not being parties to the provisions of the Washington Naval Agreement or the London Conference. They immediately laid down the *Bismarck* and the *Tirpitz*, and, while Britain, France, and the United States were all bound by the 35,000-tons limitation, these two great vessels were being designed with a displacement of over 45,000 tons, which made them, when completed, certainly the strongest vessels afloat in the world.

It was also at this moment a great diplomatic advantage to Hitler to divide the Allies, to have one of them ready to condone breaches of the Treaty of Versailles, and to invest the regaining of full freedom to rearm with the sanction of agreement with Britain. The effect of the announcement was another blow at the League of Nations. The French had every right to complain that their vital interests were affected by the permission accorded by Great Britain for the building of U-boats. Mussolini saw in this episode evidence that Great Britain was not acting in good faith with her other allies, and that, so long as her special naval interests were secured, she would apparently go to any length in accommodation with Germany, regardless of the detriment to friendly Powers menaced by the growth of the German land forces. He was encouraged by what seemed the cynical and selfish attitude of Great Britain to press on with his plans against Abyssinia. The Scandinavian Powers, who only a fortnight before had courageously sustained the protest against Hitler's introduction of compulsory service in the German Army, now found that Great Britain had behind the scenes agreed to a German Navy which, though only a third of the British, would within this limit be master of the Baltic.

Great play was made by British Ministers with a German offer to co-operate with us in abolishing the submarine. Considering that the condition attached to it was that all other countries should agree at the same time, and that it was well known there was not the slightest chance of other countries agreeing, this was a very safe offer for the Germans to make. This also applied to the German agreement to restrict the use of submarines so as to strip submarine warfare against commerce of inhumanity. Who could suppose that the Germans, possessing a great fleet of U-boats and watching their women and children being starved by a British blockade, would abstain from the fullest use of that arm? I described this view as "the acme of gullibility."

Far from being a step towards disarmament, the agreement, had it been carried out over a period of years, would inevitably have provoked a world-wide development of new warship building. The French Navy, except for its latest vessels, would require reconstruction. This again would react upon Italy. For ourselves, it was evident that we should have to rebuild the British Fleet on a very large scale in order to maintain our three-to-one superiority in modern ships. It may be that the idea of the German Navy being one-third of the British also presented itself to our Admiralty as the British Navy being three times the German. This perhaps might clear the path to a reasonable and overdue rebuilding of our Fleet. But where were the statesmen?

This agreement was announced to Parliament by the First Lord of the Admiralty on June 21, 1935. On the first opportunity, I condemned it: What had in fact been done was to authorise Germany to build to her utmost capacity for five or six years to come.

*   *   *

Meanwhile in the military sphere the formal establishment of conscription in Germany on March 16, 1935, marked the fundamental challenge to Versailles. But the steps by which the German Army was now magnified and reorganised are not of technical interest only. The name Reichswehr was changed to that of Wehrmacht. The Army was to be subordinated to the supreme leadership of the Fuehrer. Every soldier took the oath, not, as formerly, to the Constitution, but to the person of Adolf Hitler. The War Ministry was directly subordinated to the orders of the Fuehrer. A new kind of formation was planned — the Armoured, or "Panzer," Division, of which three were soon in being. Detailed arrangements were also made for the regimentation of German youth. Starting in the ranks of the Hitler Youth, the boyhood of Germany passed at the age of eighteen on a voluntary basis into the S.A. for two years. Service in the work battalions or Arbeitsdienst became a compulsory duty on every male German reaching the age of twenty. For six months he would have to serve his country, constructing roads, building barracks, or draining marshes, thus fitting him physically and morally for the crowning duty of a German citizen, service with the armed forces. In the work battalions the emphasis lay upon the abolition of class and the stressing of the social unity of the German people; in the Army it was put upon discipline and the territorial unity of the nation.

The gigantic task of training the new body and of expanding the cadres now began. On October 15, 1935, again in defiance of the clauses of Versailles, the German Staff College was reopened with formal ceremony by Hitler, accompanied by the chiefs of the armed services. Here was the apex of the pyramid, whose base was now already constituted by the myriad formations of the work battalions. On November 7, the first class, born in 1914, was called up for service: 596,-000 young men to be trained in the profession of arms. Thus at one stroke, on paper at least, the German Army was raised to nearly 700,000 effectives.

It was realised that after the first call-up of the 1914 class, in Germany as in France, the succeeding years would bring a diminishing number of recruits, owing to the decline in births during the period of the World War. Therefore in August 1936 the period of active military service in Germany was raised to two years. The 1915 class numbered 464,000, and with the retention of the 1914 class for another year the number of Germans under regular military training in 1936 was 1,511,000 men. The effective strength of the French Army, apart from reserves, in the same year was 623,000 men, of whom only 407,000 were in France.

The following figures, which actuaries could foresee with some precision, tell their tale:

TABLE OF THE COMPARATIVE FRENCH AND GERMAN FIGURES FOR THE
CLASSES BORN FROM 1914 TO 1920, AND CALLED UP FROM 1934 TO 1940

Class				German		French	
1914	..	..	..	596,000 men	..	279,000 men	
1915	..	..	..	464,000 "	..	184,000 "	
1916	..	..	..	351,000 "	..	165,000 "	
1917	..	..	..	314,000 "	..	171,000 "	
1918	..	..	..	326,000 "	..	197,000 "	
1919	..	..	..	485,000 "	..	218,000 "	
1920	..	..	..	636,000 "	..	360,000 "	
				3,172,000 men		1,574,000 men	

Until these figures became facts as the years unfolded, they were still but warning shadows. All that was done up to 1935 fell far short of the strength and power of the French Army and its vast reserves, apart from its numerous and vigorous allies. Even at this time a resolute decision upon the authority, which could easily have been obtained, of the League of Nations might have arrested the whole process. Germany could either have been brought to the bar at Geneva and invited to give a full explanation and allow inter-Allied missions of inquiry to examine the state of her armaments and military formations in breach of the Treaty, or, in the event of refusal, the Rhine bridgeheads could have been reoccupied until compliance with the Treaty had been secured, without there being any possibility of effective resistance or much likelihood of bloodshed. In this way the Second World War could have been at least delayed indefinitely. Many of the facts and their whole general tendency were well known to the French and British Staffs, and were to a lesser extent realised by the Governments. The French Government, which was in ceaseless flux in the fascinating game of party politics, and the British Government, which arrived at the same vices by the opposite process of general agreement to keep things quiet, were equally incapable of any drastic or clear-cut action, however justifiable both by treaty and by common prudence.

8

Sanctions Against Italy, 1935

WORLD PEACE now suffered its second heavy stroke. The loss by Britain of air parity was followed by the transference of Italy to the German side. The two events combined enabled Hitler to advance along his predetermined deadly course. We have seen how helpful Mussolini had been in the protection of Austrian independence, with all that it implied in Central and Southeastern Europe. Now he was to march over to the opposite camp. Nazi Germany was no longer to be alone. One of the principal Allies of the First World War would soon join her. The gravity of this downward turn in the balance of safety oppressed my mind.

Mussolini's designs upon Abyssinia were unsuited to the ethics of the twentieth century. They belonged to those dark ages when white men felt themselves entitled to conquer yellow, brown, black, or red men, and subjugate them by their superior strength and weapons. In our enlightened days, when crimes and cruelties have been committed from which the savages of former times would have recoiled, or of which they would at least have been incapable, such conduct was at once obsolete and reprehensible. Moreover, Abyssinia was a member of the League of Nations. By a curious inversion it was Italy who had in 1923 pressed for her inclusion, and Britain who had opposed it. The British view was that the character of the Ethiopian Government and the conditions prevailing in that wild land of tyranny, slavery, and tribal war were not consonant with membership of the League. But the Italians had had their way, and Abyssinia was a member of the League, with all its rights and such securities as it could offer. Here indeed was a testing case for the instrument of world government upon which the hopes of all good men were founded.

The Italian Dictator was not actuated solely by desire for territorial

74

gains. His rule, his safety, depended upon prestige. The humiliating defeat which Italy had suffered forty years before at Adowa, and the mockery of the world when an Italian army had not only been destroyed or captured but shamefully mutilated, rankled in the minds of all Italians. They had seen how Britain had after the passage of years avenged both Khartoum and Majuba. To proclaim their manhood by avenging Adowa meant almost as much in Italy as the recovery of Alsace-Lorraine in France. There seemed no way in which Mussolini could more easily or at less risk and cost consolidate his own power or, as he saw it, raise the authority of Italy in Europe than by wiping out the stain of bygone years and adding Abyssinia to the recently built Italian Empire. All such thoughts were wrong and evil, but since it is always wise to try to understand another country's point of view they may be recorded.

In the fearful struggle against rearming Nazi Germany which I could feel approaching with inexorable strides, I was most reluctant to see Italy estranged, and even driven into the opposite camp. There was no doubt that the attack by one member of the League of Nations upon another at this juncture, if not resented, would be finally destructive of the League as a factor for welding together the forces which could alone control the might of resurgent Germany and the awful Hitler menace. More could perhaps be got out of the vindicated majesty of the League than Italy could ever give, withhold, or transfer. If therefore the League were prepared to use the united strength of all its members to curb Mussolini's policy, it was our bounden duty to take our share and play a faithful part. There seemed in all the circumstances no obligation upon Britain to take the lead herself. She had a duty to take account of her own weakness caused by the loss of air parity, and even more of the military position of France, in the face of German rearmament. One thing was clear and certain. Half-measures were useless for the League, and pernicious to Britain if she assumed its leadership. If we thought it right and necessary for the law and welfare of Europe to quarrel mortally with Mussolini's Italy, we must also strike him down. The fall of the lesser dictator might combine and bring into action all the forces — and they were still overwhelming — which would enable us to restrain the greater dictator, and thus prevent a second German war.

These general reflections are a prelude to the narrative of this chapter.

*   *   *

Ever since the Stresa Conference, Mussolini's preparations for the conquest of Abyssinia had been apparent. It was evident that British opinion would be hostile to such an act of Italian aggression. Those of us who saw in Hitler's Germany a danger not only to peace but to survival dreaded this movement of a first-class Power, as Italy was then rated, from our side to the other. I remember a dinner at which Sir Robert Vansittart and Mr. Duff Cooper, then only an under-secretary, were present, at which this adverse change in the balance of Europe was clearly foreseen. The project was mooted of some of us going out to see Mussolini in order to explain to him the inevitable results which would be produced in Great Britain. Nothing came of this; nor would it have been of any good. Mussolini, like Hitler, regarded Britannia as a frightened, flabby old woman, who at the worst would only bluster, and was anyhow incapable of making war. Lord Lloyd, who was on friendly terms with him, noted how he had been struck by the Joad Resolution of the Oxford undergraduates in 1933 refusing to "fight for King and Country."

In August the Foreign Secretary invited me and also the Opposition Party leaders to visit him separately at the Foreign Office, and the fact of these consultations was made public by the Government. Sir Samuel Hoare told me of this growing anxiety about Italian aggression against Abyssinia, and asked me how far I should be prepared to go against it. Wishing to know more about the internal and personal situation at the Foreign Office under diarchy before replying, I asked about Eden's view. "I will get him to come," said Hoare, and in a few minutes Anthony arrived, smiling and in the best of tempers. We had an easy talk. I said I thought the Foreign Secretary was *justified in going as far with the League of Nations against Italy as he could carry France;* but I added that he ought not to put any pressure upon France, because of her military convention with Italy and her German preoccupation; and that in the circumstances I did not expect France would go very far. Generally I strongly advised the Ministers not to try to take a leading part or to put themselves forward too prominently. In this I was of course oppressed by my German fears and the conditions to which our defences had been reduced.

As the summer of 1935 drew on, the movement of Italian troopships through the Suez Canal was continuous, and considerable forces and supplies were assembled along the eastern Abyssinian frontier. Suddenly an extraordinary and to me, after my talks at the Foreign Office, a quite unexpected event occurred. On August 24 the Cabinet resolved and declared that Britain would uphold its obligation under its treaties and under the Covenant of the League. Mr. Eden, Minister for League of Nations Affairs and almost co-equal of the Foreign Secretary, had already been for some weeks at Geneva, where he had rallied the Assembly to a policy of "Sanctions" against Italy if she invaded Abyssinia. The peculiar office to which he had been appointed made him by its very nature concentrate upon the Abyssinian question with an emphasis which outweighed other aspects. "Sanctions" meant the cutting off from Italy of all financial aid and of economic supplies, and the giving of all such assistance to Abyssinia. To a country like Italy, dependent for so many commodities needed in war upon unhampered imports from overseas, this was indeed a formidable deterrent. Eden's zeal and address and the principles which he proclaimed dominated the Assembly. On September 11 the Foreign Secretary, Sir Samuel Hoare, having arrived at Geneva, himself addressed them:

> I will begin by reaffirming the support of the League by the Government I represent and the interest of the British people in collective security. . . . The ideas enshrined in the Covenant and in particular the aspiration to establish the rule of law in international affairs have become part of our national conscience. It is to the principles of the League and not to any particular manifestation that the British nation has demonstrated its adherence. Any other view is at once an underestimation of our good faith and an imputation upon our sincerity. In conformity with its precise and explicit obligations the League stands, and my country stands with it, for the collective maintenance of the Covenant in its entirety, and particularly for steady and collective resistance to all acts of unprovoked aggression.

In spite of my anxieties about Germany, and little as I liked the way our affairs were handled, I remember being stirred by this speech when I read it in Riviera sunshine. It aroused everyone, and reverberated throughout the United States. It united all those forces in Britain which stood for a fearless combination of righteousness and strength. Here at least was a policy. If only the orator had realised what tremendous powers he held unleashed in his hand at that moment he might indeed for a while have led the world.

These declarations gathered their validity from the fact that they had behind them, like many causes which in the past have proved vital to human progress and freedom, the British Navy. For the first and the last time the League of Nations seemed to have at its disposal a secular arm. Here was the international police force upon the ultimate authority of which all kinds of diplomatic and economic pressures and persuasion could be employed. When on September 12, the very next day, the battle cruisers *Hood* and *Renown,* accompanied by the Second Cruiser Squadron and a destroyer flotilla, arrived at Gibraltar, it was assumed on all sides that Britain would back her words with deeds. Policy and action alike gained immediate and overwhelming support at home. It was taken for granted, not unnaturally, that neither the declaration nor the movement of warships would have been made without careful expert calculation by the Admiralty of the fleet or fleets required in the Mediterranean to make our undertakings good.

At the end of September I had to make a speech at the City Carlton Club, an orthodox body of some influence. I tried to convey a warning to Mussolini, which I believe he read, but in October, undeterred by belated British naval movements, he launched the Italian armies upon the invasion of Abyssinia. On the 10th, by the votes of fifty sovereign states to one, the Assembly of the League resolved to take collective measures against Italy, and a Committee of Eighteen was appointed to make further efforts for a peaceful solution. Mussolini, thus confronted, made a clear-cut statement marked by deep shrewdness. Instead of saying, "Italy will meet sanctions with war," he said, "Italy will meet them with discipline, with frugality, and with sacrifice." At the same time however he intimated that *he would not tolerate the imposition of any sanctions which hampered his invasion of Abyssinia.* If that enterprise were endangered he would go to war with whoever stood in his path. "Fifty nations!" he said. "Fifty nations, led by one!" Such was the position in the weeks which preceded the dissolution of Parliament in Britain and the general election, which was now constitutionally due.

*   *   *

Bloodshed in Abyssinia, hatred of Fascism, the invocation of sanctions by the League, produced a convulsion within the British Labour Party. Trade unionists, among whom Mr. Ernest Bevin was outstanding, were by no means pacifist by temperament. A very strong desire to fight the Italian Dictator, to enforce sanctions of a decisive character, and to use the British Fleet, if need be, surged through the sturdy wage earners. Rough and harsh words were spoken at excited meetings. On one occasion Mr. Bevin complained that "he was tired of having George Lansbury's conscience carted about from conference to conference." Many members of the Parliamentary Labour Party shared the trade union mood. In a far wider sphere, all the leaders of the League of Nations Union felt themselves bound to the cause of the League. Here were principles in obedience to which lifelong humanitarians were ready to die, and if to die, also to kill. On October 8 Mr. Lansbury resigned his leadership of the Labour Parliamentary Party, and Major Attlee, who had a fine war record, reigned in his stead.

But this national awakening was not in accord with Mr. Baldwin's outlook or intentions. It was not till several months after the election that I began to understand the principles upon which "sanctions" were founded. The Prime Minister had declared that sanctions meant war; secondly, he was resolved there must be no war; and, thirdly, he decided upon sanctions. It was evidently impossible to reconcile these three conditions. Under the guidance of Britain and the pressures of Laval the League of Nations Committee, charged with devising sanctions, kept clear of any that would provoke war. A large number of commodities, some of which were war materials, were prohibited from entering Italy, and an imposing schedule was drawn up. But oil, without which the campaign in Abyssinia could not have been maintained, continued to enter freely, because it was understood that to stop it meant war. Here the attitude of the United States, not a member of the League of Nations, and the world's main oil supplier, though benevolent, was uncertain. Moreover, to stop it to Italy involved also stopping it to Germany. The export of aluminium to Italy was strictly forbidden; but aluminium was almost the only metal that Italy produced in quantities beyond her own needs. The importation of scrap iron and iron ore into Italy was sternly vetoed in the name of public justice. But as the Italian metallurgical industry made but little use of them, and as steel billets and pig iron were not interfered with, Italy suffered no hindrance. Thus the measures pressed with so great a parade were not real sanctions to paralyse the aggressor, but merely such halfhearted sanctions as the aggressor would tolerate, because in fact, though onerous, they stimulated Italian war spirit. The League of Nations therefore proceeded to the rescue of Abyssinia on the basis that nothing must be done to hamper the invading Italian armies. These facts were not known to the British public at the time of the election. They earnestly supported the policy of sanctions, and believed that this was a sure way of bringing the Italian assault upon Abyssinia to an end.

Still less did His Majesty's Government contemplate the use of the Fleet. All kinds of tales were told of Italian suicide squadrons of dive-bombers which would hurl themselves upon the decks of our ships and blow them to pieces. The British Fleet which was lying at Alexandria had now been reinforced. It could by a gesture have turned back Italian transports from the Suez Canal, and would as a consequence have had to offer battle to the Italian Navy. We were told that it was not capable of meeting such an antagonist. I had raised the question at the outset, but had been reassured. Our battleships of course were old, and it now appeared that we had no aircraft cover and very little anti-aircraft ammunition. It transpired however that the Admiral commanding resented the suggestion attributed to him that he was not strong enough to fight a fleet action. It would seem that before taking their first decision to oppose the Italian aggression His Majesty's Government should carefully have examined ways and means, and also made up their minds.

There is no doubt, on our present knowledge, that a bold decision would have cut the Italian communications with Ethiopia, and that we should have been successful in any naval battle which might have followed. I was never in favour of isolated action by Great Britain, but having gone so far it was a grievous deed to recoil. Moreover, Mussolini would never have dared to come to grips with a resolute British Government. Nearly the whole of the world was against him, and he would have had to risk his régime upon a singlehanded war with Britain in which a fleet action in the Mediterranean would be the early and decisive test. How could Italy have fought this war? Apart from a limited advantage in modern light cruisers, her navy was but a fourth the size of the British. Her numerous conscript army, which was vaunted in millions, could not come into action. Her air power was in quantity and quality far below even our modest establishments. She would instantly have been blockaded. The Italian armies in Abyssinia would have famished for supplies and ammunition. Germany could as yet give no effective help. If ever there was an opportunity of striking a decisive blow in a generous cause with the minimum of risk, it was here and now. The fact that the nerve of the British Government was not equal to the occasion can be excused only by their sincere love of peace. Actually it played a part in leading to an infinitely more terrible war. Mussolini's bluff succeeded, and an important spectator drew far-reaching conclusions from the fact. Hitler had long resolved on war for German aggrandisement. He now formed a view of Great Britain's degeneracy which was only to be changed too late for peace and too late for him. In Japan also there were pensive spectators.

*   *   *

The two opposite processes of gathering national unity on the burning issue of the hour and the clash of party interests inseparable from a general election moved forward together. This was greatly to the advantage of Mr. Baldwin and his supporters. "The League of Nations would remain, as heretofore, the keystone of British foreign policy"; so ran the Government's election manifesto. "The prevention of war and the establishment of peace in the world must always be the most vital interest of the British people, and the League is the instrument which has been framed and to which we look for the attainment of these objects. We shall therefore continue to do all in our power to uphold the Covenant and to maintain and increase the efficiency of the League. In the present unhappy dispute between Italy and Abyssinia *there will be no wavering in the policy we have hitherto pursued.*"

The Labour Party, on the other hand, was much divided. The majority was pacifist, but Mr. Bevin's active campaign commanded many supporters among the masses. The official leaders therefore tried to give general satisfaction by pointing opposite ways at once. On the one hand they clamoured for decisive action against the Italian Dictator; on the other they denounced the policy of rearmament. Thus Mr. Attlee in the House of Commons on October 22: "We want effective sanctions, effectively applied. We support economic sanctions. We support the League system." But then, later in the same speech: "We are not persuaded that the way to safety is by piling up armaments. We do not believe that in this [time] there is such a thing as national defence. We think that you have to go forward to disarmament and not to the piling up of armaments." Neither side usually has much to be proud of at election times. The Prime Minister himself was no doubt conscious of the growing strength behind the Government's foreign policy. He was however determined not to be drawn into war on any account. It seemed to me, viewing the proceedings from outside, that he was anxious to gather as much support as possible and use it to begin British rearmament on a modest scale.

At the general election Mr. Baldwin spoke in strong terms of the need for rearmament, and his principal speech was devoted to the unsatisfactory condition of the Navy. However, having gained all that there was in sight upon a programme of sanctions and rearmament, he became very anxious to comfort the professional peace-loving elements in the nation and allay any fears in their breasts which his talk about naval requirements might have caused. On October 1, six weeks before the poll, he made a speech to the Peace Society at the Guildhall. In the course of this he said, "I give you my word there will be no great armaments." In the light of the knowledge which the Government had of strenuous German preparations, this was a singular promise. Thus the votes both of those who sought to see the nation prepare itself against the dangers of the future and of those who believed that peace could be preserved by praising its virtues were gained. The result was a triumph for Mr. Baldwin. The electors accorded him a majority of 247 over all other parties combined, and after five years of office he reached a position of personal power unequalled by any Prime Minister since the close of the Great War. All who had opposed him, whether on India or on the neglect of our defences, were stultified by this renewed vote of confidence, which he had gained by his skilful and fortunate tactics in home politics and by the esteem so widely felt for his personal character. Thus an administration more disastrous than any in our history saw all its errors and shortcomings acclaimed by the nation. There was however a bill to be paid, and it took the new House of Commons nearly ten years to pay it.

It had been widely bruited that I should join the Government as First Lord of the Admiralty. But after the figures of his victory had been proclaimed Mr. Baldwin lost no time in announcing through the Central Office that there was no intention to include me in the Government.

There was much mocking in the press about my exclusion. But now one can see how lucky I was. Over me beat the invisible wings.

And I had agreeable consolations. I set out with my paintbox for more genial climes without waiting for the meeting of Parliament.

*   *   *

There was an awkward sequel to Mr. Baldwin's triumph, for the sake of which we may sacrifice chronology. His Foreign Secretary, Sir Samuel Hoare, travelling through Paris to Switzerland on a well-earned skating holiday, had a talk with Laval, still French Foreign Minister. The result of this was the Hoare-Laval Pact of December 9. It is worth while to look a little into the background of this celebrated incident.

The idea of Britain leading the League of Nations against Mussolini's Fascist invasion of Abyssinia had carried the nation in one of its big swings. But once the election was over and the Ministers found themselves in possession of a majority which might give them for five years the guidance of the State, many tiresome consequences had to be considered. At the root of them all lay Mr. Baldwin's "There must be no war," and also "There must be no large rearmaments." This remarkable party manager, having won the election on world leadership against aggression, was profoundly convinced that we must keep peace at any price.

Moreover, now from the Foreign Office came a very powerful thrust. Sir Robert Vansittart never removed his eyes for one moment from the Hitler peril. He and I were of one mind on that point. And now British policy had forced Mussolini to change sides. Germany was no longer isolated. The four Western Powers were divided two against two instead of three against one. This marked deterioration in our affairs aggravated the anxiety in France. The French Government had already made the Franco-Italian agreement of January. Following thereupon had come the military convention with Italy. It was calculated that this convention saved eighteen French divisions from the Italian front for transference to the front against Germany. In his negotiations it is certain that Laval had given more than a hint to Mussolini that France would not trouble herself about anything that might happen to Abyssinia. The French had a considerable case to argue with British Ministers. First, for several years we had tried to make them reduce their army, which was all they had to live upon. Secondly, the British had had a very good run in the leadership of the League of Nations against Mussolini. They had even won an election upon it; and in democracies elections are very important. Thirdly, we had made a naval agreement, supposed to be very good for ourselves, which made us quite comfortable upon the seas, apart from submarine warfare.

Now in December 1935 a new set of arguments marched upon the scene. Mussolini, hard pressed by sanctions, and under the very heavy threat of "fifty nations led by one," would, it was whispered, welcome a compromise on Abyssinia. Could not a peace be made which gave Italy what she had aggressively demanded and left Abyssinia four-fifths of her entire empire? Vansittart, who happened to be in Paris at the time the Foreign Secretary passed through, and was thus drawn into the affair, should not be misjudged because he thought continuously of the German threat and wished to have Britain and France organised at their strongest to face this major danger, with Italy in their rear a friend and not a foe.

But the British nation from time to time gives way to waves of crusading sentiment. More than any other country in the world, it is at rare intervals ready to fight for a cause or a theme, just because it is convinced in its heart and soul that it will not get any material advantage out of the conflict. Baldwin and his Ministers had given a great uplift to Britain in their resistance to Mussolini at Geneva. They had gone so far that their only salvation before history was to go all lengths. Unless they were prepared to back words and gestures by action, it might have been better to keep out of it all, like the United States, and let things rip and see what happened. Here was an arguable plan. But it was not the plan they had adopted. They had appealed to the millions, and the unarmed, and hitherto unconcerned, millions had answered with a loud shout, overpowering all other cries, "Yes, we will march against evil, and we will march now. Give us the weapons."

The new House of Commons was a spirited body. With all that lay before them in the next ten years, they had need to be. It was therefore with a horrible shock that, while tingling from the election, they received the news that a compromise had been made between Sir Samuel Hoare and M. Laval about Abyssinia. This crisis nearly cost Mr. Baldwin his political life. It shook Parliament and the nation to its base. Mr. Baldwin fell almost overnight from his pinnacle of acclaimed national leadership to a depth where he was derided and despised. His position in the House during these days was pitiful. He had never understood why people should worry about all these bothersome foreign affairs. They had a Conservative majority and no war. What more could they want? But the experienced pilot felt and measured the full force of the storm.

The Cabinet, on December 9, had approved the Hoare-Laval plan to partition Abyssinia between Italy and the Emperor. On the 13th the full text of the Hoare-Laval proposals was laid before the League. On the 18th the Cabinet abandoned the Hoare-Laval proposals, thus entailing the resignation of Sir Samuel Hoare. The crisis passed. On his return from Geneva Mr. Eden was summoned to 10 Downing Street by the Prime Minister to discuss the situation following Sir Samuel Hoare's resignation. Mr. Eden at once suggested that Sir Austen Chamberlain should be invited to take over the Foreign Office, and added that if desired he was prepared to serve under him in any capacity. Mr. Baldwin replied that he had already considered this and had informed Sir Austen himself that he did not feel able to offer the Foreign Office to him. This may have been due to Sir Austen's health. On December 22 Mr. Eden became Foreign Secretary.

My wife and I passed this exciting week at Barcelona. Several of my best friends advised me not to return. They said I should only do myself harm if I were mixed up in this violent conflict. Our comfortable Barcelona hotel was the rendezvous of the Spanish Left. In the excellent restaurant where we lunched and dined were always several groups of eager-faced, black-coated young men purring together with glistening eyes about Spanish politics, in which quite soon a million Spaniards were to die. Looking back, I think I ought to have come home. I might have brought an element of decision and combination to the anti-Government gatherings which would have ended the Baldwin régime. Perhaps a Government under Sir Austen Chamberlain might have been established at this moment. On the other hand, my friends cried, "Better stay away. Your return will only be regarded as a personal challenge to the Government." I did not relish the advice, which was certainly not flattering; but I yielded to the impression that I could do no good, and stayed on at Barcelona daubing canvases in the sunshine. Thereafter Frederick Lindemann joined me, and we cruised in a nice steamship around the eastern coasts of Spain and landed at Tangier. Here I found Lord Rothermere with a pleasant circle. He told me that Mr. Lloyd George was at Marrakesh, where the weather was lovely. We all motored thither. I lingered painting in delightful Morocco, and did not return till the sudden death of King George V on January 20.

*   *   *

The collapse of Abyssinian resistance and the annexation of the whole country by Italy produced unhelpful effects in German public opinion. Even those elements which did not approve of Mussolini's policy or action admired the swift, efficient, and ruthless manner in which, as it seemed, the campaign had been conducted. The general view was that Great Britain had emerged thoroughly weakened. She had earned the undying hatred of Italy; she had wrecked the Stresa front once and for all; and her loss of prestige in the world contrasted agreeably with the growing strength and repute of the new Germany. "I am impressed," wrote one of our representatives in Bavaria, "by the note of contempt in references to Great Britain in many quarters. . . . It is to be feared that Germany's attitude in the negotiations for a settlement in Western Europe and for a more general settlement of European and extra-European questions will be found to have stiffened." All this was only too true. His Majesty's Government had imprudently advanced to champion a great world cause. They had led fifty nations forward with much brave language. Confronted with brute facts Mr. Baldwin had recoiled. Their policy had for a long time been designed to give satisfaction to powerful elements of opinion at home rather than to seek the realities of the European situation. By estranging Italy they had upset the whole balance of Europe and gained nothing for Abyssinia. They had led the League of Nations into an utter fiasco, most damaging if not fatally injurious to its effective life as an institution.

9

Hitler Strikes, 1936

WHEN I RETURNED at the end of January 1936 I was conscious of a new atmosphere in England. Mussolini's conquest of Ethiopia and the brutal methods by which it had been accomplished, the shock of the Hoare-Laval negotiations, the discomfiture of the League of Nations, the obvious breakdown of "collective security," had altered the mood not only of the Labour and Liberal Parties but of a great body of well-meaning but hitherto futile opinion. All these forces were now prepared to contemplate war against Fascist or Nazi tyranny. Far from being excluded from lawful thought, the use of force gradually became a decisive point in the minds of a vast mass of peace-loving people, and even of many who had hitherto been proud to be called pacifists. But force, according to the principles which they served, could only be used on the initiative and under the authority of the League of Nations. Although both the Opposition parties continued to oppose all measures of rearmament, there was an immense measure of agreement open, and had His Majesty's Government risen to the occasion they could have led a united people forward into the whole business of preparation in an emergency spirit.

The Government adhered to their policy of moderation, half-measures, and keeping things quiet. It was astonishing to me that they did not seek to utilise all the growing harmonies that now existed in the nation. By this means they would enormously have strengthened themselves and have gained the power to strengthen the country. Mr. Baldwin had no such inclinations. He was ageing fast. He rested upon the great majority which the election had given him, and the Conservative Party lay tranquil in his hand.

*　　　　*　　　　*

Once Hitler's Germany had been allowed to rearm without active interference by the Allies and former associated Powers, a second World War was almost certain. The longer a decisive trial of strength was put off the worse would be our chances, at first of stopping Hitler without serious fighting, and as a second stage of being victorious after a terrible ordeal. In the summer of 1935 Germany had reinstituted conscription in breach of the Treaties. Great Britain had condoned this, and by a separate agreement her rebuilding of a navy, if desired with U-boats on the British scale. Nazi Germany had secretly and unlawfully created a military air force which, by the spring of 1935, openly claimed to be equal to the British. She was now in the second year of active munitions production after long covert preparations. Great Britain and all Europe, and what was then thought distant America, were faced with the organised might and will-to-war of seventy millions of the most efficient race in Europe, longing to regain their national glory, and driven, in case they faltered, by a merciless military, social, and party régime.

There was, perhaps, still time for an assertion of collective security, based upon the avowed readiness of all members concerned to enforce the decisions of the League of Nations by the sword. The democracies and their dependent states were still actually and potentially far stronger than the dictatorships, but their position relatively to their opponents was less than half as good as it had been twelve months before. Virtuous motives, trammelled by inertia and timidity, are no match for armed and resolute wickedness. A sincere love of peace is no excuse for muddling hundreds of millions of humble folk into total war. The cheers of weak, well-meaning assemblies soon cease to echo, and their votes soon cease to count. Doom marches on.

Germany had, during the course of 1935, repulsed and sabotaged the efforts of the Western Powers to negotiate an Eastern Locarno. The new Reich at this moment declared itself a bulwark against Bolshevism, and for them, they said, there could be no question of working with the Soviets. Hitler told the Polish Ambassador in Berlin on December 18 that "he was resolutely opposed to any co-operation of the West with Russia." It was in this mood that he sought to hinder and undermine the French attempts to reach direct agreement with Moscow. The Franco-Soviet Pact had been signed in May, but not ratified by either party. It became a major object of German diplomacy to prevent such a ratification. Laval was warned from Berlin that if this move took place there could be no hope of any further Franco-German rapprochement. His reluctance to persevere thereafter became marked, but did not affect the event.

On February 27 the French Chamber ratified the Pact, and the following day the French Ambassador in Berlin was instructed to approach the German Government and inquire upon what basis general negotiations for a Franco-German understanding could be initiated. Hitler, in reply, asked for a few days in which to reflect. At ten o'clock on the morning of March 7 Herr von Neurath, the German Foreign Minister, summoned the British, French, Belgian, and Italian Ambassadors to the Wilhelmstrasse to announce to them a proposal for a twenty-five-year pact, a demilitarisation on both sides of the Rhine frontier, a pact limiting air forces, and non-aggression pacts to be negotiated with Eastern and Western neighbours.

The "demilitarised zone" in the Rhineland had been established by Articles 42, 43, and 44 of the Treaty of Versailles. These articles declared that Germany should not have or establish fortifications on the left bank of the Rhine or within fifty kilometres of its right bank. Neither should Germany have in this zone any military forces, nor hold at any time any military manoeuvres, nor maintain any facilities for military mobilisation. On top of this lay the Treaty of Locarno, freely negotiated by both sides. In this treaty the signatory Powers guaranteed individually and collectively the permanence of the frontiers of Germany and Belgium and of Germany and France. Article 2 of the Treaty of Locarno promised that Germany, France, and Belgium would never invade or attack across these frontiers. Should, however, Articles 42 or 43 of the Treaty of Versailles be infringed, such a violation would constitute "an unprovoked act of aggression," and immediate action would be required from the offended signatories because of the assembling of armed forces in the demilitarised zone. Such a violation should be brought at once before the League of Nations, and the League, having established the fact of violation, must then advise the signatory Powers that they were bound to give their military aid to the Power against whom the offence had been perpetrated.

At noon on this same March 7, 1936, two hours after his proposal for a twenty-five-years pact, Hitler announced to the Reichstag that he intended to reoccupy the Rhineland, and even while he spoke, German columns streamed across the boundary and entered all the main German towns. They were everywhere received with rejoicing, tempered by the fear of Allied action. Simultaneously, in order to baffle British and American public opinion, Hitler declared that the occupation was purely symbolic. The German Ambassador in London handed Mr. Eden similar proposals to those which Neurath in Berlin had given to the Ambassadors of the other Locarno Powers in the morning. This provided comfort for everyone on both sides of the Atlantic who wished to be humbugged. Mr. Eden made a stern reply to the Ambassador. We now know of course that Hitler was merely using these conciliatory proposals as part of his design and as a cover for the violent act he had committed, the success of which was vital to his prestige and thus to the next step in his programme.

It was not only a breach of an obligation exacted by force of arms in war, and of the Treaty of Locarno, signed freely in full peace, but the taking advantage of the friendly evacuation by the Allies of the Rhineland several years before it was due. This news caused a world-wide sensation. The French Government, under M. Sarraut, in which M. Flandin was Foreign Minister, uprose in vociferous wrath and appealed to all its allies and to the League. Above all, France also had a right to look to Great Britain, having regard to the guarantee we had given for the French frontier against German aggression, and the pressure we had put upon France for the earlier evacuation of the Rhineland. Here if ever was the violation, not only of the Peace Treaty, but of the Treaty of Locarno, and an obligation binding upon all the Powers concerned.

MM. Sarraut and Flandin had the impulse to act at once by general mobilisation. If they had been equal to their task they would have done so, and thus compelled all others to come into line. But they appeared unable to move without the concurrence of Britain. This is an explanation, but no excuse. The issue was vital to France, and any French Government worthy of the name should have made up its own mind and trusted to the Treaty obligations. More than once in these fluid years French Ministers in their ever-changing Governments were content to find in British pacifism an excuse for their own. Be that as it may, they did not meet with any encouragement to resist the German aggression from the British. On the contrary, if they hesitated to act, their British allies did not hesitate to dissuade them. During the whole of Sunday there were agitated telephonic conversations between London and Paris. His Majesty's Government exhorted the French to wait in order that both countries might act jointly and after full consideration. A velvet carpet for retreat!

The unofficial responses from London were chilling. Mr. Lloyd George hastened to say, "In my judgment Herr Hitler's greatest crime was not the breach of a treaty, because there was provocation." He added that "he hoped we should keep our heads." The provocation was presumably the failure of the Allies to disarm themselves more than they had done. The Socialist Lord Snowden concentrated upon the proposed non-aggression pact, and said that Hitler's previous peace overtures had been ignored, but the peoples would not permit *this* peace offer to be neglected. These utterances may have expressed misguided British public opinion at the moment, but will not be deemed creditable to their authors. The British Cabinet, seeking the line of least resistance, felt that the easiest way out was to press France into another appeal to the League of Nations.

There was also great division in France. On the whole it was the politicians who wished to mobilise the army and send an ultimatum to Hitler, and the generals who, like their German counterparts, pleaded for calm, patience, and delay. We now know of the conflicts of opinion which arose at this time between Hitler and the German High Command. If the French Government had mobilised the French Army, with nearly a hundred divisions, and its air force (then still falsely believed to be the strongest in Europe), there is no doubt that Hitler would have been compelled by his own General Staff to withdraw, and a check would have been given to his pretensions which might well have proved fatal to his rule. It must be remembered that France alone was at this time quite strong enough to drive the Germans out of the Rhineland. Instead, the French Government were urged by Britain to cast their burden upon the League of Nations, already weakened and disheartened by the fiasco of sanctions and the Anglo-German Naval Agreement of the previous year.

On Monday, March 9, Mr. Eden went to Paris, accompanied by Lord Halifax and Ralph Wigram. The first plan had been to convene a meeting of the League in Paris, but presently Wigram, on Eden's authority, was sent to tell Flandin to come to London to have the meeting of the League in England, as he would thus get more effective support from Britain. This was an unwelcome mission for the faithful official. Immediately on his return to London on March 11 he came to see me, and told me the story. Flandin himself arrived late the same night, and at about eight-thirty on Thursday morning he came to my flat in Morpeth Mansions. He told me that he proposed to demand from the British Government simultaneous mobilisation of the land, sea, and air forces of both countries, and that he had received assurances of support from all the nations of the "Little Entente" and from other states. There was no doubt that superior strength still lay with the Allies of the former war. They had only to act to win. Although we did not know what was passing between Hitler and his generals, it was evident that overwhelming force lay on our side.

Mr. Neville Chamberlain was at this time, as Chancellor of the Exchequer, the most effective Member of the Government. His able biographer, Mr. Keith Feiling, gives the following extract from his diary: "March 12, talked to Flandin, emphasising that public opinion would not support us in sanctions of any kind. His view is that if a firm front is maintained Germany will yield without war. We cannot accept this as a reliable estimate of a mad Dictator's reaction." When Flandin urged at least an economic boycott Chamberlain replied by suggesting an international force during negotiations, agreed to a pact for mutual assistance, and declared that if by giving up a colony we could secure lasting peace he would consider it.

Meanwhile most of the British press, with the *Times* and the *Daily Herald* in the van, expressed their belief in the sincerity of Hitler's offers of a non-aggression pact. Austen Chamberlain, in a speech at Cambridge, proclaimed the opposite view. Wigram thought it was within the compass of his duty to bring Flandin into touch with everyone he could think of from the City, from the press, and from the Government, and also with Lord Lothian. To all whom Flandin met at the Wigrams' he spoke in the following terms: "The whole world and especially the small nations today turn their eyes towards England. If England will act now she can lead Europe. You will have a policy, all the world will follow you, and thus you will prevent war. It is your last chance. If you do not stop Germany now, all is over. France cannot guarantee Czechoslovakia any more, because that will become geographically impossible. If you do not maintain the Treaty of Locarno all that will remain to you is to await a rearmament by Germany, against which France can do nothing. If you do not stop Germany by force today war is inevitable, even if you make a temporary friendship with Germany. As for myself, I do not believe that friendship is possible between France and Germany; the two countries will always be in tension. Nevertheless, if you abandon Locarno I shall change my policy, for there will be nothing else to do." These were brave words; but action would have spoken louder.

Lord Lothian's contribution was: "After all, they are only going into their own back garden." This was a representative British view.

*　　　*　　　*

When I heard how ill things were going, and after a talk with Wigram, I advised M. Flandin to demand an interview with Mr. Baldwin before he left. This took place at Downing Street. The Prime Minister received M. Flandin with the utmost courtesy. Mr. Baldwin explained that although he knew little of foreign affairs he was able to interpret accurately the feelings of the British people. And they wanted peace. M. Flandin says that he rejoined that the only way to ensure this was to stop Hitlerite aggression while such action was still possible. France had no wish to drag Great Britain into war; she asked for no practical aid, and she would herself undertake what would be a simple police operation, as, according to French information, the German troops in the Rhineland had orders to withdraw if opposed in a forcible manner. Flandin asserts that he said that all that France asked of her ally was a free hand. This is certainly not true. How could Britain have restrained

France from action to which, under the Locarno Treaty, she was legally entitled? The British Prime Minister repeated that his country could not accept the risk of war. He asked what the French Government had resolved to do. To this no plain answer was returned. According to Flandin,[1] Mr. Baldwin then said: "You may be right, but if there is *even one chance in a hundred* that war would follow from your police operation I have not the right to commit England." And after a pause he added: "England is not in a state to go to war." There is no confirmation of this. M. Flandin returned to France convinced, first that his own divided country could not be united except in the presence of a strong will power in Britain, and secondly that, so far from this being forthcoming, no strong impulse could be expected from her. Quite wrongly he plunged into the dismal conclusion that the only hope for France was in an arrangement with an ever more aggressive Germany.

Nevertheless, in view of what I saw of Flandin's attitude during these anxious days, I felt it my duty, in spite of his subsequent lapses, to come to his aid, so far as I was able, in later years. I used my power in the winter of 1943–44 to protect him when he was arrested in Algeria by the de Gaulle Administration. In this I invoked and received active help from President Roosevelt. When after the war Flandin was brought to trial, my son Randolph, who had seen much of Flandin during the African campaign, was summoned as a witness, and I am glad to think that his advocacy, and also a letter which I wrote for Flandin to use in his defence, were not without influence in procuring the acquittal which he received from the French tribunal. Weakness is not treason, though it may be equally disastrous. Nothing however can relieve the French Government of their prime responsibility. Clemenceau or Poincaré would have left Mr. Baldwin no option.

The British and French submission to the violations of the Treaties of Versailles and Locarno involved in Hitler's seizure of the Rhineland was a mortal blow to Wigram. "After the French delegation had left," wrote his wife to me, "Ralph came back, and sat down in a corner of the room where he had never sat before, and said to me, 'War is now *inevitable*, and it will be the most terrible war there has ever been. I don't think I shall see it, but you will. Wait now for the bombs on this little house.'[2] I was frightened at his words, and he went on, 'All my work these many years has been no use. I am a failure. I have failed to make the people here realise what is at stake. I am not strong enough, I suppose. I have not been able to make them understand. Winston has always, always understood, and he is strong and will go on to the end.' "

[1] Pierre-Étienne Flandin, *Politique Française*, 1919–40, pp. 207–8.
[2] It was actually smitten.

My friend never seemed to recover from this shock. He took it too much to heart. After all, one can always go on doing what one believes to be his duty, and running ever greater risks till knocked out. Wigram's profound comprehension reacted on his sensitive nature unduly. His untimely death in December 1936 was an irreparable loss to the Foreign Office, and played its part in the miserable decline of our fortunes.

*   *   *

When Hitler met his generals after the successful reoccupation of the Rhineland he was able to confront them with the falsity of their fears and prove to them how superior his judgment or "intuition" was to that of ordinary military men. The generals bowed. As good Germans they were glad to see their country gaining ground so rapidly in Europe and its former adversaries so divided and tame. Undoubtedly Hitler's prestige and authority in the supreme circle of German power was sufficiently enhanced by this episode to encourage and enable him to march forward to greater tests. To the world he said: "All Germany's territorial ambitions have now been satisfied."

France was thrown into incoherency, amid which fear of war, and relief that it had been avoided, predominated. The simple English were taught by their simple press to comfort themselves with the reflection, "After all, the Germans are only going back to their own country. How should we feel if we had been kept out of, say, Yorkshire for ten or fifteen years?" No one stopped to note that the detrainment points from which the German Army could invade France had been advanced by one hundred miles. No one worried about the proof given to all the Powers of the "Little Entente" and to Europe that France would not fight, and that England would hold her back even if she would. This episode confirmed Hitler's power over the Reich, and stultified, in a manner ignominious and slurring upon their patriotism, the generals who had hitherto sought to restrain him.

IO

The Loaded Pause, 1936–1938

TWO WHOLE YEARS passed between Hitler's seizure of the
Rhineland in March 1936 and his rape of Austria in March 1938.
This was a longer interval than I had expected. During this period no
time was wasted by Germany. The fortification of the Rhineland, or
"the West Wall," proceeded apace, and an immense line of permanent
and semi-permanent fortifications grew continually. The German Army,
now on the full methodical basis of compulsory service and reinforced
by ardent volunteering, grew stronger month by month, both in numbers
and in the maturity and quality of its formations. The German Air
Force held and steadily improved the lead it had obtained over Great
Britain. The German munitions plants were working at high pressure.
The wheels revolved and the hammers descended day and night in Germany, making its whole industry an arsenal, and welding all its population into one disciplined war machine. At home in the autumn of 1936
Hitler inaugurated a Four Years' Plan to reorganise German economy
for greater self-sufficiency in war. Abroad he obtained that "strong al-
liance" which he had stated in *Mein Kampf* would be necessary for
Germany's foreign policy. He came to terms with Mussolini, and the
Rome-Berlin Axis was formed.

Up till the middle of 1936 Hitler's aggressive policy and treaty breaking had rested, not upon Germany's strength, but upon the disunion
and timidity of France and Britain and the isolation of the United
States. Each of his preliminary steps had been gambles in which he
knew he could not afford to be seriously challenged. The seizure of the
Rhineland and its subsequent fortification was the greatest gamble of
all. It had succeeded brilliantly. His opponents were too irresolute to
call his bluff. When next he moved in 1938 his bluff was bluff no more.
Aggression was backed by force, and it might well be by superior force.

When the Governments of France and Britain realised the terrible transformation which had taken place it was too late.

*   *   *

At the end of July 1936 the increasing degeneration of the Parliamentary régime in Spain and the growing strength of the movements
for a Communist, or alternately an anarchist, revolution led to a military
revolt which had long been preparing. It is part of the Communist doctrine and drillbook, laid down by Lenin himself, that Communists should
aid all movements towards the Left and help into office weak Constitutional, Radical, or Socialist Governments. These they should undermine, and from their falling hands snatch absolute power, and found the
Marxist State. In fact, a perfect reproduction of the Kerensky period
in Russia was taking place in Spain. But the strength of Spain had not
been shattered by foreign war. The Army still maintained a measure of
cohesion. Side by side with the Communist conspiracy there was elaborated in secret a deep military counterplot. Neither side could claim
with justice the title deeds of legality, and Spaniards of all classes were
bound to consider the life of Spain.

Many of the ordinary guarantees of civilised society had been already
liquidated by the Communist pervasion of the decayed Parliamentary
Government. Murders began on both sides, and the Communist pestilence had reached a point where it could take political opponents in the
streets or from their beds and kill them. Already a large number of
these assassinations had taken place in and around Madrid. The climax
was the murder of Señor Sotelo, the Conservative leader, who corresponded somewhat to the type of Sir Edward Carson in British politics
before the 1914 war. This crime was the signal for the generals of the
Army to act. General Franco had a month before written a letter to the
Spanish War Minister, making it clear that if the Spanish Government
could not maintain the normal securities of law in daily life the Army
would have to intervene. Spain had seen many *pronunciamientos* by
military chiefs in the past. When General Franco raised the standard of
revolt, he was supported by the Army, including the rank and file. The
Church, with the noteworthy exception of the Dominicans, and nearly
all the elements of the Right and Centre adhered to him, and he became
immediately the master of several important provinces. The Spanish
sailors killed their officers and joined what soon became the Communist
side. In the collapse of civilised government the Communist sect obtained control, and acted in accordance with their drill. Bitter civil war
now began. Wholesale cold-blooded massacres of their political opponents, and of the well-to-do, were perpetrated by the Communists who
had seized power. These were repaid with interest by the forces under

Franco. All Spaniards went to their deaths with remarkable composure, and great numbers on both sides were shot. The military cadets defended their college at the Alcazar in Toledo with the utmost tenacity, and Franco's troops, forcing their way up from the south, leaving a trail of vengeance behind them in every Communist village, presently achieved their relief. This episode deserves the notice of historians.

In this quarrel I was neutral. Naturally I was not in favour of the Communists. How could I be, when if I had been a Spaniard they would have murdered me and my family and friends? I was sure however that with all the rest they had on their hands the British Government were right to keep out of Spain. France proposed a plan of non-intervention, whereby both sides would be left to fight it out without any external aid. The British, German, Italian, and Russian Governments subscribed to this. In consequence the Spanish Government, now in the hands of the most extreme revolutionaries, found itself deprived of the right even to buy the arms ordered with the gold it physically possessed. It would have been more reasonable to follow the normal course and to have recognised the belligerency of both sides, as was done in the American Civil War of 1861–65. Instead, however, the policy of non-intervention was adopted and formally agreed to by all the Great Powers. This agreement was strictly observed by Great Britain; but Italy and Germany on the one side, and Soviet Russia on the other, broke their engagement constantly and threw their weight into the struggle one against the other. Germany in particular used her air power to commit such experimental horrors as the bombing of the defenceless little township of Guernica.

The Government of M. Léon Blum, which had succeeded the Ministry of M. Albert Sarraut on June 4, was under pressure from its Communist supporters in the Chamber to support the Spanish Government with war material. The Air Minister, M. Cot, without too much regard for the strength of the French Air Force, then in a state of decay, was secretly delivering planes and equipment to the Republican armies. I was perturbed at such developments, and on July 31, 1936, I wrote to the French Ambassador:

> One of the greatest difficulties I meet with in trying to hold on to the old position is the German talk that the anti-Communist countries should stand together. I am sure if France sent aeroplanes, etc., to the present Madrid Government, and the Germans and Italians pushed in from the other angle, the dominant forces here would be pleased with Germany and Italy, and estranged from France. I hope you will not mind my writing this, which I do of course entirely on my own account. *I do not like to hear people*

talking of England, Germany, and Italy forming up against European Communism.[1] It is too easy to be good.

I am sure that an absolutely rigid neutrality, with the strongest protest against any breach of it, is the only correct and safe course at the present time. A day may come, if there is a stalemate, when the League of Nations may intervene to wind up the horrors. But even that is very doubtful.

*   *   *

Advantage is gained in war and also in foreign policy and other things by selecting from many attractive or unpleasant alternatives the dominating point. American military thought has coined the expression "Overall Strategic Objective." When our officers first heard this they laughed; but later on its wisdom became apparent and accepted. Evidently this should be the rule, and other great business be set in subordinate relationship to it. Failure to adhere to this simple principle produces confusion and futility of action, and nearly always makes things much worse later on.

Personally I had no difficulty in conforming to the rule long before I heard it proclaimed. My mind was obsessed by the impression of the terrific Germany I had seen and felt in action during the years of 1914 to 1918 suddenly becoming again possessed of all her martial power, while the Allies, who had so narrowly survived, gaped idle and bewildered. Therefore I continued by every means and on every occasion to use what influence I had with the House of Commons and also with individual Ministers to urge forward our military preparations and to procure Allies and associates for what would before long become again the Common Cause.

One day a friend of mine in a high confidential position under the Government came over to Chartwell to swim with me in my pool when the sun shone bright and the water was fairly warm. We talked of nothing but the coming war, of the certainty of which he was not entirely convinced. As I saw him off he suddenly on an impulse turned and said to me, "The Germans are spending a thousand million pounds sterling a year on their armaments." I thought Parliament and the British public ought to know the facts. I therefore set to work to examine German finance. Budgets were produced and still published every year in Germany; but from their wealth of figures it was very difficult to tell what was happening. However, in April 1936 I privately instituted two separate lines of scrutiny. The first rested upon two German refugees of high ability and inflexible purpose. They understood all the details of the presentation of German budgets, the value of the

[1] My subsequent italics.—W. S. C.

mark, and so forth. At the same time I asked my friend Sir Henry Strakosch whether he could not find out what was actually happening. Strakosch was the head of the firm called Union Corporation, with great resources, and a highly skilled, devoted personnel. The brains of this City company were turned for several weeks onto the problem. Presently they reported with precise and lengthy detail that the German war expenditure was certainly round about a thousand million pounds sterling a year. At the same time the German refugees, by a totally different series of arguments, arrived independently at the same conclusion. One thousand million pounds sterling per annum at the money values of 1936!

I had therefore two separate structures of fact on which to base a public assertion. So I accosted Mr. Neville Chamberlain, still Chancellor of the Exchequer, in the lobby the day before a debate and said to him, "Tomorrow I shall ask you whether it is not a fact that the Germans are spending a thousand million pounds a year on warlike preparations, and I shall ask you to confirm or deny." Chamberlain said, "I cannot deny it, and if you put the point I shall confirm it."

I substituted the figure of eight hundred million for one thousand million pounds to cover my secret information, and also to be on the safe side, and Mr. Chamberlain admitted in Parliament that my estimate was "not excessive."

I sought by several means to bring the relative state of British and German armaments to a clear-cut issue. I asked for a debate in secret session. This was refused. "It would cause needless alarm." I got little support. All secret sessions are unpopular with the press. Then on July 20 I asked the Prime Minister whether he would receive a deputation of Privy Councillors and a few others who would lay before him the facts so far as they knew them. Lord Salisbury requested that a similar deputation from the House of Lords should also come. This was agreed. Although I made personal appeals both to Mr. Attlee and Sir Archibald Sinclair, the Labour and Liberal Parties declined to be represented. Accordingly, on July 28 we were received in the Prime Minister's House of Commons room by Mr. Baldwin, Lord Halifax, and Sir Thomas Inskip, an able lawyer who had the advantage of being little known himself and knowing nothing about military subjects, whom Mr. Baldwin had made Minister for the Co-ordination of Defence. A group of Conservative and non-party notables came with me. Sir Austen Chamberlain introduced us. This was a great occasion. I cannot recall anything like it in what I have seen of British public life. The group of eminent men, with no thought of personal advantage, but whose lives had been centred upon public affairs, represented a weight of Conservative opinion which could not easily be disregarded. If the leaders of the Labour and Liberal Oppositions had come with us there might have been a political situation so tense as to enforce remedial action. The proceedings occupied three or four hours on each of two successive days. I have always said Mr. Baldwin was a good listener. He certainly seemed to listen with the greatest interest and attention. With him were various members of the staff of the Committee of Imperial Defence. On the first day I opened the case in a statement of an hour and a quarter, and I ended as follows:

> First, we are facing the greatest danger and emergency of our history. Second, we have no hope of solving our problem except in conjunction with the French Republic. The union of the British Fleet and the French Army, together with their combined Air Forces operating from close behind the French and Belgian frontiers, together with all that Britain and France stand for, constitutes a deterrent in which salvation may reside. Anyhow it is the best hope. Coming down to detail, we must lay aside every impediment in raising our own strength. We cannot possibly provide against all possible dangers. We must concentrate upon what is vital and take our punishment elsewhere. Coming to still more definite propositions, we must increase the development of our air power in priority over every other consideration. At all costs we must draw the flower of our youth into piloting aeroplanes. Never mind what inducements must be offered; we must draw from every source, by every means. We must accelerate and simplify our aeroplane production and push it to the largest scale, and not hesitate to make contracts with the United States and elsewhere for the largest possible quantities of aviation material and equipment of all kinds. We are in danger, as we have never been in danger before — no, not even at the height of the submarine campaign [1917].
>
> This thought preys upon me: *The months slip by rapidly. If we delay too long in repairing our defences we may be forbidden by superior power to complete the process.*

We were much disappointed that the Chancellor of the Exchequer could not be present. It was evident that Mr. Baldwin's health was failing, and it was well known that he would soon seek rest from his burdens. There could be no doubt who would be his successor. Unhappily, Mr. Neville Chamberlain was absent upon a well-deserved holiday, and did not have the opportunity of this direct confrontation with the facts from members of the Conservative Party, who included his brother and so many of his most valued personal friends.

Most earnest consideration was given by Ministers to our formidable representations, but it was not till after the recess, on November 23,

1936, that we were all invited by Mr. Baldwin to receive a more fully considered statement on the whole position. Sir Thomas Inskip then gave a frank and able account, in which he did not conceal from us the gravity of the plight into which we had come. In substance this was to the effect that our estimates, and in particular my statements, took a too gloomy view of our prospects; that great efforts were being made (as indeed they were) to recover the lost ground; but that no case existed which would justify the Government in adopting emergency measures; that these would necessarily be of a character to upset the whole industrial life of this country, would cause widespread alarm, and advertise any deficiencies that existed, and that within these limits everything possible was being done. On this Sir Austen Chamberlain recorded our general impression that our anxieties were not relieved and that we were by no means satisfied. Thus we took our leave.

During the whole of 1936 the anxiety of the nation and Parliament continued to mount, and was concentrated in particular upon our air defences. In the debate on the Address on November 12 I severely reproached Mr. Baldwin for having failed to keep his pledge that "any Government of this country — a National Government more than any, and this Government — will see to it that in air strength and air power this country shall no longer be in a position inferior to any country within striking distance of its shores." I said, "The Government simply cannot make up their minds, or they cannot get the Prime Minister to make up his mind. So they go on in strange paradox, decided only to be undecided, resolved to be irresolute, adamant for drift, solid for fluidity, all-powerful to be impotent. So we go on preparing more months and years — precious, perhaps vital, to the greatness of Britain — for the locusts to eat."

Mr. Baldwin replied to me in a remarkable speech, in which he said:

> I would remind the House that not once but on many occasions in speeches and in various places, when I have been speaking and advocating as far as I am able the democratic principle, I have stated that a *democracy is always two years behind the dictator*. I believe that to be true. It has been true in this case. I put before the whole House my own views with an appalling frankness. You will remember at that time the Disarmament Conference was sitting in Geneva. You will remember at that time [1931–32] there was probably a stronger pacifist feeling running through this country than at any time since the war. You will remember *the election at Fulham in the autumn of 1933, when a seat which the National Government held was lost by about 7000 votes on no issue but the pacifist*. . . . My position as the leader of a great party was not alto-

gether a comfortable one. I asked myself what chance was there —
when that feeling that was given expression to in Fulham was common throughout the country — what chance was there within the
next year or two of that feeling being so changed that the country
would give a mandate for rearmament? Supposing I had gone to
the country and said that Germany was rearming, and that we must
rearm, does anybody think that this pacific democracy would have
rallied to that cry at that moment? *I cannot think of anything that
would have made the loss of the election from my point of view
more certain.*

This was indeed appalling frankness. It carried naked truth about his
motives into indecency. That a Prime Minister should avow that he
had not done his duty in regard to national safety because he was afraid
of losing the election was an incident without parallel in our Parliamentary history. Mr. Baldwin was of course not moved by any ignoble wish
to remain in office. He was in fact in 1936 earnestly desirous of retiring.
His policy was dictated by the fear that if the Socialists came into power
even less would be done than his Government intended. All their declarations and votes against defence measures are upon record. But this was
no complete defence, and less than justice to the spirit of the British
people. The success which had attended the naïve confession of miscalculation in air parity the previous year was not repeated on this occasion.
The House was shocked. Indeed, the impression produced was so painful that it might well have been fatal to Mr. Baldwin, who was also at
that time in failing health, had not the unexpected intervened.

At this time there was a great drawing together of men and women
of all parties in England who saw the perils of the future, and were
resolute upon practical measures to secure our safety and the cause of
freedom, equally menaced by both the totalitarian impulsions and our
Government's complacency. Our plan was the most rapid large-scale
rearmament of Britain combined with the complete acceptance and
employment of the authority of the League of Nations. I called this
policy "Arms and the Covenant." Mr. Baldwin's performance in the
House of Commons was viewed among us all with disdain. The culmination of this campaign was to be a meeting at the Albert Hall. Here on
December 3 we gathered many of the leading men in all the parties —
strong Tories of the right wing earnestly convinced of the national peril;
the leaders of the League of Nations Union; the representatives of many
great trade unions, including in the chair my old opponent of the General Strike, Sir Walter Citrine; the Liberal Party and its leader, Sir
Archibald Sinclair. We had the feeling that we were upon the threshold
of not only gaining respect for our views but of making them dominant.

It was at this moment that the King's passion to marry the woman he loved caused the casting of all else into the background. The abdication crisis was at hand.

Before I replied to the vote of thanks there was a cry, "God Save the King," and this excited prolonged cheering. I explained therefore on the spur of the moment my personal position:

> There is another grave matter which overshadows our minds to-night. In a few minutes we are going to sing "God Save the King." I shall sing it with more heartfelt fervour than I have ever sung it in my life. I hope and pray that no irrevocable decision will be taken in haste, but that time and public opinion will be allowed to play their part, and that a cherished and unique personality may not be incontinently severed from the people he loves so well. I hope that Parliament will be allowed to discharge its function in these high constitutional questions. I trust that our King may be guided by the opinions that are now for the first time being expressed by the British nation and the British Empire, and that the British people will not in their turn be found wanting in generous consideration for the occupant of the throne.

It is not relevant to this account to describe the brief but intensely violent controversy that followed. I had known King Edward VIII since he was a child, and had in 1910 as Home Secretary read out to a wonderful assembly the proclamation creating him Prince of Wales at Carnarvon Castle. I felt bound to place my personal loyalty to him upon the highest plane. Although during the summer I had been made fully aware of what was going forward, I in no way interfered or communicated with him at any time. However, presently in his distress he asked the Prime Minister for permission to consult me. Mr. Baldwin gave formal consent, and on this being conveyed to me I went to the King at Fort Belvedere. I remained in contact with him till his abdication, and did my utmost to plead both to the King and to the public for patience and delay. I have never repented of this — indeed, I could do no other.

The Prime Minister proved himself to be a shrewd judge of British national feeling. Undoubtedly he perceived and expressed the profound will of the nation. His deft and skilful handling of the abdication issue raised him in a fortnight from the depths to the pinnacle. There were several moments when I seemed to be entirely alone against a wrathful House of Commons. I am not, when in action, unduly affected by hostile currents of feeling; but it was on more than one occasion almost physically impossible to make myself heard. All the forces I had gathered together on "Arms and the Covenant," of which I conceived myself to be the mainspring, were estranged or dissolved, and I was myself so smitten in public opinion that it was the almost universal view that my political life was at last ended. How strange it is that this very House of Commons which had regarded me with so much hostility should have been the same instrument which hearkened to my guidance and upheld me through the long adverse years of war till victory over all our foes was gained! What a proof is here offered that the only wise and safe course is to act from day to day in accordance with what one's own conscience seems to decree!

From the abdication of one king we passed to the coronation of another, and until the end of May 1937 the ceremonial and pageantry of a solemn national act of allegiance and the consecration of British loyalties at home and throughout the Empire to the new Sovereign filled all minds. Foreign affairs and the state of our defences lost all claim upon the public mood. Our Island might have been ten thousand miles away from Europe. However, I am permitted to record that on May 18, 1937, on the morrow of the Coronation, I received from the new King a letter in his own handwriting:

> THE ROYAL LODGE,
> THE GREAT PARK,
> WINDSOR, BERKS.
> 18.v.37

My dear Mr. Churchill,

I am writing to thank you for your very nice letter to me. I know how devoted you have been, and still are, to my dear brother, and I feel touched beyond words by your sympathy and understanding in the very difficult problems that have arisen since he left us in December. I fully realise the great responsibilities and cares that I have taken on as King, and I feel most encouraged to receive your good wishes, as one of our great statesmen, and from one who has served his country so faithfully. I can only hope and trust that the good feeling and hope that exists in the Country and Empire now will prove a good example to other Nations in the world.

> Believe me,
> Yours very sincerely,
> GEORGE R.I.

This gesture of magnanimity towards one whose influence at that time had fallen to zero will ever be a cherished experience in my life.

*　　　*　　　*

On May 28, 1937, after King George VI had been crowned, Mr. Baldwin retired. His long public services were suitably rewarded by an earldom and the Garter. He laid down the wide authority he had gathered and carefully maintained, but had used as little as possible. He departed in a glow of public gratitude and esteem. There was no doubt who his successor should be. Mr. Neville Chamberlain had, as Chancellor of the Exchequer, not only done the main work of the Government for five years past, but was the ablest and most forceful Minister, with high abilities and an historic name. I had described him a year earlier at Birmingham in Shakespeare's words as the "packhorse in our great affairs," and he had accepted this description as a compliment. I had no expectation that he would wish to work with me, nor would he have been wise to do so at such a time. His ideas were far different from mine on the treatment of the dominant issues of the day. But I welcomed the accession to power of a live, competent, executive figure. Our relations continued to be cool, easy, and polite both in public and in private.

I may here set down a comparative appreciation of these two Prime Ministers, Baldwin and Chamberlain, whom I had known so long and under whom I had served or was to serve. Stanley Baldwin was the wiser, more comprehending personality, but without detailed executive capacity. He was largely detached from foreign and military affairs. He knew little of Europe, and disliked what he knew. He had a deep knowledge of British party politics, and represented in a broad way some of the strengths and many of the infirmities of our island race. He had fought five general elections as leader of the Conservative Party and had won three of them. He had a genius for waiting upon events and an imperturbability under adverse criticism. He was singularly adroit in letting events work for him, and capable of seizing the ripe moment when it came. He seemed to me to revive the impressions history gives us of Sir Robert Walpole, without of course the eighteenth-century corruption, and he was master of British politics for nearly as long.

Neville Chamberlain, on the other hand, was alert, businesslike, opinionated and self-confident in a very high degree. Unlike Baldwin, he conceived himself able to comprehend the whole field of Europe, and indeed the world. Instead of a vague but none the less deep-seated intuition, we had now a narrow, sharp-edged efficiency within the limits of the policy in which he believed. Both as Chancellor of the Exchequer and as Prime Minister he kept the tightest and most rigid control upon military expenditure. He was throughout this period the masterful opponent of all emergency measures. He had formed decided judgments about all the political figures of the day, both at home and abroad, and felt himself capable of dealing with them. His all-pervading hope was to go down to history as the great peacemaker, and for this he was prepared to strive continually in the teeth of facts, and face great risks for himself and his country. Unhappily he ran into tides the force of which he could not measure, and met hurricanes from which he did not flinch, but with which he could not cope. In these closing years before the war I should have found it easier to work with Baldwin, as I knew him, than with Chamberlain; but neither of them had any wish to work with me except in the last resort.

*   *   *

One day in 1937 I had a meeting with Herr von Ribbentrop, German Ambassador to Britain. In one of my fortnightly articles I had noted that he had been misrepresented in some speech he had made. I had of course met him several times in society. He now asked me whether I would come to see him and have a talk. He received me in the large upstairs room at the German Embassy. We had a conversation lasting for more than two hours. Ribbentrop was most polite, and we ranged over the European scene, both on respect of armaments and policy. The gist of his statement to me was that Germany sought the friendship of England (on the Continent we are still often called "England"). He said he could have been Foreign Minister of Germany, but he had asked Hitler to let him come over to London in order to make the full case for an Anglo-German entente or even alliance. Germany would stand guard for the British Empire in all its greatness and extent. They might ask for the return of the German colonies, but this was evidently not cardinal. What was required was that Britain should give Germany a free hand in the East of Europe. She must have her *Lebensraum,* or living space, for her increasing population. Therefore Poland and the Danzig Corridor must be absorbed. White Russia and the Ukraine were indispensable to the future life of the German Reich of some seventy million souls. Nothing less would suffice. All that was asked of the British Commonwealth and Empire was not to interfere. There was a large map on the wall, and the Ambassador several times led me to it to illustrate his projects.

After hearing all this I said at once that I was sure the British Government would not agree to give Germany a free hand in Eastern Europe. It was true we were on bad terms with Soviet Russia and that we hated Communism as much as Hitler did, but he might be sure that even if France were safeguarded Great Britain would never disinterest herself in the fortunes of the Continent to an extent which would enable Germany to gain the domination of Central and Eastern Europe. We were actually standing before the map when I said this. Ribbentrop turned abruptly away. He then said, "In that case, war is inevitable. There is no way out. The Fuehrer is resolved. Nothing will stop him and nothing will stop us." We then returned to our chairs. I was only a private member of Parliament, but of some prominence. I thought it right to say to the German Ambassador — in fact, I remember the words well, "When you talk of war, which no doubt would be general war, you must not underrate England. She is a curious country, and few foreigners can understand her mind. Do not judge by the attitude of the present administration. Once a great cause is presented to the people all kinds of unexpected actions might be taken by this very Government and by the British nation." And I repeated, "Do not underrate England. She is very clever. If you plunge us all into another Great War she will bring the whole world against you, like last time." At this the Ambassador rose in heat and said, "Ah, England may be very clever, but this time she will not bring the world against Germany." We turned the conversation onto easier lines, and nothing more of note occurred. The incident however remains in my memory, and as I reported it at the time to the Foreign Office I feel it right to put it on record.

When he was on his trial for his life by the conquerors Ribbentrop gave a distorted version of this conversation and claimed that I should be summoned as a witness. What I have set down about it is what I should have said had I been called.

II

Mr. Eden at the Foreign Office: His Resignation

THE FOREIGN SECRETARY has a special position in a British Cabinet. He is treated with marked respect in his high and responsible office, but he usually conducts his affairs under the continuous scrutiny, if not of the whole Cabinet, at least of its principal members. He is under an obligation to keep them informed. He circulates to his colleagues, as a matter of custom and routine, all his executive telegrams, the reports from our embassies abroad, the records of his interviews with foreign Ambassadors or other notables. At least this has been the case during my experience of Cabinet life. This supervision is of course especially maintained by the Prime Minister, who personally or through his Cabinet is responsible for controlling, and has the power to control the main course of foreign policy. From him at least there must be no secrets. No Foreign Secretary can do his work unless he is supported constantly by his chief. To make things go smoothly, there must not only be agreement between them on fundamentals, but also a harmony of outlook and even to some extent of temperament. This is all the more important if the Prime Minister himself devotes special attention to foreign affairs.

Eden was the Foreign Secretary of Mr. Baldwin, who, apart from his main well-known desire for peace and a quiet life, took no active share in foreign policy. Mr. Chamberlain, on the other hand, sought to exercise a masterful control in many departments. He had strong views about foreign affairs, and from the beginning asserted his undoubted right to discuss them with foreign Ambassadors. His assumption of the Premiership therefore implied a delicate but perceptible change in the position of the Foreign Minister.

To this was added a profound, though at first latent, difference of spirit and opinion. The Prime Minister wished to get on good terms

with the two European dictators, and believed that conciliation and
the avoidance of anything likely to offend them was the best method.
Eden, on the other hand, had won his reputation at Geneva by rallying
the nations of Europe against one dictator; and, left to himself, might
well have carried sanctions to the verge of war, and perhaps beyond.
He was a devoted adherent of the French Entente. He was anxious
to have more intimate relations with Soviet Russia. He felt and feared
the Hitler peril. He was alarmed by the weakness of our armaments,
and its reaction on foreign affairs. It might almost be said that there
was not much difference of view between him and me, except of course
that he was in harness. It seemed therefore to me from the beginning
that differences would be likely to arise between these two leading
ministerial figures as the world situation became more acute.

Moreover, in Lord Halifax the Prime Minister had a colleague who
seemed to share his views on foreign affairs with sympathy and conviction. My long and intimate associations with Edward Halifax dated
from 1922, when, in the days of Lloyd George, he became my Under-
Secretary at the Dominions and Colonial Office. Political differences —
even as serious and prolonged as those which arose between us about
his policy as Viceroy of India — had never destroyed our personal
relations. I thought I knew him very well, and I was sure that there
was a gulf between us. I felt also that this same gulf, or one like it, was
open between him and Anthony Eden. It would have been wiser, on
the whole, for Mr. Chamberlain to have made Lord Halifax his Foreign
Secretary when he formed his Government. Eden would have been
far more happily placed in the War Office or the Admiralty, and the
Prime Minister would have had a kindred spirit and his own man at
the Foreign Office. Between the summer of 1937 and the end of that
year divergence, both in method and aim, grew between the Prime
Minister and his Foreign Secretary. The sequence of events which led
to Mr. Eden's resignation in February 1938 followed a logical
course.

The original points of difference arose about our relations with Germany and Italy. Mr. Chamberlain was determined to press his suit
with the two dictators. In July 1937 he invited the Italian Ambassador, Count Grandi, to Downing Street. The conversation took place
with the knowledge but not in the presence of Mr. Eden. Mr. Chamberlain spoke of his desire for an improvement in Anglo-Italian relations.
Count Grandi suggested to him that as a preliminary move it might be
well if the Prime Minister were to write a personal appeal to Mussolini. Mr. Chamberlain sat down and wrote such a letter during the
interview. It was dispatched without reference to the Foreign Secretary, who was in the Foreign Office a few yards away. The letter pro-

duced no apparent results, and our relations with Italy, because of her increasing intervention in Spain, got steadily worse.

Mr. Chamberlain was imbued with a sense of a special and personal mission to come to friendly terms with the Dictators of Italy and Germany, and he conceived himself capable of achieving this relationship. To Mussolini he wished to accord recognition of the Italian conquest of Abyssinia as a prelude to a general settlement of differences. To Hitler he was prepared to offer colonial concessions. At the same time he was disinclined to consider in a conspicuous manner the improvements of British armaments or the necessity of close collaboration with France, both on the staff and political levels. Mr. Eden, on the other hand, was convinced that any arrangement with Italy must be part of a general Mediterranean settlement, which must include Spain, and be reached in close understanding with France. In the negotiation of such a settlement our recognition of Italy's position in Abyssinia would clearly be an important bargaining counter. To throw this away in the prelude and appear eager to initiate negotiations was, in the Foreign Secretary's view, unwise.

During the autumn of 1937 these differences became more severe. Mr. Chamberlain considered that the Foreign Office was obstructing his attempts to open discussions with Germany and Italy, and Mr. Eden felt that his chief was displaying immoderate haste in approaching the Dictators, particularly while British armaments were so weak. There was in fact a profound practical and psychological divergence of view.

*   *   *

In spite of my differences with the Government, I was in close sympathy with their Foreign Secretary. He seemed to me the most resolute and courageous figure in the Administration, and although as a Private Secretary and later as an Under-Secretary of State in the Foreign Office he had had to adapt himself to many things I had attacked and still condemn, I felt sure his heart was in the right place and that he had the root of the matter in him. For his part, he made a point of inviting me to Foreign Office functions, and we corresponded freely. There was of course no impropriety in this practice, and Mr. Eden held to the well-established precedent whereby the Foreign Secretary is accustomed to keep in contact with the prominent political figures of the day on all broad international issues.

In the autumn of 1937 Eden and I had reached, though by somewhat different paths, a similar standpoint against active Axis intervention in the Spanish Civil War. I always supported him in the House when he took resolute action, even though it was upon a very limited scale. I knew well what his difficulties were with some of his senior colleagues in the Cabinet and with his chief, and that he would act more boldly if he were not enmeshed. Soon in the Mediterranean a crisis arose which he handled with firmness and skill, and which was accordingly solved in a manner reflecting a gleam of credit upon our course. A number of merchant ships had been sunk by so-called Spanish submarines. Actually there was no doubt that they were not Spanish but Italian. This was sheer piracy, and it stirred all who knew about it to action. A conference of the Mediterranean Powers was convened at Nyon for September 10. To this the Foreign Secretary, accompanied by Vansittart and Lord Chatfield, the First Sea Lord, proceeded. The conference was brief and successful. It was agreed to establish British and French anti-submarine patrols, with orders which left no doubt as to the fate of any submarine encountered. This was acquiesced in by Italy, and the outrages stopped at once.

Although an incident, here is a proof of how powerful the combined influence of Britain and France, if expressed with conviction and a readiness to use force, would have been upon the mood and policy of the Dictators. That such a policy would have prevented war at this stage cannot be asserted. It might easily have delayed it. It is the fact that whereas "appeasement" in all its forms only encouraged their aggression and gave the Dictators more power with their own peoples, any sign of a positive counteroffensive by the Western Democracies immediately produced an abatement of tension. This rule prevailed during the whole of 1937. After that the scene and conditions were different.

During November Eden became increasingly concerned about our slow rearmament. On the 11th he had an interview with the Prime Minister and tried to convey his misgivings. Mr. Neville Chamberlain after a while refused to listen to him. He advised him to "go home and take an aspirin." By February 1938 the Foreign Secretary conceived himself to be almost isolated in the Cabinet. The Prime Minister had strong support against him and his outlook. A whole band of important Ministers thought the Foreign Office policy dangerous and even provocative. On the other hand, a number of the younger Ministers were very ready to understand his point of view. Some of them later complained that he did not take them into his confidence. He did not however contemplate anything like forming a group against his leader. The Chiefs of Staff could give him no help. Indeed, they enjoined caution and dwelt upon the dangers of the situation. They were reluctant to draw too close to the French lest we should enter into engagements beyond our power to fulfil. They took a gloomy view of Russian military strength after Stalin's purge, of which more later. They believed it necessary to deal with our problems as though we had three enemies — Germany, Italy, and Japan — who might all attack us together, and few to help us. We might ask for air bases in France, but we were not able to send an army in the first instance. Even this modest suggestion encountered strong resistance in the Cabinet.

But the actual breach came over a new and separate issue. On the evening of January 11, 1938, Mr. Sumner Welles, the American Under-Secretary of State, called upon the British Ambassador in Washington. He was the bearer of a secret and confidential message from President Roosevelt to Mr. Chamberlain. The President was deeply anxious at the deterioration of the international situation, and proposed to take the initiative by inviting the representatives of certain Governments to Washington to discuss the underlying causes of present difficulties. Before taking this step however he wished to consult the British Government on their view of such a plan, and stipulated that no other Government should be informed either of the nature or the existence of such a proposal. He asked that not later than January 17 he should be given a reply to his message, and intimated that only if his suggestion met with "the cordial approval and wholehearted support of His Majesty's Government" would he then approach the Governments of France, Germany, and Italy. Here was a formidable and measureless step.

In forwarding this most secret proposal to London the British Ambassador, Sir Ronald Lindsay, urged its acceptance in the most earnest manner. The Foreign Office received the Washington telegram on January 12, and copies were sent to the Prime Minister in the country that evening. On the following morning he came to London, and on his instructions a reply was sent to the President's message. Mr. Eden was at this time on a brief holiday in the South of France. Mr. Chamberlain's reply was to the effect that he appreciated the confidence of President Roosevelt in consulting him in this fashion upon his proposed plan to alleviate the existing tension in Europe, but he wished to explain the position of his own efforts to reach agreement with Germany and Italy, particularly in the case of the latter. "His Majesty's Government would be prepared, for their part, if possible with the authority of the League of Nations, to recognise *de jure* the Italian occupation of Abyssinia, if they found that the Italian Government on their side were ready to give evidence of their desire to contribute to the restoration of confidence and friendly relations." The Prime Minister mentioned these facts, the message continued, so that the President might consider whether his present proposal might not cut across the British efforts. Would is not therefore be wiser to postpone the launching of the American plan?

This reply was received by Mr. Roosevelt with some disappointment.

He intimated that he would reply by letter to Mr. Chamberlain on January 17. On the evening of January 15 the Foreign Secretary returned to England. He had been urged to come back, not by his chief, who was content to act without him, but by his devoted officials at the Foreign Office. The vigilant Alexander Cadogan awaited him upon the pier at Dover. Mr. Eden, who had worked long and hard to improve Anglo-American relations, was deeply perturbed. He immediately sent a telegram to Sir Ronald Lindsay attempting to minimise the effects of Mr. Chamberlain's chilling answer. The President's letter reached London on the morning of January 18. In it he agreed to postpone making his proposal in view of the fact that the British Government were contemplating direct negotiations, but he added that he was gravely concerned at the suggestion that His Majesty's Government might accord recognition to the Italian position in Abyssinia. He thought that this would have a most harmful effect upon Japanese policy in the Far East and upon American public opinion. Mr. Cordell Hull, in delivering this letter to the British Ambassador in Washington, expressed himself even more emphatically. He said that such a recognition would "rouse a feeling of disgust, would revive and multiply all fears of pulling the chestnuts out of the fire; it would be represented as a corrupt bargain completed in Europe at the expense of interests in the Far East in which America was intimately concerned."

The President's letter was considered at a series of meetings of the Foreign Affairs Committee of the Cabinet. Mr. Eden succeeded in procuring a considerable modification of the previous attitude. Most of the Ministers thought he was satisfied. He did not make it clear to them that he was not. Following these discussions two messages were sent to Washington on the evening of January 21. The substance of these replies was that the Prime Minister warmly welcomed the President's initiative, but was not anxious to bear any responsibility for its failure if American overtures were badly received. Mr. Chamberlain wished to point out that we did not accept in an unqualified manner the President's suggested procedure, which would clearly irritate both the Dictators and Japan. Nor did His Majesty's Government feel that the President had fully understood our position in regard to *de jure* recognition. The second message was in fact an explanation of our attitude in this matter. We intended to accord such recognition only as part of a general settlement with Italy.

The British Ambassador reported his conversation with Mr. Sumner Welles when he handed these messages to the President on January 22. He stated that Mr. Welles told him that "the President regarded recognition as an unpleasant pill which we should both have to swallow, and he wished that we should both swallow it together."

Thus it was that President Roosevelt's proposal to use American influence for the purpose of bringing together the leading European Powers to discuss the chances of a general settlement, this of course involving however tentatively the mighty power of the United States, was repulsed by Mr. Chamberlain.

*   *   *

It was plain that no resignation by the Foreign Secretary could be founded upon the rebuff administered by Mr. Chamberlain to the President's overture. Mr. Roosevelt was indeed running great risks in his own domestic politics by deliberately involving the United States in the darkening European scene. All the forces of isolationism would have been aroused if any part of these interchanges had transpired. On the other hand, no event could have been more likely to stave off, or even prevent, war than the arrival of the United States in the circle of European hates and fears. To Britain it was a matter almost of life and death. No one can measure in retrospect its effect upon the course of events in Austria and later at Munich. We must regard its rejection — for such it was — as the loss of the last frail chance to save the world from tyranny otherwise than by war. That Mr. Chamberlain, with his limited outlook and inexperience of the European scene, should have possessed the self-sufficiency to wave away the proffered hand stretched out across the Atlantic leaves one, even at this date, breathless with amazement. The lack of all sense of proportion, and even of self-preservation, which this episode reveals in an upright, competent, well-meaning man, charged with the destinies of our country and all who depended upon it, is appalling. One cannot today even reconstruct the state of mind which would render such gestures possible.

It must have been with declining confidence in the future that Mr. Eden went to Paris on January 25 to consult with the French. Everything now turned upon the success of the approach to Italy, of which we had made such a point in our replies to the President. The French Ministers impressed upon Mr. Eden the necessity of the inclusion of Spain in any general settlement with the Italians; on this he needed little convincing. On February 10 the Prime Minister and the Foreign Secretary met Count Grandi, who declared that the Italians were ready in principle to open the conversations.

On February 15 the news came of the submission of the Austrian Chancellor, Schuschnigg, to the German demand for the introduction into the Austrian Cabinet of the chief Nazi agent, Seyss-Inquart, as Minister of the Interior and Head of the Austrian Police. This grave event did not avert the personal crisis between Mr. Chamberlain and Mr. Eden. On February 18 they saw Count Grandi again. This was the last business they conducted together. The Ambassador refused either to discuss the Italian position towards Austria or to consider the British plan for the withdrawal of volunteers, or so-called volunteers — in this case five divisions of the regular Italian Army — from Spain. Grandi asked however for general conversations to be opened in Rome. The Prime Minister wished for these, and the Foreign Secretary was strongly opposed to such a step.

There were prolonged parleyings and Cabinet meetings. At the end Mr. Eden briefly tendered his resignation on the issue of the Italian conversations taking place at this stage and in these circumstances. At this his colleagues were astonished. They had not realised that the differences between the Foreign Secretary and the Prime Minister had reached breaking-point. Evidently if Mr. Eden's resignation was involved a new question raising larger and more general issues was posed. However, they had all committed themselves on the merits of the matter in dispute. The rest of the long day was spent in efforts to induce the Foreign Secretary to change his mind. Mr. Chamberlain was impressed by the distress of the Cabinet. "Seeing how my colleagues had been taken aback, I proposed an adjournment until next day." But Eden saw no use in continuing a search for formulas, and by midnight, on the 20th, his resignation became final. "Greatly to his credit, as I see it," noted the Prime Minister. Lord Halifax was at once appointed Foreign Secretary in his place.

*   *   *

It had of course become known that there were serious differences in the Cabinet, though the cause was obscure. I had heard something of this, but carefully abstained from any communication with Mr. Eden. I hoped that he would not on any account resign without building up his case beforehand, and giving his many friends in Parliament a chance to draw out the issues. But the Government at this time was so powerful and aloof that the struggle was fought out inside the Ministerial conclave, and mainly between the two men.

Late in the night of February 20 a telephone message reached me as I sat in my old room at Chartwell (as I often sit now) that Eden had resigned. I must confess that my heart sank, and for a while the dark waters of despair overwhelmed me. In a long life I have had many ups and downs. During all the war soon to come and in its darkest times I never had any trouble in sleeping. In the crisis of 1940, when so much responsibility lay upon me, and also at many very anxious, awkward moments in the following five years, I could always flop into bed and go to sleep after the day's work was done — subject of course to any emergency call. I slept sound and awoke refreshed, and had no feelings except appetite to grapple with whatever the morning's boxes might bring. But now on this night of February 20, 1938, and on this occasion only, sleep deserted me. From midnight till dawn I lay in my bed consumed by emotions of sorrow and fear. There seemed one strong young figure standing up against long, dismal, drawling tides of drift and surrender, of wrong measurements and feeble impulses. My conduct of affairs would have been different from his in various ways; but he seemed to me at this moment to embody the life-hope of the British nation, the grand old British race that had done so much for men, and had yet some more to give. Now he was gone. I watched the daylight slowly creep in through the windows, and saw before me in mental gaze the vision of Death.

12

The Rape of Austria,
February 1938

USUALLY IN MODERN TIMES when states have been defeated in war they have preserved their structure, their identity, and the secrecy of their archives. On this occasion, the war being fought to an utter finish, we have come into full possession of the inside story of the enemy. From this we can check up with some exactness our own information and performances. In July 1936 Hitler had instructed the German General Staff to draw up military plans for the occupation of Austria when the hour should strike. This operation was labelled "Case Otto." Now, on November 5, 1937, he unfolded his future designs to the chiefs of his armed forces. Germany must have more "living space." This could best be found in Eastern Europe — Poland, White Russia, and the Ukraine. To obtain this would involve a major war, and incidentally the extermination of the people then living in those parts. Germany would have to reckon with her two "hateful enemies," England and France, to whom "a German Colossus in the centre of Europe would be intolerable." In order to profit by the lead she had gained in munitions production and by the patriotic fervour aroused and represented by the Nazi Party, she must therefore make war at the first promising opportunity, and deal with her two obvious opponents before they were ready to fight.

Neurath, Fritsch, and even Blomberg, all of them influenced by the views of the German Foreign Office, General Staff, and Officer Corps, were alarmed by this policy. They thought that the risks to be run were too high. They recognised that by the audacity of the Fuehrer they were definitely ahead of the Allies in every form of rearmament. The Army was maturing month by month; the internal decay of France and the lack of will power in Britain were favourable factors which

116

might well run their full course. What was a year or two when all was moving so well? They must have time to complete the war machine, and a conciliatory speech now and again from the Fuehrer would keep these futile and degenerate democracies chattering. But Hitler was not sure of this. His genius taught him that victory would not be achieved by processes of certainty. Risks had to be run. The leap had to be made. He was flushed with his successes, first in rearmament, second in conscription, third in the Rhineland, fourth by the accession of Mussolini's Italy. To wait till everything was ready was probably to wait till all was too late. It is very easy for historians and other people, who do not have to live and act from day to day, to say that he would have had the whole fortunes of the world in his hand if he had gone on growing in strength for another two or three years before striking. However, this does not follow. There are no certainties in human life or in the life of states. Hitler was resolved to hurry, and have the war while he was in his prime.

Blomberg, weakened with the Officer Corps by an inappropriate marriage, was first removed; and then, on February 4, 1938, Hitler dismissed Fritsch, and himself assumed supreme command of the armed forces. So far as it is possible for one man, however gifted and powerful, however terrible the penalties he can inflict, to make his will effective over spheres so vast, the Fuehrer assumed direct control, not only of the policy of the State, but of the military machine. He had at this time something like the power of Napoleon after Austerlitz and Jena, without of course the glory of winning great battles by his personal direction on horseback, but with triumphs in the political and diplomatic field which all his circle and followers knew were due alone to him and to his judgment and daring.

*   *   *

Apart from his resolve, so plainly described in *Mein Kampf,* to bring all Teutonic races into the Reich, Hitler had two reasons for wishing to absorb the Austrian Republic. It opened to Germany both the door of Czechoslovakia and the more spacious portals of Southeastern Europe. Since the murder of Chancellor Dollfuss in July 1934 by the Austrian section of the Nazi Party the process of subverting the independent Austrian Government by money, intrigue, and force had never ceased. The Nazi movement in Austria grew with every success that Hitler reaped elsewhere, whether inside Germany or against the Allies. It had been necessary to proceed step by step. Officially Papen was instructed to maintain the most cordial relations with the Austrian Government, and to procure the official recognition by them of the Austrian Nazi Party as a legal body. At that time the attitude of

Mussolini had imposed restraint. After the murder of Dr. Dollfuss the Italian Dictator had flown to Venice to receive and comfort the widow, who had taken refuge there, and considerable Italian forces had been concentrated on the southern frontier of Austria. But now in the dawn of 1938 decisive changes in European groupings and values had taken place. The Siegfried Line confronted France with a growing barrier of steel and concrete, requiring as it seemed an enormous sacrifice of French manhood to pierce. The door from the West was shut. Mussolini had been driven into the German system by sanctions so ineffectual that they had angered him without weakening his power. He might well have pondered with relish on Machiavelli's celebrated remark, "Men avenge slight injuries, but not grave ones." Above all the Western Democracies had seemed to give repeated proofs that they would bow to violence so long as they were not themselves directly assailed. Papen was working skilfully inside the Austrian political structure. Many Austrian notables had yielded to his pressure and intrigues. The tourist trade, so important to Vienna, was impeded by the prevailing uncertainty. In the background terrorist activity and bomb outrages shook the frail life of the Austrian Republic.

It was thought that the hour had now come to obtain control of Austrian policy by procuring the entry into the Vienna Cabinet of the leaders of the lately legalised Austrian Nazi Party. On February 12, 1938, eight days after assuming the supreme command, Hitler had summoned the Austrian Chancellor, Herr von Schuschnigg, to Berchtesgaden. He had obeyed, and was accompanied by his Foreign Minister, Guido Schmidt. We now have Schuschnigg's record, in which the following dialogue occurs.[1] Hitler had mentioned the defences of the Austrian frontier. These were no more than might be required to make a military operation necessary to overcome them, and thus raise major issues of peace and war.

> *Hitler:* I only need to give an order, and overnight all the ridiculous scarecrows on the frontier will vanish. You don't really believe that you could hold me up for half an hour? Who knows — perhaps I shall be suddenly overnight in Vienna: like a spring storm. Then you will really experience something. I would willingly spare the Austrians this; it will cost many victims. *After the troops will follow the S.A. and the Legion!* No one will be able to hinder their vengeance, not even myself. Do you want to turn Austria into another Spain? All this I would like if possible to avoid.
>
> *Schuschnigg:* I will obtain the necessary information and put

[1] Schuschnigg, *Ein Requiem in Rot-Weiss-Rot*, p. 37 ff.

a stop to the building of any defence works on the German frontier. Naturally I realise that you can march into Austria, but, Mr. Chancellor, whether we wish it or not, that would lead to the shedding of blood. We are not alone in the world. That probably means war.

Hitler: That is very easy to say at this moment as we sit here in club armchairs, but behind it all there lies a sum of suffering and blood. Will you take the responsibility for that, Herr Schuschnigg? Don't believe that anyone in the world will hinder me in my decisions! Italy? I am quite clear with Mussolini: with Italy I am on the closest possible terms. England? England will not lift a finger for Austria. . . . And France? Well, two years ago when we marched into the Rhineland with a handful of battalions — at that moment I risked a great deal. If France had marched then we should have been forced to withdraw. . . . But for France it is now too late!

This first interview took place at eleven in the morning. After a formal lunch the Austrians were summoned into a small room, and there confronted by Ribbentrop and Papan with a written ultimatum. The terms were not open to discussion. They included the appointment of the Austrian Nazi Seyss-Inquart as Minister of Security in the Austrian Cabinet, a general amnesty for all Austrian Nazis under detention, and the official incorporation of the Austrian Nazi Party in the Government-sponsored Fatherland Front.

Later Hitler received the Austrian Chancellor. "I repeat to you, this is the very last chance. Within three days I expect the execution of this agreement." In Jodl's diary the entry reads, "Von Schuschnigg together with Guido Schmidt are again being put under heaviest political and military pressure. At 11 P.M. Schuschnigg signs the 'protocol.' " [2] As Papen drove back with Schuschnigg in the sledge which conveyed them over the snow-covered roads to Salzburg he commented, "Yes, that is how the Fuehrer can be; now you have experienced it for yourself. But when you next come you will have a much easier time. The Fuehrer can be really charming."

The drama ran its course. Mussolini now sent a verbal message to Schuschnigg saying that he considered the Austrian attitude at Berchtesgaden to be both right and adroit. He assured him both of the unalterable attitude of Italy towards the Austrian question and of his personal friendship. On February 24 the Austrian Chancellor himself spoke to the Austrian Parliament, welcoming the settlement with Germany, but emphasising, with some sharpness, that beyond the specific

[2] *Nuremberg Documents* (H.M. Stationery Office), Pt. I, p. 249.

terms of the agreement Austria would never go. On March 3 he sent
a confidential message to Mussolini through the Austrian Military Attaché in Rome informing the Duce that he intended to strengthen the
political position in Austria by holding a plebiscite. Twenty-four hours
later he received a message from the Attaché describing his interview
with Mussolini. In this the Duce expressed himself optimistically.
The situation would improve. An imminent *détente* between Rome and
London would ensure a lightening of the existing pressure. . . . As to
the plebiscite Mussolini uttered a warning: *"E un errore"* (It's a mistake). "If the result is satisfactory, people will say that it is not
genuine. If it is bad, the situation of the Government will be unbear-
able; and if it is indecisive, then it is worthless." But Schuschnigg
was determined. On March 9 he announced officially that a plebiscite
would be held throughout Austria on the following Sunday, March 13.

At first nothing happened. Seyss-Inquart seemed to accept the idea
without demur. At 5.30 A.M. however on the morning of the 11th
Schuschnigg was rung up on the telephone from Police Headquarters
in Vienna. He was told: "The German frontier at Salzburg was closed
an hour ago. The German customs officials have been withdrawn.
Railway communications have been cut." The next message to reach
the Austrian Chancellor was from his Consul-General in Munich saying that the German Army Corps there had been mobilised: supposed
destination — Austria!

Later in the morning Seyss-Inquart came to announce that Goering
had just telephoned to him that the plebiscite must be called off within
an hour. If no reply was received within that time Goering would assume that Seyss-Inquart had been hindered from telephoning, and
would act accordingly. After being informed by responsible officials
that the police and Army were not entirely reliable, Schuschnigg informed Seyss-Inquart that the plebiscite would be postoned. A quarter
of an hour later the latter returned with a reply from Goering scribbled
on a message pad:

> The situation can only be saved if the Chancellor resigns im-
> mediately and if within two hours Dr. Seyss-Inquart is nomin-
> ated Chancellor. If nothing is done within this period the German
> invasion of Austria will follow.[3]

Schuschnigg waited on President Miklas to tender his resignation.
While in the President's room he received a deciphered message from
the Italian Government that they could offer no counsel. The old
President was obstinate: "So in the decisive hour I am left alone." He

[3] Schuschnigg, *op. cit.*, pp. 51–52, 66, 72.

steadfastly refused to nominate a Nazi Chancellor. He was determined to force the Germans into a shameful and violent deed. But for this they were well prepared. Orders were issued by Hitler to the German armed forces for the military occupation of Austria. Operation "Otto," so long studied, so carefully prepared, began. President Miklas confronted Seyss-Inquart and the Austrian Nazi leaders in Vienna with firmness throughout a hectic day. The telephone conversation between Hitler and Prince Philip of Hesse, his special envoy to the Duce, was quoted in evidence at Nuremberg, and is of interest:

> *Hesse:* I have just come back from the Palazzo Venezia. The Duce accepted the whole thing in a very friendly manner. He sends you his regards. He had been informed from Austria; von Schuschnigg gave him the news. He had then said it [*i.e.,* Italian intervention] would be a complete impossibility; it would be a bluff; such a thing could not be done. So he [Schuschnigg] was told that it was unfortunately arranged thus, and it could not be changed any more. Then Mussolini said that Austria would be immaterial to him.
>
> *Hitler:* Then please tell Mussolini I will never forget him for this.
>
> *Hesse:* Yes.
>
> *Hitler:* Never, never, never, whatever happens. I am still ready to make a quite different agreement with him.
>
> *Hesse:* Yes, I told him that too.
>
> *Hitler:* As soon as the Austrian affair has been settled I shall be ready to go with him through thick and thin; nothing matters.
>
> *Hesse:* Yes, my Fuehrer.
>
> *Hitler:* Listen. I will make any agreement — I am no longer in fear of the terrible position which would have existed militarily in case we had become involved in a conflict. You may tell him that I do thank him ever so much; never, never shall I forget that.
>
> *Hesse:* Yes, my Fuehrer.
>
> *Hitler:* I will never forget it, whatever may happen. If he should ever need any help or be in any danger he can be convinced that I shall stick to him whatever might happen, even if the whole world were against him.
>
> *Hesse:* Yes, my Fuehrer.[4]

Certainly when he rescued Mussolini from the Italian Provisional Government in 1943 Hitler kept his word.

*   *   *

[4] Schuschnigg, *op. cit.,* pp. 102–3, and *Nuremberg Documents,* I, pp. 258–9.

A triumphal entry into Vienna had been the Austrian Corporal's dream. On the night of Saturday, March 12, the Nazi Party in the capital had planned a torchlight procession to welcome the conquering hero. But nobody arrived. Three bewildered Bavarians of the supply services who had come by train to make billeting arrangements for the invading army had therefore to be carried shoulder-high through the streets. The cause of this hitch leaked out slowly. The German war machine had lumbered falteringly over the frontier and come to a standstill near Linz. In spite of perfect weather and road conditions the majority of the tanks broke down. Defects appeared in the motorised heavy artillery. The road from Linz to Vienna was blocked with heavy vehicles at a standstill. General von Reichenau, Hitler's special favourite, Commander of Army Group IV, was deemed responsible for a breakdown which exposed the unripe condition of the German Army at this stage in its reconstruction.

Hitler himself, motoring through Linz, saw the traffic jam, and was infuriated. The light tanks were disengaged from confusion and straggled into Vienna in the early hours of Sunday morning. The armoured vehicles and motorised heavy artillery were loaded onto railway trucks, and only thus arrived in time for the ceremony. The pictures of Hitler driving through Vienna amid exultant or terrified crowds are well known. But this moment of mystic glory had an unquiet background. The Fuehrer was in fact convulsed with anger at the obvious shortcomings of his military machine. He rated his generals, and they answered back. They reminded him of his refusal to listen to Fritsch and his warnings that Germany was not in a position to undertake the risk of a major conflict. Appearances were preserved. The official celebrations and parades took place. On the Sunday, after large numbers of German troops and Austrian Nazis had taken possession of Vienna, Hitler declared the dissolution of the Austrian Republic and the annexation of its territory to the German Reich.

*   *   *

Herr von Ribbentrop was at this time about to leave London to take up his duties as Foreign Secretary in Germany. Mr. Chamberlain gave a farewell luncheon in his honour at No. 10 Downing Street. My wife and I accepted the Prime Minister's invitation to attend. There were perhaps sixteen people present. My wife sat next to Sir Alexander Cadogan, near one end of the table. About halfway through the meal a Foreign Office messenger brought him an envelope. He opened it and was absorbed in the contents. Then he got up, walked round to where the Prime Minister was sitting, and gave him the message. Although Cadogan's demeanour would not have indicated that anything had happened, I could not help noticing the Prime Minister's evident preoccupation. Presently Cadogan came back with the paper and resumed his seat. Later I was told its contents. It said that Hitler had invaded Austria and that the German mechanised forces were advancing fast upon Vienna. The meal proceeded without the slightest interruption, but quite soon Mrs. Chamberlain, who had received some signal from her husband, got up, saying, "Let us *all* have coffee in the drawing-room." We trooped in there, and it was evident to me and perhaps to some others that Mr. and Mrs. Chamberlain wished to bring the proceedings to an end. A kind of general restlessness pervaded the company, and everyone stood about ready to say good-bye to the guests of honour.

However, Herr von Ribbentrop and his wife did not seem at all conscious of this atmosphere. On the contrary, they tarried for nearly half an hour engaging their host and hostess in voluble conversation. At one moment I came in contact with Frau von Ribbentrop, and in a valedictory vein I said, "I hope England and Germany will preserve their friendship." "Be careful you don't spoil it," was her graceful rejoinder. I am sure they both knew perfectly well what had happened, but thought it was a good manoeuvre to keep the Prime Minister away from his work and the telephone. At length Mr. Chamberlain said to the Ambassador, "I am sorry I have to go now to attend to urgent business," and without more ado he left the room. The Ribbentrops lingered on, so that most of us made our excuses and our way home. Eventually I suppose they left. This was the last time I saw Herr von Ribbentrop before he was hanged.

It was the Russians who now sounded the alarm, and on March 18 proposed a conference on the situation. They wished to discuss, if only in outline, ways and means of implementing the Franco-Soviet Pact within the frame of League action in the event of a major threat to peace by Germany. This met with little warmth in Paris and London. The French Government was distracted by other preoccupations. There were serious strikes in the aircraft factories. Franco's armies were driving deep into the territory of Communist Spain. Chamberlain was both sceptical and depressed. He profoundly disagreed with my interpretation of the dangers ahead and the means of combating them. I had been urging the prospects of a Franco-British-Russian alliance as the only hope of checking the Nazi onrush.

Mr. Feiling tells us that the Prime Minister expressed his mood in a letter to his sister on March 20:

> The plan of the "Grand Alliance," as Winston calls it, had occurred to me long before he mentioned it. . . . I talked about it

to Halifax, and we submitted it to the Chiefs of Staff and F.O. experts. It is a very attractive idea; indeed, there is almost everything to be said for it until you come to examine its practicability. From that moment its attraction vanishes. You have only to look at the map to see that nothing that France or we could do could possibly save Czechoslovakia from being overrun by the Germans, if they wanted to do it. I have therefore abandoned any idea of giving guarantees to Czechoslovakia, or to the French in connection with her obligations to that country.[5]

Here was at any rate a decision. It was taken on wrong arguments. In modern wars of great nations or alliances particular areas are not defended only by local exertions. The whole vast balance of the war front is involved. This is still more true of policy before war begins and while it may still be averted. It surely did not take much thought from the "Chiefs of Staff and F.O. experts" to tell the Prime Minister that the British Navy and the French Army could not be deployed on the Bohemian mountain front to stand between the Czechoslovak Republic and Hitler's invading army. This was indeed evident from the map. But the certainty that the crossing of the Bohemian frontier line would have involved a general European war might well even at that date have deterred or delayed Hitler's next assault. How erroneous Mr. Chamberlain's private and earnest reasoning appears when we cast our minds forward to the guarantee he was to give to Poland *within a year,* after all the strategic value of Czechoslovakia had been cast away, and Hitler's power and prestige had almost doubled!

*   *   *

The reader is now invited to move westward to the Emerald Isle. "It's a long way to Tipperary," but a visit there is sometimes irresistible. In the interval between Hitler's seizure of Austria and his unfolding design upon Czechoslovakia we must turn to a wholly different kind of misfortune which befell us.

Since the beginning of 1938 there had been negotiations between the British Government and that of Mr. de Valera in Southern Ireland, and on April 25 an agreement was signed whereby, among other matters, Great Britain renounced all rights to occupy for naval purposes the two Southern Irish ports of Queenstown and Berehaven, and the base in Lough Swilly. The two southern ports were a vital feature in the naval defence of our food supply. When in 1922 as Colonial and Dominions Secretary I had dealt with the details of the Irish Settle-

[5] Keith Feiling, *Life of Neville Chamberlain,* pp. 347–8.

ment which the Cabinet of those days had made, I brought Admiral Beatty to the Colonial Office to explain to Michael Collins the importance of these ports to our whole system of bringing supplies into Britain. Collins was immediately convinced. "Of course you must have the ports," he said; "they are necessary for your life." Thus the matter was arranged, and everything had worked smoothly in the sixteen years that had passed. The reason why Queenstown and Berehaven were necessary to our safety is easy to understand. They were the fuelling bases from which our destroyer flotillas ranged westward into the Atlantic to hunt U-boats and protect incoming convoys as they reached the throat of the narrow seas. Lough Swilly was similarly needed to protect the approaches to the Clyde and Mersey. To abandon these meant that our flotillas would have to start in the north from Lamlash and in the south from Pembroke Dock or Falmouth, thus decreasing their radius of action and the protection they could afford by more than 400 miles out and home.

It was incredible to me that the Chiefs of Staff should have agreed to throw away this major security, and to the last moment I thought that at least we had safeguarded our right to occupy these Irish ports in the event of war. However, Mr. de Valera announced in the Dail that no conditions of any kind were attached to the cession. I was later assured that Mr. de Valera was surprised at the readiness with which the British Government had deferred to his request. He had included it in his proposals as a bargaining counter which could be dispensed with when other points were satisfactorily settled.

Lord Chatfield has in his last book devoted a chapter to explaining the course he and the other Chiefs of Staff took.[6] This should certainly be read by those who wish to pursue the subject. Personally I remain convinced that the gratuitous surrender of our right to use the Irish ports in war was a major injury to British national life and safety. A more feckless act can hardly be imagined — and at such a time. It is true that in the end we survived without the ports. It is also true that if we had not been able to do without them we should have retaken them by force rather than perish by famine. But this is no excuse. Many a ship and many a life were soon to be lost as the result of this improvident example of appeasement.

[6] Lord Chatfield, *It Might Happen Again*, Chapter XVIII.

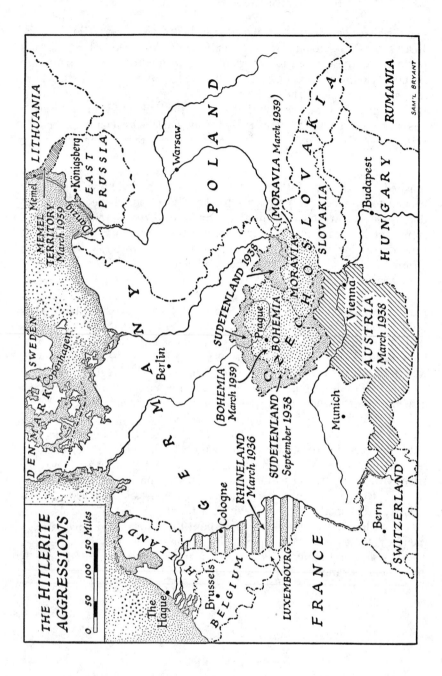

THE *HITLERITE*
AGGRESSIONS

0   50   100   150 *Miles*

DENMARK

Copenhagen

SWEDEN

LITHUANIA

Memel

MEMEL
TERRITORY
March 1939

Königsberg

EAST
PRUSSIA

Danzig

Warsaw

P O L A N D

G   E   R   M   A   N   Y

Berlin

SUDETENLAND 1938

Prague

(BOHEMIA
March 1939)

BOHEMIA

MORAVIA

(MORAVIA *March 1939*)

C Z E C H O S L O V A K I A

SLOVAKIA

Budapest

H U N G A R Y

RUMANIA

SAM'L BRYANT

RHINELAND
March 1936

Cologne

SUDETENLAND
September 1938

Munich

Vienna

AUSTRIA
March 1938

HOLLAND

The
Hague

Brussels

B E L G I U M

LUXEMBOURG

F R A N C E

Bern

SWITZERLAND

13

Czechoslovakia

WHILE THE INVASION of Austria was in full swing Hitler said in the motorcar to General von Halder, "This will be very inconvenient to the Czechs." Halder saw immediately the significance of this remark. To him it lighted up the future. It showed him Hitler's intentions, and at the same time, as he viewed it, Hitler's military ignorance. "It was practically impossible," he has explained, "for a German army to attack Czechoslovakia from the south. The single railway line through Linz was completely exposed, and surprise was out of the question." But Hitler's main political strategic conception was correct. The West Wall was growing, and, although far from complete, already confronted the French Army with horrible memories of the Somme and Passchendaele. He was convinced that neither France nor Britain would fight.

On the day of the march of the German armies into Austria we heard that Goering had given a solemn assurance to the Czech Minister in Berlin that Germany had *"no evil intentions towards Czechoslovakia."* On March 14 the French Premier, M. Blum, solemnly declared to the Czech Minister in Paris that France would unreservedly honour her engagement to Czechoslovakia. These diplomatic reassurances could not conceal the grim reality. The whole strategic position on the Continent had changed. The German arguments and armies could now concentrate directly upon the western frontiers of Czechoslovakia, whose border districts were German in racial character, with an aggressive and active German Nationalist Party eager to act as a fifth column in the event of trouble.

*　　　*　　　*

In the hope of deterring Germany the British Government, in accordance with Mr. Chamberlain's resolve, sought a settlement with

Italy in the Mediterranean. This would strengthen the position of France, and would enable both the French and British to concentrate upon events in Central Europe. Mussolini, to some extent placated by the fall of Eden, and feeling himself in a strong bargaining position, did not repulse the British repentance. On April 16, 1938, an Anglo-Italian agreement was signed, giving Italy in effect a free hand in Abyssinia and Spain in return for the imponderable value of Italian goodwill in Central Europe. The Foreign Office was sceptical of this transaction. Mr. Chamberlain's biographer tells us that he wrote in a personal and private letter, "You should have seen the draft put up to me by the F.O.; it would have frozen a Polar bear." [1]

I shared the misgivings of the Foreign Office at this move.

Hitler was watching the scene with vigilance. To him also the ultimate alignment of Italy in a European crisis was important. In conference with his Chiefs of Staff at the end of April he was considering how to force the pace. Mussolini wanted a free hand in Abyssinia. In spite of the acquiescence of the British Government, he might ultimately need German support in this venture. If so, he should accept German action against Czechoslovakia. This issue must be brought to a head, and in the settling of the Czech question Italy would be involved on Germany's side. The declarations of British and French statesmen were of course studied in Berlin. The intention of these Western Powers to persuade the Czechs to be reasonable in the interests of European peace was noted with satisfaction. The Nazi Party of the Sudetenland, led by Henlein, now formulated their demands for autonomy in the German-border regions of that country, and the British and French Ministers in Prague called on the Czech Foreign Minister shortly afterwards to "express the hope that the Czech Government will go to the furthest limit in order to settle the question."

During May the Germans in Czechoslovakia were ordered to increase their agitation. Municipal elections were due, and the German Government began a calculated war of nerves. Persistent rumours already circulated of German troop movements towards the Czech frontier. German denials did not reassure the Czechs, who on the night of May 20–21 decreed a partial mobilisation of their Army.

Hitler had for some time been convinced that neither France nor Britain would fight for Czechoslovakia. On May 28 he called a meeting of his principal advisers and gave instructions for the preparations to attack Czechoslovakia. His service advisers, however, did not share unanimously his overwhelming confidence. The German generals could not be persuaded, considering the still enormous preponderance of Allied strength except in the air, that France and Britain would submit to the Fuehrer's challenge. To break the Czech Army and pierce or

---

[1] Keith Feiling, *Life of Neville Chamberlain*, p. 350.

turn the Bohemian fortress line would require practically the whole of thirty-five divisions. The German Chiefs of Staff informed Hitler that the Czech Army must be considered efficient and up to date in arms and equipment. The fortifications of the West Wall or Siegfried Line, though already in existence as field works, were far from completed: and at the moment of attacking the Czechs only five effective and eight Reserve divisions would be available to protect the whole of Germany's western frontier against the French Army, which could mobilise a hundred divisions. The generals were aghast at running such risks, when by waiting a few years the German Army would again be master. Although Hitler's political judgment had been proved correct by the pacifism and weakness of the Allies about conscription, the Rhineland, and Austria, the German High Command could not believe that Hitler's bluff would succeed a fourth time. It seemed so much beyond the bounds of reason that great victorious nations, possessing evident military superiority, would once again abandon the path of duty and honour, which was also for them the path of common sense and prudence. Besides all this there was Russia, with her Slav affinities with Czechoslovakia, and whose attitude towards Germany at this juncture was full of menace.

The relations of Soviet Russia with Czechoslovakia as a state, and personally with President Beneš, were those of intimate and solid friendship. The roots of this lay in a certain racial affinity, and also in comparatively recent events which require a brief digression. When President Beneš visited me at Marrakesh in January 1944 he told me this story. In 1935 he had received an offer from Hitler to respect in all circumstances the integrity of Czechoslovakia in return for a guarantee that she would remain neutral in the event of a Franco-German war. When Beneš pointed to his treaty obliging him to act with France in such a case, the German Ambassador replied that there was no need to denounce the treaty. It would be sufficient to break it, if and when the time came, by simply failing to mobilise or march. The small republic was not in a position to indulge in indignation at such a suggestion. Their fear of Germany was already very grave, more especially as the question of the Sudeten Germans might at any time be raised and fomented by Germany, to their extreme embarrassment and growing peril. They therefore let the matter drop without comment or commitment, and it did not stir for more than a year. In the autumn of 1936 a message from a high military source in Germany was conveyed to President Beneš to the effect that if he wanted to take advantage of the Fuehrer's offer he had better be quick because events would shortly take place in Russia rendering any help he could give to Germany insignificant.

While Beneš was pondering over this disturbing hint he became aware that communications were passing through the Soviet Embassy in Prague between important personages in Russia and the German Government. This was a part of the so-called military and Old Guard Communist conspiracy to overthrow Stalin and introduce a new régime based on a pro-German policy. President Beneš lost no time in communicating all he could find out to Stalin.[2] Thereafter there followed the merciless, but perhaps not needless, military and political purge in Soviet Russia, and the series of trials in January 1937, in which Vyshinsky, the Public Prosecutor, played so masterful a part.

Although it is highly improbable that the Old-Guard Communists had made common cause with the military leaders, or vice versa, they were certainly filled with jealousy of Stalin, who had ousted them. It may therefore have been convenient to get rid of them at the same time, according to the standards maintained in a totalitarian state. Zinoviev, Bukharin, and others of the original leaders of the Revolution, Marshal Tukachevsky, who had been invited to represent the Soviet Union at the Coronation of King George VI, and many other high officers of the Army, were shot. In all not less than five thousand officers and officials above the rank of captain were "liquidated." The Russian Army was purged of its pro-German elements at a heavy cost to its military efficiency. The bias of the Soviet Government was turned in a marked manner against Germany. Stalin was conscious of a personal debt to President Beneš, and a very strong desire to help him and his threatened country against the Nazi peril animated the Soviet Government. The situation was of course thoroughly understood by Hitler; but I am not aware that the British and French Governments were equally enlightened. To Mr. Chamberlain and the British and French General Staffs the purge of 1937 presented itself mainly as a tearing to pieces internally of the Russian Army, and a picture of the Soviet Union as riven asunder by ferocious hatreds and vengeance. This was perhaps an excessive view; for a system of government founded on terror may well be strengthened by a ruthless and successful assertion of its power. The salient fact for the purposes of this account is the close association of Russia and Czechoslovakia, and of Stalin and Beneš.

But neither the internal stresses in Germany nor the ties between Beneš and Stalin were known to the outside world, or appreciated by the British and French Ministers. The Siegfried Line, albeit unperfected, seemed a fearful deterrent. The exact strength and fighting power of the German Army, new though it was, could not be accu-

[2] There is however some evidence that Beneš's information had previously been imparted to the Czech police by the Ogpu, who wished it to reach Stalin from a friendly foreign source. This did not detract from Beneš's service to Stalin, and is therefore irrelevant.

rately estimated and was certainly exaggerated. There were also the unmeasured dangers of air attack on undefended cities. Above all there was the hatred of war in the hearts of the democracies.

Nevertheless on June 12 M. Daladier, now Premier of France, renewed his predecessor's pledge of March 14, and declared that France's engagements towards Czechoslovakia "are sacred, and cannot be evaded." This considerable statement sweeps away all chatter about the Treaty of Locarno thirteen years before having by implication left everything in the East vague pending an Eastern Locarno. There can be no doubt before history that the treaty between France and Czechoslovakia of 1924 had complete validity not only in law but in fact, and that this was reaffirmed by successive heads of the French Government in all the circumstances of 1938.

But on this subject Hitler was convinced that his judgment alone was sound, and on June 18 he issued a final directive for the attack on Czechoslovakia, in the course of which he sought to reassure his anxious generals. "I will decide," he told Keitel, "to take action against Czechoslovakia only if I am firmly convinced, as in the case of the demilitarised zone [of the Rhineland] and the entry into Austria, that France will not march, and that therefore England will not intervene." [3]

On July 26, 1938, Chamberlain announced to Parliament the mission of Lord Runciman to Prague with the object of seeking a solution there by arrangements between the Czech Government and Herr Henlein. On the following day the Czechs issued a draft statute for national minorities to form a basis of negotiations. On the same day Lord Halifax stated in Parliament: *"I do not believe that those responsible for the Government of any country in Europe today want war."* On August 3 Lord Runciman reached Prague, and a series of interminable and complicated discussions took place with the various interested parties. Within a fortnight these negotiations broke down, and from this point events moved rapidly.

On August 27 Ribbentrop, now Foreign Minister, reported a visit which he had received from the Italian Ambassador in Berlin, who "had received another written instruction from Mussolini asking that Germany would communicate in time the probable date of action against Czechoslovakia." Mussolini asked for such notification in order "to be able to take in due time the necessary measures on the French frontier."

*   *   *

Anxiety grew steadily during August, and in the afternoon of September 2 I received a message from the Soviet Ambassador that he

---

[3] *Nuremberg Documents*, Pt. II, No. 10.

would like to come down to Chartwell and see me at once upon a matter of urgency. I had for some time had friendly personal relations with M. Maisky, who also saw a good deal of my son Randolph. I thereupon received the Ambassador, and after a few preliminaries he told me in precise and formal detail the story here set out. Before he had got very far I realised that he was making a declaration to me, a private person, because the Soviet Government preferred this channel to a direct offer to the Foreign Office, which might have encountered a rebuff. It was clearly intended that I should report what I was told to His Majesty's Government. This was not actually stated by the Ambassador, but it was implied by the fact that no request for secrecy was made. As the matter struck me at once as being of the first importance, I was careful not to prejudice its consideration by Halifax and Chamberlain by proceeding to commit myself in any way, or use language which would excite controversy between us.

The substance of what he told me was as follows:

The French Chargé d' Affaires in Moscow (the Ambassador being on leave) had that same day called upon M. Litvinov and, in the name of the French Government, asked what aid Russia would give to Czechoslovakia against a German attack, having regard particularly to the difficulties which might be created by the neutrality of Poland or Rumania. Litvinov replied that the Soviet Union had resolved to fulfil their obligations. He recognised the difficulties created by the attitude of Poland and Rumania, but thought that in the case of Rumania these could be overcome. If, for instance, the League decided that Czechoslovakia was the victim of aggression and that Germany was the aggressor, that would probably persuade Rumania to allow Russian troops and air forces to pass through her territory.

Even if the Council of the League were not unanimous, M. Litvinov thought a majority decision would be sufficient, and that Rumania would probably associate herself with it. He therefore advised that the Council should be invoked under Article 11, on the ground that there was danger of war, and that the League Powers should consult together. The sooner this was done the better, as time might be very short. Staff conversations ought to take place between Russia, France, and Czechoslovakia immediately about ways and means of giving assistance.

He also advocated consultation among the peaceful Powers about the best method of preserving peace, with a view, perhaps, to a joint declaration including France, Russia, and Great Britain. He believed that the United States would give moral support to such a declaration.

Thus M. Maisky. I said it was unlikely the British Government would consider any further step until or unless there was a fresh breakdown in the Henlein-Beneš negotiations, in which the fault could not on any account be attributed to the Government of Czechoslovakia. We should not want to irritate Hitler, if his mind was really turning towards a peaceful solution.

I sent a report of all this to Lord Halifax as soon as I had dictated it. He replied on September 5 in a guarded manner, that he did not at present feel that action of the kind proposed under Article 11 would be helpful, but that he would keep it in his mind. "For the present, I think, as you indicated, we must review the situation in the light of the report with which Henlein has returned from Berchtesgaden." He added that the situation remained very anxious.

In its leading article of September 7 the *Times* stated:

> If the Sudetens now ask for more than the Czech Government are ready to give in their latest set of proposals, it can only be inferred that the Germans are going beyond the mere removal of disabilities for those who do not find themselves at ease within the Czechoslovak Republic. In that case it might be worth while for the Czechoslovak Government to consider whether they should exclude altogether the project, which has found favour in some quarters, of making Czechoslovakia a more homogeneous state by the cession of that fringe of alien populations who are contiguous to the nation to which they are united by race.

This of course involved the surrender of the whole of the Bohemian fortress line. Although the British Government stated at once that this *Times* article did not represent their views, public opinion abroad, particularly in France, was far from reassured. M. Bonnet, then French Foreign Minister, declares that on September 10, 1938, he put the following question to our Ambassador in Paris, Sir Eric Phipps: "To-morrow Hitler may attack Czechoslovakia. If he does France will mobilise at once. She will turn to you, saying, 'We march: do you march with us?' What will be the answer of Great Britain?"

The following was the answer approved by the Cabinet, sent by Lord Halifax through Sir Eric Phipps on the 12th:

> I naturally recognise of what importance it would be to the French Government to have a plain answer to such a question. But, as you pointed out to Bonnet, the question itself, though plain in form, cannot be dissociated from the circumstances in which it might be posed, which are necessarily at this stage completely hypothetical.

Moreover, in this matter it is impossible for His Majesty's Government to have regard only to their own position, inasmuch as in any decision they may reach or action they may take they would, in fact, be committing the Dominions. Their Governments would quite certainly be unwilling to have their position in any way decided for them in advance of the actual circumstances, of which they would desire themselves to judge.

So far therefore as I am in a position to give any answer at this stage to M. Bonnet's question, it would have to be that while His Majesty's Government would never allow the security of France to be threatened, they are unable to make precise statements of the character of their future action, or the time at which it would be taken, in circumstances that they cannot at present foresee.[4]

Upon the statement that "His Majesty's Government would never allow the security of France to be threatened" the French asked what aid they could expect if it were. The reply from London was, according to Bonnet, two divisions, not motorised, and one hundred and fifty aeroplanes during the first six months of the war. If M. Bonnet was seeking for an excuse for leaving the Czechs to their fate, it must be admitted that his search had met with some success.

On this same September 12 Hitler delivered at a Nuremberg Party rally a violent attack on the Czechs, who replied on the following day by the establishment of martial law in certain districts of the republic. On September 14 negotiations with Henlein were definitely broken off, and on the 15th the Sudeten leader fled to Germany.

The summit of the crisis was now reached.

[4] Printed in Georges Bonnet, *De Washington au Quai d'Orsay*, pp. 360–61.

14

The Tragedy of Munich

M A N Y   V O L U M E S have been written, and will be written, upon
the crisis that was ended at Munich by the sacrifice of Czechoslovakia, and it is only intended here to give a few of the cardinal facts
and establish the main proportions of events. At the Assembly of the
League of Nations on September 21 an official warning was given by
Litvinov:

> . . . At the present time Czechoslovakia is suffering interference
> in its internal affairs at the hands of a neighbouring state, and
> is publicly and loudly menaced with attack. One of the oldest,
> most cultured, most hard-working of European peoples, who
> acquired their independence after centuries of oppression, today
> or tomorrow may decide to take up arms in defence of that inde-
> pendence. . . . When, a few days before I left for Geneva, the
> French Government for the first time inquired as to our attitude
> in the event of an attack on Czechoslovakia, I gave in the name
> of my Government the following perfectly clear and unambiguous
> reply:
> "We intend to fulfil our obligations under the Pact, and to-
> gether with France to afford assistance to Czechoslovakia by the
> ways open to us. Our War Department is ready immediately
> to participate in a conference with representatives of the French
> and Czechoslovak War Departments, in order to discuss the
> measures appropriate to the moment. . . ."
> It was only two days ago that the Czechoslovak Government
> addressed a formal inquiry to my Government as to whether the
> Soviet Union is prepared, in accordance with the Soviet-Czech
> Pact, to render Czechoslovakia immediate and effective aid if

France, loyal to her obligations, will render similar assistance, to which my Government gave a clear answer in the affirmative.

This public, and unqualified, declaration by one of the greatest Powers concerned played no part in Mr. Chamberlain's negotiations, or in the French conduct of the crisis. The Soviet offer was in effect ignored. They were not brought into the scale against Hitler, and were treated with an indifference — not to say disdain — which left a mark in Stalin's mind. Events took their course as if Soviet Russia did not exist. For this we afterwards paid dearly.

On the evening of the 26th Hitler spoke in Berlin. He referred to England and France in accommodating phrases, launching at the same time a coarse and brutal attack on Beneš and the Czechs. He said categorically that the Czechs must clear out of the Sudetenland, but once this was settled he had no more interest in what happened to Czechoslovakia. *"This is the last territorial claim I have to make in Europe."* About eight o'clock that night Mr. Leeper, then Head of the Foreign Office Press Department, presented to the Foreign Secretary a communiqué of which the following is the pith:

> If, in spite of the efforts made by the British Prime Minister, a German attack is made upon Czechoslovakia, the immediate result must be that France will be bound to come to her assistance, and Great Britain *and Russia* will certainly stand by France.

This was approved by Lord Halifax and immediately issued. It seemed that the moment of clash had arrived and that the opposing forces were aligned. The Czechs had a million and a half men armed behind the strongest fortress line in Europe, and equipped by a highly organised and powerful industrial machine. The French Army was partly mobilised, and, albeit reluctantly, the French Ministers were prepared to honour their obligations to Czechoslovakia. At 11.20 A.M. on September 28 orders to the British to mobilise were issued from the Admiralty.

*   *   *

There had already begun an intense, unceasing struggle between the Fuehrer and his expert advisers. The crisis seemed to provide all the circumstances which the German generals dreaded. Between thirty and forty Czech divisions were deploying upon Germany's eastern frontiers, and the weight of the French Army, at odds of nearly eight to one, began to lie heavy on the Western Wall. A hostile Russia might operate from Czech airfields, and Soviet armies might wend their way forward through Poland or Rumania. Some of them made a plot to arrest Hitler and "immunize Germany from this madman." Others declared that the low morale of the German population was incapable of sustaining a European war, and the German armed forces were not yet ready for it. Admiral Raeder, Chief of the German Admiralty, made a vehement appeal to the Fuehrer which was emphasised a few hours later by the news that the British Fleet was being mobilised. And Hitler wavered. At 2 A.M. the German radio broadcast an official denial that Germany intended to mobilise on the 29th, and at 11.45 A.M. the same morning a similar statement of the German official news agency was given to the British press. The strain upon this one man and upon his astounding will power must at this moment have been most severe. Evidently he had brought himself to the brink of a general war. Could he take the plunge in the face of an unfavourable public opinion and of the solemn warning of the chiefs of his army, navy, and air force? Could he, on the other hand, afford to retreat after living so long upon prestige?

But Mr. Chamberlain was also active, and he was now in complete control of Britain's foreign policy. Lord Halifax, in spite of increasing doubts derived from the atmosphere of his department, followed the guidance of his chief. The Cabinet was deeply perturbed, but obeyed. The Government majority in the House of Commons was skilfully handled by the Whips. One man and one man only conducted our affairs. He did not shrink either from the responsibility which he incurred or from the personal exertions required. On September 14 he had telegraphed to Hitler on his own initiative proposing to come to see him. Three times in all the British Prime Minister flew to Germany, both he and Lord Runciman being convinced that only the cession of the Sudeten areas would dissuade Hitler from invading Czechoslovakia. The last occasion was at Munich, M. Daladier, Premier of France, and Mussolini being present. No invitation was extended to Russia. Nor were the Czechs themselves allowed to be present at the meetings. The Czech Government were informed in bald terms on the evening of the 28th that a conference of the representatives of the four European Powers would take place the following day. Agreement was reached between "the Big Four" with speed. The conversations began at noon and lasted till two o'clock the next morning. A memorandum was drawn up and signed at 2 A.M. on September 30. It was in essentials the acceptance of the German demands. The Sudetenland was to be evacuated in five stages beginning on October 1, and to be completed within ten days. An international commission was to determine the final frontiers.

The document was placed before the Czech delegates. They bowed to the decisions. "They wished," they said, "to register their protest before the world against a decision in which they had no part." President Beneš resigned because "he might now prove a hindrance to the developments to which our new State must adapt itself." He departed from Czechoslovakia and found shelter in England. The dismemberment of Czechoslovakia proceeded. The Germans were not the only vultures upon the carcase. The Polish Government sent a twenty-four-hour ultimatum to the Czechs demanding the immediate handing over of the frontier district of Teschen. There was no means of resisting this harsh demand. The Hungarians also arrived with their claims.

*   *   *

While the four statesmen were waiting for the experts to draft the final document the Prime Minister asked Hitler whether he would care for a private talk. Hitler "jumped at the idea." The two leaders met in Hitler's Munich flat on the morning of September 30, and were alone except for the interpreter. Chamberlain produced a draft declaration which he had prepared, proclaiming that "the question of Anglo-German relations is of the first importance for the two countries and for Europe," and that "We regard the Agreement signed last night, and the Anglo-German Naval Agreement, as symbolic of the desire of our two peoples never to go to war with one another again."

Hitler read it and signed it without demur.

Chamberlain returned to England. At Heston, where he landed, he waved the joint declaration which he had got Hitler to sign, and read it to the crowd of notables and others who welcomed him. As his car drove through cheering crowds from the airport, he said to Halifax, sitting beside him, "All this will be over in three months"; but from the windows of Downing Street he waved his piece of paper again and used these words, "This is the second time in our history that there has come back from Germany to Downing Street peace with honour. I believe it is peace for our time." [1]

Hitler's judgment had been once more decisively vindicated. The German General Staff was utterly abashed. Once again the Fuehrer had been right after all. He with his genius and intuition alone had truly measured all the circumstances, military and political. Once again, as in the Rhineland, the Fuehrer's leadership had triumphed over the obstruction of the German military chiefs. All these generals were patriotic men. They longed to see the Fatherland regain its position in the world. They were devoting themselves night and day to every process that could strengthen the German forces. They therefore felt smitten in their hearts at having been found so much below the level

---

[1] See Keith Feiling, *Life of Neville Chamberlain*, pp. 376, 381

of the event, and in many cases their dislike and their distrust of Hitler were overpowered by admiration for his commanding gifts and miraculous luck. Surely here was a star to follow, surely here was a guide to obey. Thus did Hitler finally become the undisputed master of Germany, and the path was clear for the great design. The conspirators lay low, and were not betrayed by their military comrades.

* * *

It is not easy in these latter days, when we have all passed through years of intense moral and physical stress and exertion, to portray for another generation the passions which raged in Britain about the Munich Agreement. Among the Conservatives families and friends in intimate contact were divided to a degree the like of which I have never seen. Men and women, long bound together by party ties, social amenities, and family connections, glared upon one another in scorn and anger. The issue was not one to be settled by the cheering crowds which had welcomed Mr. Chamberlain back from the airport or blocked Downing Street and its approaches, nor by the redoubtable exertions of the Ministerial Whips and partisans. We who were in a minority at the moment cared nothing for the jokes or scowls of the Government supporters. The Cabinet was shaken to its foundations, but the event had happened and they held together. One Minister alone stood forth. The First Lord of the Admiralty, Mr. Duff Cooper, resigned his great office, which he had dignified by the mobilisation of the Fleet. At the moment of Mr. Chamberlain's overwhelming mastery of public opinion he thrust his way through the exulting throng to declare his total disagreement with its leader.

At the opening of the three days' debate on Munich he made his resignation speech. This was a vivid incident in our parliamentary life. Speaking with ease and without a note, for forty minutes he held the hostile majority of his party under his spell. It was easy for Labour men and Liberals in hot opposition to the Government of the day to applaud him. This was a rending quarrel within the Tory Party.

The debate which took place was not unworthy of the emotions aroused and the issues at stake. I well remember that when I said "We have sustained a total and unmitigated defeat" the storm which met me made it necessary to pause for a while before resuming. There was widespread and sincere admiration for Mr. Chamberlain's persevering and unflinching efforts to maintain peace, and for the personal exertions which he had made. It is impossible in this account to avoid marking the long series of miscalculations, and misjudgments of men and facts, on which he based himself; but the motives which inspired him have never been impugned, and the course he followed required the highest degree of moral courage. To this I paid tribute two years later in my speech after his death.

There was also a serious and practical line of argument, albeit not to their credit, on which the Government could rest themselves. No one could deny that we were hideously unprepared for war. Who had been more forward in proving this than I and my friends? Great Britain had allowed herself to be far surpassed by the strength of the German Air Force. All our vulnerable points were unprotected. Barely a hundred anti-aircraft guns could be found for the defence of the largest city and centre of population in the world; and these were largely in the hands of untrained men. If Hitler was honest and lasting peace had in fact been achieved, Chamberlain was right. If, unhappily, he had been deceived, at least we should gain a breathing space to repair the worst of our neglects. These considerations, and the general relief and rejoicing that the horrors of war had been temporarily averted, commanded the loyal assent of the mass of Government supporters. The House approved the policy of His Majesty's Government "by which war was averted in the recent crisis" by 366 to 144. The thirty or forty dissentient Conservatives could do no more than register their disapproval by abstention. This we did as a formal and united act.

On November 1 a nonentity, Dr. Hacha, was elected to the vacant Presidency of the remnants of Czechoslovakia. A new Government took office in Prague. "Conditions in Europe and the world in general," said the Foreign Minister of this forlorn administration, "are not such that we should hope for a period of calm in the near future." Hitler thought so too. A formal division of the spoils was made by Germany at the beginning of November. Poland was not disturbed in her occupation of Teschen. The Slovaks, who had been used as a pawn by Germany, obtained a precarious autonomy. Hungary received a piece of flesh at the expense of Slovakia. When these consequences of Munich were raised in the House of Commons Mr. Chamberlain explained that the French and British offer of an international guarantee to Czechoslovakia which had been given after the Munich Pact did not affect the existing frontiers of that State, but referred only to the hypothetical question of unprovoked aggression. "What we are doing now," he said, with much detachment, "is witnessing the readjustment of frontiers laid down in the Treaty of Versailles. I do not know whether the people who were responsible for those frontiers thought they would remain permanently as they were laid down. I doubt very much whether they did. They probably expected that from time to time the frontiers would have to be adjusted. . . . I think I have said enough about Czechoslovakia. . . ." There was, however, to be a later occasion.

*   *   *

The question has been debated whether Hitler or the Allies gained the more in strength in the year that followed Munich. Many persons in Britain who knew our nakedness felt a sense of relief as each month our air force developed and the Hurricane and Spitfire types approached issue. The number of formed squadrons grew and the anti-aircraft guns multiplied. Also the general pressure of industrial preparation for war continued to quicken. But these improvements, invaluable though they seemed, were petty compared with the mighty advance in German armaments. As has been explained, munitions production on a nation-wide plan is a four years' task. The first year yields nothing, the second very little, the third a lot, and the fourth a flood. Hitler's Germany in this period was already in the third or fourth year of intense preparation under conditions of grip and drive which were almost the same as those of war. Britain, on the other hand, had only been moving on a non-emergency basis, with a weaker impulse and on a far smaller scale. In 1938-39 British military expenditure of all kinds reached £304 millions,[2] and German was at least £1500 millions. It is probable that in this last year before the outbreak Germany manufactured at least double, and possibly treble, the munitions of Britain and France put together, and also that her great plants for tank production reached full capacity. They were therefore getting weapons at a far higher rate than we.

The subjugation of Czechoslovakia robbed the Allies of the Czech Army of twenty-one regular divisions, fifteen or sixteen second-line divisions already mobilised, and also their mountain fortress line, which in the days of Munich had required the deployment of thirty German divisions, or the main strength of the mobile and fully trained German Army. According to Generals Halder and Jodl, there were but thirteen German divisions, of which only five were composed of first-line troops, left in the West at the time of the Munich arrangement. We certainly suffered a loss through the fall of Czechoslovakia equivalent to some thirty-five divisions. Besides this the Skoda works, the second most important arsenal in Central Europe, the production of which between August 1938 and September 1939 was in itself nearly equal to the actual output of British arms factories in that period, was made to change sides adversely. While all Germany was working under intense and almost war pressure, French labour had achieved as early as 1936 the long-desired forty-hour week.

Even more disastrous was the alteration in the relative strength of the French and German Armies. With every month that passed, from 1938 onward, the German Army not only increased in numbers and formations and in the accumulation of reserves, but in quality and

[2] 1937-38, £234 millions; 1938-39, £304 millions; 1939-40, £367 millions.

maturity. The advance in training and general proficiency kept pace
with the ever-augmenting equipment. No similar improvement or expansion was open to the French Army. It was being overtaken along
every path. In 1935 France, unaided by her previous allies, could have
invaded and reoccupied Germany almost without serious fighting. In
1936 there could still be no doubt of her overwhelmingly superior
strength. We now know, from the German revelations, that this continued in 1938, and it was the knowledge of their weakness which led
the German High Command to do their utmost to restrain Hitler from
every one of the successful strokes by which his fame was enhanced. In
the year after Munich, which we are now examining, the German Army,
though still weaker in trained reserves than the French, approached
its full efficiency. As it was based upon a population double as large
as that of France, it was only a question of time when it would become
by every test the stronger. In morale also the Germans had the advantage. The desertion of an ally, especially from fear of war, saps the
spirit of any army. The sense of being forced to yield depresses both
officers and men. While on the German side confidence, success, and
the sense of growing power inflamed the martial instincts of the race, the
admission of weakness discouraged the French soldiers of every rank.

*   *   *

There was however one vital sphere in which we began to overtake
Germany and improve our own position. In 1938 the process of replacing British biplane fighters, like the Gladiators, by the then modern
types of Hurricanes and later Spitfires had only just begun. In
September of 1938 we had but five squadrons remounted on Hurricanes. Moreover, reserves and spares for the older aircraft had been
allowed to drop, since they were going out of use. The Germans were
well ahead of us in remounting with modern fighter types. They already had good numbers of the Me.109, against which our old aircraft
would have fared very ill. Throughout 1939 our position improved
as more squadrons were remounted. In July of that year we had
twenty-six squadrons of modern eight-gun fighters, though there had
been little time to build up a full scale of reserves and spares. By July
1940, at the time of the Battle of Britain, we had on the average
forty-seven squadrons of modern fighters available.

The Germans on their side had in fact done most of their air expansion both in quantity and quality before the war began. Our effort
was later than theirs by nearly two years. Between 1939 and 1940 they
made a twenty per cent increase only, whereas our increase in modern
fighter aircraft was eighty per cent. The year 1938 in fact found us
sadly deficient in quality, and although by 1939 we had gone some way

towards meeting the disparity we were still relatively worse off than in 1940, when the test came.

We might in 1938 have had air raids on London, for which we were lamentably unprepared. There was however no possibility of a decisive Air Battle of Britain until the Germans had occupied France and the Low Countries, and thus obtained the necessary bases in close striking distance of our shores. Without these bases they could not have escorted their bombers with the fighter aircraft of those days. The German armies were not capable of defeating the French in 1938 or 1939.

The vast tank production with which they broke the French front did not come into existence till 1940, and, in the face of the French superiority in the West and an unconquered Poland in the East, they could certainly not have concentrated the whole of their air power against England as they were able to do when France had been forced to surrender. This takes no account either of the attitude of Russia or of whatever resistance Czechoslovakia might have made. For all the above reasons, the year's breathing space said to be "gained" by Munich left Britain and France in a much worse position compared with Hitler's Germany than they had been at the Munich crisis.

*   *   *

Finally there is this staggering fact that in the single year 1938 Hitler had annexed to the Reich and brought under his absolute rule 6,750,000 Austrians and 3,500,000 Sudetens, a total of over ten millions of subjects, toilers, and soldiers. Indeed the dread balance had turned in his favour.

15

Prague, Albania, and the Polish Guarantee

A FTER THE SENSE OF RELIEF springing from the Munich agreement had worn off, Mr. Chamberlain and his Government found themselves confronted by a sharp dilemma. The Prime Minister had said, "I believe it is peace for our time." But the majority of his colleagues wished to utilise "our time" to rearm as rapidly as possible. Here a division arose in the Cabinet. The sensations of alarm which the Munich crisis had aroused, the flagrant exposure of our deficiencies, especially in anti-aircraft guns, dictated vehement rearmament. And this of course was criticised by the German Government and its inspired Press. However, there was no doubt of the opinion of the British nation. While rejoicing at being delivered from war by the Prime Minister and cheering peace slogans to the echo, they felt the need of weapons acutely. All the Service departments put in their claims and referred to the alarming shortages which the crisis had exposed. The Cabinet reached an agreeable compromise on the basis of all possible preparations without disturbing the trade of the country or irritating the Germans and Italians by large-scale measures.

Mr. Chamberlain continued to believe that he had only to form a personal contact with the Dictators to effect a marked improvement in the world situation. He little knew that their decisions were taken. In a hopeful spirit he proposed that he and Lord Halifax should visit Italy in January. After some delay an invitation was extended, and on January 11, 1939, the meeting took place. It makes one flush to read in Ciano's diary the comments which were made behind the Italian scene about our country and its representatives. "Essentially," writes Ciano, "the visit was kept in a minor key. . . . Effective contact has not been made. How far apart we are from these people! It is another world. We were talking about it after dinner to the Duce. 'These men,' said

Mussolini, 'are not made of the same stuff as Francis Drake and the other magnificent adventurers who created the Empire. They are after all the tired sons of a long line of rich men.'" "The British," noted Ciano, " do not want to fight. They try to draw back as slowly as possible, but they do not want to fight. . . . Our conversations with the British have ended. Nothing was accomplished. I have telephoned to Ribbentrop saying it was a fiacso, absolutely innocuous. . . ." And then a fortnight later "Lord Perth [the British Ambassador] has submitted for our approval the outlines of the speech that Chamberlain will make in the House of Commons in order that we may suggest changes if necessary." The Duce approved it, and commented: "I believe this is the first time that the head of the British Government has submitted to a foreign Government the outlines of one of his speeches. It's a bad sign for them." [1] However, in the end it was Ciano and Mussolini who went to their doom.

Meanwhile, in this same January 1939 Ribbentrop was at Warsaw to continue the diplomatic offensive against Poland. The absorption of Czechoslovakia was to be followed by the encirclement of Poland. The first stage in this operation would be the cutting off of Poland from the sea by the assertion of German sovereignty in Danzig and by the prolongation of the German control of the Baltic to the vital Lithuanian port of Memel. The Polish Government displayed strong resistance to this pressure, and for a while Hitler watched and waited for the campaigning season.

During the second week of March rumours gathered of troop movements in Germany and Austria, particularly in the Vienna-Salzburg region. Forty German divisions were reported to be mobilised on a war footing. Confident of German support, the Slovaks were planning the separation of their territory from the Czechoslovak Republic. The Polish Foreign Minister, Colonel Beck, relieved to see the Teutonic wind blowing in another direction, declared publicly in Warsaw that his Government had full sympathy with the aspirations of the Slovaks. Father Tiso, the Slovak leader, was received by Hitler in Berlin with the honours due to a Prime Minister. On the 12th Mr. Chamberlain, questioned in Parliament about the guarantee of the Czechoslovak frontier, reminded the House that this proposal had been directed against unprovoked aggression. No such aggression had yet taken place. He did not have long to wait.

*   *   *

A wave of perverse optimism had swept across the British scene during these March days of 1939. In spite of the growing stresses in

---

[1] *Ciano's Diary*, 1939–43 (ed. Malcolm Muggeridge), pp. 9, 10.

Czechoslovakia under intense German pressure from without and from within, the Ministers and newspapers identified with the Munich Agreement did not lose faith in the policy into which they had drawn the nation. On March 10 the Home Secretary addressed his constituents about his hopes of a Five Years' Peace Plan which would lead in time to the creation of "a Golden Age." A plan for a commercial treaty with Germany was still being hopefully discussed. The famous periodical *Punch* produced a cartoon showing John Bull waking with a gasp of relief from a nightmare, while all the evil rumours, fancies, and suspicions of the night were flying away out of the window. On the very day when this appeared Hitler launched his ultimatum to the tottering Czech Government, bereft of their fortified line by the Munich decisions. German troops, marching into Prague, assumed absolute control of the unresisting State. I remember sitting with Mr. Eden in the smoking-room of the House of Commons when the editions of the evening papers recording these events came in. Even those who like us had no illusions and had testified earnestly were surprised at the sudden violence of this outrage. One could hardly believe that with all their secret information His Majesty's Government could be so far adrift. March 14 witnessed the dissolution and subjugation of the Czechoslovak Republic. The Slovaks formally declared their independence. Hungarian troops, backed surreptitiously by Poland, crossed into the eastern province of Czechoslovakia, or the Carpatho-Ukraine, which they demanded. Hitler, having arrived in Prague, proclaimed a German protectorate over Czechoslovakia, which was thereby incorporated in the Reich.

On the 15th Mr. Chamberlain had to say to the House, "The occupation of Bohemia by German military forces began at six o'clock this morning. The Czech people have been ordered by their Government not to offer resistance." He then proceeded to state that the guarantee he had given Czechoslovakia no longer in his opinion had validity: " . . . the position has altered since the Slovak Diet declared the independence of Slovakia. The effect of this declaration put an end by internal disruption to the State whose frontiers we had proposed to guarantee, and His Majesty's Government cannot accordingly hold themselves bound by this obligation."

This seemed decisive. "It is natural," he said in conclusion, "that I should bitterly regret what has now occurred, but do not let us on that account be deflected from our course. Let us remember that the desire of all the peoples of the world still remains concentrated on the hopes of peace."

Mr. Chamberlain was due to speak at Birmingham two days later. I fully expected that he would accept what had happened with the best grace possible. The Prime Minister's reaction surprised me. He had conceived himself as having a special insight into Hitler's character, and the power to measure with shrewdness the limits of German action. He believed, with hope, that there had been a true meeting of minds at Munich, and that he, Hitler, and Mussolini had together saved the world from the infinite horrors of war. Suddenly as by an explosion his faith and all that had followed from his actions and his arguments was shattered. Responsible as he was for grave misjudgments of facts, having deluded himself and imposed his errors on his subservient colleagues and upon the unhappy British public opinion, he none the less between night and morning turned his back abruptly upon his past. If Chamberlain failed to understand Hitler, Hitler completely underrated the nature of the British Prime Minister. He mistook his civilian aspect and passionate desire for peace for a complete explanation of his personality, and thought that his umbrella was his symbol. He did not realise that Neville Chamberlain had a very hard core, and that he did not like being cheated.

The Birmingham speech struck a new note. He reproached Hitler with a flagrant personal breach of faith about the Munich Agreement. He quoted all the assurances Hitler had given. "This is the last territorial claim which I have to make in Europe." "I shall not be interested in the Czech State any more, and I can guarantee it. We don't want any more Czechs." "I am convinced," said the Prime Minister, "that after Munich the great majority of the British people shared my honest desire that that policy should be carried further, but today I share their disappointment, their indignation, that those hopes have been so wantonly shattered. How can these events this week be reconciled with those assurances which I have read out to you? . . . Is this the last attack upon a small state, or is it to be followed by another? Is this in fact a step in the direction of an attempt to dominate the world by force?"

It is not easy to imagine a greater contradiction to the mood and policy of the Prime Minister's statement two days earlier in the House of Commons. He must have been through a period of intense stress. Moreover, Chamberlain's change of heart did not stop at words. The next "small state" on Hitler's list was Poland. When the gravity of the decision and all those who had to be consulted are borne in mind, the period must have been busy. Within a fortnight (March 31) the Prime Minister said in Parliament:

> . . . In the event of any action which clearly threatened Polish independence and which the Polish Government accordingly considered it vital to resist with their national forces, His Majesty's Government would feel themselves bound at once to lend the Polish Government all support in their power. They have given the Polish Government an assurance to this effect.

I may add that the French Government have authorised me to make it plain that they stand in the same position in this matter as do His Majesty's Government. . . . [And later] The Dominions have been kept fully informed.

This was no time for recriminations about the past. The guarantee to Poland was supported by the leaders of all parties and groups in the House. "God helping, we can do no other," was what I said. At the point we had reached it was a necessary action. But no one who understood the situation could doubt that it meant in all human probability a major war, in which we should be involved.

In this sad tale of wrong judgments formed by well-meaning and capable people we now reach our climax. That we should all have come to this pass makes those responsible, however honourable their motives, blameworthy before history. Look back and see what we had successively accepted or thrown away: a Germany disarmed by solemn treaty; a Germany rearmed in violation of a solemn treaty; air superiority or even air parity cast away; the Rhineland forcibly occupied and the Siegfried Line built or building; the Berlin-Rome Axis established; Austria devoured and digested by the Reich; Czechoslovakia deserted and ruined by the Munich Pact, its fortress line in German hands, its mighty arsenal of Skoda henceforward making munitions for the German armies; President Roosevelt's effort to stabilise or bring to a head the European situation by the intervention of the United States waved aside with one hand, and Soviet Russia's undoubted willingness to join the Western Powers and go all lengths to save Czechoslovakia ignored on the other; the services of thirty-five Czech divisions against the still unripened German Army cast away, when Great Britain could herself supply only two to strengthen the front in France; all gone with the wind.

And now, when every one of these aids and advantages has been squandered and thrown away, Great Britain advances, leading France by the hand, to guarantee the integrity of Poland — of that very Poland which with hyena appetite had only six months before joined in the pillage and destruction of the Czechoslovak State. There was sense in fighting for Czechoslovakia in 1938, when the German Army could scarcely put half a dozen trained divisions on the Western Front, when the French with nearly sixty or seventy divisions could most certainly have rolled forward across the Rhine or into the Ruhr. But this had been judged unreasonable, rash, below the level of modern intellectual thought and morality. Yet now at last the two Western democracies declared themselves ready to stake their lives upon the territorial integrity of Poland. History, which, we are told, is mainly the record of the crimes, follies, and miseries of mankind, may be scoured and ransacked to find a parallel to this sudden and complete reversal of five or six years' policy of easygoing placatory appeasement, and its transformation almost overnight into a readiness to accept an obviously imminent war on far worse conditions and on the greatest scale.

Moreover, how could we protect Poland and make good our guarantee? Only by declaring war upon Germany and attacking a stronger Western Wall and a more powerful German Army than those from which we had recoiled in September 1938. Here is a line of milestones to disaster. Here is a catalogue of surrenders, at first when all was easy and later when things were harder, to the ever-growing German power. But now at last was the end of British and French submission. Here was decision at last, taken at the worst possible moment and on the least satisfactory ground, which must surely lead to the slaughter of tens of millions of people. Here was the righteous cause deliberately and with a refinement of inverted artistry committed to mortal battle after its assets and advantages had been so improvidently squandered. Still, if you will not fight for the right when you can easily win without bloodshed, if you will not fight when your victory will be sure and not too costly, you may come to the moment when you will have to fight with all the odds against you and only a precarious chance of survival. There may even be a worse case. You may have to fight when there is no hope of victory, because it is better to perish than live as slaves.

*   *   *

The Poles had gained Teschen by their shameful attitude towards the liquidation of the Czechoslovak State. They were soon to pay their own forfeits. On March 21, when Ribbentrop saw the Polish Ambassador in Berlin, he adopted a sharper tone than in previous discussions. The occupation of Bohemia and the creation of satellite Slovakia brought the German Army to the southern frontiers of Poland. The Ambassador explained that the Polish man-in-the-street could not understand why the Reich had assumed the protection of Slovakia, that protection being directed against Poland. He also inquired about the recent conversations between Ribbentrop and the Lithuanian Foreign Minister. Did they affect Memel? He received his answer two days later (March 23). German troops occupied Memel.

The means of organising any resistance to German aggression in Eastern Europe were now almost exhausted. Hungary was in the German camp. Poland had stood aside from the Czechs, and was unwilling to work closely with Rumania. Neither Poland nor Rumania would accept Russian intervention against Germany across their territories. The key to a Grand Alliance was an understanding with Russia. On March 19 the Russian Government, which was profoundly affected by all that was taking place, and in spite of having been left outside the door in the Munich crisis, proposed a Six-Power Conference. On this subject also Mr. Chamberlain had decided views. In a private letter he "confessed to the most profound distrust of Russia. I have no belief whatever in her ability to maintain an effective offensive, even if she wanted to. And I distrust her motives, which seem to me to have little connection with our ideas of liberty, and to be concerned only with setting everyone else by the ears. Moreover, she is both hated and suspected by many of the smaller states, notably by Poland, Rumania, and Finland."

The Soviet proposal for a Six-Power Conference was therefore coldly received and allowed to drop.

The possibilities of weaning Italy from the Axis, which had loomed so large in British official calculations, were also vanishing. On March 26 Mussolini made a violent speech asserting Italian claims against France in the Mediterranean. At dawn on April 7, 1939, Italian forces landed in Albania, and after a brief scuffle took over the country. As Czechoslovakia was to be the base for aggression against Poland, so Albania would be the springboard for Italian action against Greece and for the neutralising of Yugoslavia. The British Government had already undertaken a commitment in the interests of peace in Northeastern Europe. What about the threat developing in the Southeast? The British Mediterranean Fleet, which might have checked the Italian move, had been allowed to disperse. The vessel of peace was springing a leak from every seam. On April 15, after the declaration of the German protectorate of Bohemia and Moravia, Goering met Mussolini and Ciano in Rome to explain the progress of German preparations for war. On the same day President Roosevelt sent a personal message to Hitler and Mussolini urging them to give a guarantee not to undertake any further aggression for ten "or even twenty-five years, if we are to look that far ahead." The Duce at first refused to read the document, and then remarked: "A result of infantile paralysis!" He little thought he was himself to suffer worse afflictions.

*   *   *

On April 27 the Prime Minister took the serious decision to introduce conscription, although repeated pledges had been given by him against such a step. To Mr. Hore-Belisha, the Secretary of State for War, belongs the credit of forcing this belated awakening. He certainly took his political life in his hands, and several of his interviews with his chief were of a formidable character. I saw something of him in this ordeal, and he was never sure that each day in office would not be his last.

Of course the introduction of conscription at his stage did not give us an army. It only applied to the men of twenty years of age; they had still to be trained; and after they had been trained they had still to be armed. It was however a symbolic gesture of the utmost consequence to France and Poland, and to other nations on whom we had lavished our guarantees. In the debate the Opposition failed in their duty. Both Labour and Liberal Parties shrank from facing the ancient and deep-rooted prejudice which has always existed in England against compulsory military service, and their leaders found reasons for opposing this step. Both these men were distressed at the course they felt bound on party grounds to take. But they both took it, and adduced a wealth of reasons. The division was on party lines, and the Conservatives carried their policy by 380 to 143 votes. In my speech I tried my best to persuade the Opposition to support this indispensable measure; but my efforts were in vain. I understood fully their difficulties, especially when confronted with a Government to which they were opposed. I must record the event, because it deprives Liberal and Labour partisans of any right to censure the Government of the day. They showed their own measure in relation to events only too plainly. Presently they were to show a truer measure.

In March I had joined Mr. Eden and some thirty Conservative Members in tabling a resolution for a National Government. During the summer there arose a very considerable stir in the country in favour of this, or at least for my and Mr. Eden's inclusion in the Cabinet. Sir Stafford Cripps, in his independent position, became deeply distressed about the national danger. He visited me and various Ministers to urge the formation of what he called an "All-in Government." I could do nothing; but Mr. Stanley, President of the Board of Trade, was deeply moved. He wrote to the Prime Minister offering his own office if it would facilitate a reconstruction. Mr. Chamberlain contented himself with a formal acknowledgment.

As the weeks passed by, almost all the newspapers, led by the *Daily Telegraph* and emphasised by the *Manchester Guardian,* reflected this surge of opinion. I was surprised to see its daily recurrent and repeated expression. Thousands of enormous posters were displayed for weeks on end on Metropolitan hoardings, "Churchill Must Come Back." Scores of young volunteer men and women carried sandwich-board placards with similar slogans up and down before the House of Commons. I had nothing to do with such methods of agitation, but I should certainly have joined the Government had I been invited. Here again my personal good fortune held, and all else flowed out in its logical, natural, and horrible sequence.

16

On the Verge

WE HAVE REACHED THE PERIOD when all relations between Britain and Germany were at an end. We now know of course that there never had been any true relationship between our two countries since Hitler came into power. He had only hoped to persuade or frighten Britain into giving him a free hand in Eastern Europe, and Mr. Chamberlain had cherished the hope of appeasing and reforming him and leading him to grace. However, the time had come when the last illusions of the British Government had been dispelled. The Cabinet was at length convinced that Nazi Germany meant war, and the Prime Minister offered guarantees and contracted alliances in every direction still open, regardless of whether we could give any effective help to the countries concerned. To the Polish guarantee was added a Greek and Rumanian guarantee, and to these an alliance with Turkey.

We must now recall the sad piece of paper which Mr. Chamberlain had got Hitler to sign at Munich and which he waved triumphantly to the crowd when he quitted his aeroplane at Heston. In this he had invoked the two bonds which he assumed existed between him and Hitler and between Britain and Germany, namely, the Munich Agreement and the Anglo-German Naval Treaty. The subjugation of Czechoslovakia had destroyed the first; on April 28 Hitler brushed away the second. He also denounced the German-Polish Non-Aggression Pact. He gave as his direct reason the Anglo-Polish Guarantee.

The British Government had to consider urgently the practical implications of the guarantees given to Poland and to Rumania. Neither set of assurances had any military value except within the framework of a general agreement with Russia. It was therefore with this object that talks at last began in Moscow on April 15 between the British Ambassador and Mr. Litvinov. Considering how the Soviet Government had hitherto been treated, there was not much to be expected from them now. However, on April 16 they made a formal offer, the text of which was not published, for the creation of a united front of mutual assistance between Great Britain, France, and the U.S.S.R. The three Powers, with Poland added if possible, were furthermore to guarantee those states in Central and Eastern Europe which lay under the menace of German aggression. The obstacle to such an agreement was the terror of these same border countries of receiving Soviet help in the shape of Soviet armies marching through their territories to defend them from the Germans, and incidentally incorporating them in the Soviet-Communist system, of which they were the most vehement opponents. Poland, Rumania, Finland, and the three Baltic States did not know whether it was German aggression or Russian rescue that they dreaded more. It was this hideous choice that paralysed British and French policy.

There can however be no doubt, even in the after-light, that Britain and France should have accepted the Russian offer, proclaimed the Triple Alliance, and left the method by which it could be made effective in case of war to be adjusted between allies engaged against a common foe. In such circumstances a different temper prevails. Allies in war are inclined to defer a great deal to each other's wishes; the flail of battle beats upon the front, and all kinds of expedients are welcomed which in peace would be abhorrent. It would not be easy in a Grand Alliance, such as might have been developed, for one ally to enter the territory of another unless invited.

But Mr. Chamberlain and the Foreign Office were baffled by this riddle of the Sphinx. When events are moving at such speed and in such tremendous mass as at this juncture it is wise to take one step at a time. The alliance of Britain, France, and Russia would have struck deep alarm into the heart of Germany in 1939, and no one can prove that war might not even then have been averted. The next step could have been taken with superior power on the side of the allies. The initiative would have been regained by their diplomacy. Hitler could afford neither to embark upon the war on two fronts, which he himself had so deeply condemned, nor to sustain a check. It was a pity not to have placed him in this awkward position, which might well have cost him his life. Statesmen are not called upon only to settle easy questions. These often settle themselves. It is where the balance quivers, and the proportions are veiled in mist, that the opportunity for world-saving decisions presents itself. Having got ourselves into this awful plight of 1939 it was vital to grasp the larger hope. If, for instance, Mr. Chamberlain on receipt of the Russian offer had replied, "Yes. Let us three band together and break Hitler's neck," or words to that effect, Parliament would have approved, Stalin would have understood, and history might have taken a different course. At least it could not have taken a worse.

Instead, there was a long silence while half-measures and judicious compromises were being prepared. This delay was fatal to Litvinov. His last attempt to bring matters to a clear-cut decision with the Western Powers was deemed to have failed. Our credit stood very low. A wholly different foreign policy was required for the safety of Russia, and a new exponent must be foud. On May 3 an official communiqué from Moscow announced that M. Litvinov had been released from the office of Foreign Commissar at his request and that his duties would be assumed by the Premier, M. Molotov. The eminent Jew, the target of German antagonism, was flung aside for the time being like a broken tool, and, without being allowed a word of explanation, was bundled off the world stage to obscurity, a pittance, and police supervision. Molotov, little known outside Russia, became Commissar for Foreign Affairs, in the closest confederacy with Stalin. He was free from all encumbrance of previous declarations, free from the League of Nations atmosphere, and able to move in any direction which the self-preservation of Russia might seem to require. There was in fact only one way in which he was now likely to move. He had always been favourable to an arrangement with Hitler. The Soviet Government were convinced by Munich and much else that neither Britain nor France would fight till they were attacked, and would not be much good then. The gathering storm was about to break. Russia must look after herself.

This violent and unnatural reversal of Russian policy was a transmogrification of which only totalitarian states are capable. It was barely two years since the leaders of the Russian Army, and several thousands of its most accomplished officers, had been slaughtered for the very inclinations which now became acceptable to the handful of anxious masters in the Kremlin. Then pro-Germanism had been heresy and treason. Now, overnight, it was the policy of the State, and woe was mechanically meted out to any who dared dispute it, and often to those not quick enough on the turn-about.

For the task in hand no one was better fitted or equipped than the new Foreign Commissar.

*   *   *

The figure whom Stalin had now moved to the pulpit of Soviet foreign policy deserves some description, not available to the British or French Governments at the time. Vyacheslav Molotov was a man of outstanding ability and cold-blooded ruthlessness. He had survived the fearful hazards and ordeals to which all the Bolshevik leaders had been subjected in the years of triumphant revolution. He had lived and thrived in a society where ever-varying intrigue was accompanied by the constant menace of personal liquidation. His cannon-ball head, black moustache, and comprehending eyes, his slab face, his verbal adroitness and imperturbable demeanour, were appropriate manifestations of his qualities and skill. He was above all men fitted to be the agent and instrument of the policy of an incalculable machine. I have only met him on equal terms, in parleys where sometimes a strain of humour appeared, or at banquets where he genially proposed a long succession of conventional and meaningless toasts. I have never seen a human being who more perfectly represented the modern conception of a robot. And yet with all this there was an apparently reasonable and keenly polished diplomatist. What he was to his inferiors I cannot tell. What he was to the Japanese Ambassador during the years when after the Teheran Conference Stalin had promised to attack Japan once the German Army was beaten can be deduced from his recorded conversations. One delicate, searching, awkward interview after another was conducted with perfect poise, impenetrable purpose, and bland, official correctitude. Never a chink was opened. Never a needless jar was made. His smile of Siberian winter, his carefully measured and often wise words, his affable demeanour, combined to make him the perfect agent of Soviet policy in a deadly world.

Correspondence with him upon disputed matters was always useless, and, if pushed far, ended in lies and insults, of which this work will presently contain some examples. Only once did I seem to get a natural, human reaction. This was in the spring of 1942, when he alighted in Britain on his way back from the United States. We had signed the Anglo-Soviet Treaty, and he was about to make his dangerous flight home. At the garden gate of Downing Street, which we used for secrecy, I gripped his arm and we looked each other in the face. Suddenly he appeared deeply moved. Inside the image there appeared the man. He responded with an equal pressure. Silently we wrung each other's hands. But then we were all together, and it was life or death for the lot. Havoc and ruin had been around him all his days, either impending on himself or dealt by him to others. Certainly in Molotov the Soviet machine had found a capable and in many ways a characteristic representative — always the faithful Party man and Communist disciple. How glad I am at the end of my life not to have had to endure the stresses which he has suffered; better never be born. In the conduct of foreign affairs Mazarin, Talleyrand, Metternich, would welcome him to their company, if there be another world to which Bolsheviks allow themselves to go.

From the moment when Molotov became Foreign Commissar he pursued the policy of an arrangement with Germany at the expense of Poland. The Russian negotiations with Britain proceeded languidly, and on May 19 the whole issue was raised in the House of Commons. The debate, which was short and serious, was practically confined to the leaders of parties and to prominent ex-Ministers. Mr. Lloyd George, Mr. Eden, and I all pressed upon the Government the vital need of an immediate arrangement with Russia of the most far-reaching character and on equal terms. The Prime Minister replied, and for the first time revealed to us his views on the Soviet offer. His reception of it was certainly cool, and indeed disdainful; and seemed to show the same lack of proportion as we have seen in the rebuff to the Roosevelt proposals a year before. Attlee, Sinclair, and Eden spoke on the general line of the imminence of the danger and the need of the Russian alliance. There can be little doubt that all this was now too late. Our efforts had come to a seemingly unbreakable deadlock. The Polish and Rumanian Governments, while accepting the British guarantee, were not prepared to accept a similar undertaking in the same form from the Russian Government. A similar attitude prevailed in another vital strategic quarter — the Baltic States. The Soviet Government made it clear that they would only adhere to a pact of mutual assistance if Finland and the Baltic States were included in a general guarantee. All four countries now refused, and perhaps in their terror would for a long time have refused such a condition. Finland and Esthonia even asserted that they would consider a guarantee extended to them without their assent as an act of aggression. On June 7 Esthonia and Latvia signed non-aggression pacts with Germany. Thus Hitler penetrated with ease into the final defences of the tardy, irresolute coalition against him.

Summer advanced, preparations for war continued throughout Europe, and the attitudes of diplomatists, the speeches of politicians, and the wishes of mankind counted each day for less. German military movements seemed to portend the forcible settlement of the dispute with Poland over Danzig as a preliminary to the assault on Poland itself. Mr. Chamberlain expressed his anxieties to Parliament on June 10, and repeated his intention to stand by Poland if her indendence were threatened. In a spirit of detachment from the facts the Belgian Government, largely under the influence of their King, announced on the 23rd that they were opposed to Staff talks with England and France and that Belgium intended to maintain a strict neutrality. The tide of events brought with it a closing of the ranks between England and France, and also at home. There was much coming and going between Paris and London during the month of July. The celebrations of the Fourteenth of July were an occasion for a display of Anglo-French union. I was invited by the French Government to attend this brilliant spectacle.

As I was leaving Le Bourget after the parade General Gamelin suggested that I should visit the French front. "You have never seen the

Rhine sector," he said. "Come then in August; we will show you every-thing." Accordingly a plan was made, and on August 15 General Spears and I were welcomed by his close friend, General Georges, Commander-in-Chief of the armies on the northeastern front and *successeur eventuel* to the Supreme Command. I was delighted to meet this most agreeable and competent officer, and we passed the next ten days in his company, revolving military problems and making contacts with Gamelin, who was also inspecting certain points on this part of the front.

Beginning at the angle of the Rhine near Lauterbourg, we traversed the whole section to the Swiss frontier. In England, as in 1914, the care-free people were enjoying their holidays and playing with their children on the sands. But here along the Rhine a different light glared. All the temporary bridges across the river had been removed to one side or the other. The permanent bridges were heavily guarded and mined. Trusty officers were stationed night and day to press at a signal the buttons which would blow them up. The great river, swollen by the melting Alpine snows, streamed along in sullen, turgid flow. The French out-posts crouched in their rifle pits amid the brushwood. Two or three of us could stroll together to the water's edge, but nothing like a target, we were told, must be presented. Three hundred yards away on the farther side, here and there among the bushes, German figures could be seen working rather leisurely with pick and shovel at their defences. All the riverside quarter of Strasbourg had already been cleared of civilians. I stood on its bridge some time and watched one or two motorcars pass over it. Prolonged examination of passports and character took place at either end. Here the German post was little more than a hundred yards away from the French. There was no intercourse between them. Yet Europe was at peace. There was no dispute between Germany and France. The Rhine flowed on, swirling and eddying, at six or seven miles an hour. One or two canoes with boys in them sped past on the current. I did not see the Rhine again until more than five years later, in March 1945, when I crossed it in a small boat with Field Marshal Montgomery. But that was near Wesel, far to the north.

What was remarkable about all I learned on my visit was the com-plete acceptance of the defensive which dominated my most responsible French hosts, and imposed itself irresistibly upon me. In talking to all these highly competent French officers one had the sense that the Ger-mans were the stronger, and that France had no longer the life-thrust to mount a great offensive. She would fight for her existence — *voilà tout!* There was the fortified Siegfried Line, with all the increased fire-power of modern weapons. In my own bones, too, was the horror of the Somme and Passchendaele offensives. The Germans were of course far stronger than in the days of Munich. We did not know the deep anxieties which rent their High Command. We had allowed ourselves to get into such a condition, physically and psychologically, that no responsible person — and up to this point I had no responsibilities — could act on the assumption, which was true, that only forty-two half-equipped and half-trained German divisions guarded their long front from the North Sea to Switzerland. This compared with thirteen at the time of Munich.

*   *   *

In these final weeks my fear was that His Majesty's Government, in spite of our guarantee, would recoil from waging war upon Germany if she attacked Poland. There is no doubt that at this time Mr. Chamberlain had resolved to take the plunge, bitter though it was to him. But I did not know him so well as I did a year later. I feared that Hitler might try a bluff about some novel agency or secret weapon which would baffle or puzzle the overburdened Cabinet. From time to time Professor Lindemann had talked to me about atomic energy. I therefore asked him to let me know how things stood in this sphere, and after a conversation I wrote the following letter to Kingsley Wood, the Secretary of State for Air, with whom my relations were fairly intimate:

> Some weeks ago one of the Sunday papers splashed the story of the immense amount of energy which might be released from uranium by the recently discovered chain of processes which take place when this particular type of atom is split by neutrons. At first sight this might seem to portend the appearance of new explosives of devastating power. *In view of this it is essential to realise that there is no danger that this discovery, however great its scientific interest, and perhaps ultimately its practical importance, will lead to results capable of being put into operation on a large scale for several years.*
>
> There are indications that tales will be deliberately circulated when international tension becomes acute about the adaptation of this process to produce some terrible new secret explosive, capable of wiping out London. Attempts will no doubt be made by the Fifth Column to induce us by means of this threat to accept another surrender. For this reason it is imperative to state the true position.
>
> . . . The fear that his new discovery has provided the Nazis with some sinister, new, secret explosive with which to destroy their enemies is clearly without foundation. Dark hints will no doubt be dropped and terrifying whispers will be assiduously circulated, but it is hoped that nobody will be taken in by them.

It is remarkable how accurate this forecast was. Nor was it the Germans who found the path. Indeed they followed the wrong trail, and had actually abandoned the search for the atomic bomb in favour of rockets or pilotless aeroplanes at the moment when President Roosevelt and I were taking the decisions and reaching the memorable agreements, which will be described in their proper place, for the large-scale manufacture of atomic bombs.

*   *   *

"Tell Chamberlain," said Mussolini to the British Ambassador on July 7, "that if England is ready to fight in defence of Poland, Italy will take up arms with her ally, Germany." But behind the scenes his attitude was the opposite. He sought at this time no more than to consolidate his interests in the Mediterranean and North Africa, to cull the fruits of his intervention in Spain, and to digest his Albanian conquest. He did not like being dragged into a European war for Germany to conquer Poland. For all his public boastings, he knew the military and political fragility of Italy better than anyone. He was willing to talk about a war in 1942, if Germany would give him the munitions; but in 1939 — no!

As the pressure upon Poland sharpened during the summer, Mussolini turned his thoughts upon repeating his Munich role of mediator, and he suggested a World Peace Conference. Hitler curtly dispelled such ideas. In August he made it clear to Ciano that he intended to settle with Poland, that he would be forced to fight England and France as well, and that he wanted Italy to come in. He said, "If England keeps the necessary troops in her own country, she can send to France at the most two infantry divisions and one armoured division. For the rest she could supply a few bomber squadrons, but hardly any fighters, because the German Air Force would at once attack England, and the English fighters would be urgently needed for its defence." About France he said that after the destruction of Poland — which would not take long — Germany would be able to assemble hundreds of divisions along the West Wall, and France would thus be compelled to concentrate all her available forces from the colonies and from the Italian frontier and elsewhere on her Maginot Line for the life-and-death struggle. After these interchanges Ciano returned gloomily to report to his master, whom he found more deeply convinced that the Democracies would fight, and even more resolved to keep out of the struggle himself.

*   *   *

A renewed effort to come to an arrangement with Soviet Russia was now made by the British and French Governments. It was decided to send a special envoy to Moscow. Mr. Eden, who had made useful contacts with Stalin some years before, volunteered to go. This generous offer was declined by the Prime Minister. Instead on June 12 Mr. Strang, an able official but without any special standing outside the Foreign Office, was entrusted with this momentous mission. This was another mistake. The sending of so subordinate a figure gave actual offence. It is doubtful whether he was able to pierce the outer crust of the Soviet organism. In any case all was now too late. Much had happened since M. Maisky had been sent to see me at Chartwell in September 1938. Munich had happened. Hitler's armies had had a year more to mature. His munitions factories, reinforced by the Skoda works, were all in full blast. The Soviet Government cared much for Czechoslovakia; but Czechoslovakia was gone. Beneš was in exile. A German Gauleiter ruled in Prague.

On the other hand, Poland presented to Russia an entirely different set of age-long political and strategic problems. Their last major contact had been the Battle of Warsaw in 1920, when the Bolshevik armies under Kamieniev had been hurled back from their invasion by Pilsudski, aided by the advice of General Weygand and the British Mission under Lord D'Abernon, and thereafter pursued with bloody vengeance. All these years Poland had been a spear point of anti-Bolshevism. With her left hand she joined and sustained the anti-Soviet Baltic States. But with her right hand, at Munich-time, she had helped to despoil Czechoslovakia. The Soviet Government were sure that Poland hated them, and also that Poland had no power to withstand a German onslaught. They were however very conscious of their own perils and of their need for time to repair the havoc in the High Commands of their armies. In these circumstances the prospects of Mr. Strang's mission were not exuberant.

The negotiations wandered around the question of the reluctance of Poland and the Baltic States to be rescued from Germany by the Soviets; and here they made no progress. All through July the discussions continued fitfully, and eventually the Soviet Government proposed that conversations should be continued on a military basis with both French and British representatives. The British Government therefore dispatched Admiral Drax with a mission to Moscow on August 10. These officers possessed no written authority to negotiate. The French Mission was headed by General Doumenc. On the Russian side Marshal Voroshilov officiated. We now know that at this same time the Soviet Government agreed to the journey of a German negotiator to Moscow. The military conference soon foundered upon the refusal of Poland and Rumania to allow the transit of Russian troops. The Polish attitude was, "With the Germans we risk losing our liberty; with the Russians our soul." [1]

[1] Quoted in Paul Reynaud, *La France a Sauvé l'Europe*, I, p. 587.

At the Kremlin in August 1942 Stalin, in the early hours of the morning, gave me one aspect of the Soviet position. "We formed the impression," said Stalin, "that the British and French Governments were not resolved to go to war if Poland were attacked, but that they hoped the diplomatic line-up of Britain, France, and Russia would deter Hitler. We were sure it would not." "How many divisions," Stalin had asked, "will France send against Germany on mobilisation?" The answer was, "About a hundred." He then asked, "How many will England send?" The answer was, "Two and two more later." "Ah, two, and two more later," Stalin had repeated. "Do you know," he asked, "how many divisions we shall have to put on the Russian front if we go to war with Germany?" There was a pause. "More than three hundred." I was not told with whom this conversation took place or its date. It must be recognised that this was solid ground, but not favourable for Mr. Strang of the Foreign Office.

It was judged necessary by Stalin and Molotov for bargaining purposes to conceal their true intentions till the last possible moment. Remarkable skill in duplicity was shown by Molotov and his subordinates in all their contacts with both sides. On the evening of August 19 Stalin announced to the Politburo his intention to sign a pact with Germany. On August 22 Marshal Voroshilov was not to be found by the Allied missions until evening. The next day Ribbentrop arrived in Moscow. In a secret agreement Germany declared herself politically disinterested in Latvia, Esthonia, and Finland, but considered Lithuania to be in her sphere of influence. A demarcation line was drawn for the Polish partition. In the Baltic countries Germany claimed only economic interests. The Non-Aggression Pact and the secret agreement were signed rather late on the night of August 23.[2]

Despite all that has been dispassionately recorded in this chapter, only totalitarian despotism in both countries could have faced the odium of such an unnatural act. It is a question whether Hitler or Stalin loathed it most. Both were aware that it could only be a temporary expedient. The antagonisms between the two empires and systems were mortal. Stalin no doubt felt that Hitler would be a less deadly foe to Russia after a year of war with the Western Powers. Hitler followed his method of "One at a time." The fact that such an agreement could be made marks the culminating failure of British and French foreign policy and diplomacy over several years.

On the Soviet side it must be said that their vital need was to hold the deployment positions of the German armies as far to the west as possible, so as to give the Russians more time for assembling their forces from all parts of their immense empire. They had burnt in their

[2] *Nuremberg Documents*, Pt. X, p. 210 ff.

minds the disasters which had come upon their armies in 1914, when they had hurled themselves forward to attack the Germans while still themselves only partly mobilised. But now their frontiers lay far to the east of those of the previous war. They must be in occupation of the Baltic States and a large part of Poland by force or fraud before they were attacked. If their policy was cold-blooded, it was also at the moment realistic in a high degree.

It is still worth while to record the terms of the Pact.

> Both High Contracting Parties obligate themselves to desist from any act of violence, any aggressive action, and any attack on each other, either individually or jointly with other Powers.

This treaty was to last ten years, and if not denounced by either side one year before the expiration of that period would be automatically extended for another five years. There was much jubilation and many toasts around the conference table. Stalin spontaneously proposed the toast of the Fuehrer, as follows: "I know how much the German nation loves its Fuehrer; I should therefore like to drink his health." A moral may be drawn from all this, which is of homely simplicity. "Honesty is the best policy." Several examples of this will be shown in these pages. Crafty men and statesmen will be shown misled by all their elaborate calculations. But this is the signal instance. Only twenty-two months were to pass before Stalin and the Russian nation in its scores of millions were to pay a frightful forfeit. If a Government has no moral scruples it often seems to gain great advantages and liberties of action, but "All comes out even at the end of the day, and all will come out yet more even when all the days are ended."

*   *   *

The sinister news broke upon the world like an explosion. Whatever emotions the British Government may have experienced, fear was not among them. They lost no time in declaring that "such an event would in no way affect their obligations, which they were determined to fulfill." They at once took precautionary measures. Orders were issued for key parties of the coast and anti-aircraft defences to assemble, and for the protection of vulnerable points. Warning telegrams were sent to Dominion Governments and to the Colonies. All leave was stopped throughout the fighting services. The Admiralty issued warnings to merchant shipping. Many other steps were taken. On August 25 the British Government proclaimed a formal treaty with Poland, confirming the guarantees already given. It was hoped by this step to give the best chance of a settlement by direct negotiation between Germany and Poland in the face of the fact that if this failed Britain would stand by Poland. In fact Hitler postponed D-Day from August 25 to September 1, and entered into direct negotiation with Poland, as Chamberlain desired. His object was not however to reach an agreement with Poland, but to give His Majesty's Government every opportunity to escape from their guarantee. Their thoughts, like those of Parliament and the nations, were upon a different plane. It is a curious fact about the British Islanders, who hate drill and have not been invaded for nearly a thousand years, that as danger comes nearer and grows they become progressively less nervous; when it is imminent they are fierce; when it is mortal they are fearless. These habits have led them into some very narrow escapes.

From correspondence with Mussolini at this point Hitler now learnt, if he had not divined it already, that he could not count upon the armed intervention of Italy if war came. It seems to have been from English rather than from German sources that the Duce learnt of the final moves. Ciano records in his diary on August 27, "The English communicate to us the text of the German proposals to London, about which we are kept entirely in the dark." [3] Mussolini's only need now was Hitler's acquiescence in Italy's neutrality. This was accorded to him.

On August 31 Hitler issued his "Directive Number 1 for the Conduct of the War."

> 1. Now that all the political possibilities of disposing by peaceful means of a situation on the Eastern frontier which is intolerable for Germany are exhausted, I have determined on a solution by force.
>
> 2. The attack on Poland is to be carried out in accordance with the preparation made. . . . The date of attack — September 1, 1939. Time of attack — 04.45 [inserted in red pencil].
>
> 3. In the West it is important that the responsibility for the opening of hostilities should rest unequivocally with England and France. At first purely local action should be taken against insignificant frontier violations.[4]

*   *   *

On my return from the Rhine front I passed some sunshine days at Mme. Balsan's place, with a pleasant but deeply anxious company, in the old chateau where King Henry of Navarre had slept the night before the Battle of Ivry. One could feel the deep apprehension brooding over all, and even the light of this lovely valley of the Eure seemed robbed of its genial ray. I found painting hard work in this uncertainty.

[3] *Ciano's Diary*, p. 136.
[4] *Nuremberg Documents*, Pt. II, p. 172.

On August 26 I decided to go home, where at least I could find out what was going on. I told my wife I would send her word in good time. On my way through Paris I gave General Georges luncheon. He produced all the figures of the French and German Armies, and classified the divisions in quality. The result impressed me so much that for the first time I said, "But you are the masters." He replied, "The Germans have a very strong Army, and we shall never be allowed to strike first. If they attack, both our countries will rally to their duty."

That night I slept at Chartwell, where I had asked General Ironside to stay with me next day. He had just returned from Poland, and the reports he gave of the Polish Army were most favourable. He had seen a divisional attack exercise under a live barrage, not without casualties. Polish morale was high. He stayed three days with me, and we tried hard to measure the unknowable. Also at this time I completed bricklaying the kitchen of the cottage which during the year past I had prepared for our family home in the years which were to come. My wife, on my signal, came over via Dunkirk on August 30.

There were known to be twenty thousand organised German Nazis in England at this time, and it would only have been in accord with their procedure in other friendly countries that the outbreak of war should be preceded by a sharp prelude of sabotage and murder. I had at that time no official protection, and I did not wish to ask for any: but I thought myself sufficiently prominent to take precautions. I had enough information to convince me that Hitler recognised me as a foe. My former Scotland Yard detective, Inspector Thompson, was in retirement. I told him to come along and bring his pistol with him. I got out my own weapons, which were good. While one slept the other watched. Thus nobody would have had a walkover. In these hours I knew that if war came — and who could doubt it coming? — a major burden would fall upon me.

17

Twilight War

POLAND WAS ATTACKED by Germany at dawn on September 1. The mobilisation of all our forces was ordered during the morning. The Prime Minister asked me to visit him in the afternoon at Downing Street. He told me that he saw no hope of averting a war with Germany, and that he proposed to form a small War Cabinet of Ministers without departments to conduct it. He mentioned that the Labour Party were not, he understood, willing to share in a national coalition. He still had hopes that the Liberals would join him. He invited me to become a member of the War Cabinet. I agreed to his proposal without comment, and on this basis we had a long talk on men and measures.

I was surprised to hear nothing from Mr. Chamberlain during the whole of September 2, which was a day of intense crisis. I thought it probable that a last-minute effort was being made to preserve peace; and this proved true. However, when Parliament met in the evening a short but very fierce debate occurred, in which the Prime Minister's temporising statement was ill received by the House. When Mr. Greenwood rose to speak on behalf of the Labour Opposition Mr. Amery from the Conservative benches cried out to him, "Speak for England." This was received with loud cheers. There was no doubt that the temper of the House was for war. I even deemed it more resolute and united than in the similar scene on August 3, 1914, in which I had also taken part. I learnt later that a British ultimatum had been given to Germany at 9.30 P.M. on September 1, and that this had been followed by a second and final ultimatum at 9 A.M. on September 3. The early broadcast of the 3rd announced that the Prime Minister would speak on the radio at 11.15 A.M.

The Prime Minister's broadcast informed us that we were already at war, and he had scarcely ceased speaking when a strange, prolonged,

165

wailing noise, afterwards to become familiar, broke upon the ear. My wife came into the room braced by the crisis and commented favourably upon the German promptitude and precision, and we went up to the flat top of the house to see what was going on. Around us on every side, in the clear, cool September light, rose the roofs and spires of London. Above them were already slowly rising thirty or forty cylindrical balloons. We gave the Government a good mark for this evident sign of preparation, and as the quarter of an hour's notice which we had been led to expect we should receive was now running out we made our way to the shelter assigned to us, armed with a bottle of brandy and other appropriate medical comforts.

Our shelter was a hundred yards down the street, and consisted merely of an open basement, not even sandbagged, in which the tenants of half a dozen flats were already assembled. Everyone was cheerful and jocular, as is the English manner when about to encounter the unknown. As I gazed from the doorway along the empty street and at the crowded room below, my imagination drew pictures of ruin and carnage and vast explosions shaking the ground; of buildings clattering down in dust and rubble, of fire brigades and ambulances scurrying through the smoke, beneath the drone of hostile aeroplanes. For had we not all been taught how terrible air raids would be? The Air Ministry had, in natural self-importance, greatly exaggerated their power. The pacifists had sought to play on public fears, and those of us who had so long pressed for preparation and a superior air force, while not accepting the most lurid forecasts, had been content that they should act as a spur. I knew that the Government were prepared, in the first few days of the war, with over 250,000 beds for air-raid casualties. Here at least there had been no underestimation. Now we should see what were the facts.

After about ten minutes had passed, the wailing broke out again. I was myself not sure that this was not a reiteration of the previous warning, but a man came running along the street shouting "All clear," and we dispersed to our dwellings and went about our business. Mine was to go to the House of Commons, which duly met at noon with its unhurried procedure and brief, stately prayers. There I received a note from the Prime Minister asking me to come to his room as soon as the debate died down. As I sat in my place, listening to the speeches, a very strong sense of calm came over me, after the intense passions and excitements of the last few days. I felt a serenity of mind and was conscious of a kind of uplifted detachment from human and personal affairs. The glory of Old England, peace-loving and ill prepared as she was, but instant and fearless at the call of honour, thrilled my being and seemed to lift our fate to those spheres far removed from earthly facts and physical sensation. I tried to convey some of this mood to the House when I spoke, not without acceptance.

Mr. Chamberlain told me that it was now possible for him to offer me the Admiralty as well as a seat in the War Cabinet. I was very glad of this, because, though I had not raised the point, I naturally preferred a definite task to that exalted brooding over the work done by others which may well be the lot of a Minister, however influential, who has no department. It is easier to give directions than advice, and more agreeable to have the right to act, even in a limited sphere, than the privilege to talk at large. Had the Prime Minister in the first instance given me the choice between the War Cabinet and the Admiralty, I should of course have chosen the Admiralty. Now I was to have both.

Nothing had been said about when I should formally receive my office from the King, and in fact I did not kiss hands till the 5th. But the opening hours of war may be vital with navies. I therefore sent word to the Admiralty that I would take charge forthwith and arrive at six o'clock. On this the Board were kind enough to signal to the Fleet, "Winston is back." So it was that I came again to the room I had quitted in pain and sorrow almost exactly a quarter of a century before, when Lord Fisher's resignation had led to my removal from my post as First Lord and ruined irretrievably, as it proved, the important conception of forcing the Dardanelles. A few feet behind me, as I sat in my old chair, was the wooden map case I had had fixed in 1911, and inside it still remained the chart of the North Sea on which each day, in order to focus attention on the supreme objective, I had made the Naval Intelligence Branch record the movements and dispositions of the German High Seas Fleet. Since 1911 much more than a quarter of a century had passed, and still mortal peril threatened us at the hands of the same nation. Once again defence of the rights of a weak state, outraged and invaded by unprovoked aggression, forced us to draw the sword. Once again we must fight for life and honour against all the might and fury of the valiant, disciplined, and ruthless German race. Once again! So be it.

Presently the First Sea Lord came to me. I had known Dudley Pound slightly in my previous tenure of the Admiralty as one of Lord Fisher's trusted staff officers. I had strongly condemned in Parliament the dispositions of the Mediterranean Fleet when he commanded it, at the moment of the Italian descent upon Albania. Now we met as colleagues upon whose intimate relations and fundamental agreement the smooth working of the vast Admiralty machine would depend. We eyed each other amicably if doubtfully. But from the earliest days our friendship and mutual confidence grew and ripened. I measured

and respected the great professional and personal qualities of Admiral
Pound. As the war, with all its shifts and fortunes, beat upon us with
clanging blows we became ever truer comrades and friends. And when,
four years later, he died at the moment of the general victory over Italy,
I mourned with a personal pang for all the Navy and the nation had lost.

*   *   *

I had, as the reader may be aware, a considerable knowledge of the
Admiralty and of the Royal Navy. The four years from 1911 to 1915,
when I had the duty of preparing the Fleet for war and the task of
directing the Admiralty during the first ten critical months, had been
the most vivid of my life. I had amassed an immense amount of detailed information and had learned many lessons about the Fleet and
war at sea. In the interval I had studied and written much about naval
affairs. I had spoken repeatedly upon them in the House of Commons.
I had always preserved a close contact with the Admiralty, and although their foremost critic in these years, I had been made privy to
many of their secrets. My four years' work on the Air Defence Research Committee had given me access to all the most modern development in radar, which now vitally affected the naval service. In June
1938 Lord Chatfield, then the First Sea Lord, had himself shown me
over the Anti-Submarine School at Portland, and we had gone to sea
in destroyers on an exercise in submarine detection by the use of the
Asdic apparatus. My intimacy with the late Admiral Henderson, Con-
troller of the Navy till 1938, and the discussions which the First Lord
of those days had encouraged me to have with Lord Chatfied upon the
design of new battleships and cruisers, gave me a full view over the
sphere of new construction. I was of course familiar from the published
records with the strength, composition, and structure of the Fleet, actual
and prospective, and with those of the German, Italian, and Japanese
Navies.

*   *   *

One of the first steps I took on taking charge of the Admiralty and
becoming a member of the War Cabinet was to form a statistical department of my own. For this purpose I relied on Professor Lindemann,
my friend and confidant of so many years. Together we had formed
our views and estimates about the whole story. I now installed him at
the Admiralty with half a dozen statisticians and economists whom we
could trust to pay no attention to anything but realities. This group
of capable men, with access to all official information, was able, under
Lindemann's guidance, to present me continually with tables and diagrams, illustrating the whole war so far as it came within our knowl-

edge. They examined and analysed with relentless pertinacity all the departmental papers which were circulated to the War Cabinet, and also pursued all the inquiries which I wished to make myself.

At this time there was no general Government statistical organisation. Each department presented its tale on its own figures and data. The Air Ministry counted one way, the War Office another. The Ministry of Supply and the Board of Trade, though meaning the same thing, talked different dialects. This led sometimes to misunderstandings and waste of time when some point or other came to a crunch in the Cabinet. I had however from the beginning my own sure, steady source of information, every part of which was integrally related to all the rest. Although at first this covered only a portion of the field, it was most helpful to me in forming a just and comprehensible view of the innumerable facts and figures which flowed out upon us.

The tremendous naval situation of 1914 in no way repeated itself. Then we had entered the war with a ratio of sixteen to ten in capital ships and two to one in cruisers. In those days we had mobilised eight battle squadrons of eight battleships, with a cruiser squadron and a flotilla assigned to each, together with important detached cruiser forces, and I looked forward to a general action with a weaker but still formidable fleet. Now the German Navy had only begun their rebuilding and had no power even to form a line of battle. Their two great battleships, *Bismarck* and *Tirpitz*, both of which, it must be assumed, had transgressed the agreed Treaty limits in tonnage, were at least a year from completion. The light battle cruisers, *Scharnhorst* and *Gneisenau,* which had been fraudulently increased by the Germans from 10,000 tons to 26,000 tons, had been completed in 1938. Besides this Germany had available the three "pocket battleships" of 10,000 tons, *Admiral Graf Spee, Admiral Scheer,* and *Deutschland,* together with two fast 8-inch-gun cruisers of 10,000 tons, six light cruisers, and sixty destroyers and smaller vessels. Thus there was no challenge in surface craft to our command of the seas. There was no doubt that the British Navy was overwhelmingly superior to the German in strength and in numbers, and no reason to assume that its science, training or skill was in any way defective. Apart from the shortage of cruisers and destroyers, the Fleet had been maintained at its customary high standard. It had to face enormous and innumerable duties, rather than an antagonist.

Italy had not declared war, and it was already clear that Mussolini was waiting upon events. In this uncertainty and as a measure of precaution till all our arrangements were complete we thought it best to divert our shipping round the Cape. We had however already on our side, in addition to our own preponderance over Germany and Italy combined, the powerful fleet of France, which by the remarkable capacity and long administration of Admiral Darlan had been brought to the highest strength and degree of efficiency ever attained by the French Navy since the days of the Monarchy. Should Italy become hostile our first battlefield must be the Mediterranean. I was entirely opposed, except as a temporary convenience, to all plans for quitting the centre and merely sealing up the ends of the great inland sea. Our forces alone, even without the aid of the French Navy and its fortified harbours, were sufficient to drive the Italian ships from the sea, and should secure complete naval command of the Mediterranean within two months, and possibly sooner.

*   *   *

Newspaper opinion, headed by the *Times,* favoured the principle of a War Cabinet of not more than five or six Ministers, all of whom should be free from departmental duties. Thus alone, it was argued, could a broad and concerted view be taken upon war policy, especially in its larger aspects. Put shortly, "Five men with nothing to do but to run the war" was deemed the ideal. There are however many practical objections to such a course. A group of detached statesmen, however high their nominal authority, are at a serious disadvantage in dealing with the Ministers at the head of the great departments vitally concerned. This is especially true of the Service departments. The War Cabinet personages can have no direct responsibility for day-to-day events. They may take major decisions, they may advise in general terms beforehand or criticise afterwards, but they are no match, for instance, for a First Lord of the Admiralty or a Secretary of State for War or Air, who, knowing every detail of the subject and supported by his professional colleagues, bears the burden of action. United, there is little they cannot settle, but usually there are several opinions among them. Words and arguments are interminable, and meanwhile the torrent of war takes its headlong course. The War Cabinet Ministers themselves would naturally be diffident of challenging the responsible Minister, armed with all his facts and figures. They feel a compunction in adding to the strain upon those actually in executive control. They tend therefore to become more and more theoretical supervisors and commentators, reading an immense amount of material every day, but doubtful how to use their knowledge without doing more harm than good. Often they can do little more than arbitrate or find a compromise in inter-departmental disputes. It is therefore necessary that the Ministers in charge of the Foreign Office and the fighting departments should be integral members of the supreme body. Usually some at least of the "Big Five" are chosen for their political influence, rather than for their knowledge of and aptitude for warlike operations. The numbers therefore begin to grow far beyond the limited circle originally conceived. Of course, where the Prime Minister himself becomes Minister of Defence a strong compression is obtained. Personally, when I was placed in charge I did not like having unharnessed Ministers around me. I preferred to deal with chiefs of organisations rather than counsellors. Everyone should do a good day's work and be accountable for some definite task, and then they do not make trouble for trouble's sake or to cut a figure.

Mr. Chamberlain's original War Cabinet plan was almost immediately expanded, by the force of circumstances, to include Lord Halifax, Foreign Secretary; Sir Samuel Hoare, Lord Privy Seal; Sir John Simon, Chancellor of the Exchequer; Lord Chatfield, Minister for the Co-ordination of Defence; and Lord Hankey, Minister without Portfolio. To these were added the Service Ministers, of whom I was now one, with Mr. Hore Belisha, Secretary of State for War, and Sir Kingsley Wood, Secretary of State for Air. In addition it was necessary that Mr. Eden, who had now rejoined the Government as Dominions Secretary, and Sir John Anderson, the Home Secretary and Minister of Home Security, though not actual members of the War Cabinet, should be present on all occasions. Thus our total was eleven.

Apart from myself all the other Ministers had directed our affairs for a good many recent years or were involved in the situation we now had to face both in diplomacy and war. I had not held public office for nearly eleven years. I had therefore no responsibility for the past or for any want of preparation now apparent. On the contrary, I had for the last six or seven years been a continual prophet of evils which had now in large measure come to pass. Thus, armed as I now was with the mighty machine of the Navy, on which fell in this phase the sole burden of active fighting, I did not feel myself at any disadvantage, and had I done so it would have been removed by the courtesy and loyalty of the Prime Minister and his colleagues. All these men I knew very well. Most of us had served together for five years in Mr. Baldwin's Cabinet, and we had of course been constantly in contact, friendly or controversial, though the changing scenes of Parliamentary life. Sir John Simon and I however represented an older political generation. I had served, off and on, in British Governments for fifteen years, and he for almost as long, before any of the others had gained public office. I had been at the head of the Admiralty or Ministry of Munitions through the stresses of the First World War. Although the Prime Minister was my senior by some years in age, I was almost the only antediluvian. This might well have been a matter of reproach in a time of crisis, when it was natural and popular to demand the force of young men and new ideas. I saw therefore that I should have to strive my ut-

most to keep pace with the generation now in power and with fresh
young giants who might at any time appear. In this I relied upon knowledge as well as upon all possible zeal and mental energy.

For this purpose I had recourse to a method of life which had been
forced upon me at the Admiralty in 1914 and 1915, and which I found
greatly extended my daily capacity for work. I always went to bed at
least for one hour as early as possible in the afternoon, and exploited to
the full my happy gift of falling almost immediately into deep sleep.
By this means I was able to press a day and a half's work into one.
Nature had not intended mankind to work from eight in the morning
until midnight without that refreshment of blessed oblivion which, even
if it only lasts twenty minutes, is sufficient to renew all the vital forces.
I regretted having to send myself to bed like a child every afternoon,
but I was rewarded by being able to work through the night until two
or even later — sometimes much later — in the morning, and begin the
new day between eight and nine o'clock. This routine I observed
throughout the war, and I commend it to others if and when they find it
necessary for a long spell to get the last scrap out of the human structure. The First Sea Lord, Admiral Pound, as soon as he had realised
my technique, adopted it himself, except that he did not actually go to
bed, but dozed off in his armchair. He even carried the policy so far as
often to go to sleep during the Cabinet meetings. One word about the
Navy was however sufficient to awaken him to the fullest activity.
Nothing slipped past his vigilant ear, or his comprehending mind.

*   *   *

Meanwhile around the Cabinet table we were witnessing the swift
and almost mechanical destruction of a weaker state according to
Hitler's method and long design. Over fifteen hundred modern aircraft
were hurled on Poland, and fifty-six divisions, including all his nine
armoured and motorised divisions, composed the invading armies. In
numbers and equipment the Poles were no match for their assailants,
nor were their dispositions wise. They spread all their forces along the
frontiers of their native land. They had no central reserve. While taking
a proud and haughty line against German ambitions, they had nevertheless feared to be accused of provocation by mobilising in good time
against the masses gathering around them. Thirty divisions, representing only two-thirds of their active army, were ready or nearly ready to
meet the first shock. The speed of events and the violent intervention
of the German Air Force prevented the rest from reaching the forward
positions till all was broken, and they were only involved in the final
disasters. Thus the Poles faced nearly double their numbers around
a long perimeter with nothing behind them. Nor was it in numbers alone

that they were inferior. They were heavily outclassed in artillery, and had but a single armoured brigade to meet the nine German Panzers, as they were already called. Their horse cavalry, of which they had twelve brigades, charged valiantly against the swarming tanks and armoured cars, but could not harm them with their swords and lances. Their nine hundred first-line aircraft, of which perhaps half were modern types, were taken by surprise, and many were destroyed before they even got into the air. In two days the Polish air power was virtually annihilated. Within a week the German armies had bitten deep into Poland. Resistance everywhere was brave but vain, and by the end of a fortnight the Polish Army, nominally of about two million men, ceased to exist as an organised force.

It was now the turn of the Soviets. What they now call "Democracy" came into action. On September 17 the Russian armies swarmed across the almost undefended Polish eastern frontier and rolled westward on a broad front. On the 18th they met their German collaborators at Brest-Litovsk. Here in the previous war the Bolsheviks, in breach of their solemn agreements with the Western Allies, had made their separate peace with the Kaiser's Germany and had bowed to its harsh terms. Now in Brest-Litovsk it was with Hitler's Germany that the Russian Communists grinned and shook hands. The ruin of Poland and its entire subjugation proceeded apace. The resistance of Warsaw, largely arising from the surge of its citizens, was magnificent and forlorn. After many days of violent bombardment from the air and by heavy artillery, much of which was rapidly transported across the great lateral highways from the idle Western Front, the Warsaw radio ceased to play the Polish National Anthem, and Hitler entered the ruins of the city. In one month all was over, and a nation of thirty-five millions fell into the merciless grip of those who sought not only conquest but enslavement and indeed extinction for vast numbers.

We had seen a perfect specimen of the modern Blitzkrieg; the close interaction on the battlefield of army and air force; the violent bombardment of all communications and of any town that seemed an attractive target; the arming of an active Fifth Column; the free use of spies and parachutists; and above all the irresistible forward thrusts of great masses of armour. The Poles were not to be the last to endure this ordeal.

18

The Admiralty Task

ASTONISHMENT WAS WORLD-WIDE when Hitler's crashing onslaught upon Poland and the declarations of war upon Germany by Britain and France were followed only by a prolonged and oppressive pause. Mr. Chamberlain in a private letter published by his biographer described this phase as "twilight war"; [1] and I find the expression so just and expressive that I have adopted it as the title for this period. The French armies made no attack upon Germany. Their mobilisation completed, they remained in contact motionless along the whole front. No air action, except reconnaissance, was taken against Britain; nor was any air attack made upon France by the Germans. The French Government requested us to abstain from air attack on Germany, stating that it would provoke retaliation upon their war factories, which were unprotected. We contented ourselves with dropping pamphlets to rouse the Germans to a higher morality. This strange phase of the war on land and in the air astounded everyone. France and Britain remained impassive while Poland was in a few weeks destroyed or subjugated by the whole might of the German war machine. Hitler had no reason to complain of this.

The war at sea, on the contrary, began from the first hour with full intensity, and the Admiralty therefore became the active centre of events. On September 3 all our ships were sailing about the world on their normal business. Suddenly they were set upon by U-boats carefully posted beforehand, especially in the western approaches. At 9 P.M. that very night the outward-bound passenger liner *Athenia*, of 13,500 tons, was torpedoed, and foundered with a loss of one hundred and twelve lives, including twenty-eight American citizens. This outrage broke upon the world within a few hours. The German Government, to prevent

---

[1] Keith Feiling, *Life of Neville Chamberlain*, p. 424.

174

any misunderstanding in the United States, immediately issued a statement that I personally had ordered a bomb to be placed on board this vessel in order by its destruction to prejudice German-American relations. This falsehood received some credence in unfriendly quarters. On the 5th and 6th the *Bosnia, Royal Sceptre,* and *Rio Claro* were sunk off the coast of Spain. All these were important vessels.

Comprehensive plans existed at the Admiralty for multiplying our anti-submarine craft, and a wartime building programme of destroyers, both large and small, and of cruisers, with many ancillary vessels, was also ready in every detail, and came into operation automatically with the declaration of war. The previous conflict had proved the sovereign merits of convoy, and we adopted it in the North Atlantic forthwith. Before the end of the month regular ocean convoys were in operation, outward from the Thames and Liverpool and homeward from Halifax, Gibraltar, and Freetown. Upon all the vital need of feeding the island and developing our power to wage war there now at once fell the numbing loss of the southern Irish ports. This imposed a grievous restriction on the radius of action of our already scarce destroyers.

*   *   *

After the institution of the convoy system the next vital naval need was a safe base for the Fleet. In a war with Germany Scapa Flow is the true strategic point from which the British Navy can control the exits from the North Sea and enforce blockade and I felt it my duty to visit Scapa at the earliest moment. I therefore obtained leave from our daily Cabinets, and started for Wick with a small personal staff on the night of September 14. I spent most of the next two days inspecting the harbour and the entrances, with their booms and nets. I was assured that they were as good as in the last war, and that important additions and improvements were being made or were on the way. I stayed with Sir Charles Forbes, the Commander-in-Chief, in his flagship, *Nelson,* and discussed not only Scapa but the whole naval problem with him and his principal officers. The rest of the Fleet was hiding in Loch Ewe, and on the 17th the Admiral took me to them in the *Nelson.* The narrow entry into the loch was closed by several lines of indicator nets, and patrolling craft with Asdics and depth charges, as well as picket boats, were numerous and busy. On every side rose the purple hills of Scotland in all their splendour. My thoughts went back a quarter of a century to that other September when I had last visited Sir John Jellicoe and his captains in this very bay, and had found them with their long lines of battleships and cruisers drawn out at anchor, a prey to the same uncertainties as now afflicted us. Most of the captains and admirals of those days were dead, or had long passed into retirement. The responsible senior officers who were now presented to me as I visited the various ships had been young lieutenants or even midshipmen in those far-off days. Before the former war I had had three years' preparation in which to make the acquaintance and approve the appointments of most of the high personnel, but now all these were new figures and new faces. The perfect discipline, style, and bearing, the ceremonial routine — all were unchanged. But an entirely different generation filled the uniforms and the posts. Only the ships had most of them been laid down in my tenure. None of them was new. It was a strange experience, like suddenly resuming a previous incarnation. It seemed that I was all that survived in the same position I had held so long ago. But no; the dangers had survived too. Danger from beneath the waves, more serious with more powerful U-boats; danger from the air, not merely of being spotted in your hiding-place, but of heavy and perhaps destructive attack!

No one had ever been over the same terrible course twice with such an interval between. No one had felt its dangers and responsibilities from the summit as I had, or, to descend to a small point, understood how First Lords of the Admiralty are treated when great ships are sunk and things go wrong. If we were in fact going over the same cycle a second time, should I have once again to endure the pangs of dismissal? Fisher, Wilson, Battenburg, Jellicoe, Beatty, Pakenham, Sturdee, all gone!

> *I feel like one*
> *Who treads alone*
> *Some banquet-hall deserted,*
> *Whose lights are fled,*
> *Whose garlands dead,*
> *And all but he departed.*

And what of the supreme, measureless ordeal in which we were again irrevocably plunged? Poland in its agony; France but a pale reflection of her former warlike ardour; the Russian Colossus no longer an ally, not even neutral, possibly to become a foe. Italy no friend. Japan no ally. Would America ever come in again? The British Empire remained intact and gloriously united, but ill prepared, unready. We still had command of the sea. We were woefully outmatched in numbers in this new mortal weapon of the air. Somehow the light faded out of the landscape.

We joined our train at Inverness and travelled through the afternoon and night to London. As we got out at Euston the next morning I was surprised to see the First Sea Lord on the platform. Admiral Pound's look was grave. "I have bad news for you, First Lord. The *Courageous* was sunk yesterday evening in the Bristol Channel." The *Courageous* was one of our oldest aircraft carriers, but a very necessary ship at this time. I thanked him for coming to break it to me himself, and said, "We can't expect to carry on a war like this without that sort of thing happening from time to time, I have seen lots of it before." And so to bath and the toil of another day.

                    *   *   *

By the end of September we had little cause for dissatisfaction with the results of the first impact of the war at sea. I could feel that I had effectively taken over the great department which I knew so well and loved with a discriminating eye. I now knew what there was in hand and on the way. I knew where everything was. I had visited all the principal naval ports and met all the Commanders-in-Chief. By the letters patent constituting the Board, the First Lord is "responsible to Crown and Parliament for all the business of the Admiralty," and I certainly felt prepared to discharge that duty in fact as well as in form.

We had made the immense, delicate, and hazardous transition from peace to war. Forfeits had to be paid in the first few weeks by a world-wide commerce suddenly attacked contrary to formal international agreement by indiscriminate U-boat warfare; but the convoy system was now in full flow, and merchant ships were leaving our ports every day by scores with a gun and a nucleus of trained gunners. The Asdic-equipped trawlers and other small craft armed with depth charges, all well prepared by the Admiralty before the outbreak, were now coming into commission in a growing stream. We all felt sure that the first attack of the U-boat on British trade had been broken and that the menace was in thorough and hardening control. It was obvious that the Germans would build submarines by hundreds, and no doubt numerous shoals were upon the slips in various stages of completion. In twelve months, certainly in eighteen, we must expect the main U-boat war to begin. But by that time we hoped that our mass of new flotillas and anti-U-boat craft, which was our first priority, would be ready to meet it with a proportionate and effective predominance.

Meanwhile the transport of the Expeditionary Force to France was proceeding smoothly, and the blockade of Germany was being enforced by similar methods to those employed in the previous war. Overseas our cruisers were hunting down German ships, while at the same time providing cover against attack on our shipping by raiders. German shipping had come to a standstill and 325 German ships, totalling nearly 750,000 tons, were immobilised in foreign ports. Our Allies also played their part. The French took an important share in the control of the Mediterranean. In home waters and the Bay of Biscay they helped in the battle against the U-boats, and in the central Atlantic a powerful force based on Dakar formed part of the Allied plans against surface raiders.

In this same month I was delighted to receive a personal letter from President Roosevelt. I had only met him once in the previous war. It was at a dinner at Gray's Inn, and I had been struck by his magnificent presence in all his youth and strength. There had been no opportunity for anything but salutations. "It is because you and I," he wrote on the 11th, "occupied similar positions in the World War that I want you to know how glad I am that you are back again in the Admiralty. Your problems are, I realise, complicated by new factors, but the essential is not very different. What I want you and the Prime Minister to know is that I shall at all times welcome it if you will keep me in touch personally with anything you want me to know about. You can always send sealed letters through your pouch or my pouch."

I responded with alacrity, using the signature of "Naval Person," and thus began that long and memorable correspondence — covering nearly a thousand communications on each side, and lasting till his death more than five years later.

*   *   *

In October there burst upon us suddenly an event which touched the Admiralty in a most sensitive spot.

A report that a U-boat was *inside Scapa Flow* had driven the Grand Fleet to sea on the night of October 17, 1914. The alarm was premature. Now, after exactly a quarter of a century almost to a day, it came true. At 1.30 A.M. on October 14, 1939, a German U-boat braved the tides and currents, penetrated our defences, and sank the battleship *Royal Oak* as she lay at anchor. At first, out of a salvo of torpedoes, only one hit the bow, and caused a muffled explosion. So incredible was it to the Admiral and captain on board that a torpedo could have struck them, safe in Scapa Flow, that they attributed the explosion to some internal cause. Twenty minutes passed before the U-boat, for such she was, had reloaded her tubes and fired a second salvo. Then three or four torpedoes, striking in quick succession, ripped the bottom out of the ship. In ten minutes she capsized and sank. Most of the men were at action stations, but the rate at which the ship turned over made it almost impossible for anyone below to escape.

This episode, which must be regarded as a feat of arms on the part of the German U-boat commander, Captain Prien, gave a shock to public opinion. It might well have been politically fatal to any Minister who had been responsible for the prewar precautions. Being a newcomer I was immune from such reproaches in these early months, and moreover, the Opposition did not attempt to make capital out of the misfortune. I promised the strictest inquiry. The event showed how necessary it was to perfect the defences of Scapa against all forms of attack before allowing it to be used. It was nearly six months before we were able to enjoy its commanding advantages.

Presently a new and formidable danger threatened our life. During September and October nearly a dozen merchant ships were sunk at the entrance of our harbours, although these had been properly swept for mines. The Admiralty at once suspected that a magnetic mine had been used. This was no novelty to us; we had even begun to use it on a small scale at the end of the previous war, but the terrible damage that could be done by large ground mines laid in considerable depth by ships or aircraft had not been fully realised. Without a specimen of the mine it was impossible to devise the remedy. Losses by mines, largely Allied and neutral, in September and October had amounted to 56,000 tons, and in November Hitler was encouraged to hint darkly at his new "secret weapon" to which there was no counter. One night when I was at Chartwell Admiral Pound came down to see me in serious anxiety. Six ships had been sunk in the approaches to the Thames. Every day hundreds of ships went in and out of British harbours, and our survival depended on their movement. Hitler's experts may well have told him that this form of attack would compass our ruin. Luckily he began on a small scale, and with limited stocks and manufacturing capacity.

Fortune also favoured us more directly. On November 22 between 9 and 10 P.M. a German aircraft was observed to drop a large object attached to a parachute into the sea near Shoeburyness. The coast here is girdled with great areas of mud which uncover with the tide, and it was immediately obvious that whatever the object was it could be examined and possibly recovered at low water. Here was our golden opportunity. Before midnight that same night two highly skilled officers, Lieutenant Commanders Ouvry and Lewis, from H.M.S. *Vernon*, the naval establishment responsible for developing underwater weapons, were called to the Admiralty, where the First Sea Lord and I interviewed them and heard their plans. By one-thirty in the morning they were on their way by car to Southend to undertake the hazardous task of recovery. Before daylight on the 23rd, in pitch-darkness, aided only by a signal lamp, they found the mine some five hundred yards below high-water mark, but as the tide was then rising they could only inspect it and make their preparations for attacking it after the next high water.

The critical operation began early in the afternoon, by which time it had been discovered that a second mine was also on the mud near the first. Ouvry with Chief Petty Officer Baldwin tackled the first, whilst their colleagues, Lewis and Able Seaman Vearncombe, waited at a safe distance in case of accidents. After each prearranged operation Ouvry would signal to Lewis, so that the knowledge gained would be available when the second mine came to be dismantled. Eventually the combined efforts of all four men were required on the first, and their skill and devotion were amply rewarded. That evening some of the party came to the Admiralty to report that the mine had been recovered intact and was on its way to Portsmouth for detailed examination. I received them with enthusiasm. I gathered together eighty or a hundred officers and officials in our largest room, and a thrilled audience listened to the tale, deeply conscious of all that was at stake.

The whole power and science of the Navy were now applied; and it was not long before trial and experiment began to yield practical results. We worked all ways at once, devising first active means of attacking the mine by new methods of minesweeping and fuze-provocation, and, secondly, passive means of defence for all ships against possible mines in unswept, or ineffectually swept, channels. For this second purpose a most effective system of demagnetising ships by girdling them with an electric cable was developed. This was called "degaussing," and was at once applied to ships of all types. But serious casualties continued. The new cruiser *Belfast* was mined in the Firth of Forth on November 21, and on December 4 the battleship *Nelson* was mined whilst entering Loch Ewe. Both ships were however able to reach a dockyard port. It is remarkable that German Intelligence failed to pierce our security measures covering the injury to the *Nelson* until the ship had been repaired and was again in service. Yet from the first many thousands in England had to know the true facts.

Experience soon gave us new and simpler methods of degaussing. The moral effect of its success was tremendous, but it was on the faithful, courageous, and persistent work of the mine sweepers and the patient skill of the technical experts, who devised and provided the equipment they used, that we relied chiefly to defeat the enemy's efforts. From this time onward, despite many anxious periods, the mine menace was always under control, and eventually the danger began to recede.

It is well to ponder this side of the naval war. In the event a significant proportion of our whole war effort had to be devoted to combating the mine. A vast output of material and money was diverted from other tasks, and many thousands of men risked their lives night and day in the mine sweepers alone. The peak figure was reached in June 1944, when nearly sixty thousand were thus employed. Nothing daunted the ardour of the Merchant Navy, and their spirits rose with the deadly complications of the mining attack and our effective measures for countering it. Their toils and tireless courage were our salvation. In the wider sphere of naval operations no definite challenge had yet been made to our position. This was still to come, and a description of two major conflicts with German surface raiders may conclude my account of the war at sea in the year 1939.

*   *   *

Our long, tenuous blockade line north of the Orkneys, largely composed of armed merchant cruisers with supporting warships at intervals, was of course always liable to a sudden attack by German capital ships, and particular by their two fast and most powerful battle cruisers, the *Scharnhorst* and the *Gneisenau*. We could not prevent such a stroke being made. Our hope was to bring the intruders to decisive action.

Late in the afternoon of November 23 the armed merchant cruiser *Rawalpindi*, on patrol between Iceland and the Faroes, sighted an enemy warship which closed her rapidly. She believed the stranger to be the pocket battleship *Deutschland*, and reported accordingly. Her commanding officer, Captain Kennedy, could have had no illusions about the outcome of such an encounter. His ship was but a converted passenger liner with a broadside of four old 6-inch guns, and his presumed antagonist mounted six 11-inch guns, besides a powerful secondary armament. Nevertheless he accepted the odds, determined to fight his ship to the last. The enemy opened fire at 10,000 yards, and the *Rawalpindi* struck back. Such a one-sided action could not last long, but the fight continued until, with all her guns out of action, the *Rawalpindi* was reduced to a blazing wreck. She sank some time after dark, with the loss of her captain and 270 of her gallant crew.

In fact it was not the *Deutschland* but the two battle cruisers *Scharnhorst* and *Gneisenau* which were engaged. These ships had left Germany two days before to attack our Atlantic convoys, but having encountered and sunk the *Rawalpindi,* and fearing the consequences of the exposure, they abandoned the rest of their mission and returned at once to Germany. The *Rawalpindi*'s heroic fight was not therefore in vain. The cruiser *Newcastle,* nearby on patrol, saw the gun flashes, and responded at once to the *Rawalpindi*'s first report, arriving on the scene with the cruiser *Delhi* to find the burning ship still afloat. She pursued the enemy, and at 6.15 P.M. sighted two ships in gathering darkness and heavy rain. One of these she recognised as a battle cruiser, but lost contact in the gloom, and the enemy made good his escape.

The hope of bringing these two vital German ships to battle dominated all concerned, and the Commander-in-Chief put to sea at once with his whole fleet. By the 25th fourteen British cruisers were combing the North Sea, with destroyers and submarines co-operating and with the battle fleet in support. But fortune was adverse; nothing was found, nor was there any indication of an enemy move to the west. Despite very severe weather the arduous search was maintained for seven days, and we eventually learnt that the *Scharnhorst* and the *Gneisenau* had safely re-entered the Baltic. It is now known that they passed through our cruiser line patrolling near the Norwegian coast on the morning of November 26. The weather was thick and neither saw the other. Modern radar would have ensured contact, but then it was not available. Public impressions were unfavourable to the Admiralty. We could not bring home to the outside world the vastness of the seas or the intense exertions which the Navy was making in so many areas. After more than two months of war and several serious losses we had nothing to show on the other side. Nor could we yet answer the question, "What is the Navy doing?"

*   *   *

The attack on our ocean commerce by surface raiders would have been even more formidable could it have been sustained. The three German pocket battleships permitted by the Treaty of Versailles had been designed with profound thought as commerce destroyers. Their six 11-inch guns, their 26-knot speed, and the armour they carried had been compressed with masterly skill into the limits of a 10,000-ton displacement. No single British cruiser could match them. The German 8-inch-gun cruisers were more modern than ours, and if employed as commerce raiders would also be a formidable threat. Besides this the enemy might use disguised heavily armed merchantmen. We had vivid memories of the depredations of the *Emden* and *Koenigsberg* in 1914, and of the thirty or more warships and armed merchantmen they had forced us to combine for their destruction.

There were rumours and reports before the outbreak of the new war that one or more pocket battleships had already sailed from Germany. The Home Fleet searched but found nothing. We now know that both the *Deutschland* and the *Graf Spee* sailed from Germany between August 21 and 24, and were already through the danger zone and loose in the oceans before our blockade and northern patrols were organised. On September 3 the *Deutschland,* having passed through the Denmark Straits, was lurking near Greenland. The *Graf Spee* had crossed the North Atlantic trade route unseen and was already far south of the Azores. Each was accompanied by an auxiliary vessel to replenish fuel and stores. Both at first remained inactive and lost in the ocean spaces. Unless they struck they won no prizes. Until they struck they were in no danger.

On September 30 the British liner *Clement*, of 5000 tons, sailing independently, was sunk by the *Graf Spee* off Pernambuco. The news electrified the Admiralty. It was the signal for which we had been waiting. A number of hunting groups were immediately formed, comprising all our available aircraft carriers, supported by battleships, battle cruisers, and cruisers. Each group of two or more ships was judged to be capable of catching and destroying a pocket battleship.

In all, during the ensuing months the search for two raiders entailed the formation of nine hunting groups, comprising twenty-three powerful ships. Working from widely dispersed bases in the Atlantic and Indian Oceans, these groups could cover the main focal areas traversed by our shipping. To attack our trade the enemy must place himself within reach of at least one of them.

The *Deutschland*, which was to have harassed our lifeline across the Northwest Atlantic, interpreted her orders with comprehending caution. At no time during her two and a half months' cruise did she approach a convoy. Her determined efforts to avoid British forces prevented her from making more than two kills, one being a small Norwegian ship. Early in November she slunk back to Germany, passing again through Arctic waters. The mere presence of this powerful ship upon our main trade route had however imposed, as was intended, a serious strain upon our escorts and hunting groups in the North Atlantic. We should in fact have preferred her activity to the vague menace she embodied.

The *Graf Spee* was more daring and imaginative, and soon became the centre of attention in the South Atlantic. Her practice was to make a brief appearance at some point, claim a victim, and vanish again into the trackless ocean wastes. After a second appearance farther south on the Cape route, in which she sank only one ship, there was no further sign of her for nearly a month, during which our hunting groups were searching far and wide in all areas, and special vigilance was enjoined in the Indian Ocean. This was in fact her destination, and on November 15 she sank a small British tanker in the Mozambique Channel, between Madagascar and the mainland. Having thus registered her appearance as a feint in the Indian Ocean, in order to draw the hunt in that direction, her captain — Langsdorff, a high-class person — promptly doubled back and, keeping well south of the Cape, re-entered the Atlantic. This move had not been unforeseen; but our plans to intercept him were foiled by the quickness of his withdrawal. It was by no means clear to the Admiralty whether in fact one raider was on the prowl or two, and exertions were made both in the Indian and Atlantic Oceans. We also thought that the *Spee* was her sister ship, the *Scheer*. The disproportion between the strength of the enemy and the countermeasures forced upon us was vexatious. It recalled to me the anxious weeks before the action at Coronel and later at the Falkland Islands in December 1914, when we had to be prepared at seven or eight different points, in the Pacific and South Atlantic, for the arrival of Admiral von Spee with the earlier edition of the *Scharnhorst* and *Gneisenau*. A quarter of a century had passed, but the puzzle was the same. It was with a definite sense of relief that we learnt that the *Spee* had appeared once more on the Cape-Freetown route, sinking the *Doric Star* and another ship on December 2 and one more on the 7th.

*   *   *

From the beginning of the war Commodore Harwood's special care and duty had been to cover British shipping off the River Plate and Rio de Janeiro. He was convinced that sooner or later the *Spee* would come towards the Plate, where the richest prizes were offered to her. He had carefully thought out the tactics which he would adopt in an encounter. Together, his 8-inch cruisers *Cumberland* and *Exeter,* and his 6-inch cruisers *Ajax* and *Achilles,* the latter being a New Zealand ship manned mainly by New Zealanders, could not only catch but kill. However, the needs of fuel and refit made it unlikely that all four would be present "on the day." If they were not, the issue was disputable. On hearing that the *Doric Star* had been sunk on December 2, Harwood guessed right. Although she was over three thousand miles away he assumed that the *Spee* would come towards the Plate. He estimated with luck and wisdom that she might arrive by the 13th. He ordered all his available forces to concentrate there by December 12. Alas, the *Cumberland* was refitting at the Falklands; but on the morning of the 13th *Exeter, Ajax,* and *Achilles* were in company at the centre of the shipping routes off the mouth of the river. Sure enough, at 6.14 A.M. smoke was sighted to the east. The longed-for collision had come.

Harwood, in the *Ajax,* disposing his forces so as to attack the pocket battleship from widely divergent quarters and thus confuse her fire, advanced at the utmost speed of his small squadron. Captain Langsdorff thought at first glance that he had only to deal with one light cruiser and two destroyers, and he too went full speed ahead; but a few moments later he recognized the quality of his opponents, and knew that a mortal action impended. The two forces were now closing at nearly fifty miles an hour. Langsdorff had but a minute to make up his mind. His right course would have been to turn away immediately so as to keep his assailants as long as possible under the superior range and weight of his 11-inch guns, to which the British could not at first have replied. He would thus have gained for his undisturbed firing the difference between adding speeds and subtracting them. He might well have crippled one of his foes before any could fire at him. He decided, on the contrary, to hold on his course and make for the *Exeter.* The action therefore began almost simultaneously on both sides.

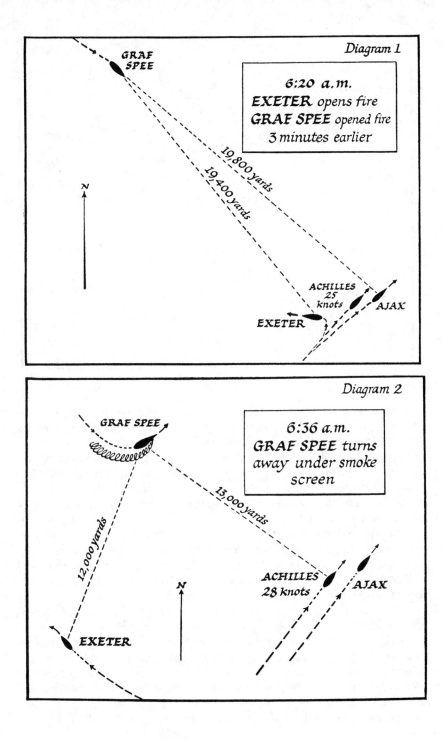

Diagram 1

6:20 a.m.
EXETER *opens fire*
GRAF SPEE *opened fire*
3 minutes earlier

GRAF SPEE

19,800 yards

19,400 yards

N

ACHILLES
25
knots

AJAX

EXETER

Diagram 2

6:36 a.m.
GRAF SPEE *turns*
away under smoke
screen

GRAF SPEE

13,000 yards

12,000 yards

ACHILLES
28 knots

AJAX

N

EXETER

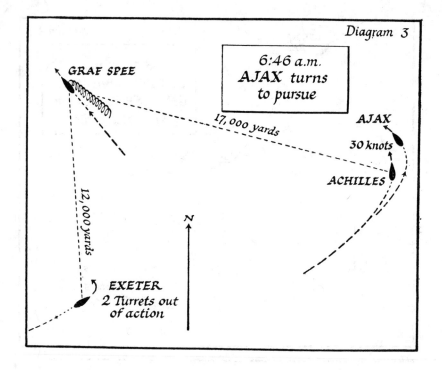

Diagram 3

6:46 a.m.
AJAX turns
to pursue

GRAF SPEE

17,000 yards

AJAX

30 knots

ACHILLES

12,000 yards

N

EXETER
2 Turrets out
of action

Commodore Harwood's tactics proved advantageous. The 8-inch salvoes from the *Exeter* struck the *Spee* from the earliest stages of the fight. Meanwhile the 6-inch cruisers were also hitting hard and effectively. Soon the *Exeter* received a hit which, besides knocking out B turret, destroyed all the communications on the bridge, killed or wounded nearly all upon it, and put the ship temporarily out of control. By this time however the 6-inch cruisers could no longer be neglected by the enemy, and the *Spee* shifted her main armament to them, thus giving respite to the *Exeter* at a critical moment. The German battleship, plastered from three directions, found the British attack too hot, and soon afterwards turned away under a smoke screen with the apparent intention of making for the River Plate. Langsdorff had better have done this earlier.

After this turn the *Spee* once more engaged the *Exeter*, hard hit by the 11-inch shells. All her forward guns were out of action. She was burning fiercely amidships and had a heavy list. Captain Bell, unscathed by the explosion on the bridge, gathered two or three officers round him in the after control-station, and kept his ship in action with her sole remaining turret, until at seven-thirty failure of pressure put this too out of action. He could do no more. At seven-forty the *Exeter* turned away to effect repairs and took no further part in the fight.

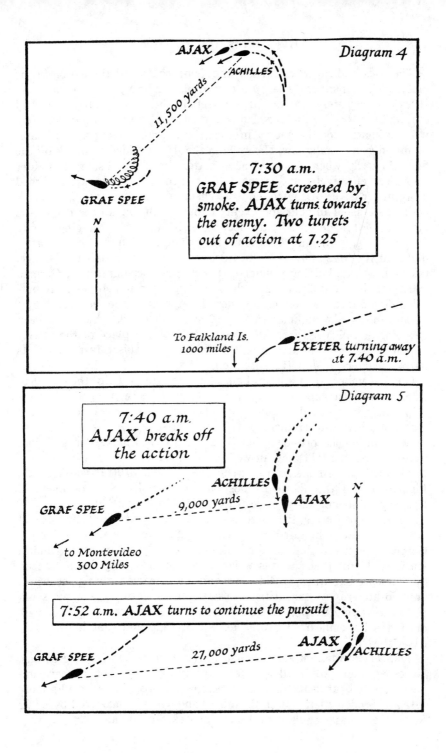

Diagram 4

AJAX   ACHILLES

11,500 yards

GRAF SPEE

N

7:30 a.m.
GRAF SPEE screened by
smoke. AJAX turns towards
the enemy. Two turrets
out of action at 7.25

To Falkland Is.
1000 miles

EXETER turning away
at 7.40 a.m.

Diagram 5

7:40 a.m.
AJAX breaks off
the action

ACHILLES   AJAX

GRAF SPEE   9,000 yards

N

to Montevideo
300 Miles

7:52 a.m. AJAX turns to continue the pursuit

AJAX

GRAF SPEE   27,000 yards   ACHILLES

The *Ajax* and *Achilles*, already in pursuit, continued the action in the most spirited manner. The *Spee* turned all her heavy guns upon them. By seven twenty-five the two after turrets in the *Ajax* had been knocked out, and the *Achilles* had also suffered damage. These two light cruisers were no match for the enemy in gun power, and, finding that his ammunition was running low, Harwood in the *Ajax* decided to break off the fight till dark, when he would have better chances of using his lighter armament effectively, and perhaps his torpedoes. He therefore turned away under cover of smoke, and the enemy did not follow. This fierce action had lasted an hour and twenty minutes. During all the rest of the day the *Spee* made for Montevideo, the British cruisers hanging grimly on her heels, with only occasional interchanges of fire. Shortly after midnight the *Spee* entered Montevideo, and lay there repairing damage, taking in stores, landing wounded, transhipping personnel to a German merchant ship, and reporting to the Fuehrer. *Ajax* and *Achilles* lay outside, determined to dog her to her doom should she venture forth. Meanwhile on the night of the 14th the *Cumberland*, which had been steaming at full speed from the Falklands, took the place of the utterly crippled *Exeter*. The arrival of this 8-inch-gun cruiser restored to its narrow balance a doubtful situation.

On December 16 Captain Langsdorff telegraphed to the German Admiralty that escape was hopeless. "Request decision on whether the ship should be scuttled in spite of insufficient depth in the estuary of the Plate, or whether internment is to be preferred."

At a conference presided over by Hitler, at which Raeder and Jodl were present, the following answer was decided on:

"Attempt by all means to extend the time in neutral waters. . . . Fight your way through to Buenos Aires if possible. No internment in Uruguay. Attempt effective destruction if ship is scuttled."

Accordingly during the afternoon of the 17th the *Spee* transferred more than seven hundred men, with baggage and provisions, to the German merchant ship in the harbour. Shortly afterwards Admiral Harwood learnt that she was weighing anchor. At 6.15 P.M., watched by immense crowds, she left harbour and steamed slowly seaward, awaited hungrily by the British cruisers. At 8.54 P.M., as the sun sank, the *Ajax*'s aircraft reported: *"Graf Spee has blown herself up."* Langsdorff was brokenhearted by the loss of his ship and shot himself two days later.

Thus ended the first surface challenge to British trade on the oceans. No other raider appeared until the spring of 1940, when a new campaign opened, utilising disguised merchant ships. These could more easily avoid detection, but on the other hand could be mastered by lesser forces than those required to destroy a pocket battleship.

19

The Front in France

IMMEDIATELY UPON THE OUTBREAK of war the British Expeditionary Force, or "B.E.F.," began to move to France. By mid-October four British divisions, formed into two Army Corps of professional quality, were in their stations along the Franco-Belgian frontier, and by March 1940 six more divisions had joined them, making a total of ten. As our numbers grew we took over more line. We were not of course at any point in contact with the enemy.

When the B.E.F. reached their prescribed positions they found ready-prepared a fairly complete artificial anti-tank ditch along the front line, and every thousand yards or so was a large and very visible pillbox giving enfilade fire along the ditch for machine and anti-tank guns. There was also a continuous belt of wire. Much of the work of our troops during this strange autumn and winter was directed to improving the French-made defences and organising a kind of Siegfried Line. In spite of frost progress was rapid. Air photographs showed the rate at which the Germans were extending their own Siegfried Line northwards from the Moselle. Despite the many advantages they enjoyed in home resources and forced labour, we seemed to be keeping pace with them. Large base installations were created, roads improved, a hundred miles of broad-gauge railway line laid. Nearly fifty new airfields and satellites were developed or improved. Behind our front immense masses of stores and ammunition were accumulated in the depots all along the communications. Ten days' supply was gathered between the Seine and the Somme, *and seven days' additional north of the Somme.* This latter provision saved the Army after the German break-through. Gradually, in view of the prevailing tranquillity, many ports north of Havre were brought into use in succession and in the end we were making use in all of thirteen French harbours.

*　　　*　　　*

In 1914 the spirit of the French Army and nation, burning from sire to son since 1870, was vehemently offensive. Their doctrine was that the numerically weaker Power could only meet invasion by the counter-offensive, not only strategic but tactical at every point. It was now a very different France from that which had hurled itself upon its ancient foe in August 1914. The spirit of *revanche* had exhausted its mission and itself in victory. The chiefs who had nursed it were long dead. The French people had undergone the frightful slaughter of a million and a half of their manhood. Offensive action was associated in the great majority of French minds with the initial failures of the French on-slaught of 1914, with General Nivelle's repulse in 1917, with the long agonies of the Somme and Passchendaele, and above all with the sense that the fire power of modern weapons was devastating to the attacker. Neither in France nor in Britain had there been any effective compre-hension of the consequences of the new fact that armoured vehicles could be made capable of withstanding artillery fire, and could ad-vance a hundred miles a day. An illuminating book on this subject, published some years before by a Commander de Gaulle, had met with no response. The authority of the aged Marshal Pétain in the Conseil Supérieur de la Guerre had weighed heavily upon French military thought in closing the door to new ideas, and especially in discouraging what had been quaintly called "offensive weapons."

In the afterlight the policy of the Maginot Line has often been con-demned. It certainly engendered a defensive mentality. Yet it is always a wise precaution in defending a frontier of hundreds of miles to bar off as much as possible by fortifications, and thus economise in the use of troops in sedentary roles and "canalise" potential invasion. Properly used in the French scheme of war, the Maginot Line would have been of immense service to France. It could have been viewed as presenting a long succession of invaluable sally ports, and above all as blocking off large sections of the front as a means of accumulating the general reserves or "mass manoeuvre." Having regard to the disparity of the population of France to that of Germany, the Maginot Line must be regarded as a wise and prudent measure. Indeed, it was extraordinary that it should not have been carried forward at least along the river Meuse. It could then have served as a trusty shield, freeing a heavy, sharp, offensive French sword. But Marshal Pétain had opposed this extension. He held strongly that the Ardennes could be ruled out as a channel of invasion on account of the nature of the ground. Ruled out accordingly it was. The offensive conceptions of the Maginot Line were explained to me by General Giraud when I visited Metz in 1937. They were however not carried into effect, and the Line not only absorbed very large numbers of highly trained regular soldiers and technicians, but exercised an enervating effect both upon military strategy and national vigilance.

The new air power was justly esteemed a revolutionary factor in all operations. Considering the comparatively small numbers of aircraft available on either side at this time, its effects were even exaggerated, and were held in the main to favour the defensive by hampering the concentrations and communications of great armies once launched in attack. Even the period of the French mobilisation was regarded by the French High Command as most critical on account of the possible destruction of railway centres, although the numbers of German aircraft, like those of the Allies, were far too few for such a task. These thoughts expressed by air chiefs followed correct lines, and were justified in the later years of the war, when the air strength had grown ten- or twenty-fold. At the outbreak they were premature.

*   *   *

It is a joke in Britain to say that the War Office is always preparing for the last war. But this is probably true of other departments and of other countries, and it was certainly true of the French Army. I also rested under the impression of the superior power of the defensive provided it were actively conducted. I had neither the responsibility nor the continuous information to make a new measurement. I knew that the carnage of the previous war had bitten deeply into the soul of the French people. The Germans had been given time to build the Siegfried Line. How frightful to hurl the remaining manhood of France against this wall of fire and concrete! In my mind's outlook in the opening months of this Second World War I did not dissent from the general view about the defensive, and I believed that anti-tank obstacles and field guns, cleverly posted and with suitable ammunition, could frustrate or break up tanks except in darkness or fog, real or artificial.

In the problems which the Almighty sets his humble servants things hardly ever happen the same way twice over, or if they seem to do so there is some variant which stultifies undue generalisation. The human mind, except when guided by extraordinary genius, cannot surmount the established conclusions amid which it has been reared. Yet we are to see, after eight months of inactivity on both sides, the Hitler inrush of a vast offensive, led by spear-point masses of cannon-proof or heavily armoured vehicles, breaking up all defensive opposition, and for the first time for centuries, and even perhaps since the invention of gunpowder, making artillery for a while almost impotent on the battlefield.

We are also to see that the increase of fire power made the actual battles less bloody by enabling the necessary ground to be held with very small numbers of men, thus offering a far smaller human target.

Anyway, the earliest date at which the French could have mounted a big attack was perhaps at the end of the third week of September. But by that time the Polish campaign had ended. By mid-October the Germans had seventy divisions on the Western Front. The fleeting French numerical superiority in the West was passing. A French offensive from their eastern frontier would have denuded their far more vital northern front. Even if an initial success had been gained by the French armies at the outset, within a month they would have had extreme difficulty in maintaining their conquests in the East, and would have been exposed to the whole force of the German counterstroke to the north.

This is the answer to the question "Why remain passive till Poland was destroyed?" But this battle had been lost some years before. In 1938 there was a good chance of victory while Czechoslovakia still existed. In 1936 there could have been no effective opposition. In 1933 a rescript from Geneva would have procured bloodless compliance. General Gamelin cannot be the only one to blame because in 1939 he did not run the risks which had so enormously increased since the previous crisis, from which both the French and British Governments had recoiled.

*   *   *

What then were the probabilities of a German offensive against France? There were of course three methods open. First, invasion through Switzerland. This might turn the southern flank of the Maginot Line, but had many geographical and strategic difficulties. Secondly, invasion of France across the common frontier. This appeared unlikely, as the German Army was not believed to be fully equipped or armed for a heavy attack on the Maginot Line. And, thirdly, invasion of France through Holland and Belgium. This would turn the Maginot Line, and would not entail the losses likely to be sustained in a frontal attack against permanent fortifications. We could not meet an onslaught through the Low Countries so far forward as Holland, but it would be in the Allied interest to stem it, if possible, in Belgium, and at this period there were two lines to which the Allies could advance if they chose to come to her succour, or which they could occupy by a well-planned secret and sudden scheme, if so invited. The first of these lines was what may be called the line of the Scheldt. This was no great march from the French frontier and involved little serious risk. At the worst it would do no harm to hold it as a "false front." At the best it

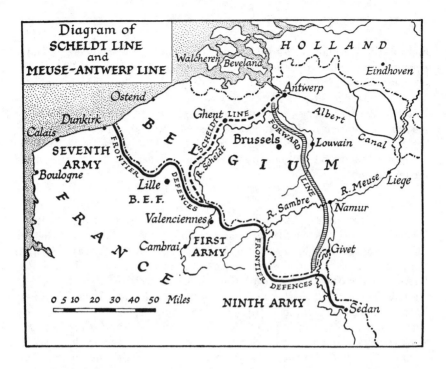

Diagram of SCHELDT LINE and MEUSE-ANTWERP LINE

might be built up according to events. The second line was far more ambitious. It followed the Meuse through Givet, Dinant, and Namur by Louvain to Antwerp. If this adventurous line was seized by the Allies and held in hard battles the German right-handed swing of invasion would be heavily checked; and if their armies were proved inferior it would be an admirable prelude to the entry and control of the vital centre of Germany's munitions production in the Ruhr.

"We understand," wrote the Chiefs of Staff, "that the French idea [1] is that, provided the Belgians are still holding out on the Meuse, the French and British Armies should occupy the line Givet-Namur, the British Expeditionary Force operating on the left. *We consider it would be unsound to adopt this plan unless plans are concerted with the Belgians for the occupation of this line in sufficient time before the Germans advance. . . . Unless the present Belgian attitude alters and plans can be prepared for early occupation of the Givet-Namur* [also called Meuse-Antwerp] *line, we are strongly of opinion that the German advance should be met in prepared positions on the French frontier.*"

The Allied Supreme Council met in Paris on November 17. Mr. Chamberlain took with him Lord Halifax, Lord Chatfield, and Sir Kingsley Wood. The decision was taken: "Given the importance of

[1] Known as Plan D.

holding the German forces as far east as possible, *it is essential to make every endeavour to hold the line Meuse-Antwerp in the event of a German invasion of Belgium.*" At this meeting Mr. Chamberlain and M. Daladier insisted on the importance which they attached to this resolution and thereafter it governed action. In this posture therefore we passed the winter and awaited the spring. No new decisions of strategic principle were taken by the French and British Staffs or by their Governments in the six months which lay between us and the German onslaught.

*   *   *

During the winter and spring the B.E.F were extremely busy setting themselves to rights, fortifying their line and preparing for war, whether offensive or defensive. From the highest rank to the lowest all were hard at it, and the good showing that they eventually made was due largely to the full use made of the opportunities provided during the winter. The British was a far better army at the end of the "Twilight War." It was also larger. But the awful gap, reflecting on our pre-war arrangements, was *the absence of even one armoured division in the British Expeditionary Force.* Britain, the cradle of the tank in all its variants, had between the wars so far neglected the development of this weapon, soon to dominate the battlefields, that eight months after the declaration of war our small but good Army had only with it, when the hour of trial arrived, the 1st Tank Brigade, comprising seventeen light two-pounder tanks and "Infantry" tanks. Only twenty-three of the latter carried even the two-pounder gun, the rest machine guns only. There were also seven cavalry and Yeomanry regiments equipped with carriers and light tanks which were in process of being formed into two light armoured brigades.

Developments on the French front were less satisfactory. In a great national conscript force the mood of the people is closely reflected in its Army, the more so when that Army is quartered in the homeland and contacts are close. It cannot be said that France in 1939–40 viewed the war with uprising spirit, or even with much confidence. The restless internal politics of the past decade had bred disunity and discontents. Important elements, in reaction to growing Communism, had swung towards Fascism, lending a ready ear to Goebbels' skilful propaganda and passing it on in gossip and rumour. So also in the Army the disintegration influences of both Communism and Fascism were at work; the long winter months of waiting gave time and opportunity for the poisons to be established.

Very many factors go to the building up of sound morale in an army, but one of the greatest is that the men be fully employed at useful and interesting work. Idleness is a dangerous breeding ground. Through-out the winter there were many tasks that needed doing: training demanded continuous attention; defences were far from satisfactory or complete — even the Maginot Line lacked many supplementary field works; physical fitness demands exercise. Yet visitors to the French front were often struck by the prevailing atmosphere of calm aloofness, by the seemingly poor quality of the work in hand, by the lack of visible activity of any kind. The emptiness of the roads behind the line was in great contrast to the continual coming and going which extended for miles behind the British sector.

There can be no doubt that the quality of the French Army was allowed to deteriorate during the winter, and that it would have fought better in the autumn than in the spring. Soon it was to be stunned by the swiftness and violence of the German assault. It was not until the last phases of that brief campaign that the true fighting qualities of the French soldier rose uppermost in defence of his country against the age-long enemy. But then it was too late.

On January 10, 1940, anxieties about the Western Front received confirmation. A German staff major of the 7th Air Division had been ordered to take some documents to headquarters in Cologne. He missed his train and decided to fly. His machine overshot the mark and made a forced landing in Belgium, where Belgian troops arrested him and impounded his papers, which he tried desperately to destroy. Those contained the entire and actual scheme for the invasion of Belgium, Holland, and France on which Hitler had resolved. Shortly the German major was released to explain matters to his superiors. I was told about all this at the time, and it seemed to me incredible that the Belgians would not make a plan to invite us in. But they did nothing about it. It was argued in all three countries concerned that probably it was a plant. But this could not be true. There could be no sense in the Germans trying to make the Belgians believe that they were going to attack them in the near future. This might make them do the very last thing the Germans wanted, namely, make a plan with the French and British Armies to come forward privily and quickly one fine night. I therefore believed in the impending attack.

We appealed to Belgium, but the Belgian King and his Army staff merely waited, hoping that all would turn out well. In spite of the German major's papers no further action of any kind was taken by the Allies or the threatened States. Hitler, on the other hand, as we know, summoned Goering to his presence, and on being told that the captured papers were in fact the complete plans for invasion, ordered, after venting his anger, new variants to be prepared. Of course, if British and French policy during the five years preceding the war had been of a manly and resolute character, within the sanctity of treaties and the approval of the League of Nations, Belgium might have adhered to her old allies and allowed a common front to be formed. Such an alliance properly organised would have erected a shield along the Belgian frontier to the sea against that terrible turning movement which had nearly compassed our destruction in 1914 and was to play its part in the ruin of France in 1940. At the worst Belgium could have suffered no harder fate than actually befell her. When we recall the aloofness of the United States; Mr. Ramsay MacDonald's campaign for the disarmament of France; the repeated rebuffs and humiliations which we had accepted in the various German breaches of the disarmament clauses of the Treaty; our submission to the German violation of the Rhineland; our acquiescence in the absorption of Austria; our pact at Munich and acceptance of the German occupation of Prague — when we recall all this, no man in Britain or France who in those years was responsible for public action has a right to blame Belgium. In a period of vacillation and appeasement the Belgians clung to neutrality, and vainly comforted themselves with the belief that they could hold the German invader on their fortified frontiers until the British and French Armies could come to their aid.

20

Scandinavia, Finland

THE THOUSAND-MILE-LONG PENINSULA stretching from the mouth of the Baltic to the Arctic Circle had an immense strategic significance. The Norwegian mountains run into the ocean in a continuous fringe of islands. Between these islands and the mainland there was a corridor in territorial waters through which Germany could communicate with the outer sea to the grievous injury of our blockade. German war industry was mainly based upon supplies of Swedish iron ore, which in the summer were drawn from the Swedish port of Luleå, at the head of the Gulf of Bothnia, and in the winter, when this was frozen, from Narvik, on the west coast of Norway. To respect the "Leads," as these sheltered waters were called, would be to allow the whole of this traffic to proceed under the shield of neutrality in the face of our superior sea power. The Admiralty Staff were seriously perturbed at this important advantage being presented to Germany, and at the earliest opportunity I raised the issue in the Cabinet.

At first the reception of my case was favourable. All my colleagues were deeply impressed with the evil; but strict respect for the neutrality of small states was a principle of conduct to which we all adhered. In September, at the invitation of my colleagues, and after the whole subject had been minutely examined at the Admiralty, I drafted a paper for the Cabinet upon this subject, and on the chartering of neutral tonnage, which was linked with it. Again there was general agreement upon the need; but I was unable to obtain assent to action. The Foreign Office arguments about neutrality were weighty, and I could not prevail. I continued, as will be seen, to press my point by every means and on all occasions. It was not however until April 1940 that the decision that I asked for in September 1939 was taken. By that time it was too late.

Almost at this very moment, as we now know, German eyes were turned in the same direction. On October 3 Admiral Raeder, Chief of the Naval Staff, submitted a proposal to Hitler headed "Gaining of Bases in Norway." He asked, "That the Fuehrer be informed as soon as possible of the opinions of the Naval War Staff on the possibilities of extending the operational base to the north. It must be ascertained whether it is possible to gain bases in Norway under the combined pressure of Russia and Germany, with the aim of improving our strategic and operational position." He framed therefore a series of notes, which he placed before Hitler on October 10. "In these notes," he wrote, "I stressed the disadvantages which an occupation of Norway by the British would have for us: the control of the approaches to the Baltic, the outflanking of our naval operations and of our air attacks on Britain, the end of our pressure on Sweden. I also stressed the advantages for us of the occupation of the Norwegian coast: outlet to the North Atlantic, no possibility of a British mine barrier, as in the year 1917–18."

Rosenberg, the foreign affairs expert of the Nazi Party, and in charge of a special bureau to deal with propaganda activities in foreign countries, shared the Admiral's view. He dreamed of "converting Scandinavia to the idea of a Nordic community embracing the northern peoples under the natural leadership of Germany." Early in 1939 he thought he had discovered an instrument in the extreme Nationalist Party in Norway, which was led by a former Norwegian Minister of War named Vidkun Quisling. Contacts were established, and Quisling's activity was linked with the plans of the German Naval Staff through Rosenberg's organisation and the German Naval Attaché in Oslo. Quisling and his assistant, Hagelin, went to Berlin on December 14, and were taken by Raeder to Hitler, to discuss a political stroke in Norway. Quisling arrived with a detailed plan. Hitler, careful of secrecy, affected reluctance to increase his commitments, and said he would prefer a neutral Scandinavia. Nevertheless, according to Raeder, it was on this very day that he gave the order to the Supreme Commander to prepare for a Norwegian operation.

Of all this we of course knew nothing.

*   *   *

Meanwhile the Scandinavian peninsula became the scene of an unexpected conflict which aroused strong feeling in Britain and France and powerfully affected the discussion about Norway. Stalin's "Mutual Assistance Pacts" with Esthonia, Latvia, and Lithuania had already led to the occupation and ruin of those countries, and the Red Army and Air Force now blocked the lines of entry into the Soviet Union from the

North Cape

N O R W A Y

Narvik

Petsamo

Murmansk

Kola
Peninsula

S W E D E N

Kemijarvi

White
Sea

Vaasa

F I N L A N D

Kajaani

U.
S.

S.

Aland Is.
Turku

MANNERHEIM
LINE
Viipuri

Lake
Ladoga

R.

Helsingfors

Hango

Stockholm

Leningrad

ESTHONIA

B a l t i c   S e a

**RUSSIAN ATTACK
ON FINLAND**
December 1939

━ ━ ━ ━ ━ *December 1939*

0   50   100   200 Miles

west, so far, at any rate, as the Baltic route was concerned. There remained only the approach through Finland.

Early in October Mr. Paasikivi, one of the Finnish statesmen who
had signed the Peace of 1921 with the Soviet Union, went to Moscow.
The Soviet demands were sweeping: the Finnish frontier on the Karelian Isthmus must be moved back a considerable distance so as to remove Leningrad from the range of hostile artillery. The cession of certain Finnish islands in the Gulf of Finland; the lease of Finland's only
ice-free port in the Arctic Sea, Petsamo; and, above all, the leasing of
the port of Hango, at the entrance of the Gulf of Finland, as a Russian
naval and air base. The Finns were prepared to make concessions on
every point except the last. With the keys of the Gulf in Russian hands
the strategic and national security of Finland seemed to them to vanish.
The negotiations broke down on November 13, and the Finnish Government began to mobilise. On November 28 Molotov denounced the
Non-Aggression Pact between Finland and Russia; two days later the
Russians attacked at eight points along Finland's thousand-mile frontier, and on the same morning the capital, Helsingfors, was bombed by
the Red Air Force.

The brunt of the Russian attack fell at first upon the frontier defences
of the Finns in the Karelian Isthmus. These comprised a fortified zone
about twenty miles in depth running north and south through forest
country, deep in snow. This was called the "Mannerheim Line," after
the Finnish Commander-in-Chief and saviour of Finland from Bolshevik subjugation in 1917. The indignation excited in Britain, France and
even more vehemently in the United States, at the unprovoked attack
by the enormous Soviet Power upon a small, spirited, and highly
civilised nation was soon followed by astonishment and relief. The
early weeks of fighting brought no success to the Soviet forces. The
Finnish Army, whose total fighting strength was only about 200,000
men, gave a good account of themselves. The Russian tanks were encountered with audacity and a new type of hand grenade, soon nicknamed "The Molotov Cocktail."

It is probable that the Soviet Government had counted on a walkover. Their early air raids on Helsingfors and elsewhere, though not on
a heavy scale, were expected to strike terror. The troops they used at
first, though numerically much stronger, were inferior in quality and
ill-trained. The effect of the air raids and of the invasion of their land
roused the Finns, who rallied to a man against the aggressor and fought
with absolute determination and the utmost skill. The attack on the
"Waist" of Finland proved disastrous to the invaders. The country
here is almost entirely pine forests, gently undulating and at the time

covered with a foot of snow. The cold was intense. The Finns were well equipped with skis and warm clothing, of which the Russians had neither. Moreover, the Finns proved themselves aggressive individual fighters, highly trained in reconnaissance and forest warfare. The Russians relied in vain on numbers and heavier weapons. All along this front the Finnish frontier posts withdrew slowly down the roads, followed by the Russian columns. After these had penetrated about thirty miles they were set upon by the Finns. Held in front at Finnish defence lines constructed in the forests, violently attacked in flank by day and night, their communications severed behind them, the columns were cut to pieces, or, if lucky, got back after heavy loss whence they came. By the end of December the whole Russian plan for driving in across the "Waist" had broken down.

Meanwhile the assault on the Mannerheim Line in the Karelian Isthmus fared no better. A series of mass attacks by nearly twelve divisions was launched in early December, and continued throughout the month. By the end of the year failure all along the front convinced the Soviet Government that they had to deal with a very different enemy from what they had expected. They determined upon a major effort. This required preparation on a large scale, and from the end of the year fighting died down all along the Finnish front, leaving the Finns so far victorious over their mighty assailant. This surprising event was received with equal satisfaction in all countries, belligerent or neutral, throughout the world. It was a pretty bad advertisement for the Soviet Army. In British circles many people congratulated themselves that we had not gone out of our way to bring the Soviets in on our side, and preened themselves on their foresight. The conclusion was drawn too hastily that the Russian Army had been ruined by the purge, and that the inherent rottenness and degradation of their system of government and society was now proved. It was not only in England that this view was taken. There is no doubt that Hitler and his generals meditated profoundly upon the Finnish exposure, and that it played a potent part in influencing the Fuehrer's thought.

All the resentment felt against the Soviet Government for the Ribbentrop-Molotov pact was fanned into flame by this latest exhibition of brutal bullying and aggression. With this was also mingled scorn for the inefficiency displayed by the Soviet troops and enthusiasm for the gallant Finns. In spite of the Great War which had been declared, there was a keen desire to help the Finns by aircraft and other precious war material and by volunteers from Britain, from the United States, and still more from France. Alike for the munitions supplies and the volunteers there was only one possible route to Finland. The iron ore port of Narvik, with its railroad over the mountains to the Swedish iron ore mines, acquired a new sentimental if not strategic significance. Its use as a line of supply for the Finnish armies affected the neutrality both of Norway and Sweden. These two states, in equal fear of Germany and Russia, had no aim but to keep out of the wars by which they were encircled and might be engulfed. For them this seemed the only chance of survival. But whereas the British Government were naturally reluctant to commit even a technical infringement of Norwegian territorial waters by laying mines in the Leads for their own advantage against Germany, they moved upon a generous emotion, only indirectly connected with our war problem, towards a far more serious demand upon both Norway and Sweden for the passage of men and supplies to Finland.

I sympathised ardently with the Finns and supported all proposals for their aid; and I welcomed this new and favourable breeze as a means of achieving the major strategic advantage of cutting off the vital iron ore supplies of Germany. If Narvik was to become a kind of Allied base to supply the Finns, it would certainly be easy to prevent the German ships loading ore at the port and sailing safely down the Leads to Germany. Once Norwegian and Swedish protestations were overborne, for whatever reason, the greater measures would include the less. On December 16 I therefore renewed my efforts to win consent to the simple and bloodless operation of mining the Leads.

My memorandum was considered by the Cabinet on December 22, and I pleaded the case to the best of my ability. I could not obtain any decision for action. Diplomatic protest might be made to Norway about the misuse of her territorial waters by Germany, and the Chiefs of Staff were instructed to "consider the military consequences of commitments on Scandinavian soil." They were authorised "to plan for landing a force at Narvik for the sake of Finland, and also a possible German occupation of Southern Norway." But no executive orders could be issued to the Admiralty. In a paper which I circulated on December 21 I summarised the Intelligence reports which showed the possibilities of a Russian design upon Norway. The Soviet were said to have three divisions concentrated at Murmansk preparing for a seaborne expedition. "It may be," I concluded, "that this theatre will become the scene of early activities." This proved only too true; but from a different quarter.

*   *   *

I had long been concerned that the *Altmark*, the auxiliary of the *Spee*, should be captured. This vessel was also a floating prison for the crews of our sunk merchant ships. British captives released by Captain Langsdorff according to international law in Montevideo harbour told us that nearly three hundred British merchant seamen were on board the *Altmark*. This vessel hid in the South Atlantic for nearly two months, and then, hoping that the search had died down, her captain made a bid to return to Germany. Luck and the weather favoured her, and not until February 14, after passing between Iceland and the Faroes, was she sighted by our aircraft in Norwegian territorial waters.

In the words of an Admiralty communiqué, "certain of His Majesty's ships which were conveniently disposed were set in motion." A destroyer flotilla, under the command of Captain Philip Vian, of H.M.S. *Cossack*, intercepted the *Altmark*, but did not immediately molest her. She took refuge in Jösing Fiord, a narrow inlet about a mile and a half long surrounded by high snow-clad cliffs. Two British destroyers were told to board her for examination. At the entrance to the fiord they were met by two Norwegian gunboats, who informed them that the ship was unarmed, had been examined the previous day, and had received permission to proceed to Germany, making use of Norwegian territorial waters. Our destroyers thereupon withdrew.

When this information reached the Admiralty I intervened, and, with the concurrence of the Foreign Secretary, ordered our ships to enter the fiord. Vian did the rest. That night in the *Cossack*, with searchlights burning, he entered the fiord through the ice floes. He first went on board the Norwegian gunboat *Kjell* and requested that the *Altmark* should be taken to Bergen under a joint escort, for inquiry according to international law. The Norwegian captain repeated his assurance that the *Altmark* had been twice searched, that she was unarmed, and that no British prisoners had been found. Vian then stated that he was going to board her, and invited the Norwegian officer to join him. This offer was eventually declined.

Meanwhile the *Altmark* got under way, and in trying to ram the *Cossack* ran herself aground. The *Cossack* forced her way alongside and a boarding party sprang across, after grappling the two ships together. A sharp hand-to-hand fight followed, in which four Germans were killed and five wounded; part of the crew fled ashore and the rest surrendered. The search began for the British prisoners. They were soon found in their hundreds, battened down, locked in storerooms, and even in an empty oil-tank. Then came the cry, "The Navy's here!" The doors were broken in and the captives rushed on deck. It was also found that the *Altmark* carried two pom-poms and four machine guns, and that, despite having been boarded twice by the Norwegians, she had not been searched. The Norwegian gunboats remained passive observers throughout. By midnight Vian was clear of the fiord, and making for the Forth.

Admiral Pound and I sat up together in some anxiety in the Admiralty War Room. I had put a good screw on the Foreign Office, and was fully aware of the technical gravity of the measures taken. But what mattered at home and in the Cabinet was whether British prisoners were found on board or not. We were delighted when at three o'clock in the morning news came that three hundred had been found and rescued. This was a dominating fact.

Hitler's decision to invade Norway had, as we have seen, been taken on December 14, and the staff work was proceeding under Keitel. The incident of the *Altmark* no doubt gave a spur to action. At Keitel's suggestion, on February 20 Hitler summoned urgently to Berlin General von Falkenhorst, who was at that time in command of an army corps at Coblenz. Falkenhorst had taken part in the German campaign in Finland in 1918, and he discussed that afternoon with Hitler, Keitel, and Jodl the detailed operational plans for the Norwegian expedition which Falkenhorst was now to command. The question of priorities was of supreme importance. Would Hitler commit himself in Norway before or after the execution of "Case Yellow" — the attack on France? On March 1 he made his decision: Norway was to come first. The Fuehrer held a military conference on the afternoon of March 16, and D-Day was provisionally fixed, apparently for April 9.

*   *   *

Meanwhile the Soviets had brought their main power to bear on the Finns. They redoubled their efforts to pierce the Mannerheim Line before the melting of the snows. Alas, this year the spring and its thaw, on which the hard-pressed Finns based their hopes, came nearly six weeks late. The great Soviet offensive on the Isthmus, which was to last forty-two days, opened on February 1, combined with heavy air bombing of base depots and railway junctions behind the lines. Ten days of heavy bombardment from Soviet guns, massed wheel to wheel, heralded the main infantry attack. After a fortnight's fighting the line was breached. The air attacks on the key fort and base of Viipuri increased in intensity. By the end of the month the Mannerheim defence system had been disorganised and the Russians were able to concentrate against the Gulf of Viipuri. The Finns were short of ammunition and their troops exhausted.

The honourable correctitude which had deprived us of any strategic initiative equally hampered all effective measures for sending munitions to Finland. In France however a warmer and deeper sentiment prevailed, and this was strongly fostered by M. Daladier. On March 2, without consulting the British Government, he agreed to send fifty thousand volunteers and a hundred bombers to Finland. We could certainly not act on this scale, and in view of the documents found on the German major in Belgium, and of the ceaseless Intelligence reports of the steady massing of German troops on the Western Front, it went far beyond what prudence would allow. However, it was agreed to send fifty British bombers. On March 12 the Cabinet again decided to revive the plans for military landings at Narvik and Trondheim, to be followed at Stavanger and Bergen, as a part of the extended help to Finland into which we had been drawn by the French. These plans were to be available for action on March 20, although the need of Norwegian and Swedish permission had not been met. Meanwhile on March 7 Mr. Paasikivi had gone again to Moscow, this time to discuss armistice terms. On the 12th the Russian terms were accepted by the Finns. All our plans for military landings were again shelved, and the forces which were being collected were to some extent dispersed. The two divisions which had been held back in England were now allowed to proceed to France, and our striking power towards Norway was reduced to eleven battalions.

*   *   *

The military collapse of Finland led to further repercussions. On March 18 Hitler met Mussolini at the Brenner Pass. Hitler deliberately gave the impression to his Italian host that there was no question of Germany launching a land offensive in the West. On the 19th Mr. Chamberlain spoke in the House of Commons. In view of growing criticism he reviewed in some detail the story of British aid to Finland. He rightly emphasised that our main consideration had been the desire to respect the neutrality of Norway and Sweden, and he also defended the Government for not being hustled into attempts to succour the Finns, which had offered little chance of success. The defeat of Finland was fatal to the Daladier Government, whose chief had taken such marked, if tardy, action, and who had personally given disproportionate prominence to this part of our anxieties. On March 21 a new Cabinet was formed, under M. Reynaud, pledged to an increasingly vigorous conduct of the war.

My relations with M. Reynaud stood on a different footing from any I had established with M. Daladier. Reynaud, Mandel, and I had felt the same emotions about Munich. Daladier had been on the other side. I therefore welcomed the change. The French Ministers came to London for a meeting of the Supreme War Council on March 28. Mr.

Chamberlain opened with a full and clear description of the scene as he saw it. He said that Germany had two weaknesses: her supplies of iron ore and of oil. The main sources of supply of these were situated at the opposite ends of Europe. The iron ore came from the North. He unfolded with precision the case for intercepting the German iron ore supplies from Sweden. He dealt also with the Rumanian and Baku oilfields, which ought to be denied to Germany, if possible by diplomacy. I listened to this powerful argument with increasing pleasure. I had not realised how fully Mr. Chamberlain and I were agreed.

M. Reynaud spoke of the impact of German propaganda upon French morale. The German radio blared each night that the Reich had no quarrel with France; that the origin of the war was to be found in the blank cheque given by Britain to Poland; that France had been dragged into war at the heels of the British, and even that she was not in a position to sustain the struggle. Goebbels' policy towards France seemed to be to let the war run on at the present reduced tempo, counting upon growing discouragement among the five million Frenchmen now called up and upon the emergence of a French Government willing to come to compromise terms with Germany at the expense of Great Britain.

The question, he said, was widely asked in France, "How can the Allies win the war?" The number of divisions, "despite British efforts," was increasing faster on the German side than on ours. When therefore could we hope to secure that superiority in manpower required for successful action in the West? We had no knowledge of what was going on in Germany in material equipment. There was a general feeling in France that the war had reached a deadlock, and that Germany had only to wait. Unless some action were taken to cut the enemy's supply of oil and other raw material "the feeling might grow that blockade was not a weapon strong enough to secure victory for the Allied cause." He was far more responsive about cutting off supplies of Swedish iron ore, and he stated that there was an exact relation between the supplies of Swedish iron ore to Germany and the output of the German iron and steel industry. His conclusion was that the Allies should lay mines in the territorial waters along the Norwegian coast and later obstruct by similar action ore being carried from the port of Luleå to Germany. He emphasised the importance of hampering German supplies of Rumanian oil.

It was at last decided that, after addressing communications in general terms to Norway and Sweden, we should lay minefields in Norwegian territorial waters on April 5. It was also agreed that if Germany invaded Belgium the Allies should immediately move into that country without waiting for a formal invitation; and that if

Germany invaded Holland, and Belgium did not go to her assistance, the Allies should consider themselves free to enter Belgium for the purpose of helping Holland.

Finally, as an obvious point on which all were at one, the communiqué stated that the British and French Governments had agreed on the following solemn declaration: *That during the present war they would neither negotiate nor conclude an armistice or treaty of peace except by mutual agreement.*

This pact later acquired high importance.

*   *   *

On April 3 the British Cabinet implemented the resolve of the Supreme War Council, and the Admiralty was authorised to mine the Norwegian Leads on April 8. I called the actual mining operation "Wilfred," because by itself it was so small and innocent. As our mining of Norwegian waters might provoke a German retort, it was also agreed that a British brigade and a French contingent should be sent to Narvik to clear the port and advance to the Swedish frontier. Other forces should be dispatched to Stavanger, Bergen, and Trondheim, in order to deny these bases to the enemy.

Ominous items of news of varying credibility now began to come in. At this same meeting of the War Cabinet on April 3 the Secretary of State for War told us that a report had been received at the War Office that the Germans had been collecting strong forces of troops at Rostock with the intention of taking Scandinavia if necessary. The Foreign Secretary said that the news from Stockholm tended to confirm this report. According to the Swedish Legation in Berlin, 200,000 tons of German shipping were now concentrated at Stettin and Swinemünde, with troops on board which rumour placed at 400,000. It was suggested that these forces were in readiness to deliver a counterstroke against a possible attack by us upon Narvik or other Norwegian ports, about which the Germans were said to be still nervous.

On Thursday, April 4, Mr. Chamberlain delivered a speech of unusual optimism. Hitler, he declared, had "missed the bus." Seven months had enabled us to remove our weaknesses and add enormously to our fighting strength. Germany, on the other hand, had prepared so completely that she had very little margin of strength to call upon.

This proved an ill-judged utterance. Its main assumption that we and the French were relatively stronger than at the beginning of the war was not reasonable. As has been previously explained, the Germans were now in the fourth year of vehement munitions manufacture, whereas we were at a much earlier stage, probably comparable in fruitfulness to the second year. Moreover, with every month that had passed, the German Army, now four years old, was becoming a mature and perfected weapon, and the former advantage of the French Army in training and cohesion was steadily passing away. All lay in suspense. The various minor expedients I had been able to suggest had gained acceptance; but nothing of a major character had been done by either side. Our plans, such as they were, rested upon enforcing the blockade by the mining of the Norwegian corridor in the north and by hampering German oil supplies from the southeast. Complete immobility and silence reigned behind the German front. Suddenly the passive or small-scale policy of the Allies was swept away by a cataract of violent surprises. We were to learn what total war means.

21

Norway

BEFORE RESUMING THE NARRATIVE I must explain the alterations in my position which occurred during the month of April 1940.

Lord Chatfield's office as Minister for the Co-ordination of Defence had become redundant, and on the 3rd Mr. Chamberlain accepted his resignation, which he proffered freely. On the 4th a statement was issued from No. 10 Downing Street that it was not proposed to fill the vacant post, but that arrangements were being made for the First Lord of the Admiralty, as the senior Service Minister concerned, to preside over the Military Co-ordination Committee. Accordingly I took the chair at its meetings, which were held daily, and sometimes twice daily, from the 8th to 15th of April. I had therefore an exceptional measure of responsibility, but no power of effective direction. Among the other Service Ministers who were also members of the War Cabinet I was "first among equals." I had however no power to take or to enforce decisions. I had to carry with me both the Service Ministers and their professional chiefs. Thus many important and able men had a right and duty to express their views on the swiftly changing phases of the battle — for battle it was — which now began.

The Chiefs of Staff sat daily together after discussing the whole situation with their respective Ministers. They then arrived at their own decisions, which obviously became of dominant importance. I learned about these either from the First Sea Lord, who kept nothing from me, or by the various memoranda or *aide-mémoires* which the Chiefs of Staff Committee issued. If I wished to question any of these opinions I could of course raise them in the first instance at my Co-ordinating Committee, where the Chiefs of Staff, supported by their departmental Ministers, whom they had usually carried along with them, were all present as individual members. There was a copious flow of polite conversation, at the end of which a tactful report was drawn up by the secretary in attendance and checked by the three Service departments to make sure there were no discrepancies. Thus we had arrived at those broad, happy uplands where everything is settled for the greatest good of the greatest number by the common sense of most after the consultation of all. But in war of the kind we were now to feel the conditions were different. Alas, I must write it: the actual conflict had to be more like one ruffian bashing the other on the snout with a club, a hammer, or something better. All this is deplorable, and it is one of the many good reasons for avoiding war, and having everything settled by agreement in a friendly manner, with full consideration for the rights of minorities and the faithful recording of dissentient opinions.

The Defence Committee of the War Cabinet sat almost every day to discuss the reports of the Military Co-ordination Committee and those of the Chiefs of Staff; and their conclusions or divergences were again referred to frequent Cabinets. All had to be explained and re-explained; and by the time this process was completed the whole scene had often changed. At the Admiralty, which is of necessity in wartime a battle headquarters, decisions affecting the Fleet were taken on the instant, and only in the gravest cases referred to the Prime Minister, who supported us on every occasion. Where the action of the other Services was involved the procedure could not possibly keep pace with events. However, at the beginning of the Norway campaign the Admiralty in the nature of things had three-quarters of the executive business in its own hands.

I do not pretend that, whatever my powers, I should have been able to take better decisions or reach good solutions of the problems with which we were now confronted. The impact of the events about to be described was so violent and the conditions so chaotic that I soon perceived that only the authority of the Prime Minister could reign over the Military Co-ordination Committee. Accordingly on the 15th I requested Mr. Chamberlain to take the chair, and he presided at practically every one of our subsequent meetings during the campaign in Norway. He and I continued in close agreement, and he gave his supreme authority to the views which I expressed.

Loyalty and goodwill were forthcoming from all concerned. Nevertheless both the Prime Minister and I were acutely conscious of the formlessness of our system, especially when in contact with the surprising course of events. Although the Admiralty was at this time inevitably the prime mover, obvious objections could be raised to an organisation in which one of the Service Ministers attempted to concert all the operations of the other Services, while at the same time man-

aging the whole business of the Admiralty and having a special responsibility for the naval movements. These difficulties were not removed
by the fact that the Prime Minister himself took the chair and backed
me up. But while one stroke of misfortune after another, the results
of want of means or of indifferent management, fell upon us almost
daily, I nevertheless continued to hold my position in this fluid, friendly,
but unfocused circle.

Eventually, but not until many disasters had fallen upon us in
Scandinavia, I was authorised to convene and preside over the meetings
of the Chiefs of Staff Committee, without whom nothing could be
done, and I was made responsible formally "for giving guidance and
direction" to them. General Ismay, the Senior Staff Officer in charge of
the Central Staff, was placed at my disposal *as my staff officer and
representative,* and in this capacity was made a full member of the
Chiefs of Staff Committee. I had known Ismay for many years, but
now for the first time we became hand-in-glove, and much more. Thus
the Chiefs of Staff were to large extent made responsible to me in their
collective capacity, and as a deputy of the Prime Minister I could
nominally influence with authority their decisions and policies. On the
other hand, it was only natural that their primary loyalties should be
to their own Service Ministers, who would have been less than human
if they had not felt some resentment at the delegation of a part of their
authority to one of their colleagues. Moreover, it was expressly laid
down that my responsibilities were to be discharged *on behalf* of the
Military Co-ordination Committee. I was thus to have immense responsibilities, without effective power in my own hands to discharge
them. Nevertheless I had a feeling that I might be able to make the
new organisation work. It was destined to last only a week. But my
personal and official connection with General Ismay and his relation
to the Chiefs of Staff Committee was preserved unbroken and un-
weakened from May 1, 1940, to July 26, 1945, when I laid down my
charge.

*   *   *

On the evening of Friday, April 5, the German Minister in Oslo
invited distinguished guests, including members of the Government,
to a film show at the Legation. The film depicted the German conquest
of Poland, and culminated in a crescendo of horror scenes during the
German bombing of Warsaw. The caption read: "For this they could
thank their English and French friends." The party broke up in silence
and dismay. The Norwegian Government was however chiefly concerned with the activities of the British. Between 4.30 and 5 A.M. on
April 8 four British destroyers laid our minefield off the entrance to

West Fiord, the channel to the port of Narvik. At 5 A.M. the news
was broadcast from London, and at 5.30 a note from His Majesty's
Government was handed to the Norwegian Foreign Minister. The
morning in Oslo was spent in drafting protests to London. But later
that afternoon the Admiralty informed the Norwegian Legation in
London that German warships had been sighted off the Norwegian coast
proceeding northward, and presumably bound for Narvik. About the
same time reports reached the Norwegian capital that a German troop-
ship, the *Rio de Janeiro*, had been sunk off the south coast of Norway
by the Polish submarine *Orzel*, that large numbers of German soldiers
had been rescued by the local fishermen, and that they said they were
bound for Bergen to help the Norwegians defend their country against
the British and French. More was to come. Germany had broken into
Denmark, but the news did not reach Norway until after she herself
was invaded. Thus she received no formal warning. Denmark was
easily overrun after a resistance in which a few faithful soldiers were
killed.

That night German warships approached Oslo. The outer batteries
opened fire. The Norwegian defending force consisted of a minelayer,
the *Olav Tryggvason*, and two minesweepers. After dawn two German
minesweepers entered the mouth of the fiord to disembark troops in
the neighbourhood of the shore batteries. One was sunk by the *Olav
Tryggvason*, but the German troops were landed and the batteries taken.
The gallant minelayer however held off two German destroyers at the
mouth of the fiord and damaged the cruiser *Emden*. An armed Norwegian whaler mounting a single gun also went into action at once and
without special orders against the invaders. Her gun was smashed and
the commander had both legs shot off. To avoid unnerving his men,
he rolled himself overboard and died nobly. The main German force,
led by the heavy cruiser *Bluecher*, now entered the fiord, making for
the narrows defended by the fortress of Oscarsborg. The Norwegian
batteries opened, and two torpedoes fired from the shore at five hundred
yards scored a decisive strike. The *Bluecher* sank rapidly, taking with
her the senior officers of the German administrative staff and detachments of the Gestapo. The other German ships, including the *Luetzow*,
retired. The damaged *Emden* took no further part in the fighting at
sea. Oslo was ultimately taken, not from the sea, but by troop-carrying aeroplanes and by landings in the fiord.

Hitler's plan immediately flashed into its full scope. German forces
descended at Kristiansand, at Stavanger, and to the north at Bergen
and Trondheim.

The most daring stroke was at Narvik. For a week supposedly empty
German ore ships returning to that port in the ordinary course had

been moving up the corridor sanctioned by Norwegian neutrality, filled with supplies and ammunition. Ten German destroyers, each carrying two hundred soldiers and supported by the *Scharnhorst* and *Gneisenau*, had left Germany some days before, and reached Narvik early on the 9th.

Two Norwegian warships, *Norge* and *Eidsvold*, lay in the fiord. They were prepared to fight to the last. At dawn destroyers were sighted approaching the harbour at high speed, but in the prevailing snow-squalls their identity was not at first established. Soon a German officer appeared in a motor launch and demanded the surrender of the *Eidsvold*. On receiving from the commanding officer the curt reply, "I attack," he withdrew, but almost at once the ship was destroyed with nearly all hands by a volley of torpedoes. Meanwhile the *Norge* opened fire, but in a few minutes she too was torpedoed and sank instantly. In this gallant but hopeless resistance 287 Norwegian seamen perished, less than a hundred being saved from the two ships. Thereafter the capture of Narvik was easy. It was a strategic key — for ever to be denied us.

That morning Admiral Forbes, with the main fleet, was abreast of Bergen. The situation at Narvik was obscure. Hoping to forestall a German seizure of the port, the Commander-in-Chief directed our destroyers to enter the fiord and prevent any landing. Accordingly Captain Warburton-Lee, with the five destroyers of his own flotilla, *Hardy, Hunter, Havock, Hotspur,* and *Hostile,* entered West Fiord. He was told by Norwegian pilots at Tranoy that six ships larger than his own and a U-boat had passed in and that the entrance to the harbour was mined. He signalled this information and added: "Intend attacking at dawn." In the mist and snowstorms of April 10 the five British destroyers steamed up the fiord, and at dawn stood off Narvik. Inside the harbour were five enemy destroyers. In the first attack the *Hardy* torpedoed the ship bearing the pennant of the German Commodore, who was killed; another destroyer was sunk by two torpedoes, and the remaining three were so smothered by gunfire that they could offer no effective resistance. There were also in the harbour twenty-three merchant ships of various nations, including five British: six German were destroyed. Only three of our five destroyers had hitherto attacked. The *Hotspur* and *Hostile* had been left in reserve to guard against any shore batteries or against fresh German ships approaching. They now joined in a second attack, and the *Hotspur* sank two more merchantmen with torpedoes. Captain Warburton-Lee's ships were unscathed; the enemy's fire was apparently silenced, and after an hour's fighting no ships had come out from any of the inlets against him.

But now fortune turned. As he was coming back from a third attack

Captain Warburton-Lee sighted three fresh ships approaching. They showed no signs of wishing to close the range, and action began at seven thousand yards. Suddenly out of the mist ahead appeared two more warships. They were not, as was at first hoped, British reinforcements, but German destroyers which had been anchored in a nearby fiord. Soon the heavier guns of the German ships began to tell; the bridge of the *Hardy* was shattered, Warburton-Lee mortally stricken, and all his officers and companions killed or wounded except Lieutenant Stanning, his secretary, who took the wheel. A shell then exploded in the engine room, and under heavy fire the destroyer was beached. The last signal from the *Hardy's* captain to his flotilla was "Continue to engage the enemy."

Meanwhile *Hunter* had been sunk, and *Hotspur* and *Hostile*, which were both damaged, with the *Havock,* made for the open sea. The enemy who had barred their passage was by now in no condition to stop them. Half an hour later they encountered a large ship coming in from the sea, which proved to be the *Rauenfels*, carrying the German reserve ammunition. She was fired upon by the *Havock,* and soon blew up. The survivors of the *Hardy* struggled ashore with the body of their commander, who was awarded posthumously the Victoria Cross. He and they had left their mark on the enemy and in our naval records.

Surprise, ruthlessness, and precision were the characteristics of the onslaught upon innocent and naked Norway. Seven army divisions were employed. Eight hundred operational aircraft and 250 to 300 transport planes were the salient and vital feature of the design. Within forty-eight hours all the main ports of Norway were in the German grip. The King, the Government, the Army and the people, as soon as they realised what was happening, flamed into furious anger. But it was all too late. German infiltration and propaganda had hitherto clouded their vision, and now sapped their powers of resistance. Major Quisling presented himself at the radio, now in German hands, as the pro-German ruler of the conquered land. Almost all Norwegian officials refused to serve him. The Army was mobilised, and at once began to fight the invaders pressing northward from Oslo. Patriots who could find arms took to the mountains and the forests. The King, the Ministry, and the Parliament withdrew first to Hamar, a hundred miles from Oslo. They were hotly pursued by German armoured cars, and ferocious attempts were made to exterminate them by bombing and machine-gunning from the air. They continued however to issue proclamations to the whole country urging the most strenuous resistance. The rest of the population was overpowered and terrorised by bloody examples into stupefied or sullen submission. The peninsula of

Norway is nearly a thousand miles long. It is sparsely inhabited, and roads and railways are few, especially to the northward. The rapidity with which Hitler effected the domination of the country was a remarkable feat of war and policy, and an enduring example of German thoroughness, wickedness, and brutality.

*   *   *

The Norwegian Government, hitherto in their fear of Germany so frigid to us, now made vehement appeals for succour. It was from the beginning obviously impossible for us to rescue Southern Norway. Almost all our trained troops, and many only half trained, were in France. Our modest but growing air force was fully assigned to supporting the British Expeditionary Force, to Home Defence, and vigorous training. All our anti-aircraft guns were demanded ten times over for vulnerable points of the highest importance. Still, we felt bound to do our utmost to go to their aid, even at violent derangement of our own preparations and interests. Narvik, it seemed, could certainly be seized and defended with benefit to the whole Allied cause. Here the King of Norway might fly his flag unconquered. Trondheim might be fought for, at any rate as a means of delaying the northward advance of the invader until Narvik could be regained and made the base of an army. This, it seemed, could be maintained from the sea at a strength superior to anything which could be brought against it by land through five hundred miles of mountain country. The Cabinet heartily approved all possible measures for the rescue and defence of Narvik and Trondheim. The troops which had been released from the Finnish project, and a nucleus kept in hand for Narvik, could soon be ready. They lacked aircraft, anti-aircraft guns, anti-tank guns, tanks, transport, and training. The whole of Northern Norway was covered with snow to depths which none of our soldiers had ever seen, felt, or imagined. There were neither snowshoes nor skis — still less skiers. We must do our best. Thus began a ramshackle campaign.

We landed, or tried to land, at Narvik, Trondheim, and other places. The superiority of the Germans in design, management, and energy was plain. They put into ruthless execution a carefully prepared plan of action. They comprehended perfectly the use of the air arm on a great scale in all its aspects. Moreover, their individual ascendancy was marked, especially in small parties. At Narvik a mixed and improvised German force barely six thousand strong held at bay for six weeks some twenty thousand Allied troops, and, though driven out of the town, lived to see them depart. The sea-borne attack, brilliantly opened by the Navy, was paralysed by the refusal of the military

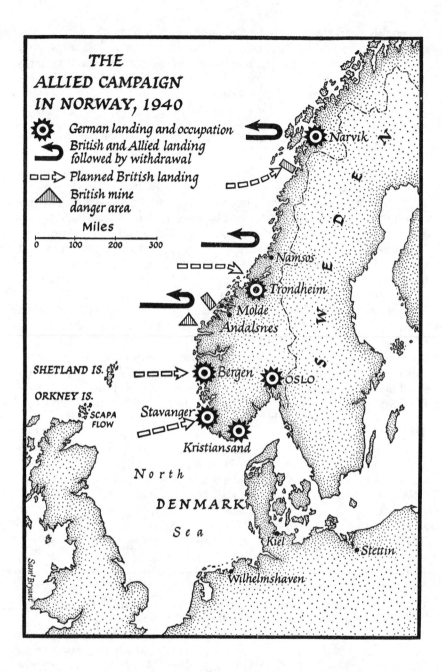

THE
ALLIED CAMPAIGN
IN NORWAY, 1940

German landing and occupation
British and Allied landing
followed by withdrawal
Planned British landing
British mine
danger area

Miles

0   100   200   300

Narvik

Namsos

Trondheim

Molde
Andalsnes

SHETLAND IS.

ORKNEY IS.

SCAPA
FLOW

Bergen

OSLO

Stavanger

Kristiansand

North

DENMARK

Sea

Kiel

Stettin

Wilhelmshaven

Sam¹ Bryant

commander to run what was admittedly a desperate risk. We divided our resources between Narvik and Trondheim and injured both our plans. At Namsos there was a muddy waddle forward and back. Only in an expedition to Andalsnes did we bite. The Germans, although they had to overcome hundreds of miles of rugged, snow-clogged country, drove us back in spite of gallant episodes. We, who had the command of the sea and could pounce anywhere on an undefended coast, were outpaced by the enemy moving by land across very large distances in the face of every obstacle.

We tried hard at the call of duty to entangle and imbed ourselves in Norway. We thought fortune had been cruelly against us. We can now see that we were well out of it. Meanwhile by early May we had to comfort ourselves as best we might by a series of successful evacuations. Considering the prominent part I played in these events and the impossibility of explaining the difficulties by which we had been overcome, or the defects of our staff and governmental organisation and our methods of conducting war, it was a marvel that I survived and maintained my position in public esteem and Parliamentary confidence. This was due to the fact that for six or seven years I had predicted with truth the course of events, and had given ceaseless warnings, then unheeded but now remembered.

The aircraft carrier *Glorious*, attacked on June 8 by the German battle cruisers *Scharnhorst* and *Gneisenau*, perished in an hour and a half. In the death of one of her escorting destroyers, the *Acasta*, told by her sole survivor, Leading Seaman C. Carter, we have a vivid and typical picture of the clash at sea:

On board our ship, what a deathly calm, hardly a word spoken, the ship was now steaming full speed away from the enemy, then came a host of orders, prepare all smoke floats, hose-pipes connected up, various other jobs were prepared, we were still stealing away from the enemy, and making smoke, and all our smoke floats had been set going. The Captain then had this message passed to all positions: "You may think we are running away from the enemy, we are not, our chummy ship *[Ardent]* has sunk, the *Glorious* is sinking, the least we can do is make a show, good luck to you all." We then altered course into our own smoke screen. I had the order stand by to fire tubes 6 and 7, we then came out of the smoke screen, altered course to starboard firing our torpedoes from port side. It was then I had my first glimpse of the enemy, to be honest it appeared to me to be a large one [ship] and a small one, and we were very close. I fired my two torpedoes from my tubes [aft], the foremost tubes fired theirs, we were all watching results. I'll never forget that cheer that went up; on the port bow of one of the ships a yellow flash and a great column of smoke and water shot up from her. We knew we had hit, personally I could not see how we could have missed so close as we were. The enemy never fired a shot at us, I feel they must have been very surprised. After we had fired our torpedoes we went back into our own smoke screen, altered course again to starboard. "Stand by to fire remaining torpedoes"; and this time as soon as we poked our nose out of the smoke screen, the enemy let us have it. A shell hit the engine room, killed my tubes' crew, I was blown to the after end of the tubes, I must have been knocked out for a while, because when I came to, my arm hurt me; the ship had stopped with a list to port. Here is something, believe it or believe it not, I climbed back into the control seat, I see those two ships, I fired the remaining torpedoes, no one told me to, I guess I was raving mad. God alone knows why I fired them, but I did. The *Acasta*'s guns were firing the whole time, even firing with a list on the ship. The enemy then hit us several times, but one big explosion took place right aft, and I have often wondered whether the enemy hit us with a torpedo, in any case it seemed to lift the ship out of the water. At last the Captain gave orders to abandon ship. I will always remember the Surgeon Lieutenant,[1] his first ship, his first action. Before I jumped over the side, I saw him still attending to the wounded, a hopeless task, and when I was in the water I saw the Captain leaning over the bridge, take a cigarette from a case and light it. We shouted to him to come on our raft, he waved "Good-bye and good luck" — the end of a gallant man.

But from all the wreckage and confusion there emerged one fact of major importance potentially affecting the future of the war. In a desperate grapple with the British Navy the Germans ruined their own, such as it was, for the impending climax. The Allied losses in all the sea fighting off Norway amounted to one aircraft carrier, two cruisers, one sloop, and nine destroyers. Six cruisers, two sloops, and eight destroyers were disabled, but could be repaired within our margin of sea power. On the other hand, at the end of June 1940, a momentous date, the effective German Fleet consisted of no more than *one 8-inch cruiser, two light cruisers, and four destroyers*. Although many of their damaged ships, like ours, could be repaired, the German Navy was no factor in the supreme issue of the invasion of Britain.

*   *   *

[1] Temporary Surgeon-Lieutenant H. J. Stammers, R.N.V.R.

Twilight War ended with Hitler's assault on Norway. It broke into the glare of the most fearful military explosion so far known to man. I have described the trance in which for eight months France and Britain had been held while all the world wondered. This phase proved most harmful to the Allies. From the moment when Stalin made terms with Hitler the Communists in France took their cue from Moscow and denounced the war as "an imperialist and capitalist crime against democracy." They did what they could to undermine morale in the Army and impede production in the workshops. The morale of France, both of her soldiers and her people, was now in May markedly lower than at the outbreak of war.

Nothing like this happened in Britain, where Soviet-directed Communism, though busy, was weak. Nevertheless we were still a party Government, under a Prime Minister from whom the Opposition was bitterly estranged, and without the ardent and positive help of the trade union movement. The sedate, sincere, but routine character of the Administration did not evoke that intense effort, either in the government circles or in the munitions factories, which was vital. The stroke of catastrophe and the spur of peril were needed to call forth the dormant might of the British nation. The tocsin was about to sound.

22

The Fall of the Government

THE MANY DISAPPOINTMENTS and disasters of the brief
campaign in Norway caused profound perturbation at home, and
the currents of passion mounted even in the breasts of some of those
who had been most slothful and purblind in the years before the war.
The Opposition asked for a debate on the war situation, and this was
arranged for May 7. The House was filled with Members in a high
state of irritation and distress. Mr. Chamberlain's opening statement
did not stem the hostile tide. He was mockingly interrupted, and
reminded of his speech of April 4, when in quite another connection
he had incautiously said, "Hitler missed the bus." He defined my new
position and my relationship with the Chiefs of Staff, and in reply to
Mr. Herbert Morrison made it clear that I had not held those powers
during the Norwegian operations. One speaker after another from
both sides of the House attacked the Government, and especially its
chief, with unusual bitterness and vehemence, and found themselves
sustained by growing applause from all quarters. Sir Roger Keyes,
burning for distinction in the new war, sharply criticised the Naval
Staff for their failure to attempt the capture of Trondheim. "When
I saw," he said, "how badly things were going I never ceased im-
portuning the Admiralty and War Cabinet to let me take all responsibility and lead the attack." Wearing his uniform as Admiral of the
Fleet, he supported the complaints of the Opposition with technical
details and his own professional authority in a manner very agreeable
to the mood of the House. From the benches behind the Government
Mr. Amery quoted, amid ringing cheers, Cromwell's imperious words
to the Long Parliament: "You have sat too long here for any good
you have been doing. Depart, I say, and let us have done with you.
In the name of God, go!" These were terrible words, coming from a

friend and colleague of many years, a fellow Birmingham Member, and a Privy Counsellor of distinction and experience.

On the second day, May 8, the debate, although continuing upon an adjournment motion, assumed the character of a vote of censure, and Mr. Herbert Morrison, in the name of the Opposition, declared their intention to have a vote. The Prime Minister rose again, accepted the challenge, and in an unfortunate passage appealed to his friends to stand by him. He had a right to do this, as these friends had sustained his action, or inaction, and thus shared his responsibility in "the years which the locusts had eaten" before the war. But today they sat abashed and silenced, and some of them had joined the hostile demonstrations. This day saw the last decisive intervention of Mr. Lloyd George in the House of Commons. In a speech of not more than twenty minutes he struck a deeply wounding blow at the head of the Government. He endeavoured to exculpate me: "I do not think that the First Lord was entirely responsible for all the things which happened in Norway." I immediately interposed, "I take complete responsibility for everything that has been done by the Admiralty, and I take my full share of the burden." After warning me not to allow myself to be converted into an air-raid shelter to keep the splinters from hitting my colleages, Mr. Lloyd George turned upon Mr. Chamberlain. "It is not a question of who are the Prime Minister's friends. It is a far bigger issue. He has appealed for sacrifice. The nation is prepared for every sacrifice so long as it has leadership, so long as the Government show clearly what they are aiming at, and so long as the nation is confident that those who are leading it are doing their best." He ended, "I say solemnly that the Prime Minister should give an example of sacrifice, because there is nothing which can contribute more to victory in this war than that he should sacrifice the seals of office."

As Ministers we all stood together. The Secretaries of State for War and Air had already spoken. I had volunteered to wind up the debate, which was no more than my duty, not only in loyalty to the chief under whom I served but also because of the exceptionally prominent part I had played in the use of our inadequate forces during our forlorn attempt to succour Norway. I did my very best to regain control of the House for the Government in the teeth of continuous interruption, coming chiefly from the Labour Opposition benches. I did this with good heart when I thought of their mistakes and dangerous pacifism in former years, and how only four months before the outbreak of the war they had voted solidly against conscription. I felt that I, and a few friends who had acted with me, had the right to inflict these censures, but they had not. When they broke in upon me I retorted upon them and defied them, and several times the clamour was such that I could not make myself heard. Yet all the time it was clear that their anger was not directed against me but at the Prime Minister, whom I was defending to the utmost of my ability and without regard for any other considerations. When I sat down at eleven o'clock the House divided. The Government had a majority of 81, but over thirty Conservatives voted with the Labour and Liberal Oppositions, and a further sixty abstained. There was no doubt that in effect, though not in form, both the debate and the division were a violent manifestation of want of confidence in Mr. Chamberlain and his Administration.

After the debate was over he asked me to go to his room, and I saw at once that he took the most serious view of the sentiment of the House towards himself. He felt he could not go on. There ought to be a National Government. One party alone could not carry the burden. Someone must form a Government in which all parties would serve, or we could not get through. Aroused by the antagonisms of the debate, and being sure of my own past record on the issues at stake, I was strongly disposed to fight on. "This has been a damaging debate, but you have a good majority. Do not take the matter grievously to heart. We have a better case about Norway than it has been possible to convey to the House. Strengthen your Government from every quarter, and let us go on until our majority deserts us." To this effect I spoke. But Chamberlain was neither convinced nor comforted, and I left him about midnight with the feeling that he would persist in his resolve to sacrifice himself if there was no other way, rather than attempt to carry the war further with a one-party Government.

I do not remember exactly how things happened during the morning of May 9, but the following occurred. Sir Kingsley Wood was very close to the Prime Minister as a colleague and a friend. They had long worked together in complete confidence. From him I learned that Mr. Chamberlain was resolved upon the formation of a National Government, and if he could not be the head he would give way to anyone commanding his confidence who could. Thus by the afternoon I became aware that I might well be called upon to take the lead. The prospect neither excited nor alarmed me. I thought it would be by far the best plan. I was content to let events unfold. In the afternoon the Prime Minister summoned me to Downing Street, where I found Lord Halifax, and after a talk about the situation in general we were told that Mr. Attlee and Mr. Greenwood would visit us in a few minutes for a consultation.

When they arrived we three Ministers sat on one side of the table and the Opposition leaders on the other. Mr. Chamberlain declared the paramount need of a National Government, and sought to ascertain whether the Labour Party would serve under him. The Conference of their party was in session at Bournemouth. The conversation was most polite, but it was clear that the Labour leaders would not commit themselves without consulting their people, and they hinted, not obscurely, that they thought the response would be unfavourable. They then withdrew. It was a bright, sunny afternoon, and Lord Halifax and I sat for a while on a seat in the garden of Number 10 and talked about nothing in particular. I then returned to the Admiralty, and was occupied during the evening and a large part of the night in heavy business.

*   *   *

The morning of the 10th of May dawned, and with it came tremendous news. Boxes with telegrams poured in from the Admiralty, the War Office, and the Foreign Office. The Germans had struck their long-awaited blow. Holland and Belgium were both invaded. Their frontiers had been crossed at numerous points. The whole movement of the German Army upon the invasion of the Low Countries and of France had begun.

At about ten o'clock Sir Kingsley Wood came to see me, having just been with the Prime Minister. He told me that Mr. Chamberlain was inclined to feel that the great battle which had broken upon us made it necessary for him to remain at his post. Kingsley Wood had told him that, on the contrary, the new crisis made it all the more necessary to have a National Government, which alone could confront it, and he added that Mr. Chamberlain had accepted this view. At eleven o'clock I was again summoned to Downing Street by the Prime Minister. There once more I found Lord Halifax. We took our seats at the table opposite Mr. Chamberlain. He told us that he was satisfied that it was beyond his power to form a National Government. The response he had received from the Labour leaders left him in no doubt of this. The question therefore was whom he should advise the King to send for after his own resignation had been accepted. His demeanour was cool, unruffled, and seemingly quite detached from the personal aspect of the affair. He looked at us both across the table.

I have had many important interviews in my public life, and this was certainly the most important. Usually I talked a great deal, but on this occasion I was silent. Mr. Chamberlain evidently had in his mind the stormy scene in the House of Commons two nights before, when I had semed to be in such heated controversy with the Labour Party. Although this had been in his support and defence, he nevertheless felt that it might be an obstacle to my obtaining their adherence at this juncture. I do not recall the actual words he used, but this was the implication. His biographer, Mr. Feiling, states definitely that he preferred Lord Halifax. As I remained silent a very long pause ensued. It certainly seemed longer than the two minutes which one observes in the commemorations of Armistice Day. Then at length Halifax spoke. He said that he felt that his position as a peer, out of the House of Commons, would make it very difficult for him to discharge the duties of Prime Minister in a war like this. He would be held responsible for everything, but would not have the power to guide the assembly upon whose confidence the life of every Government depended. He spoke for some minutes in this sense, and by the time he had finished it was clear that the duty would fall upon me — had in fact fallen upon me. Then for the first time I spoke. I said I would have no communication with either of the Opposition parties until I had the King's Commission to form a Government. On this the momentous conversation came to an end, and we reverted to our ordinary easy and familiar manners of men who had worked for years together and whose lives in and out of office had been spent in all the friendliness of British politics. I then went back to the Admiralty, where, as may well be imagined, much awaited me.

The Dutch Ministers were in my room. Haggard and worn, with horror in their eyes, they had just flown over from Amsterdam. Their country had been attacked without the slightest pretext or warning. The avalanche of fire and steel had rolled across the frontiers, and when resistance broke out and the Dutch frontier guards fired, an overwhelming onslaught was made from the air. The whole country was in a state of wild confusion. The long-prepared defence scheme had been put into operation; the dykes were opened, the waters spread far and wide. But the Germans had already crossed the outer lines, and were now streaming down the banks of the Rhine and through the inner Gravelines defences. They threatened the causeway which encloses the Zuyder Zee. Could we do anything to prevent this? Luckily, we had a flotilla not far away, and this was immediately ordered to sweep the causeway with fire and take the heaviest toll possible of the swarming invaders. The Queen was still in Holland, but it did not seem she could remain there long.

As a consequence of these discussions, a large number of orders were dispatched by the Admiralty to all our ships in the neighbourhood, and close relations were established with the Royal Dutch Navy. Even with the recent overrunning of Norway and Denmark in their minds, the Dutch Ministers seemed unable to understand how the great German nation, which up to the night before had professed nothing but friendship, should suddenly have made this frightful and brutal onslaught. Upon these proceedings and other affairs an hour or two passed. A spate of telegrams pressed in from all the frontiers affected by the forward heave of the German armies. It seemed that the old Schlieffen plan, brought up to date with its Dutch extension, was already in full operation. In 1914 the swinging right arm of the German invasion had swept through Belgium but had stopped short of Holland. It was well known then that had that war been delayed for three or four years the extra army group would have been ready and the railway terminals and communications adapted for a movement through Holland. Now the famous movement had been launched with all these facilities and with every circumstance of surprise and treachery. But other developments lay ahead. The decisive stroke of the enemy was not to be a turning movement on the flank but a break through the main front. This none of us or the French, who were in responsible command, foresaw. Earlier in the year I had, in a published interview, warned these neutral countries of the fate which was impending upon them, and which was evident from the troop dispositions and road and rail development, as well as from the captured German plans. My words had been resented.

In the splintering crash of this vast battle the quiet conversations we had had in Downing Street faded or fell back in one's mind. However, I remember being told that Mr. Chamberlain had gone, or was going, to see the King, and this was naturally to be expected. Presently a message arrived summoning me to the Palace at six o'clock. It only takes two minutes to drive there from the Admiralty along the Mall. Although I suppose the evening newspapers must have been full of the terrific news from the Continent, nothing had been mentioned about the Cabinet crisis. The public had not had time to take in what was happening either abroad or at home, and there was no crowd about the Palace gates.

I was taken immediately to the King. His Majesty received me most graciously and bade me sit down. He looked at me searchingly and quizzically for some moments, and then said, "I suppose you don't know why I have sent for you?" Adopting his mood, I replied, "Sir, I simply couldn't imagine why." He laughed and said, "I want to ask you to form a Government." I said I would certainly do so.

The King had made no stipulation about the Government being national in character, and I felt that my commission was in no formal way dependent upon this point. But in view of what had happened, and the conditions which had led to Mr. Chamberlain's resignation, a Government of national character was obviously inherent in the situation. If I found it impossible to come to terms with the Opposition parties, I should not have been constitutionally debarred from trying to form the strongest Government possible of all who would stand by the country in the hour of peril, provided that such a Government could command a majority in the House of Commons. I told the King that I would immediately send for the leaders of the Labour and Liberal Parties, that I proposed to form a War Cabinet of five or six Ministers, and that I hoped to let him have at least five names before midnight. On this I took my leave and returned to the Admiralty.

Between seven and eight, at my request, Mr. Attlee called upon me. He brought with him Mr. Greenwood. I told him of the authority I had to form a Government, and asked if the Labour Party would join. He said they would. I proposed that they should take rather more than a third of the places, having two seats in the War Cabinet of five, or it might be six, and I asked Mr. Attlee to let me have a list of men so that we could discuss particular offices. I mentioned Mr. Ernest Bevin, Mr. Alexander, Mr. Morrison, and Mr. Dalton as men whose services in high office were immediately required. I had, of course, known both Attlee and Greenwood for a long time in the House of Commons. During the ten years before the outbreak of war I had in my more or less independent position come far more often into collision with the Conservative and National Governments than with the Labour and Liberal Oppositions. We had a pleasant talk for a little while, and they went off to report by telephone to their friends and followers at Bournemouth, with whom of course they had been in the closest contact during the previous forty-eight hours.

I invited Mr. Chamberlain to lead the House of Commons as Lord President of the Council, and he replied by telephone that he accepted, and had arranged to broadcast at nine that night, stating that he had resigned, and urging everyone to support and aid his successor. This he did in magnanimous terms. I asked Lord Halifax to join the War Cabinet while remaining Foreign Secretary. At about ten I sent the King a list of the five names, as I had promised. The appointment of the three Service Ministers was vitally urgent. I had already made up my mind who they should be. Mr. Eden should go to the War Office, Mr. Alexander should come to the Admiralty, and Sir Archibald Sinclair, Leader of the Liberal Party, should take the Air Ministry. At the same time I assumed the office of Minister of Defense, without however attempting to define its scope and powers.

*   *   *

Thus, then, on the night of the 10th of May, at the outset of this mighty battle, I acquired the chief power in the State, which henceforth I wielded in ever-growing measure for five years and three months of world war, at the end of which time, all our enemies having surrendered unconditionally or being about to do so, I was immediately dismissed by the British electorate from all further conduct of their affairs.

During these last crowded days of the political crisis my pulse had not quickened at any moment. I took it all as it came. But I cannot conceal from the reader of this truthful account that as I went to bed at about 3 A.M. I was conscious of a profound sense of relief. At last I had the authority to give directions over the whole scene. I felt as if I were walking with destiny, and that all my past life had been but a preparation for this hour and for this trial. Ten years in the political wilderness had freed me from ordinary party antagonisms. My warnings over the last six years had been so numerous, so detailed, and were now so terribly vindicated, that no one could gainsay me. I could not be reproached either for making the war or with want of preparation for it. I thought I knew a good deal about it all, and I was sure I should not fail. Therefore, although impatient for the morning, I slept soundly and had no need for cheering dreams. Facts are better than dreams.

BOOK II

Alone

May 10, 1940–June 22, 1941

After the first forty days we were alone, with victorious Germany and Italy engaged in a mortal attack upon us, with Soviet Russia a hostile neutral actively aiding Hitler, and Japan an unknowable menace.

☆

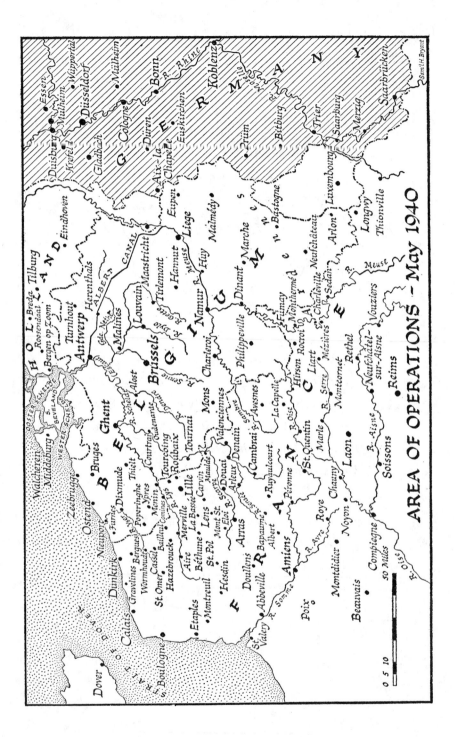

AREA OF OPERATIONS – MAY 1940

I

The National Coalition

N OW AT LAST the slowly gathered, long-pent-up fury of the
storm broke upon us. Four or five millions of men met each
other in the first shock of the most merciless of all the wars of which
record has been kept. Within a week the front in France, behind which
we had been accustomed to dwell through the hard years of the former
war and the opening phase of this, was to be irretrievably broken.
Within three weeks the long-famed French Army was to collapse in
rout and ruin, and our only British Army to be hurled into the sea
with all its equipment lost. Within six weeks we were to find ourselves
alone, almost disarmed, with triumphant Germany and Italy at our
throats, with the whole of Europe open to Hitler's power, and Japan
glowering on the other side of the globe. It was amid these facts and
looming prospects that I entered upon my duties as Prime Minister and
Minister of Defence and addressed myself to the first task of forming
a Government of all parties to conduct His Majesty's business at home
and abroad by whatever means might be deemed best suited to the national interest.

Five years later almost to a day it was possible to take a more
favorable view of our circumstances. Italy was conquered and Mussolini
slain. The mighty Germany Army had surrendered unconditionally.
Hitler had committed suicide. In addition to the immense captures by
General Eisenhower, nearly three million German soldiers were taken
prisoners in twenty-four hours by Field Marshal Alexander in Italy and
Field Marshal Montgomery in Germany. France was liberated, rallied,
and revived. Hand in hand with our Allies, the two mightiest empires
in the world, we advanced to the swift annihilation of Japanese resistance. The contrast was certainly remarkable. The road across these
five years was long, hard, and perilous. Those who perished upon it did

not give their lives in vain. Those who marched forward to the end will always be proud to have trodden it with honour.

*   *   *

In giving an account of my stewardship and in telling the tale of the famous National Coalition Government it is my first duty to make plain the scale and force of the contribution which Great Britain and her Empire, whom danger only united more tensely, made to what eventually became the common cause of so many states and nations. I do this with no desire to make invidious comparisons or rouse purposeless rivalries with our greatest ally, the United States, to whom we owe immeasurable and enduring gratitude. But it is to the combined interest of the English-speaking world that the magnitude of the British war-making effort should be known and realised. I have therefore had a table made which I print on the following page, which covers the whole period of the war. This shows that until July 1944 Britain and her Empire had a substantially larger number of divisions *in contact with the enemy* than the United States. This general figure includes not only the European and African spheres but also all the war in Asia against Japan. Till the arrival in Normandy in the autumn of 1944 of the great mass of the American Army, we had always the right to speak at least as an equal and usually as the predominant partner in every theatre of war except the Pacific and the Australasian; and this remains also true, up to the time mentioned, of the aggregation of all divisions in all theatres for any given month. From July 1944 the fighting front of the United States, as represented by divisions in contact with the enemy, became increasingly predominant, and so continued, mounting and triumphant, till the final victory ten months later.

Another comparison which I have made shows that the British and Empire sacrifice in loss of life was even greater than that of our valiant ally. The British total dead, and missing, presumed dead, of the armed forces amounted to 303,240, to which should be added over 109,000 from the Dominions, India, and the Colonies, a total of over 412,240. This figure does not include 60,500 civilians killed in the air raids on the United Kingdom, nor the losses of our Merchant Navy and fishermen, which amounted to about 30,000. Against this figure the United States mourn the deaths in the Army and Air Force, the Navy, Marines, and Coast Guard, of 322,188.[1] I cite these sombre rolls of honour in the confident faith that the equal comradeship sanctified by so much precious blood will continue to command the reverence and inspire the conduct of the English-speaking world.

On the seas the United States naturally bore almost the entire weight

---

[1] Eisenhower, *Crusade in Europe*, p. 1.

LAND FORCES IN FIGHTING CONTACT WITH THE ENEMY
"EQUIVALENT DIVISIONS"

	BRITISH EMPIRE			U.S.A.		
	Western Theatre	Eastern Theatre	Total	Western Theatre	Eastern Theatre	Total
Jan. 1, 1940	5⅓	—	5⅓ *	—	—	—
July 1, 1940	6	—	6	—	—	—
Jan. 1, 1941	10⅓	—	10⅓ †	—	—	—
July 1, 1941	13	—	13 †	—	—	—
Jan. 1, 1942	7⅔	7	14⅔	—	2⅔	2⅔ ‡
July 1, 1942	10	4⅔	14⅔	—	8⅓	8⅓
Jan. 1, 1943	10⅓	8⅔	19	5	10	15
July 1, 1943	16⅔	7⅔	24⅓	10	12⅓	22⅓
Jan. 1, 1944	11⅓	12⅓	23⅔	6⅔	9⅓	16
July 1, 1944	22⅔	16	38⅔	25	17	42
Jan. 1, 1945	30⅓	18⅔	49	55⅔	23⅓	79

NOTES AND ASSUMPTIONS

* B.E.F. in France.
† Excludes guerrillas in Abyssinia.
‡ Excludes Filipino troops.
The dividing line between the Eastern and Western theatres is taken as a north-south line through Karachi.
The following are NOT taken as operational theatres:
Northwest Frontier of India; Gibraltar; West Africa; Iceland; Hawaii; Palestine; Iraq; Syria (except on July 1, 1941).
Malta is taken as an operational theatre; also Alaska from January 1942 to July 1943.
Foreign contingents — e.g., Free French, Poles, Czechs — are NOT included.

of the war in the Pacific, and the decisive battles which they fought near Midway Island, at Guadalcanal, and in the Coral Sea in 1942 gained for them the whole initiative in that vast ocean domain, and opened to them the assault of all the Japanese conquests, and eventually of Japan herself. The American Navy could not at the same time carry the main burden in the Atlantic and the Mediterranean. Here again it is a duty to set down the facts. Out of 781 German and 85 Italian U-boats destroyed in the European theatre, the Atlantic and Indian Oceans, 594 were accounted for by British sea and air forces, who also disposed of all the German battleships, cruisers, and destroyers, besides destroying or capturing the whole Italian Fleet.

The table of U-boat losses is as follows:

Grand Total of U-Boats Destroyed: 996

Destroyed by	German	Italian	Japanese
British Forces *	525	69	9½
United States Forces *	174	5	110½
Other and unknown causes	82	11	10
Totals	781	85	130

* The terms British and United States Forces include Allied Forces under their operational control. Where fractional losses are shown the "kill" was shared. There were many cases of shared "kills," but in the German totals the fractions add up to whole numbers.

In the air superb efforts were made by the United States to come into action, especially with their daylight Fortress bombers, on the greatest scale from the earliest moment after Pearl Harbour, and their power was used both against Japan and from the British Isles against Germany. However, when we reached Casablanca in January 1943 it is a fact that no single American bomber plane had cast a daylight bomb on Germany. Very soon the fruition of the great exertions they were making was to come, but up till the end of 1943 the British discharge of bombs upon Germany had in the aggregate exceeded by eight tons to one those cast from American machines by day or night, and it was only in the spring of 1944 that the preponderance of discharge was achieved by the United States. Here, as in the armies and on the sea, we ran the full course from the beginning, and it was not until 1944 that we were overtaken and surpassed by the tremendous war effort of the United States.

It must be remembered that our munitions effort from the beginning of Lend-Lease in January 1941 was increased by over one-fifth through the generosity of the United States. With the materials and weapons which they gave us we were actually able to wage war as *if we were a nation of fifty-eight millions instead of forty-eight*. In shipping also the marvellous production of Liberty Ships enabled the flow of supplies to be maintained across the Atlantic. On the other hand, the analysis of shipping losses by enemy action suffered by all nations through the war should be borne in mind. Here are the figures:

Nationality	Losses in Gross Tons	Percentage
British	11,357,000	54
United States	3,334,000	16
All other nations (outside enemy control)	6,503,000	30
	21,194,000	100

Of these losses eighty per cent were suffered in the Atlantic Ocean, including British coastal waters and the North Sea. Only five per cent were lost in the Pacific.

This is all set down, not to claim undue credit, but to establish on a footing capable of commanding fair-minded respect the intense output in every form of war activity of the people of this small island, upon whom in the crisis of the world's history the brunt fell.

*   *   *

It is probably easier to form a cabinet, especially a coalition cabinet, in the heat of battle than in quiet times. The sense of duty dominates all else, and personal claims recede. Once the main arrangements had been settled with the leaders of the other parties, with the formal authority of their organisations, the attitude of all those I sent for was like that of soldiers in action, who go to the places assigned to them at once without question. The party basis being officially established, it seemed to me that no sense of Self entered into the minds of any of the very large number of gentlemen I had to see. If some few hesitated it was only because of public considerations. Even more did this high standard of behaviour apply to the large number of Conservative and National Liberal Ministers who had to leave their offices and break their careers, and at this moment of surpassing interest and excitement to step out of official life, in many cases for ever.

The Conservatives had a majority of more than 120 over all other parties in the House combined. Mr. Chamberlain was their chosen leader. I could not but realise that his supersession by me must be very unpleasant to many of them, after all my long years of criticism and often fierce reproach. Besides this, it must be evident to the majority of them how my life had been passed in friction or actual strife with the Conservative Party, that I had left them on Free Trade and had later returned to them as Chancellor of the Exchequer. After that I had been for many years their leading opponent on India, on foreign policy, and on the lack of preparation for war. To accept me as Prime Minister was to them very difficult. It caused pain to many honourable men. Moreover, loyalty to the chosen leader of the party is a prime characteristic of the Conservatives. If they had on some questions fallen short of their duty to the nation in the years before the war, it was because of this sense of loyalty to their appointed chief. None of these considerations caused me the slightest anxiety. I knew they were all drowned by the cannonade.

In the first instance I had offered to Mr. Chamberlain, and he had accepted, the leadership of the House of Commons, as well as the Lord Presidency. Nothing had been published. Mr. Atlee informed me that the Labour Party would not work easily under this arrangement. In a coalition the leadership of the House must be generally acceptable. I put this point to Mr. Chamberlain, and, with his ready agreement, I took the leadership myself, and held it till February 1942. During this time Mr. Attlee acted as my deputy and did the daily work. His long experience in Opposition was of great value. I came down only on the most serious occasions. These were, however, recurrent. Many Conservatives felt that their party leader had been slighted. Everyone admired his personal conduct. On his first entry into the House in his new capacity (May 13) the whole of his party — the large majority of the House — rose and received him in a vehement demonstration of sympathy and regard. In the early weeks it was from the Labour benches that I was mainly greeted. But Mr. Chamberlain's loyalty and support was steadfast, and I was sure of myself.

There was considerable pressure by elements of the Labour Party, and by some of those many able and ardent figures who had not been included in the new Government, for a purge of the "guilty men" and of Ministers who had been responsible for Munich or could be criticised for the many shortcomings in our war preparation. But this was no time for proscriptions of able, patriotic men of long experience in high office. If the censorious people could have had their way at least a third of the Conservative Ministers would have been forced to resign. Considering that Mr. Chamberlain was the leader of the Conservative Party, it was plain that this movement would be destructive of the national unity. Moreover, I had no need to ask myself whether all the blame lay on one side. Official responsibility rested upon the Government of the time. But moral responsibilities were more widely spread. A long, formidable list of quotations from speeches and votes recorded by Labour, and not less by Liberal, Ministers, all of which had been stultified by events, was in my mind and available in detail. No one had more right than I to pass a sponge across the past. I therefore resisted these disruptive tendencies. "If the present," I said a few weeks later, "tries to sit in judgment on the past it will lose the future." This argument and the awful weight of the hour quelled the would-be heresy hunters.

*   *   *

My experiences in those first days were peculiar. One lived with the battle, upon which all thoughts were centred, and about which nothing could be done. All the time there was the Government to form and the gentlemen to see and the party balances to be adjusted. I cannot remember, nor do my records show, how all the hours were spent. A British Ministry at that time contained between sixty and seventy Ministers of the Crown, and all these had to be fitted in like a jigsaw puzzle, in this case having regard to the claims of three parties. It was necessary for me to see not only all the principal figures but, for a few minutes at least, the crowd of able men who were to be chosen for important tasks. In forming a coalition government the Prime Minister has to attach due weight to the wishes of the party leaders as to who among their followers shall have the offices allotted to the party. By this principle I was mainly governed. If any who deserved better were left out on the advice of their party authorities, or even in spite of that advice, I can only express regret. On the whole however the difficulties were few.

In Clement Attlee I had a colleague of war experience long versed in the House of Commons. Our only differences in outlook were about Socialism, but these were swamped by a war soon to involve the almost complete subordination of the individual to the State. We worked together with perfect ease and confidence during the whole period of the Government. Mr. Arthur Greenwood was a wise counsellor of high courage and a good and helpful friend.

Sir Archibald Sinclair, as official leader of the Liberal Party, found it embarrassing to accept the office of Air Minister because his followers felt he should instead have a seat in the War Cabinet. But this ran contrary to the principle of a small War Cabinet. I therefore proposed that he should join the War Cabinet when any matter affecting fundamental political issues or party union was involved. He was my friend, and had been my second-in-command when in 1916 I commanded the 6th Royal Scots Fusiliers at Ploegsteert ("Plug Street"), and personally longed to enter upon the great sphere of action I had reserved for him. After no little intercourse this had been amicably settled. Mr. Ernest Bevin, with whom I had made acquaintance at the beginning of the war, in trying to mitigate the severe Admiralty demands for trawlers, had to consult the Transport and General Workers' Union, of which he was secretary, before he could join the team in the most important office of Minister of Labour. This took two or three days, but it was worth it. The Union, the largest of all in Britain, said unanimously that he was to do it, and stuck solid for five years till we won.

The greatest difficulty was with Lord Beaverbrook. I believed he had services to render of a very high quality. I had resolved, as the result of my experiences in the previous war, to remove the supply and design of aircraft from the Air Ministry, and I wished him to become the Minister of Aircraft Production. He seemed at first reluctant to undertake the task, and of course the Air Ministry did not like having their Supply Branch separated from them. There were other resistances to his appointment. I felt sure however that our life depended upon the flow of new aircraft; I needed his vital and vibrant energy, and I persisted in my view.

In deference to prevailing opinions expressed in Parliament and the press it was necessary that the War Cabinet should be small. I therefore began by having only five members, of whom one only, the Foreign Secretary, had a department. These were naturally the leading party politicians of the day. For the convenient conduct of business it was necessary that the Chancellor of the Exchequer and the leader of the Liberal Party should usually be present, and as time passed the number of "constant attenders" grew. But all the responsibility was laid upon the five War Cabinet Ministers. They were the only ones who had the right to have their heads cut off on Tower Hill if we did not win. The rest could suffer for departmental shortcomings, but not on account of the policy of the State. Apart from the War Cabinet, no one could say "I cannot take the responsibility for this or that." The burden of policy was borne at a higher level. This saved many people a lot of worry in the days which were immediately to fall upon us.

*   *   *

In my long political experience I had held most of the great offices of State, but I readily admit that the post which had now fallen to me was the one I like the best. Power, for the sake of lording it over fellow creatures or adding to personal pomp, is rightly judged base. But power in a national crisis, when a man believes he knows what orders should be given, is a blessing. In any sphere of action there can be no comparison between the positions of number one and number two, three, or four. The duties and the problems of all persons other than number one are quite different and in many ways more difficult. It is always a misfortune when number two or three has to initiate a dominant plan or policy. He has to consider not only the merits of the policy but the mind of his chief; not only what to advise but what it is proper for him in his station to advise; not only what to do but how to get it agreed, and how to get it done. Moreover, number two or three will have to reckon with numbers four, five, and six, or maybe some bright outsider, number twenty. Ambition, not so much for vulgar ends, but for fame, glints in every mind. There are always several points of view which may be right, and many which are plausible. I was ruined for the time being in 1915 over the Dardanelles, and a supreme enterprise was cast away, through my trying to carry out a major and cardinal operation of war from a subordinate position. Men are ill-advised to try such ventures. This lesson had sunk into my nature.

At the top there are great simplifications. An accepted leader has only to be sure of what it is best to do, or at least to have made up his mind about it. The loyalties which centre upon number one are enormous. If he trips he must be sustained. If he makes mistakes they must be covered. If he sleeps he must not be wantonly disturbed. If he is no good he must be pole-axed. But this last extreme process cannot be carried out every day; and certainly not in the days just after he has been chosen.

The fundamental changes in the machinery of war direction were more real than apparent. "A Constitution," said Napoleon, "should be short and obscure." The existing organisms remained intact. No official personalities were changed. The War Cabinet and the Chiefs of Staff Committee at first continued to meet every day as they had done before. In calling myself, with the King's approval, Minister of Defence I had made no legal or constitutional change. I had been careful not to define my rights and duties. I asked for no special powers either from the Crown or Parliament. It was however understood and accepted that I should assume the general direction of the war, subject to the support of the War Cabinet and of the House of Commons. The key change which occurred on my taking over was of course the supervision and direction of the Chiefs of Staff Committee by a Minister of Defence with undefined powers. As this Minister was also the Prime Minister, he had all the rights inherent in that office, including very wide powers of selection and removal of all professional and political personages. Thus for the first time the Chiefs of Staff Committee assumed its due and proper place in direct daily contact with the executive head of the Government, and in accord with him had full control over the conduct of the war and the armed forces.

The position of the First Lord of the Admiralty and of the Secretaries of State for War and Air was decisively affected in fact though not in form. They were not members of the War Cabinet, nor did they attend the meetings of the Chiefs of Staff Committee. They remained entirely responsible for their departments, but rapidly and almost imperceptibly ceased to be responsible for the formulation of strategic plans and the day-to-day conduct of operations. These were settled by the Chiefs of Staff Committee acting directly under the Minister of Defence and Prime Minister, and thus with the authority of the War Cabinet. The three Service Ministers, very able and trusted friends of mine whom I had picked for these duties, stood on no ceremony. They organised and administered the ever-growing forces, and helped all they could in the easy, practical English fashion. They had the fullest information by virtue of their membership of the Defence Committee, and constant access to me. Their professional subordinates, the Chiefs of Staff, discussed everything with them and treated them with the utmost respect. But there was an integral direction of the war to which they loyally submitted. There never was an occasion when their powers were abrogated or challenged, and anyone in this circle could always speak his mind; but the actual war direction soon settled into a very few hands, and what had seemed so difficult before became much more simple — apart of course from Hitler. In spite of the turbulence of events and the many disasters we had to endure the machinery worked almost automatically, and one lived in a stream of coherent thought capable of being translated with great rapidity into executive action.

*   *   *

Although the awful battle was now going on across the Channel, and the reader is no doubt impatient to get there, it may be well at this point to describe the system and machinery for conducting military and other affairs which I set on foot and practised from my earliest days of power. I am a strong believer in transacting official business by the *Written Word*. No doubt, surveyed in the after-time, much that is set down from hour to hour under the impact of events may be lacking in proportion or may not come true. I am willing to take my chance of that. It is always better, except in the hierarchy of military discipline, to express opinions and wishes rather than to give orders. Still, written directives coming personally from the lawfully constituted Head of the Government and Minister specially charged with defence counted to such an extent that, though not expressed as orders, they very often found their fruition in action.

To make sure that my name was not used loosely, I issued during the crisis of July the following minute:

> Let it be very clearly understood that all directions emanating from me are made in writing, or should be immediately afterwards confirmed in writing, and that I do not accept any responsibility for matters relating to national defence on which I am alleged to have given decisions unless they are recorded in writing.

When I woke about 8 A.M. I read all the telegrams, and from my bed dictated a continuous flow of minutes and directives to the departments and to the Chiefs of Staff Committee. These were typed in relays as they were done, and handed at once to General Ismay, Deputy Secretary (Military) to the War Cabinet, and my representative on the Chiefs of Staff Committee, who came to see me early each morning. Thus he usually had a good deal in writing to bring before the Chiefs of Staff Committee when they met at ten-thirty. They gave all consideration to my views at the same time as they discussed the general situation. Thus between three and five o'clock in the afternoon, unless there were some difficulties between us requiring further consultation, there was ready a whole series of orders and telegrams sent by me or by the Chiefs of Staff and agreed between us, usually giving all the decisions immediately required.

In total war it is quite impossible to draw any precise line between military and non-military problems. That no such friction occurred between the military staff and the War Cabinet staff was due primarily to the personality of Sir Edward Bridges, Secretary to the War Cabinet. Not only was this son of a former Poet Laureate an extremely competent and tireless worker, but he was also a man of exceptional force, ability, and personal charm, without a trace of jealousy in his nature. All that mattered to him was that the War Cabinet Secretariat as a whole should serve the Prime Minister and War Cabinet to the very best of their ability. No thought of his own personal position ever entered his mind, and never a cross word passed between the civil and military officers of the Secretariat.

In larger questions, or if there were any differences of view, I called a meeting of the War Cabinet Defence Committee, which at the outset comprised Mr. Chamberlain, Mr. Attlee, and the three Service Ministers, with the Chiefs of Staff in attendance. These formal meetings got fewer after 1941.[2] As the machine began to work more smoothly I came to the conclusion that the daily meetings of the War Cabinet with the Chiefs of Staff present were no longer necessary. I therefore eventually instituted what came to be known among ourselves as the "Monday Cabinet Parade." Every Monday there was a considerable gathering — all the War Cabinet, the Service Ministers, and the Minister of Home Security, the Chancellor of the Exchequer, the Secretaries of State for the Dominions and for India, the Minister of Information, the Chiefs of Staff, and the official head of the Foreign Office. At these meetings each Chief of Staff in turn unfolded his account of all that had happened during the previous seven days; and the Foreign Secretary followed them with his story of any important developments in foreign affairs. On other days of the week the War Cabinet sat alone, and all important matters requiring decision were brought before them. Other ministers primarily concerned with the subjects to be discussed attended for their own particular problems. The members of the War Cabinet had the fullest circulation of all papers affecting the war, and saw all important telegrams sent by me. As confidence grew, the War Cabinet intervened less actively in operational matters, though they watched them with close attention and full knowledge. They took almost the whole weight of home and party affairs off my shoulders, thus setting me free to concentrate upon the

[2] The Defence Committee met 40 times in 1940, 76 in 1941, 20 in 1942, 14 in 1943, and 10 in 1944.

main theme. With regard to all future operations of importance I always consulted them in good time; but, while they gave careful consideration to the issues involved, they frequently asked not to be informed of dates and details, and indeed on several occasions stopped me when I was about to unfold these to them.

I had never intended to embody the office of Minister of Defence in a department. This would have required legislation, and all the delicate adjustments I have described, most of which settled themselves by personal goodwill, would have had to be thrashed out in a process of ill-timed constitution-making. There was however in existence and activity under the personal direction of the Prime Minister the Military Wing of the War Cabinet Secretariat, which had in pre-war days been the Secretariat of the Committee of Imperial Defence. At the head of this stood General Ismay, with Colonel Hollis and Colonel Jacob as his two principals, and a group of specially selected younger officers drawn from all three Services. This Secretariat became the staff of the office of the Minister of Defence. My debt to its members is immeasurable. General Ismay, Colonel Hollis, and Colonel Jacob rose steadily in rank and repute as the war proceeded, and none of them was changed. Displacements in a sphere so intimate and so concerned with secret matters are detrimental to continuous and efficient dispatch of business.

After some early changes almost equal stability was preserved in the Chiefs of Staff Committee. On the expiry of his term as Chief of the Air Staff, in September 1940, Air Marshal Newall became Governor-General of New Zealand, and was succeeded by Air Marshal Portal, who was the accepted star of the air force. Portal remained with me throughout the war. Sir John Dill, who had succeeded General Ironside in May 1940, remained C.I.G.S. until he accompanied me to Washington in December 1941. I then made him my personal Military Representative with the President and head of our Joint Staff Mission. His relations with General Marshall, Chief of Staff of the United States Army, became a priceless link in all our business, and when he died in harness some two years later he was accorded the unique honour of a resting place in Arlington Cemetery, the Valhalla hitherto reserved exclusively for American warriors. He was succeeded as C.I.G.S. by Sir Alan Brooke, who stayed with me till the end.

From 1941, for nearly four years, the early part of which was passed in much misfortune and disappointment, the only change made in this small band either among the Chiefs or in the Defence staff was due to the death in harness of Admiral Pound. This may well be a record in British military history. A similar degree of continuity was achieved by President Roosevelt in his own circle. The United States Chiefs of Staff — General Marshall, Admiral King, and General Arnold, subsequently joined by Admiral Leahy — started together on the American entry into the war, and were never changed. As both the British and Americans presently formed the Combined Chiefs of Staff Committee this was an inestimable advantage for all. Nothing like it between allies has ever been known before.

I cannot say that we never differed among ourselves even at home, but a kind of understanding grew up between me and the British Chiefs of Staff that we should convince and persuade rather than try to overrule each other. This was of course helped by the fact that we spoke the same technical language, and possessed a large common body of military doctrine and war experience. In this ever-changing scene we moved as one, and the War Cabinet clothed us with ever more discretion, and sustained us with unwearied and unflinching constancy. There was no division, as in the previous war, between politicians and soldiers, between the "Frocks" and the "Brass Hats" — odious terms which darkened counsel. We came very close together indeed, and friendships were formed which I believe were deeply valued.

The efficiency of a war administration depends mainly upon whether decisions emanating from the highest approved authority are in fact strictly, faithfully, and punctually obeyed. This was achieved in Britain in this time of crisis owing to the intense fidelity, comprehension, and whole-hearted resolve of the War Cabinet upon the essential purpose to which we had devoted ourselves. According to the directions given, ships, troops, and aeroplanes moved, and the wheels of factories spun. By all these processes, and by the confidence, indulgence, and loyalty by which I was upborne, I was soon able to give an integral direction to almost every aspect of the war. This was really necessary, because times were so very bad. The method was accepted because everyone realised how near were death and ruin. Not only individual death, which is the universal experience, stood near, but, incomparably more commanding, the life of Britain, her message, and her glory.

*   *   *

Any account of the methods of government which developed under the National Coalition would be incomplete without an explanation of the series of personal messages which I sent to the President of the United States and the heads of other foreign countries and the Dominions Governments. This correspondence must be described. Having obtained from the Cabinet any specific decisions required on policy, I composed and dictated these documents myself, for the most part on the basis that they were intimate and informal correspondence with friends and fellow workers. One can usually put one's thoughts better in one's own words. It was only occasionally that I read the text to the Cabinet beforehand. Knowing their views, I used the ease and freedom needed for the doing of my work. I was of course hand-in-glove with the Foreign Secretary and his department, and any differences of view were settled together. I circulated these telegrams, in some cases after they had been sent, to the principal members of the War Cabinet, and, where he was concerned, to the Dominions Secretary. Before dispatching them I of course had my points and facts checked departmentally, and nearly all military messages passed through Ismay's hands to the Chiefs of Staff. This correspondence in no way ran counter to the official communications or the work of the Ambassadors. It became however in fact the channel of much vital business, and played a part in my conduct of the war not less, and sometimes even more, important than my duties as Minister of Defence.

The very select circle, who were entirely free to express their opinion, were almost invariably content with the drafts and gave me an increasing measure of confidence. Differences with American authorities for instance, insuperable at the second level, were settled often in a few hours by direct contact at the top. Indeed, as time went on, the efficacy of this top-level transaction of business was so apparent that I had to be careful not to let it become a vehicle for ordinary departmental affairs. I had repeatedly to refuse the requests of my colleagues to address the President personally on important matters of detail. Had these intruded unduly upon the personal correspondence they would soon have destroyed its privacy and consequently its value.

My relations with the President gradually became so close that the chief business between our two countries was virtually conducted by these personal interchanges between him and me. In this way our perfect understanding was gained. As Head of the State as well as Head of the Government, Roosevelt spoke and acted with authority in every sphere; and, carrying the War Cabinet with me, I represented Great Britain with almost equal latitude. Thus a very high degree of concert was obtained, and the saving in time and the reduction in the number of people informed were both invaluable. I sent my cables to the American Embassy in London, which was in direct touch with the President at the White House through special coding machines. The speed with which answers were received and things settled was aided by clock-time. Any message which I prepared in the evening, at night, or even up to two o'clock in the morning, would reach the President before he went to bed, and very often his answer would come back to me when I woke the next morning. In all I sent him nine hundred and fifty messages, and received about eight hundred in reply. I felt

I was in contact with a very great man, who was also a warm-hearted friend and the foremost champion of the high causes which we served.

*   *   *

On Monday, May 13, 1940, I asked the House of Commons, which had been specially summoned, for a vote of confidence in the new Administration. After reporting the progress which had been made in filling the various offices, I said: "I have nothing to offer but blood, toil, tears, and sweat." In all our long history no Prime Minister had ever been able to present to Parliament and the nation a programme at once so short and so popular. I ended:

> You ask, What is our policy? I will say: It is to wage war, by sea, land, and air, with all our might and with all the strength that God can give us: to wage war against a monstrous tyranny, never surpassed in the dark, lamentable catalogue of human crime. That is our policy. You ask, What is our aim? I can answer in one word: Victory — victory at all costs, victory in spite of all terror; victory, however long and hard the road may be; for without victory there is no survival. Let that be realised: no survival for the British Empire; no survival for all that the British Empire has stood for, no survival for the urge and impulse of the ages, that mankind will move forward towards its goal. But I take up my task with buoyancy and hope. I feel sure that our cause will not be suffered to fail among men. At this time I feel entitled to claim the aid of all, and I say: Come, then, let us go forward together with our united strength.

Upon these simple issues the House voted unanimously, and adjourned till May 21.

Thus, then, we all started on our common task. Never did a British Prime Minister receive from Cabinet colleagues the loyal and true aid which I enjoyed during the next five years from these men of all parties in the State. Parliament, while maintaining free and active criticism, gave continuous, overwhelming support to all measures proposed by the Government, and the nation was united and ardent as never before. It was well indeed that this should be so, because events were to come upon us of an order more terrible than anyone had foreseen.

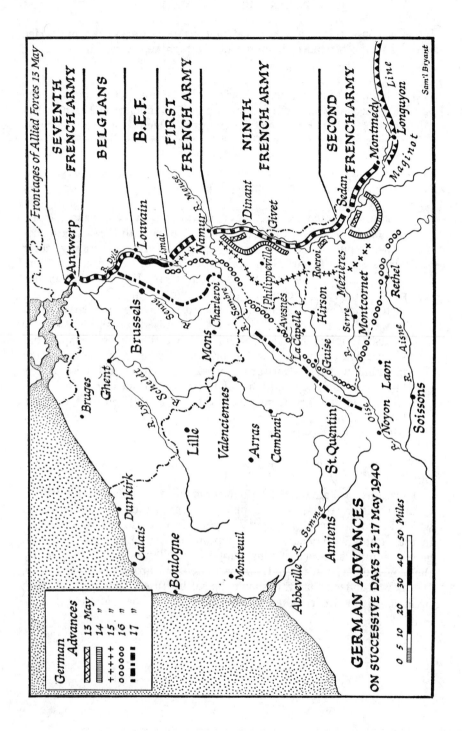

GERMAN ADVANCES
ON SUCCESSIVE DAYS 13–17 MAY 1940

Sam.¹ Bryant

Frontages of Allied Forces 15 May

SEVENTH FRENCH ARMY

BELGIANS

B.E.F.

FIRST FRENCH ARMY

NINTH FRENCH ARMY

SECOND FRENCH ARMY

German
Advances

13 May
14   "
15.   "
16   "
17   "

0 5 10   20   30   40   50 Miles

Antwerp
R. Dyle
Louvain
Limal
Namur
R. Meuse
Dinant
Givet
Sedan
Montmédy
Longuyon
Maginot Line

Brussels
Ghent
Bruges
R. Lys
R. Scheldt
R. Senne
Mons
Charleroi
R. Sambre
Philippeville
Avesnes
Rocroi
Mézières
Hirson
R. Serre
Montcornet
Rethel
R. Aisne

Dunkirk
Calais
Boulogne
Montreuil
Lille
Valenciennes
Arras
Cambrai
La Capelle
Guise
St. Quentin
R. Oise
Laon
Noyon
Soissons

Abbeville
R. Somme
Amiens

2

The Battle of France

O N T H E O U T B R E A K of the war in September 1939, the main power of the German Army and Air Force had been concentrated on the invasion and conquest of Poland. Along the whole of the Western front, from Aix-la-Chapelle to the Swiss frontier, there had stood forty-two German divisions without armour. After the French mobilisation France could deploy the equivalent of seventy divisions opposite to them. For reasons which have been explained, it was not deemed possible to attack the Germans then. Very different was the situation on May 10, 1940. The enemy, profiting by the eight months' delay and by the destruction of Poland, had armed, equipped, and trained about 155 divisions, of which ten were armoured ("Panzer"). Hitler's agreement with Stalin had enabled him to reduce the German forces in the East to the smallest proportions. Opposite Russia, according to General Halder, the German Chief of Staff, there was "no more than a light covering force, scarcely fit for collecting customs duties." Without premonition of their own future, the Soviet Government watched the destruction of that "Second Front" in the West for which they were soon to call so vehemently and to wait in agony so long. Hitler was therefore in a position to deliver his onslaught on France with 126 divisions, and the whole of the immense armour weapon of ten Panzer divisions, comprising nearly three thousand armoured vehicles, of which a thousand at least were heavy tanks.

Opposite this array, the exact strength and disposition of which was of course unknown to us, the French had the equivalent total, including the British, of 103 divisions. If the armies of Belgium and Holland became involved, this number would be increased by twenty-two Belgian and ten Dutch divisions. As both these countries were immediately attacked, the grand total of Allied divisions of all qualities nominally available on May 10 was therefore 135, or practically the same number as we now know the enemy possessed. Properly organised and equipped, well trained and led, this force should, according to the standards of the previous war, have had a good chance of bringing the invasion to a stop.

However, the Germans had full freedom to choose the moment, the direction, and the strength of their attack. More than half of the French Army stood on the southern and eastern sectors of France, and the fifty-one French and British divisions of General Billotte's Army Group No. 1, with whatever Belgian and Dutch aid was forthcoming, had to face the onslaught of upwards of seventy hostile divisions under Bock and Rundstedt between Longwy and the sea. The combination of the almost cannon-proof tank and dive-bomber aircraft which had proved so successful in Poland on a smaller scale was again to form the spearhead of the main attack, and a group of five Panzer and three motorised divisions under Kleist was directed through the Ardennes on Sedan and Monthermé.

To meet such modern forms of war the French deployed about 2300 tanks, mostly light. Their armoured formations included some powerful modern types, but more than half their total armoured strength was held in dispersed battalions of light tanks, for co-operation with the infantry. Their six armoured divisions,[1] with which alone they could have countered the massed Panzer assault, were widely distributed over the front, and could not be collected together to operate in coherent action. Britain, the birthplace of the tank, had only just completed the formation and training of her first armoured division (328 tanks), which was still in England.

The German fighter aircraft now concentrated in the West were far superior to the French in numbers and quality. The British Air Force in France comprised the ten fighter squadrons of Hurricanes which could be spared from vital Home Defence and nineteen squadrons of other types. Neither the French nor the British air authorities had equipped themselves with dive bombers, which at this time, as in Poland, became prominent, and were to play an important part in the demoralisation of the French infantry and particularly of their coloured troops.

*   *   *

During the night of May 9–10, heralded by widespread air attacks against airfields, communications, headquarters, and magazines, all the German forces sprang forward towards France across the frontiers of Belgium, Holland, and Luxembourg. Complete tactical surprise was

---

[1] This figure includes the so-called light motorised divisions, which possessed tanks.

achieved in nearly every case. Out of the darkness came suddenly innumerable parties of well-armed, ardent storm troops, often with light artillery, and long before daybreak a hundred and fifty miles of front were aflame. Holland and Belgium, assaulted without the slightest pretext or warning, cried aloud for help. The Dutch had trusted to their waterline; all the sluices not seized or betrayed were opened, and the Dutch frontier guards fired upon the invaders.

Mr. Colijn, when as Dutch Prime Minister he visited me in 1937, had explained to me the marvellous efficiency of the Dutch inundations. He could, he explained, by a telephone message from the luncheon table at Chartwell press a button which would confront an invader with impassable water obstacles. But all this was nonsense. The power of a great state against a small one under modern conditions is overwhelming. The Germans broke through at every point, bridging the canals or seizing the locks and water controls. In a single day all the outer line of the Dutch defences was mastered. At the same time the German Air Force began to use its might upon a defenceless country. Rotterdam was reduced to a blazing ruin. The Hague, Utrecht, and Amsterdam were threatened with the same fate. The Dutch hope that they would be by-passed by the German right-handed swing as in the former war was vain.

During the 14th the bad news began to come in. At first all was vague. At 7 P.M. I read to the Cabinet a message received from M. Reynaud stating that the Germans had broken through at Sedan, that the French were unable to resist the combination of tanks and dive-bombing, and asking for ten more squadrons of fighters to re-establish the line. Other messages received by the Chiefs of Staff gave similar information, and added that both Generals Gamelin and Georges took a serious view of the situation and that General Gamelin was surprised at the rapidity of the enemy's advance. At almost all points where the armies had come in contact the weight and fury of the German attack was overpowering. All the British air squadrons fought continuously, their principal effort being against the pontoon bridges in the Sedan area. Several of these were destroyed and others damaged in desperate and devoted attacks. The losses in the low-level attacks on the bridges from the German anti-aircraft artillery were cruel. In one case, of six aircraft only one returned from the successful task. On this day alone we lost a total of sixty-seven machines, and, being engaged principally with the enemy's anti-aircraft forces, accounted for only fifty-three German aircraft. That night there remained in France of the Royal Air Force only 206 serviceable aircraft out of 474.

This detailed information came only gradually to hand. But it was already clear that the continuance of fighting on this scale would soon completely consume the British Air Force in spite of its individual ascendancy. The hard question of how much we could send from Britain without leaving ourselves defenceless and thus losing the power to continue the war pressed itself henceforward upon us. Our own natural promptings and many weighty military arguments lent force to the incessant, vehement French appeals. On the other hand, there was a limit, and that limit if transgressed would cost us our life.

At this time all these issues were discussed by the whole War Cabinet, which met several times a day. Air Chief Marshal Dowding, at the head of our metropolitan fighter command, had declared to me that with twenty-five squadrons of fighters he would defend the island against the whole might of the German Air Force, but that with less he would be overpowered. Defeat would have entailed not only the destruction of all our airfields and our air power but of the aircraft factories on which our whole future hung. My colleagues and I were resolved to run all risks for the sake of the battle up to that limit — and those risks were very great — but not to go beyond it, no matter what the consequences might be.

About half past seven on the morning of the 15th I was woken up with the news that M. Reynaud was on the telephone at my bedside. He spoke in English, and evidently under stress. "We have been defeated." As I did not immediately respond he said again, "We are beaten; we have lost the battle." I said, "Surely it can't have happened so soon?" But he replied, "The front is broken near Sedan; they are pouring through in great numbers with tanks and armoured cars" — or words to that effect. I then said, "All experience shows that the offensive will come to an end after a while. I remember the 21st of March, 1918. After five or six days they have to halt for supplies, and the opportunity for counter-attack is presented. I learned all this at the time from the lips of Marshal Foch himself." Certainly this was what we had always seen in the past and what we ought to have seen now. However, the French Premier came back to the sentence with which he had begun, which proved indeed only too true: "We are defeated; we have lost the battle." I said I was willing to come over and have a talk.

A gap of some fifty miles had in fact been punched in the French line, through which the vast mass of enemy armour was pouring, and the French Ninth Army was in a state of complete dissolution. By the evening of the 15th, German armoured cars were reported to be sixty miles behind the original front. On this day also the struggle in Holland came to an end. Owing to the capitulation of the Dutch High Command at 11 A.M., only a very few Dutch troops could be evacuated.

Of course this picture presented a general impression of defeat. I had seen a good deal of this sort of thing in the previous war, and the idea of the line being broken, even on a broad front, did not convey to my mind the appalling consequences that now flowed from it. Not having had access to official information for so many years, I did not comprehend the violence of the revolution effected since the last war by the incursion of a mass of fast-moving heavy armour. I knew about it, but it had not altered my inward convictions as it should have done. There was nothing I could have done if it had. I rang up General Georges, who seemed quite cool, and reported that the breach at Sedan was being plugged. A telegram from General Gamelin also stated that although the position between Namur and Sedan was serious he viewed the situation with calm. I reported Reynaud's message and other news to the Cabinet at 11 A.M.

But on the 16th the penetration of over sixty miles inward upon us from the frontier near Sedan was confirmed. Although few details were available even to the War Office, and no clear view could be formed of what was happening, the gravity of the crisis was obvious. I felt it imperative to go to Paris that afternoon.

*   *   *

At about 3 P.M. I boarded a Flamingo, a Government passenger plane, of which there were three. General Dill, Vice-Chief of the Imperial General Staff, came with me, and Ismay.

It was a good machine, very comfortable, and making about a hundred and sixty miles an hour. As it was unarmed, an escort was provided, but we soared off into a rain-cloud and reached Le Bourget in little more than an hour. From the moment we got out of the Flamingo it was obvious that the situation was incomparably worse than we had imagined. The officers who met us told General Ismay that the Germans were expected in Paris in a few days at most. After hearing at the Embassy about the position, I drove to the Quai d'Orsay, arriving at 5.30 o'clock. I was conducted into one of its fine rooms. Reynaud was there, Daladier, Minister of National Defence and War, and General Gamelin. Everybody was standing. At no time did we sit down around a table. Utter dejection was written on every face. In front of Gamelin on a student's easel was a map, about two yards square, with a black ink line purporting to show the Allied front. In this line there was drawn a small but sinister bulge at Sedan.

The Commander-in-Chief briefly explained what had happened. North and south of Sedan, on a front of fifty or sixty miles, the Germans had broken through. The French army in front of them was destroyed or scattered. A heavy onrush of armoured vehicles was advancing with unheard-of speed towards Amiens and Arras, with the

intention, apparently, of reaching the coast at Abbeville or thereabouts.
Alternatively they might make for Paris. Behind the armour, he said,
eight or ten German divisions, all motorised, were driving onward,
making flanks for themselves as they advanced against the two dis-
connected French armies on either side. The General talked perhaps
five minutes without anyone saying a word. When he stopped there
was a considerable silence. I then asked, "Where is the strategic reserve?" and, breaking into French, which I used indifferently (in
every sense): *"Où est la masse de manoeuvre?"* General Gamelin turned
to me and, with a shake of the head and a shrug, said, *"Aucune."*

There was another long pause. Outside in the garden of the Quai
d'Orsay clouds of smoke arose from large bonfires, and I saw from the
window venerable officials pushing wheelbarrows of archives onto
them. Already therefore the evacuation of Paris was being prepared.

Past experience carries with its advantages the drawback that things
never happen the same way again. Otherwise I suppose life would be
too easy. After all, we had often had our fronts broken before; always we had been able to pull things together and wear down the
momentum of the assault. But here were two new factors that I had
never expected to have to face. First, the overrunning of the whole
of the communications and countryside by an irresistible incursion of
armoured vehicles, and secondly NO STRATEGIC RESERVE. *"Aucune."*
I was dumbfounded. What were we to think of the great French Army
and its highest chiefs? It had never occurred to me that any commanders having to defend five hundred miles of engaged front would have
left themselves unprovided with a mass of manoeuvre. No one can
defend with certainty so wide a front; but when the enemy has committed himself to a major thrust which breaks the line one can always
have, one *must* always have, a mass of divisions which marches up in
vehement counterattack at the moment when the first fury of the
offensive has spent its force.

What was the Maginot Line for? It should have economised troops
upon a large sector of the frontier, not only offering many sally ports
for local counterstrokes, but also enabling large forces to be held in
reserve; and this is the only way these things can be done. But now
there was no reserve. I admit this was one of the greatest surprises I
have had in my life. Why had I not known more about it, even though
I had been so busy at the Admiralty? Why had the British Government, and the War Office above all, not known more about it? It was
no excuse that the French High Command would not impart their
dispositions to us or to Lord Gort except in vague outline. We had a
right to know. We ought to have insisted. Both armies were fighting
in the line together. I went back again to the window and the curling

wreaths of smoke from the bonfires of the State documents of the French Republic. Still the old gentlemen were bringing up their wheelbarrows, and industriously casting their contents into the flames.

Presently General Gamelin was speaking again. He was discussing whether forces should now be gathered to strike at the flanks of the penetration, or "Bulge," as we called such things later on. Eight or nine divisions were being withdrawn from quiet parts of the front, the Maginot Line; there were two or three armoured divisions which had not been engaged; eight or nine more divisions were being brought from Africa and would arrive in the battle zone during the next fortnight or three weeks. The Germans would advance henceforward through a corridor between two fronts on which warfare in the fashion of 1917 and 1918 could be waged. Perhaps the Germans could not maintain the corridor, with its ever-increasing double flank guards to be built up, and at the same time nourish their armoured incursion. Something in this sense Gamelin seemed to say, and all this was quite sound. I was conscious however that it carried no conviction in this small but hitherto influential and responsible company. Presently I asked General Gamelin when and where he proposed to attack the flanks of the Bulge. His reply was: "Inferiority of numbers, inferiority of equipment, inferiority of method" — and then a hopeless shrug of the shoulders. There was no argument; there was no need of argument. And where were we British anyway, having regard to our tiny contribution — ten divisions after eight months of war, and not even one modern tank division in action?

*   *   *

The burden of General Gamelin's, and indeed of all the French High Command's subsequent remarks, was insistence on their inferiority in the air and earnest entreaties for more squadrons of the Royal Air Force, bomber as well as fighter, but chiefly the latter. This prayer for fighter support was destined to be repeated at every subsequent conference until France fell. In the course of his appeal General Gamelin said that fighters were needed not only to give cover to the French Army, but also to stop the German tanks. At this I said, "No. It is the business of the artillery to stop the tanks. The business of the fighters is to cleanse the skies [nettoyer le ciel] over the battle." It was vital that our metropolitan fighter air force should not be drawn out of Britain on any account. Our existence turned on this. Nevertheless it was necessary to cut to the bone. In the morning, before I started, the Cabinet had given me authority to move four more squadrons of fighters to France. On our return to the Embassy, and after talking it over with Dill, I decided to ask sanction for the dispatch of six more. This would leave us with only the twenty-five fighter squadrons at home, and that was the final limit. It was a rending decision either way. I told General Ismay to telephone to London that the Cabinet should assemble at once to consider an urgent telegram which would be sent over in the course of the next hour or so.

The reply came at about eleven-thirty. The Cabinet said "Yes." I immediately took Ismay off with me in a car to M. Reynaud's flat. We found it more or less in darkness. After an interval M. Reynaud emerged from his bedroom in his dressing gown and I told him the favourable news. Ten fighter squadrons! I then persuaded him to send for M. Daladier, who was duly summoned and brought to the flat to hear the decision of the British Cabinet. In this way I hoped to revive the spirits of our French friends, as much as our limited means allowed. Daladier never spoke a word. He rose slowly from his chair and wrung my hand. I got back to the Embassy about 2 A.M., and slept well, though the cannon fire in petty aeroplane raids made one roll over from time to time. In the morning I flew home, and, in spite of other preoccupations, pressed on with the construction of the second level of the new Government.

The War Cabinet met at 10 A.M. on the 17th, and I gave them an account of my visit to Paris, and of the situation so far as I could measure it.

I said I had told the French that unless they made a supreme effort we should not be justified in accepting the grave risk to the safety of our country that we were incurring by the dispatch of the additional fighter squadrons to France. I felt that the question of air reinforcements was one of the gravest that a British Cabinet had ever had to face. It was claimed that the German air losses had been four or five times our own, but I had been told that the French had only a quarter of their fighter aircraft left. On this day Gamelin thought the situation "lost," and is reported to have said, "I will guarantee the safety of Paris only for today, tomorrow [the 18th], and the night following." The battle crisis grew hourly in intensity. That afternoon the Germans entered Brussels. The next day they reached Cambrai, passed St. Quentin, and brushed our small parties out of Péronne. The Belgian, the British, and the French Armies concerned continued their withdrawal to the Scheldt.

At midnight (May 18–19) Lord Gort was visited at his headquarters by General Billotte. Neither the personality of this French General nor his proposals, such as they were, inspired confidence in his allies. From this moment the possibility of a withdrawal to the coast began to present itself to the British Commander-in-Chief. In his dispatch published in March 1941 he wrote: "The picture was now [night of the

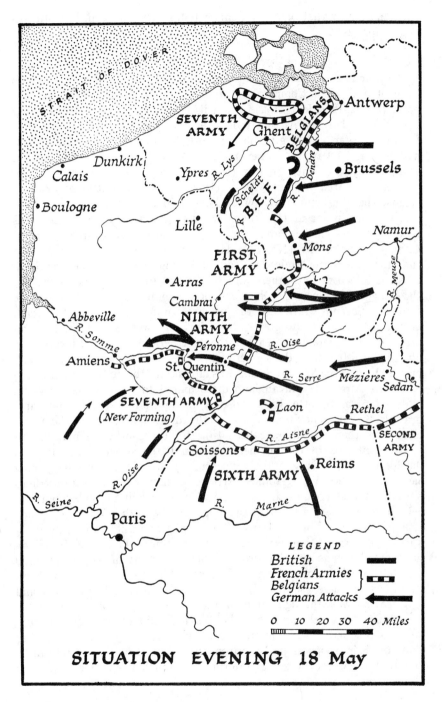

STRAIT OF DOVER

SEVENTH ARMY

Ghent

BELGIANS

Antwerp

Dunkirk

R. Lys

R. Scheldt

R. Dendre

Brussels

Calais

Ypres

B.E.F.

Boulogne

Lille

Mons

R. Meuse

Namur

FIRST ARMY

Arras

Cambrai

Abbeville

NINTH ARMY

R. Somme

Péronne

R. Oise

Amiens

St. Quentin

R. Serre

Mézières

Sedan

SEVENTH ARMY
(New Forming)

Laon

Rethel

R. Aisne

SECOND ARMY

R. Oise

Soissons

SIXTH ARMY

Reims

R. Seine

R. Marne

Paris

LEGEND

British
French Armies
Belgians }
German Attacks

0   10   20   30   40 Miles

SITUATION EVENING 18 May

19th] no longer that of a line bent or temporarily broken, but of a be-sieged fortress."

*   *   *

Far-reaching changes were now made by M. Reynaud in the French Cabinet and High Command. On the 18th Marshal Pétain was appointed Vice-President of the Council. Reynaud himself, transferring Daladier to Foreign Affairs, took over the Ministry of National Defence and War. At 7 P.M. on the 19th he appointed Weygand, who had just arrived from the Levant, to replace General Gamelin. I had known Weygand when he was the right-hand man of Marshal Foch, and had admired his masterly intervention in the Battle of Warsaw against the Bolshevik invasion of Poland in August 1920 — an event decisive for Europe at that time. He was now seventy-three, but was reported to be efficient and vigorous in a very high degree. General Gamelin's final Order (No. 12), dated 9.45 A.M. on May 19, prescribed that the Northern Armies, instead of letting themselves be encircled, must at all costs force their way southward to the Somme, attacking the Panzer divisions which had cut their communications. At the same time the Second Army and the newly forming Sixth were to attack northward towards Mézières. These decisions were sound. Indeed, an order for the general retreat of the Northern Armies southward was already at least four days overdue. Once the gravity of the breach in the French centre at Sedan was apparent, the only hope for the Northern Armies lay in an immediate march to the Somme. Instead, under General Billotte, they had only made gradual and partial withdrawals to the Scheldt and formed the defensive flank to the right. Even now there might have been time for the southward march.

The confusion of the northern command, the apparent paralysis of the First French Army, and the uncertainty about what was happening had caused the War Cabinet extreme anxiety. All our proceedings were quiet and composed, but we had a united and decided opinion, behind which there was silent passion. On the 19th we were informed at 4.30 P.M. that Lord Gort was "examining a possible withdrawal towards Dunkirk if that were forced upon him." The C.I.G.S. (Ironside) could not accept this proposal, as, like most of us, he favoured the southward march. We therefore sent him to Lord Gort with instructions to move the British Army in a southwesterly direction and to force his way through all opposition in order to join up with the French in the south, and that the Belgians should be urged to conform to this movement, or, alternatively, that we would evacuate as many of their troops as possible from the Channel ports. He was to be told that we would ourselves inform the French Government of what

had been resolved. At the same Cabinet we sent Dill to General
Georges' Headquarters, with which we had a direct telephone line. He
was to stay there for four days, and tell us all he could find out. Contacts even with Lord Gort were intermittent and difficult, but it was
reported that only four days' supplies and ammunition for one battle
were available.

At the morning War Cabinet of May 20 we again discussed the
situation of our Army. Even on the assumption of a successful fighting
retreat to the Somme, I thought it likely that considerable numbers
might be cut off or driven back on the sea. It is recorded in the minutes
of the meeting: "The Prime Minister thought that as a precautionary
measure the Admiralty should assemble a large number of small vessels
in readiness to proceed to ports and inlets on the French coast." On this
the Admiralty acted immediately and with ever-increasing vigour as
the days passed and darkened. Operational control had been delegated on the 19th to Admiral Ramsay, commanding at Dover, and on
the afternoon of the 20th, in consequence of the orders from London,
the first conference of all concerned, including representatives of the
Shipping Ministry, was held at Dover to consider *"the emergency
evacuation across the Channel of very large forces."* It was planned if
necessary to evacuate from Calais, Boulogne, and Dunkirk, at a rate of
ten thousand men from each port every twenty-four hours. From
Harwich round to Weymouth sea-transport officers were directed to
list all suitable ships up to a thousand tons, and a complete survey was
made of all shipping in British harbours. These plans for what was
called "Operation Dynamo" proved the salvation of the Army ten days
later.

*   *   *

The direction of the German thrust had now become more obvious.
Armoured vehicles and mechanised divisions continued to pour through
the gap towards Amiens and Arras, curling westwards along the
Somme towards the sea. On the night of the 20th they entered Abbeville, having traversed and cut the whole communications of the Northern Armies. These hideous, fatal scythes encountered little or no resistance once the front had been broken. The German tanks — the
dreaded *"chars allemands"* — ranged freely through the open country,
and, aided and supplied by mechanised transport, advanced thirty or
forty miles a day. They had passed through scores of towns and hundreds of villages without the slightest opposition, their officers looking
out of the open cupolas and waving jauntily to the inhabitants. Eyewitnesses spoke of crowds of French prisoners marching along with
them, many still carrying their rifles, which were from time to time

collected and broken under the tanks. I was shocked by the utter failure to grapple with the German armour, which, with a few thousand vehicles, was compassing the entire destruction of mighty armies, and by the swift collapse of all French resistance once the fighting front had been pierced. The whole German movement was proceeding along the main roads, which at no point seemed to be blocked.

Weygand's first act was to confer with his senior commanders. It was not unnatural that he should wish to see the situation in the north for himself, and to make contact with the commanders there. Allowances must be made for a general who takes over the command in the crisis of a losing battle. But now there was no time. He should not have left the summit of the remaining controls and have become involved in the delays and strains of personal movement. We may note in detail what followed. On the morning of the 20th Weygand, installed in Gamelin's place, made arrangements to visit the Northern Armies on the 21st. After learning that the roads to the north were cut by the Germans he decided to fly. His plane was attacked, and forced to land at Calais. The hour appointed for his conference at Ypres had to be altered to 3 P.M. on the 21st. Here he met King Leopold of Belgium and General Billotte. Lord Gort, who had not been notified of time and place, was not present, and no British officers were there. The King described this conference as "four hours of confused talking." It discussed the co-ordination of the three armies, the execution of the Weygand plan, and if that failed the retirement of the British and French to the Lys, and the Belgians to the Yser. At 7 P.M. General Weygand had to leave. Lord Gort did not arrive till eight, when he received an account of the proceedings from General Billotte. Weygand drove back to Calais, embarked on a submarine for Dieppe, and returned to Paris. Billotte drove off in his car to deal with the crisis, and within the hour was killed in a motor collision. Thus all was again in suspense.

On the 21st Ironside returned and reported that Lord Gort, on receiving the Cabinet instructions, had seemed adverse to a southward march. It would involve a rearguard action from the Scheldt at the same time as an attack into an area already strongly held by the enemy armoured and mobile formations. During such a movement both flanks would have to be protected, and neither the French First Army nor the Belgians were likely to be able to conform to such a manoeuvre if attempted. Ironside added that confusion reigned in the French High Command in the north, that General Billotte had failed to carry out his duties of co-ordination for the past eight days and appeared to have no plans, that the British Expeditionary Force were in good heart and had so far had only about five hundred battle casualties. He gave a vivid description of the state of the roads, crowded with refugees, scourged by the fire of German aircraft. He had had a rough time himself.

Two fearsome alternatives therefore presented themselves to the War Cabinet. The first, the British Army at all costs, with or without French and Belgian co-operation, to cut its way to the south and the Somme, a task which Lord Gort doubted its ability to perform; the second, to fall back on Dunkirk and face a sea evacuation under hostile air attack, with the certainty of losing all the artillery and equipment, then so scarce and precious. Obviously great risks should be run to achieve the first, but there was no reason why all possible precautions and preparations should not be taken for the sea evacuation if the southern plan failed. I proposed to my colleagues that I should go to France to meet Reynaud and Weygand and come to a decision. Dill was to meet me there from General Georges' Headquarters.

*   *   *

When I arrived in Paris on May 22 there was a new setting. Gamelin was gone; Daladier was gone from the war scene. Reynaud was both Prime Minister and Minister of War. As the German thrust had definitely turned seaward, Paris was not immediately threatened. Grand Quartier Général was still at Vincennes. Reynaud drove me down there about noon. In the garden some of those figures I had seen round Gamelin — one a very tall cavalry officer — were pacing moodily up and down. *"C'est l'ancien régime,"* remarked the aide-de-camp. Reynaud and I were brought into Weygand's room, and afterwards to the map room, where we had the great maps of the Supreme Command. Weygand met us. In spite of his physical exertions and a night of travel, he was brisk, buoyant, and incisive. He made an excellent impression upon all. He unfolded his plan of war. He was not content with a southward march or retreat for the Northern Armies. They should strike southeast from around Cambrai and Arras in the general direction of St. Quentin, thus taking in flank the German armoured divisions at present engaged in what he called the St. Quentin–Amiens pocket. Their rear, he thought, would be protected by the Belgian Army, which would cover them towards the east, and if necessary towards the north. Meanwhile a new French army under General Frère, composed of eighteen to twenty divisions drawn from Alsace, from the Maginot Line, from Africa, and from every other quarter, were to form a front along the Somme. Their left hand would push forward through Amiens to Arras, and thus by their utmost efforts establish contact with the armies of the north. The enemy armour must be kept under constant pressure. "The Panzer divisions

S.B.

Antwerp

BELGIANS

Ghent

Brussels

Dunkirk

Calais

R. Yser

Ypres

R. Lys

Oudenarde

Cassel

Menin

St. Omer

B.E.F.

Namur

Boulogne

Hazebrouck

Lille

Béthune

Maulde

Montreuil

FIRST
ARMY

Hesdin

St. Pol

Douai

Valenciennes

Arras

Abbeville

Doullens

Cambrai

(10 GERMAN PZ. DIVISIONS)

R. Somme

Péronne

R. Oise

Amiens

St. Quentin

R. Serre

Sedan

SEVENTH ARMY
(Concentrating)

Chauny

Laon

Beauvais

R. Oise

R. Aisne

Soissons
SIXTH ARMY

Reims

R. Seine

Ch. Thierry

Marne

Epernay

R.

Paris

Châlons-
sur-Marne

LEGEND
British
French Armies }
Belgians
Germans
German Attacks

SITUATION · EVENING
22 May

0   10   20   30   40 Miles

must not," said Weygand, "be allowed to keep the initiative." All necessary orders had been given so far as it was possible to give orders at all. We were now told that General Billotte, to whom he had imparted his whole plan, had just been killed in the motor accident. Dill and I were agreed that we had no choice, and indeed no inclination, except to welcome the plan. I emphasised that "it was indispensable to reopen communications between the armies of the north and those of the south by way of Arras." I explained that Lord Gort, while striking southward, must also guard his path to the coast. To make sure there was no mistake about what was settled, I myself dictated a *résumé* of the decision and showed it to Weygand, who agreed. I reported accordingly to the Cabinet, and passed the news to Lord Gort.

It will be seen that Weygand's new plan did not differ except in emphasis from the cancelled Instruction No. 12 of General Gamelin. Nor was it out of harmony with the vehement opinion which the War Cabinet had expressed on the 19th. The Northern Armies were to shoulder their way southward by offensive action, destroying, if possible, the armoured incursion. They were to be met by a helpful thrust through Amiens by the new French Army Group under General Frère. This would be most important if it came true. In private I complained to M. Reynaud that Gort had been left entirely without orders for four consecutive days. Even since Weygand had assumed command three days had been lost in taking decisions. The change in the Supreme Command was right. The resultant delay was evil.

In the absence of any supreme war direction events and the enemy had taken control. A small forlorn battle was fought by the British around Arras between the 21st and the 23rd, but the enemy armour, some of it commanded by a general named Rommel, was too strong. Up to this time General Weygand had been counting on General Frère's army advancing northward on Amiens, Albert, and Péronne. They had, in fact, made no noticeable progress, and were still forming and assembling.

*   *   *

There was a very strong feeling in Cabinet and high military circles that the abilities and strategic knowledge of Sir John Dill, who had been since April 23 Vice-Chief of the Imperial General Staff, should find their full scope in his appointment as our principal Army adviser. No one could doubt that his professional standing was in many ways superior to that of Ironside.

As the adverse battle drew to its climax I and my colleagues greatly desired that Sir John Dill should become C.I.G.S. We had also to choose a Commander-in-Chief for the British Island, if we were invaded. Late at night on May 25 Ironside, Dill, Ismay, myself, and one or two others in my room at Admiralty House were trying to measure the position. General Ironside volunteered the proposal that he should cease to be C.I.G.S., but declared himself quite willing to command the British Home Forces. Considering the unpromising task that such a command was at the time thought to involve, this was a spirited and selfless offer. I therefore accepted General Ironside's proposal; and the high dignities and honours which were later conferred upon him arose from my appreciation of his bearing at this moment in our affairs. Sir John Dill became C.I.G.S. on May 27. The changes were generally judged appropriate for the time being.

3

The March to the Sea

WE MAY NOW REVIEW up to this point the course of this memorable battle.

Only Hitler was prepared to violate the neutrality of Belgium and Holland. Belgium would not invite the Allies in until she was herself attacked. Therefore the military initiative rested with Hitler. On May 10 he struck his blow. The First Army Group, with the British in the centre, instead of standing behind their fortifications, leaped forward into Belgium on a vain, because belated, mission of rescue, in accordance with General Gamelin's Plan D.[1] The French had left the gap opposite the Ardennes ill fortified and weakly guarded. An armoured inroad on a scale never known in war broke the centre of the French line of armies, and in forty-eight hours threatened to cut all the Northern Armies alike from their southern communications and from the sea. By the 14th at the latest the French High Command should have given imperative orders to these armies to make a general retreat at full speed, accepting not only risks but heavy losses of *matériel*. This issue was not faced in its brutal realism by General Gamelin. The French commander of the northern group, Billotte, was incapable of taking the necessary decisions himself. Confusion reigned throughout the armies of the threatened left wing.

As the superior power of the enemy was felt they fell back. As the turning movement swung round their right they formed a defensive flank. If they had started back on the 14th they could have been on their old line by the 17th and would have had a good chance of fighting their way out. At least three mortal days were lost. From the 17th onward the British War Cabinet saw clearly that an immediate fighting march southward would alone save the British Army. They were

[1] See p. 193.

resolved to press their view upon the French Government and General Gamelin, but their own commander, Lord Gort, was doubtful whether it was possible to disengage the fighting fronts, and still more to break through at the same time. On the 19th General Gamelin was dismissed, and Weygand reigned in his stead. Gamelin's "Instruction No. 12," his last order, though five days late, was sound in principle, and also in conformity with the main conclusions of the British War Cabinet and Chiefs of Staff. The change in the Supreme Command, or want of command, led to another three days' delay. The spirited plan which General Weygand proposed after visiting the Northern Armies was never more than a paper scheme. In the main it was the Gamelin plan, rendered still more hopeless by further delay.

In the hideous dilemma which now presented itself we accepted the Weygand plan, and made loyal and persistent, though now ineffectual, efforts to carry it out until the 25th, when, all the communications being cut, our weak counterattack being repulsed, with the loss of Arras, the Belgian front breaking and King Leopold about to capitulate, all hope of escape to the southward vanished. There remained only the sea. Could we reach it, or must we be surrounded and broken up in the open field? In any case the whole artillery and equipment of our Army, irreplaceable for many months, must be lost. But what was that compared with saving the Army, the nucleus and structure upon which alone Britain could build her armies of the future? Lord Gort, who had from the 25th onward felt that evacuation by sea was our only chance, now proceeded to form a bridgehead around Dunkirk and to fight his way into it with what strength remained. All the discipline of the British, and the qualities of their commanders, who included Brooke, Alexander, and Montgomery, were to be needed. Much more was to be needed. All that man could do was done. Would it be enough?

*   *   *

A much disputed episode must now be examined. General Halder, Chief of the German General Staff, has declared that at this moment Hitler made his only effective direct personal intervention in the battle. He became, according to this authority, "alarmed about the armoured formations, because they were in considerable danger in a difficult country, honeycombed with canals, without being able to attain any vital results." He felt he could not sacrifice armoured formations uselessly, as they were essential to the second stage of the campaign. He believed, no doubt, that his air superiority would be sufficient to prevent a large-scale evacuation by sea. He therefore, according to Halder, sent a message through Brauchitsch ordering "the armoured formations to be stopped, the points even taken back." Thus, says Halder, the way to Dunkirk was cleared for the British Army. At any rate we intercepted a German message sent in clear at 11.42 A.M. on May 24, to the effect that the attack on the Dunkirk line was to be discontinued for the present. Halder states that he refused, on behalf of Supreme Army Headquarters (O.K.H.), to interfere in the movement of Army Group Rundstedt, which had clear orders to prevent the enemy from reaching the coast. The quicker and more complete the success here, he argued, the easier it would be later to repair the loss of some tanks.

The controversy finished with a definite order by Hitler, to which he added that he would ensure execution of his order by sending personal liaison officers to the front. "I have never been able," said General Halder, "to figure how Hitler conceived the idea of the useless endangering of the armoured formations. It is most likely that Keitel, who was for a considerable time in Flanders in the First World War, had originated these ideas by his tales."

Other German generals have told much the same story, and have even suggested that Hitler's order was inspired by a political motive, to improve the chances of peace with England after France was beaten. Authentic documentary evidence has now come to light in the shape of the actual diary of Rundstedt's headquarters *written at the time*. This tells a different tale. At midnight on the 23rd orders came from Brauchitsch at O.K.H., confirming that the Fourth Army was to remain under Rundstedt for "the last act" of "the encirclement battle." Next morning Hitler visited Rundstedt, who represented to him that his armour, which had come so far and so fast, was much reduced in strength and needed a pause wherein to reorganise and regain its balance for the final blow against an enemy who, his staff diary says, was "fighting with extraordinary tenacity." Moreover, Rundstedt foresaw the possibility of attacks on his widely dispersed forces from north and south; in fact, the Weygand plan, which, if it had been feasible, was the obvious Allied counterstroke. Hitler "agreed entirely." He also dwelt on the paramount necessity of conserving the armoured forces for further operations. However, very early on the 25th a fresh directive was sent from Brauchitsch as the Commander-in-Chief ordering the continuation of the advance by the armour. Rundstedt, fortified by Hitler's verbal agreement, would have none of it. He did not pass on the order to the Fourth Army Commander, Kluge, who was told to continue to husband the Panzer divisions. Kluge protested at the delay, but it was not till next day, the 26th, that Rundstedt released them, although even then he enjoined that Dunkirk was not yet itself to be directly assaulted. The diary records that the

Fourth Army protested at this restriction, and its Chief of Staff telephoned on the 27th: "The picture in the Channel ports is as follows.
Big ships come up the quayside, boards are put down, and the men
crowd on the ships. All material is left behind. But we are not keen
on finding these men, newly equipped, up against us later."

It is therefore certain that the armour was halted; that this was done
on the initiative not of Hitler but of Rundstedt. Rundstedt no doubt
had reasons for his view both in the condition of the armour and in the
general battle, but he ought to have obeyed the formal orders of the
Army Command, or at least told them what Hitler had said in conversation. There is general agreement among the German commanders
that a great opportunity was lost.

*   *   *

There was however a separate cause which affected the movements
of the German armour at the decisive point.

After reaching the sea beyond Abbeville on the night of the 20th,
the leading German armoured and motorised columns had moved
northward along the coast towards Boulogne, Calais, and Dunkirk, with
the evident intention of cutting off all escape by sea. This region was
lighted in my mind from the previous war, when I had maintained
the mobile Marine Brigade operating from Dunkirk against the flanks
and rear of the German armies marching on Paris. I did not therefore have to learn about the inundation system between Calais and
Dunkirk, or the significance of the Gravelines waterline. The sluices
had already been opened, and with every day the floods were spreading, thus giving southerly protection to our line of retreat. The defence of Boulogne, but still more of Calais, to the latest hour, stood
forth upon the confused scene, and garrisons were immediately sent
there from England. Boulogne, isolated and attacked on May 22, was
defended by two battalions of the Guards and one of our few anti-
tank batteries, with some French troops. After thirty-six hours' resistance it was reported to be untenable, and I consented to the remainder
of the garrison, including the French, being taken off by sea. The
Guards were embarked by eight destroyers on the night of May 23–24,
with a loss of only two hundred men. The French continued to fight in
the Citadel until the morning of the 25th. I regretted our evacuation.

Some days earlier I had placed the conduct of the defence of the
Channel ports directly under the Chief of the Imperial General Staff,
with whom I was in constant touch. I now resolved that Calais should
be fought to the death, and that no evacuation by sea could be allowed
to the garrison, which consisted of one battalion of the Rifle Brigade,
one of the 60th Rifles, the Queen Victoria Rifles, the 229th Anti-tank

Battery, R.A., and a battalion of the Royal Tank Regiment, with twenty-one light and twenty-seven cruiser tanks, and an equal number of Frenchmen. It was painful thus to sacrifice these splendid, trained troops, of which we had so few, for the doubtful advantage of gaining two or perhaps three days, and the unknown uses that could be made of these days. The Secretary of State for War and the C.I.G.S. agreed to this hard measure.

The final decision not to relieve the garrison was taken on the evening of May 26. Till then the destroyers were held ready. Eden and Ironside were with me at the Admiralty. We three came out from dinner and at 9 P.M. did the deed. It involved Eden's own regiment, in which he had long served and fought in the previous struggle. One has to eat and drink in war, but I could not help feeling physically sick as we afterwards sat silent at the table.

Calais was the crux. Many other causes might have prevented the deliverance of Dunkirk, but it is certain that the three days gained by the defence of Calais enabled the Gravelines waterline to be held, and that without this, even in spite of Hitler's vacillations and Rundstedt's orders, all would have been cut off and lost.

*   *   *

Upon all this there now descended a simplifying catastrophe. The Germans, who had hitherto not pressed the Belgian front severely, on May 24 broke the Belgian line on either side of Courtrai, which is but thirty miles from Ostend and Dunkirk. The King of the Belgians soon considered the situation hopeless, and prepared himself for capitulation.

In the evening of the 25th Lord Gort took a vital decision. His orders still were to pursue the Weygand plan of a southerly attack towards Cambrai, in which the 5th and 50th Divisions, in conjunction with the French, were to be employed. The promised French attack northward from the Somme showed no sign of reality. The last defenders of Boulogne had been evacuated. Calais still held out. Gort now abandoned the Weygand plan. There was in his view no longer hope of a march to the south and to the Somme. Moreover, at the same time the crumbling of the Belgian defence and the gap opening to the north created a new peril, dominating in itself. Confident in his military virtue, and convinced of the complete breakdown of all control, either by the British and French Governments or by the French Supreme Command, Gort resolved to abandon the attack to the southward, to plug the gap which a Belgian capitulation was about to open in the north, and to march to the sea. At this moment here was the only hope of saving anything from destruction or surrender. At 6 P.M. he ordered the 5th and 50th Divisions to join the Second British Corps to fill the impending Belgian gap. He informed General Blanchard, who had succeeded Billotte in command of the First Army Group, of his action; and this officer, acknowledging the force of events, gave orders at 11.30 P.M. for a withdrawal on the 26th to a line behind the Lys canal west of Lille, with a view of forming a bridgehead around Dunkirk.

Early on May 26 Gort and Blanchard drew up their plan for withdrawal to the coast. As the First French Army had farther to go, the first movements of the B.E.F. on the night of the 26th–27th were to be preparatory, and rearguards of the British First and Second Corps remained on the frontier defences till the night of the 27th–28th. In all this Lord Gort had acted upon his own responsibility. But by now we at home, with a somewhat different angle of information, had already reached the same conclusions. On the 26th a telegram from the War Office approved his conduct, and authorised him "to operate towards the coast forthwith in conjunction with the French and Belgian armies." The emergency gathering on a vast scale of naval vessels of all kinds and sizes was already in full swing.

Meanwhile the organisation of the bridgeheads around Dunkirk was proceeding. The French were to hold from Gravelines to Bergues, and the British thence along the canal by Furnes to Nieuport and the sea. The various groups and parties of all arms which were arriving from both directions were woven into this line. Confirming the orders of the 26th, Lord Gort received from the War Office a telegram, dispatched at 1 P.M. on the 27th, telling him that his task henceforward was "to evacuate the maximum force possible." I had informed M. Reynaud the day before that the policy was to evacuate the British Expeditionary Force, and had requested him to issue corresponding orders. Such was the breakdown in communications that at 2 P.M. on the 27th the commander of the First French Army issued an order to his corps: *"La bataille sera livrée sans esprit de recul sur la position de la Lys."*

Four British divisions and the whole of the First French Army were now in dire peril of being cut off around Lille. The two arms of the German encircling movement strove to close the pincers upon them. This however was one of those rare but decisive moments when mechanical transport exercises its rights. When Gort gave the order all these four divisions came back with surprising rapidity almost in a night. Meanwhile, by fierce battles on either side of the corridor, the rest of the British Army kept the path open to the sea. The pincer-claws, which were delayed by the 2nd Division, and checked for three days by the 5th Division, eventually met on the night of May 29 in a manner similar to the great Russian operation around Stalingrad in 1942. The trap had taken two and a half days to close, and in that time

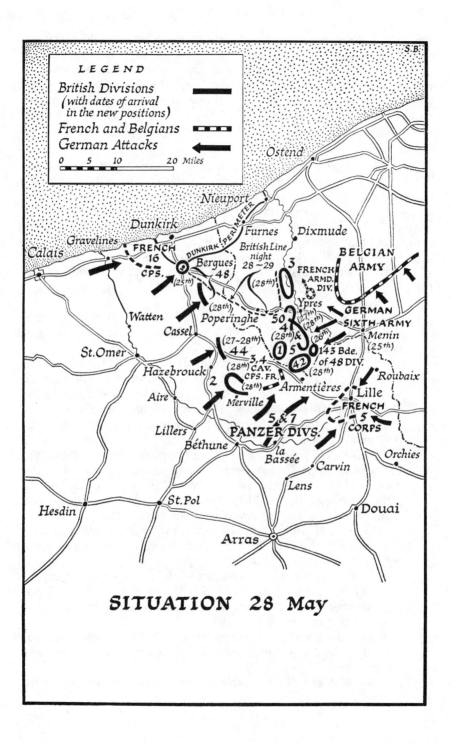

LEGEND

British Divisions
(with dates of arrival
in the new positions)

French and Belgians

German Attacks

0   5   10   20 Miles

Ostend

Nieuport

Dunkirk

Gravelines

Calais

FRENCH
16
CPS.

Bergues
48

(25th)

Furnes

British Line
night
28-29

Dixmude

BELGIAN
ARMY

3

FRENCH
ARMD.
DIV.

GERMAN
SIXTH ARMY

Menin
(25th)

Ypres

(27th)
(28th)

(28th)

Watten

(28th)

Poperinghe

50
4
(28th) &

(26th)

Cassel

(27-28th)
44

1 5

143 Bde.
of 48 DIV.
(28th)

St. Omer

(28th)
3, 4
CAV.
CPS. FR.
(28th)

4 2

Roubaix

Hazebrouck

2

Armentières

Lille

FRENCH
5
CORPS

Aire

Merville

5 & 7

Lillers

PANZER DIVS.

Orchies

Béthune

la
Bassée

Carvin

Hesdin

St. Pol

Lens

Douai

Arras

SITUATION 28 May

the British divisions and a great part of the First French Army, except the Fifth Corps, which was lost, withdrew in good order through the gap, in spite of the French having only horse transport, and the main road to Dunkirk being already cut and the secondary roads filled with retiring troops, long trains of transport, and many thousands of refugees.

*   *   *

The question about our ability to go on alone, which I had asked Mr. Chamberlain to examine with other Ministers ten days before, was now put formally by me to our military advisers. I drafted the reference purposely in terms which, while giving a lead, left freedom to the Chiefs of Staff to express their view, whatever it might be. I knew beforehand that they were absolutely determined; but it is wise to have written records of such decisions. I wished moreover to be able to assure Parliament that our resolve was backed by professional opinion. Here it is, with the answer:

1. We have reviewed our report on "British Strategy in a Certain Eventuality" in the light of the following terms of reference remitted to us by the Prime Minister.

"In the event of France being unable to continue in the war and becoming neutral, with the Germans holding their present position and the Belgian Army being forced to capitulate after assisting the British Expeditionary Force to reach the coast; in the event of terms being offered to Britain which would place her entirely at the mercy of Germany through disarmament, cession of naval bases in the Orkneys, etc.; what are the prospects of our continuing the war alone against Germany and probably Italy? Can the Navy and the Air Force hold out reasonable hopes of preventing serious invasion, and could the forces gathered in this Island cope with raids from the air involving detachments not greater than 10,000 men; it being observed that a prolongation of British resistance might be very dangerous for Germany, engaged in holding down the greater part of Europe?"

2. Our conclusions are contained in the following paragraphs:

3. While our Air Force is in being, our Navy and Air Force together should be able to prevent Germany carrying out a serious seaborne invasion of this country.

4. Supposing Germany gained complete air superiority, we consider that the Navy could hold up an invasion for a time, but not for an indefinite period.

5. If, with our Navy unable to prevent it, and our Air Force

gone, Germany attempted an invasion, our coast and beach defences could not prevent German tanks and infantry getting a firm footing on our shores. In the circumstances envisaged above our land forces would be insufficient to deal with a serious invasion.

6. The crux of the matter is air superiority. Once Germany had attained this she might attempt to subjugate this country by air attack alone.

7. Germany could not gain complete air superiority unless she could knock out our Air Force, and the aircraft industries, some vital portions of which are concentrated at Coventry and Birmingham.

8. Air attacks on the aircraft factories would be made by day or by night. We consider that we should be able to inflict such casualties on the enemy by day as to prevent serious damage. Whatever we do however by way of defensive measures — and we are pressing on with these with all dispatch — we cannot be sure of protecting the large industrial centres, upon which our aircraft industries depend, from serious material damage by night attack. The enemy would not have to employ precision bombing to achieve this effect.

9. Whether the attacks succeed in eliminating the aircraft industry depends not only on the material damage by bombs, but on the moral effect on the workpeople and their determination to carry on in the face of wholesale havoc and destruction.

10. If therefore the enemy presses home night attacks on our aircraft industry, he is likely to achieve such material and moral damage within the industrial area concerned as to bring all work to a standstill.

11. It must be remembered that numerically the Germans have a superiority of four to one. Moreover, the Germany aircraft factories are well dispersed and relatively inaccessible.

12. On the other hand, so long as we have a counteroffensive bomber force we can carry out similar attacks on German industrial centres and by moral and material effect bring a proportion of them to a standstill.

13. To sum up, our conclusion is that *prima facie* Germany has most of the cards; but the real test is whether the morale of our fighting personnel and civil population will counterbalance the numerical and material advantages which Germany enjoys. We believe it will.

This report, which of course was written at the darkest moment before the Dunkirk Deliverance, was signed not only by the three Chiefs of Staff, Newall, Pound, and Ironside, but by the three Vice-Chiefs, Dill, Phillips, and Peirse. Reading it in after years, I must admit it was grave and grim. But the War Cabinet and the few other Ministers who saw it were all of one mind. There was no discussion. Heart and soul we were together.

At home I issued the following general injunction:

(Strictly Confidential)   28.v.40

In these dark days the Prime Minister would be grateful if all his colleagues in the Government, as well as important officials, would maintain a high morale in their circles; not minimising the gravity of events, but showing confidence in our ability and inflexible resolve to continue the war till we have broken the will of the enemy to bring all Europe under his domination.

No tolerance should be given to the idea that France will make a separate peace; but whatever may happen on the Continent, we cannot doubt our duty, and we shall certainly use all our power to defend the Island, the Empire, and our Cause.

*   *   *

In the early hours of the 28th the Belgian Army surrendered. Lord Gort received the formal intimation of this only one hour before the event, but the collapse had been foreseen three days earlier, and in one fashion or another the gap was plugged. All that day the escape of the British Army hung in the balance. On the front from Comines to Ypres and thence to the sea, facing east and attempting to fill the Belgian gap, General Brooke and his Second Corps fought a magnificent battle, but as the Belgians withdrew northward, and then capitulated, the gap widened beyond repair. The German thrust between the British and Belgian Armies was not to be prevented, but its fatal consequence, an inward turn across the Yser, which would have brought the enemy on to the beaches behind our fighting troops, was foreseen and everywhere forestalled.

The Germans sustained a bloody repulse. All the time, only about four miles behind Brooke's struggling front, vast masses of transport and troops poured back into the developing bridgehead of Dunkirk, and were fitted with skillful improvisation into its defences. By the 29th a large part of the B.E.F had arrived within the perimeter, and by this time the naval measures for evacuation were beginning to attain their full effect. On May 30 G.H.Q. reported that all British divisions, or the remains of them, had come in.

More than half the First French Army found their way to Dunkirk where the great majority were safely embarked. But the line of retreat of at least five divisions was cut by the German pincers movement west of Lille. The French in Lille fought on gradually contracting fronts against increasing pressure, until on the evening of the 31st, short of food and with their ammunition exhausted, they were forced to surrender. About fifty thousand men thus fell into German hands. These Frenchmen, under the gallant leadership of General Molinié, had for four critical days contained no less than seven German divisions which otherwise could have joined in the assaults on the Dunkirk perimeter. This was a splendid contribution to the escape of their more fortunate comrades and of the B.E.F.

*   *   *

It was a severe experience for me, bearing so heavy an overall responsibility, to watch during these days in flickering glimpses this drama in which control was impossible, and intervention more likely to do harm than good. There is no doubt that by pressing in all loyalty the Weygand plan of retirement to the Somme as long as we did, our dangers, already so grave, were increased. But Gort's decision, in which we speedily concurred, to abandon the Weygand plan and march to the sea was executed by him and his staff with masterly skill, and will ever be regarded as a brilliant episode in British military annals.

4

The Deliverance of Dunkirk

IT WAS TUESDAY, MAY 28, and I did not again attend the House until that day week. There was no advantage to be gained by a further statement in the interval, nor did Members express a wish for one. But everyone realized that the fate of our Army and perhaps much else might well be decided before the week was out. "The House," I said, "should prepare itself for hard and heavy tidings. I have only to add that nothing which may happen in this battle can in any way relieve us of our duty to defend the world cause to which we have vowed ourselves; nor should it destroy our confidence in our power to make our way, as on former occasions in our history, through disaster and through grief to the ultimate defeat of our enemies." I had not seen many of my colleagues outside the War Cabinet, except individually, since the formation of the Government, and I thought it right to have a meeting in my room at the House of Commons of all Ministers of Cabinet rank other than the War Cabinet Members. We were perhaps twenty-five round the table. I described the course of events, and I showed them plainly where we were, and all that was in the balance. Then I said quite casually, and not treating it as a point of special significance: "Of course, whatever happens at Dunkirk, we shall fight on."

There occurred a demonstration which, considering the character of the gathering — twenty-five experienced politicians and Parliament men, who represented all the different points of view, whether right or wrong, before the war — surprised me. Quite a number seemed to jump up from the table and come running to my chair, shouting and patting me on the back. There is no doubt that had I at this juncture faltered at all in the leading of the nation I should have been hurled out of office. I was sure that every Minister was ready to be killed quite soon, and have all his family and possessions destroyed, rather than give in. In this they represented the House of Commons and almost all the people. It fell to me in these coming days and months to express their sentiments on suitable occasions. This I was able to do because they were mine also. There was a white glow, overpowering, sublime, which ran through our island from end to end.

* * *

Accurate and excellent accounts have been written of the evacuation of the British and French armies from Dunkirk. Ever since the 20th the gathering of shipping and small craft had been proceeding under the control of Admiral Ramsay, who commanded at Dover. On the evening of the 26th an Admiralty signal put "Operation Dynamo" into play, and the first troops were brought home that night. After the loss of Boulogne and Calais only the remains of the port of Dunkirk and the open beaches next to the Belgian frontier were in our hands. At this time it was thought that the most we could rescue was about forty-five thousand men in two days. Early the next morning, May 27, emergency measures were taken to find additional small craft "for a special requirement." This was no less than the full evacuation of the British Expeditionary Force. It was plain that large numbers of such craft would be required for work on the beaches, in addition to bigger ships which could load in Dunkirk harbour. On the suggestion of Mr. H. C. Riggs, of the Ministry of Shipping, the various boatyards, from Teddington to Brightlingsea, were searched by Admiralty officers, and yielded upwards of forty serviceable motor-boats or launches, which were assembled at Sheerness on the following day. At the same time lifeboats from liners in the London docks, tugs from the Thames, yachts, fishing craft, lighters, barges, and pleasure boats — anything that could be of use along the beaches — were called into service. By the night of the 27th a great tide of small vessels began to flow towards the sea, first to our Channel ports, and thence to the beaches of Dunkirk and the beloved Army.

Once the need for secrecy was relaxed the Admiralty did not hesitate to give full rein to the spontaneous movement which swept the seafaring population of our south and southeastern shores. Everyone who had a boat of any kind, steam or sail, put out for Dunkirk, and the preparations, fortunately begun a week earlier, were now aided by the brilliant improvisation of volunteers on an amazing scale. The numbers arriving on the 29th were small, but they were the forerunners of nearly four hundred small craft which from the 31st were destined to play a vital part in ferrying from the beaches to the off-lying ships almost a hundred thousand men. In these days I missed the head of my Admiralty map room, Captain Pim, and one or two other familiar faces. They had got hold of a Dutch *schuit,* which in four days brought off eight hundred soldiers. Altogether there came to the rescue of the Army under the ceaseless air bombardment of the enemy about eight hundred and sixty vessels, of which nearly seven hundred were British and the rest Allied.

*   *   *

Meanwhile ashore around Dunkirk the occupation of the perimeter was effected with precision. The troops arrived out of chaos and were formed in order along the defences, which even in two days had grown. Those men who were in best shape turned about to form the line. Divisions like the 2nd and 5th, which had suffered most, were held in reserve on the beaches and were then embarked early. In the first instance there were to be three corps on the front, but by the 29th, with the French taking a greater share in the defences, two sufficed. The enemy had closely followed the withdrawal, and hard fighting was incessant, especially on the flanks near Nieuport and Bergues. As the evacuation went on the steady decrease in the number of troops, both British and French, was accompanied by a corresponding contraction of the defence. On the beaches, among the sand dunes, for three, four, or five days scores of thousands of men dwelt under unrelenting air attack. Hitler's belief that the German Air Force would render escape impossible, and that therefore he should keep his armoured formations for the final stroke of the campaign, was a mistaken but not unreasonable one.

Three factors falsified his expectations. First, the incessant air bombing of the masses of troops along the seashore did them very little harm. The bombs plunged into the soft sand, which muffled their explosions. In the early stages, after a crashing air raid, the troops were astonished to find that hardly anybody had been killed or wounded. Everywhere there had been explosions, but scarcely anyone was the worse. A rocky shore would have produced far more deadly results. Presently the soldiers regarded the air attacks with contempt. They crouched in the sand dunes with composure and growing hope. Before them lay the grey but not unfriendly sea. Beyond, the rescuing ships and — Home.

The second factor which Hitler had not foreseen was the slaughter of his airmen. British and German air quality was put directly to the test. By intense effort Fighter Command maintained successive patrols over the scene, and fought the enemy at long odds. Hour after hour they bit into the German fighter and bomber squadrons, taking a heavy toll, scattering them and driving them away. Day after day this went on, till the glorious victory of the Royal Air Force was gained. Wherever German aircraft were encountered, sometimes in forties and fifties, they were instantly attacked, often by single squadrons or less, and shot down in scores, which presently added up into hundreds. The whole Metropolitan Air Force, our last sacred reserve, was used. Sometimes the fighter pilots made four sorties a day. A clear result was obtained. The superior enemy were beaten or killed, and for all their bravery mastered, or even cowed. This was a decisive clash. Unhappily, the troops on the beaches saw very little of this epic conflict in the air, often miles away or above the clouds. They knew nothing of the loss inflicted on the enemy. All they felt was the bombs scourging the beaches, cast by the foes who had got through, but did not perhaps return. There was even a bitter anger in the Army against the Air Force, and some of the troops landing at Dover or at Thames ports in their ignorance insulted men in Air Force uniform. They should have clasped their hands; but how could they know? In Parliament I took pains to spread the truth.

But all the aid of the sand and all the prowess in the air would have been vain without the sea. The instructions given ten or twelve days before had, under the pressure and emotion of events, borne amazing fruit. Perfect discipline prevailed ashore and afloat. The sea was calm. To and fro between the shore and the ships plied the little boats, gathering the men from the beaches as they waded out or picking them from the water, with total indifference to the air bombardment, which often claimed its victims. Their numbers alone defied air attack. The Mosquito Armada as a whole was unsinkable. In the midst of our defeat glory came to the island people, united and unconquerable; and the tale of the Dunkirk beaches will shine in whatever records are preserved of our affairs.

Notwithstanding the valiant work of the small craft it must not be forgotten that the heaviest burden fell on the ships plying from Dunkirk harbour, where two-thirds of the men were embarked. The destroyers played the predominant part, as the casualty lists show. Nor must the great part played by the personnel ships with their mercantile crews be overlooked.

*　　　*　　　*

The progress of the evacuation was watched with anxious eyes and growing hope. On the evening of the 27th Lord Gort's position appeared critical to the naval authorities, and Captain Tennant, R.N., from the Admiralty, who had assumed the duties of Senior Naval Officer at Dunkirk, signalled for all available craft to be sent to the beaches immediately, as "evacuation tomorrow night is problematical." The picture presented was grim, even desperate. Extreme efforts were made to meet the call, and a cruiser, eight destroyers, and twenty-six other vessels were sent. The 28th was a day of tension, which gradually eased as the position on land was stabilised with the powerful help of the Royal Air Force. The naval plans were carried through despite severe losses on the 29th, when three destroyers and twenty-one other vessels were sunk and many others damaged.

On the 30th I held a meeting of the three Service Ministers and the Chiefs of Staff in the Admiralty War Room. We considered the events of the day on the Belgian coast. The total number of troops brought off had risen to 120,000, including only 6000 French; 860 vessels of all kinds were at work. A message from Admiral Wake-Walker at Dunkirk said that, in spite of intense bombardment and air attack, 4000 men had been embarked in the previous hour. He also thought that Dunkirk itself would probably be untenable by the next day. I emphasised the urgent need of getting off more French troops. To fail to do so might do irreparable harm to the relations between ourselves and our ally. I also said that when the British strength was reduced to that of a corps we ought to tell Lord Gort to embark and return to England, leaving a corps commander in charge. The British Army would have to stick it out as long as possible so that the evacuation of the French could continue.

Knowing well the character of Lord Gort, I wrote out in my own hand the following order to him, which was sent officially by the War Office at 2 P.M. on the 30th:

> Continue to defend the present perimeter to the utmost in order to cover maximum evacuation now proceeding well. Report every three hours through La Panne. If we can still communicate we shall send you an order to return to England with such officers as you may choose at the moment when we deem your command so reduced that it can be handed over to a corps commander. You should now nominate this commander. If communications are broken you are to hand over and return as specified when your effective fighting force does not exceed the equivalent of three divisions. This is in accordance with correct military procedure, and no personal discretion is left you in the matter. On political grounds it would be a needless triumph to the enemy to capture you when only a small force remained under your orders. The corps commander chosen by you should be ordered to carry on the defence in conjunction with the French, and evacuation whether from Dunkirk or the beaches, but when in his judgment no further organized evacuation is possible and no further proportionate damage can be inflicted on the enemy he is authorized in consultation with the senior French commander to capitulate formally to avoid useless slaughter.

It is possible that this last message influenced other great events and the fortunes of another valiant commander. When I was at the White House at the end of December 1941 I learned from the President and Mr. Stimson of the approaching fate of General MacArthur and the American garrison at Corregidor. I thought it right to show them the way in which we had dealt with the position of a Commander-in-Chief whose force was reduced to a small fraction of his original command. The President and Mr. Stimson both read the telegram with profound attention, and I was struck by the impression it seemed to make upon them. A little later in the day Mr. Stimson came back and asked for a copy of it, which I immediately gave him. It may be (for I do not know) that this influenced them in the right decision which they took in ordering General MacArthur to hand over his command to one of his subordinate generals, and thus saved for all his future glorious services the great Commander who would otherwise have perished or passed the war as a Japanese captive. I should like to think this was true.

On this same 30th day of May, 1940, members of Lord Gort's staff in conference with Admiral Ramsay at Dover informed him that daylight on June 1 was the latest time up to which the eastern perimeter might be expected to hold. Evacuation was therefore pressed on with the utmost urgency to ensure, so far as possible, that a British rearguard of no more than about four thousand men would then remain ashore. Later it was found that this number would be insufficient to defend the final covering positions, and it was decided to hold the British sector until midnight June 1–2, evacuation proceeding meanwhile on the basis of full equality between French and British forces.

Such was the situation when on the evening of the 31st Lord Gort in accordance with his orders handed over his command to Major General Alexander and returned to England.

*   *   *

To avoid misunderstandings by keeping personal contact it was necessary for me to fly to Paris on May 31 for a meeting of the Supreme War Council. With me in the plane came Mr. Attlee and Generals Dill and Ismay. I also took General Spears, who had flown over on the 30th with the latest news from Paris. This brilliant officer and Member of Parliament was a friend of mine from the First Great War. Liaison officer between the left of the French and the right of the British Armies, he had taken me round the Vimy Ridge in 1916. Speaking French with a perfect accent and bearing five wound stripes on his sleeve, he was a personality at this moment fitted to our anxious relations. When Frenchmen and Englishmen are in trouble together and arguments break out, the Frenchman is often voluble and vehement, and the Englishman unresponsive or even rude. But Spears could say things to the high French personnel with an ease and force which I have never seen equalled.

This time we did not go to the Quai d'Orsay, but to M. Reynaud's room at the War Office in the Rue Saint-Dominique. Attlee and I found Reynaud and Marshal Pétain opposite to us as the only French Ministers. This was the first appearance of Pétain, now Vice-President of the Council, at any of our meetings. He wore plain clothes. Our Ambassador, Dill, Ismay, and Spears were with us, and Weygand and Darlan, Captain de Margerie, head of Reynaud's private office, and M. Baudouin, Secretary of the French War Cabinet, represented the French.

The French seemed to have no more idea of what was happening to the Northern Armies than we had about the main French front. When I told them that 165,000 men, of whom 15,000 were French, had been taken off they were astonished. They naturally drew attention to the marked British preponderance. I explained that this was due largely to the fact that there had been many British administrative units in the back area who had been able to embark before fighting troops could be spared from the front. Moreover, the French up to the present had had no orders to evacuate. One of the chief reasons why I had come to Paris was to make sure that the same orders were given to the French troops as to the British. His Majesty's Government had felt it necessary in the dire circumstances to order Lord Gort to take off fighting men and leave the wounded behind. If present hopes were confirmed, 200,000 able-bodied troops might be got away. This would be almost a miracle. Four days ago I would not have wagered on more than 50,000 as a maximum. I dwelt upon our terrible losses in equipment. Reynaud paid a handsome tribute to the work of the British Navy and Air Force, for which I thanked him. We then spoke at some length upon what could be done to rebuild the British forces in France.

Meanwhile Admiral Darlan had drafted a telegram to Admiral Abrial at Dunkirk:

> (1) A bridgehead shall be held round Dunkirk with the divisions under your command and those under British command.
> (2) As soon as you are convinced that no troops outside the bridgehead can make their way to the points of embarkation the troops holding the bridgehead shall withdraw and embark, *the British forces embarking first.*

I intervened at once to say that the British would not embark first, but that the evacuation should proceed on equal terms between the British and the French — *"Bras dessus bras dessous."* The British would form the rearguard. This was agreed.

The conversation next turned to Italy. I expressed the British view that if Italy came in we should strike at her at once in the most effective manner. Many Italians were opposed to war, and all should be made to realise its severity. I proposed that we should strike by air-bombing at the northwestern industrial triangle enclosed by the three cities of Milan, Turin, and Genoa. Reynaud agreed that the Allies must strike at once; and Admiral Darlan said he had a plan ready for the naval and aerial bombardment of Italy's oil supplies, largely stored along the coast between the frontier and Naples. The necessary technical discussions were arranged.

After some talk about the importance of keeping Spain out of the war, I spoke on the general outlook. The Allies, I said, must maintain an unflinching front against all their enemies. The United States had been roused by recent events, and, even if they did not enter the war, would soon be prepared to give us powerful aid. An invasion of England, if it took place, would have a still more profound effect on the United States. England did not fear invasion, and would resist it most fiercely in every village and hamlet. It was only after her essential need of troops had been met that the balance of her armed forces could be put at the disposal of her French ally. I was absolutely convinced we had only to carry on the fight to conquer. Even if one of us should be struck down, the other must not abandon the struggle. The British Government were prepared to wage war from the New World, if through some disaster England herself were laid waste. If Germany defeated either ally or both, she would give no mercy; we should be reduced to the status of vassals and slaves forever. It would be better far that the civilisation of Western Europe with all its achievements should come to a tragic but splendid end than that the two great democracies should linger on, stripped of all that made life worth living.

Mr. Attlee then said that he entirely agreed with my view. "The British people now realize the danger with which they are faced, and know that in the event of a German victory everything they have built up will be destroyed. The Germans kill not only men but ideas. Our people are resolved as never before in their history." Reynaud thanked us for what we had said. He was sure that the morale of the German people was not up to the level of the momentary triumph of their Army. If France could hold the Somme with the help of Britain and if American industry came in to make good the disparity in arms, then we could be sure of victory. He was most grateful, he said, for my renewed assurance that if one country went under the other would not abandon the struggle.

The formal meeting then ended.

After we rose from the table some of the principals talked together in the bay window in a somewhat different atmosphere. Chief among these was Marshal Pétain. Spears was with me, helping me out with my French and speaking himself. The young Frenchman, Captain de Margerie, had already spoken about fighting it out in Africa. But Marshal Pétain's attitude, detached and sombre, gave me the feeling that he would face a separate peace. The influence of his personality, his reputation, his serene acceptance of the march of adverse events, apart from any words he used, was almost overpowering to those under his spell. One of the Frenchmen, I cannot remember who, said in their polished way that a continuance of military reverses might in certain eventualities enforce a modification of foreign policy upon France. Here Spears rose to the occasion, and, addressing himself particularly to Marshal Pétain, said in perfect French: "I suppose you understand, M. le Maréchal, that that would mean blockade?" Someone else said: "That would perhaps be inevitable." But then Spears to Pétain's face: "That would not only mean blockade but *bombardment* of all French ports in German hands." I was glad to have this said. I sang my usual song: we would fight on whatever happened or whoever fell out. Again we had a night of petty raids, and in the morning I departed.

*   *   *

May 31 and June 1 saw the climax though not the end at Dunkirk. On these two days over 132,000 men were safely landed in England, nearly one-third of them having been brought from the beaches in small craft under fierce air attack and shell fire. On June 1 from early dawn onward the enemy bombers made their greatest efforts, often timed when our own fighters had withdrawn to refuel. These attacks took heavy toll of the crowded shipping, which suffered almost as much as in all the previous week. On this single day our losses by air attack, by mines, E-boats, or other misadventures were thirty-one ships sunk and eleven damaged. On land the enemy increased their pressure on the bridgehead, doing their utmost to break through. They were held at bay by the desperate resistance of the Allied rearguards.

The final phase was carried through with much skill and precision. For the first time it became possible to plan ahead instead of being forced to rely on hourly improvisations. At dawn on June 2 about four thousand British with seven anti-aircraft guns and twelve anti-tank guns remained on the outskirts of Dunkirk with the still considerable French forces holding the contracting perimeter. Evacuation was now possible only in darkness, and Admiral Ramsay determined to make a massed descent on the harbour that night with all his available resources. Besides tugs and small craft, forty-four ships were sent that evening from England, including eleven destroyers and fourteen mine sweepers. Forty French and Belgian vessels also participated. Before midnight the British rearguard was embarked.

This was not however the end of the Dunkirk story. We had been prepared to carry considerably greater numbers of French that night than had offered themselves. The result was that when our ships, many of them still empty, had to withdraw at dawn, great numbers of French troops, many still in contact with the enemy, remained ashore. One more effort had to be made. Despite the exhaustion of ships' companies after so many days without rest or respite, the call was answered. On June 4 26,175 Frenchmen were landed in England, over 21,000 of them in British ships. Unfortunately several thousands remained, who continued the fight in the contracting bridgehead until the morning of the 4th, when the enemy was in the outskirts of the town and they had come to an end of their powers. They had fought gallantly for many days to cover the evacuation of their British and French comrades. They were to spend the next years in captivity. Let us remember that but for the endurance of the Dunkirk rearguard the re-creation of an army in Britain for home defence and final victory would have been gravely prejudiced.

Finally, at 2.23 P.M. on June 4 the Admiralty, in agreement with the French, announced that "Operation Dynamo" was now completed. More than 338,000 British and Allied troops had been landed in England.

*   *   *

Parliament assembled on June 4, and it was my duty to lay the story fully before them both in public and later in secret session. The narrative requires only a few extracts from my speech. It was imperative to explain not only to our own people but to the world that our resolve to fight on was based on serious grounds, and was no mere despairing effort. It was also right to lay bare my own reasons for confidence.

We must be very careful not to assign to this deliverance the attributes of a victory. Wars are not won by evacuations. But there was a victory inside this deliverance, which should be noted. It was gained by the Air Force. Many of our soldiers coming back have not seen the Air Force at work; they saw only the bombers which escaped its protective attack. They underrate its achievements. I have heard much talk of this; that is why I go out of my way to say this. I will tell you about it.

This was a great trial of strength between the British and German Air Forces. Can you conceive a greater objective for the Germans in the air than to make evacuation from these beaches impossible, and to sink all these ships which were displayed, almost to the extent of thousands? Could there have been an objective of greater military importance and significance for the whole purpose of the war than this? They tried hard, and they were beaten back; they were frustrated in their task. We got the Army away; and they have paid fourfold for any losses which they have inflicted. . . . All of our types and all our pilots have been vindicated as superior to what they have at present to face.

When we consider how much greater would be our advantage in defending the air above this Island against an overseas attack, I must say that I find in these facts a sure basis upon which practical and reassuring thoughts may rest. I will pay my tribute to these young airmen. The great French Army was very largely, for the time being, cast back and disturbed by the onrush of a few thousands of armoured vehicles. May it not also be that the cause of civilisation itself will be defended by the skill and devotion of a few thousand airmen?

We are told that Herr Hitler has a plan for invading the British Isles. This has often been thought of before. When Napoleon lay at Boulogne for a year with his flat-bottomed boats and his Grand Army he was told by someone: "There are bitter weeds in England." There are certainly a great many more of them since the British Expeditionary Force returned.

The whole question of Home Defence against invasion is, of course, powerfully affected by the fact that we have for the time being in this island incomparably stronger military forces than we have ever had at any moment in this war or the last. But this will not continue. We shall not be content with a defensive war. We have our duty to our Ally. We have to reconstitute and build up the British Expeditionary Force once again, under its gallant Commander-in-Chief, Lord Gort. All this is in train; but in the interval we must put our defences in this Island into such a high state of organisation that the fewest possible numbers will be required to give effective security and that the largest possible potential of offensive effort may be realised. On this we are now engaged.

I ended in a passage which was to prove, as will be seen, a timely and important factor in United States decisions.

Even though large tracts of Europe and many old and famous States have fallen or may fall into the grip of the Gestapo and all the odious apparatus of Nazi rule, we shall not flag or fail. We shall go on to the end. We shall fight in France, we shall fight in the seas and oceans, we shall fight with growing confidence and growing strength in the air; we shall defend our Island, whatever the cost may be. We shall fight on the beaches, we shall fight on the landing-grounds, we shall fight in the fields and in the streets, we shall fight in the hills; we shall never surrender; and even if, which I do not for a moment believe, this island or a large part of it were subjugated and starving, then our Empire beyond the seas, armed and guarded by the British Fleet, would carry on the struggle, until, in God's good time, the New World, with all its power and might, steps forth to the rescue and the liberation of the Old.

5

The Rush for the Spoils

THE FRIENDSHIP between the British and Italian peoples
sprang from the days of Garibaldi and Cavour. Every stage in the
liberation of Northern Italy from Austria and every step towards Italian
unity and independence had commanded the sympathies of Victorian
Liberalism. British influence had powerfully contributed to the Italian
accession to the Allied cause in the First World War. The rise of Mussolini and the establishment of Fascism as a counter to Bolshevism had
in its early phases divided British opinion on party lines, but had not affected the broad foundations of goodwill between the peoples. We have
seen that until Mussolini's designs against Abyssinia had raised grave
issues he had ranged himself with Great Britain in opposition to Hitlerism and German ambitions. I have told the sad tale of how the Baldwin-
Chamberlain policy about Abyssinia brought us the worst of both
worlds, how we estranged the Italian dictator without breaking his
power, and how the League of Nations was injured without Abyssinia
being saved. We have also seen the earnest but futile efforts made by
Mr. Chamberlain, Sir Samuel Hoare, and Lord Halifax to win back
during the period of appeasement Mussolini's lost favour. And finally
there was the growth of Mussolini's conviction that Britain's sun had
set, and that Italy's future could, with German help, be founded on the
ruins of the British Empire. This had been followed by the creation
of the Berlin-Rome Axis, in accordance with which Italy might well
have been expected to enter the war against Britain and France on its
very first day.

It was certainly only common prudence for Mussolini to see how
the war would go before committing himself and his country irrevocably. The process of waiting was by no means unprofitable. Italy
was courted by both sides, and gained much consideration for her

interests, many profitable contracts, and time to improve her armaments. Thus the twilight months had passed. It is an interesting speculation what the Italian fortunes would have been if this policy had
been maintained. The United States with its large Italian vote might
well have made it clear to Hitler that an attempt to rally Italy to his
side by force of arms would raise the gravest issues. Peace, prosperity,
and growing power would have been the prize of a persistent neutrality.
Once Hitler was embroiled with Russia this happy state might have
been almost indefinitely prolonged, with ever-growing benefits, and
Mussolini might have stood forth in the peace or in the closing year
of the war as the wisest statesman the sunny peninsula and its industrious and prolific people had known. This was a more agreeable
situation than that which in fact awaited him.

On the two occasions in 1927 when I met Mussolini our personal
relations had been intimate and easy. I would never have encouraged
Britain to make a breach with him about Abyssinia or roused the
League of Nations against him unless we were prepared to go to war
in the last extreme. He, like Hitler, understood and in a way respected
my campaign for British rearmament, though he was very glad British
public opinion did not support my view.

In the crisis we had now reached of the disastrous Battle of France
it was clearly my duty as Prime Minister to do my utmost to keep Italy
out of the conflict, and though I did not indulge in vain hopes I at
once used what resources and influence I might possess. Six days after
becoming head of the Government I wrote at the Cabinet's desire the
appeal to Mussolini which, together with his answer, was published
two years later in very different circumstances. It was dated 16 May,
1940.

Now that I have taken up my office as Prime Minister and
Minister of Defence I look back to our meetings in Rome and feel
a desire to speak words of goodwill to you as Chief of the Italian
nation across what seems to be a swift-widening gulf. Is it too late
to stop a river of blood from flowing between the British and
Italian peoples? We can no doubt inflict grievous injuries upon
one another and maul each other cruelly, and darken the Mediterranean with our strife. If you so decree, it must be so; but I declare that I have never been the enemy of Italian greatness, nor
ever at heart the foe of the Italian lawgiver. It is idle to predict
the course of the great battles now raging in Europe, but I am sure
that whatever may happen on the Continent, England will go
on to the end, even quite alone, as we have done before, and I believe with some assurance that we shall be aided in increasing

measure by the United States, and, indeed, by all the Americas.

I beg you to believe that it is in no spirit of weakness or of fear that I make this solemn appeal, which will remain on record. Down the ages above all other calls comes the cry that the joint heirs of Latin and Christian civilisation must not be ranged against one another in mortal strife. Hearken to it, I beseech you in all honour and respect, before the dread signal is given. It will never be given by us.

The response was hard. It had at least the merit of candour.

I reply to the message which you have sent me in order to tell you that you are certainly aware of grave reasons of an historical and contingent character which have ranged our two countries in opposite camps. Without going back very far in time I remind you of the initiative taken in 1935 by your Government to organise at Geneva sanctions against Italy, engaged in securing for herself a small space of the African sun without causing the slightest injury to your interests and territories or those of others. I remind you also of the real and actual state of servitude in which Italy finds herself in her own sea. If it was to honour your signature that your Government declared war on Germany, you will understand that the same sense of honour and of respect for engagements assumed in the Italian-German Treaty guides Italian policy today and tomorrow in the face of any event whatsoever.

From this moment we could have no doubt of Mussolini's intention to enter the war at his most favourable opportunity. His resolve had in fact been made as soon as the defeat of the French armies was obvious. On May 13 he had told Ciano that he would declare war on France and Britain within a month. His official decision to declare war on any date suitable after June 5 was imparted to the Italian Chiefs of Staff on May 29. At Hitler's request the date was postponed to June 10.

*   *   *

On May 26, while the fate of the Northern Armies hung in the balance and no one could be sure that any would escape, Reynaud flew over to England to have a talk with us about this topic which had not been absent from our minds. The Italian declaration of war must be expected at any moment. Thus France would burn upon another front, and a new foe would march hungrily upon her in the South. Could anything be done to buy off Mussolini? That was the question posed. I did not think there was the slightest chance, and every fact that the French

Premier used as an argument for trying only made me surer there was no hope. However, Reynaud was under strong pressure at home, and we on our side wished to give full consideration to our Ally, whose one vital weapon, her Army, was breaking in her hand. Although there was no need to marshal the grave facts, M. Reynaud dwelt not obscurely upon the possible French withdrawal from the war. He himself would fight on, but there was always the possibility that he might soon be replaced by others of a different temper.

We had already on May 25 at the instance of the French Government made a joint request to President Roosevelt to intervene. In this message Britain and France authorised him to state that we understood Italy had territorial grievances against us in the Mediterranean, that we were disposed to consider at once any reasonable claims, that the Allies would admit Italy to the Peace Conference with a status equal to that of any belligerent, and that we would invite the President to see that any agreement reached now was carried out. The President acted accordingly; but his addresses were repulsed by the Italian dictator in the most abrupt manner. At our meeting with Reynaud we had already this answer before us. The French Premier now suggested more precise proposals. Obviously, if these were to remedy Italy's "state of servitude in her own sea," they must affect the status both of Gibraltar and Suez. France was prepared to make similar concessions about Tunis.

We were not able to show any favour to these ideas. This was not because it was wrong to examine them or because it did not seem worth while at this moment to pay a heavy price to keep Italy out of the war. My own feeling was that at the pitch in which our affairs lay we had nothing to offer which Mussolini could not take for himself or be given by Hitler if we were defeated. One cannot easily make a bargain at the last gasp. Once we started negotiating for the friendly mediation of the Duce we should destroy our power of fighting on. I found my colleagues very stiff and tough. All our minds ran much more on bombing Milan and Turin the moment Mussolini declared war, and seeing how he liked that. Reynaud, who did not at heart disagree, seemed convinced, or at least content. This did not prevent the French Government from making a few days later a direct offer of their own to Italy of territorial concessions, which Mussolini treated with disdain. "He was not interested," said Ciano to the French Ambassador on June 3, "in recovering any French territories by peaceful negotiation. He had decided to make war on France." This was only what we had expected.

In spite of extreme efforts made by the United States, nothing could turn Mussolini from his course. On June 10 at 4.45 P.M. the Italian Minister for Foreign Affairs informed the British Ambassador that Italy would consider herself at war with the United Kingdom from midnight that day. A similar communication was made to the French Government. When Ciano delivered his note to the French Ambassador, M. François-Poncet remarked as he reached the door: "You too will find the Germans are hard masters." From his balcony in Rome Mussolini announced to well-organised crowds that Italy was at war with France and Britain. It was, as Ciano is said to have apologetically remarked later on, "a chance which comes only once in five thousand years." Such chances, though rare, are not necessarily good.

Forthwith the Italians attacked the French troops on the Alpine front and Great Britain reciprocally declared war on Italy. Five Italian ships detained at Gibraltar were seized, and orders were given to the Navy to intercept and bring into controlled ports all Italian vessels at sea. On the night of the 12th our bomber squadrons, after a long flight from England, which meant light loads, dropped their first bombs upon Turin and Milan. We looked forward however to a much heavier delivery as soon as we could use the French airfields at Marseilles.

The French could only muster three divisions, with fortress troops equivalent to three more, to meet invasion over the Alpine passes and along the Riviera coast by the western group of Italian armies. These comprised thirty-two divisions, under Prince Umberto. Moreover, strong German armour, rapidly descending the Rhone Valley, soon began to traverse the French rear. Nevertheless the Italians were still confronted, and even pinned down, at every point on the new front by the French Alpine units, even after Paris had fallen and Lyons was in German hands. When on June 18 Hitler and Mussolini met at Munich the Duce had little cause to boast. A new Italian offensive was launched on June 21. The French Alpine positions however proved impregnable, and the major Italian effort towards Nice was halted in the outskirts of Mentone. But although the French Army on the southeastern borders saved its honour, the German march to the south behind them made further fighting impossible, and the conclusion of the armistice with Germany was linked with a French request to Italy for the cessation of hostilities.

A speech from President Roosevelt had been announced for the night of the 10th. About midnight I listened to it with a group of officers in the Admiralty War Room, where I still worked. When he uttered the scathing words about Italy, "On this tenth day of June, 1940, the hand that held the dagger has struck it into the back of its neighbour," there was a deep growl of satisfaction. I wondered about the Italian vote in the approaching presidential election; but I knew that Roosevelt was a most experienced American party politician, although never afraid to run risks for the sake of his resolves. It was a magnificent speech, instinct with passion and carrying to us a message of hope. While the impression was strong upon me, and before going to bed, I expressed my gratitude.

* * *

The rush for the spoils had begun. But Mussolini was not the only hungry animal seeking prey. To join the Jackal came the Bear.

I have already recorded the course of Anglo-Soviet relations up till the outbreak of war, and the hostility, verging upon an actual breach with Britain and France, which arose during the Russian invasion of Finland. Germany and Russia now worked together as closely as their deep divergences of interest permitted. Hitler and Stalin had much in common as totalitarians, and their systems of government were akin. Molotov beamed on the German Ambassador, Count Schulenburg, on every important occasion, and was forward and fulsome in his approval of German policy and praise for Hitler's military measures. When the German assault had been made upon Norway he had said that the Soviet Government understood the measures which were forced upon Germany. The English had certainly gone much too far. They had disregarded completely the rights of neutral nations. *"We wish Germany complete success in her defensive measures."* Hitler had taken pains to inform Stalin on the morning of May 10 of the onslaught he had begun upon France and the neutral Low Countries. "I called on Molotov," wrote Schulenburg. "He appreciated the news, and added that he understood that Germany had to protect herself against Anglo-French attack. He had no doubt of our success."

Although these expressions of their opinion were of course unknown till after the war, we were under no illusions about the Russian attitude. We nonetheless pursued a patient policy of trying to re-establish relations of a confidential character with Russia, trusting to the march of events and to their fundamental antagonisms to Germany. It was thought wise to use the abilities of Sir Stafford Cripps as Ambassador to Moscow. He willingly accepted this bleak and unpromising task. We did not at that time realise sufficiently that Soviet Communists hate extreme left-wing politicians even more than they do Tories or Liberals. The nearer a man is to Communism in sentiment the more obnoxious he is to the Soviets unless he joins the Party. The Soviet Government agreed to receive Cripps as Ambassador, and explained this step to their Nazi confederates. "The Soviet Union," wrote Schulenburg to Berlin on May 29, "is interested in obtaining rubber and tin from England in exchange for lumber. There is no reason for apprehension concerning Cripps's mission, since there is no reason to doubt the loyal attitude of the Soviet Union towards us, and since the unchanged direction of Soviet policy towards England precludes damage to Germany or vital German interests. There are no indications of any kind here for belief that the latest German successes cause alarm or fear of Germany in the Soviet Government."

The collapse of France and the destruction of the French armies and of all counter-poise in the West ought to have produced some reaction in Stalin's mind, but nothing seemed to warn the Soviet leaders of the gravity of their own peril. On June 18, when the French defeat was total, Schulenburg reported: "Molotov summoned me this evening to his office and expressed the warmest congratulations of the Soviet Government on *the splendid success* of the German armed forces." This was almost exactly a year from the date when these same armed forces, taking the Soviet Government by complete surprise, fell upon Russia in cataracts of fire and steel. We now know that only four months later in 1940 Hitler definitely decided upon a war of extermination against the Soviets, and began the long, vast, stealthy movements of these much-congratulated German armies to the East. No recollection of their miscalculation and former conduct ever prevented the Soviet Government and its Communist agents and associates all over the world from screaming for a Second Front, in which Britain, whom they had consigned to ruin and servitude, was to play a leading part. However, we comprehended the future more truly than these cold-blooded calculators, and understood their dangers and their interests better than they did themselves.

On June 14, the day Paris fell, Moscow sent an ultimatum to Lithuania accusing her and the other Baltic States of military conspiracy against the U.S.S.R. and demanding radical changes of government and military concessions. On June 15, Red Army troops invaded the country. Latvia and Esthonia were exposed to the same treatment. Pro-Soviet Governments must be set up forthwith and Soviet garrisons admitted into these small countries. Resistance was out of the question. The President of Latvia was deported to Russia, and Mr. Vyshinsky arrived to nominate a Provisional Government to manage new elections. In Esthonia the pattern was identical. On June 19 Zhdanov arrived to install a similar regime. On August 3–6 the pretence of pro-Soviet friendly and democratic Governments was swept away, and the Kremlin annexed the Baltic States to the Soviet Union.

A Russian ultimatum to Rumania was delivered to the Rumanian Minister in Moscow at 10 P.M. on June 26. The cession of Bessarabia and the northern part of the province of Bukovina was demanded, and an immediate reply requested by the following day. Germany, though annoyed by this precipitate action of Russia, which threatened her economic interests in Rumania, was bound by the terms of the Ribbentrop-

Molotov pact of August 1939, which recognised the exclusive political interest of Russia in these areas of Southeast Europe. The German Government therefore counselled Rumania to yield. On June 27, Rumanian troops were withdrawn from the two provinces concerned, and the territories passed into Russian hands. The armed forces of the Soviet Union were now firmly planted on the shores of the Baltic and at the mouths of the Danube.

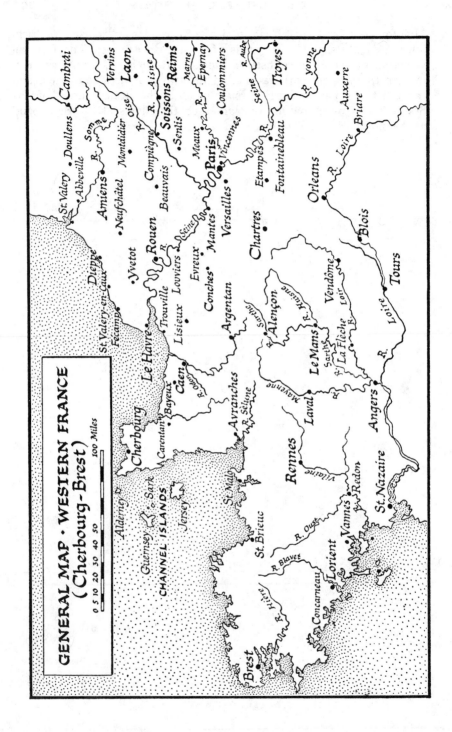

GENERAL MAP · WESTERN FRANCE
(Cherbourg - Brest)

0 5 10 20 30 40 50   100 Miles

6

Back to France
June 4 to June 12

W HEN IT WAS KNOWN how many had been rescued from Dunkirk, a sense of deliverance spread in the island and throughout the Empire. There was a feeling of intense relief, melting almost into triumph. The safe homecoming of a quarter of a million men, the flower of our Army, was a milestone in our pilgrimage through years of defeat. The troops returned with nothing but rifles and bayonets and a few hundred machine guns, and were forthwith sent to their homes for seven days' leave. Their joy at being once again united with their families did not overcome a stern desire to engage the enemy at the earliest moment. Those who had actually fought the Germans in the field had the belief that, given a fair chance, they could beat them. Their morale was high, and they rejoined their regiments and batteries with alacrity.

There was of course a darker side to Dunkirk. We had lost the whole equipment of the Army to which all the first fruits of our factories had hitherto been given. Many months must elapse, even if the existing programmes were fulfilled without interruption by the enemy, before this loss could be repaired.

However, across the Atlantic in the United States strong emotions were already stirring in the breasts of its leading men. It was at once realised that the bulk of the British Army had got away only with the loss of all their equipment. As early as June 1 the President sent out orders to the War and Navy Departments to report what weapons they could spare for Britain and France. At the head of the American Army as Chief of Staff was General Marshall, not only a soldier of proved quality but a man of commanding vision. He instantly directed his Chief of Ordnance and his Assistant Chief of Staff to survey the entire list of the American reserve ordnance and munitions stocks. In forty-

eight hours the answers were given, and on June 3 Marshall approved
the lists. The first list comprised half a million .30 calibre rifles out of
two million manufactured in 1917 and 1918 and stored in grease for
more than twenty years. For these there were about 250 cartridges
apiece. There were 900 *"soixante-quinze"* field guns, with a million
rounds, 80,000 machine guns, and various other items. The Chief of
Ordnance, Major General Wesson, was told to handle the matter, and
immediately all the American Army depots and arsenals started packing
the material for shipment. By the end of the week more than six hundred heavily loaded freight cars were rolling towards the Army docks at
Raritan, New Jersey, up the river from Gravesend Bay. By June 11 a
dozen British merchant ships moved into the bay and anchored, and
loading from lighters began.

By these extraordinary measures the United States left themselves
with the equipment for only 1,800,000 men, the minimum figure stipulated by the American Army's mobilisation plan. All this reads easily
now, but at that time it was a supreme act of faith and leadership for
the United States to deprive themselves of this very considerable mass
of arms for the sake of a country which many deemed already beaten.
They never had need to repent of it. As will presently be recounted, we
ferried these precious weapons safely across the Atlantic during July, and
they formed not only a material gain, but an important factor in all calculations made by friend or foe about invasion.

*   *   *

The month of June was particularly trying to all of us, because of the
dual and opposite stresses to which in our naked condition we were subjected by our duty to France on the one hand and the need to create an
effective army at home and to fortify the island on the other. The double tension of antagonistic but vital needs was most severe. Nevertheless
we followed a firm and steady policy without undue excitement. First
priority continued to be given to sending whatever trained and equipped
troops we had in order to reconstitute the B.E.F. in France. After that
our efforts were devoted to the defence of the island — first, by reforming and re-equipping the Regular Army; secondly, by fortifying the
likely landing places; thirdly, by arming and organising the population,
so far as was possible; and of course by bringing home whatever forces
could be gathered from the Empire. It was not men that were lacking
but arms. Over 80,000 rifles were retrieved from the communications
and bases south of the Seine, and by the middle of June every fighting
man in the Regular forces had at least a personal weapon in his hand.
We had very little field artillery, even for the Regular Army. Nearly all

the new 25-pounders had been lost in France. There remained about 500 guns and only 103 cruiser, 114 infantry, and 252 light tanks. Never has a great nation been so naked before her foes.

Apart from our last twenty-five fighter squadrons, on which we were adamant, we regarded the duty of sending aid to the French Army as paramount. The movement of the 52nd Lowland Division to France, under previous orders, was due to begin on June 7. These orders were confirmed. The leading division of the Canadian Army, which had concentrated in England early in the year and was well armed, was directed, with the full assent of the Dominion Government, to Brest, to begin arriving there on June 11 for what might by this time already be deemed a forlorn hope. That we should have sent our only two formed divisions, the 52nd Lowland Division and 1st Canadian Division, over to our failing French ally in this mortal crisis, when the whole fury of Germany must soon fall upon us, must be set to our credit against the very limited forces we had been able to put into France in the first eight months of war. Looking back on it, I wonder how, when we were resolved to continue the war to the death, and under the threat of invasion, and France was evidently falling, we had the nerve to strip ourselves of the remaining effective military formations we possessed. This was only possible because we understood the difficulties of the Channel crossing without the command of the sea or the air, or the necessary landing craft.

*   *   *

We had still in France, behind the Somme, the 51st Highland Division, which had been withdrawn from the Maginot Line and was in good condition. There was also our 1st (and only) Armoured Division, less the tank battalion and the support group which had been sent to Calais. This however had lost heavily in attempts to cross the Somme as part of Weygand's plan. By June 1 it was reduced to one-third of its strength, and was sent back across the Seine to refit. At the same time nine infantry battalions, armed mainly with rifles, were scraped together from the bases and lines of communication in France. They had very few anti-tank weapons, and neither transport nor signals.

On June 5 the final phase of the Battle of France began. We have seen how the German armour had been hobbled and held back in the Dunkirk battle, in order to save it for the final phase in France. All this armour now rolled forward upon the weak and improvised or quivering French front between Paris and the sea. It is here only possible to record the battle on the coastal flank, in which we played a part. The Tenth French Army tried to hold the line of the Somme. On June 7 two German armoured divisions drove towards Rouen. The French left, including the 51st Highland Division, were separated from the rest of the front, and, with the remnants of the French Ninth Corps, was cut off in the Rouen–Dieppe *cul-de-sac*.

We had been intensely concerned lest this division should be driven back to the Havre peninsula and thus be separated from the main armies, and its commander, Major General Fortune, had been told to fall back if necessary in the direction of Rouen. This movement was forbidden by the already disintegrating French command. Repeated urgent representations were made by us, but they were of no avail. It was a case of gross mismanagement, for this very danger was visible a full three days before.

On June 10, after sharp fighting, the division fell back, together with the French Ninth Corps, to the perimeter of St. Valéry, expecting to be evacuated by sea. During the night of the 11th–12th fog prevented the ships from evacuating the troops. By the morning on the 12th the Germans had reached the sea cliffs to the south and the beach was under direct fire. White flags appeared in the town. The French corps capitulated at eight o'clock, and the remains of the Highland Division were forced to do so at 10.30 A.M. Eight thousand British and four thousand French fell into the hands of the 7th Panzer Division, commanded by General Rommel. I was vexed that the French had not allowed our division to retire on Rouen in good time, but had kept it waiting till it could neither reach Havre nor retreat southward, and thus forced it to surrender with their own troops. The fate of the Highland Division was hard, but in after years not unavenged by those Scots who filled their places, re-created the division by merging it with the 9th Scottish, and marched across all the battlefields from Alamein to final victory beyond the Rhine.

                    *   *   *

About eleven o'clock on the morning of June 11 there was a message from Reynaud, who had also cabled to the President. The French tragedy had moved and slid downward. For several days past I had pressed for a meeting of the Supreme Council. We could no longer meet in Paris. We were not told what were the conditions there. Certainly the German spearheads were very close. I had had some difficulty in obtaining a rendezvous, but this was no time to stand on ceremony. We must know what the French were going to do. Reynaud now told me that he could receive us at Briare, near Orléans. The seat of government was moving from Paris to Tours. Grand Quartier Général was near Briare. Nothing loth, I ordered the Flamingo to be ready at Hendon after luncheon, and, having obtained the approval of my colleagues at the morning Cabinet, we started about two o'clock.

This was my fourth journey to France; and since military conditions evidently predominated, I asked Mr. Eden, now Secretary of State for War, to come with me, as well as General Dill, the C.I.G.S., and of course Ismay. The German aircraft were now reaching far down into the Channel, and we had to make a still wider sweep. As before, the Flamingo had an escort of twelve Spitfires. After a couple of hours we alighted at a small landing-ground. There were a few Frenchmen about, and soon a colonel arrived in a motorcar. I displayed the smiling countenance and confident air which are thought suitable when things are very bad, but the Frenchman was dull and unresponsive. I realised immediately how very far things had fallen even since we were in Paris a week before. After an interval we were conducted to the château, where we found Reynaud, Marshal Pétain, General Weygand, the Air General Vuillemin, and some others, including the relatively junior General de Gaulle, who had just been appointed Under-Secretary for National Defence. Hard by on the railway was the Headquarters train, in which some of our party were accommodated. The château possessed but one telephone, in the lavatory. It was kept very busy, with long delays and endless shouted repetitions.

At seven o'clock we entered into conference. There were no reproaches or recriminations. We were all up against brute facts. In effect, the discussion ran on the following lines: I urged the French Government to defend Paris. I emphasised the enormous absorbing power of the house-to-house defence of a great city upon an invading army. I recalled to Marshal Pétain the nights we had spent together in his train at Beauvais after the British Fifth Army disaster in 1918, and how he, as I put it, not mentioning Marshal Foch, had restored the situation. I also reminded him how Clemenceau had said: "I will fight in front of Paris, in Paris, and behind Paris." The Marshal replied very quietly and with dignity that in those days he had a mass of manoeuvre of upwards of sixty divisions; now there was none. He mentioned that there were then sixty British divisions in the line. Making Paris into a ruin would not affect the final event.

Then General Weygand exposed the military position, so far as he knew it, in the fluid battle proceeding fifty or sixty miles away, and he paid a high tribute to the prowess of the French Army. He requested that every reinforcement should be sent — above all, that every British fighter air squadron should immediately be thrown into the battle. "Here," he said, "is the decisive point. Now is the decisive moment. It is therefore wrong to keep *any* squadrons back in England." But in accordance with the Cabinet decision, taken in the presence of Air Chief Marshal Dowding, whom I had brought specially to a Cabinet meeting, I replied: "This is not the decisive point and this is not the decisive moment. That moment will come when Hitler hurls his Luftwaffe against Great Britain. If we can keep command of the air, and if we can keep the seas open, as we certainly shall keep them open, we will win it all back for you." [1] Twenty-five fighter squadrons must be maintained at all costs for the defence of Britain and the Channel, and nothing would make us give up these. We intended to continue the war whatever happened, and we believed we could do so for an indefinite time, but to give up these squadrons would destroy our chance of life.

Presently General Georges, the Commander-in-Chief of the North-western Front, arrived. After being apprised of what had passed, he confirmed the account of the French front which had been given by Weygand. I again urged my guerrilla plan. The German Army was not so strong as might appear at their points of impact. If all the French armies, every division and brigade, fought the troops on their front with the utmost vigour a general standstill might be achieved. I was answered by statements of the frightful conditions on the roads, crowded with refugees harried by unresisted machine-gun fire from the German aeroplanes, and of the wholesale flight of vast numbers of inhabitants and the increasing breakdown of the machinery of government and of military control. At one point General Weygand mentioned that the French might have to ask for an armistice. Reynaud at once snapped at him: "That is a political affair." According to Ismay, I said: "If it is thought best for France in her agony that her Army should capitulate, let there be no hesitation on our account, because whatever you may do we shall fight on forever and ever and ever." When I said that the French Army, fighting on, wherever it might be, could hold or wear out a hundred German divisions, General Weygand replied: "Even if that were so, they would still have another hundred to invade and conquer you. What would you do then?" On this I said that I was not a military expert, but that my technical advisers were of the opinion that the best method of dealing with a German invasion of the island of Britain was to drown as many as possible on the way over and knock the others on the head as they crawled ashore. Weygand answered with a sad smile: "At any rate I must admit you have a very good anti-tank obstacle." These were the last striking words I remember to have heard from him. In all this miserable discussion it must be borne in mind that I was haunted and undermined by the grief I felt that Britain, with her forty-eight million population, had not been able to make a greater contribution to the land war against Germany, and that so far nine-tenths of the slaughter and ninety-nine-hundredths of the suffering had fallen upon France and upon France alone.

After another hour or so we got up and washed our hands while a

_____
[1] I am obliged to General Ismay for his recollection of these words.

meal was brought to the conference table. In this interval I talked to General Georges privately, and suggested first the continuance of fighting everywhere on the home front and a prolonged guerrilla in the mountainous regions, and secondly the move to Africa, which a week before I had regarded as "defeatist." My respected friend, who, although charged with much direct responsibility, had never had a free hand to lead the French armies, did not seem to think there was much hope in either of these.

I have written lightly of the happenings of these days, but here to all of us was real agony of mind and soul.

*   *   *

At about ten o'clock everyone took his place at the dinner. I sat on Reynaud's right and General de Gaulle was on my other side. There was soup, an omelette or something, coffee and light wine. Even at this point in our awful tribulation under the German scourge we were quite friendly. But presently there was a jarring interlude. The reader will recall the importance I had attached to striking hard at Italy the moment she entered the war, and arrangements had been made with full French concurrence to move a force of British heavy bombers to the French airfields near Marseilles in order to attack Turin and Milan. All was now in readiness to strike. Scarcely had we sat down when Air Marshal Barratt, commanding the British Air Force in France, rang up Ismay on the telephone to say that the local authorities objected to the British bombers taking off, on the grounds that an attack on Italy would only bring reprisals upon the South of France, which the British were in no position to resist or prevent. Reynaud, Weygand, Eden, Dill, and I left the table, and, after some parleying, Reynaud agreed that orders should be sent to the French authorities concerned that the bombers were not to be stopped. But later that night Air Marshal Barratt reported that the French people near the airfields had dragged all kinds of country carts and lorries onto them, and that it had been impossible for the bombers to start on their mission.

Presently, when we left the dinner table and sat with some coffee and brandy, M. Reynaud told me that Marshal Pétain had informed him that it would be necessary for France to seek an armistice, and that he had written a paper upon the subject which he wished him to read. "He has not," said Reynaud, "handed it to me yet. He is still ashamed to do it." He ought also to have been ashamed to support even tacitly Weygand's demand for our last twenty-five squadrons of fighters, when he had made up his mind that all was lost and that France should give in. Thus we all went unhappily to bed in this disordered château or in the military train a few miles away. The Germans entered Paris on the 14th.

Early in the morning we resumed our conference. Air Marshal Barratt was present. Reynaud renewed his appeal for five more squadrons of fighters to be based in France, and General Weygand said that he was badly in need of day bombers to make up for his lack of troops. I gave them an assurance that the whole question of increased air support for France would be examined carefully and sympathetically by the War Cabinet immediately I got back to London; but I again emphasised that it would be a vital mistake to denude the United Kingdom of its essential Home defences.

After some fruitless discussion about a counterattack on the Lower Seine, I expressed in the most formal manner my hope that if there was any change in the situation the French Government would let the British Government know at once, in order that we might come over and see them at any convenient spot, before they took any final decisions which would govern their action in the second phase of the war.

We then took leave of Pétain, Weygand, and their staff, and this was the last we saw of them. Finally I took Admiral Darlan apart and spoke to him alone: "Darlan, you must never let them get the French Fleet." He promised solemnly that he would never do so.

                    *   *   *

Lack of suitable petrol made it impossible for the twelve Spitfires to escort us. We had to choose between waiting till it turned up or taking a chance in the Flamingo. We were assured that it would be cloudy all the way. It was urgently necessary to get back home. Accordingly we started alone, calling for an escort to meet us, if possible, over the Channel. As we approached the coast the skies cleared and presently became cloudless. Eight thousand feet below us on our right hand was Havre, burning. The smoke drifted away to the eastward. No new escort was to be seen. Presently I noticed some consultations going on with the captain, and immediately after we dived to a hundred feet or so above the calm sea, where aeroplanes are often invisible. What had happened? I learned later that they had seen two German aircraft below us firing at fishing boats. We were lucky that their pilots did not look upward. The new escort met us as we approached the English shore, and the faithful Flamingo alighted safely at Hendon.

At five o'clock that evening I reported to the War Cabinet the results of my mission. I described the condition of the French armies as it had been reported to the conference by General Weygand. For six days they had been fighting night and day, and they were now almost wholly exhausted. The enemy attack, launched by a hundred and twenty divisions, with supporting armour, had fallen on forty French divisions. The French armies were now on the last line on which they could attempt to offer an organised resistance. This line had already been penetrated in two or three places. General Weygand evidently saw no prospect of the French going on fighting, and Marshal Pétain had quite made up his mind that peace must be made. He believed that France was being systematically destroyed by the Germans, and that it was his duty to save the rest of the country from this fate. I mentioned his memorandum to this effect, which he had shown to Reynaud but had not left with him. "There can be no doubt," I said, "that Pétain is a dangerous man at this juncture: he was always a defeatist, even in the last war." On the other hand, M. Reynaud had seemed quite determined to fight on, and General de Gaulle, who had attended the conference with him, was in favour of carrying on a guerrilla warfare. He was young and energetic and had made a very favourable impression on me. I thought it probable that if the present line collapsed Reynaud would turn to him to take command. Admiral Darlan also had declared that he would never surrender the French Navy to the enemy: in the last resort, he had said, he would send it over to Canada; but in this he might be overruled by the French politicians.

It was clear that France was near the end of organised resistance, and a chapter in the war was now closing. The French might by some means continue the struggle. There might even be two French Governments, one which made peace, and one which organised resistance from the French colonies, carrying on the war at sea through the French Fleet and in France through guerrillas. It was too early yet to tell. Though for a period we might still have to send some support to France, we must now concentrate our main efforts on the defence of our Island.

7

Home Defence and the
Apparatus of Counterattack

T HE READER OF THESE PAGES in future years should
realise how dense and baffling is the veil of the Unknown. Now in
the full light of the after-time it is easy to see where we were ignorant or
too much alarmed, where we were careless or clumsy. Twice in two
months we had been taken completely by surprise. The overrunning of
Norway and the break-through at Sedan, with all that followed from
these, proved the deadly power of the German initiative. What else had
they got ready — prepared and organised to the last inch? Would they
suddenly pounce out of the blue with new weapons, perfect planning,
and overwhelming force upon our almost totally unequipped and disarmed island at any one of a dozen or score of possible landing-places?
Or would they go to Ireland? He would have been a very foolish man
who allowed his reasoning, however clean-cut and seemingly sure, to blot
out any possibility against which provision could be made. "Depend
upon it," said Dr. Johnson, "when a man knows he is going to be hanged
in a fortnight, it concentrates his mind wonderfully." I was always sure
we should win, but nevertheless I was highly geared up by the situation,
and very thankful to be able to make my views effective.

My colleagues had already felt it right to obtain from Parliament the
extraordinary powers for which a bill had been prepared during the last
few days. This measure would give the Government practically unlimited power over the life, liberty, and property of all His Majesty's
subjects in Great Britain. In general terms of law the powers granted
by Parliament were absolute. The Act was to "include power by Order
in Council to make such Defence Regulations making provision for requiring persons to place themselves, their services and their property at
the disposal of His Majesty as appear to him to be necessary or expedient for securing the public safety, the defence of the realm, the main-

tenance of public order, or the efficient prosecution of any war in which His Majesty may be engaged, or for maintaining supplies or services essential to the life of the community."

In regard to persons, the Minister of Labour was empowered to direct anyone to perform any service required. The regulation giving him this power included a Fair Wages clause, which was inserted in the Act to regulate wage conditions. Labour supply committees were to be set up in important centres. The control of property in the widest sense was imposed in equal manner. Control of all establishments, including banks, was imposed under the authority of Government orders. Employers could be required to produce their books, and excess profits were to be taxed at one hundred per cent. A Production Council, to be presided over by Mr. Greenwood, was to be formed, and a Director of Labour Supply to be appointed.

This bill had been presented to Parliament on the afternoon of May 22 by Mr. Chamberlain and Mr. Attlee, the latter himself moving the Second Reading. Both the Commons and the Lords with their immense Conservative majorities passed it unanimously through all its stages in a single afternoon, and it received the Royal Assent that night.

> *For Romans in Rome's quarrel*
> *Spared neither land nor gold,*
> *Nor son nor wife, nor limb nor life,*
> *In the brave days of old.*

Such was the temper of the hour.

This was a time when all Britain worked and strove to the utmost limit and was united as never before. Men and women toiled at the lathes and machines in the factories till they fell exhausted on the floor and had to be dragged away and ordered home, while their places were occupied by newcomers ahead of time. The one desire of all the males and many women was to have a weapon. The Cabinet and Government were locked together by bonds the memory of which is still cherished by all. The sense of fear seemed entirely lacking in the people, and their representatives in Parliament were not unworthy of their mood. We had not suffered like France under the German flail. Nothing moves an Englishman so much as the threat of invasion, the reality unknown for a thousand years. Vast numbers of people were resolved to conquer or die. There was no need to rouse their spirit by oratory. They were glad to hear me express their sentiments and give them good reasons for what they meant to do, or try to do. The only possible divergence was from people who wished to do even more than was possible, and had the idea that frenzy might sharpen action.

*   *   *

Our decision to send our only two well-armed divisions back to France made it all the more necessary to take every possible measure to defend the island against direct assault. The swift fate of Holland was in all our minds. Mr. Eden had already proposed to the War Cabinet the formation of Local Defence Volunteers or "Home Guards," and this plan was energetically pressed. All over the country, in every town and village, bands of determined men came together armed with shotguns, sporting rifles, clubs, and spears. From this a vast organisation was soon to spring. It presently approached one and a half million men and gradually acquired good weapons.

My principal fear was of German tanks coming ashore. Since my mind was attracted to landing tanks on their coast, I naturally thought they might have the same idea. We had hardly any anti-tank guns or ammunition, or even ordinary field artillery. The plight to which we were reduced in dealing with this danger may be measured from the following incident. I visited our beaches in St. Margaret's Bay, near Dover. The Brigadier informed me that he had only three anti-tank guns in his brigade, covering four or five miles of this highly menaced coastline. He declared that he had only six rounds of ammunition for each gun, and he asked me with a slight air of challenge whether he was justified in letting his men fire one single round for practice in order that they might at least know how the weapon worked. I replied that we could not afford practice rounds, and that fire should be held for the last moment at the closest range.

This was therefore no time to proceed by ordinary channels in devising expedients. In order to secure quick action, free from departmental processes, upon any bright idea or gadget, I decided to keep under my own hand as Minister of Defence the experimental establishment formed by Major Jefferis at Whitchurch. Since 1939 I had had useful contacts with this brilliant officer, whose ingenious, inventive mind proved, as will be seen, fruitful during the whole war. Lindemann was in close touch with him and me. I used their brains and my power. Jefferis and others connected with him were at work upon a bomb which could be thrown at a tank, perhaps from a window, and would stick upon it. The impact of a very high explosive in actual contact with a steel plate is particularly effective. We had the picture in mind that devoted soldiers or civilians would run close up to the tank and even thrust the bomb upon it, though its explosion cost them their lives. There were undoubtedly many who would have done it. I thought also that the bomb fixed on a rod might be fired with a reduced charge from rifles. In the end the "sticky" bomb was accepted as one of our best emergency weapons. We never had to use it at home; but in Syria, where equally primitive conditions prevailed, it proved its value.

*   *   *

For the first time in a hundred and twenty-five years a powerful enemy was now established across the narrow waters of the English Channel. Our re-formed Regular Army, and the larger but less well trained Territorials, had to be organized and deployed to create an elaborate system of defences, and to stand ready, if the invader came, to destroy him — for there could be no escape. It was for both sides "Kill or Cure." Already the Home Guard could be included in the general framework of defence. On June 25 General Ironside, Commander-in-Chief Home Forces, exposed his plans to the Chiefs of Staff. They were of course scrutinised with anxious care by the experts, and I examined them myself with no little attention. On the whole they stood approved. There were three main elements in this early outline of a great future plan: first, an entrenched "crust" on the probable invasion beaches of the coast, whose defenders should fight where they stood, supported by mobile reserves for immediate counterattack; secondly, a line of anti-tank obstacles, manned by the Home Guard, running down the east centre of England and protecting London and the great industrial centres from inroads by armoured vehicles; thirdly, behind that line, the main reserves for major counter-offensive action.

Ceaseless additions and refinements to this first plan were effected as the weeks and months passed; but the general conception remained. All troops, if attacked, should stand firm, not in linear only but *in all-round defence,* while others moved rapidly to destroy the attackers, whether they came from sea or air. Men who had been cut off from immediate help would not have merely remained in position. Active measures were prepared to harass the enemy from behind; to interfere with his communications and to destroy material, as the Russians did with great results when the German tide flowed over their country a year later. Many people must have been bewildered by the innumerable activities all around them. They could understand the necessity for wiring and mining the beaches, the anti-tank obstacles at the defiles, the concrete pillboxes at the crossroads, the intrusions into their houses to fill an attic with sandbags, onto their golf courses or most fertile fields and gardens to burrow out some wide anti-tank ditch. All these inconveniences, and much more, they accepted in good part. But sometimes they must have wondered if there was a general scheme, or whether lesser individuals were not running amok in their energetic use of newly granted powers of interference with the property of the citizen.

There was however a central plan, elaborate, co-ordinated, and all-embracing. As it grew it shaped itself thus: the overall command was maintained at General Headquarters in London. All Great Britain and Northern Ireland were divided into seven commands; these again into areas of corps and divisional commands. Commands, corps, and divisions were each required to hold a proportion of their resources in mobile reserve, only the minimum being detailed to hold their own particular defences. Gradually there were built up in rear of the beaches zones of defence in each divisional area; behind these were similar corps zones and command zones, the whole system amounting in depth to a hundred miles or more. And behind these was established the main anti-tank obstacle running across Southern England and northward into Nottinghamshire. Above all was the final reserve directly under the Commander-in-Chief of the Home Forces. This it was our policy to keep as large and mobile as possible.

Within this general structure were many variations. Each of our ports on the east and south coasts was a special study. Direct frontal attack upon a defended port seemed an unlikely contingency, and all were made into strong points equally capable of defence from the landward or the seaward side. Obstacles were placed on many thousand square miles of Britain to impede the landing of air-borne troops. All our aerodromes, radar stations, and fuel depots, of which even in the summer of 1940 there were three hundred and seventy-five, needed defence by special garrisons and by their own airmen. Many thousands of "vulnerable points" — bridges, power stations, depots, vital factories, and the like — had to be guarded day and night from sabotage or sudden onset. Schemes were ready for the immediate demolition of resources helpful to the enemy if captured. The destruction of port facilities, the cratering of key roads, the paralysis of motor transport and of telephones and telegraph stations, of rolling stock or permanent way, before they passed out of our control, were planned to the last detail. Yet despite all these wise and necessary precautions, in which the civilian departments gave unstinted help to the military, there was no question of a "scorched earth policy"; England was to be defended by its people, not destroyed.

*   *   *

There was also another side to all this. My first reaction to the "Miracle of Dunkirk" had been to turn it to proper use by mounting a counter-offensive. When so much was uncertain, the need to recover the initiative glared forth. June 4 was much occupied for me by the need to prepare and deliver the long and serious speech to the House of Commons of which some account has been given, but as soon as this was over I made haste to strike the note which I thought should rule our minds and inspire our actions at this moment, and I accordingly sent the following minute to General Ismay:

> We are greatly concerned — and it is certainly wise to be so — with the dangers of the Germans landing in England in spite of our possessing the command of the seas and having very strong defence by fighters in the air. Every creek, every beach, every harbour has become to us a source of anxiety. Besides this the parachutists may sweep over and take Liverpool or Ireland, and so forth. All this mood is very good if it engenders energy. But if it is so easy for the Germans to invade us, in spite of sea power, some may feel inclined to ask the question, why should it be thought impossible for us to do anything of the same kind to them? The completely defensive habit of mind which has ruined the French must not be allowed to ruin all our initiative. It is of the highest consequence to keep the largest numbers of German forces all along the coasts of the countries they have conquered, and we should immediately set to work to organise raiding forces on these coasts where the populations are friendly. Such forces might be composed of self-contained, thoroughly equipped units of say one thousand up to not more than ten thousand when combined. Surprise would be ensured by the fact that the destination would be concealed until the last moment. What we have seen at Dunkirk shows how quickly troops can be moved off (and I suppose on to) selected points if need be. How wonderful it would be if the Germans could be made to wonder where they were going to be struck next, instead of forcing us to try to wall in the island and roof it over! An effort must be made to shake off the mental and moral prostration to the will and initiative of the enemy from which we suffer.

Ismay conveyed this to the Chiefs of Staff, and in principle it received their cordial approval and was reflected in many of the decisions which we took. Out of it gradually sprang a policy. My thought was at this time firmly fixed on tank warfare, not merely defensive but offensive. This required the construction of large numbers of tank-landing vessels, which henceforward became one of my constant cares. As all this was destined to become of major importance in the future I must now make a retrogression into a subject which had long ago lain in my mind and was now revived.

*   *   *

I had always been fascinated by amphibious warfare, and the idea of using tanks to run ashore from specially constructed landing craft

on beaches where they were not expected had long been in my mind.
Ten days before I joined Mr. Lloyd George's Government as Minister
of Munitions on July 17, 1917, I had prepared, without expert assistance, a scheme for the capture of the two Frisian islands Borkum and
Sylt. It contained the following paragraphs which have never yet seen
the light of day:

> The landing of the troops upon the island [of Borkum or Sylt]
> under cover of the guns of the Fleet [should be] aided by gas and
> smoke from torpedo-proof transports by means of *bullet-proof
> lighters.* Approximately one hundred should be provided for land-
> ing a division. In addition a number — say fifty — [of] *tank-
> landing lighters should be provided, each carrying a tank or tanks*
> [and] fitted for wire-cutting in its bow. By means of a drawbridge
> or shelving bow [the tanks] would land under [their] own power,
> and prevent the infantry from being held up by wire when attack-
> ing the gorges of the forts and batteries. This is a new feature,
> and removes one of the very great previous difficulties, namely, the
> rapid landing of [our] field artillery to cut wire.

And further:

> There is always the danger of the enemy getting wind of our
> intentions and reinforcing his garrisons with good troops before-
> hand, at any rate so far as Borkum, about which he must always
> be very sensitive, is concerned. On the other hand, *the landing
> could be effected under the shields of lighters, proof against ma-
> chine-gun bullets,* and too numerous to be seriously affected by
> heavy gunfire [*i.e.*, the fire of heavy guns]; *and tanks employed in
> even larger numbers than are here suggested, especially the quick-
> moving tank and lighter varieties,* would operate in an area where
> no preparations could have been made to receive them. These may
> be thought new and important favourable considerations.

In this paper also I had an alternative plan for making an artificial
island in the shallow waters of Horn Reef (to the northward).

> One of the methods suggested for investigation is as follows:
> *A number of flat-bottomed barges or caissons, made not of steel,
> but of concrete,* should be prepared in the Humber, at Harwich, and
> in the Wash, the Medway, and the Thames. These structures would
> be adapted to the depths in which they were to be sunk, according
> to a general plan. They would float when empty of water, and
> thus could be towed across to the site of the artificial island. On

arrival at the buoys marking the island sea cocks would be opened, and they would settle down on the bottom. They could subsequently be gradually filled with sand, as opportunity served, by suction dredgers. These structures would range in size from 50′ × 40′ × 20′ to 120′ × 80′ × 40′. *By this means a torpedo- and weather-proof harbour, like an atoll, would be created in the open sea, with regular pens for the destroyers and submarines, and alighting platforms for aeroplanes.*

This project, if feasible, is capable of great elaboration, and it might be applied in various places. Concrete vessels can perhaps be made to carry a complete heavy-gun turret, and these, on the admission of water to their outer chambers, would sit on the sea floor, like the Solent forts, at the desired points. Other sinkable structures could be made to contain storerooms, oil tanks, or living chambers. It is not possible, without an expert inquiry, to do more here than indicate the possibilities, which embrace nothing less than the creation, transportation in pieces, assemblement and posing of an artificial island and destroyer base.

Such a scheme, if found mechanically sound, avoids the need of employing troops and all the risks of storming a fortified island. *It could be applied as a surprise, for although the construction of these concrete vessels would probably be known in Germany, the natural conclusion would be that they were intended for an attempt to block up the river mouths, which indeed is an idea not to be excluded.* Thus, until the island or system of breakwaters actually began to grow the enemy would not penetrate the design.

For nearly a quarter of a century this paper had slumbered in the archives of the Committee of Imperial Defence. I did not print it in *The World Crisis,* of which it was to have been a chapter, for reasons of space, and because it was never put into effect. This was fortunate, because the ideas expressed were in this war more than ever vital; and the Germans certainly read my war books with attention. The underlying conceptions of this old paper were deeply imprinted in my mind, and in the new emergency formed the foundation of action which, after a long interval, found memorable expression in the vast fleet of tank-landing craft of 1943 and in the "Mulberry" harbours of 1944.

Henceforth intense energy was imparted to the development of all types of landing craft, and a special department was formed in the Admiralty to deal with these matters. By October 1940 the trials of the first Landing-Craft Tank (L.C.T.) were in progress. An improved design followed, many of which were built in sections for more convenient transport by sea to the Middle East, where they began to arrive in the summer of 1941. These proved their worth, and as we gained experience the capabilities of later editions of these strange crafts steadily improved. Fortunately it proved that the building of L.C.T.'s could be delegated to constructional engineering firms not engaged in shipbuilding, and thus the labour and plant of the larger shipyards need not be disturbed. This rendered possible the large-scale programme which we contemplated but also placed a limit on the size of the craft.

The L.C.T. was suitable for cross-Channel raiding operations or for more extended work in the Mediterranean, but not for long voyages in the open sea. The need arose for a larger, more seaworthy craft which besides transporting tanks and other vehicles on ocean voyages could also land them over beaches like the L.C.T. I gave directions for the design of such a vessel, which was called "Landing Ship Tank" (L.S.T.). In due course it was taken to the United States, where the details were jointly worked out. It was put into production in America on a massive scale and figured prominently in all our later operations, making perhaps the greatest single contribution to the solution of the stubborn problem of landing heavy vehicles over beaches. Ultimately over a thousand of these were built.

By the end of 1940 we had a sound conception of the physical expression of amphibious warfare. The production of specialised craft and equipment of many kinds was gathering momentum, and the necessary formations to handle all this new material were being developed and trained under the Combined Operations Command. Special training centres for this purpose were established both at home and in the Middle East. All these ideas and their practical manifestation we presented to our American friends as they took shape. The results grew steadily across the years of struggle, and thus in good time they formed the apparatus which eventually played an indispensable part in our greatest plans and deeds. In 1940 and 1941 our efforts in this field were limited by the demands of the U-boat struggle. Not more than seven thousand men could be spared for landing-craft production up to the end of 1940, nor was this number greatly exceeded in the following year. However, by 1944 no less than seventy thousand men in Britain alone were dedicated to this stupendous task, besides much larger numbers in the United States.

In view of the many accounts which are extant and multiplying of my supposed aversion from any kind of large-scale opposed landing, such as took place in Normandy in 1944, it may be convenient if I make it clear that from the very beginning I provided a great deal of the impulse and authority for creating the immense apparatus and armada for the landing of armour on beaches, without which it is now universally recognised that all such major operations would have been impossible.

8

The French Agony

FUTURE GENERATIONS may deem it noteworthy that the supreme question of whether we should fight on alone never found a place upon the War Cabinet agenda. It was taken for granted and as a matter of course by these men of all parties in the State, and we were much too busy to waste time upon such unreal, academic issues. We were united also in viewing the new phase with good confidence.

On June 13 I made my last visit to France for four years almost to a day. The French Government had now withdrawn to Tours, and tension had mounted steadily. I took Edward Halifax and General Ismay with me, and Max Beaverbrook volunteered to come too. In trouble he is always buoyant. This time the weather was cloudless, and we sailed over in the midst of our Spitfire squadron, making however a rather wider sweep to the southward than before. Arrived over Tours, we found the airport had been heavily bombed the night before, but we and all our escort landed smoothly in spite of the craters. Immediately one sensed the increasing degeneration of affairs. No one came to meet us or seemed to expect us. We borrowed a service car from the station commander and motored into the city, making for the Prefecture, where it was said the French Government had their headquarters. No one of consequence was there, but Reynaud was reported to be motoring in from the country.

It being already nearly two o'clock, I insisted upon luncheon, and after some parleyings we drove through streets crowded with refugees' cars, most of them with a mattress on top and crammed with luggage. We found a café, which was closed, but after explanations we obtained a meal. During luncheon I was visited by M. Baudouin, whose influence had risen in these latter days. He began at once in his soft, silky manner about the hopelessness of the French resistance. If the

United States would declare war on Germany it might be possible for France to continue. What did I think about this? I did not discuss the question further than to say that I hoped America would come in, and that we should certainly fight on. He afterwards, I was told, spread it about that I had agreed that France should surrender unless the United States came in.

We then returned to the Prefecture, where Mandel, Minister of the Interior, awaited us. This faithful former secretary of Clemenceau, and a bearer forward of his life's message, seemed in the best of spirits. He was energy and defiance personified. His luncheon, an attractive chicken, was uneaten on the tray before him. He was a ray of sunshine. He had a telephone in each hand, through which he was constantly giving orders and decisions. His ideas were simple: fight on to the end in France, in order to cover the largest possible movement into Africa. This was the last time I saw this valiant Frenchman. The restored French Republic rightly shot to death the hirelings who murdered him. His memory is honoured by his countrymen and their allies.

Presently Reynaud arrived. At first he seemed depressed. General Weygand had reported to him that the French armies were exhausted. The line was pierced in many places; refugees were pouring along all the roads through the country, and many of the troops were in disorder. The Generalissimo felt it was necessary to ask for an armistice while there were still enough French troops to keep order until peace could be made. Such was the military advice. He would send that day a further message to Mr. Roosevelt saying that the last hour had come and that the fate of the Allied cause lay in America's hand. Hence arose the alternative of armistice and peace.

M. Reynaud proceeded to say that the Council of Ministers had on the previous day instructed him to inquire what would be Britain's attitude should the worst come. He himself was well aware of the solemn pledge that no separate peace would be entered into by either ally. General Weygand and others pointed out that France had already sacrificed everything in the common cause. She had nothing left, but she had succeeded in greatly weakening the common foe. It would in those circumstances be a shock if Britain failed to concede that France was physically unable to carry on, if France was still expected to fight on and thus deliver up her people to the certainty of corruption and evil transformation at the hands of ruthless specialists in the art of bringing conquered peoples to heel. That then was the question which he had to put. Would Great Britain realise the hard facts with which France was faced?

I thought the issue was so serious that I asked to withdraw with my colleagues before answering it. So Lords Halifax and Beaverbrook and the rest of our party went out into a dripping but sunlit garden and talked things over for half an hour. On our return I restated our position. We could not agree to a separate peace however it might come. Our war aim remained the total defeat of Hitler, and we felt that we could still bring this about. We were therefore not in a position to release France from her obligation. Whatever happened, we would level no reproaches against France; but that was a different matter from consenting to release her from her pledge. I urged that the French should now send a new, final appeal to President Roosevelt, which we would support from London. M. Reynaud agreed to do this, and promised that the French would hold on until the result was known.

At the end of our talk he took us into the adjoining room, where MM. Herriot and Jeanneney, the Presidents of the Chamber and Senate respectively, were seated. Both these French patriots spoke with passionate emotion about fighting on to the death. As we went down the crowded passage into the courtyard I saw General de Gaulle standing stolid and expressionless at the doorway. Greeting him, I said in a low tone, in French: *"L'homme du destin."* He remained impassive. In the courtyard there must have been more than a hundred leading Frenchmen in frightful misery. Clemenceau's son was brought up to me. I wrung his hand. The Spitfires were already in the air, and I slept sound on our swift and uneventful journey home. This was wise, for there was a long way to go before bedtime.

*   *   *

At 10.15 P.M. I made my new report to the Cabinet. My account was endorsed by my two companions. While we were still sitting Ambassador Kennedy arrived with President Roosevelt's reply to an earlier appeal which Reynaud had made to him on June 10.

> Your message [he cabled] has moved me very deeply. As I have already stated to you and to Mr. Churchill, this Government is doing everything in its power to make available to the Allied Governments the material they so urgently require, and our efforts to do still more are being redoubled. This is so because of our faith in and our support of the ideals for which the Allies are fighting.
>
> The magnificent resistance of the French and British Armies has profoundly impressed the American people.
>
> I am, personally, particularly impressed by your declaration that France will continue to fight on behalf of Democracy, even if it means slow withdrawal, even to North Africa and the Atlantic. It is most important to remember that the French and British

Fleets continue [in] mastery of the Atlantic and other oceans; also to remember that vital materials from the outside world are necessary to maintain all armies.

I am also greatly heartened by what Prime Minister Churchill said a few days ago about the continued resistance of the British Empire, and that determination would seem to apply equally to the great French Empire all over the world. Naval power in world affairs still carries the lessons of history, as Admiral Darlan well knows.

We all thought the President had gone a very long way. He had authorised Reynaud to publish his message of June 10, with all that that implied, and now he had sent this formidable answer. If, upon this, France decided to endure the further torture of the war, the United States would be deeply committed to enter it. At any rate, it contained two points which were tantamount to belligerence: first, a promise of all material aid, which implied active assistance; secondly, a call to go on fighting even if the Government were driven right out of France. I sent our thanks to the President immediately, and I also sought to commend the President's message to Reynaud in the most favourable terms. Perhaps these points were stressed unduly; but it was necessary to make the most of everything we had or could get.

The next day arrived a telegram from the President explaining that he could not agree to the publication of his message to Reynaud. He himself, according to Mr. Kennedy, had wished to do so, but the State Department, while in full sympathy with him, saw the gravest dangers. The President complimented the British and French Governments on the courage of their troops. He renewed the assurances about furnishing all possible material and supplies; but he then said that his message was in no sense intended to commit and did not commit the Government of the United States to military participation. There was no authority under the American Constitution except Congress which could make any commitment of that nature. He bore particularly in mind the question of the French Fleet. Congress, at his desire, had appropriated fifty million dollars for the purpose of supplying food and clothing to civilian refugees in France.

This was a disappointing telegram.

Around our table we all fully understood the risks the President ran of being charged with exceeding his constitutional authority, and consequently of being defeated on this issue at the approaching election, on which our fate, and much more, depended. I was convinced that he would give up life itself, to say nothing of public office, for the cause of world freedom now in such awful peril. But what would have been the good of that? Across the Atlantic I could feel his suffering. In the White House the torment was of a different character from that of Bordeaux or London. But the degree of personal stress was not unequal.

In my reply I tried to arm Mr. Roosevelt with some arguments which he could use to others about the danger to the United States if Europe fell and Britain failed. This was no matter of sentiment, but of life and death. "The fate of the British Fleet," I cabled, "as I have already mentioned to you, would be decisive on the future of the United States, because if it were joined to the Fleets of Japan, France, and Italy and the great resources of German industry, overwhelming sea power would be in Hitler's hands. He might of course use it with a merciful moderation. On the other hand, he might not. This revolution in sea power might happen very quickly, and certainly long before the United States would be able to prepare against it. If we go down you may have a United States of Europe under the Nazi command for more numerous, far stronger, far better armed than the New World. . . ."

*   *   *

Meanwhile the situation on the French front went from bad to worse. The German operations northwest of Paris, in which our 51st Division had been lost, had brought the enemy to the lower reaches of the Seine and the Oise. On the southern banks the dispersed remnants of the Tenth and Seventh French Armies were hastily organising a defence; they had been riven asunder, and to close the gap the garrison of the capital, the so-called Armée de Paris, had been marched out and interposed.

Farther to the east, along the Aisne, the Sixth, Fourth, and Second Armies were in far better shape. They had had three weeks in which to establish themselves and to absorb such reinforcements as had been sent. During all the period of Dunkirk and of the drive to Rouen they had been left comparatively undisturbed, but their strength was small for the hundred miles they had to hold, and the enemy had used the time to concentrate against them a great mass of divisions to deliver the final blow. On June 9 it fell. Despite a dogged resistance, for the French were now fighting with great resolution, bridgeheads were established south of the river from Soissons to Rethel, and in the next two days these were expanded until the Marne was reached. German Panzer divisions, which had played so decisive a part in the drive down the coast, were brought across to join the new battle. Eight of these, in two great thrusts, turned the French defeat into a rout. The French armies, decimated and in confusion, were quite unable to withstand this powerful assembly of superior numbers, equipment, and technique. In four days, by June 16, the enemy had reached Orléans and the Loire; while to the east the other thrust had passed through Dijon and Besançon, almost to the Swiss frontier.

West of Paris the remains of the Tenth Army, the equivalent of no more than two divisions, had been pressed back southwestward from the Seine towards Alençon. The capital fell on the 14th; its defending armies, the Seventh and the Armée de Paris, were scattered; a great gap now separated the exiguous French and British forces in the west from the rest and the remains of the once proud Army of France.

And what of the Maginot Line, the shield of France, and its defenders? Until June 14 no direct attack was made, and already some of the active formations, leaving behind the garrison troops, had started to join, if they could, the fast-withdrawing armies of the centre. But it was too late. On that day the Maginot Line was penetrated before Saarbrucken and across the Rhine by Colmar; the retreating French were caught up in the battle and unable to extricate themselves. Two days later the German penetration to Besançon had cut off their retreat. More than four hundred thousand men were surrounded without hope of escape. Many encircled garrisons held out desperately; they refused to surrender until after the armistice, when French officers were dispatched to give them the order. The last forts obeyed on June 30, the commander protesting that his defences were still intact at every point.

Thus the vast disorganised battle drew to its conclusion all along the French front. It remains only to recount the slender part which the British were able to play.

*   *   *

General Brooke had won distinction in the retreat to Dunkirk, and especially by his battle in the gap opened by the Belgian surrender. We had therefore chosen him to command the British troops which remained in France and all reinforcements until they should reach sufficient numbers to require the presence of Lord Gort as an Army Commander. Brooke had now arrived in France, and on the 14th he met Generals Weygand and Georges. Weygand stated that the French forces were no longer capable of organised resistance or concerted action. The French Army was broken into four groups, of which its Tenth Army was the westernmost. Weygand also told him that the Allied Governments had agreed that a bridgehead should be created in the Brittany peninsula, to be held jointly by the French and British troops on a line running roughly north and south through Rennes. He ordered him to deploy his forces on a defensive line running through this town. Brooke pointed out that this line of defence was 150 kilometres long and required at least fifteen divisions. He was told that the instructions he was receiving must be regarded as an order.

It is true that on June 11, at Briare, Reynaud and I had agreed to try to draw a kind of "Torres Vedras line" across the foot of the Brittany peninsula. Everything however was dissolving at the same time, and the plan, for what it was worth, never reached the domain of action. In itself the idea was sound, but there were no facts to clothe it with reality. Once the main French armies were broken or destroyed, this bridgehead, precious though it was, could not have been held for long against concentrated German attack. But even a few weeks' resistance here would have maintained contact with Britain and enabled large French withdrawals to Africa from other parts of the immense front, now torn to shreds. If the battle in France was to continue, it could be only in the Brest peninsula and in wooded or mountainous regions like the Vosges. The alternative for the French was surrender. Let none, therefore, mock at the conception of a bridgehead in Brittany. The Allied armies under Eisenhower, then an unknown American colonel, bought it back for us later at a high price.

General Brooke, after his talk with the French commanders, and having measured from his own headquarters a scene which was getting worse every hour, reported to the War Office and by telephone to Mr. Eden that the position was hopeless. All further reinforcements should be stopped, and the remainder of the British Expeditionary Force, now amounting to a hundred and fifty thousand men, should be re-embarked at once. On the night of June 14, as I was thought to be obdurate, he rang me up on a telephone line which by luck and effort was open, and pressed this view upon me. I could hear quite well, and after ten minutes I was convinced that he was right and we must go. Orders were given accordingly. He was released from French command. The back-loading of great quantities of stores, equipment, and men began. The leading elements of the Canadian Division which had landed got back into their ships, and the 52nd Lowland Division, of which the larger part had not yet been committed to action, retreated on Brest. On June 15 the rest of our troops were released from the orders of the Tenth French Army, and next day moved towards Cherbourg. On June 17 it was announced that the Pétain Government had asked for an armistice, ordering all French forces to cease fighting, without even communicating this information to our troops. General Brooke was consequently told to come away with all men he could embark and any equipment he could save.

We repeated now on a considerable scale, though with larger vessels, the Dunkirk evacuation. Over twenty thousand Polish troops who refused to capitulate cut their way to the sea and were carried by our ships to Britain. The Germans pursued our forces at all points. In the Cherbourg peninsula they were in contact with our rearguard ten miles south of the harbour on the morning of the 18th. The last ship left at 4 P.M., when the enemy, led by Rommel's 7th Panzer Division, were within three miles of the port. Very few of our men were taken prisoners. In all there were evacuated from all French harbours 136,000 British troops and 310 guns; a total, with the Poles, of 156,000 men.

The German air attack on the transports was heavy. One frightful incident occurred on the 17th at St. Nazaire. The 20,000-ton liner *Lancastria,* with five thousand men on board, was bombed just as she was about to leave. Upwards of three thousand men perished. The rest were rescued under continued air attack by the devotion of the small craft. When this news came to me in the quiet Cabinet Room during the afternoon I forbade its publication, saying: "The newspapers have got quite enough disaster for today at least." I had intended to release the news a few days later, but events crowded upon us so black and so quickly that I forgot to lift the ban, and it was some time before the knowledge of this horror became public.

*   *   *

We must now quit the field of military disaster for the convulsions in the French Cabinet and the personages who surrounded it at Bordeaux.

On the afternoon of June 16 M. Monnet and General de Gaulle visited me in the Cabinet Room. The General in his capacity of Under-Secretary of State for National Defence had just ordered the French ship *Pasteur,* which was carrying weapons to Bordeaux from America, to proceed instead to a British port. Monnet was very active upon a plan to transfer all French contracts for munitions in America to Britain if France made a separate peace. He evidently expected this, and wished to save as much as possible from what seemed to him to be the wreck of the world. His whole attitude in this respect was most helpful. Then he turned to our sending all our remaining fighter air squadrons to share in the final battle in France, which was of course already over. I told him that there was no possibility of this being done. Even at this stage he used the usual arguments — "the decisive battle," "now or never," "if France falls all falls," and so forth. But I could not do anything to oblige him in this field. My two French visitors then got up and moved towards the door, Monnet leading. As they reached it, de Gaulle, who had hitherto scarcely uttered a single word, turned back, and, taking two or three paces towards me, said in English: "I think you are quite right." Under an impassive, imperturbable demeanour he seemed to me to have a remarkable capacity for feeling pain. I preserved the impression, in contact with this very tall, phleg-

matic man: "Here is the Constable of France." He returned that afternoon in a British aeroplane, which I had placed at his disposal, to
Bordeaux. But not for long.

The War Cabinet sat until six o'clock that evening. They were in a
state of unusual emotion. The fall and the fate of France dominated
their minds. Our own plight, and what we should have to face and face
alone, seemed to take a second place. Grief for our ally in her agony,
and desire to do anything in human power to aid her, was the prevailing mood. There was also the overpowering importance of making
sure of the French Fleet. Some days beforehand we had evolved a
Declaration for a Franco-British Union, common citizenship, joint
organs for defence, foreign, financial and economic policy, and so forth,
with the object, apart from its general merits, of giving M. Reynaud
some new fact of a vivid and stimulating nature with which to carry a
majority of his Cabinet into a move to Africa and the continuance of
the war. Armed with this document, and accompanied by the leaders
of the Labour and Liberal Parties, the three Chiefs of Staff, and various
important officers and officials, I now set out on yet another mission to
France. A special train was waiting at Waterloo. We could reach
Southampton in two hours, and a night of steaming at thirty knots in
a cruiser would bring us to our rendezvous by noon on the 17th.
We had taken our seats in the train. My wife had come to see me off.
There was an odd delay in starting. Evidently some hitch had occurred.
Presently my private secretary arrived from Downing Street breathless
with the following message from Sir Ronald Campbell, our Ambassador
at Bordeaux: "Ministerial crisis has opened. . . . Hope to have news by
midnight. Meanwhile meeting arranged for tomorrow impossible."

On this I returned to Downing Street with a heavy heart.

*   *   *

The final scene in the Reynaud Cabinet was as follows.

The hopes which M. Reynaud had founded upon the Declaration of
Union were soon dispelled. Rarely has so generous a proposal encountered such a hostile reception. The Premier read the document
twice to the Council. He declared himself strongly for it, and added
that he was arranging a meeting with me for the next day to discuss the
details. But the agitated Ministers, some famous, some nobodies, torn
by division and under the terrible hammer of defeat, were staggered.
Most were wholly unprepared to receive such far-reaching themes. The
overwhelming feeling of the Council was to reject the whole plan. Surprise and mistrust dominated the majority, and even the most friendly
and resolute were baffled. The Council had met expecting to receive
the answer to the French request, on which they had all agreed, that

Britain should release France from her obligations in order that the French might ask the Germans what their terms of armistice would be. It is possible, even probable, that if our formal answer had been laid before them the majority would have accepted our primary condition about sending their Fleet to Britain, or at least would have made some other suitable proposal and thus have freed them to open negotiations with the enemy, while reserving to themselves a final option of retirement to Africa if the German conditions were too severe. But now there was a classic example of "Order, counterorder, disorder."

Paul Reynaud was quite unable to overcome the unfavourable impression which the proposal of Anglo-French Union created. The defeatist section, led by Marshal Pétain, refused even to examine it. Violent charges were made. It was "a last-minute plan," "a surprise," "a scheme to put France in tutelage, or to carry off her colonial empire." It relegated France, so they said, to the position of a Dominion. Others complained that not even equality of status was offered to the French, because Frenchmen were to receive only the citizenship of the British Empire instead of that of Great Britain, while the British were to be citizens of France. This suggestion is contradicted by the text.

Beyond these came other arguments. Weygand had convinced Pétain without much difficulty that England was lost. High French military authorities had advised: "In three weeks England will have her neck wrung like a chicken." To make a union with Great Britain was, according to Pétain, "fusion with a corpse." Ybarnegaray, who had been so stout in the previous war, exclaimed: "Better be a Nazi province. At least we know what that means." Senator Reibel, a personal friend of General Weygand's, declared that this scheme meant complete destruction for France, and anyhow definite subordination to England. In vain did Reynaud reply: "I prefer to collaborate with my allies rather than with my enemies." And Mandel: "Would you rather be a German district than a British Dominion?" All was in vain.

We are assured that Reynaud's statement of our proposal was never put to a vote in the Council. It collapsed of itself. This was a personal and fatal reverse for the struggling Premier which marked the end of his influence and authority upon the Council. All further discussion turned upon the armistice and asking the Germans what terms they would give, and in this M. Chautemps was cool and steadfast. Two telegrams we had sent about the Fleet were never presented to the Council. The demand that it should be sailed to British ports as a prelude to the negotiations with the Germans was never considered by the Reynaud Cabinet, which was now in complete decomposition. At about eight o'clock Reynaud, utterly exhausted by the physical and mental strain to which he had for so many days been subjected, sent his resignation to the President, and advised him to send for Marshal Pétain. This action must be judged precipitate. He still seems to have cherished the hope that he could keep his rendezvous with me the next day, and spoke of this to General Spears. "Tomorrow there will be another Government, and you will no longer speak for anyone," said Spears.

Forthwith Marshal Pétain formed a French Government with the main purpose of seeking an immediate armistice from Germany. Late on the night of June 16 the defeatist group of which he was the head was already so shaped and knit together that the process did not take long. M. Chautemps ("to ask for terms is not necessarily to accept them") was Vice-President of the Council. General Weygand, whose view was that all was over, held the Ministry of National Defence. Admiral Darlan was Minister of Marine, and M. Baudouin Minister for Foreign Affairs.

The only hitch apparently arose over M. Laval. The Marshal's first thought had been to offer him the post of Minister of Justice. Laval brushed this aside with disdain. He demanded the Ministry of Foreign Affairs, from which position alone he conceived it possible to carry out his plan of reversing the alliances of France, finishing up England, and joining as a minor partner the New Nazi Europe. Marshal Pétain surrendered at once to the vehemence of this formidable personality. M. Baudouin, who had already undertaken the Foreign Office, for which he knew himself to be utterly inadequate, was quite ready to give it up. But when he mentioned the fact to M. Charles-Roux, Permanent Under-Secretary to the Ministry of Foreign Affairs, the latter was indignant. He enlisted the support of Weygand. When Weygand entered the room and addressed the illustrious Marshal, Laval became so furious that both military chiefs were overwhelmed. The permanent official however refused point-blank to serve under Laval. Confronted with this, the Marshal again subsided, and after a violent scene Laval departed in wrath and dudgeon.

This was a critical moment. When, four months later, on October 28, Laval eventually became Foreign Minister there was a new consciousness of military values. British resistance to Germany was by then a factor. Apparently the Island could not be entirely discounted. Anyhow, its neck had not been "wrung like a chicken's in three weeks." This was a new fact; and a fact at which the whole French nation rejoiced.

*   *   *

At the desire of the Cabinet I had broadcast the following statement on the evening of June 17:

> The news from France is very bad, and I grieve for the gallant French people who have fallen into this terrible misfortune. Noth-

ing will alter our feelings towards them or our **faith** that the genius
of France will rise again. What has happened in France makes no
difference to our actions and purpose. We have become the sole
champions now in arms to defend the world cause. We shall do our
best to be worthy of this high honour. We shall defend our island
home, and with the British Empire we shall fight on unconquerable until the curse of Hitler is lifted from the brows of mankind.
We are sure that in the end all will come right.

That morning I had mentioned to my colleagues in the Cabinet a
telephone conversation which I had had during the night with General
Spears, who said he did not think he could perform any useful service
in the new structure at Bordeaux. He spoke with some anxiety about
the safety of General de Gaulle. Spears had apparently been warned
that as things were shaping it might be well for de Gaulle to leave
France. I readily assented to a good plan being made for this. So that
very morning — the 17th — de Gaulle went to his office in Bordeaux,
made a number of engagements for the afternoon, as a blind, and then
drove to the airfield with his friend Spears to see him off. They shook
hands and said good-bye, and as the plane began to move, de Gaulle
stepped in and slammed the door. The machine soared off into the air,
while the French police and officials gaped. De Gaulle carried with him,
in this small aeroplane, the honour of France.

9

Admiral Darlan and
the French Fleet:
Oran

A FTER THE COLLAPSE of France the question which arose
in the minds of all our friends and foes was: "Will Britain surrender too?" So far as public statements count in the teeth of events,
I had in the name of His Majesty's Government repeatedly declared our
resolve to fight on alone. After Dunkirk on June 4 I had used the expression "if necessary for years, *if necessary alone.*" This was not inserted without design, and the French Ambassador in London had been
instructed the next day to inquire what I actually meant. He was told
"exactly what was said." I could remind the House of my remark when
I addressed it on June 18, the morrow of the Bordeaux collapse. I then
gave "some indication of the solid practical grounds on which we based
our inflexible resolve to continue the war." I was able to assure Parliament that our professional advisers of the three Services were confident that there were good and reasonable hopes of ultimate victory. I
told them that I had received from all the four Dominion Prime Ministers messages in which they endorsed our decision to fight on and
declared themselves ready to share our fortunes. "In casting up this
dread balance sheet and contemplating our dangers with a disillusioned
eye I see great reasons for vigilance and exertion, but none whatever
for panic or fear." I added: "During the first four years of the last war
the Allies experienced nothing but disaster and disappointment. . . .
We repeatedly asked ourselves the question 'How are we going to
win?' and no one was ever able to answer it with much precision, until
at the end, quite suddenly, quite unexpectedly, our terrible foe collapsed before us, and we were so glutted with victory that in our folly
we threw it away."

I ended: "What General Weygand called the Battle of France is over. I expect that the Battle of Britain is about to begin. Upon this battle depends the survival of Christian civilisation. Upon it depends our own British life, and the long continuity of our institutions and our Empire. The whole fury and might of the enemy must very soon be turned on us. Hitler knows that he will have to break us in this island or lose the war. If we can stand up to him, all Europe may be free and the life of the world may move forward into broad, sunlit uplands. But if we fail, then the whole world, including the United States, including all that we have known and cared for, will sink into the abyss of a new Dark Age, made more sinister, and perhaps more protracted, by the lights of perverted science. Let us therefore brace ourselves to our duties, and so bear ourselves that, if the British Empire and its Commonwealth last for a thousand years, men will still say: 'This was their finest hour.' "

All these often-quoted words were made good in the hour of victory. But now they were only words. Foreigners who do not understand the temper of the British race all over the globe when its blood is up might have supposed that they were only a bold front, set up as a good prelude for peace negotiations. Hitler's need to finish the war in the West was obvious. He was in a position to offer the most tempting terms. To those who like myself had studied his moves it did not seem impossible that he would consent to leave Britain and her Empire and Fleet intact and make a peace which would have secured him that free hand in the East of which Ribbentrop had talked to me in 1937, and which was his heart's main desire. So far we had not done him much harm. We had indeed only added our own defeat to his triumph over France. Can one wonder that astute calculators in many countries, ignorant as they mostly were of the problems of overseas invasion, and of the quality of our air force, and who dwelt under the overwhelming impression of German might and terror, were not convinced? Not every Government called into being by Democracy or by Despotism, and not every nation, while quite alone, and as it seemed abandoned, would have courted the horrors of invasion and disdained a fair chance of peace for which many plausible excuses could be presented. Rhetoric was no guarantee. Another administration might come into being. "The warmongers have had their chance and failed." America had stood aloof. No one was under any obligation to Soviet Russia. Why should not Britain join the spectators who in Japan and in the United States, in Sweden, and in Spain, might watch with detached interest, or even relish, a mutually destructive struggle between the Nazi and Communist Empires? Future generations will find it hard to believe that the issues I have summarised here were never thought worth a place upon the Cabinet agenda, or even mentioned in our most private conclaves. Doubts could be swept away only by deeds. The deeds were to come.

*   *   *

In the closing days at Bordeaux Admiral Darlan became very important. My contacts with him had been few and formal. I respected him for the work he had done in re-creating the French Navy, which after ten years of his professional control was more efficient than at any time since the French Revolution. When in December 1939 he had visited England we gave him an official dinner at the Admiralty. In response to the toast, he began by reminding us that his great-grandfather had been killed at the Battle of Trafalgar. I therefore thought of him as one of those good Frenchmen who hate England. Anglo-French naval discussions in January had also shown how very jealous the Admiral was of his professional position in relation to whoever was the political Minister of Marine. This had become a positive obsession, and, I believe, played a definite part in his action.

For the rest, Darlan had been present at most of the conferences which I have described, and as the end of the French resistance approached he had repeatedly assured me that whatever happened the French Fleet should never fall into German hands. Now at Bordeaux came the fateful moment in the career of this ambitious, self-seeking, and capable Admiral. His authority over the Fleet was for all practical purposes absolute. He had only to order the ships to British, American, or French colonial harbours — some had already started — to be obeyed. In the morning of June 17, after the fall of M. Reynaud's Cabinet, he declared to General Georges that he was resolved to give the order. The next day Georges met him in the afternoon and asked him what had happened. Darlan replied that he had changed his mind. When asked why, he answered simply: "I am now Minister of Marine." This did not mean that he had changed his mind in order to become Minister of Marine; but that being Minister of Marine he had a different point of view.

How vain are human calculations of self-interest! Rarely has there been a more convincing example. Darlan had but to sail in any one of his ships to any port outside France to become the master of all French interests beyond German control. He would not have come, like General de Gaulle, with only on unconquerable heart and a few kindred spirits. He would have carried with him outside the German reach the fourth Navy in the world, whose officers and men were personally devoted to him. Acting thus, Darlan would have become the chief of the French Resistance with a mighty weapon in his hand. British and American dockyards and arsenals would have been at his disposal

for the maintenance of his fleet. The French gold reserve in the United
States would have assured him, once recognised, of ample resources.
The whole French Empire would have rallied to him. Nothing could
have prevented him from being the Liberator of France. The fame
and power which he so ardently desired were in his grasp. Instead, he
went forward through two years of worrying and ignominious office to a
violent death, a dishonoured grave, and a name long to be execrated
by the French Navy and the nation he had hitherto served so well.

There is a final note which should be struck at this point. In a letter
which Darlan wrote to me on December 4, 1942, just three weeks
before his assassination, he vehemently claimed that he had kept his
word. This letter states his case and has been printed by me elsewhere.[1]
It cannot be disputed that no French ship was ever manned by the
Germans or used against us by them in the war. This was not entirely
due to Admiral Darlan's measures; but he had certainly built up in the
minds of the officers and men of the French Navy that at all costs their
ships should be destroyed before being seized by the Germans, whom
he disliked as much as he did the English.

But in June 1940, the addition of the French Navy to the German
and Italian Fleets, with the menace of Japan measureless upon the
horizon, confronted Great Britain with mortal dangers and gravely affected the safety of the United States. Article 8 of the Armistice prescribed that the French Fleet, except that part left free for safeguarding French colonial interests, "shall be collected in ports to be specified
and there demobilised and disarmed under German or Italian control."
It was therefore clear that the French war vessels would pass into that
control while fully armed. It was true that in the same article the
German Government solemnly declared that they had no intention of
using them for their own purposes during the war. But who in his
senses would trust the word of Hitler after his shameful record and the
facts of the hour? Furthermore, the article excepted from this assurance
"those units necessary for coast surveillance and mine-sweeping." The
interpretation of this lay with the Germans. Finally, the Armistice
could at any time be voided on any pretext of non-observance. There
was in fact no security for us at all. At all costs, at all risks, in one way
or another, we must make sure that the Navy of France did not fall
into wrong hands, and then perhaps bring us and others to ruin.

*   *   *

The War Cabinet never hesitated. Those Ministers who, the week
before, had given their whole hearts to France and offered common
nationhood resolved that all necessary measures should be taken. This

[1] *Their Finest Hour*, Chapter 11.

was a hateful decision, the most unnatural and painful in which I have ever been concerned. It recalled the episode of the seizure by the Royal Navy of the Danish fleet at Copenhagen in 1807; but now the French had been only yesterday our dear allies, and our sympathy for the misery of France was sincere. On the other hand, the life of the State and the salvation of our cause were at stake. It was Greek tragedy. But no act was ever more necessary for the life of Britain and for all that depended upon it. I thought of Danton in 1793: "The coalesced Kings threaten us, and we hurl at their feet as a gage of battle the head of a King." The whole event was in this order of ideas.

The French Navy was disposed in the following manner: Two battleships, four light cruisers (or *contre-torpilleurs*), some submarines, including a very large one, the *Surcouf*, eight destroyers, and about two hundred smaller but valuable mine-sweeping and anti-submarine craft lay for the most part at Portsmouth and Plymouth. These were in our power. At Alexandria there were a French battleship, four French cruisers, three of them modern 8-inch-gun cruisers, and a number of smaller ships. These were covered by a strong British battle squadron. At Oran, at the other end of the Mediterranean, and at its adjacent military port of Mers-el-Kebir, were two of the finest vessels of the French fleet, the *Dunkerque* and the *Strasbourg*, modern battle cruisers much superior to the *Scharnhorst* and *Gneisenau*, and built for the express purpose of being superior to them. These vessels in German hands on our trade routes would have been most disagreeable. With them were two French battleships, several light cruisers, and a number of destroyers, submarines, and other vessels. At Algiers were seven cruisers, of which four were 8-inch armed, and at Martinique an aircraft carrier and two light cruisers. At Casablanca lay the *Jean Bart*, newly arrived from St. Nazaire, but without her guns. This was one of the key ships in the computation of world naval strength. She was unfinished, and could not be finished at Casablanca. She must not go elsewhere. The *Richelieu*, which was far nearer completion, had reached Dakar. She could steam, and her 15-inch guns could fire. There were many other French ships of minor importance in various ports. Finally, at Toulon a number of warships were beyond our reach. "Operation Catapult" comprised the simultaneous seizure, control, or effective disablement or destruction of all the accessible French Fleet.

In the early morning of July 3 all the French vessels at Portsmouth and Plymouth were taken under British control. The action was sudden and necessarily a surprise. Overwhelming force was employed, and the whole transaction showed how easily the Germans could have taken possession of any French warships lying in ports which they controlled. In Britain the transfer, except in the *Surcouf*, was amicable, and the crews came willingly ashore. In the *Surcouf* two gallant British officers and a leading seaman were killed,[2] and another seaman wounded. One French seaman also was killed, but many hundreds volunteered to join us. The *Surcouf*, after rendering distinguished service, perished on February 19, 1942, with all her gallant French crew.

The deadly stroke was in the Western Mediterranean. Here, at Gibraltar, Vice-Admiral Somerville with "Force H'" consisting of the battle cruiser *Hood,* the battleships *Valiant* and *Resolution,* the aircraft carrier *Ark Royal,* two cruisers, and eleven destroyers, received orders sent from the Admiralty at 2.25 A.M. on July 1:

> Be prepared for "Catapult" July 3.

Among Somerville's officers was Captain Holland, a gallant and distinguished officer, lately Naval Attaché in Paris and with keen French sympathies, who was influential. In the early afternoon of July 1 the Vice-Admiral telegraphed:

> After talk with Holland and others Vice-Admiral "Force H" is impressed with their view that the use of force should be avoided at all costs. Holland considers offensive action on our part would alienate all French wherever they are.

To this the Admiralty replied at 6.20 P.M.

> Firm intention of His Majesty's Government that if French will not accept any of your alternatives they are to be destroyed.

Shortly after midnight (1.08 A.M., July 2) Admiral Somerville was sent a carefully conceived communication to be made to the French Admiral. The crucial portion was as follows:

> (*a*) Sail with us and continue to fight for victory against the Germans and Italians.
> (*b*) Sail with reduced crews under our control to a British port. The reduced crews will be repatriated at the earliest moment.
> If either of these courses is adopted by you, we will restore your ships to France at the conclusion of the war, or pay full compensation if they are damaged meanwhile.
> (*c*) Alternatively, if you feel bound to stipulate that your ships should not be used against the Germans or Italians unless these break the Armistice, then sail them with us with reduced crews to some French port in the West Indies — Martinique, for instance — where they can be demilitarised to our satisfaction, or perhaps

---

[2] Commander D. V. Sprague, R.N., Lieutenant P. M. K. Griffiths, R.N., and Leading Seaman A. Webb, R.N.

be entrusted to the United States and remain safe until the end of the war, the crews being repatriated.

If you refuse these fair offers, I must, with profound regret, require you to sink your ships within six hours.

Finally, failing the above, I have the orders of His Majesty's Government to use whatever force may be necessary to prevent your ships from falling into German or Italian hands.

The Admiral sailed at daylight and was off Oran at about nine-thirty. He sent Captain Holland himself in a destroyer to wait upon the French Admiral Gensoul. After being refused an interview Holland sent by messengers the document already quoted. Admiral Gensoul replied in writing, that in no case would the French warships be allowed to fall intact into German and Italian hands, and that force would be met with force.

All day negotiations continued. At 4.15 P.M. Captain Holland was at last permitted to board the *Dunkerque,* but the ensuing meeting with the French Admiral was frigid. Admiral Gensoul had meanwhile sent two messages to the French Admiralty, and at 3 P.M. the French Council of Ministers had met to consider the British terms. General Weygand was present at this meeting, and what transpired has now been recorded by his biographer. From this it seems that the third alternative, namely, the removal of the French Fleet to the West Indies, was never mentioned. He states, " . . . It would appear that Admiral Darlan, whether deliberately or not, or whether he was aware of them or not, I do not know, *did not in fact inform us of all the details of the matter at the time.* It now appears that the terms of the British ultimatum were less crude than we were led to believe, and suggested a third and far more acceptable alternative, namely, the departure of the Fleet for West Indian waters." [3] No explanation of this omission, if it were an omission, has so far been seen. [4]

The distress of the British Admiral and his principal officers was evident to us from the signals which had passed. Nothing but the most direct orders compelled them to open fire on those who had been so lately their comrades. At the Admiralty also there was manifest emotion. But there was no weakening in the resolve of the War Cabinet. I sat all the afternoon in the Cabinet Room in frequent contact with my principal colleagues and the First Lord and First Sea Lord. A final signal was dispatched at 6.26 P.M.:

French ships must comply with our terms or sink themselves or be sunk by you before dark.

[3] *The Rôle of General Weygand,* by Jacques Weygand.
[4] Written in 1950.

But the action had already begun. At 5.45 P.M. Admiral Somerville opened fire upon this powerful French fleet, which was also protected by its shore batteries. At 6 P.M. he reported that he was heavily engaged. The bombardment lasted for some ten minutes. The battleship *Bretagne* was blown up. The *Dunkerque* ran aground. The battleship *Provence* was beached. The *Strasbourg* escaped, and, though attacked by torpedo aircraft from the *Ark Royal,* reached Toulon, as did also the cruisers from Algiers.

At Alexandria, after protracted negotiations with Admiral Cunningham, the French Admiral Godefroy agreed to discharge his oil fuel, to remove important parts of his gun mechanisms, and to repatriate some of his crews. At Dakar on July 8 an attack was made on the battleship *Richelieu* by the aircraft carrier *Hermes,* and most gallantly by a motorboat. The *Richelieu* was hit by an air torpedo and seriously damaged. The French aircraft carrier and two light cruisers in the French West Indies were immobilised after long-drawn-out discussions under an agreement with the United States.

*   *   *

On July 4 I reported at length to the House of Commons what we had done. Although the battle cruiser *Strasbourg* had escaped from Oran and the effective disablement of the *Richelieu* had not then been reported, the measures we had taken had removed the French Navy from major German calculations. I spoke for an hour or more that afternoon, and gave a detailed account of all these sombre events as they were known to me. I have nothing to add to the account which I then gave to Parliament and to the world. I thought it better, for the sake of proportion, to end upon a note which placed this mournful episode in true relation with the plight in which we stood. I therefore read to the House the admonition which I had, with Cabinet approval, circulated through the inner circles of the governing machine the day before:

> On what may be the eve of an attempted invasion or battle for our native land, the Prime Minister desires to impress upon all persons holding responsible positions in the Government, in the Fighting Services, or in the Civil departments their duty to maintain a spirit of alert and confident energy. While every precaution must be taken that time and means afford, there are no grounds for supposing that more German troops can be landed in this country, either from the air or across the sea, than can be destroyed or captured by the strong forces at present under arms. The Royal Air Force is in excellent order and at the highest strength yet attained. The German Navy was never so weak nor the British Army at home so

strong as now. The Prime Minister expects all His Majesty's servants in high places to set an example of steadiness and resolution. They should check and rebuke the expression of loose and ill-digested opinions in their circles, or by their subordinates. They should not hesitate to report, or if necessary remove, any persons, officers, or officials who are found to be consciously exercising a disturbing or depressing influence, and whose talk is calculated to spread alarm and despondency. Thus alone will they be worthy of the fighting men who, in the air, on the sea, and on land, have already met the enemy without any sense of being outmatched in martial qualities.

The House was very silent during the recital, but at the end there occurred a scene unique in my own experience. Everybody seemed to stand up all around, cheering, for what seemed a long time. Up till this moment the Conservative Party had treated me with some reserve, and it was from the Labour benches that I received the warmest welcome when I entered the House or rose on serious occasions. But now all joined in solemn stentorian accord.

The elimination of the French Navy as an important factor almost at a single stroke by violent action produced a profound impression in every country. Here was this Britain which so many had counted down and out, which strangers had supposed to be quivering on the brink of surrender to the mighty power arrayed against her, striking ruthlessly at her dearest friends of yesterday and securing for a while to herself the undisputed command of the sea. It was made plain that the British War Cabinet feared nothing and would stop at nothing. This was true.

The genius of France enabled her people to comprehend the whole significance of Oran, and in her agony to draw new hope and strength from this additional bitter pang. General de Gaulle, whom I did not consult beforehand, was magnificent in his demeanour, and France liberated and restored has ratified his conduct. I am indebted to M. Teitgen, a prominent member of the Resistance Movement, afterwards French Minister of Defence, for a tale which should be told. In a village near Toulon dwelt two peasant families, each of whom had lost their sailor son by British fire at Oran. A funeral service was arranged to which all their neighbors sought to go. Both families requested that the Union Jack should lie upon the coffins side by side with the Tricolour, and their wishes were respectfully observed. In this we may see how the comprehending spirit of simple folk touches the sublime.

IO

At Bay

IN THESE SUMMER DAYS OF 1940 after the fall of France we were all alone. None of the British Dominions or India or the Colonies could send decisive aid, or send what they had in time. The victorious, enormous German armies, thoroughly equipped and with large reserves of captured weapons and arsenals behind them, were gathering for the final stroke. Italy, with numerous and imposing forces, had declared war upon us, and eagerly sought our destruction in the Mediterranean and in Egypt. In the Far East, Japan glared inscrutably, and pointedly requested the closing of the Burma Road against supplies for China. Soviet Russia was bound to Nazi Germany by her pact, and lent important aid to Hitler in raw materials. Spain, which had already occupied the International Zone of Tangier, might turn against us at any moment and demand Gibraltar, or invite the Germans to help her attack it, or mount batteries to hamper passage through the Straits. The France of Pétain and Bordeaux, soon moved to Vichy, might any day be forced to declare war upon us. What was left at Toulon of the French Fleet seemed to be in German power. Certainly we had no lack of foes.

After Oran it became clear to all countries that the British Government and nation were resolved to fight on to the last. But even if there were no moral weakness in Britain, how could the appalling physical facts be overcome? Our armies at home were known to be almost unarmed except for rifles. Months must pass before our factories could make good even the munitions lost at Dunkirk. Can one wonder that the world at large was convinced that our hour of doom had struck?

Deep alarm spread through the United States, and indeed through all the surviving free countries. Americans gravely asked themselves

whether it was right to cast away any of their own severely limited resources to indulge a generous though hopeless sentiment. Ought they
not to strain every nerve and nurse every weapon to remedy their own
unpreparedness? It needed a very sure judgment to rise above these
cogent, matter-of-fact arguments. The gratitude of the British nation
is due to the noble President and his great officers and high advisers for
never, even in the advent of the Third Term Presidential Election,
losing their confidence in our fortunes or our will.

The buoyant and imperturbable temper of Britain, which I had the
honour to express, may well have turned the scale. Here was this
people, who in the years before the war had gone to the extreme bounds
of pacifism and improvidence, who had indulged in the sport of party
politics, and who, though so weakly armed, had advanced lightheartedly
into the centre of European affairs, now confronted with the reckoning
alike of their virtuous impulses and neglectful arrangements. They
were not even dismayed. They defied the conquerors of Europe. They
seemed willing to have their island reduced to a shambles rather than
give in. This would make a fine page in history. But there were other
tales of this kind. Athens had been conquered by Sparta. The Cartha-
ginians made a forlorn resistance to Rome. Not seldom in the annals
of the past — and how much more often in tragedies never recorded or
long-forgotten — had brave, proud, easygoing states, and even entire
races, been wiped out, so that only their name or even no mention of
them remains.

Few British and very few foreigners understood the peculiar technical advantages of our insular position; nor was it generally known how
even in the irresolute years before the war the essentials of sea and
latterly air defence had been maintained. It was nearly a thousand years
since Britain had seen the fires of a foreign camp on English soil. At
the summit of British resistance everyone remained calm, content to
set their lives upon the cast. That this was our mood was gradually
recognised by friends and foes throughout the whole world. What
was there behind the mood? That could be settled only by brute
force.

There was also another aspect. One of our greatest dangers during
June lay in having our last reserves drawn away from us into a wasting,
futile French resistance in France, and the strength of our air forces
gradually worn down by their flights or transference to the Continent.
If Hitler had been gifted with supernatural wisdom he would have
slowed down the attack on the French front, making perhaps a pause
of three or four weeks after Dunkirk on the line of the Seine, and meanwhile developing his preparations to invade England. Thus he would

have had a deadly option, and could have tortured us with the hooks of either deserting France in her agony or squandering the last resources for our future existence. The more we urged the French to fight on, the greater was our obligation to aid them, and the more difficult it would have become to make any preparations for defence in England, and above all to keep in reserve the twenty-five squadrons of fighter aircraft on which all depended. On this point we should never have given way, but the refusal would have been bitterly resented by our struggling ally, and would have poisoned all our relations. It was even with an actual sense of relief that some of our high commanders addressed themselves to our new and grimly simplified problem. As the commissionaire at one of the Service clubs in London said to a rather downcast member: "Anyhow, sir, we're in the Final, and it's to be played on the Home Ground."

*   *   *

The strength of our position was not, even at this date, underrated by the German High Command. Ciano tells how, when he visited Hitler in Berlin on July 7, 1940, he had a long conversation with General von Keitel. Keitel, like Hitler, spoke to him about the attack on England. He repeated that up to the present nothing definite had been decided. He regarded the landing as possible, but considered it an "extremely difficult operation, which must be approached with the utmost caution, in view of the fact that the intelligence available on the military preparedness of the island and on the coastal defences is meagre and not very reliable."[1] What would appear to be easy and also essential was a major air attack upon the airfields, factories, and the principal communication centres in Great Britain. It was necessary however to bear in mind that the British Air Force was extremely efficient. Keitel calculated that the British had about fifteen hundred machines ready for defence and counterattack. He admitted that recently the offensive action of the British Air Force had been greatly intensified. Bombing missions were carried out with noteworthy accuracy, and the groups of aircraft which appeared numbered up to eighty machines at a time. There was however in England a great shortage of pilots, and those who were now attacking the German cities could not be replaced by the new pilots, who were completely untrained. Keitel also insisted upon the necessity of striking at Gibraltar in order to disrupt the British imperial system. Neither Keitel nor Hitler made any reference to the duration of the war. Only Himmler said incidentally that the war ought to be finished by the beginning of October.

Such was Ciano's report. He also offered Hitler, at "the earnest wish

[1] Ciano, *Diplomatic Papers*. p. 378.

of the Duce," an army of ten divisions and an air component of thirty squadrons to take part in the invasion. The army was politely declined. Some of the air squadrons came, but, as will be presently related, fared ill.

*   *   *

On July 19 Hitler delivered a triumphant speech in the Reichstag, in which, after predicting that I would shortly take refuge in Canada, he made what has been called his Peace Offer. This gesture was accompanied during the following days by diplomatic representations through Sweden, the United States, and at the Vatican. Naturally Hitler would have been very glad, after having subjugated Europe to his will, to bring the war to an end by procuring British acceptance of what he had done. It was in fact an offer not of peace but of readiness to accept the surrender by Britain of all she had entered the war to maintain.

My first thought was a solemn, formal debate in both Houses of Parliament, but my colleagues thought that this would be making too much of the matter, upon which we were all of one mind. It was decided instead that the Foreign Secretary should dismiss Hitler's gesture in a broadcast. On the night of the 22nd he "brushed aside" Hitler's "summons to capitulate to his will." He contrasted Hitler's picture of Europe with the picture of the Europe for which we were fighting, and declared that "we shall not stop fighting until Freedom is secure." In fact however the rejection of any idea of a parley had already been given in the British press and by the B.B.C., without any prompting from His Majesty's Government, as soon as Hitler's speech was heard over the radio.

Ciano records in his diaries that "late in the evening of the 19th, when the first cold British reaction to the speech arrived, a sense of ill-concealed disappointment spread among the Germans." Hitler "would like an understanding with Great Britain. He knows that war with the British will be hard and bloody, and knows also that people everywhere are averse from bloodshed." Mussolini, on the other hand, "fears that the English may find in Hitler's much too cunning speech a pretext to begin negotiations." "That," remarks Ciano "would be sad for Mussolini, because now more than ever he wants war." [2] He need not have fretted himself. He was not to be denied all the war he wanted.

*   *   *

At the end of June the Chiefs of Staff through General Ismay had suggested to me at the Cabinet that I should visit the threatened sectors

[2] *Ciano's Diaries*, pp. 277-78.

of the east and south coasts. Accordingly I devoted a day or two every week to this agreeable task, sleeping when necessary in my train, where I had every facility for carrying on my regular work and was in constant contact with Whitehall. I inspected the Tyne and the Humber and many possible landing places. The Canadian Division did an exercise for me in Kent. I examined the landward defences of Harwich and Dover. One of my earliest visits was to the 3rd Division, commanded by General Montgomery, an officer whom I had not met before. My wife came with me. The 3rd Division was stationed near Brighton. It had been given the highest priority in re-equipment, and had been about to sail for France when the French resistance ended. General Montgomery's headquarters were near Steyning, and he showed me a small exercise of which the central feature was a flanking movement of Bren-gun carriers, of which he could at that moment muster only seven or eight. After this we drove together along the coast through Shoreham and Hove till we came to the familiar Brighton front, of which I had so many schoolboy memories. We dined in the Royal Albion Hotel, which stands opposite the end of the pier. The hotel was entirely empty, a great deal of evacuation having taken place; but there were still a number of people airing themselves on the beaches or the parade. I was amused to see a platoon of the Grenadier Guards making a sand-bag machine-gun post in one of the kiosks of the pier, like those where in my childhood I had often admired the antics of the performing fleas. It was lovely weather. I had very good talks with the General, and enjoyed my outing thoroughly.

In mid-July the Secretary of State for War recommended that General Brooke should replace General Ironside in command of our Home Forces. On July 19, in the course of my continuous inspection of the invasion sectors, I visited the Southern Command. Some sort of tactical exercise was presented to me in which no fewer than twelve tanks were able to participate. All the afternoon I drove with General Brooke, who commanded this front. His record stood high. Not only had he fought the decisive flank battle near Ypres during the retirement to Dunkirk, but he had acquitted himself with singular firmness and dexterity, in circumstances of unimaginable difficulty and confusion, when in command of the new forces we had sent to France during the first three weeks of June. I also had a personal link with Alan Brooke through his two gallant brothers — the friends of my early military life.

These connections and memories did not decide my opinion on the grave matters of selection; but they formed a personal foundation upon which my unbroken wartime association with Alan Brooke was maintained and ripened. We were four hours together in the motorcar on this July afternoon of 1940, and we seemed to be in agreement on the methods of Home Defence. After the necessary consultations with others, I approved the Secretary of State for War's proposal to place Brooke in command of the Home Forces in succession to General Ironside. Ironside accepted his retirement with soldierly dignity which on all occasions characterised his actions.

During the invasion menace for a year and a half Brooke organised and commanded the Home Forces, and thereafter when he had become C.I.G.S. we continued together for three and a half years until victory was won. I shall presently narrate the benefits which I derived from his advice in the decisive changes of command in Egypt and the Middle East in August 1942, and also the heavy disappointment which I had to inflict upon him about the command of the cross-Channel invasion "Operation Overlord" in 1944. His long tenure as chairman of the Chiefs of Staff Committee during the greater part of the war and his work as C.I.G.S. enabled him to render services of the highest order, not only to the British Empire, but also to the Allied Cause. This tale will record occasional differences between us, but also an overwhelming measure of agreement, and will witness to a friendship which I cherish.

*   *   *

During this same month of July, American weapons in considerable quantities were safely brought across the Atlantic. When the ships approached our shores with their priceless arms, special trains were waiting in all the ports to receive their cargoes. The Home Guard in every country, in every town, in every village, sat up all through the night to receive them. Men and women worked night and day making them fit for use. By the end of July we were an armed nation, so far as parachute or air-borne landings were concerned. We had become a "hornet's nest." Anyhow, if we had to go down fighting (which I did not anticipate) a lot of our men and some women had weapons in their hands. The arrival of the first instalment of the half-million .300 rifles for the Home Guard (albeit with only about fifty cartridges apiece, of which we dared issue only ten, and no factories yet set in motion) enabled us to transfer three hundred thousand .303 British-type rifles to the rapidly expanding formations of the Regular Army.

At the seventy-fives, with their thousand rounds apiece, some fastidious experts presently turned their noses up. There were no limbers and no immediate means of procuring more ammunition. Mixed calibres complicate operations. But I would have none of this, and during all 1940 and 1941 these nine hundred seventy-fives were a great addition to our military strength for Home Defence. Arrangements were devised and men were drilled to run them up on planks into lorries for movement. When you are fighting for existence any cannon is better than no cannon at all, and the French seventy-five, although outdated by the British 25-pounder and the German field-gun howitzer, was still a splendid weapon.

As July and August passed without any disaster we settled ourselves down with increasing assurance that we could make a long and hard fight. Our gains of strength were borne in upon us from day to day. The entire population laboured to the last limit of its strength, and felt rewarded when they fell asleep after their toil or vigil by a growing sense that we should have time and that we should win. All the beaches now bristled with defences of various kinds. The whole country was organised in defensive localities. The factories poured out their weapons. By the end of August we had over two hundred and fifty new tanks! The fruits of the American "Act of Faith" had been gathered. The whole trained professional British Army and its Territorial comrades drilled and exercised from morn till night, and longed to meet the foe. The Home Guard overtopped the million mark, and when rifles were lacking grasped lustily the shotgun, the sporting rifle, the private pistol, or, when there was no firearm, the pike and the club. No Fifth Column existed in Britain, though a few spies were carefully rounded up and examined. What few Communists there were lay low. Everyone else gave all they had to give.

When Ribbentrop visited Rome in September he said to Ciano: "The English territorial defence is nonexistent. A single German division will suffice to bring about a complete collapse." This merely shows his ignorance. I have often wondered however what would have happened if two hundred thousand German storm troops had actually established themselves ashore. The massacre would have been on both sides grim and great. There would have been neither mercy nor quarter. They would have used terror, and we were prepared to go all lengths. I intended to use the slogan "You can always take one with you." I even calculated that the horrors of such a scene would in the last resort turn the scale in the United States. But none of these emotions was put to the proof. Far out on the grey waters of the North Sea and the Channel coursed and patrolled the faithful, eager flotillas peering through the night. High in the air soared the fighter pilots, or waited serene at a moment's notice around their excellent machines. This was a time when it was equally good to live or die.

*   *   *

Sea power, when properly understood, is a wonderful thing. The passage of an army across salt water in the face of superior fleets and flotillas is an almost impossible feat. Steam had added enormously to the power of the Navy to defend Great Britain. In Napoleon's day the same wind which would carry his flat-bottomed boats across the Channel from Boulogne would drive away our blockading squadrons. But everything that had happened since then had magnified the power of the superior navy to destroy the invaders in transit. Every complication which modern apparatus had added to armies made their voyage more cumbrous and perilous, and the difficulties of their maintenance when landed probably insuperable. At that former crisis in our island fortunes we possessed superior and, as it proved, ample sea power. The enemy was unable to gain a major sea battle against us. He could not face our cruiser forces. In flotillas and light craft we outnumbered him tenfold. Against this must be set the incalculable chances of weather, particularly fog. But even if this were adverse and a descent were effected at one or more points, the problem of maintaining a hostile line of communications and of nourishing any lodgments remained unsolved. Such was the position in the First Great War.

But now there was the air. What effect had this sovereign development produced upon the invasion problem? Evidently if the enemy could dominate the narrow seas, on both sides of the Straits of Dover, by superior air power, the losses of our flotillas would be very heavy and might eventually be fatal. No one would wish, except on a supreme occasion, to bring heavy battleships or large cruisers into waters commanded by the German bombers. We did not in fact station any capital ships south of the Forth or east of Plymouth. But from Harwich, the Nore, Dover, Portsmouth, and Portland we maintained a tireless, vigilant patrol of light fighting vessels which steadily increased in number. By September they exceeded eight hundred, which only a hostile air power could destroy, and then only by degrees.

But who had the power in the air? In the Battle of France we had fought the Germans against odds of two and three to one and inflicted losses in similar proportions. Over Dunkirk, where we had to maintain continuous patrol to cover the escape of the Army, we had fought at four or five to one with success and profit. Over our own waters and exposed coasts and counties Air Chief Marshall Dowding contemplated profitable fighting at seven or eight to one. The strength of the German Air Force at this time, taken as a whole, so far as we knew — and we were well informed — apart from particular concentrations, was about three to one. Although these were heavy odds at which to fight the brave and efficient German foe, I rested upon the conclusion that in our own air, over our own country and its waters, we could beat the German Air Force. And if this were true our naval power would continue to rule the seas and oceans, and would destroy all enemies who set their course towards us.

There was of course a third potential factor. Had the Germans with their renowned thoroughness and foresight secretly prepared a vast armada of special landing craft, which needed no harbours or quays, but could land tanks, cannon, and motor vehicles anywhere on the beaches, and which thereafter could supply the landed troops? As has been shown, such ideas had risen in my mind long ago in 1917, and were now being actually developed as the result of my directions. We had however no reason to believe that anything of this kind existed in Germany, though it is always best when counting the cost not to exclude the worst. It took us four years of intense effort and experiment and immense material aid from the United States to provide such equipment on a scale equal to the Normandy landing. Much less would have sufficed the Germans at this moment. But they had only a few ferries.

Thus the invasion of England in the summer and autumn of 1940 required from Germany local naval superiority and air superiority and immense special fleets and landing craft. But it was we who had the naval superiority; it was we who conquered the mastery in the air; and finally we believed, as we now know rightly, that they had not built or conceived any special craft. These were the foundations of my thought about invasion in 1940. In July there was growing talk and anxiety on the subject both inside the British Government and at large. In spite of ceaseless reconnaissance and all the advantages of air photography, no evidence had yet reached us of large assemblies of transport in the Baltic or in the Rhine or Scheldt harbours, and we were sure that no movement either of shipping or self-propelled barges through the Straits into the Channel had taken place. Nevertheless preparation to resist invasion was the supreme task before us all, and intense thought was devoted to it throughout our war circle and Home Command.

As will presently be described, the German plan was to invade across the Channel with medium ships (4000 to 5000 tons) and small craft, and we now know that they never had any hope or intention of moving an army from the Baltic and North Sea ports in large transports; still less did they make any plans for an invasion from the Biscay ports. This does not mean that in choosing the south coast as their target they were thinking rightly and we wrongly. The east coast invasion was by far the more formidable if the enemy had had the means to attempt it. There could of course be no south coast invasion unless or until the necessary shipping had passed southward through the Straits of Dover and had been assembled in the French Channel ports. Of this, during July, there was no sign.

We had none the less to prepare against all variants, and yet at the same time avoid the dispersion of our mobile forces, and to gather reserves. This nice and difficult problem could only be solved in relation to the news and events from week to week. The British coastline, indented with innumerable inlets, is over two thousand miles in circumference, without including Ireland. The only way of defending so vast a perimeter, any part or parts of which may be simultaneously or successively attacked, is by lines of observation and resistance around the coast or frontiers with the object of delaying an enemy, and meanwhile creating the largest possible reserves of highly trained mobile troops so disposed as to be able to reach any point assailed in the shortest time for strong counterattack. When in the last phases of the war Hitler found himself encircled and confronted with a similar problem he made, as we shall see, the gravest possible mistakes in handling it. He had created a spider's web of communications, *but he forgot the spider.* With the example of the unsound French dispositions for which such a fatal penalty had just been exacted fresh in our memories, we did not forget the "mass of manoeuvre"; and I ceaselessly inculcated this policy to the utmost extent that our growing resources would allow.

My views were in general harmony with Admiralty thought, and on July 12 Admiral Pound sent me a full and careful statement which he and the Naval Staff had drawn up in pursuance of it. Naturally and properly, the dangers we had to meet were forcefully stated. But in summing up, Admiral Pound said: *"It appears probable that a total of some hundred thousand men might reach these shores without being intercepted by naval forces . . .* but the maintenance of their line of supply, unless the German Air Force had overcome both our Air Force and our Navy, seems practically impossible. . . . If the enemy undertook this operation he would do so in the hope that he could make a quick rush on London, living on the country as he went, and force the Government to capitulate." I was content with this estimate.

Then in August the situation began to change in a decisive manner. Our excellent Intelligence confirmed that the operation "Sea Lion" had been definitely ordered by Hitler and was in active preparation. It seemed certain that the man was going to try. Moreover, the front to be attacked was altogether different from *or additional* to the east coast, on which the Chiefs of Staff, the Admiralty and I, in full agreement, still laid the major emphasis. A large number of self-propelled barges and motorboats began to pass by night through the Straits of Dover, creeping along the French coast and gradually assembling in all the French Channel ports from Calais to Brest. Our daily photographs showed this movement with precision. It had not been found possible to re-lay our minefields close to the French shore. We immediately began to attack the vessels in transit with our small craft, and Bomber Command was concentrated upon the new set of invasion ports now opening upon us. At the same time a great deal of information came to hand about the assembly of a German Army or Armies of Invasion along this stretch of the hostile coast, of movement on the railways, and of large concentrations in the Pas de Calais and Normandy. Large numbers of powerful long-range batteries all along the French Channel coast came into existence.

In response to the new menace we began to shift our weight from one leg to the other and to improve all our facilities for moving our increasingly large mobile reserves towards the southern front. All the time our forces were increasing in numbers, efficiency, mobility, and equipment, and in the last half of September we were able to bring into action on the south coast front sixteen divisions of high quality, of which three were armoured divisions or their equivalent in brigades, all of which were additional to the local coastal defence and could come into action with great speed against any invasion landing. This provided us with a punch or series of punches which General Brooke was well poised to deliver as might be required; and no one more capable.

                    *   *   *

All this while we could not feel any assurance that the inlets and river mouths from Calais to Terschelling and Heligoland, with all that swarm of islands off the Dutch and German coasts (the "Riddle of the Sands" of the previous war), might not conceal other large hostile forces with small or moderate-sized ships. An attack from Harwich right round to Portsmouth, Portland, or even Plymouth, centring upon the Kent promontory, seemed to impend. We had nothing but negative evidence that a third wave of invasion harmonised with the others might not be launched from the Baltic through the Skagerrak in large ships. This was indeed essential to a German success, because in no other way could heavy weapons reach the landed armies or large depots of supply be established.

We now entered upon a period of extreme tension and vigilance. We had of course all this time to maintain heavy forces north of the Wash, right up to Cromarty; and arrangements were perfected to draw from these should the assault declare itself decidedly in the south. The abundant intricate railway system of the island and our continued mastery of our home air would have enabled us to move with certainty another four or five divisions to reinforce the southern defence if it were necessary on the fourth, fifth, and sixth days after the enemy's full effort had been exposed.

A very careful study was made of the moon and the tides. We thought that the enemy would like to cross by night and land at dawn; and we now know the German Army Command felt like this too.

They would also be glad of a half-moonlight on the way over, so as to keep their order and make their true landfall. Measuring it all with precision, the Admiralty thought the most favourable conditions for the enemy would arise between the 15th and 30th of September. Here also we now find that we were in agreement with our foes. We had little doubt of our ability to destroy anything that got ashore on the Dover promontory or on the sector of coast from Dover to Portsmouth, or even Portland. As all our thoughts at the summit moved together in harmonious and detailed agreement, one could not help liking the picture which presented itself with growing definition. Here perhaps was the chance of striking a blow at the mighty enemy which would resound throughout the world. One could not help being inwardly excited alike by the atmosphere and the evidence of Hitler's intentions which streamed in upon us. There were indeed some who on purely technical grounds, and for the sake of the effect the total defeat and destruction of his expedition would have on the general war, were quite content to see him try.

In July and August we had asserted air mastery over Great Britain, and were especially powerful and dominant over the home counties of the southeast. Vast intricate systems of fortifications, defended localities, anti-tank obstacles, blockhouses, pillboxes, and the like laced the whole area. The coastline bristled with defences and batteries, and at the cost of heavier losses through reduced escorts in the Atlantic, and also by new construction coming into commission, the flotillas grew substantially in numbers and quality. We had brought the battleship *Revenge,* and the old target ship and dummy battleship *Centurion,* and a cruiser to Plymouth. The Home Fleet was at its maximum strength and could operate without much risk to the Humber and even to the Wash. In all respects therefore we were fully prepared.

Finally, we were already not far from the equinoctial gales customary in October. Evidently September was the month for Hitler to strike if he dared, and the tides and the moon phase were favourable in the middle of that month.

It is time to go over to the other camp and set forth the enemy's preparations and plans as we now know them.

II

Operation Sea Lion

SOON AFTER WAR BROKE OUT on September 3, 1939, the German Admiralty, as we have learned from their captured archives, began their staff study of the invasion of Britain. Unlike us, they had no doubt that the only way was across the narrow waters of the English Channel. They never considered any other alternative. If we had known this it would have been an important relief. An invasion across the Channel came upon our best-defended coast, the old sea front against France, where all the ports were fortified and our main flotilla bases and in later times most of our airfields and air-control stations for the defence of London were established. There was no part of the island where we could come into action more quickly or in such great strength with all three Services. Admiral Raeder was anxious not to be found wanting should the demand to invade Britain be made upon the German Navy. At the same time he asked for a lot of conditions. The first of these was the entire control of the French, Belgian, and Dutch coasts, harbours, and river mouths. Therefore the project slumbered during the Twilight War.

Suddenly all these conditions were surprisingly fulfilled, and it must have been with some misgivings but also satisfaction that on the morrow of Dunkirk and the French surrender he could present himself to the Fuehrer with a plan. On May 21 and again on June 20 he spoke to Hitler on the subject, not with a view to proposing an invasion, but in order to make sure that if it were ordered the planning in detail should not be rushed. Hitler was sceptical, saying that "he fully appreciated the exceptional difficulties of such an undertaking." He also nursed the hope that England would sue for peace. It was not until the last week in June that the Supreme Headquarters turned to this idea, nor till July 2 that the first directive was issued for planning the invasion of Britain as a possible event. "The Fuehrer has decided that under certain conditions — the most important of which is achieving air superiority — a landing in England may take place." On July 16 Hitler issued his directive: "Since England in spite of her militarily hopeless position shows no sign of coming to terms, I have decided to prepare a landing operation against England, and if necessary to carry it out. . . . The preparations for the entire operation must be completed by mid-August." Active measures in every direction were already in progress.

*   *   *

The German Navy plan was essentially mechanical. Under the cover of heavy-gun batteries firing from Gris-Nez towards Dover, and a very strong artillery protection along the French coast in the Straits, they proposed to make a narrow corridor across the Channel on the shortest convenient line and to wall this in by minefields on either side, with outlying U-boat protection. Through this the Army was to be ferried over and supplied in a large number of successive waves. There the Navy stopped, and on this the German Army chiefs were left to address themselves to the problem.

Considering that we could, with our overwhelming naval superiority, tear these minefields to pieces with small craft under superior air power and also destroy the dozen or score of U-boats concentrated to protect them, this was at the outset a bleak proposition. Nevertheless, after the fall of France anyone could see that the only hope of avoiding a long war, with all that it might entail, was to bring Britain to her knees. The German Navy itself had been, as we have recorded, knocked about in a most serious manner in the fighting off Norway; and in their crippled condition they could not offer more than minor support to the Army. Still, they had their plan, and no one could say that they had been caught unawares by good fortune.

The German Army Command had from the first regarded the invasion of England with considerable qualms. They had made no plans or preparations for it; and there had been no training. As the weeks of prodigious, delirious victory succeeded one another they were emboldened. The responsibility for the safe crossing was not departmentally theirs, and, once landed in strength, they felt that the task was within their power. Indeed, already in August Admiral Raeder felt it necessary to draw their attention to the dangers of the passage, during which perhaps the whole of the army forces employed might be lost. Once the responsibility for putting the Army across was definitely thrust upon the Navy, the German Admiralty became consistently pessimistic.

On July 21 the heads of the three Services met the Fuehrer. He informed them that the decisive stage of the war had already been reached, but that England had not yet recognised it and still hoped for a turn of fate. He spoke of the support of England by the United States and of a possible change in German political relations with Soviet Russia. The execution of "Sea Lion," he said, must be regarded as the most effective means of bringing about a rapid conclusion of the war. After his long talks with Admiral Raeder, Hitler had begun to realise what the crossing of the channel, with its tides and currents, and all the mysteries of the sea, involved. He described "Sea Lion" as "an exceptionally bold and daring undertaking." "Even if the way is short, this is not just a river crossing, but the crossing of a sea which is dominated by the enemy. This is not a case of a single-crossing operation, as in Norway; operational surprise cannot be expected; a defensively prepared *and utterly determined enemy* faces us and dominates the sea area which we must use. For the Army operation forty divisions will be required. The most difficult part will be the material reinforcements and stores. We cannot count on supplies of any kind being available to us in England." The prerequisites were complete mastery of the air, the operational use of powerful artillery in the Dover Straits, and protection by minefields. "The time of year," he said, "is an important factor, since the weather in the North Sea and in the Channel during the second half of September is very bad, and the fogs begin in the middle of October. The main operation must therefore be completed by September 15, for after that date co-operation between the Luftwaffe and the heavy weapons becomes too unreliable. But as air co-operation is decisive it must be regarded as the principal factor in fixing the date."

A vehement controversy, conducted with no little asperity, arose in the German staffs about the width of the front and the number of points to be attacked. The Army demanded a series of landings along the whole English southern coast from Dover to Lyme Regis, west of Portland. They also desired an ancillary landing north of Dover at Ramsgate. The German Naval Staff now stated that the most suitable area for the safe crossing of the English Channel was between the North Foreland and the western end of the Isle of Wight. On this the Army Staff developed a plan for a landing of 100,000 men, followed almost immediately by 160,000 more at various points from Dover westward to Lyme Bay. Colonel-General Halder, Chief of the Army Staff, declared that it was necessary to land at least four divisions in the Brighton area. He also required landings in the area Deal–Ramsgate; at least thirteen divisions must be deployed, as far as possible simultaneously, at points along the whole front. In addition, the Luftwaffe demanded shipping to transport fifty-two anti-aircraft batteries with the first wave.

The Chief of the Naval Staff however made it clear that nothing like

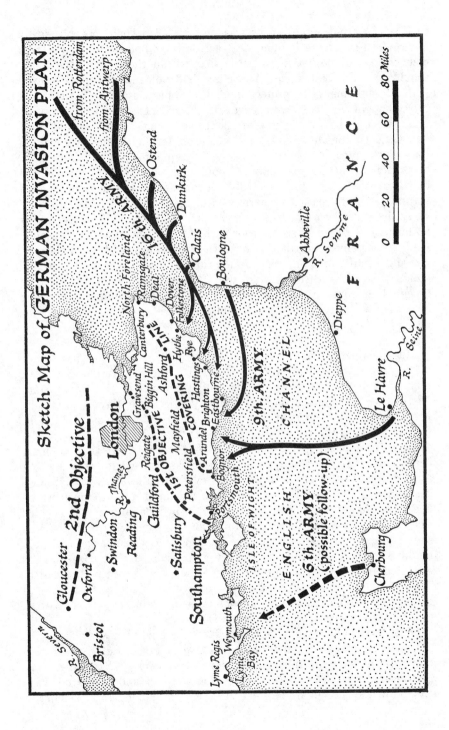

Sketch Map of GERMAN INVASION PLAN

— — — 2nd Objective

— · — · — 1st OBJECTIVE

— — — COVERING LINE

from Rotterdam

from Antwerp

Ostend

Dunkirk

Calais

Boulogne

16 th ARMY

North Foreland
Ramsgate
Deal
Dover
Folkestone
Hythe
Rye
Hastings
Eastbourne
Brighton
Bognor

Canterbury
Biggin Hill
Ashford
Mayfield
Arundel
Petersfield
Portsmouth

Gravesend
Reigate
Guildford
Salisbury
Southampton

London
R. Thames
Reading
Swindon
Oxford
Gloucester
Bristol
R. Severn

9th ARMY

ENGLISH CHANNEL

ISLE OF WIGHT

6th ARMY
(possible follow-up)

Lyme Regis
Lyme Bay
Weymouth

Cherbourg

Le Havre

Dieppe

Abbeville

R. Somme

R. Seine

F R A N C E

0   20   40   60   80 Miles

so large or rapid a movement was possible. He could not physically undertake to escort a landing fleet across the whole width of the area mentioned. All he had meant was that within these limits the Army should pick the best place. The Navy had not enough strength, even with air supremacy, to protect more than one passage at a time, and they thought the narrowest parts of the Straits of Dover the least difficult. To carry the whole of the 160,000 men of the second wave and their equipment in a single operation would require two million tons of shipping. Even if this fantastic requirement could have been met, such quantities of shipping could not have been accommodated in the area of embarkation. Only the first échelons could be thrown across for the formation of narrow bridgeheads, and at least two days would be needed to land the second échelons of these divisions, to say nothing of the second six divisions which were thought indispensable. He further pointed out that a broad-front landing would mean three to five and a half hours' difference in the times of high water at the various points selected. Either therefore unfavourable tide conditions must be accepted at some places, or simultaneous landings renounced. This objection must have been very difficult to answer.

*   *   *

Much valuable time had been consumed in these exchanges of memoranda. It was not until August 7 that the first verbal discussion took place between General Halder and the Chief of the Naval Staff. At this meeting Halder said : "I utterly reject the Navy's proposals. From the Army viewpoint I regard it as complete suicide. I might just as well put the troops that have been landed straight through the sausage machine." The Naval Chief of Staff rejoined that he must equally reject the landing on a broad front, as that would lead only to a sacrifice of the troops on the passage over. In the end a compromise decision was given by Hitler which satisfied neither the Army nor the Navy. A Supreme Command Directive, issued on August 27, decided that "the Army operations must allow for the facts regarding available shipping space and security of the crossing and disembarkation." All landings in the Deal–Ramsgate area were abandoned, but the front was extended from Folkestone to Bognor. Thus it was nearly the end of August before even this measure of agreement was reached; and of course everything was subject to victory being gained in the air battle, which had now been raging for six weeks.

On the basis of the frontage at last fixed, the final plan was made. The military command was entrusted to Rundstedt, but shortage of shipping reduced his force to thirteen divisions with twelve in reserve. The Sixteenth Army, from ports between Rotterdam and Boulogne, were to land in the neighbourhood of Hythe, Rye, Hastings, and Eastbourne.

the Ninth Army, from ports between Boulogne and Havre, attacking between Brighton and Worthing. Dover was to be captured from the landward side; then both armies would advance to the covering line of Canterbury–Ashford–Mayfield–Arundel. In all, eleven divisions were to be landed in the first waves. A week after the landing it was hoped, optimistically, to advance yet farther, to Gravesend, Reigate, Petersfield, Portsmouth. In reserve lay the Sixth Army, with divisions ready to reinforce, or, if circumstances allowed, to extend the frontage of attack to Weymouth. There was indeed no lack of fierce and well-armed troops, but they required shipping and safe conveyance.

On the Naval Staff fell the heaviest initial task. Germany had about 1,200,000 tons of seagoing shipping available to meet all her needs. To embark the invasion force would require more than half this amount, and would involve great economic disturbance. By the beginning of September the Naval Staff were able to report that the following had been requisitioned:

> 168 transports (of 700,000 tons)
> 1910 barges
> 419 tugs and trawlers
> 1600 motorboats

All this armada had to be manned, and brought to the assembly ports by sea and canal. When on September 1 the great southward flow of invasion shipping began it was watched, reported, and violently assailed by the Royal Air Force along the whole front from Antwerp to Havre. The German Naval Staff recorded: "The enemy's continuous fighting defence off the coast, his concentration of bombers on the 'Sea Lion' embarkation ports, and his coastal reconnaissance activities indicate that he is now expecting an immediate landing."

And again: "The English bombers, however, and the minelaying forces of the British Air Force . . . are still at full operational strength, and it must be confirmed that the activity of the British forces has undoubtedly been successful even if no decisive hindrance has yet been caused to German transport movement."

Yet, despite delays and damage, the German Navy completed the first part of its task. The ten per cent margin for accidents and losses it had provided was fully expended. What survived however did not fall short of the minimum it had planned to have for the first stage.

*   *   *

Both Navy and Army now cast their burden on the German Air Force. All this plan of the corridor, with its balustrades of minefields to be laid and maintained under the German Air Force canopy against the overwhelming superiority of the British flotillas and small craft, depended upon the defeat of the British Air Force and the complete mastery of the air by Germany over the Channel and Southeast England, and not only over the crossing but over the landing points. Both the older services passed the buck to Reichsmarschall Goering.

Goering was by no means unwilling to accept this responsibility, because he believed that the German Air Force, with its large numerical superiority, would, after some weeks of hard fighting, beat down the British air defence, destroy their airfields in Kent and Sussex, and establish a complete domination of the Channel. But apart from this he felt assured that the bombing of England, and particularly of London, would reduce the decadent, peace-loving British to a condition in which they would sue for peace, more especially if the threat of invasion grew steadily upon their horizon. The German Admiralty were by no means convinced; indeed their misgivings were profound. They considered "Sea Lion" should be launched only in the last resort, and in July they had recommended the postponement of the operation till the spring of 1941, unless *the unrestricted air attack and the unlimited U-boat warfare* should "cause the enemy to negotiate with the Fuehrer on his own terms." But Feldmarschall Keitel and General Jodl were glad to find the Air Supreme Commander so confident.

These were great days for Nazi Germany. Hitler had danced his jig of joy before enforcing the humiliation of the French Armistice at Compiègne. The German Army marched triumphantly through the Arc de Triomphe and down the Champs Elysées. What was there they could not do? Why hesitate to play out a winning hand? Thus each of the three Services involved in the operation "Sea Lion" worked upon the hopeful factors in their own theme and left the ugly side to their companions.

As the days passed, doubts and delays appeared and multiplied. Hitler's directive of July 16 had laid down that all preparations were to be completed by the middle of August. All three Services saw that this was impossible. And at the end of July Hitler accepted September 15 as the earliest D-Day, reserving his decision for action until the results of the projected intensified air battle could be known.

On August 30 the Naval Staff reported that owing to British counter-action against the invasion fleet preparations could not be completed by September 15. At their request D-Day was postponed to September 21, with a proviso of ten days' previous warning. This meant that the preliminary order had to be issued on September 11. On September 10 the Naval Staff again reported their various difficulties from the weather, which is always tiresome, and from British counterbombing. They pointed out that although the necessary naval preparations could in fact be completed by the 21st, the stipulated operational condition of undisputed air superiority over the Channel had not been achieved. On the 11th therefore Hitler postponed the preliminary order by three days, thus setting back the earliest D-Day to the 24th; on the 14th he further put it off. On the 17th the postponement became indefinite, and for good reason, in their view as in ours.

On September 7 the information before us showed that the westerly and southerly movement of barges and small ships to ports between Ostend and Havre was in progress, and as these assembly harbours were under heavy British air attack it was not likely the ships would be brought to them until shortly before the actual attempt. The striking strength of the German Air Force between Amsterdam and Brest had been increased by the transfer of one hundred and sixty bomber aircraft from Norway; and short-range dive-bomber units were observed on the forward airfields in the Pas de Calais area. Four Germans captured a few days earlier after landing from a rowboat on the southeast coast had confessed to being spies, and said that they were to be ready at any time during the next fortnight to report the movement of British reserve formations in the area Ipswich–London–Reading–Oxford. Moon and tide conditions between the 8th and 10th of September were favourable for invasion on the southeast coast. On this the Chiefs of Staff concluded that the possibility of invasion had become imminent and that the defence forces should stand by at immediate notice.

There was however at that time no machinery at General Headquarters, Home Forces, by which the existing eight hours' notice for readiness could be brought to "readiness for immediate action" by intermediate stages. The code word "Cromwell," which meant "invasion imminent," was therefore issued by Home Forces at 8 P.M., September 7, to the Eastern and Southern Commands, implying action stations for the forward coastal divisions. It was also sent to all formations in the London area and to the IVth and VIIth Corps in G.H.Q. Reserve. It was repeated for information to all other commands in the United Kingdom. On this, in some parts of the country, the Home Guard commanders, acting on their own initiative, called out the Home Guard by ringing the church bells. Neither I nor the Chiefs of Staff were aware that the decisive code word "Cromwell" had been used, and the next morning instructions were given to devise intermediate stages by which vigilance could be increased on future occasions without declaring an invasion imminent. As may be imagined, this incident caused a great deal of talk and stir, but no mention of it was made in the newspapers or in Parliament. It served as a useful tonic and rehearsal for all concerned.

*   *   *

Having traced the German invasion preparations steadily mounting to a climax, we have seen how the early mood of triumph changed gradually to one of doubt and finally to complete loss of confidence in the outcome. During the fateful months of July and August we see the Naval Commander, Raeder, endeavouring to teach his military and air colleagues about the grave difficulties attending large-scale amphibious war. He realized his own weakness and the lack of time for adequate preparation, and sought to impose limits on the grandiose plans advanced by Halder for landing immense forces simultaneously over a wide front. Meanwhile Goering with soaring ambition was determined to achieve spectacular victory with his Air Force alone and was disinclined to play the humbler role of working to a combined plan for the systematic reduction of opposing sea and air forces in the invasion area.

It is apparent from the records that the German High Command were very far from being a co-ordinated team working together with a common purpose and with a proper understanding of each other's capabilities and limitations. Each wished to be the brightest star in the firmament. Friction was apparent from the outset, and so long as Halder could thrust responsibility onto Raeder he did little to bring his own plans into line with practical possibilities. Intervention by the Fuehrer was necessary, but seems to have done little to improve the relations between the Services. In Germany the prestige of the Army was paramount and the military leaders regarded their naval colleagues with some condescension. It is impossible to resist the conclusion that the German Army was reluctant to place itself in the hands of its sister service in a major operation. When questioned after the war about these plans, General Jodl impatiently remarked, "Our arrangements were much the same as those of Julius Caesar." Here speaks the authentic German soldier in relation to the sea affair, having little conception of the problems involved in landing and deploying large military forces on a defended coast exposed to all the hazards of the sea.

In Britain, whatever our shortcomings, we understood the sea affair very thoroughly. For centuries it has been in our blood, and its traditions stir not only our sailors but the whole race. It was this above all things which enabled us to regard the menace of invasion with a steady gaze. The system of control of operations by the three Chiefs of Staff concerted under a Minister of Defence produced a standard of teamwork, mutual understanding, and ready co-operation unrivalled in the past. When in course of time our opportunity came to undertake great invasions from the sea it was upon a foundation of solid achievement in preparation for the task and with a full understanding of the technical needs of such vast and hazardous undertakings. Had the Germans possessed in 1940 well-trained amphibious forces equipped with all the apparatus of modern amphibious war their task would still have been a forlorn hope in the face of our sea and air power. In fact they had neither the tools nor the training.

The more the German High Command and the Fuehrer looked at the venture the less they liked it. We could not of course know each other's moods and valuations: but with every week from the middle of July to the middle of September the unknown identity of views upon the problem between the German and British Admiralties, between the German Supreme Command and the British Chiefs of Staff, and also between the Fuehrer and the author of this book, became more definitely pronounced. If we could have agreed equally well about other matters there need have been no war. It was of course common ground between us that all depended upon the battle in the air. The question was how this would end between the combatants; and in addition the Germans wondered whether the British people would stand up to the air bombardment, the effect of which in these days was greatly exaggerated, or whether they would crumple and force His Majesty's Government to capitulate. About this Reichsmarschall Goering had high hopes, and we had no fears.

12

The Battle of Britain

OUR FATE NOW DEPENDED upon victory in the air. The German leaders had recognised that all their plans for the invasion of Britain depended on winning air supremacy above the Channel and the chosen landing places on our south coast. The preparation of the embarkation ports, the assembly of the transports, the minesweeping of the passages, and the laying of the new minefields were impossible without protection from British air attack. For the actual crossing and landings complete mastery of the air over the transports and the beaches was the decisive condition. The result therefore turned upon the destruction of the Royal Air Force and the system of airfields between London and the sea. We now know that Hitler said to Admiral Raeder on July 31: "If after eight days of intensive air war the Luftwaffe has not achieved considerable destruction of the enemy's air force, harbours, and naval forces, the operation will have to be put off till May 1941." This was the battle that had now to be fought.

I did not myself at all shrink mentally from the impending trial of strength. I had told Parliament on June 4: "The great French Army was very largely, for the time being, cast back and disturbed by the onrush of a few thousand armoured vehicles. May it not also be that the cause of civilisation itself will be defended by the skill and devotion of a few thousand airmen?" And to Smuts, on June 9: "I see only one sure way through now — to wit, that Hitler should attack this country, and in so doing break his air weapon." The occasion had now arrived.

Admirable accounts have been written of the struggle between the British and German Air Forces which constitutes the Battle of Britain. We have now also access to the views of the German High Command and to their inner reactions in the various phases. It appears that the German losses in some of the principal combats were a good deal less

356

than we thought at the time, and that reports on both sides were materially exaggerated. But the main features and the outline of this famous conflict, upon which the life of Britain and the freedom of the world depended, are not in dispute.

The German Air Force had been engaged to the utmost limit in the Battle of France, and, like the German Navy after the Norway campaign, they required a period of weeks or months for recovery. This pause was convenient to us too, for all but three of our fighter squadrons had at one time or another been engaged in the Continental operations. Hitler could not conceive that Britain would not accept a peace offer after the collapse of France. Like Marshal Pétain, Weygand, and many of the French generals and politicians, he did not understand the separate, aloof resources of an island state, and like these Frenchmen he misjudged our will power. We had travelled a long way and learned a lot since Munich. During the month of June he had addressed himself to the new situation as it gradually dawned upon him, and meanwhile the German Air Force recuperated and redeployed for their next task. There could be no doubt what this would be. Either Hitler must invade and conquer England, or he must face an indefinite prolongation of the war, with all its incalculable hazards and complications. There was always the possibility that victory over Britain in the air would bring about the end of the British resistance, and that actual invasion, even if it became practicable, would also become unnecessary except for the occupying of a defeated country.

During June and early July the German Air Force revived and regrouped its formations and established itself on all the French and Belgian airfields from which the assault had to be launched, and by reconnaissance and tentative forays sought to measure the character and scale of the opposition which would be encountered. It was not until July 10 that the first heavy onslaught began, and this date is usually taken as the opening of the battle. Two other dates of supreme consequence stand out, August 15 and September 15. There were also three successive but overlapping phases in the German attack. First, from July 10 to August 18, the harrying of British convoys in the Channel and of our southern ports from Dover to Plymouth, whereby our Air Force should be tested, drawn into battle, and depleted; whereby also damage should be done to those seaside towns marked as objectives for the forthcoming invasion. In the second phase, August 24 to September 27, a way to London was to be forced by the elimination of the Royal Air Force and its installations, leading to the violent and continuous bombing of the capital. This would also cut communications with the threatened shores. But in Goering's view there was good reason to believe that a greater prize was here in sight, no less than throwing the world's largest city into confusion and paralysis, the cowing of the Government and the people, and their consequent submission to the German will. Their Navy and Army staffs devoutly hoped that Goering was right. As the situation developed they saw that the R.A.F. was not being eliminated, and meanwhile their own urgent needs for the "Sea Lion" adventure were neglected for the sake of destruction in London. And then, when all were disappointed, when invasion was indefinitely postponed for lack of the vital need, air supremacy, there followed the third and last phase. The hope of daylight victory had faded, the Royal Air Force remained vexatiously alive, and Goering in October resigned himself to the indiscriminate bombing of London and the centres of industrial production.

*   *   *

In the quality of the fighter aircraft there was little to choose. The Germans' were faster, with a better rate of climb; ours more manoeuvrable, better armed. Their airmen, well aware of their great numbers, were also the proud victors of Poland, Norway, the Low Countries, France; ours had supreme confidence in themselves as individuals and that determination which the British race displays in fullest measure when in supreme adversity. One important strategical advantage the Germans enjoyed and skilfully used: their forces were deployed on many and widely spread bases, whence they could concentrate upon us in great strengths and with feints and deceptions as to the true points of attack. By August the Luftwaffe had gathered 2669 operational aircraft, comprising 1015 bombers, 346 dive bombers, 933 fighters, and 375 heavy fighters. The Fuehrer's Directive No. 17 authorized the intensified air war against England on August 5. Goering never set much store by "Sea Lion"; his heart was in the "absolute" air war. His consequent distortion of the arrangements disturbed the German Naval Staff. The destruction of the Royal Air Force and our aircraft industry was to them but a means to an end: when this was accomplished the air war should be turned against the enemy's warships and shipping. They regretted the lower priority assigned by Goering to the naval targets, and they were irked by the delays. On August 6 they reported to the Supreme Command that the preparations for German minelaying in the Channel area could not proceed because of the constant British threat from the air.

The continuous heavy air fighting of July and early August had been directed upon the Kent promontory and the Channel coast. Goering and his skilled advisers formed the opinion that they must have drawn nearly all our fighter squadrons into this southern struggle. They therefore decided to make a daylight raid on the manufacturing cities north of the Wash. The distance was too great for their first-class fighters, the Me. 109's. They would have to risk their bombers with only escorts from the Me. 110's, which, though they had the range, had nothing like the quality, which was what mattered now. This was nevertheless a reasonable step for them to take, and the risk was well run.

Accordingly, on August 15, about a hundred bombers, with an escort of forty Me. 110's, were launched against Tyneside. At the same time a raid of more than eight hundred planes was sent to pin down our forces in the South, where it was thought they were already all gathered. But now the dispositions which Dowding had made of the Fighter Command were signally vindicated. The danger had been foreseen. Seven Hurricane or Spitfire squadrons had been withdrawn from the intense struggle in the South to rest in and at the same time to guard the North. They had suffered severely, but were nonetheless deeply grieved to leave the battle. The pilots respectfully represented that they were not at all tired. Now came an unexpected consolation. These squadrons were able to welcome the assailants as they crossed the coast. Thirty German planes were shot down, most of them heavy bombers (Heinkel 111's, with four trained men in each crew), for a British loss of only two pilots injured. The foresight of Air Marshal Dowding in his direction of Fighter Command deserves high praise, but even more remarkable had been the restraint and the exact measurement of formidable stresses which had reserved a fighter force in the North through all these long weeks of mortal conflict in the South. We must regard the generalship here shown as an example of genius in the art of war. Henceforth everything north of the Wash was safe by day.

August 15 was the largest air battle of this period of the war; five major actions were fought, on a front of five hundred miles. It was indeed a crucial day. In the South all our twenty-two squadrons were engaged, many twice, some three times, and the German losses, added to those in the North, were seventy-six to our thirty-four. This was a recognisable disaster to the German Air Force.

It must have been with anxious minds that the German Air Chiefs measured the consequences of this defeat, which boded ill for the future. The German Air Force however had still as their target the Port of London, all that immense line of docks with their masses of shipping, and the largest city in the world, which did not require much accuracy to hit.

*   *   *

During these weeks of intense struggle and ceaseless anxiety Lord Beaverbrook rendered signal service. At all costs the fighter squadrons must be replenished with trustworthy machines. This was no time for red tape and circumlocution, although these have their place in a well-ordered, placid system. All his remarkable qualities fitted the need. His personal buoyancy and vigour were a tonic. I was glad to be able sometimes to lean on him. He did not fail. This was his hour. His personal force and genius, combined with so much persuasion and contrivance, swept aside many obstacles. Everything in the supply pipeline was drawn forward to the battle. New or repaired aeroplanes streamed to the delighted squadrons in numbers they had never known before. All the services of maintenance and repair were driven to an intense degree. I felt so much his value that on August 2, with the King's approval, I invited him to join the War Cabinet. At this time also his eldest son, Max Aitken, gained high distinction and at least six victories as a fighter pilot.

Another Minister I consorted with at this time was Ernest Bevin, Minister of Labour and National Service, with the whole man-power of the nation to manage and animate. All the workers in the munitions factories were ready to take his direction. In September he too joined the War Cabinet. The trade unionists cast their slowly framed, jealously guarded rules and privileges upon the altar where wealth, rank, privilege, and property had already been laid. I was much in harmony with both Beaverbrook and Bevin in the white-hot weeks. Afterwards they quarrelled, which was a pity, and caused much friction. But at this climax we were all together. I cannot speak too highly of the loyalty of Mr. Chamberlain, or of the resolution and efficiency of all my Cabinet colleagues. Let me give them my salute.

*   *   *

Up till the end of August Goering did not take an unfavourable view of the air conflict. He and his circle believed that the English ground organisation and aircraft industry and the fighting strength of the R.A.F. had already been severely damaged. There was a spell of fine weather in September, and the Luftwaffe hoped for decisive results. Heavy attacks fell upon our aerodrome installations round London, and on the night of the 6th sixty-eight aircraft attacked London, followed on the 7th by the first large-scale attack of about three hundred. On this and succeeding days, during which our anti-aircraft guns were doubled in numbers, very hard and continuous air fighting took place over the capital, and the Luftwaffe were still confident through their over-estimation of our losses.

Indeed in the fighting between August 24 and September 6 the scales had tilted against Fighter Command. During these crucial days the Germans had continuously applied powerful forces against the airfields of South and Southeast England. Their object was to break down the day fighter defence of the capital, which they were impatient to attack. Far more important to us than the protection of London from terror bombing was the functioning and articulation of these airfields and the squadrons working from them. In the life-and-death struggle of the two air forces this was a decisive phase. We never thought of the struggle in terms of the defence of London or any other place, but only who won in the air. There was much anxiety at Fighter Headquarters at Stanmore, and particularly at the headquarters of No. 11 Fighter Group at Uxbridge. Extensive damage had been done to five of the group's forward airfields, and also to the six sector stations. Biggin Hill Sector Station, to the south of London, was so severely damaged that for a week only one fighter squadron could operate from it. If the enemy had persisted in heavy attacks against the adjacent sectors and damaged their operations rooms or telephone communications the whole intricate organisation of Fighter Command might have been broken down. This would have meant not only merely the maltreatment of London, but the loss to us of the perfected control of our own air in the decisive area. I was led to visit several of these stations, particularly Manston (August 28) and Biggin Hill, which is quite near my home. They were getting terribly knocked about, and their runways were ruined by craters. It was therefore with a sense of relief that Fighter Command felt the German attack turn onto London on September 7, and concluded that the enemy had changed his plan. Goering should certainly have persevered against the airfields, on whose organisation and combination the whole fighting power of our air force at this moment depended. By departing from the classical principles of war, as well as from the hitherto accepted dictates of humanity, he made a foolish mistake.

This same period (August 24–September 6) had seriously drained the strength of Fighter Command as a whole. The Command had lost in this fortnight 103 pilots killed and 128 seriously wounded, while 466 Spitfires and Hurricanes had been destroyed or seriously damaged. Out of a total pilot strength of about a thousand nearly a quarter had been lost. Their places could only be filled by 260 new, ardent, but inexperienced pilots drawn from training units, in many cases before their full courses were complete. The night attacks on London for ten days after September 7 struck at the London docks and railway centres, and killed and wounded many civilians, but they were in effect for us a breathing space of which we had the utmost need.

We must take September 15 as the culminating date. On this day the Luftwaffe, after two heavy attacks on the 14th, made its greatest concentrated effort in a resumed daylight attack on London.

It was one of the decisive battles of the war, and, like the Battle of Waterloo, it was on a Sunday. I was at Chequers. I had already on several occasions visited the headquarters of No. 11 Fighter Group in order to witness the conduct of an air battle, when not much had happened. However, the weather on this day seemed suitable to the enemy, and accordingly I drove over to Uxbridge and arrived at the Group Headquarters. No. 11 Group comprised no fewer than twenty-five squadrons covering the whole of Essex, Kent, Sussex, and Hampshire, and all the approaches across them to London. Air Vice-Marshal Park had for six months commanded this group, on which our fate largely depended. From the beginning of Dunkirk all the daylight actions in the South of England had already been conducted by him, and all his arrangements and apparatus had been brought to the highest perfection. My wife and I were taken down to the bomb-proof Operations Room, fifty feet below ground. All the ascendancy of the Hurricanes and Spitfires would have been fruitless but for this system of underground control centres and telephone cables, which had been devised and built before the war by the Air Ministry under Dowding's advice and impulse. The Supreme Command was exercised from the Fighter Headquarters at Stanmore, but the actual handling of the direction of the squadrons was wisely left to No. 11 Group, which controlled the units through its fighter stations located in each county.

The Group Operations Room was like a small theatre, about sixty feet across, and with two storeys. We took our seats in the dress circle. Below us was the large-scale map table, around which perhaps twenty highly trained young men and women, with their telephone assistants, were assembled. Opposite to us, covering the entire wall, where the theatre curtain would be, was a gigantic blackboard divided into six columns with electric bulbs, for the six fighter stations, each of their squadrons having a subcolumn of its own, and also divided by lateral lines. Thus the lowest row of bulbs showed as they were lighted the squadrons which were "Standing By" at two minutes' notice, the next row those at "Readiness," five minutes, then at "Available," twenty minutes, then those which had taken off, the next row those which had reported having seen the enemy, the next — with red lights — those which were in action, and the top row those which were returning home. On the left-hand side, in a kind of glass stage box, were the four or five officers whose duty it was to weigh and measure the information received from our Observer Corps, which at this time numbered upwards of fifty thousand men, women, and youths. Radar was still in its infancy, but it gave warning of raids approaching our coast, and the observers, with fieldglasses and portable telephones, were our main sources of information about raiders flying overland.

Thousands of messages were therefore received during an action. Several roomfuls of experienced people in other parts of the underground headquarters sifted them with great rapidity, and transmitted the results from minute to minute directly to the plotters seated around the table on the floor and to the officer supervising from the glass stage box.

On the right hand was another glass stage box containing Army officers who reported the action of our anti-aircraft batteries, of which at this time in the Command there were two hundred. At night it was of vital importance to stop these batteries firing over certain areas in which our fighters would be closing with the enemy. I was not unacquainted with the general outlines of this system, having had it explained to me a year before the war by Dowding when I visited him at Stanmore. It had been shaped and refined in constant action, and all was now fused together into a most elaborate instrument of war, the like of which existed nowhere in the world.

"I don't know," said Park, as we went down, "whether anything will happen today. At present all is quiet." However, after a quarter of an hour the raid plotters began to move about. An attack of "40 plus" was reported to be coming from the German stations in the Dieppe area. The bulbs along the bottom of the wall display panel began to glow as various squadrons came to "Stand By." Then in quick succession "20 plus," "40 plus" signals were received, and in another ten minutes it was evident that a serious battle impended. On both sides the air began to fill.

One after another signals came in, "40 plus," "60 plus"; there was even an "80 plus." On the floor table below us the movement of all the waves of attack was marked by pushing discs forward from minute to minute along different lines of approach, while on the blackboard facing us the rising lights showed our fighter squadrons getting into the air, till there were only four or five left at "Readiness." These air battles, on which so much depended, lasted little more than an hour from the first encounter. The enemy had ample strength to send out new waves of attack, and our squadrons, having gone all out to gain the upper air, would have to refuel after seventy or eighty minutes, or land to rearm after a five-minute engagement. If at this moment of refuelling or rearming the enemy were able to arrive with fresh unchallenged squadrons some of our fighters could be destroyed on the ground. It was therefore one of our principal objects to direct our squadrons so as not to have too many on the ground refuelling or rearming simultanteously during daylight.

Presently the red bulbs showed that the majority of our squadrons were engaged. A subdued hum arose from the floor, where the busy

plotters pushed their discs to and fro in accordance with the swiftly
changing situation. Air Vice-Marshal Park gave general directions for
the disposition of his fighter force, which were translated into detailed
orders to each fighter station by a youngish officer in the centre of the
dress circle, at whose side I sat. Some years after I asked his name.
He was Lord Willoughby de Broke. (I met him next in 1947, when the
Jockey Club, of which he was a steward, invited me to see the Derby.
He was surprised that I remembered the occasion.) He now gave the
orders for the individual squadrons to ascend and patrol as the results of
the final information which appeared on the map table. The Air
Marshal himself walked up and down behind, watching with vigilant
eye every move in the game, supervising his junior executive hand, and
only occasionally intervening with some decisive order, usually to
reinforce a threatened area. In a little while all our squadrons were
fighting, and some had already begun to return for fuel. All were in
the air. The lower line of bulbs was out. There was not one squadron
left in reserve. At this moment Park spoke to Dowding at Stanmore,
asking for three squadrons from No. 12 Group to be put at his disposal
in case of another major attack while his squadrons were rearming and
refuelling. This was done. They were specially needed to cover
London and our fighter aerodromes, because No. 11 Group had already
shot their bolt.

The young officer, to whom this seemed a matter of routine, continued to give his orders, in accordance with the general directions of
his Group Commander, in a calm, low monotone, and the three reinforcing squadrons were soon absorbed. I became conscious of the
anxiety of the Commander, who now stood still behind his subordinate's chair. Hitherto I had watched in silence. I now asked: "What
other reserves have we?" "There are none," said Air Vice-Marshal
Park. In an account which he wrote about it afterwards he said that at
this I "looked grave." Well I might. What losses should we not suffer
if our refuelling planes were caught on the ground by further raids of
"40 plus" or "50 plus"! The odds were great; our margins small; the
stakes infinite.

Another five minutes passed, and most of our squadrons had now
descended to refuel. In many cases our resources could not give them
overhead protection. Then it appeared that the enemy were going
home. The shifting of the discs on the table showed a continuous eastward movement of German bombers and fighters. No new attack appeared. In another ten minutes the action was ended. We climbed
again the stairways which led to the surface, and almost as we emerged
the "All Clear" sounded.

"We are very glad, sir, you have seen this," said Park. "Of course, during the last twenty minutes we were so choked with information that we couldn't handle it. This shows you the limitation of our present resources. They have been strained far beyond their limits today." I asked whether any results had come to hand, and remarked that the attack appeared to have been repelled satisfactorily. Park replied that he was not satisfied that we had intercepted as many raiders as he had hoped we should. It was evident that the enemy had everywhere pierced our defences. Many scores of German bombers, with their fighter escort, had been reported over London. About a dozen had been brought down while I was below, but no picture of the results of the battle or of the damage or losses could be obtained.

It was 4.30 P.M. before I got back to Chequers, and I immediately went to bed for my afternoon sleep. I must have been tired by the drama of No. 11 Group, for I did not wake till eight. When I rang, John Martin, my principal private secretary, came in with the evening budget of news from all over the world. It was repellent. This had gone wrong here; that had been delayed there; an unsatisfactory answer had been received from so-and-so; there had been bad sinkings in the Atlantic. "However," said Martin, as he finished this account, "all is redeemed by the air. We have shot down one hundred and eighty-three for a loss of under forty."

*　　　*　　　*

Although postwar information has shown that the enemy's losses on this day were only fifty-six, September 15 was the crux of the Battle of Britain. That same night our Bomber Command attacked in strength the shipping in the ports from Boulogne to Antwerp. At Antwerp particularly heavy losses were inflicted. On September 17, as we now know, the Fuehrer decided to postpone "Sea Lion" indefinitely. It was not till October 12 that the invasion was formally called off till the following spring. In July 1941 it was postponed again by Hitler till the spring of 1942, "by which time the Russian campaign will be completed." This was a vain but an important imagining. On February 13, 1942, Admiral Raeder had his final interview on "Sea Lion" and got Hitler to agree to a complete "stand-down." Thus perished "Operation Sea Lion." And September 15 may stand as the date of its demise.

No doubt we were always oversanguine in our estimates of enemy scalps. In the upshot we got two to one of the German assailants, instead of three to one, as we believed and declared. But this was enough. The Royal Air Force, far from being destroyed, was triumphant. A strong flow of fresh pilots was provided. The aircraft factories, upon which not only our immediate need but our power to wage a long war depended, were mauled but not paralysed. The workers, skilled and unskilled, men and women alike, stood to their lathes and manned the workshops under fire as if they were batteries in action — which indeed they were. At the Ministry of Supply, Herbert Morrison spurred all in his wide sphere. "Go to it," he adjured, and to it they went. Skilful and ever-ready support was given to the air fighting by the Anti-Aircraft Command under General Pile. Their main contribution came later. The Observer Corps, devoted and tireless, were hourly at their posts. The carefully wrought organisation of Fighter Command, without which all might have been in vain, proved equal to months of continuous strain. All played their part.

At the summit the stamina and valour of our fighter pilots remained unconquerable and supreme. Thus Britain was saved. Well might I say in the House of Commons: "Never in the field of human conflict was so much owed by so many to so few."

13

"London Can Take It"

THE GERMAN AIR ASSAULT on Britain is a tale of divided counsels, conflicting purposes, and never fully accomplished plans. Three or four times in these months the enemy abandoned a method of attack which was causing us severe stress, and turned to something new. But all these stages overlapped one another, and cannot be readily distinguished by precise dates. Each one merged into the next. The early operations sought to engage our air forces in battle over the Channel and the south coast; next the struggle was continued over our southern counties, principally Kent and Sussex, the enemy aiming to destroy our air-power organisation; then nearer to and over London; then London became the supreme target; and finally, when London triumphed, there was a renewed dispersion to the provincial cities and to our sole Atlantic lifeline by the Mersey and the Clyde.

We have seen how very hard they had run us in the attack on the south coast airfields in the last week of August and the first week of September. But on September 7 Goering publicly assumed command of the air battle, and turned from daylight to night attack and from the fighter airfields of Kent and Sussex to the vast build-up areas of London. Minor raids by daylight were frequent, indeed constant, and one great daylight attack was still to come; but in the main the whole character of the German offensive was altered. For fifty-seven nights the bombing of London was unceasing. This constituted an ordeal for the world's largest city, the results of which no one could measure beforehand. Never before was so wide an expanse of houses subjected to such bombardment or so many families required to face its problems and its terrors.

The sporadic raiding of London towards the end of August was promptly answered by us in a retaliatory attack on Berlin. Because of the distance we had to travel, this could only be on a very small scale compared with attacks on London from nearby French and Belgian airfields. The War Cabinet were much in the mood to hit back, to raise the stakes, and to defy the enemy. I was sure they were right, and believed that nothing impressed or disturbed Hitler so much as his realisation of British wrath and will power. In his heart he was one of our admirers. He took of course full advantage of our reprisal on Berlin, and publicly announced the previously settled German policy of reducing London and other British cities to chaos and ruin. "If they attack our cities," he declared on September 4, " we will simply erase theirs." He tried his best.

From September 7 to November 3 an average of two hundred German bombers attacked London every night. The various preliminary raids which had been made on our provincial cities in the previous three weeks had led to a considerable dispersion of our anti-aircraft artillery, and when London first became the main target there were but ninety-two guns in position. It was thought better to leave the air free for our night fighters, working under No. 11 Group. Of these there were six squadrons of Blenheims and Defiants. Night fighting was in its infancy, and very few casualties were inflicted on the enemy. Our batteries therefore remained silent for three nights in succession. Their own technique was at this time woefully imperfect. Nevertheless, in view of the weakness of our night fighters and of their unsolved problems it was decided that the anti-aircraft gunners should be given a free hand to fire at unseen targets, using any methods of control they liked. In forty-eight hours General Pile, commanding the Air Defence Artillery, had more than doubled the number of guns in the capital by withdrawals from the provincial cities. Our own aircraft were kept out of the way, and the batteries were given their chance.

For three nights Londoners had sat in their houses or inadequate shelters enduring what seemed to be an utterly unresisted attack. Suddenly, on September 10, the whole barrage opened, accompanied by a blaze of searchlights. This roaring cannonade did not do much harm to the enemy, but gave enormous satisfaction to the population. Everyone was cheered by the feeling that we were hitting back. From that time onward the batteries fired regularly, and of course practice, ingenuity, and grinding need steadily improved the shooting. A slowly increasing toll was taken of the German raiders. Upon occasions the batteries were silent and the night fighters, whose methods were also progressing, came on the scene. The night raids were accompanied by more or less continuous daylight attacks by small groups or even single enemy planes, and the sirens often sounded at brief intervals throughout the whole twenty-four hours. To this curious existence the seven million inhabitants of London accustomed themselves.

*　　　*　　　*

In the hope that it may lighten the hard course of this narrative I record a few personal notes about the "Blitz," well knowing how many thousands have far more exciting tales to tell.

When the bombardment first began, the idea was to treat it with disdain. In the West End everybody went about their business and pleasure and dined and slept as they usually did. The theatres were full, and the darkened streets were crowded with casual traffic. All this was perhaps a healthy reaction from the frightful squawk which the defeatist elements in Paris had put up on the occasion when they were first seriously raided in May. I remember dining in a small company when very lively and continuous raids were going on. The large windows of Stornoway House opened upon the Green Park, which flickered with the flashes of the guns and was occasionally lit by the glare of an exploding bomb. I felt that we were taking unnecessary risks. After dinner we went to the Imperial Chemicals building overlooking the Embankment. From these high stone balconies there was a splendid view of the river. At least a dozen fires were burning on the south side, and while we were there several heavy bombs fell, one near enough for my friends to pull me back behind a substantial stone pillar. This certainly confirmed my opinion that we should have to accept many restrictions upon the ordinary amenities of life.

The group of Government buildings around Whitehall were repeatedly hit. Downing Street consists of houses two hundred and fifty years old, shaky and lightly built by the profiteering contractor whose name they bear. At the time of the Munich alarm, shelters had been constructed for the occupants of No. 10 and No. 11, and the rooms on the garden level had had their ceilings propped up with a wooden under-ceiling and strong timbers. It was believed that this would support the ruins if the building was blown or shaken down; but of course neither these rooms nor the shelters were effective against a direct hit. During the last fortnight of September preparations were made to transfer my Ministerial headquarters to the more modern and solid Government offices looking over St. James's Park by Storey's Gate. These quarters we called "the Annexe." Here during the rest of the war my wife and I lived comfortably. We felt confidence in this solid stone building, and only on very rare occasions went down below the armour. My wife even hung up our few pictures in the sitting room, which I had thought it better to keep bare. Her view prevailed and was justified by the event. From the roof near the cupola of the Annexe there was a splendid view of

London on clear nights. They made a place for me with light overhead cover from splinters, and one could walk in the moonlight and watch the fireworks. Below was the War Room and a certain amount of bomb-proof sleeping accommodation. The bombs at this time were of course smaller than those of the later phases. Still, in the interval before the new apartments were ready life at Downing Street was exciting. One might as well have been at a battalion headquarters in the line.

One evening (October 17) stands out in my mind. We were dining in the garden room of No. 10 when the usual night raid began. My companions were Archie Sinclair, Oliver Lyttelton, and Moore-Brabazon. The steel shutters had been closed. Several loud explosions occurred around us at no great distance, and presently a bomb fell, perhaps a hundred yards away, on the Horse Guards Parade, making a great deal of noise. Suddenly I had a providential impulse. The kitchen at No. 10 Downing Street is lofty and spacious, and looks out through a large plate-glass window about twenty-five feet high. The butler and parlour-maid continued to serve the dinner with complete detachment, but I became acutely aware of this big window, behind which Mrs. Landemare, the cook, and the kitchen maid, never turning a hair, were at work. I got up abruptly, went into the kitchen, told the butler to put the dinner on the hot plate in the dining room, and ordered the cook and the other servants into the shelter, such as it was. I had been seated again at table only about three minutes when a really very loud crash, close at hand, and a violent shock showed that the house had been struck. My detective came into the room and said much damage had been done. The kitchen, the pantry, and the offices on the Treasury side were shattered.

We went into the kitchen to view the scene. The devastation was complete. The bomb had fallen fifty yards away on the Treasury, and the blast had smitten the large, tidy kitchen, with all its bright sauce-pans and crockery, into a heap of black dust and rubble. The big plate-glass window had been hurled in fragments and splinters across the room, and would of course have cut its occupants, if there had been any, to pieces. But my fortunate inspiration, which I might so easily have neglected, had come in the nick of time. The underground Treasury shelter across the court had been blown to pieces by a direct hit, and the four civil servants who were doing Home Guard night duty there were killed. All however were buried under tons of brick rubble, and we did not know who was missing.

As the raid continued and seemed to grow in intensity we put on our tin hats and went out to view the scene from the top of the Annexe building. Before doing so, however, I could not resist taking Mrs. Landemare and the others from the shelter to see their kitchen. They were upset at the sight of the wreck, but principally on account of the general untidiness!

Archie and I went up to the cupola of the Annexe building. The night was clear and there was a wide view of London. It seemed that the greater part of Pall Mall was in flames. At least five fierce fires were burning there, and others in St. James's Street and Piccadilly. Farther back over the river in the opposite direction there were many conflagrations. But Pall Mall was the vivid flame picture. Gradually, the attack died down, and presently the "All Clear" sounded, leaving only the blazing fires. We went downstairs to my new apartments on the first floor of the Annexe, and there found Captain David Margesson, the Chief Whip, who was accustomed to live at the Carlton Club. He told us the club had been blown to bits, and indeed we had thought, by the situation of the fires, that it must have been hit. He was in the club with about two hundred and fifty members and staff. It had been struck by a heavy bomb. The whole of the façade and the massive coping on the Pall Mall side had fallen into the street, obliterating his motorcar, which was parked near the front door. The smoking room had been full of members, and the whole ceiling had come down upon them. When I looked at the ruins next day it seemed incredible that most of them should not have been killed. However, by what seemed a miracle, they all crawled out of the dust, smoke, and rubble, and though many were injured not a single life was lost. When in due course these facts came to the notice of the Cabinet our Labour colleagues facetiously remarked: "The devil looks after his own." Mr. Quintin Hogg had carried his father, a former Lord Chancellor, on his shoulders from the wreck, as Aeneas had borne Pater Anchises from the ruins of Troy. Margesson had nowhere to sleep, and we found him blankets and a bed in the basement of the Annexe. Altogether it was a lurid evening, and considering the damage to buildings it was remarkable that there were not more than five hundred people killed and about a couple of thousand injured.

Another time I visited Ramsgate. An air raid came upon us, and I was conducted into their big tunnel, where quite large numbers of people lived permanently. When we came out, after a quarter of an hour, we looked at the still-smoking damage. A small hotel had been hit. Nobody had been hurt, but the place had been reduced to a litter of crockery, utensils, and splintered furniture. The proprietor, his wife, and the cooks and waitresses were in tears. Where was their home? Where was their livelihood? Here is a privilege of power. I formed an immediate resolve. On the way back in my train I dictated a letter to the Chancellor of the Exchequer, Kingsley Wood, laying down the principle that all damage from the fire of the enemy must be a charge upon the State

and compensation be paid in full and at once. Thus the burden would
not fall alone on those whose homes or business premises were hit, but
would be borne evenly on the shoulders of the nation. Kingsley Wood
was naturally a little worried by the indefinite character of this obligation. But I pressed hard, and an insurance scheme was devised in a
fortnight which afterwards played a substantial part in our affairs. The
Treasury went through various emotions about this insurance scheme.
First they thought it was going to be their ruin; but when, after May
1941, the air raids ceased for over three years they began to make a
great deal of money, and considered the plan provident and statesman-
like. However, later on in the war, when the "doodle bugs" and rockets
began, the accounts swung the other way, and eight hundred and ninety
millions were soon paid out. I am very glad this was so.

*   *   *

In this new phase of warfare it became important to extract the
optimum of work not only from the factories but even more from the
departments in London which were under frequent bombardment during both the day and night. At first, whenever the sirens gave the alarm,
all the occupants of a score of Ministries were promptly collected and
led down to the basements, for what these were worth. Pride even was
being taken in the efficiency and thoroughness with which this evolu-
tion was performed. In many cases it was only half a dozen aeroplanes
which approached — sometimes only one. Often they did not arrive. A
petty raid might bring to a standstill for over an hour the whole executive and administrative machine in London.

I therefore proposed the stage "Alert," operative on the siren warning, as distinct from the "Alarm," which should be enforced only when
the spotters on the roof, or Jim Crows, as they came to be called, reported "Imminent danger," which meant that the enemy was actually
overhead or very near. Schemes were worked out accordingly. Parliament also required guidance about the conduct of its work in these
dangerous days. Members felt that it was their duty to set an example.
This was right, but it might have been pushed too far; I had to reason
with the Commons to make them observe ordinary prudence and conform to the peculiar conditions of the time. I convinced them in secret
session of the need to take necessary and well-considered precautions.
They agreed that their days and hours of sitting should not be adver-
tised, and to suspend their debates when the Jim Crow reported to the
Speaker "Imminent danger." Then they all trooped down dutifully to
the crowded, ineffectual shelters that had been provided. It will always
add to the renown of the British Parliament that its Members continued
to sit and discharge their duties through all this period. The Commons

are very touchy in such matters, and it would have been easy to mis-judge their mood. When one Chamber was damaged they moved to another, and I did my utmost to persuade them to follow wise advice with good grace. In short, everyone behaved with sense and dignity. It was also lucky that when the Chamber was blown to pieces a few months later it was by night and not by day, when empty and not when full. With our mastery of the daylight raids there came considerable relief in personal convenience. But during the first few months I was never free from anxiety about the safety of the Members. After all, a free sovereign Parliament, fairly chosen by universal suffrage, able to turn out the Government any day, but proud to uphold it in the darkest days, was one of the points which were in dispute with the enemy. Parliament won.

I doubt whether any of the dictators had as much effective power throughout his whole nation as the British War Cabinet. When we expressed our desires we were sustained by the people's representatives, and cheerfully obeyed by all. Yet at no time was the right of criticism impaired. Nearly always the critics respected the national interest. When on occasions they challenged us, the Houses voted them down by overwhelming majorities, and this, in contrast with totalitarian methods, without the slightest coercion, intervention, or use of the police or Secret Service. It was a proud thought that Parliamentary Democracy, or whatever our British public life can be called, can endure, surmount, and survive all trials. Even the threat of annihilation did not daunt our Members, but this fortunately did not come to pass.

*　　　　*　　　　*

In the middle of September a new and damaging form of attack was used against us. Large numbers of delayed-action bombs were now widely and plentifully cast upon us and became an awkward problem. Long stretches of railway line, important junctions, the approaches to vital factories, airfields, main thoroughfares, had scores of times to be blocked off and denied to us in our need. These bombs had to be dug out and exploded or rendered harmless. This was a task of the utmost peril, especially at the beginning, when the means and methods had all to be learned by a series of decisive experiences. I have already recounted the drama of dismantling the magnetic mine, but this form of self-devotion now became commonplace while remaining sublime. I had always taken an interest in the delayed-action fuze, which had first impressed itself on me in 1918, when the Germans had used it on a large scale to deny us the use of the railways by which we planned to advance into Germany. I had urged its use by us both in Norway and in the Kiel Canal and the Rhine. There is no doubt that it is a most effective agent in warfare, on account of the prolonged uncertainty which it creates. We were now to taste it ourselves. A special organisation to deal with it was set up. Special companies were formed in every city, town, and district. Volunteers pressed forward for the deadly game. Teams were formed which had good or bad luck. Some survived this phase of our ordeal. Others ran twenty, thirty, or even forty courses before they met their fate. The Unexploded Bomb detachments presented themselves wherever I went on my tours. Somehow or other their faces seemed different from those of ordinary men, however brave and faithful. They were gaunt, they were haggard, their faces had a bluish look, with bright gleaming eyes and exceptional compression of the lips; withal a perfect demeanour. In writing about our hard times we are apt to overuse the word "grim." It should have been reserved for the U.X.B. disposal squads.

One squad I remember which may be taken as symbolic of many others. It consisted of three people — the Earl of Suffolk, his lady private secretary, and his rather aged chauffeur. They called themselves "the Holy Trinity." Their prowess and continued existence got around among all who knew. Thirty-four unexploded bombs did they tackle with urbane and smiling efficiency. But the thirty-fifth claimed its forfeit. Up went the Earl of Suffolk in his Holy Trinity. But we may be sure that, as for Mr. Valiant-for-truth, "all the trumpets sounded for them on the other side."

Very quickly, but at heavy sacrifice of our noblest, the devotion of the U.X.B. detachments mastered the peril.

*   *   *

It is difficult to compare the ordeal of the Londoners in the winter of 1940–41 with that of the Germans in the last three years of the war. In this latter phase the bombs were much more powerful and the raids far more intense. On the other hand, long preparation and German thoroughness had enabled a complete system of bomb-proof shelters to be built, into which all were forced to go by iron routine. When eventually we got into Germany we found cities completely wrecked, but strong buildings standing up above the ground, and spacious subterranean galleries where the inhabitants slept night after night, although their houses and property were being destroyed above. In many cases only the rubble heaps were stirred. But in London, although the attack was less overpowering, the security arrangements were far less developed. Apart from the Tubes there were no really safe places. There were very few basements or cellars which could withstand a direct hit. Virtually the whole mass of the London population lived and slept in their homes or in their Anderson shelters under the fire of the enemy, taking their chance with British phlegm after a hard day's work. Not one in a thousand had any protection except against blast and splinters. But there was as little psychological weakening as there was physical pestilence. Of course, if the bombs of 1943 had been applied to the London of 1940 we should have passed into conditions which might have pulverised all human organisation. However, everything happens in its turn and in its relation, and no one has a right to say that London, which was certainly unconquered, was not also unconquerable.

Little or nothing had been done before the war or during the passive period to provide bomb-proof strongholds from which the central government could be carried on. Elaborate plans had been made to move the seat of government from London. Complete branches of many departments had already been moved to Harrogate, Bath, Cheltenham, and elsewhere. Accommodation had been requisitioned over a wide area, providing for all Ministers and important functionaries in the event of an evacuation of London. But now under the bombardment the desire and resolve of the Government and of Parliament to remain in London was unmistakable, and I shared this feeling to the full. I, like others, had often pictured the destruction becoming so overpowering that a general move and dispersal would have to be made. But under the impact of the event all our reactions were in the contrary sense.

In these months we held our evening Cabinets in the War Room in the Annexe basement. To get there from Downing Street it was necessary to walk through the Foreign Office quadrangle and then clamber through the working parties who were pouring in the concrete to make the War Room and basement offices safer. I did not realise what a trial this was to Mr. Chamberlain, with all the consequences of his major operation upon him. Nothing deterred him, and he was never more spick and span or cool and determined than at the last Cabinets which he attended.

One evening in late September 1940 I looked out of the Downing Street front door and saw workmen piling sandbags in front of the low basement windows of the Foreign Office opposite. I asked what they were doing. I was told that after his operation Mr. Neville Chamberlain had to have special periodical treatment, and that it was embarrassing to carry this out in the shelter of No. 11 where at least twenty people were gathered during the constant raids, so a small private place was being prepared over there for him. Every day he kept all his appointments, reserved, efficient, faultlessly attired. But here was the background. It was too much. I used my authority. I walked through the passage between No. 10 and No. 11 and found Mrs. Chamberlain. I said: "He ought not to be here in this condition. You must take him

away till he is well again. I will send all the telegrams to him each day."
She went off to see her husband. In an hour she sent me word. "He will
do what you wish. We are leaving tonight." I never saw him again. I
am sure he wanted to die in harness. This was not to be.

*   *   *

The retirement of Mr. Chamberlain led to important Ministerial
changes. Mr. Herbert Morrison had been an efficient and vigorous
Minister of Supply, and Sir John Anderson had faced the Blitz of
London with firm and competent management. By the early days of
October the continuous attack on the largest city in the world was so
severe and raised so many problems of a social and political character
in its vast harassed population that I thought it would be a help to have
a long-trained Parliamentarian at the Home Office, which was now also
the Ministry of Home Security. London was bearing the brunt. Herbert
Morrison was a Londoner, versed in every aspect of metropolitan administration. He had unrivalled experience of London government, having been leader of the County Council, and in many ways the principal
figure in its affairs. At the same time I needed John Anderson, whose
work at the Home Office had been excellent, as Lord President of the
Council in the wider sphere of the Home Affairs Committee, to which
an immense mass of business was referred, with great relief to the
Cabinet. This also lightened my own burden and enabled me to concentrate upon the military conduct of the war, in which my colleagues
seemed increasingly disposed to give me latitude.

I therefore invited these two high Ministers to change their offices.
It was no bed of roses which I offered Herbert Morrison. These pages
certainly cannot attempt to describe the problems of London government, when often night after night ten or twenty thousand people were
made homeless, and when nothing but the ceaseless vigil of the citizens
as fire guards on the roofs prevented uncontrollable conflagrations;
when hospitals, filled with mutilated men and women, were themselves
struck by the enemy's bombs; when hundreds of thousands of weary
people crowded together in unsafe and insanitary shelters; when communications by road and rail were ceaselessly broken down; when drains
were smashed and light, power, and gas paralysed; and when nevertheless the whole fighting, toiling life of London had to go forward, and
nearly a million people be moved in and out for their work every night
and morning. We did not know how long it would last. We had no
reason to suppose that it would not go on getting worse. When I made
the proposal to Mr. Morrison he knew too much about it to treat it
lightly. He asked for a few hours' consideration; but in a short time
he returned and said he would be proud to shoulder the job. I highly
approved his manly decision.

Quite soon after the Ministerial movements a change in the enemy's method affected our general policy. Till now the hostile attack had been confined almost exclusively to high-explosive bombs; but with the full moon of October 15, when the heaviest attack of the month fell upon us, German aircraft dropped in addition 70,000 incendiary bombs. Hitherto we had encouraged the Londoners to take cover, and every effort was being made to improve their protection. But now "To the basements" must be replaced by "To the roofs." It fell to the new Minister of Home Security to institute this policy. An organisation of fire watchers and fire services on a gigantic scale and covering the whole of London (apart from measures taken in provincial cities) was rapidly brought into being. At first the fire watchers were volunteers; but the numbers required were so great, and the feeling that every man should take his turn upon the roster so strong, that fire watching soon became compulsory. This form of service had a bracing and buoyant effect upon all classes. Women pressed forward to take their share. Large-scale systems of training were developed to teach the fire watchers how to deal with the various kinds of incendiaries which were used against us. Many became adept, and thousands of fires were extinguished before they took hold. The experience of remaining on the roof night after night under fire, with no protection but a tin hat, soon became habitual.

Mr. Morrison presently decided to consolidate the fourteen hundred local fire brigades into a single National Fire Service, and to supplement this with a great fire guard of civilians trained and working in their spare time. The fire guard, like the roof watchers, was at first recruited on a voluntary basis, but like them it became by general consent compulsory. The National Fire Service gave us the advantages of greater mobility, a universal standard of training and equipment, and formally recognised ranks. The other Civil Defence forces produced regional columns ready at a minute's notice to go anywhere. The name Civil Defence Service was substituted for the prewar title of Air Raid Precautions (A.R.P.). Good uniforms were provided for large numbers, and they became conscious of being a fourth arm of the Crown.

I was glad that, if any of our cities were to be attacked, the brunt should fall on London. London was like some huge prehistoric animal, capable of enduring terrible injuries, mangled and bleeding from many wounds, and yet preserving its life and movement. The Anderson shelters were widespread in the working-class districts of two-storey houses, and everything was done to make them habitable and to drain them in wet weather. Later the Morrison shelter was developed, which was no more than a heavy kitchen table made of steel with strong wire sides, capable of holding up the ruins of a small house and thus giving a measure of protection. Many owed their lives to it. For the rest, "London could take it." They took all they got, and could have taken more.

Indeed, at this time we saw no end but the demolition of the whole metropolis. Still, as I pointed out to the House of Commons at the time, the law of diminishing returns operates in the case of the demolition of large cities. Soon many of the bombs would only fall upon houses already ruined and only make the rubble jump. Over large areas there would be nothing more to burn or destroy, and yet human beings might make their homes here and there, and carry on their work with infinite resource and fortitude.

*   *   *

On the night of November 3 for the first time in nearly two months no alarm sounded in London. The silence seemed quite odd to many. They wondered what was wrong. On the following night the enemy's attacks were widely dispersed throughout the island; and this continued for a while. There had been another change in the policy of the German offensive. Although London was still regarded as the principal target, a major effort was now to be made to cripple the industrial centres of Britain. Special squadrons had been trained, with new navigational devices, to attack specific key centres. For instance, one formation was trained solely for the destruction of the Rolls-Royce aero-engine works at Hillington, Glasgow. All this was a makeshift and interim plan. The invasion of Britain had been temporarily abandoned, and the attack upon Russia had not yet been mounted, nor was expected outside Hitler's intimate circle. The remaining winter months were therefore to be for the German Air Force a period of experiment, both in technical devices in night bombing and in attacks upon British sea-borne trade, together with an attempt to break down our production, military and civil. They would have done much better to have stuck to one thing at a time and pressed it to a conclusion. But they were already baffled and for the time being unsure of themselves.

These new bombing tactics began with the blitz on Coventry on the night of November 14. London seemed too large and vague a target for decisive results, but Goering hoped that provincial cities or munitions centres might be effectively obliterated. The raid started early in the dark hours of the 14th, and by dawn nearly five hundred German aircraft had dropped six hundred tons of high explosives and thousands of incendiaries. On the whole this was the most devastating raid which we sustained. The centre of Coventry was shattered, and its life for a spell completely disrupted. Four hundred people were killed and many more seriously injured. The German radio proclaimed that our other cities would be similarly "Coventrated." Nevertheless the all-important aero-engine and machine-tool factories were not brought to a standstill; nor was the population, hitherto untried in the ordeal of bombing, put out of action. In less than a week an emergency reconstruction committee did wonderful work in restoring the life of the city.

On November 15 the enemy switched back to London with a very heavy raid in full moonlight. Much damage was done, especially to churches and other monuments. The next target was Birmingham, and three successive raids from the 19th to the 22nd of November inflicted much destruction and loss of life. Nearly eight hundred people were killed and over two thousand injured; but the life and spirit of Birmingham survived this ordeal, and its million inhabitants, highly organised, conscious and comprehending, rode high above their physical suffering. During the last week of November and the beginning of December the weight of the attack shifted to the ports. Bristol, Southampton, and above all Liverpool, were heavily bombed. Later on, Plymouth, Sheffield, Manchester, Leeds, Glasgow, and other munitions centres passed through the fire undaunted. It did not matter where the blow struck, the nation was as sound as the sea is salt.

The climax raid of these weeks came once more to London, on Sunday, December 29. All the painfully gathered German experience was expressed on this occasion. It was an incendiary classic. The weight of the attack was concentrated upon the City of London itself. It was timed to meet the dead-low-water hour. The water mains were broken at the outset by very heavy high-explosive parachute mines. Nearly fifteen hundred fires had to be fought. The damage to railway stations and docks was serious. Eight Wren churches were destroyed or damaged. The Guildhall was smitten by fire and blast, and St. Paul's Cathedral was only saved by heroic exertions. A void of ruin at the very centre of the British world gaped upon us, but when the King and Queen visited the scene they were received with enthusiasm far exceeding any Royal festival.

During this prolonged ordeal, of which several months were still to come, the King was constantly at Buckingham Palace. Proper shelters were being constructed in the basement, but all this took time. Also it happened several times that His Majesty arrived from Windsor in the middle of an air raid. Once he and the Queen had a very narrow escape. His Majesty had a shooting range made in the Buckingham Palace garden, at which he and other members of his family and his equerries practised assiduously with pistols and tommy guns. Presently I brought the King an American short-range carbine, from a number which had been sent to me. This was a very good weapon.

About this time the King changed his practice of receiving me in a formal weekly audience at about five o'clock which had prevailed during my first two months of office. It was now arranged that I should lunch with him every Tuesday. This was certainly a very agreeable

method of transacting State business, and sometimes the Queen was
present. On several occasions we all had to take our plates and glasses
in our hands and go down to the shelter, which was making progress,
to finish our meal. The weekly luncheons became a regular institution.
After the first few months His Majesty decided that all servants should
be excluded, and that we should help ourselves and help each other.
During the four and a half years that this continued I became aware
of the extraordinary diligence with which the King read all the telegrams
and public documents submitted to him. Under the British constitutional system the Sovereign has a right to be made acquainted with
everything for which his Ministers are responsible, and has an unlimited
right of giving counsel to his Government. I was most careful that
everything should be laid before the King, and at our weekly meetings
he frequently showed that he had mastered papers which I had not yet
dealt with. It was a great help to Britain to have so good a King and
Queen in those fateful years, and as a convinced upholder of constitutional monarchy I valued as a signal honour the gracious intimacy with
which I, as first Minister, was treated, for which I suppose there had
been no precedent since the days of Queen Anne and Marlborough during his years of power.

*   *   *

This brings us to the end of the year, and for the sake of continuity
I have gone ahead of the general war. The reader will realise that all
this clatter and storm was but an accompaniment to the cool processes
by which our war effort was maintained and our policy and diplomacy
conducted. Indeed, I must record that at the summit these injuries,
failing to be mortal, were a positive stimulant to clarity of view, faithful comradeship, and judicious action. It would be unwise however to
suppose that if the attack had been ten or twenty times as severe — or
even perhaps two or three times as severe — the healthy reactions I
have described would have followed.

14

Lend-Lease

ABOVE THE ROAR AND CLASH of arms there now loomed upon us a world-fateful event of a different order. The presidential election took place on November 5. In spite of the tenacity and vigour with which these four-yearly contests are conducted, and the bitter differences on domestic issues which at this time divided the two main parties, the Supreme Cause was respected by the responsible leaders, Republicans and Democrats alike. At Cleveland on November 2 Mr. Roosevelt said: "Our policy is to give all possible material aid to the nations which still resist aggression across the Atlantic and Pacific Oceans." His opponent, Mr. Wendell Willkie, declared the same day at Madison Square Garden: "All of us — Republicans, Democrats, and Independents — believe in giving aid to the heroic British people. We must make available to them the products of our industry."

This larger patriotism guarded both the safety of the American Union and our life. Still, it was with profound anxiety that I awaited the result. No newcomer into power could possess or soon acquire the knowledge and experience of Franklin Roosevelt. None could equal his commanding gifts. My own relations with him had been most carefully fostered by me, and seemed already to have reached a degree of confidence and friendship which was a vital factor in all my thought. To close the slowly built-up comradeship, to break the continuity of all our discussions, to begin again with a new mind and personality, seemed to me a repellent prospect. Since Dunkirk I had not been conscious of the same sense of strain. It was with indescribable relief that I received the news that President Roosevelt had been re-elected.

*   *   *

Up till this time we had placed our orders for munitions in the

United States separately from, though in consultation with, the American Army, Navy, and Air Services. The ever-increasing volume of our several needs had led to overlapping at numerous points, with possibilities of friction arising at lower levels in spite of general goodwill. "Only a single, unified Government procurement policy for all defence purposes," writes Mr. Stettinius,[1] "could do the tremendous job that was now ahead." This meant that the United States Government should place all the orders for weapons in America. Three days after his re-election the President publicly announced a "rule of thumb" for the division of American arms output. As weapons came off the production line they were to be divided roughly fifty-fifty between the United States forces and the British and Canadian forces. That same day the Priorities Board approved a British request to order twelve thousand more aeroplanes in the United States in addition to the eleven thousand we had already booked. But how was all this to be paid for?

In mid-November Lord Lothian, who had recently flown home from Washington, spent two days with me at Ditchley. I had been advised not to make a habit of staying at Chequers every week end, especially when the moon was full, in case the enemy should pay me special attention. Mr. Ronald Tree and his wife made me and my staff very welcome many times at their large and charming house near Oxford. Ditchley is only four or five miles away from Blenheim. In these agreeable surroundings I received the Ambassador. He was primed with every aspect and detail of the American attitude. He had won nothing but goodwill and confidence in Washington. He was fresh from intimate contact with the President, with whom he had established a warm personal friendship. His mind was now set upon the dollar problem; this was grim indeed.

Britain entered the war with about 4,500,000,000 in dollars, or in gold and in United States investments that could be turned into dollars. The only way in which these resources could be increased was by new gold-production in the British Empire, mainly of course in South Africa, and by vigorous efforts to export goods, principally luxury goods, such as whisky, fine woollens, and pottery, to the United States. By these means an additional two thousand millions were procured during the first sixteen months of the war. During the period of the "Twilight War" we were torn between a vehement desire to order munitions in America and gnawing fear as our dollar resources dwindled. Always in Mr. Chamberlain's day the Chancellor of the Exchequer, Sir John Simon, would tell us of the lamentable state of our dollar resources and emphasise the need for conserving them. It was more or less accepted that we should have to reckon with a rigorous limitation of pur-

[1] Edward R. Stettinius, *Lend-Lease*, p. 62.

chases from the United States. We acted, as Mr. Purvis, the head of
our Purchasing Commission and a man of outstanding ability, once said
to Stettinius, "as if we were on a desert island on short rations which
we must stretch as far as we could." [2]

This had meant elaborate arrangements for eking out our money. In
peace we imported freely and made payments as we liked. When war
came we had to create a machine which mobilised gold and dollars and
other private assets, which stopped the ill-disposed from remitting their
funds to countries where they felt things were safer, and which cut out
wasteful imports and other expenditures. On top of making sure that
we did not waste our money, we had to see that others went on taking
it. The countries of the sterling area were with us: they adopted the
same kind of exchange control policy as we did and were willing takers
and holders of sterling. With others we made special arrangements by
which we paid them in sterling, which could be used anywhere in the
sterling area, and they undertook to hold any sterling for which they
had no immediate use and to keep dealings at the official rates of exchange. Such arrangements were originally made with the Argentine
and Sweden, but were extended to a number of other countries on the
Continent and in South America. These arrangements were completed
after the spring of 1940, and it was a matter of satisfaction — and a
tribute to sterling — that we were able to achieve and maintain them in
circumstances of such difficulty. In this way we were able to go on dealing with most parts of the world in sterling, and to conserve most of our
precious gold and dollars for our vital purchases in the United States.

When the war exploded into hideous reality, in May 1940, we were
conscious that a new era had dawned in Anglo-American relations.
From the time I formed the new Government, and Sir Kingsley Wood
became Chancellor of the Exchequer, we followed a simpler plan, namely,
to order everything we possibly could and leave future financial problems on the lap of the Eternal Gods. Fighting for life and presently
alone, under ceaseless bombardment, with invasion glaring upon us, it
would have been false economy and misdirected prudence to worry too
much about what would happen when our dollars ran out. We were
conscious of the tremendous changes taking place in American opinion,
and of the growing belief, not only in Washington but throughout the
Union, that their fate was bound up with ours. Moreover, at this time
an intense wave of sympathy and admiration for Britain surged across
the American nation. Very friendly signals were made to us from
Washington direct, and also through Canada, encouraging our boldness
and indicating that somehow or other a way would be found. In Mr.
Morgenthau, Secretary of the Treasury, the cause of the Allies had a

[2] Stettinius, *op. cit.*, p. 60

tireless champion. The taking over of the French contracts in June had almost doubled our rate of spending across the Exchange. Besides this, we placed new orders for aeroplanes, tanks, and merchant ships in every direction, and promoted the building of great new factories both in the United States and Canada.

*   *   *

Up till November we had paid for everything we had received. We had already sold $335,000,000 worth of American shares requisitioned for sterling from private owners in Britain. We had paid out over $4,500,000,000 in cash. We had only two thousand millions left, the greater part in investments, many of which were not readily marketable. It was plain that we could not go on any longer in this way. Even if we divested ourselves of all our gold and foreign assets, we could not pay for half we had ordered, and the extension of the war made it necessary for us to have ten times as much. We must keep something in hand to carry on our daily affairs.

Lothian was confident that the President and his advisers were earnestly seeking the best way to help us. Now that the election was over, the moment to act had come. Ceaseless discussions on behalf of the Treasury were proceeding in Washington between their representative, Sir Frederick Phillips, and Mr. Morgenthau. The Ambassador urged me to write a full statement of our position to the President. Accordingly that Sunday at Ditchley I drew up, in consultation with him, a personal letter. As the document had to be checked and rechecked by the Chiefs of Staff and the Treasury, and approved by the War Cabinet, it was not completed before Lothian's return to Washington. In its final form it was dated December 8, and was immediately sent to Mr. Roosevelt. The letter, which was one of the most important I ever wrote, reached our great friend when he was cruising, on board an American warship, the *Tuscaloosa*, in the sunlight of the Caribbean Sea. He had only his own intimates around him. Harry Hopkins, then unknown to me, told me later that Mr. Roosevelt read and reread this letter as he sat alone in his deck chair, and that for two days he did not seem to have reached any clear conclusion. He was plunged in intense thought, and brooded silently.

From all this there sprang a wonderful decision. It was never a question of the President not knowing what he wanted to do. His problem was how to carry his country with him and to persuade Congress to follow his guidance. According to Stettinius, the President, as early as the late summer, had suggested at a meeting of the Defence Advisory Commission on Shipping Resources that "It should not be necessary for the British to take their own funds and have ships built in the United

States, or for us to loan them money for this purpose. There is no reason why we should not take a finished vessel and lease it to them for the duration of the emergency." It appeared that by a Statute of 1892 the Secretary for War, "when in his discretion it will be for the public good," could lease Army property if not required for public use for a period of not longer than five years. Precedents for the use of this Statute, by the *lease* of various Army items, from time to time were on record.

Thus the word "lease" and the idea of applying the lease principle to meeting British needs had been in President Roosevelt's mind for some time as an alternative to a policy of indefinite loans which would soon far outstrip all possibilities of repayment. Now suddenly all this sprang into decisive action, and the glorious conception of Lend-Lease was proclaimed.

The President returned from the Caribbean on December 16, and broached his plan at his press conference next day. He used a simple illustration. "Suppose my neighbour's house catches fire and I have a length of garden hose four or five hundred feet away. If he can take my garden hose and connect it up with his hydrant, I may help him to put out the fire. Now what do I do? I don't say to him before that operation, 'Neighbour, my garden hose cost me fifteen dollars; you have to pay me fifteen dollars for it.' No! What is the transaction that goes on? I don't want fifteen dollars — I want my garden hose back after the fire is over." And again: "There is absolutely no doubt in the mind of a very overwhelming number of Americans that the best immediate defence of the United States is the success of Great Britain defending itself; and that therefore, quite aside from our historic and current interest in the survival of Democracy in the world as a whole, it is equally important from a selfish point of view and of American defence that we should do everything possible to help the British Empire to defend itself."

Finally: "I am trying to eliminate the dollar mark."

On this foundation the ever-famous Lend-Lease Bill was at once prepared for submission to Congress. I described this to Parliament later as "the most unsordid act in the history of any nation." Once it was accepted by Congress it transformed immediately the whole position. It made us free to shape by agreement long-term plans of vast extent for all our needs. There was no provision for repayment. There was not even to be a formal account kept in dollars or sterling. What we had was lent or leased to us because our continued resistance to the Hitler tyranny was deemed to be of vital interest to the great Republic. According to President Roosevelt, the defence of the United States and not dollars was henceforth to determine where American weapons were to go.

* * *

It was at this moment, the most important in his public career, that Philip Lothian was taken from us. Shortly after his return to Washington he fell suddenly and gravely ill. He worked unremittingly to the end. On December 12, in the full tide of success, he died. This was a loss to the nation and to the Cause. He was mourned by wide circles of friends on both sides of the ocean. To me, who had been in such intimate contact with him a fortnight before, it was a personal shock. I paid my tribute to him in a House of Commons united in deep respect for his work and memory.

I had to turn immediately to the choice of his successor. It seemed that our relations with the United States at this time required as Ambassador an outstanding national figure and a statesman versed in every aspect of world politics. Having ascertained from the President that my suggestion would be acceptable, I invited Mr. Lloyd George to take the post. He had not felt able to join the War Cabinet in July, and was not happily circumstanced in British politics. His outlook on the war and the events leading up to it was from a different angle from mine. There could be no doubt however that he was our foremost citizen, and that his incomparable gifts and experience would be devoted to the success of his mission. I had a long talk with him in the Cabinet Room, and also at luncheon on a second day. He showed genuine pleasure at having been invited. "I tell my friends," he said, "I have had honourable offers made to me by the Prime Minister." He was sure that at the age of seventy-seven he ought not to undertake so exacting a task. As a result of my long conversations with him I was conscious that he had aged even in the months which had passed since I had asked him to join the War Cabinet, and with regret but also with conviction I abandoned my plan.

I next turned to Lord Halifax, whose prestige in the Conservative Party stood high, and was enhanced by his being at the Foreign Office. For a Foreign Secretary to become an Ambassador marks in a unique manner the importance of the mission. His high character was everywhere respected, yet at the same time his record in the years before the war and the way in which events had moved left him exposed to much disapprobation and even hostility from the Labour side of our National Coalition. I knew that he was conscious of this himself.

When I made him this proposal, which was certainly not a personal advancement, he contented himself with saying in a simple and dignified manner that he would serve wherever he was thought to be most useful. In order to emphasise still further the importance of his duties, I arranged that he should resume his function as a member of the War Cabinet whenever he came home on leave. This arrangement worked without the slightest inconvenience, owing to the qualities and experience of the personalities involved, and for six years thereafter, both under the National Coalition and the Labour-Socialist Government, Halifax discharged the work of Ambassador to the United States with conspicuous and ever-growing influencing and success.

President Roosevelt, Mr. Hull, and other high personalities in Washigton were extremely pleased with the selection of Lord Halifax. Indeed it was at once apparent to me that the President greatly preferred it to my first proposal. The appointment of the new Ambassador was received with marked approval both in America and at home, and was judged in every way adequate and appropriate to the scale of events.

*   *   *

I had no doubt who should fill the vacancy at the Foreign Office. On all the great issues of the past four years I had, as these pages have shown, dwelt in close agreement with Anthony Eden. I have described my anxieties and emotions when he parted company with Mr. Chamberlain in the spring of 1938. Together we had abstained from the vote on Munich. Together we had resisted the party pressures brought to bear upon us in our constituencies during the winter of that melancholy year. We had been united in thought and sentiment at the outbreak of the war and as colleagues during its progress. The greater part of Eden's public life had been devoted to the study of foreign affairs. He had held the splendid office of Foreign Secretary with distinction, and had resigned it when only forty-two years of age for reasons which are in retrospect, and at this time, viewed with the approval of all parties in the State. He had played a fine part as Secretary of State for War during this terrific year, and his conduct of Army affairs had brought us very close together. We thought alike, even without consultation, on a very great number of practical issues as they arose from day to day. I looked forward to an agreeable and harmonious comradeship between the Prime Minister and the Foreign Secretary, and this hope was certainly fulfilled during the four and a half years of war and policy which lay before us. Eden was sorry to leave the War Office, in all the stresses and excitements of which he was absorbed; but he returned to the Foreign Office like a man going home.

15

Desert Victory

IN SPITE OF THE ARMISTICE and Oran and the ending of our diplomatic relations with Vichy, whither the French Government had moved under Marshal Pétain, I never ceased to feel a unity with France. People who have not been subjected to the personal stresses which fell upon prominent Frenchmen in the awful ruin of their country should be careful in their judgments of individuals. It is beyond the scope of this story to enter the maze of French politics. But I felt sure that the French nation would do its best for the common cause according to the facts presented to it. When they were told that their only salvation lay in following the advice of the illustrious Marshal, and that England, which had given them so little help, would soon be conquered or give in, very little choice was offered to the masses. But I was sure they wanted us to win, and that nothing would give them more joy than to see us continue the struggle with vigour. It was our first duty to give loyal support to General de Gaulle in his valiant constancy. On August 7 I signed a military agreement with him which dealt with practical needs. His stirring addresses were made known to France and the world by the British broadcast. The sentence of death which the Pétain Government passed upon him glorified his name. We did everything in our power to aid him and magnify his movement.

At the same time it was necessary to keep in touch not only with France but even with Vichy. I therefore always tried to make the best of them. I was very glad when at the end of 1940 the United States sent an Ambassador to Vichy of so much influence and character as Admiral Leahy, who was himself so close to the President. I repeatedly encouraged the Canadian Premier, Mr. Mackenzie King, to keep his representative, the skilful and accomplished M. Dupuy, at Vichy. Here at least was a window upon a courtyard to which we had no other access. On July 25 I sent a minute to the Foreign Secretary in which I said: "I want to promote a kind of collusive conspiracy in the Vichy Government whereby certain members of that Government, perhaps with the consent of those who remain will levant to North Africa in order to make a better bargain for France from the North African shore and from a position of independence. For this purpose I would use both food and other inducements, as well as the obvious arguments." Our consistent policy was to make the Vichy Government and its members feel that, so far as we were concerned, it was never too late to mend. Whatever had happened in the past, France was our comrade in tribulation, and nothing but actual war between us should prevent her being our partner in victory.

This mood was hard upon de Gaulle, who had risked all and kept the flag flying, but whose handful of followers outside France could never claim to be an effective alternative French Government. Nevertheless we did our utmost to increase his influence, authority, and power. He for his part naturally resented any kind of truck on our part with Vichy, and thought we ought to be exclusively loyal to him. He also felt it to be essential to his position before the French people that he should maintain a proud and haughty demeanour towards "perfidious Albion," although an exile, dependent upon our protection and dwelling in our midst. He had to be rude to the British to prove to French eyes that he was not a British puppet. He certainly carried out this policy with perseverance. He even one day explained this technique to me, and I fully comprehended the extraordinary difficulties of his problem. I always admired his massive strength. Whatever Vichy might do for good or ill, we would not abandon him or discourage accessions to his growing colonial domain. Above all we would not allow any portion of the French Fleet, now immobilised in French colonial harbours, to return to France. There were times when the Admiralty were deeply concerned lest France should declare war upon us and thus add to our many cares. I always believed that once we had proved our resolve and ability to fight on indefinitely the spirit of the French people would never allow the Vichy Government to take so unnatural a step. Indeed, there was by now a strong enthusiasm and comradeship for Britain, and French hopes grew as the months passed. This was recognised even by M. Laval when he presently became Foreign Minister to Pétain.

*   *   *

It was otherwise with Italy. With the disappearance of France as a combatant and with Britain set on her struggle for life at home, Mussolini might well feel that his dream of dominating the Mediterranean and rebuilding the former Roman Empire would come true. Relieved from any need to guard against the French in Tunis, he could still further reinforce the numerous army he had gathered for the invasion of Egypt. Nevertheless the War Cabinet were determined to defend Egypt against all comers with whatever resources could be spared from the decisive struggle at home. All the more was this difficult when the Admiralty declared themselves unable to pass even military convoys through the Mediterranean on account of the air dangers. All must go round the Cape. Thus we might easily rob the Battle of Britain without helping the Battle of Egypt. It is odd that, while at the time everyone concerned was quite calm and cheerful, writing about it afterwards makes one shiver.

When Italy declared war on June 10, 1940, the British Intelligence estimated — we now know correctly — that, apart from her garrisons in Abyssinia, Eritrea, and Somaliland, there were about 215,000 Italian troops in North African coastal provinces. The British forces in Egypt amounted to perhaps fifty thousand men. From these both the defence of the western frontier and the internal security of Egypt had to be provided. We therefore had heavy odds against us in the field, and the Italians had also many more aircraft.

During July and August the Italians became active at many points. There was a threat from Kassala westward towards Khartoum. Alarm was spread in Kenya by the fear of an Italian expedition marching four hundred miles south from Abyssinia towards the Tana River and Nairobi. Considerable Italian forces advanced into British Somaliland. But all these anxieties were petty compared with the Italian invasion of Egypt, which was obviously being prepared on the greatest scale. Even before the war a magnificent road had been made along the coast from the main base at Tripoli, through Tripolitania and Cyrenaica, to the Egyptian frontier. Along this road there had been for many months a swelling stream of military traffic. Large magazines were slowly established and filled at Benghazi, Derna, Tobruk, Bardia, and Sollum. The length of this road was over a thousand miles, and all these swarming Italian garrisons and supply depots were strung along it like beads on a string.

At the head of the road and near the Egyptian frontier an Italian army of seventy or eighty thousand men, with a good deal of modern equipment, had been patiently gathered and organised. Before this army glittered the prize of Egypt. Behind it stretched the long road back to Tripoli; and after that the sea! If this force, built up in driblets week by week for years, could advance continually eastward, conquering all who sought to bar the path, its fortunes would be bright. If it could gain the fertile regions of the Delta all worry about the long road back would vanish. On the other hand, if ill-fortune befell it only a few would ever get home. In the field army and in the series of great supply depots all along the coast there were by the autumn at least three hundred thousand Italians, who could, even if unmolested, retreat westward along the road only gradually or piecemeal. For this they required many months. And if the battle were lost on the Egyptian border, if the army's front were broken, and if time were not given to them, all were doomed to capture or death. However, in July 1940 it was not known who was going to win the battle.

Our foremost defended position at that time was the railhead at Mersa Matruh. There was a good road westward to Sidi Barrani, but thence to the frontier at Sollum there was no road capable of maintaining any considerable strength for long near the frontier. A small covering mechanised force had been formed of some of our finest Regular troops, and orders had been given to attack the Italian frontier posts immediately on the outbreak of war. Accordingly, within twenty-four hours they crossed the frontier, took the Italians, who had not heard that war had been declared, by surprise, and captured prisoners. The next night, June 12, they had a similar success, and on June 14 captured the frontier forts at Capuzzo and Maddalena, taking two hundred twenty prisoners. On the 16th they raided deeper, destroyed twelve tanks, intercepted a convoy on the Tobruk–Bardia road, and captured a general.

In this small but lively warfare our troops felt they had the advantage, and soon conceived themselves to be masters of the desert. Until they came up against large formed bodies or fortified posts they could go where they liked, collecting trophies from sharp encounters. When armies approach each other it makes all the difference which owns only the ground on which it stands or sleeps and which one owns all the rest. I saw this in the Boer War, where we owned nothing beyond the fires of our camps and bivouacs, whereas the Boers rode where they pleased all over the country. The published Italian casualties for the first three months were nearly three thousand five hundred men, of whom seven hundred were prisoners. Our own losses barely exceeded one hundred and fifty. Thus the first phase in the war which Italy had declared upon the British Empire opened favourably for us.

*   *   *

I felt an acute need of talking over the serious events impending in the Libyan desert with General Wavell himself. I had not met this distinguished officer, on whom so much was resting, and I asked the Secretary of State for War to invite him over for a week for consultation when an opportunity could be found. He arrived on August 8. He toiled with the Staffs and had several long conversations with me and

Mr. Eden. The command in the Middle East at that time comprised an extraordinary amalgam of military, political, diplomatic, and administrative problems of extreme complexity. It took nearly a year of ups and downs for me and my colleagues to learn the need of dividing the responsibilities of the Middle East among a Commander-in-Chief, a Minister of State, and an Intendant-General to cope with the supply problem. While not in full agreement with General Wavell's use of the resources at his disposal, I thought it best to leave him in command. I admired his fine qualities, and was impressed with the confidence so many people had in him.

As a result of the Staff discussions Dill, with Eden's ardent approval, wrote me that the War Office were arranging to send immediately to Egypt over one hundred and fifty tanks and many guns. The only question open was whether they should go round the Cape or take a chance through the Mediterranean. I pressed the Admiralty hard for direct convoy through the Mediterranean. Much discussion proceeded on this latter point. Meanwhile the Cabinet approved the embarkation and dispatch of the armoured force, leaving the final decision about which way they should go till the convoy approached Gibraltar. This option remained open to us till August 26, by which time we should know a good deal more about the imminence of any Italian attack. No time was lost. The decision to give this blood transfusion while we braced ourselves to meet a mortal danger was at once awful and right. No one faltered.

*   *   *

Until the French collapse the control of the Mediterranean had been shared between the British and French Fleets. Now France was out and Italy in. The numerically powerful Italian Fleet and a strong Italian Air Force were ranged against us. So formidable did the situation appear that Admiralty first thoughts contemplated the abandonment of the Eastern Mediterranean and concentration at Gibraltar. I resisted this policy, which, though justified on paper by the strength of the Italian Fleet, did not correspond to my impressions of the fighting values, and also seemed to spell the doom of Malta. It was resolved to fight it out at both ends. The burdens which lay upon the Admiralty at this time were however heavy in the extreme. The invasion danger required a high concentration of flotillas and small craft in the Channel and North Sea. The U-boats, which had by August begun to work from Biscayan ports, took severe toll of our Atlantic convoys without suffering many losses themselves. Until now the Italian Fleet had never been tested. The possibility of a Japanese declaration of war, with all that it would bring upon our Eastern Empire, could never be excluded from our thoughts. It is therefore not strange that the Admiralty viewed with the deepest anxiety all risking of warships in the Mediterranean, and were sorely tempted to adopt the strictest defensive at Gibraltar and Alexandria. I, on the other hand, did not see why the large number of ships assigned to the Mediterranean should not play an active part from the outset. Malta had to be reinforced both with air squadrons and troops. Although all commercial traffic was rightly suspended, and all large troop convoys to Egypt must go round the Cape, I could not bring myself to accept the absolute closure of the inland sea. Indeed I hoped that by running a few special convoys we might arrange and provoke a trial of strength with the Italian Fleet. I hoped that this might happen and Malta be properly garrisoned and equipped with aeroplanes and A.A. guns before the appearance, which I already dreaded, of the Germans in this theatre. All through the summer and autumn months I engaged in friendly though tense discussion with the Admiralty upon this part of our war effort.

However I was not able to induce the Admiralty to send the armoured force, or at least their vehicles, through the Mediterranean, and the whole convoy continued on its way round the Cape.

I was both grieved and vexed at this. No serious disaster did in fact occur in Egypt. Everywhere, despite the Italian air strength, we held the initiative, and Malta remained in the foreground of events as an advanced base for offensive operations against the Italian communications with their forces in Africa.

*   *   *

Our anxieties about the Italian invasion of Egypt were, it now appears, far surpassed by those of Marshal Graziani, who commanded it. A few days before it was due to start he asked for a month's postponement. Mussolini replied that if he did not attack on Monday he would be replaced. The Marshal answered that he would obey. "Never," says Ciano, "has a military operation been undertaken so much against the will of the commanders."

On September 13 the main Italian army began its long-expected advance across the Egyptian frontier. Their forces amounted to six infantry divisions and eight battalions of tanks. Our covering troops consisted of three battalions of infantry, one battalion of tanks, three batteries, and two squadrons of armoured cars. They were ordered to make a fighting withdrawal, an operation for which their quality and desert-worthiness fitted them. The Italian attack opened with a heavy barrage on our positions near the frontier town of Sollum. When the dust and smoke cleared the Italian forces were seen ranged in a remarkable order. In front were motorcyclists in precise formation from flank to flank and front to rear; behind them were light tanks and many rows of mechanical vehicles. In the words of a British colonel, the spectacle resembled "a birthday party in the Long Valley at Aldershot." The 3rd Coldstream Guards, who confronted this imposing array, withdrew slowly, and our artillery took its toll of the generous targets presented to them.

Farther south two large enemy columns moved across the open desert south of the long ridge that runs parallel to the sea and could be crossed only at Halfaya — the "Hellfire Pass" which played its part in all our later battles. Each Italian column consisted of many hundreds of vehicles, with tanks, anti-tanks guns, and artillery in front, and with lorried infantry in the centre. This formation, which was several times adopted, we called the "Hedgehog." Our forces fell back before these great numbers, taking every opportunity to harass the enemy, whose movements seemed erratic and indecisive. Graziani afterwards explained that at the last moment he decided to change his plan of an enveloping desert movement and "concentrate all my forces on the left to make a lightning movement along the coast to Sidi Barrani." Accordingly the great Italian mass moved slowly forward along the coast road by two parallel tracks. They attacked in waves of infantry carried in lorries, sent forward in fifties. The Coldstream Guards fell back skilfully at their convenience from Sollum to successive positions for four days, inflicting severe punishments as they went.

On the 17th the Italian army reached Sidi Barrani. Our casualties were forty killed and wounded, and the enemy's about ten times as many, including one hundred and fifty vehicles destroyed. Here, with their communications lengthened by sixty miles, the Italians settled down to spend the next three months. They were continually harassed by our small mobile columns, and suffered serious maintenance difficulties. Mussolini at first was "radiant with joy." As the weeks lengthened into months his satisfaction diminished. It seemed however certain to us in London that in two or three months an Italian army far larger than any we could gather would renew the advance to capture the Delta. And then there were always the Germans who might appear! We could not of course expect the long halt which followed Graziani's advance. It was reasonable to suppose that a major battle would be fought at Mersa Matruh. The weeks that had already passed had enabled our precious armour to come round the Cape without the time lag so far causing disadvantage.

When I look back on all these worries I remember the story of the old man who said on his deathbed that he had had a lot of trouble in his life, most of which had never happened. Certainly this is true of my life in September 1940. The Germans were beaten in the Air Battle of Britain. The overseas invasion of Britain was not attempted. In fact, by this date Hitler had already turned his glare upon the East. The Italians did not press their attack upon Egypt. The Tank Brigade sent

all round the Cape arrived in good time, not indeed for a defensive
battle of Mersa Matruh in September, but for a later operation incomparably more advantageous. We found means to reinforce Malta
before any serious attack from the air was made upon it, and no one
dared to try a landing upon the island fortress at any time. Thus
September passed.

*   *   *

A fresh though not entirely unexpected outrage by Mussolini, with
baffling problems and far-reaching consequences to all our harassed
affairs, now broke upon the Mediterranean scene.

The Duce took the final decision to attack Greece on October 15,
1940, and before dawn on the 28th the Italian Minister in Athens presented an ultimatum to General Metaxas, the Premier of Greece. Mussolini demanded that the whole of Greece should be opened to Italian
troops. At the same time the Italian army in Albania invaded Greece
at various points. The Greek Government, whose forces were by no
means unready on the frontier, rejected the ultimatum. They also invoked the guarantee given by Mr. Chamberlain on April 13, 1939. This
we were bound to honour. By the advice of the War Cabinet, and
from his own heart, His Majesty replied to the King of the Hellenes:
"Your cause is our cause; we shall be fighting against a common foe."
I responded to the appeal of General Metaxas: "We will give you all
the help in our power. We will fight a common foe and we will share
a united victory." This undertaking was during a long story made good.

Apart from a few air squadrons, a British mission, and perhaps some
token troops, we had nothing to give; and even these trifles were a
painful subtraction from ardent projects already lighting in the Libyan
theatre. One salient strategic fact leaped out upon us — CRETE! The
Italians must not have it. We must get it first — and at once. It was
fortunate that at this moment Mr. Eden was in the Middle East, and
that I thus had a ministerial colleague on the spot with whom to deal.
I telegraphed to him, and at the invitation of the Greek Government,
Suda Bay, the best harbour in Crete, was occupied by our forces a few
days later.

The story of Suda Bay is sad. The tragedy was not reached until
1941. I believe I had as much direct control over the conduct of the
war as any public man had in any country at this time. The knowledge
I possessed, the fidelity and active aid of the War Cabinet, the loyalty
of all my colleagues, the ever-growing efficiency of our war machine,
all enabled an intense focusing of constitutional authority to be achieved.
Yet how far short was the action taken by the Middle East Command of
what was ordered and what we all desired! In order to appreciate the
limitations of human action, it must be remembered how much was

going on in every direction at the same time. Nevertheless it remains astonishing to me that we should have failed to make Suda Bay the amphibious citadel of which all Crete was the fortress.

*　　　　　*　　　　　*

The Italian invasion of Greece from Albania was another heavy rebuff to Mussolini. The first assault was repulsed with heavy loss, and the Greeks immediately counterattacked. The Greek Army, under General Papagos, showed superior skill in mountain warfare, outmanoeuvring and outflanking their enemy. By the end of the year their prowess had forced the Italians thirty miles behind the Albanian frontier along the whole front. For several months twenty-seven Italian divisions were pinned in Albania by sixteen Greek divisions. The remarkable Greek resistance did much to hearten the other Balkan countries and Mussolini's prestige sank low.

There was more to follow. Mr. Eden got back home on November 8, and came that evening after the usual raid had begun to see me. He brought with him a carefully guarded secret which I wished I had known earlier. Nevertheless no harm had been done. Eden unfolded in considerable detail to a select circle, including the C.I.G.S. and General Ismay, the offensive plan which General Wavell and General Wilson had conceived and prepared. No longer were we to await in our fortified lines at Mersa Matruh an Italian assault, for which defensive battle such long and artful preparations had been made. On the contrary, within a month or so we were ourselves to attack.

We were all delighted. I purred like six cats. Here was something worth doing. It was decided there and then, subject to the agreement of the Chiefs of Staff and the War Cabinet, to give immediate sanction and all possible support to this splendid enterprise. In due course the proposals were brought before the War Cabinet. I was ready to state the case or have it stated. But when my colleagues learned that the Generals on the spot and the Chiefs of Staff were in full agreement with me and Mr. Eden, they declared that they did not wish to know the details of the plan, that the fewer who knew them the better, and that they whole-heartedly approved the general policy of the offensive. This was the attitude which the War Cabinet adopted on several important occasions, and I record it here that it may be a model, should similar dangers and difficulties arise in future times.

*　　　　　*　　　　　*

Although we were still heavily outnumbered on paper by the Italian Fleet, marked improvements had now been made in our Mediterranean strength. During September the *Valiant*, the armoured-deck aircraft carrier *Illustrious*, and two anti-aircraft cruisers had come safely through the Mediterranean to join Admiral Cunningham at Alexandria. Hitherto his ships had always been observed and usually bombed by the greatly superior Italian Air Force. The *Illustrious*, with her modern fighters and latest radar equipment, by striking down patrols and assailants gave a new secrecy to our movements. This advantage was timely.

He had long been anxious to strike a blow at the Italian Fleet as they lay in their main base at Taranto. The attack was delivered on November 11 as the climax of a well-concerted series of operations. Taranto lies in the heel of Italy three hundred and twenty miles from Malta. Its magnificent harbour was heavily defended against all modern forms of attack. The arrival at Malta of some fast reconnaissance machines enabled us to discern our prey. The *Illustrious* released her aircraft shortly after dark from a point about a hundred and seventy miles from Taranto. For an hour the battle raged amid fire and destruction among the Italian ships. Despite the heavy flak only two of our aircraft were shot down. The rest flew safely back.

By this single stroke the balance of naval power in the Mediterranean was decisively altered. The air photographs showed that three battleships, one of them the new *Littorio*, had been torpedoed, and in addition one cruiser was reported hit and much damage inflicted on the dockyard. Half the Italian battle fleet was disabled for at least six months, and the Fleet Air Arm could rejoice at having seized by their gallant exploit one of the rare opportunities presented to them.

An ironic touch is imparted to this event by the fact that on this very day the Italian Air Force at the express wish of Mussolini had taken part in the air attack on Great Britain. An Italian bomber force, escorted by about sixty fighters, attempted to bomb Allied convoys in the Medway. They were intercepted by our fighters, eight bombers and five fighters being shot down. This was their first and last intervention in our domestic affairs. They might have found better employment defending their fleet at Taranto.

*   *   *

For a month or more all the troops to be used in our desert offensive practised the special parts they had to play in the extremely complicated attack. Only a small circle of officers knew the full scope of the plan, and practically nothing was put on paper. On December 6 our lean, bronzed, desert-hardened, and completely mechanised army of about twenty-five thousand men leaped forward more than forty miles, and all next day lay motionless on the desert sand unseen by the Italian Air Force. They swept forward again on December 8, and that evening, for the first time, the troops were told that this was no desert exercise, but the "real thing." At dawn on the 9th the battle of Sidi Barrani began.

It is not my purpose to describe the complicated and dispersed fighting which occupied the next four days over a region as large as Yorkshire. Everything went smoothly. Fighting continued all through the 10th, and by ten o'clock the Coldstream battalion headquarters signalled that it was impossible to count the prisoners on account of their numbers, but that "there were about five acres of officers and two hundred acres of other ranks." At home in Downing Street they brought me hour-to-hour signals from the battlefield. It was difficult to understand exactly what was happening, but the general impression was favourable, and I remember being struck by a message from a young officer in a tank of the 7th Armoured Division: "Have arrived at the second B in Buq Buq." Sidi Barrani was captured on the afternoon of the 10th and by December 15 all enemy troops had been driven from Egypt.

Bardia was our next objective. Within its perimeter, seventeen miles in extent, was the greater part of four more Italian divisions. The defences comprised a continuous anti-tank ditch and wire obstacles with concrete blockhouses at intervals, and behind this was a second line of fortifications. The storming of this considerable stronghold required preparation, and to complete this episode of desert victory I shall intrude upon the New Year. The attack opened early on January 3. One Australian battalion, covered by a strong artillery concentration, seized and held a lodgment in the western perimeter. Behind them engineers filled in the anti-tank ditch. Two Australian brigades carried on the attack and swept east and southeastward. They sang at that time a song from an American film, which soon became popular also in Britain:

> *We're off to see the Wizard,*
> *The wonderful Wizard of Oz.*
> *We hear he is a Whiz of a Wiz,*
> *If ever a Wiz there was.*

This tune always reminds me of these buoyant days. By the afternoon of the 4th, British tanks — "Matildas," as they were named — supported by infantry, entered Bardia, and by the 5th all the defenders had surrendered. Forty-five thousand prisoners and 462 guns were taken.

By next day, January 6, Tobruk in its turn had been isolated. It was not possible to launch the assault till January 21. By early next morning all resistance ceased. The prisoners amounted to nearly 30,000 with 236 guns. The Desert Army had in six weeks advanced over two hundred miles of waterless and foodless space, had taken by assault two strongly fortified seaports with permanent air and marine defences, and captured 113,000 prisoners and over 700 guns. The great Italian Army which had invaded and hoped to conquer Egypt scarcely existed as a military

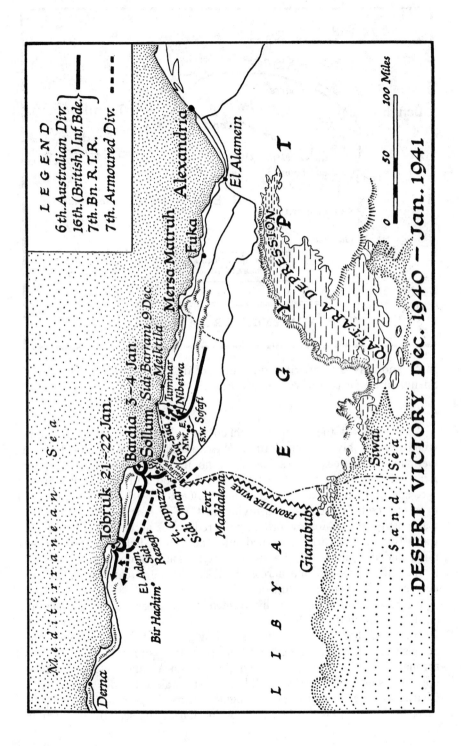

LEGEND

6th. Australian Div.
16th. (British) Inf. Bde. ——
7th. Bn. R.I.R.
7th. Armoured Div. - - - - -

Mediterranean Sea

Derna

Tobruk 21–22 Jan.

El Adem
Bir Hachim
Sidi Rezegh
Ft. Capuzzo
Sidi Omar
Fort Maddalena

Bardia 3–4 Jan.
Sollum
Sidi Barrani 9 Dec.
Meiktila
Buq Buq
S.W. E.
N.W. E.
Nibeiwa
Tummar
S.W. Sofafi

Mersa Matruh
Fuka
El Alamein
Alexandria

FRONTIER WIRE

Giarabub

Siwa

Sand Sea

L I B Y A

E G Y P T

QATTARA DEPRESSION

100 Miles
50
0

DESERT VICTORY Dec. 1940 – Jan. 1941

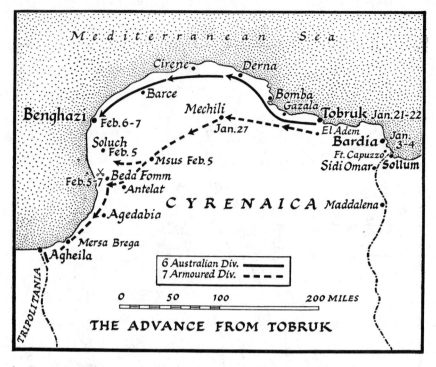

6 Australian Div. ——————
7 Armoured Div. — — — —

0   50   100   200 MILES

THE ADVANCE FROM TOBRUK

force, and only the imperious difficulties of distance and supplies delayed an indefinite British advance to the west.

*   *   *

As the end of the year approached both its lights and its shadows stood out harshly on the picture. We were alive. We had beaten the German Air Force. There had been no invasion of the island. The Army at home was now very powerful. London had stood triumphant through all her ordeals. Everything connected with our air mastery over our own island was improving fast. The smear of Communists who obeyed their Moscow orders gibbered about a capitalist-imperialist war. But the factories hummed and the whole British nation toiled night and day, uplifted by a surge of relief and pride. Victory sparkled in the Libyan desert, and across the Atlantic the Great Republic drew ever nearer to her duty and our aid.

We may, I am sure, rate this tremendous year as the most splendid, as it was the most deadly, year in our long English and British story. It was a great, quaintly organised England that had destroyed the Spanish Armada. A strong flame of conviction and resolve carried us through the twenty-five years' conflict which William III and Marlborough waged against Louis XIV. There was a famous period with Chatham. There was the long struggle against Napoleon, in which our survival was secured through the domination of the seas by the British

Navy under the classic leadership of Nelson and his associates. A million Britons died in the first World War. But nothing surpasses 1940. By the end of that year this small and ancient island, with its devoted Commonwealth, Dominions, and attachments under every sky, had proved itself capable of bearing the whole impact and weight of world destiny. We had not flinched or wavered. We had not failed. The soul of the British people and race had proved invincible. The citadel of the Commonwealth and Empire could not be stormed. Alone, but upborne by every generous heartbeat of mankind, we had defied the tyrant in the height of his triumph.

All our latent strength was now alive. The air terror had been measured. The island was intangible, inviolate. Henceforward we too would have weapons with which to fight. Henceforward we too would be a highly organised war machine. We had shown the world that we could hold our own. There were two sides to the question of Hitler's world domination. Britain, whom so many had counted out, was still in the ring, far stronger than she had ever been, and gathering strength with every day. Time had once again come over to our side. And not only to our national side. The United States was arming fast and drawing ever nearer to the conflict. Soviet Russia, who with callous miscalculation had adjudged us worthless at the outbreak of the war, and had bought from Germany fleeting immunity and a share of the booty, had also become much stronger and had secured advanced positions for her own defence. Japan seemed for the moment to be overawed by the evident prospect of a prolonged world war, and, anxiously watching Russia and the United States, meditated profoundly what it would be wise and profitable to do.

And now this Britain, and its far-spread association of states and dependencies, which had seemed on the verge of ruin, whose very heart was about to be pierced, had been for fifteen months concentrated upon the war problem, training its men and devoting all its infinitely varied vitalities to the struggle. With a gasp of astonishment and relief the smaller neutrals and the subjugated states saw that the stars still shone in the sky. Hope, and within it passion, burned anew in the hearts of hundreds of millions of men. The good cause would triumph. Right would not be trampled down. The flag of Freedom, which in this fateful hour was the Union Jack, would still fly in all the winds that blew.

But I and my faithful colleagues who brooded with accurate information at the summit of the scene had no lack of cares. The shadow of the U-boat blockade already cast its chill upon us. All our plans depended upon the defeat of this menace. The Battle of France was lost. The Battle of Britain was won. The Battle of the Atlantic had now to be fought.

16

The Widening War

W I T H   T H E   N E W   Y E A R   more intimate contacts developed with President Roosevelt. I had already sent him the compliments of the season, and on January 10, 1941, a gentleman arrived to see me at Downing Street with the highest credentials. Telegrams had been received from Washington stating that he was the closest confidant and personal agent of the President. I therefore arranged that he should be met by Mr. Brendan Bracken on his arrival at Poole Airport, and that we should lunch together alone the next day. Thus I met Harry Hopkins, that extraordinary man, who played, and was to play, a sometimes decisive part in the whole movement of the war. His was a soul that flamed out of a frail and failing body. He was a crumbling lighthouse from which there shone the beams that led great fleets to harbour. He had also a gift of sardonic humour. I always enjoyed his company, especially when things went ill. He could also be very disagreeable and say hard and sour things. My experiences were teaching me to be able to do this too, if need be.

At our first meeting we were about three hours together, and I soon comprehended his personal dynamism and the outstanding importance of his mission. This was the height of the London bombing, and many local worries imposed themselves upon us. But it was evident to me that here was an envoy from the President of supreme importance to our life. With gleaming eye and quiet, constrained passion he said: "The President is determined that we shall win the war together. Make no mistake about it.

"He has sent me here to tell you that at all costs and by all means he will carry you through, no matter what happens to him — there is nothing that he will not do so far as he has human power."

Everyone who came in contact with Harry Hopkins in the long

struggle will confirm what I have set down about his remarkable personality. And from this hour began a friendship between us which sailed serenely over all earthquakes and convulsions. He was the most faithful and perfect channel of communication between the President and me. But far more than that, he was for several years the main prop and animator of Roosevelt himself. Together these two men, the one a subordinate without public office, the other commanding the mighty Republic, were capable of taking decisions of the highest consequence over the whole area of the English-speaking world. Hopkins was of course jealous about his personal influence with his Chief and did not encourage American competitors. He therefore in some ways bore out the poet Gray's line, "A favourite has no friend." But this was not my affair. There he sat. slim, frail, ill, but absolutely glowing with refined comprehension of the Cause. It was to be the defeat, ruin. and slaughter of Hitler, to the exclusion of all other purposes, loyalties, or aims. In the history of the United States few brighter flames have burned.

Harry Hopkins always went to the root of the matter. I have been present at several great conferences, where twenty or more of the most important executive personages were gathered together. When the discussion flagged and all seemed baffled, it was on these occasions he would rap out the deadly question, "Surely, Mr. President, here is the point we have got to settle. Are we going to face it or not?" Faced it always was, and, being faced, was conquered. He was a true leader of men, and alike in ardour and in wisdom in times of crisis he has rarely been excelled. His love for the causes of the weak and poor was matched by his passion against tyranny, especially when tyranny was, for the time, triumphant.

*   *   *

Meanwhile the Blitz continued. But with a difference. At the end of 1940 Hitler had realized that Britain could not be destroyed by direct air assault. The Battle of Britain had been his first defeat, and the malignant bombing of the cities had not cowed the nation or its Government. The preparations to invade Russia in the early summer of 1941 absorbed much of the German air power. The many very severe raids which we suffered till the end of May no longer represented the full strength of the enemy. To us they were most grievous, but they were no longer the prime thought either of the German High Command or of the Fuehrer. To Hitler the continuance of the air attack on Great Britain was a necessary and convenient cover to the concentration against Russia. His optimistic timetable assumed that the Soviets, like the French, would be overthrown in a six-weeks campaign and that all German forces would then be free for the final overthrow of Britain in the autumn of 1941. Meanwhile the obstinate nation was to be worn down, first by the combination of the U-boat blockade sustained by the long-range air, and secondly by air attacks upon her cities and especially her ports. For the German Army "Sea Lion" (against Britain) was now replaced by "Barbarossa" (against Russia). The German Navy was instructed to concentrate on our Atlantic traffic and the German Air Force on our harbours and their approaches. This was a far more deadly plan than the indiscriminate bombing of London and the civil population, and it was fortunate for us that it was not pursued with all available forces and greater persistence.

During January and February, the enemy were frustrated by bad weather, and, apart from attacks on Cardiff, Portsmouth, and Swansea, our Civil Defence Services gained a well-deserved breathing space, by which they did not fail to profit. But when better weather came, the Blitz started in earnest over again. What was sometimes called "the Luftwaffe's tour of the ports" began in early March. It consisted of single or double attacks, which, though serious, failed to cripple our harbours. On the 8th and for three succeeding nights Portsmouth was heavily attacked and the dockyards damaged. Manchester and Salford were attacked on the 11th. On the ensuing nights it was the turn of Merseyside. On the 13th and 14th the Luftwaffe fell for the first time heavily on the Clyde, killing or injuring over two thousand people and putting the shipyards out of action, some till June and others till November. The heaviest blows did not fall till April. On the 8th the concentration was on Coventry, and in the rest of the country the sharpest impact was at Portsmouth. London had heavy attacks on the 16th and 17th; over two thousand three hundred people were killed, more than three thousand seriously injured. The enemy went on trying to destroy most of our principal ports by attacks prolonged in some cases over a whole week. Bristol was mauled. Plymouth was attacked from April 21 to 29, and though decoy fires helped to save the dockyards this was only at the expense of the city. The climax came on May 1, when Liverpool and the Mersey were attacked for seven successive nights. Seventy-six thousand people were made homeless and three thousand killed or injured. Sixty-nine out of a hundred and forty-four berths were put out of action, and the tonnage landed for a while was cut to a quarter. Had the enemy persisted, the Battle of the Atlantic would have been even more closely run than it was. But as usual he turned away. For two nights he battered Hull heavily, where forty thousand people had their dwellings destroyed, the food stores were wrecked, and the marine engineering works were crippled for nearly two months. In that month he struck again at Belfast, already twice raided.

The worst attack was the last. On May 10 the enemy returned to London with incendiary bombs. He lit more than two thousand fires, and by the smashing of nearly a hundred and fifty water mains, coupled with the low tide in the Thames, he stopped us putting them out. At six o'clock next morning hundreds were reported as out of control, and four were still glowing on the night of the 13th. It was the most destructive attack of the whole night Blitz. Five docks and seventy-one key points, half of which were factories, had been hit. All but one of the main railway stations were blocked for weeks, and the through routes were not fully opened till early June. Over three thousand people were killed or injured. In other respects also it was historic. It destroyed the House of Commons. One single bomb created ruin for years. We were however thankful that the Chamber was empty. On the other hand, our batteries and night fighters destroyed sixteen enemy planes, the maximum we had yet attained in night fighting.

This, though we did not know it, was the enemy's parting fling. On May 22 Kesselring shifted the headquarters of his air fleet to Posen, and at the beginning of June the whole force was moved to the east. Nearly three years were to pass before our Civil Defence organisation in London had to deal with the "baby Blitz" of February 1944 and the later onslaught of the rockets and the flying bombs. In the twelve months from June 1940 to June 1941 our civilian casualties were 43,381 killed and 50,856 seriously injured, a total of 94,237.

*   *   *

It is not possible in a major war to divide military from political affairs. At the summit they are one. It is natural that soldiers should regard the military aspects as single and supreme, and even that they should speak of political considerations with a certain amount of disdain. Also the word "politics" has been confused, and even tarnished, by its association with party politics. Thus much of the literature of this tragic century is biased by the idea that in war only military considerations count and that soldiers are obstructed in their clear, professional view by the intrusion of politicians, who for personal or party advantage tilt the dread balances of battle. The extremely close, intimate contacts which prevailed between the War Cabinet, the Chiefs of Staff, and myself, and the total absence of party feeling in Britain at this time, reduced these discords to a minimum.

While the war with the Italians in Northeast Africa continued to prosper, and the Greeks battled valiantly in Albania, all the news we got about the German movements and intentions proved every day more plainly that Hitler was about to intervene upon a large scale in the Balkans and the Mediterranean. From the beginning of January I had apprehended the arrival of German air power in Sicily, with the consequent menace to Malta and to all our hopes of resuming traffic through the inland sea. I also feared a movement of German troops, presumably armoured, into Tripoli. We could not doubt that their plans were progressing to establish a north-and-south passage through Italy to Africa, and at the same time and by the same measures to interrupt all our movements east and west in the Mediterranean.

On top of this now came the menace to the Balkan States, including Greece and Turkey, of being enticed or coerced into the Hitler empire, or conquered if they did not comply. Was the same hideous process we had witnessed in Norway, Denmark, Holland, Belgium, and France to be reproduced in Southeast Europe? Were all the Balkan States, including heroic Greece, to be subjugated one by one, and Turkey, isolated, to be compelled to open for the German legions the road to Palestine, Egypt, Iraq, and Persia? Was there no chance of creating a Balkan unity and Balkan front which would make this new German aggression too costly to be worth while? Might not the fact of Balkan resistance to Germany produce serious and helpful reactions in Soviet Russia? Certainly this was a sphere in which the Balkan States were affected by interest, and even, so far as they allowed it to influence their calculations, by sentiment. Could we from our strained but growing resources find the extra outside contribution which might galvanise all these States, whose interests were largely the same, into action for a common cause? Or ought we, on the other hand, to mind our own business and make a success of our campaign in Northeast Africa, let Greece, the Balkans, and it might be Turkey and all else in the Middle East, slide to ruin?

There would have been much mental relief in such a clear-cut decision; and it has found its adherents in the books of various officers occupying subordinate positions who have given us their views. These writers certainly have the advantage of pointing to the misfortunes which we sustained, but they had not the knowledge to consider sufficiently what the results of the opposite policy might have been. If Hitler had been able, with hardly any fighting, to bring Greece to her knees and the whole of the Balkans into his system and then force Turkey to allow the passage of his armies to the south and east, might he not have made terms with the Soviets upon the conquest and partition of these vast regions and postponed his ultimate, inevitable quarrel with them to a later part of his programme? Or, as is more likely, would he not have been able to attack Russia in greater strength at an earlier date? The main question which the ensuing chapters will probe and expose is whether His Majesty's Government by their action influenced in a decisive, or even in an appreciable manner, Hitler's movements in Southeast Europe, and moreover whether that action did not produce consequences first upon the behaviour of Russia and next upon her fortunes.

*　　　　*　　　　*

Throughout January and February good news continued to reach us from the Middle East. Malta had been reinforced and survived by the skin of its teeth the first fierce onslaught of the German Air Force in Sicily. The conquest of the Italian Empire in Eritrea, Somaliland, and Abyssinia was in process of completion. The Desert Army had advanced five hundred miles in two months, destroyed an Italian army of more than nine divisions, and seized Benghazi and all Cyrenaica. But in spite of these victories, so grave and complex were the issues, both diplomatic and military, which were at stake, and General Wavell had so much on his hands, that at the meeting of the Defence Committee on February 11 it was resolved to send the Foreign Secretary and General Dill, the Chief of the Imperial General Staff, to join him in Cairo. From there Eden, with Wavell, Dill, and other officers, flew to Athens, to confer with the Greek King and Government. At the meeting the Prime Minister, M. Korysis, read him a statement setting forth the outcome of the Greek Cabinet discussions in the past day or two. As this statement formed the basis of our action, I set forth the material part in full: "I desire to repeat most categorically that Greece, as a faithful ally, is determined to go on fighting with all her forces until final victory. This determination is not limited to the case of Italy, but will apply to any German aggression . . . whatever the outcome and whether Greece has or has not any hope of repulsing the enemy in Macedonia, she will defend her national territory, even if she can only count on her own forces." The Greek Government made it clear that their decision had been taken before they knew whether we could give them any help or not. Mr. Eden then explained that we in London, in full agreement with the Commanders-in-Chief in the Middle East, were resolved to give Greece the fullest help in our power. Military conferences and staff meetings were held all night and the next day, and on the 24th Eden sent us the following most important telegrams:

> We were all impressed by frankness and fair dealing of Greek representatives on all subjects discussed. I am quite sure that it is their determination to resist to the utmost of their strength, and that His Majesty's Government have no alternative but to back them whatever the ultimate consequences. . . . We are all convinced that we have chosen the right course, and as the eleventh hour has already struck felt sure that you would not wish us to

delay for detailed reference home. The risks are great, but there is a chance of success. . . .

On these messages, which carried with them the assent of both Dill and Wavell, it was decided in the Cabinet to give full approval to the proposals.

Mr. Eden now went on to Angora and had long discussions with the Turks. His account was not encouraging. They realised their own dangers as acutely as we did, but were convinced that the forces we could offer them would not be sufficient to make any real difference to an actual battle. As they had no offensive power they considered the common cause would be better served by Turkey remaining out of the war until her deficiencies had been remedied and she could be employed with the maximum effect. If attacked she would of course come in. I well understood how perilous the position of Turkey had become. It was obviously impossible to consider the treaty we had made with her before the war as binding upon her in the altered circumstances. When war had broken out in 1939 the Turks had mobilised their strong, good, brave army. But this was all based upon the conditions of the First Great War. The Turkish infantry were as fine as they had ever been, and their field artillery was presentable. But they had none of the modern weapons which from May 1940 were proved to be decisive. Aviation was lamentably weak and primitive. They had no tanks or armoured cars, and neither the workshops to make and maintain them nor the trained men and staffs to handle them. They had hardly any anti-aircraft or anti-tank artillery. Their signals service was rudimentary. Radar was unknown to them. Nor did their warlike qualities include any aptitude for all these modern developments.

On the other hand, Bulgaria had been largely armed by Germany out of the immense quantities of equipment of all kinds taken from France and the Low Countries as a result of the battles of 1940. The Germans had therefore plenty of modern weapons with which to arm their allies. We, for our part, having lost so much at Dunkirk, having to build up our home army against invasion and to face all the continuous pressure of the Blitz on our cities as well as maintain the war in the Middle East, could only give very sparingly and at the cost of other clamant needs. The Turkish army in Thrace was, under these conditions, at a serious and almost hopeless disadvantage compared with the Bulgarians. If to this danger were added even moderate detachments of German air and armour the weight upon Turkey might well prove insupportable.

The only policy or hope throughout this phase of the ever-extending war was in an organised plan of uniting the forces of Yugoslavia, Greece, and Turkey; and this we were now trying to do. Our aid to Greece had been limited in the first place to the few air squadrons which had been sent from Egypt when Mussolini first attacked her. The next stage had been an offer of technical units, which had been declined by the Greeks on grounds which were by no means unreasonable. We now reach the third phase, where it seemed possible to make a safe and secure desert flank at and beyond Benghazi and concentrate the largest army of manoeuvre or strategic reserve possible in Egypt.

So far we had not taken any steps which went beyond gathering the largest possible strategic reserve in the Delta and making plans and shipping preparations to transport an army to Greece. If the situation changed through a reversal of Greek policy or any other event we should be in the best position to deal with it. It was agreeable, after being so hard pressed, to be able to wind up satisfactorily the campaigns in Abyssinia, Somaliland, and Eritrea and bring substantial forces into our "mass of manoeuvre" in Egypt. While neither the intentions of the enemy nor the reactions of friends and neutrals could be divined or forecast, we seemed to have various important options open. The future remained inscrutable, but not a division had yet been launched, and meanwhile not a day was being lost in preparation.

17

The Battle of the Atlantic

THE ONLY THING that ever really frightened me during the war was the U-boat peril. Invasion, I thought, even before the air battle, would fail. After the air victory it was a good battle for us. It was the kind of battle which, in the cruel conditions of war, one ought to be content to fight. But now our lifeline, even across the broad oceans, and especially in the entrances to the Island, was endangered. I was even more anxious about this battle than I had been about the glorious air fight called the Battle of Britain.

The Admiralty, with whom I lived in the closest amity and contact, shared these fears, all the more because it was their prime responsibility to guard our shores from invasion and to keep the lifelines open to the outer world. This had always been accepted by the Navy as their ultimate, sacred, inescapable duty. So we poised and pondered together on this problem. It did not take the form of flaring battles and glittering achievements. It manifested itself through statistics, diagrams, and curves unknown to the nation, incomprehensible to the public.

How much would the U-boat warfare reduce our imports and shipping? Would it ever reach the point where our life would be destroyed? Here was no field for gestures or sensations; only the slow, cold drawing of lines on charts, which showed potential strangulation. Compared with this there was no value in brave armies ready to leap upon the invader, or in a good plan for desert warfare. The high and faithful spirit of the people counted for nought in this bleak domain. Either the food, supplies, and arms from the New World and from the British Empire arrived across the oceans, or they failed. With the whole French seaboard from Dunkirk to Bordeaux in their hands, the Germans lost no time in making bases for their U-boats and co-operating aircraft in the captured territory. From July onward we were compelled to divert our shipping from the approaches south of Ireland, where of course we were not allowed to station fighter aircraft. All had to come in around Northern Ireland. Here, by the grace of God, Ulster stood a faithful sentinel. The Mersey, the Clyde, were the lungs through which we breathed. On the east coast and in the English Channel small vessels continued to ply under an ever-increasing attack by air, by E-boat,[1] and by mines, and the passage of each convoy between the Forth and London became almost every day an action in itself.

The losses inflicted on our merchant shipping became most grave during the twelve months from July 1940 to July 1941, when we could claim that the British Battle of the Atlantic was won. The week ending September 22, 1940, was the worst since the beginning of the war, and sinkings were greater than any we had suffered in a similar period in 1917. The pressure grew unceasingly, and our losses were fearfully above new construction. The vast resources of the United States were only slowly coming into action. We could not expect any further large windfalls of vessels such as those which had followed the overrunning of Norway, Denmark, and the Low Countries in the spring of 1940. Twenty-seven ships were sunk, many of them in a Halifax convoy, and in October another Atlantic convoy was massacred by U-boats, twenty ships being sunk out of thirty-four. As November and December drew on, the entrances and estuaries of the Mersey and Clyde far surpassed in mortal significance all other factors in the war. We could of course at this time have descended upon de Valera's Ireland and regained the southern ports by force of modern arms. I had always declared that nothing but self-preservation would lead me to this. Even this hard measure would only have given a mitigation. The only sure remedy was to secure free exit and entrance in the Mersey and the Clyde. Every day when they met, those few who knew looked at one another. One understands the diver deep below the surface of the sea, dependent from minute to minute upon his air pipe. What would he feel if he could see a growing shoal of sharks biting at it? All the more when there was no possibility of his being hauled to the surface! For us there was no surface. The diver was forty-six millions of people in an overcrowded island, carrying on a vast business of war all over the world, anchored by nature and gravity to the bottom of the sea. What could the sharks do to his air pipe? How could he ward them off or destroy them?

There was another aspect of the U-boat attack. At the outset the Admiralty naturally thought first of bringing the ships safely to port, and judged their success by a minimum of sinkings. But now this was no longer the test. We all realised that the life and war effort of the

[1] E-boat, the German equivalent of British "light coastal craft."

country depended equally upon the weight of imports safely landed. In the week ending June 8, during the height of the battle in France, we had brought into the country about a million and a quarter tons of cargo, exclusive of oil. From this peak figure imports had declined at the end of July to less than 750,000 tons a week. Although substantial improvement was made in August, the weekly average again fell, and for the last three months of the year was little more than 800,000 tons. I became increasingly concerned about this ominous fall in imports. *"I see,"* I minuted to the First Lord in the middle of February 1941, *"that entrance of ships with cargo in January were less than half of what they were last January."*

The very magnitude and refinement of our protective measures — convoy, diversion, degaussing, mine clearance, the avoidance of the Mediterranean — the lengthening of most voyages in time and distance and the delays at the ports through bombing and the blackout, all reduced the operative fertility of our shipping to an extent even more serious than the actual losses. Every week our ports became more congested and we fell farther behind. At the beginning of March over 2,600,000 tons of damaged shipping had accumulated, of which more than half was immobilised by the need of repairs.

To the U-boat scourge was soon added air attack far out on the ocean by long-range aircraft. Of these the Focke-Wulf 200, known as the Condor, was the most formidable, though happily at the beginning there were few of them. They could start from Brest or Bordeaux, fly right round the British Island, refuel in Norway, and then make a return journey next day. On their way they would see far below them the very large convoys of forty or fifty ships to which scarcity of escort had forced us to resort, moving inward or outward on their voyages. They could attack these convoys, or individual ships, with destructive bombs, or they could signal the positions to which the waiting U-boats should be directed in order to make interceptions.

Powerful German cruisers were active. The *Scheer* was now in the South Atlantic, moving towards the Indian Ocean. In three months she destroyed ten ships, of sixty thousand tons in all, and then succeeded in making her way back to Germany. The *Hipper* was sheltering in Brest. At the end of January the battle cruisers *Scharnhorst* and *Gneisenau,* having at length repaired the damage inflicted upon them in Norway, were ordered to make a sortie into the North Atlantic, while *Hipper* raided the route from Sierra Leone. During a two months' cruise they sank or captured twenty-two ships, amounting to 115,000 tons. *Hipper* fell upon a homeward-bound convoy near the Azores which had not yet been joined by an escort, and in a savage attack lasting an hour she destroyed seven out of nineteen ships, making no attempt to rescue survivors, and regained Brest two days later. These formidable vessels compelled the employment on convoy duty of nearly every available British capital ship. At one period the Commander-in-Chief of the Home Fleet had only one battleship in hand.

The *Bismarck* was not yet on the active list. The German Admiralty should have waited for her completion, and for that of her consort, the *Tirpitz*. In no way could Hitler have used his two giant battleships more effectively than by keeping them both in full readiness in the Baltic and allowing rumours of an impending sortie to leak out from time to time. We should thus have been compelled to keep concentrated at Scapa Flow or thereabouts practically every new ship we had, and he would have had all the advantages of a selected moment without the strain of being always ready. As ships have to go for periodic refits it would have been almost beyond our power to maintain a reasonable margin of superiority, and any serious accident would have destroyed it.

*   *   *

My thought had rested day and night upon this awe-striking problem. At this time my sole and sure hope of victory depended upon our ability to wage a long and indefinite war until overwhelming air superiority was gained, and probably other Great Powers were drawn in on our side. But this mortal danger to our lifelines gnawed my bowels. Early in March exceptionally heavy sinkings were reported by Admiral Pound to the War Cabinet. I had already seen the figures, and after our meeting, which was in the Prime Minister's room at the House of Commons, I said to Pound, "We have got to lift this business to the highest plane, over everything else. I am going to proclaim 'the Battle of the Atlantic.'" This, like featuring "the Battle of Britain" nine months earlier, was a signal intended to concentrate all minds and all departments concerned upon the U-boat war.

In order to follow this matter with the closest personal attention, and to give timely directions which would clear away difficulties and obstructions and force action upon the great number of departments and branches involved, I brought into being the Battle of the Atlantic Committee. The meetings of this committee were held weekly, and were attended by all Ministers and high functionaries concerned, both from the fighting services and from the civil side. They usually lasted not less than two and a half hours. The whole field was gone over and everything thrashed out; nothing was held up for want of decision. Throughout the wide circles of our war machine, embracing thousands of able, devoted men, a new proportion was set, and from a hundred angles the gaze of searching eyes was concentrated.

The U-boats now began to use new methods, which became known as "wolf pack" tactics. These consisted of attacks from different directions by several U-boats working together. They were at this time usually made by night, on the surface, and at full speed. Only the destroyers could rapidly overhaul them, and Asdic was virtually impotent. The solution lay not only in the multiplication of fast escorts but still more in the development of effective radar, which would warn us of their approach. The scientists, sailors, and airmen did their best but the results came slowly. We also needed an air weapon which would kill the surfaced U-boat, and time to train our forces in its use. When eventually both these problems were solved, the U-boat was once more driven back to the submerged attack, in which it could be dealt with by the older and well-tried methods. This was not achieved for another two years.

Meanwhile the new wolf-pack tactics, inspired by Admiral Doenitz, the head of the U-boat service, and himself a U-boat captain of the previous war, were vigorously applied by the redoubtable Prien and the other tiptop U-boat commanders. But retribution followed. On March 8 Prien's U-47 was sunk with himself and all hands by the destroyer *Wolverine,* and nine days later U-99 and U-100 were sunk while engaged in a combined attack on a convoy. Both were commanded by outstanding officers, and the elimination of these three able men had a marked effect on the progress of the struggle. Few U-boat commanders who followed them were their equals in ruthless ability and daring. Five U-boats were sunk in March in the Western Approaches, and though we suffered grievous losses, amounting to 243,000 tons, by U-boat, and a further 113,000 tons by air attack, the first round in the Battle of the Atlantic may be said to have ended in a draw.

Finding the Western Approaches too hot, the U-boats moved farther west into waters where, since the Southern Irish ports were denied us, only a few of our flotilla escorts could reach them and air cover was impossible. Escorts from the United Kingdom could only protect our convoys over about a quarter of the route to Halifax. Early in April a "wolf pack" struck a convoy in longitude 28° West before the escort had joined it. Ten ships out of twenty-two were sunk, for the loss of a single U-boat. Somehow we had to contrive to extend our reach or our days would be numbered.

Between Canada and Great Britain are the islands of Newfoundland, Greenland, and Iceland. All these lie near the flank of the shortest, or great-circle, track between Halifax and Scotland. Forces based on these "steppingstones" could control the whole route by sectors. Greenland was entirely devoid of resources, but the other two islands could be quickly turned to good account. It has been said, "Whoever possesses Iceland holds a pistol firmly pointed at England, America, and Canada."

It was upon this thought that, with the concurrence of its people, we had occupied Iceland when Denmark was overrun in 1940, and in April 1941 we established bases there for our escort groups and aircraft. Thence we extended the range of the surface escorts to 35° West. Even so there remained an ominous gap to the westward which for the time being could not be bridged. In May a Halifax convoy was heavily attacked in 41° West and lost nine ships before help could arrive.

It was clear that nothing less than end-to-end escort from Canada to Britain would suffice, and on May 23 the Admiralty invited the Governments of Canada and Newfoundland to use St. John's, Newfoundland, as an advanced base for our joint escort forces. The response was immediate, and by the end of the month continuous escort over the whole route was at last achieved. Thereafter the Royal Canadian Navy accepted responsibility for the protection, out of its own resources, of convoys on the western section of the ocean route. From Great Britain and from Iceland we were able to cover the remainder of the passage. Even so our strength available remained perilously small, and our losses mounted steeply. In the three months ending with May U-boats alone sank 142 ships, of 818,000 tons, of which 99 were British.

*   *   *

In this growing tension the President, acting with all the powers accorded to him as Commander-in-Chief of the armed forces and enshrined in the American Constitution, began to give us armed aid. He resolved not to allow the German U-boat and raider war to come near the American coast, and to make sure that the munitions he was sending Britain at least got nearly halfway across. From plans made long before, there sprang the broad design for the joint defence of the Atlantic Ocean by the two English-speaking Powers. As we had found it necessary to develop bases in Iceland, so Mr. Roosevelt took steps to establish an air base of his own in Greenland. It was known that the Germans had already installed weather-reporting stations on the east coast and opposite Iceland. His action was therefore timely. By other decisions not only our merchant ships but our warships, damaged in the heavy fighting in the Mediterranean and elsewhere, could now be repaired in American shipyards, giving instant and much-needed relief to our strained resources at home.

Great news arrived at the beginning of April. The President cabled me on April 11 that the United States Government would extend their so-called security zone and patrol areas, which had been in effect since very early in the war, to a line covering all North Atlantic waters west of about West Longitude 26°. For this purpose he proposed to use aircraft and naval vessels working from Greenland, Newfoundland, Nova

Scotia, the United States, Bermuda, and the West Indies, with possibly a later extension to Brazil. He invited us to notify him in great secrecy of the movement of our convoys, "so that our patrol units can seek out any ships or planes of aggressor nations operating west of the new line of the security zones." The Americans for their part would immediately publish the position of possible aggressor ships or planes when located in the American patrol area. I transmitted this telegram to the Admiralty with a deep sense of relief.

On the 18th the United States Government announced the line of demarcation between the Eastern and Western Hemispheres to which the President had referred in his message of April 11. This line became thereafter the virtual sea frontier of the United States. It included within the United States' sphere all British territory in or near the American continent, Greenland, and the Azores, and was soon afterwards extended eastward to include Iceland. Under this declaration United States warships would patrol the waters of the Western hemisphere, and would incidentally keep us informed of any enemy activities therein. The United States however remained nonbelligerent and could not at this stage provide direct protection for our convoys. This remained solely a British responsibility over the whole route.

The effects of the President's policy were far-reaching, and we continued our struggle with important parts of our load taken off our backs by the Royal Canadian and the United States Navies. The United States was moving ever nearer to war, and this world tide was still further speeded by the irruption of the *Bismarck* into the Atlantic towards the end of May. In a broadcast on May 27, the very day that the *Bismarck* was sunk, the President declared, "It would be suicide to wait until they [the enemy] are in our front yard. . . . We have accordingly extended our patrol in North and South Atlantic waters." At the conclusion of this speech Mr. Roosevelt declared an "Unlimited National Emergency."

There is ample evidence to show that the Germans were greatly disturbed at all this, and Admirals Raeder and Doenitz besought the Fuehrer to grant greater latitude to the U-boats and permit them to operate towards the American coast as well as against American ships if convoyed or if proceeding without lights. Hitler however remained adamant. He always dreaded the consequences of war with the United States, and insisted that German forces should avoid provocative action.

The expansion of the enemy's efforts also brought its own correctives. By June, apart from those training, he had about thirty-five U-boats at sea, but the new craft now coming forward outstripped his resources in highly trained crews, and above all in exerienced captains. The "diluted" crews of the new U-boats, largely composed of young and unpractised men, showed a decline in pertinacity and skill, and the extension of the battle into the remoter expanses of the ocean disrupted the dangerous combination of the U-boats and the air. German aircraft in large numbers had not been equipped or trained for operations over the sea. Nonetheless, in the same three months of March, April, and May 179 ships, of 545,000 tons, were sunk by air attack, mainly in the coastal regions. Of this total 40,000 tons were destroyed in two fierce attacks on the Liverpool docks early in May. I was thankful the Germans did not persevere on this tormented target. All the while the stealthy, insidious menace of the magnetic mine had continued around our coasts, with varying but diminished success. We developed and expanded our bases in Canada and Iceland with all possible speed, and planned our convoys accordingly. We increased the fuel capacity of our older destroyers, and their consequent radius. The newly formed Combined Headquarters at Liverpool threw itself heart and soul into the struggle. As more escorts came into service and the personnel gained experience, Admiral Noble formed them into permanent groups under Group Commanders. The team spirit was fostered and men became accustomed to working in unison with a clear understanding of their commander's methods. The escort groups became ever more efficient, and as their power grew that of the U-boats declined.

By June we began once more to gain the upper hand. The utmost exertions were being made to improve the organisation of our convoy escorts and develop new weapons and devices. The chief needs were for more and faster escorts with greater fuel endurance, for more long-range aircraft, and above all for good radar. Shore-based aircraft alone were not enough, and every convoy needed ship-borne aeroplanes to detect any U-boat within striking distance in daylight, and by forcing it to dive prevent it making contact, or making a report which would draw others to the scene. Fighter aircraft discharged from catapults mounted in ordinary merchant ships, as well as in converted ships manned by the Royal Navy, soon met the thrust of the Focke-Wulf. The fighter pilot, having been tossed like a falcon against his prey, had at first to rely for his life on being retrieved from the sea by one of the escorts. The Focke-Wulf gradually became the hunted rather than the hunter. Hitler's invasion of Russia compelled him to redeploy his machines in strength, and from an April peak of nearly three hundred thousand tons, our losses had dwindled by midsummer to about a fifth.

The President now made another important move. He decided to establish a base in Iceland. It was agreed that United States forces should relieve the British garrison. They reached Iceland on July 7, and this island was included in the defence system of the Western Hemisphere. Thereafter American convoys escorted by American warships ran regularly to Reykjavik, and although the United States were still not at war they admitted foreign ships to the protection of their convoys.

*   *   *

At the height of this struggle I made one of the most important and fortunate appointments of my war administration. In 1930, when I was out of office, I accepted for the first and only time in my life a director-ship. It was in one of the subsidiary companies of Lord Inchcape's far-spreading organisation of the Peninsular and Oriental shipping lines. For eight years I regularly attended the monthly board meetings, and discharged my duties with care. At these meetings I gradually became aware of a very remarkable man. He presided over thirty or forty companies, of which the one with which I was connected was a small unit. I soon perceived that Frederick Leathers was the central brain and controlling power of this combination. He knew everything and commanded absolute confidence. Year after year I watched him from my small position at close quarters. I said to myself, "If ever there is another war, here is a man who will play the same kind of part as the great business leaders who served under me at the Ministry of Munitions in 1917 and 1918."

Leathers volunteered his services to the Ministry of Shipping on the outbreak in 1939. We did not come much into contact while I was at the Admiralty, because his functions were specialised and subordinate. But now in 1941, in the stresses of the Battle of the Atlantic, and with the need for combining the management of our shipping with all the movements of our supplies by rail and road from our harried ports, he came more and more into my mind. On May 8 I turned to him. After much discussion I remodelled the Ministries of Shipping and Transport into one integral machine. I placed Leathers at its head. To give him the necessary authority I created the office of Minister of War Transport. I was always shy of bringing people into high Ministerial positions in the House of Commons if they had not been brought up there for a good many years. Experienced Members out of office may badger the newcomer, and he will always be unduly worried by the speeches he has to prepare and deliver. I therefore made a submission to the Crown that a peerage should be conferred upon the new Minister.

Henceforward to the end of the war Lord Leathers remained in complete control of the Ministry of War Transport, and his reputation grew with every one of the four years that passed. He won the confidence of the Chiefs of Staff and of all departments at home, and established intimate and excellent relations with the leading Americans in this vital sphere. With none was he more closely in harmony than with

Mr. Lewis Douglas, of the United States Shipping Board, and later Ambassador in London. Leathers was an immense help to me in the conduct of the war. It was very rarely that he was unable to accomplish the hard tasks I set. Several times when all staff and departmental processes had failed to solve the problems of moving an extra division or transhipping it from British to American ships, or of meeting some other need, I made a personal appeal to him, and the difficulties seemed to disappear as if by magic.

Throughout these critical months the two German battle cruisers *Scharnhorst* and *Gneisenau* remained poised in Brest. At any moment it seemed that they might again break out into the Atlantic. It was due to the Royal Air Force that they continued inactive. Repeated air attacks were made on them in port, with such good effect that they remained idle throughout the year. The enemy's concern soon became to get them home; but even this they were unable to do until 1942. We shall see in due course the extent to which the Navy and the R.A.F. Coastal Command succeeded; how we became the masters of the outlets; how the Heinkel 111's were shot down by our fighters, and the U-boats choked in the very seas in which they sought to choke us, until once again with shining weapons we swept the approaches to the isle.

18

Yugoslavia and Greece

NOW THE MOMENT had come when the irrevocable decision must be taken whether or not to send the Army of the Nile to Greece. This grave step was required not only to help Greece in her peril and torment, but to form against the impending German attack a Balkan Front comprising Yugoslavia, Greece, and Turkey, with effects upon Soviet Russia which could not be measured by us. These would certainly have been all-important if the Soviet leaders had realised what was coming upon them. It was not what we could send ourselves that could decide the Balkan issue. Our limited hope was to stir and organise united action. If at the wave of our wand Yugoslavia, Greece, and Turkey would all act together, it seemed to us that Hitler might either let the Balkans off for the time being or become so heavily engaged with our combined forces as to create a major front in that theatre. We did not then know that he was already deeply set upon his gigantic invasion of Russia. If we had we should have felt more confidence in the success of our policy. We should have seen that he risked falling between two stools, and might easily impair his supreme undertaking for the sake of a Balkan preliminary. This is what actually happened, but we could not know at the time. Some may think we builded rightly; at least we builded better than we knew. It was our aim to animate and combine Yugoslavia, Greece, and Turkey. Our duty so far as possible was to aid the Greeks. For all these purposes our four divisions in the Delta were well placed.

*   *   *

On March 1 the German Army began to move into Bulgaria. The Bulgarian Army mobilised and took up positions along the Greek frontier. A general southward movement of the German forces was in progress, aided in every way by the Bulgarians. On the following day Mr. Eden and General Dill resumed their military conversations in Athens. As the result of these Mr. Eden sent a very serious message, and a marked change came over our views in London. Admiral Cunningham, though convinced our policy was right, left us in no doubt as to the considerable naval risks in the Mediterranean which were involved. The Chiefs of Staff recorded the various factors developing unfavourably against our Balkan policy, and particularly against sending an army to Greece. "The hazards of the enterprise," they reported, "have considerably increased." They did not however feel that they could as yet question the military advice of those on the spot, who described the position as not by any means hopeless.

After reflecting alone at Chequers on the Sunday night upon the trend of discussion in the War Cabinet that morning, I sent the following message to Mr. Eden, who had now left Athens for Cairo. This certainly struck a different note on my part. But I take full responsibility for the eventual decision, because I am sure I could have stopped it all if I had been convinced. It is so much easier to stop than to do.

> . . . We have done our best to promote Balkan combination against Germany. We must be careful not to urge Greece against her better judgment into a hopeless resistance alone when we have only handfuls of troops which can reach the scene in time. Grave Imperial issues are raised by committing New Zealand and Australian troops to an enterprise which, as you say, has become even more hazardous. . . . We must liberate Greeks from feeling bound to reject a German ultimatum. If on their own they resolve to fight, we must to some extent share their ordeal. But rapid German advance will probably prevent any appreciable British Imperial forces from being engaged.
>
> Loss of Greece and Balkans by no means a major catastrophe for us, provided Turkey remains honest neutral. We could take Rhodes and consider plans for a descent on Sicily or Tripoli. We are advised from many quarters that our ignominious ejection from Greece would do us more harm in Spain and Vichy than the fact of submission of Balkans, which with our scanty forces alone we have never been expected to prevent. . . .

Attached to this was the grave commentary of the Chiefs of Staff.

As soon as my warning telegram was read by our Ambassador in Athens he showed lively distress. "How," he telegraphed to the Foreign Secretary, "can we possibly abandon the King of Greece after the assurances given him by the Commander-in-Chief and Chief of the Imperial General Staff as to reasonable chances of success? This seems to me quite unthinkable. We shall be pilloried by the Greeks and the world in general as going back on our word. There is no question of 'liberating the Greeks from feeling bound to reject the ultimatum.' They have decided to fight Germany alone if necessary. The question is whether we help or abandon them."

The War Cabinet thereupon resolved to take no decision till we had a reply to all this from Mr. Eden. His answer arrived next day. The material portion ran as follows:

> . . . Collapse of Greece without further effort on our part to save her by intervention on land, after the Libyan victories had, as all the world knows, made forces available, would be the greatest calamity. Yugoslavia would then certainly be lost; nor can we feel confident that even Turkey would have the strength to remain steadfast if the Germans and Italians were established in Greece without effort on our part to resist them. No doubt our prestige will suffer if we are ignominiously ejected, but in any event to have fought and suffered in Greece would be less damaging to us than to have left Greece to her fate. . . . In the existing situation we are all agreed that the course advocated should be followed and help given to Greece.

Accompanied by the Chiefs of Staff, I brought the issue before the War Cabinet, who were fully apprised of everything as it happened, for final decision. In spite of the fact that we could not send more aircraft than were already ordered and on the way, there was no hesitation or division among us. Personally I felt that the men on the spot had been searchingly tested. There was no doubt that their hands had not been forced in any way by political pressure from home. Smuts, with all his wisdom, and from his separate angle of thought and fresh eye, had concurred. Nor could anyone suggest that we had thrust ourselves upon Greece against her wishes. No one had been overpersuaded. Certainly we had with us the highest expert authority, acting in full freedom and with all knowledge of the men and the scene. My colleagues, who were toughened by the many risks we had run successfully, had independently reached the same conclusions. Mr. Menzies, on whom a special burden rested, was full of courage. There was a strong glow for action. The Cabinet was short; the decision final; the answer brief:

> Chiefs of Staff [have] advised that, in view of steadfastly expressed opinion of Commanders-in-Chief on the spot, of the Chief of the Imperial General Staff, and commanders of the forces to be

employed, it would be right to go on. Cabinet decided to authorise you to proceed with the operation, *and by so doing Cabinet accepts for itself the fullest responsibility*.[1] We will communicate with Australian and New Zealand Governments accordingly.

* * *

The fate of Yugoslavia must now be described. The whole defence of Salonika depended on her coming in, and it was vital to know what she would do. On March 2, Mr. Campbell, our Ambassador at Belgrade, met Mr. Eden in Athens. He said that the Yugoslavs were frightened of Germany and unsettled internally by political difficulties. There was a chance however that if they knew our plans for aiding Greece they might be ready to help. On the 5th the Foreign Secretary sent Mr. Campbell back to Belgrade with a confidential letter to the Regent, Prince Paul. In this he portrayed Yugoslavia's fate at German hands, and said that Greece and Turkey intended to fight if attacked. In such a case Yugoslavia must join us. The Regent was to be told verbally that the British had decided to help Greece with land and air forces as strongly and quickly as possible, and that if a Yugoslav Staff officer could be sent to Athens we would include him in our discussions.

In this atmosphere much turned on the Regent's attitude. Prince Paul was an amiable, artistic personage, but the prestige of the monarchy had long been on the wane and he now carried the policy of neutrality to its limits. He dreaded particularly that any move by Yugoslavia or her neighbours might provoke the Germans into a southward advance into the Balkans. He declined a proposed visit from Mr. Eden. Fear reigned. The Ministers and the leading politicians did not dare to speak their minds. There was one exception. An air force general named Simovic represented the nationalist elements among the officer corps of the armed forces. Since December his office had become a clandestine centre of opposition to German penetration into the Balkans and to the inertia of the Yugoslav Government.

On March 4 Prince Paul left Belgrade on a secret visit to Berchtesgaden, and under dire pressure undertook verbally that Yugoslavia would follow the example of Bulgaria. On his return, at a meeting of the Royal Council and in separate discussion with political and military leaders, he found opposing views. Debate was violent, but the German ultimatum was real. General Simovic, when summoned to the White Palace, Prince Paul's residence on the hills above Belgrade, was firm against capitulation. Serbia would not accept such a decision, and the dynasty would be endangered. But Prince Paul had already in effect committed his country.

During the night of March 20 at a Cabinet meeting, the Yugoslav

[1] My subsequent italics.—W.S.C.

Government decided to adhere to the Tripartite Pact. Three Ministers however resigned on this issue. On March 24 the Prime Minister and the Minister for Foreign Affairs crept out of Belgrade from a suburban railway station on the Vienna train. Next day they signed the pact with Hitler in Vienna, and the ceremony was broadcast over the Belgrade radio. Rumors of imminent disaster swept through the cafés and conclaves of the Yugoslav capital.

Direct action, if the Government capitulated to Germany, had been discussed for some months in the small circle of officers round Simovic. When during March 26 the news of the return from Vienna of the Yugoslav Ministers began to circulate in Belgrade the conspirators decided to act. Few revolutions have gone more smoothly. There was no bloodshed. Certain senior officers were placed under arrest. The Prime Minister was brought by the police to Simovic's headquarters and obliged to sign a letter of resignation. Prince Paul was informed that Simovic had taken over the Government in the name of the King, and that the Council of Regency had been dissolved. He was escorted to the office of General Simovic. Together with the other two regents, he then signed the act of abdication. He was allowed a few hours to collect his effects, and, together with his family, he left the country that night for Greece.

The plan had been made and executed by a close band of Serb nationalist officers who had identified themselves with the true public mood. Their action let loose an outburst of popular enthusiasm. The streets of Belgrade were soon thronged with Serbs, chanting. "Rather war than the pact; rather death than slavery." There was dancing in the squares; English and French flags appeared everywhere; the Serb national anthem was sung with wild defiance by valiant, helpless multitudes. On March 28 King Peter, who by climbing down a rainpipe had made his own escape from Regency tutelage, attended divine service in Belgrade Cathedral, amid fervent acclamation. The German Minister was publicly insulted, and the crowd spat on his car. The military exploit had roused a surge of national vitality. A people paralysed in action, hitherto ill governed and ill led, long haunted by the sense of being ensnared, flung their reckless, heroic defiance at the tyrant and conqueror in the moment of his greatest power.

Hitler was stung to the quick. He had a burst of that convulsive anger which momentarily blotted out thought and sometimes impelled him on his most dire adventures. In a passion he summoned the German High Command. Goering, Keitel, and Jodl were present, and Ribbentrop arrived later. Hitler said that Yugoslavia was an uncertain factor in the coming action against Greece, and even more in the "Barbarossa" undertaking against Russia later on. He deemed it fortunate that the Yugoslavs had revealed their temper before "Barbarossa" was launched.

Yugoslavia must be destroyed "militarily and as a national unit." The blow must be *carried out with unmerciful harshness.* The night was spent by the generals in drafting the operation orders. Keitel confirms our view that the greatest danger to Germany was "an attack upon the Italian army from the rear." "The decision to attack Yugoslavia meant completely upsetting all military movements and arrangements made up to that time. The invasion of Greece had to be completely readjusted. New forces had to be brought through Hungary from the north. All had to be improvised."

Hungary was directly and immediately affected. Although the main German thrust against the Yugoslavs would clearly come through Rumania, all lines of communication led through Hungarian territory. Almost the first reaction of the German Government to the events in Belgrade was to send the Hungarian Minister in Berlin by air to Budapest with an urgent message to the Hungarian Regent, Admiral Horthy:

> *Yugoslavia will be annihilated,* for she has just renounced publicly the policy of understanding with the Axis. The greater part of the German armed forces must pass through Hungary. But the principal attack will not be made on the Hungarian sector. Here the Hungarian Army should intervene, and, in return for its co-operation, Hungary will be able to reoccupy all those former territories which she had been forced at one time to cede to Yugoslavia. The matter is urgent. An immediate and affirmative reply is requested.[2]

Hungary was bound by a pact of friendship to Yugoslavia signed only in December 1940. But open opposition to the German demands could only lead to the German occupation of Hungary in the course of the imminent military operations. There was also the temptation of regaining the territories on her southern frontiers which Hungary had lost to Yugoslavia after the first World War. The Hungarian Premier, Count Teleki, had been working consistently to maintain some liberty of action for his country. He was by no means convinced that Germany would win. At the time of signing the Tripartite Pact he had little confidence in the independence of Italy as an Axis partner. Hitler's ultimatum required the breach of his own Hungarian agreement with Yugoslavia. The initiative was however wrested from him by the Hungarian General Staff, whose chief, General Werth, himself of German origin, made his own arrangements with the German High Command behind the back of the Hungarian Government.

Teleki at once denounced Werth's action as treasonable. On the

---

[2] Ullein-Reviczy, *Guerre Allemande: Paix Russe,* p. 89.

evening of April 2, 1941, he received a telegram from the Hungarian
Minister in London that the British Foreign Office had stated formally
to him that if Hungary took part in any German move against Yugoslavia she must expect a declaration of war upon her by Great Britain.
Thus the choice for Hungary was either a vain resistance to the passage
of German troops or ranging herself openly against the Allies and be-
traying Yugoslavia. In this cruel position Count Teleki saw but one
means of saving his personal honour. Shortly after nine o'clock he left
the Hungarian Ministry of Foreign Affairs and retired to his apartments in the Sandor Palace. There he received a telephone call. It is
believed that this message stated that the German armies had already
crossed the Hungarian frontier. Shortly afterwards he shot himself.
His suicide was a sacrifice to absolve himself and his people from guilt
in the German attack upon Yugoslavia. It clears his name before history.
It could not stop the march of the German armies, nor the consequences.

*　　　　*　　　　*

The movement of our expedition to Greece had meanwhile begun. In
order of embarkation, it comprised the British 1st Armoured Brigade,
the New Zealand Division, and the 6th Australian Division. These were
all fully equipped at the expense of other formations in the Middle East.
They were to be followed by the Polish Brigade and the 7th Australian
Division. The plan was to hold the Aliakhmon line, which ran from the
mouth of the river of that name through Veria and Edhessa to the
Yugoslav frontier. Our forces were to join the Greek forces deployed on
this front, which were nominally the equivalent of seven divisions, and
were to come under the command of General Wilson.

The Greek troops were far less than General Papagos had originally
promised.[3] The great majority of the Greek Army, about fifteen divi-
sions, was in Albania. The remainder were in Macedonia, whence
Papagos declined to withdraw them, and where, after four days' fighting, when the Germans attacked, they ceased to be a military force.
Our air force numbered only eighty operational aircraft, against a
German air strength of over ten times as many. The weakness of the
Aliakhmon position lay on its left flank, which could be turned by a
German advance through Southern Yugoslavia. There had been little
contact with the Yugoslav General Staff, whose plan of defence and
degree of preparedness were not known to the Greeks or ourselves. It
was hoped however that in the difficult country which the enemy would
have to cross, the Yugoslavs would at least be able to impose consider-

[3] Papagos has since claimed that his first agreement to the holding of the Aliakhmon
line was contingent on a clarification of the situation with the Government of Yugoslavia,
which never was reached.

able to delay on them. This hope was to prove ill founded. General Papagos did not consider that withdrawal from Albania to meet such a turning movement was a feasible operation. Not only would it severely affect morale, but the Greek Army was so ill equipped with transport and communications were so bad that a general withdrawal in the face of the enemy was impossible. He had certainly left the decision till too late. It was in these circumstances that our 1st Armoured Brigade reached the forward area on March 27, where it was joined a few days later by the New Zealand Division.

The news of the revolution in Belgrade naturally gave us great satisfaction. Here at least was one tangible result of our desperate efforts to form an Allied front in the Balkans and prevent all falling piecemeal into Hitler's power. It was settled that Eden should remain in Athens to deal with Turkey and that General Dill should proceed to Belgrade. Anyone could see that the position of Yugoslavia was forlorn unless a common front was immediately presented by all the Powers concerned. There was however open to Yugoslavia the chance already mentioned of striking a deadly blow at the naked rear of the disorganised Italian armies in Albania. If they acted promptly they might bring about a major military event, and while their own country was being ravaged from the north might possess themselves of the masses of munitions and equipment which would give them the power of conducting the guerrilla in their mountains, which was now their only hope. It would have been a grand stroke, and would have reacted upon the whole Balkan scene. In our circle in London we all saw this together. The diagram on page 427 shows the movement which was deemed feasible.

But the mistakes of years cannot be remedied in hours. When the general excitement had subsided everyone in Belgrade realised that disaster and death approached them and that there was little they could do to avert their fate. The High Command could now at last mobilise their armies. But there was no strategic plan. Dill found only confusion and paralysis. The Yugoslav Government, mainly for fear of the effect on the internal situation, were determined to take no step which might be considered provocative to Germany. At this moment all the might of Germany within reach was descending like an avalanche upon them. One would have thought from the mood and outlook of the Yugoslav Ministers that they had months in which to take their decision about peace or war with Germany. Actually they had only seventy-two hours before the onslaught fell upon them.

On the morning of April 6 German bombers appeared over Belgrade. Flying in relays from occupied airfields in Rumania, they delivered a methodical attack lasting three days upon the Yugoslav capital. From roof-top height, without fear of resistance, they blasted the city without mercy. This was called "Operation Punishment." When the silence came at last on April 8 over seventeen thousand citizens of Belgrade lay dead in the streets or under the débris. Out of the nightmare of smoke and fire came the maddened animals released from their shattered cages in the zoological gardens. A stricken stork hobbled past the main hotel, which was a mass of flames. A bear, dazed and uncomprehending, shuffled through the inferno with slow and awkward gait down towards the Danube. He was not the only bear who did not understand.

Simultaneously with the ferocious bombardment of Belgrade, the converging German armies already poised on the frontiers invaded Yugoslavia from several directions. The Yugoslav General Staff did not attempt to strike their one deadly blow at the Italian rear. They conceived themselves bound not to abandon Croatia and Slovenia, and were therefore forced to attempt the defence of the whole frontier line. The four Yugoslav Army Corps in the north were rapidly and irresistibly bent inwards by the German armoured columns, supported by Hungarian troops which crossed the Danube, and by German and Italian forces advancing towards Zagreb. The main Yugoslav forces were thus driven in confusion southwards, and on April 13 the Germans entered

Belgrade. Meanwhile the Twelfth German Army, assembled in Bulgaria, had swung into Serbia and Macedonia. They had entered Monastir and Yannina on the 10th, and thus had prevented any contact between the Yugoslavs and Greeks and broken up the Yugoslav forces in the south.

Seven days later Yugoslavia capitulated.

This sudden collapse destroyed the main hope of the Greeks. It was another example of "One at a time." We had done our utmost to procure concerted action, but through no fault of ours we had failed. A grim prospect now gaped upon us all. Five German divisions, including three armoured, took part in the southward drive to Athens. By April 8 it was clear that Yugoslav resistance in the south was breaking down and that the left flank of the Aliakhmon position would shortly be threatened, and on April 10 the attack on our flank guard began. It was arrested during two days of stiff fighting in severe weather.

Farther west there was only one Greek cavalry division keeping touch with the forces in Albania, and General Wilson decided that his hard-pressed left flank must be pulled back. This move was completed on April 13, but in the process the Greek Divisions began to disintegrate. Henceforward our Expeditionary Force was alone. Wilson, still menaced upon his left flank, decided to withdraw to Thermopylae. He put this to Papagos, who approved, and who himself at this stage suggested British evacuation from Greece The next few days were decisive. Wavell telegraphed on the 16th that General Wilson had had a conversation with Papagos, who described the Greek Army as being severely pressed and getting into administrative difficulties owing to air action. Wavell's instructions to Wilson were to continue the fighting in co-operation with the Greeks so long as they were able to resist, but authorised any further withdrawal judged necessary. Orders had been given for all ships on the way to Greece to be turned back, for no more ships to be loaded, and for those already loading or loaded to be emptied.

To this grave but not unexpected news I replied at once that we could not remain in Greece against the wish of the Greek Commander-in-Chief, and thus expose the country to devastation, and that if the Greek Government assented, the evacuation should proceed.

"Crete," I added, "must be held in force."

On the 17th General Wilson motored from Thebes to the palace at Tatoi, and there met the King, General Papagos, and our Ambassador. It was accepted that withdrawal to the Thermopylae line had been the only possible plan. General Wilson was confident that he could hold that line for a while. The main discussion was the method and order of evacuation. The Greek Government would not leave for at least another week.

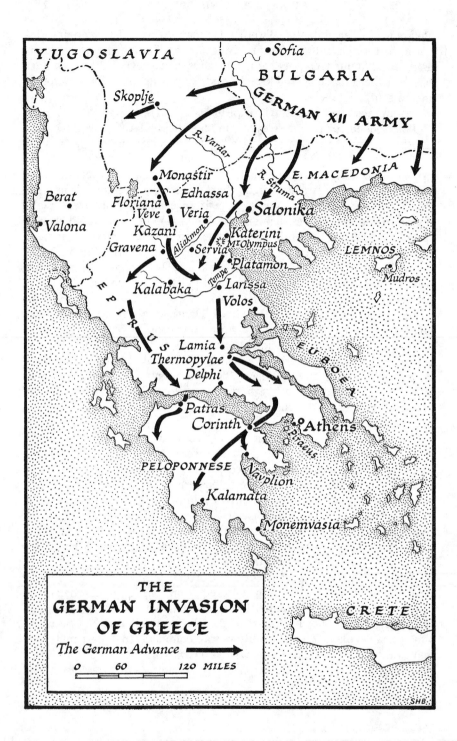

YUGOSLAVIA

• Sofia

BULGARIA

GERMAN XII ARMY

Skoplje •

R. Vardar

E. MACEDONIA

R. Struma

Monastir •

Edhassa

Berat •

Floriana

Veve

Veria

Salonika •

• Valona

Kazani

Aliakmon

Katerini •

LEMNOS

Gravena •

Servia •

Mt Olympus

Platamon

• Mudros

E P I R U S

Kalabaka •

Tempe

Larissa •

Volos

E U B O E A

Lamia •

Thermopylae

Delphi

Patras •

Corinth

• Athens

Piraeus

PELOPONNESE

Navplion

Kalamata •

Monemvasia •

C R E T E

THE
GERMAN INVASION
OF GREECE

The German Advance ⟶

0   60   120 MILES

SHB.

The Greek Prime Minister, M. Korysis, has already been mentioned. He had been chosen to fill the gap when Metaxas died. He had no claim to public office except a blameless private life and clear, resolute convictions. He could not survive the ruin, as it seemed, of his country or bear longer his own responsibilities. Like Count Teleki in Hungary, he resolved to pay the forfeit of his life. On the 18th he committed suicide. His memory should be respected.

*   *   *

The retreat to Thermopylae was a difficult manoeuvre, but stubborn and skilful rearguard actions checked the impetuous German advance at all points, inflicting severe losses. By April 20 the occupation of the Thermopylae position was complete. Frontally it was strong, but our forces were strained. The Germans made slow progress and the position was never severely tested. On this same day the Greek armies on the Albanian front surrendered. On the 21st His Majesty told General Wavell that time rendered it impossible for any organised Greek force to support the British left flank before the enemy could attack. Wavell replied that in that case he felt that it was his duty to take immediate steps for re-embarkation of such portion of his army as he could extricate. The King entirely agreed, and seemed to have expected this. He spoke with deep regret of having been the means of placing the British forces in such a position. He promised what help he could. But all was vain. The final surrender of Greece to overwhelming German might was made on April 24.

We were now confronted with another of those evacuations by sea which we had endured in 1940. The organised withdrawal of over fifty thousand men from Greece under the conditions prevailing might well have seemed an almost hopeless task. At Dunkirk on the whole we had air mastery. In Greece the Germans were in complete and undisputed control of the air and could maintain an almost continuous attack on the ports and on the retreating army. It was obvious that embarkation could only take place by night, and moreover that troops must avoid being seen near the beaches in daylight. This was Norway over again, and on ten times the scale.

Admiral Cunningham threw nearly the whole of his light forces, including six cruisers and nineteen destroyers, into the task. Working from the small ports and beaches in Southern Greece, together with transports, assault ships and many smaller craft, the work of rescue began on the night of April 24.

For five successive nights the work continued. On the 26th the enemy captured the vital bridge over the Corinth Canal by parachute attack, and thereafter German troops poured into the Peloponnese, harrying our hard-pressed soldiers as they strove to reach the southern beaches. At Nauplion there was disaster. The transport *Slamat* in a gallant but misguided effort to embark the maximum stayed too long in the anchorage. Soon after dawn, when clearing the land, she was attacked and sunk by dive bombers. Two destroyers, who rescued most of the seven hundred men on board, were both in turn sunk by air attack a few hours later. There were only fifty survivors from all three ships.

On the 28th and 29th efforts were made by two cruisers and six destroyers to rescue 8000 troops and 1400 Yugoslav refugees from the beaches near Kalamata. A destroyer sent on ahead to arrange the embarkation found the enemy in possession of the town and large fires burning, and the main operation had to be abandoned. Although a counterattack drove the Germans out of the town, only about 450 men were rescued from beaches to the eastward by four destroyers, using their own boats. These events marked the end of the main evacuation. Small isolated parties were picked up in various islands or in small craft at sea during the next two days, and 1400 officers and men, aided by the Greeks at mortal peril, made their way back to Egypt independently in later months.

In all over 11,000 of our own troops were lost and 50,662 were safely brought out, including men of the Royal Air Force and several thousand Cypriots, Palestinians, Greeks, and Yugoslavs. This figure represented about eighty per cent of the forces originally sent into Greece. These results were only made possible by the determination and skill of the seamen of the Royal and Allied Merchant Navies, who never faltered under the enemy's most ruthless efforts to halt their work. From April 21 until the end of the evacuation twenty-six ships were lost by air attack. The Royal Air Force, with a Fleet Air Arm contingent from Crete, did what they could to help, but they were overwhelmed by numbers. Nevertheless, from November onward our few squadrons had done fine service. They inflicted on the enemy confirmed losses of 231 planes and had dropped 500 tons of bombs. Their own losses of 209 machines, of which 72 were in combat, were severe, their record exemplary.

The small but efficient Greek Navy now passed under British control. A cruiser, six modern destroyers, and four submarines escaped to Alexandria, where they arrived on April 25. Thereafter the Greek Navy was represented with distinction in many of our operations in the Mediterranean.

If in telling this tale of tragedy the impression is given that the Imperial and British forces received no effective military assistance from their Greek allies, it must be remembered that these three weeks of April fighting at desperate odds were for the Greeks the culmination of the hard five months' struggle against Italy in which they had expended almost the whole life strength of their country. Attacked in October 1940 without warning by at least twice their numbers, they had first repulsed the invaders and then in counterattack had beaten them back forty miles into Albania. Throughout the bitter winter in the mountains they had been at close grips with a more numerous and better-equipped foe. The Greek Army of the Northwest had neither the transport nor the roads for a rapid manoeuvre to meet at the last moment the new overpowering German attack cutting in behind its flank and rear. Its strength had already been strained almost to the limit in a long and gallant defence of the homeland.

There were no recriminations. The friendliness and aid which the Greeks had so faithfully shown to our troops endured nobly to the end. The people of Athens and at other points of evacuation seemed more concerned for the safety of their would-be rescuers than with their own fate. Greek martial honour stands undimmed.

In a broadcast I tried not only to express the feeling of the English-speaking world but to state the dominant facts which ruled our fate.

While we naturally view with sorrow and anxiety much that is happening in Europe and in Africa, and may happen in Asia, we must not lose our sense of proportion and thus become discouraged or alarmed. When we face with a steady eye the difficulties which lie before us, we may derive new confidence from remembering those we have already overcome. Nothing that is happening now is comparable in gravity with the dangers through which we passed last year. Nothing that can happen in the East is comparable with what is happening in the West.

I have some lines which seem apt and appropriate to our fortunes tonight, and I believe they will be so judged wherever the English language is spoken or the flag of freedom flies:

> *For while the tired waves, vainly breaking,*
> *Seem here no painful inch to gain,*
> *Far back, through creeks and inlets making,*
> *Comes silent, flooding in, the main.*
>
> *And not by eastern windows only,*
> *When daylight comes, comes in the light;*
> *In front the sun climbs slow, how slowly!*
> *But westward, look, the land is bright.*

19

The Desert Flank:
Rommel: Tobruk

ALL OUR EFFORTS to form a front in the Balkans were founded upon the sure maintenance of the Desert Flank in North Africa. This might have been fixed at Tobruk; but Wavell's rapid westward advance and the capture of Benghazi had given us all Cyrenaica. To this the sea corner at Agheila was the gateway. It was common ground between all authorities in London and Cairo that this must be held at all costs and in priority over every other venture. The utter destruction of the Italian forces in Cyrenaica and the long road distances to be traversed before the enemy could gather a fresh army led Wavell to believe that for some time to come he could afford to hold this vital western flank with moderate forces and to relieve his tried troops with others less well trained. The Desert Flank was the peg on which all else hung, and there was no idea in any quarter of losing or risking that for the sake of Greece or anything in the Balkans.

But now a new figure sprang upon the world stage, a German warrior who will hold his place in their military annals. Erwin Rommel was born at Heidenheim, in Württemberg, in November 1891. He fought in the First World War in the Argonne, in Rumania, and in Italy, being twice wounded and awarded the highest classes of the Iron Cross and of the order Pour de Mérite. On the outbreak of the Second World War he was appointed commandant of the Fuehrer's field headquarters in the Polish campaign, and was then given command of the 7th Panzer Division of the XVth Corps. This division, nicknamed "the Phantoms," formed the spearhead of the German break-through across the Meuse. He narrowly escaped capture when the British counterattacked at Arras on May 21, 1940. His was the spearhead which crossed the Somme and advanced on the Seine in the direction of Rouen, rolling up the French left wing and capturing numerous French and British forces around St. Valéry. His division entered Cherbourg just after our final evacuation, where Rommel took the surrender of the port and thirty thousand prisoners.

These many services and distinctions led to his appointment early in 1941 to command the German troops sent to Libya. At that time Italian hopes were limited to holding Tripolitania, and Rommel took charge of the growing German contingent under Italian command. He strove immediately to enforce an offensive campaign. When early in April the Italian Commander-in-Chief tried to persuade him that the German Afrika Korps should not advance without his permission Rommel protested that "as a German general he had to issue orders in accordance with what the situation demanded."

Throughout the African campaign Rommel proved himself a master in handling mobile formations, especially in regrouping rapidly after an operation and following up success. He was a splendid military gambler, dominating the problems of supply and scornful of opposition. At first the German High Command, having let him loose, were astonished by his successes, and were inclined to hold him back. His ardour and daring inflicted grievous disasters upon us, but he deserves the salute which I made him — and not without some reproaches from the public — in the House of Commons in January 1942, when I said of him, "We have a very daring and skilful opponent against us, and, may I say across the havoc of war, a great general." He also deserves our respect because, although a loyal German soldier, he came to hate Hitler and all his works, and took part in the conspiracy of 1944 to rescue Germany by displacing the maniac and tyrant. For this he paid the forfeit of his life.

*   *   *

The Agheila defile was the kernel of the situation. If the enemy broke through to Agedabia, Benghazi and everything west of Tobruk were imperilled. They could choose between taking the good coast road to Benghazi and beyond or using the tracks leading straight to Mechili and Tobruk, which cut off the bulge of desert, two hundred miles long by a hundred miles broad. Taking this latter route in February, we had nipped and captured many thousands of Italians retiring through Benghazi. It should not have been a matter of surprise to us if Rommel also took the desert route to play the same trick on us. However, so long as we held the gateway at Agheila the enemy was denied the opportunity of bemusing us in this fashion.

All this depended upon a knowledge not only of the ground but of the conditions of desert warfare. A superiority in armour and in quality rather than numbers, and a reasonable parity in the air, would have enabled the better and more lively force to win in a rough-and-tumble in the desert, even if the gateway had been lost. None of these conditions were established by the arrangements which were made. We were inferior in the air; and our armour, for reasons which will appear later, was utterly inadequate, as was also the training and equipment of the troops west of Tobruk.

Rommel's attack upon Agheila began on March 31. Our armoured division, which had in fact only one armoured brigade and its support group, withdrew slowly during the next two days. In the air the enemy proved greatly superior. The Italian Air Force still counted for little, but there were about a hundred German fighters and a hundred bombers and dive bombers. Our armoured forces under the German attack became disorganised, and there were serious losses. At a single stroke, and almost in a day, the desert flank upon which all our decisions depended had crumpled.

The evacuation of Benghazi was ordered, and by the night of April 6 the retreat was in full progress. Tobruk was reinforced and held, but the headquarters of the 2nd Armoured Division and two Indian motorised regiments found themselves surrounded. A number of men fought their way out, bringing in a hundred German prisoners, but the great majority were forced to surrender. The enemy pushed on very quickly towards Bardia and Sollum, with heavy armoured cars and motorised infantry. Other troops attacked the Tobruk defences. The garrison beat off two assaults, destroying a number of enemy tanks, and for a time the position there and on the Egyptian frontier was stabilised.

*   *   *

The beating in of our Desert Flank while we were full-spread in the Greek adventure was a disaster of the first magnitude. I was for some time completely mystified about its cause, and as soon as there was a momentary lull I felt bound to ask General Wavell for some explanation of what had happened. Characteristically he took the responsibility upon himself.[1] The disaster had stripped him almost entirely of his armour.

On Sunday, April 20, I was spending the week end at Ditchley and working in bed when I received two telegrams from General Wavell to the C.I.G.S. which disclosed his plight in all its gravity. He described his tank position in detail. The picture looked dark. "It will be seen," he said, "that there are only two regiments of cruiser tanks in sight for Egypt by the end of May, and no reserves to replace casualties, *whereas there are now in Egypt, trained, an excellent personnel for six tank regiments.* I consider the provision of cruiser tanks

[1] That Rommel's early attack, with its fruitful consequences, was as great a surprise to his own superiors as to us is explained by Desmond Young in his book *Rommel*.

vital, in addition to infantry tanks, which lack speed and radius of action for desert operations. C.I.G.S., please give your personal assistance."

On reading these alarming messages I resolved not to be governed any longer by the Admiralty reluctance, but to send a convoy through the Mediterranean direct to Alexandria carrying all the tanks which General Wavell needed. We had a convoy containing large armoured reinforcements starting immediately round the Cape. I decided that the fast tank-carrying ships in this convoy should turn off at Gibraltar and take the short cut, thus saving nearly forty days. General Ismay, who was staying nearby, came over at noon to see me. I prepared a personal minute to him for the Chiefs of Staff. I asked him to go to London with it at once and make it clear that I attached supreme importance to this step being taken.

The Chiefs of Staff were assembled by the time Ismay reached London, and they discussed my minute until late into the night. Their first reactions to the proposals were unfavourable. The chances of getting the motor transport ships through the Central Mediterranean unscathed were not rated very high, since on the day before entering the Narrows and on the morning after passing Malta they would be subjected to dive-bombing attack out of range of our own shore-based fighters. The view was also expressed that we were dangerously weak in tanks at home, and that if we now suffered heavy losses in tanks abroad there would be demands for their replacement, and consequently a further diversion of tanks from the Home Forces.

However, when the Defence Committee met the next day Admiral Pound, to my great satisfaction, stood by me and agreed to pass the convoy through the Mediterranean. The Chief of the Air Staff, Air Marshal Portal, said he would try to arrange for a Beaufighter squadron to give additional protection from Malta. I then asked the Committee to consider sending a hundred additional cruiser tanks with the convoy. General Dill opposed the dispatch of these additional tanks *in view of the shortage for Home Defence.* Considering what he had agreed to ten months before, when we sent half our few tanks round the Cape to the Middle East in July 1940, I could not feel that this reason was at this time valid. As the reader is aware, I did not regard invasion as a serious danger in April 1941, since proper preparations had been made against it. We now know that this view was correct. It was settled that this operation, which I called "Tiger," should proceed.

*   *   *

While all this was on the move Tobruk lay heavily upon our minds. All Hurricanes in Greece had been lost, and many of those in Tobruk had been destroyed or damaged. Air Marshal Longmore considered that any further attempt to maintain a fighter squadron inside Tobruk would only result in heavy loss to no purpose. Thus the enemy would have complete air superiority over Tobruk until a fresh fighter force could be built up. However, the garrison had recently beaten off an attack, causing the enemy heavy casualties and taking 150 prisoners.

Soon General Wavell sent us more disquieting information about Rommel's approaching reinforcements. The disembarkation of the 15th German Armoured Division would probably be completed by April 21. There were signs that Benghazi was being regularly used, and although at least fifteen days would be required for the gathering of supplies, it seemed probable that the Armoured Division, the 5th Light Motorised Division, and the Ariete and Trento divisions would be able to move forward after the middle of June. It seemed very unsatisfactory to us at home that Benghazi, which we had failed to make a useful base, was already playing so important a part now that it had passed into German hands.

*   *   *

During the next fortnight my keen attention and anxieties were riveted upon the fortunes of "Tiger." I did not underrate the risks which the First Sea Lord had been willing to accept, and I knew that there were many misgivings in the Admiralty. The convoy, consisting of five 15-knot ships, escorted by Admiral Somerville's Force H (*Renown, Malaya, Ark Royal,* and *Sheffield*), passed Gibraltar on May 6. With it also were the reinforcements for the Mediterranean Fleet, comprising the *Queen Elizabeth* and the cruisers *Naiad* and *Fiji.* Air attacks on May 8 were beaten off without damage. During that night however two ships of the convoy struck mines when approaching the Narrows. One caught fire and sank after an explosion; the other was able to continue with the convoy. On reaching the entrance to the Skerki Channel, Admiral Somerville parted company and returned to Gibraltar. In the afternoon of the 9th Admiral Cunningham, having seized the opportunity to pass a convoy into Malta, met the "Tiger" convoy with the fleet fifty miles south of Malta. All his forces then shaped their course for Alexandria, which they reached without further loss or damage.

While this hung in the balance my thoughts turned to Crete, upon which we were now sure a heavy air-borne attack impended. It seemed to me that if the Germans could seize and use the airfields on the island, they would have the power of reinforcing almost indefinitely, and that even a dozen infantry tanks might play a decisive part in preventing their doing so. I therefore asked the Chiefs of Staff to consider turning one ship of "Tiger" to unload a few of these tanks in Crete on their way through. My expert colleagues, while agreeing that tanks would be of special value for the purpose I had in mind, deemed it inadvisable to endanger the rest of the ship's valuable cargo by such a diversion. Accordingly I suggested to them on May 9 that if it were "thought too dangerous to take the *Clan Lamont* into Suda, she should take twelve tanks, or some other ship should take them, immediately after she has discharged her cargo at Alexandria." Orders were sent accordingly. Wavell informed us on May 10 that he "had already arranged to send six infantry tanks and fifteen light tanks to Crete," and that they "should arrive within next few days if all goes well." But we had very few days left to us.

20

Crete

THE STRATEGIC IMPORTANCE of Crete in all our Mediterranean affairs has already been explained by argument and
events. British warships based on Suda Bay or able to refuel there could
give an all-important protection to Malta. If our base in Crete was
well defended against air attack the whole process of superior sea power
would come into play and ward off any sea-borne expedition. But only
a hundred miles away lay the Italian fortress of Rhodes, with its ample
airfields and well-established installations, and locally in Crete everything had proceeded in a halting manner. I had issued repeated injunctions to have Suda Bay fortified. I had even used the expression "a
second Scapa." The island had been in our possession for nearly six
months, but it would only have been possible to equip the harbour with
a more powerful outfit of anti-aircraft guns at the expense of other
still more urgent needs; nor was the Middle East Command able to
find the labour, locally or otherwise, to develop the airfields. There
could be no question of sending a large garrison to Crete or of basing
strong air forces upon its airfields while Greece was still in Allied hands.
But all should have been in readiness to receive reinforcements should
they become available and should the need arise. The responsibility
for the defective study of the problem and for the feeble execution of
the directions given must be shared between Cairo and Whitehall. It
was only after the disasters had occurred in Cyrenaica, in Crete, and in
the desert that I realised how overloaded and undersustained General
Wavell's organisation was. Wavell tried his best; but the handling
machine at his disposal was too weak to enable him to cope with the
vast mass of business which four or five simultaneous campaigns imposed upon him.

*　　　*　　　*

At no moment in the war was our Intelligence so truly and precisely informed. In the exultant confusion of their seizure of Athens the German staffs preserved less than their usual secrecy, and our agents in Greece were active and daring. In the last week of April we obtained from trustworthy sources good information about the next German stroke. The movements and excitement of the German XIth Air Corps, and also the frantic collection of small craft in Greek harbours, could not be concealed from attentive eyes and ears. In no operation did I take more personal pains to study and weigh the evidence or to make sure that the magnitude of the impending onslaught was impressed upon the Commanders-in-Chief and imparted to the general on the actual scene.

I had suggested to the C.I.G.S. that General Freyberg should be placed in command of Crete, and he proposed this to Wavell, who had immediately agreed. Bernard Freyberg and I had been friends for many years. The Victoria Cross and the D.S.O. with two bars marked his unsurpassed service, and like his only equal, Carton de Wiart, he deserved the title with which I acclaimed them of "Salamander." Both thrived in the fire, and were literally shot to pieces without being affected physically or in spirit. At the outset of the war no man was more fitted to command the New Zealand Division, for which he was eagerly chosen. In September 1940 I had toyed with the idea of giving him a far greater scope. Now at length this decisive personal command had come to him.

Freyberg and Wavell had no illusions. The geography of Crete made its defence problem difficult. There was but a single road running along the north coast, upon which were strung all the vulnerable points of the island. Each of these had to be self-supporting. There could be no central reserve free to move to a threatened point once this road was cut and firmly held by the enemy. Only tracks unfit for motor transport ran from the south coast to the north. As the impending danger began to dominate directing minds strong efforts were made to carry reinforcements and supplies of weapons, especially artillery, to the island, but it was then too late. During the second week in May the German Air Force in Greece and in the Aegean established a virtual daylight blockade, and took their toll of all traffic, especially on the northern side, where alone the harbours lay. Out of 27,000 tons of vital munitions sent in the first three weeks of May under 3000 could be landed, and the rest had to turn back. Our strength in anti-aircraft weapons was fifty guns and twenty-four searchlights. There were only twenty-five part-worn or light tanks. Our defending forces were distributed principally to protect the landing grounds, and the total of Imperial troops that took part in the defence amounted to about 28,600.

But of course it was only our weakness in the air that rendered the German attack possible. The R.A.F. strength early in May was thirty-six aircraft of which only one-half were serviceable. These were distributed between Retimo, Maleme, and Heraklion and were but a trifle compared with the overwhelming air forces about to be hurled upon the island. Our inferiority in the air was fully realised by all concerned, and on May 19, the day before the attack, all remaining aircraft were evacuated to Egypt. It was known to the War Cabinet, the Chiefs of Staff, and the Commanders-in-Chief in the Middle East that the only choice lay between fighting under this fearful disadvantage or hurrying out of the island, as might have been possible in the early days of May. But there was no difference of opinion between any of us about facing the attack; and when we see in the light and knowledge of the after-time how nearly, in spite of all our shortcomings, we won, and how far-reaching were the advantages even of our failure, we must be well content with the risks we ran and the price we paid.

The battle began on the morning of May 20, and never was a more reckless, ruthless attack launched by the Germans. In many of its aspects at the time it was unique. Nothing like it had ever been seen before. It was the first large-scale air-borne attack in the annals of war. The German Air Corps represented the flame of the Hitler Youth Movement and was an ardent embodiment of the Teutonic spirit of revenge for the defeat of 1918. The flower of German manhood was expressed in these valiant, highly trained, and completely devoted Nazi parachute troops. To lay down their lives on the altar of German glory and world power was their passionate resolve. They were destined to encounter proud soldiers many of whom had come all the way from the other side of the world to fight as volunteers for the Motherland and what they deemed the cause of right and freedom.

The Germans used the whole strength they could command. This was to be Goering's prodigious air achievement. It might have been launched upon England in 1940 if British air power had been broken. But this expectation had not been fulfilled. It might have fallen on Malta. But this stroke was spared us. The German Air Corps had waited for more than seven months to strike their blow and prove their mettle. Now at length Goering could give them the long-awaited signal. When the battle joined we did not know what were the total resources of Germany in parachute troops. The XIth Air Corps might have been only one of half a dozen such units. It was not till many months afterwards that we were sure it was the only one. It was in fact the spearpoint of the German lance. And this is the story of how it triumphed and was broken.

At Maleme the bulk of our anti-aircraft artillery was put out of ac-

tion practically at once. Before the bombardment was over gliders began
to land west of the airfield. Wherever our troops were noticed they were
subjected to tremendous bombardment. Counterattacks were impossible
in daylight. Gliders or troop-carriers landed or crashed on the beaches
and in the scrub or on the fireswept airfield. In all, around and between Maleme and Canea over 5000 Germans reached the ground on
the first day. They suffered very heavy losses from the fire and fierce

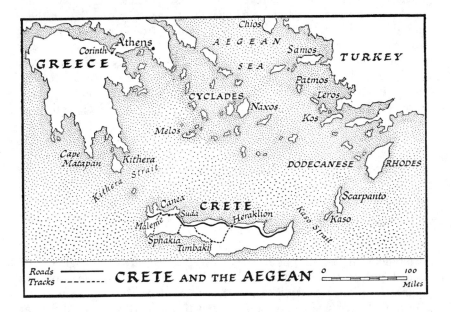

CRETE AND THE AEGEAN

hand-to-hand fighting of the New Zealanders. At the end of the day
we were still in possession of the airfield, but that evening the few who
were left of the battalion fell back on its supports.

Retimo and Heraklion were both treated to a heavy air bombardment on that morning, followed by parachute drops in the afternoon.
Heavy fighting followed, but at nightfall we remained in firm possession
of both airfields. The result of this first day's fighting was therefore
fairly satisfactory, except at Maleme; but in every sector bands of
well-armed men were now at large. The strength of the attacks far
exceeded the expectations of the British command, and the fury of our
resistance astonished the enemy.

The onslaught continued on the second day, when troop-carrying
aircraft again appeared. Although Maleme airfield remained under our
close artillery and mortar fire, troop-carriers continued to land upon it
and in the rough ground to the west. The German High Command

seemed indifferent to losses, and at least a hundred planes were wrecked by crash-landing in this area. Nevertheless the build-up continued. A counterattack made that night reached the edge of the airfield, but with daylight the German Air Force reappeared and the gains could not be held.

On the third day Maleme became an effective operational airfield for the enemy. Troop-carriers continued to arrive at a rate of more than twenty an hour. Even more decisive was the fact that they could also return for reinforcements. Altogether it was estimated that in these and the ensuing days more than six hundred troop-carriers landed or crashed more or less successfully on the airfield. Under the increasing pressure the New Zealand Brigade gradually gave way until they were nearly ten miles from Maleme. At Canea and Suda there was no change, and at Retimo the situation was well in hand. At Heraklion the enemy were landing east of the airfield, and an effective hostile lodgment there began and grew.

Next night our weary troops saw to the northward the whole sky-line alive with flashes and knew the Royal Navy was at work. The first German sea-borne convoy had started on its desperate mission. For two and a half hours the British ships hunted their prey, sinking not less than a dozen caiques and three steamers, all crowded with enemy troops. It was estimated that about four thousand men were drowned that night. Meanwhile Rear-Admiral King, with four cruisers and three destroyers, had spent the night of the 21st patrolling off Heraklion, and at daylight on the 22nd he began to sweep northwards. A single caique loaded with troops was destroyed and by ten o'clock the squadron was approaching the island of Melos. A few minutes later an enemy destroyer with five small craft was sighted to the northward, and was at once engaged. Another destroyer was then seen laying a smoke screen, and behind the smoke were a large number of caiques. We had in fact intercepted another important convoy crammed with soldiers. Our air reconnaissance had reported this fact to Admiral Cunningham, but it took more than an hour for this news to be confirmed to Admiral King. His ships had been under incessant air attack since daylight, and although they had hitherto suffered no damage all were running short of anti-aircraft ammunition. The Rear-Admiral, not fully realising the prize which was almost within his grasp, felt that to go farther north would jeopardise his whole force, and ordered a withdrawal to the west. As soon as this signal was read by the Commander-in-Chief he sent the following order: "Stick it out. Keep in visual signalling touch. Must not let Army down in Crete. It is essential no seaborne enemy force land in Crete."

It was now too late to destroy the convoy, which had turned back and scattered in all directions among the numerous islands. Thus at least five thousand German soldiers escaped the fate of their comrades. The audacity of the German authorities in ordering these practically defenceless convoys of troops across waters of which they did not possess the naval command as well as that of the air is a sample of what might have happened on a gigantic scale in the North Sea and the English Channel in September 1940. It shows the German lack of comprehension of sea power against invading forces, and also the price which may be exacted in human life as the penalty for this kind of ignorance.

Inflexibly resolved, whatever the cost, to destroy all sea-borne invaders, Admiral Cunningham threw everything into the scale. It is clear that throughout these operations he did not hesitate for this purpose to hazard not only his most precious ships but the whole naval command of the Eastern Mediterranean. His conduct on this issue was highly approved by the Admiralty. In this grim battle the German command was not alone in playing the highest stakes. The events of these forty-eight hours of sea fighting convinced the enemy, and no further attempts at sea-borne landings were made until the fate of Crete had been decided.

But May 22 and 23 were costly days for the Navy. Two cruisers and three destroyers were sunk, one battleship, the *Warspite,* was put out of action for a long time, and the *Valiant* and many other vessels were considerably damaged. Nevertheless the sea guard of Crete had been maintained. The Navy had not failed. Not a single German landed in Crete from the sea until the battle for the island was ended.

May 26 was decisive. Our troops had been under ever-growing pressure for six days. Finally they could stand it no more. Late that night the decision to evacuate Crete was taken, and we had to face once again a bitter and dismal task and the certainty of heavy losses. The harassed, overstrained Fleet had to undertake the embarking of about twenty-two thousand men, mostly from the open beach at Sphakia across three hundred and fifty miles of sea dominated by hostile air forces. It was necessary for the troops to hide near the edge until called forward for embarkation. At least fifteen thousand men lay concealed in the broken ground near Sphakia, and Freyberg's rearguard was in constant action.

A tragedy awaited the simultaneous expedition by Admiral Rawlings, which went to rescue the Heraklion garrison. Arriving before midnight, the destroyers ferried the troops to the cruisers waiting outside. By 3.20 A.M. the work was complete. Four thousand men had been embarked and the return voyage began. Fighter protection had been arranged, but partly through the change in times the aircraft did not find the ships. The dreaded bombing began at 6 A.M., and continued until 3 P.M., when the squadron was within a hundred miles of Alexandria. The destroyer *Hereward* was the first casualty At 6.25 A.M. she was hit by a bomb and could no longer keep up with the convoy. The Admiral rightly decided that he must leave the stricken ship to her fate. She was last seen approaching the coast of Crete. The majority of those on board survived, though as prisoners of war. Worse was to follow. During the next four hours the cruisers *Dido* and *Orion* and the destroyer *Decoy* were all hit. The speed of the squadron fell to 21 knots, but all kept their southerly course in company. In the *Orion* conditions were appalling. Besides her own crew, she had 1100 troops on board. On her crowded mess decks about 260 men were killed and 280 wounded by a bomb which penetrated the bridge. Her commander, Captain G. R. B. Back, was also killed, the ship heavily damaged and set on fire. At noon two Fulmars of the Fleet Air Arm appeared, and thereafter afforded a measure of relief. The fighters of the Royal Air Force, despite all efforts, could not find the tortured squadron, though they fought several engagements and destroyed at least two aircraft. When the squadron reached Alexandria at 8 P.M. on the 29th it was found that one-fifth of the garrison rescued from Heraklion had been killed, wounded, or captured.

*   *   *

After such experiences, General Wavell and his colleagues had to decide how far the effort to bring our troops off from Crete should be pursued. The Army was in mortal peril, the Air could do little, and again the task fell upon the wearied and bomb-torn Navy. To Admiral Cunningham it was against all tradition to abandon the Army in such a crisis. He declared, "It takes the Navy three years to build a new ship. It will take three hundred years to build a new tradition. The evacuation [*i.e.*, rescue] will continue." By the morning of the 29th nearly 5000 men had been brought off, but very large numbers were holding out and sheltering on all the approaches to Sphakia, and were bombed whenever they showed themselves by day. The decision to risk unlimited further naval losses was justified, not only in its impulse but by the results.

On the evening of the 28th Admiral King had sailed for Sphakia. Next night about 6000 men were embarked without interference, and, though attacked three times during the 30th, reached Alexandria safely. This good luck was due to the R.A.F. fighters, who, few though they were, broke up more than one attack before they struck home. On the morning of the 30th Captain Arliss once more sailed for Sphakia, with four destroyers. Two of these had to return, but he continued with the other pair and successfully embarked over fifteen hundred troops. Both ships were damaged by near misses on the return voyage, but reached Alexandria safely. The King of Greece, after many perils, had been brought off with the British Minister a few days earlier. That night also General Freyberg was evacuated by air on instructions from the Commanders-in-Chief.

On May 30 a final effort was ordered to bring out the remaining troops. It was thought that the numbers at Sphakia did not now exceed 3000 men, but later information showed that there were more than double that number. Admiral King sailed again on the morning of the 31st. They could not hope to carry all, but Admiral Cunningham ordered the ships to be filled to the utmost. At the same time the Admiralty were told that this would be the last night of evacuation. The embarkation went well, and the ships sailed again at 3 A.M. on June 1, carrying nearly 4000 troops safely to Alexandria.

Upward of 5000 British and Imperial troops were left somewhere in Crete, and were authorised by General Wavell to capitulate. Many individuals however dispersed in the mountainous island, which is 160 miles long. They and the Greek soldiers were succoured by the villagers and countryfolk, who were mercilessly punished whenever detected. Barbarous reprisals were made upon innocent or valiant peasants, who were shot by twenties and thirties. It was for this reason that I proposed to the Supreme War Council three years later, in 1944, that local crimes should be locally judged, and the accused persons sent back for trial on the spot. This principle was accepted, and some of the outstanding debts were paid.

*   *   *

Sixteen thousand five hundred men were brought safely back to Egypt. These were almost entirely British and Imperial troops. Nearly a thousand more were helped to escape later by various commando enterprises. Our losses were about 13,000 killed, wounded, and taken prisoner. To these must be added nearly 2000 naval casualties. Since the war more than 4000 German graves have been counted near Maleme and Suda Bay, and another thousand at Retimo and Heraklion. Besides these were the very large but unknown numbers drowned at sea, and those who later died of wounds in Greece. In all, the enemy must have suffered casualties in killed and wounded of well over 15,000. About 170 troop-carrying aircraft were lost or heavily damaged. But the price they paid for their victory cannot be measured by the slaughter.

The Battle of Crete is an example of the decisive results that may emerge from hard and well-sustained fighting apart from manoeuvring for strategic positions. We did not know how many parachute divisions the Germans had. But in fact the 7th Airborne Division was the only one which Goering possessed. This division was destroyed in the Battle of Crete. Upwards of five thousand of his bravest men were killed, and the whole structure of this organisation was irretrievably broken. It never appeared again in any effective form. The New Zealanders and other British, Imperial, and Greek troops who fought in the confused, disheartening, and vain struggle for Crete may feel that they played a definite part in an event which brought us far-reaching relief at a hingeing moment.

The German losses of their highest class fighting men removed a formidable air and parachute weapon from all further part in immediate events in the Middle East. Goering gained only a Pyrrhic victory in Crete; for the forces he expended there might easily have given him Cyprus, Iraq, Syria, and even perhaps Persia. These troops were the very kind needed to overrun large wavering regions where no serious resistance would have been encountered. He was foolish to cast away such almost measureless opportunities and irreplaceable forces in a mortal struggle, often hand to hand, with the warriors of the British Empire.

We now have in our possession the "battle report" of the XIth Air Corps, of which the 7th Airborne Division was a part. When we recall the severe critcicsm and self-criticism to which our arrangements were subjected, it is interesting to read the other side. "British land forces in Crete," said the Germans, "were about three times the strength which had been assumed. The area of operations on the island had been prepared for defence with the greatest care and by every possible means. . . . All works were camouflaged with great skill. . . . The failure, owing to lack of information, to appreciate correctly the enemy situation endangered the attack of the XIth Air Corps and resulted in exceptionally high and bloody losses."

The naval position in the Mediterranean was, on paper at least, gravely affected by our losses in the battle and evacuation of Crete. The Battle of Matapan on March 28 had for the time being driven the Italian fleet into its harbours. But now new, heavy losses had fallen upon our Fleet. On the morrow of Crete Admiral Cunningham had ready for service only two battleships, three cruisers, and seventeen destroyers. Nine other cruisers and destroyers were under repair in Egypt, but the battleships *Warspite* and *Barham* and his only aircraft carrier, the *Formidable,* besides several other vessels, would have to leave Alexandria for repair elsewhere. Three cruisers and six destroyers had been lost. Reinforcements must be sent without delay to restore the balance. But, as will presently be recorded, still further misfortunes were in store. The period which we now had to face offered to the enemy their best chance of challenging our dubious control of the Mediterranean and the Middle East, with all that this involved. We could not tell they would not seize it.

21

General Wavell's Final Effort

WHILE the struggle in Crete and the Western Desert was moving to a climax and the *Bismarck* was hunted and destroyed in the Atlantic Ocean, less sanguinary though not graver, dangers had threatened us in Syria and Iraq. Our Treaty of 1930 with Iraq provided that in time of peace Britain should, among other things, maintain air bases near Basra and at Habbaniya, and have the right of transit for military forces and supplies at all times. It also provided that in war we should have all possible facilities, including the use of railways, rivers, ports, and airfields, for the passage of our armed forces. When war came Iraq broke off diplomatic relations with Germany, but did not declare war. When Italy came in Iraq did not even sever relations, and the Italian Legation in Baghdad became the chief centre for Axis propaganda and for fomenting anti-British feeling. In this they were aided by the Mufti of Jerusalem, who had fled from Palestine shortly before the outbreak of war and later received asylum in Baghdad. With the collapse of France British prestige sank very low, and the situation gave us much anxiety. But military action had been out of the question, and we had had to carry on as best we could.

In March 1941 there was a turn for the worse. Rashid Ali, who was working with the Germans, became Prime Minister, and the pro-British Regent, Emir Abdul Ilah, fled. It became essential to make sure of Basra, the main port of Iraq on the Persian Gulf, and a brigade group sent by General Auchinleck, the Commander-in-Chief in India, disembarked there without opposition on April 18. Rashid Ali, who had been counting on the assistance of German aircraft, and even of German airborne troops, was thereupon forced into action.

His first move was against Habbaniya, our air force training base in the Iraqi desert. The cantonment held just over 2200 fighting men, and no fewer than 9000 civilians, and its Flying School thus became a point of grave importance. Air Vice-Marshal Smart, who commanded, took bold and timely precautions. The school had previously held only obsolescent or training types, but a few Gladiator fighters had arrived from Egypt, and eighty-two aircraft of all sorts were improvised into four squadrons. A British battalion, flown from India, arrived on the 29th. The ground defence of the seven miles perimeter, with its solitary wire fence, was indeed scanty. On the 30th, Iraqi troops from Baghdad appeared barely a mile away on the plateau overlooking both the airfield and the camp. They were soon reinforced until they numbered about 9000 men, with fifty guns. The next two days were spent in fruitless parleys, and at dawn on May 2 fighting began.

In Syria the threat was no less imminent and our resources no less strained. It was one of the many overseas territories of the French Empire which considered themselves bound by the surrender of the French Government, and the Vichy authorities had done their utmost to prevent anybody in the French Army of the Levant from crossing into Palestine to join us. In August 1940 an Italian Armistice Commission appeared, and German agents, who had been interned on the outbreak of war, were released and became active. By the end of the year many more Germans had arrived, and, with ample funds, proceeded to arouse anti-British and anti-Zionist feeling among the Arab peoples of the Levant. At the same moment as Rashid Ali seized power in Iraq, Syria forced itself on our attention. The Luftwaffe were already attacking the Suez Canal from bases in the Dodecanese, and they could obviously, if they chose, operate against Syria, especially with air-borne troops. If the Germans once got control, then Egypt, the Canal Zone, and the oil refineries at Abadan would come under the direct threat of continuous air attack. Our land communications between Palestine and Iraq would be in danger. There might well be political repercussions in Egypt, and our repute in Turkey and throughout the Middle East would be smitten.

Soon after Rashid Ali appealed to the Fuehrer for armed support against us in Iraq, Admiral Darlan negotiated a preliminary agreement with the Germans about Syria. Three-quarters of the war material assembled under the control of the Italian Armistice Commission was to be transported to Iraq and the German Air Force granted landing facilities. General Dentz, the Vichy High Commissioner and Commander-in-Chief, was ordered to comply, and by the end of May about a hundred German and twenty Italian aircraft landed on Syrian airfields.

*   *   *

From the outset of these new dangers General Wavell showed himself most reluctant to assume more burdens. In Syria all he could manage was a single brigade group. He said he would make preparations and do what he could to create the impression of a large force being prepared for action from Palestine, which might have some effect on the Iraqi Government, but anything he could send would be both inadequate and too late. It would leave Palestine most dangerously weak, and incitement to rebellion there was already taking place. "I have consistently warned you," he telegraphed, "that no assistance could be given to Iraq from Palestine in present circumstances, and have

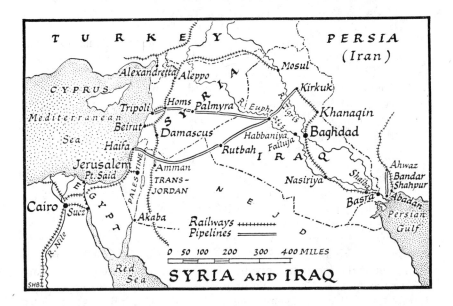

SYRIA AND IRAQ

always advised that a commitment in Iraq should be avoided. . . . My forces are stretched to the limit everywhere, and I simply cannot afford to risk part of them on what cannot produce any effect."

General Auchinleck, on the other hand, continued to offer reinforcements for Iraq up to five infantry brigades and ancillary troops if shipping could be provided. We were gratified by his forward mood. General Wavell only obeyed under protest. "I feel it my duty to warn you in the gravest terms," he cabled on May 5, "that I consider the prolongation of fighting in Iraq will seriously endanger the defence of Palestine and Egypt. The political repercussions will be incalculable, and may result in what I have spent nearly two years trying to avoid, namely, serious internal trouble in our bases. I therefore urge again most strongly that a settlement should be negotiated as early as possible."

I was not content with this, and, supported by the Chiefs of Staff, I brought the issue before the Defence Committee when it met at noon

next day. There was a resolute temper. The following orders were sent
to General Wavell at their direction:

> . . . Settlement by negotiation cannot be entertained except on
> the basis of a climb down by Iraqis, with safeguard against future
> Axis designs on Iraq. Realities of the situation are that Rashid
> Ali has all along been hand-in-glove with Axis Powers, and was
> merely waiting until they could support him before exposing his
> hand. Our arrival at Basra forced him to go off at half-cock before
> the Axis was ready. Thus there is an excellent chance of restoring
> the situation by bold action, if it is not delayed.
>
> Chiefs of Staff have therefore advised Defence Committee that
> they are prepared to accept responsibility for dispatch of the force
> specified in your telegram at the earliest possible moment. Defence
> Committee direct that Air Vice-Marshal Smart should be informed
> that he will be given assistance, and that in the meanwhile it is his
> duty to defend Habbaniya to the last. Subject to the security of
> Egypt being maintained, maximum air support possible should be
> given to operations in Iraq.

Meanwhile the squadrons of the Habbaniya Flying School, together
with Wellington bombers from Shaiba, at the head of the Persian Gulf,
attacked the Iraqi troops on the plateau. They replied by shelling the
cantonment, their aircraft joining in with bombs and machine guns.
Over forty of our men were killed or wounded in the first day, and
twenty-two aircraft destroyed or disabled. Despite the hazards of
taking off under close artillery fire, our airmen persevered. No enemy
infantry assault developed, and gradually their batteries were mastered.
It was found that the enemy gunners would not stand to their pieces
under air attack, or even if our aircraft were to be seen overhead. Full
advantage was taken of their nervousness, and after the second day we
were able to turn a proportion of our air effort against the Iraqi Air
Force and their bases. On the nights of May 3 and 4 patrols moved out
to raid the enemy lines, and by the 5th, after four days of attack from
the Royal Air Force, the enemy had had enough. That night they withdrew from the plateau. They were followed up, and a very successful
action yielded four hundred prisoners, a dozen guns, sixty machine guns,
and ten armoured cars. A reinforcing column was caught on the road
and destroyed by our aircraft. By May 7 the siege was over, and on the
18th the advance guard of the relieving force arrived from Palestine.

By now the Iraqis were not the only enemy. The first German aircraft were established on Mosul airfield on May 13, and thencefor-

ward the principal task of the R.A.F was to attack them and prevent their being supplied by railway from Syria. After a few days we succeeded in crushing them. Later an Italian fighter squadron appeared, but accomplished nothing. The German officer charged with co-ordinating the action of the Axis air squadrons with the Iraqi forces, a son of Field Marshal Blomberg, landed at Baghdad with a bullet in his head, thanks to misjudged firing by his allies. His successor, though more fortunate in his landing, could do nothing and all chance of useful Axis intervention passed away.

Our forward troops reached the outskirts of Baghdad on May 30. Although they were weak in numbers and there was an Iraqi division in the city, their presence was too much for Rashid Ali and his companions, who thereupon fled to Persia, accompanied by the German and Italian Ministers and the ex-Mufti of Jerusalem. An armistice was signed next day, the Regent was reinstated, a new Government took office, and we soon occupied all the important points in the country.

Thus the German plan for raising rebellion in Iraq and mastering cheaply this wide area was frustrated on a small margin. They had of course at their disposal an air-borne force which would have given them at this time Syria, Iraq, and Persia, with their precious oilfields. Hitler's hand might have reached out very far towards India, and beckoned to Japan. He had chosen however, as we have seen, to employ and expend his prime air organism in another direction. He certainly cast away the opportunity of taking a great prize for little cost in the Middle East.

*   *   *

The bitter need to forestall the Germans in Syria also compelled us to press Wavell hard. He said he hoped he would not be burdened with a campaign in Syria unless it was absolutely essential. The Chiefs of Staff replied that there was no option but to improvise the largest force he could without prejudice to the security of the Western Desert, and on May 21 — at the moment of the German attack on Crete — he instructed General Maitland Wilson to prepare for an advance.

Aided by Free French troops, it began on June 8, and at first met little opposition. No one could tell how much Vichy would fight. Although we could hardly achieve surprise, it was thought by some that the enemy would offer only a token resistance. But when they realised how weak we were they took heart and reacted vigorously, if only for the honour of their arms. After a week's fighting it was clear to Wavell that reinforcements were necessary. He was able to scrape up more troops, including a portion of the force which had captured Baghdad. Damascus was captured by the Australians on the 21st, after three days of severe fighting. Their advance was helped by a daring and costly raid by

Number 11 Commando, which was landed from the sea behind the
enemy lines. General Dentz realised that his limit was reached. He still
had about 24,000 men, but he could not hope to offer continued
resistance. Barely one-fifth of his air force remained. At 8:30 A.M. on
July 12 Vichy envoys arrived to sue for an armistice. This was granted,
a convention was signed, and Syria passed into Allied occupation. Our
casualties in killed and wounded were over 4600; those of the enemy
about 6500. One distasteful incident remained. British prisoners taken
during the fighting had been hurriedly shipped off to Vichy France,
whence they would certainly have passed into German keeping. When
this was discovered and no redress was offered, General Dentz and
other highly placed officers were taken into custody as hostages. This
had the desired effect, and our men were returned.

                         *   *   *

The successful campaigns in Syria and Iraq greatly improved our
strategical position in the Middle East. It closed the door to any further
attempt at enemy penetration eastward from the Mediterranean, moved
our defence of the Suez Canal northward by 250 miles, and relieved
Turkey of anxiety for her southern frontier. She could now be assured of aid from a friendly Power if she were attacked. The battle in
Crete, which cost us so dear, ruined the striking power of the German
air-borne corps. The Iraq revolt was finally crushed, and with pitifully
small and improvised forces we regained mastery of the wide regions involved. The occupation and conquest of Syria, which was undertaken
to meet a desperate need, ended, as it proved for ever, the German advance towards the Persian Gulf and India. If, under all the temptations
of prudence, the War Cabinet and Chiefs of Staff had not made every
post a winning post, and imposed their will on all commanders, we
should have been left only with the losses sustained in Crete, without
gathering the rewards which followed from the hard and glorious fighting there. If General Wavell, though exhausted, had broken under the
intense strain to which he was subjected by events and by our orders,
the whole future of the war and of Turkey might have been fatefully
altered. There is always much to be said for not attempting more than
you can do and for making a certainty of what you try. But this principle, like others in life and war, has its exceptions.

It must be remembered that the revolt in Iraq and the thrust to Syria
was but one small sector of the immense emergency in the Middle East
which lapped General Wavell on all sides simultaneously. In the same
way the whole Mediterranean scene, as viewed from London, was but
a secondary part of our world problem, in which the Invasion menace,
the U-boat war, and the attitude of Japan were dominant features.

Only the strength and cohesion of the War Cabinet, the relations of mutual respect and harmony of outlook between political and military chiefs, and the smooth working of our war machine enabled us to surmount, though sorely mauled, these trials and perils. One more operation, the battle in the Western Desert, which ranked first with me and the Chiefs of Staff, has still to be described. And this, though denied success, brought Rommel to a standstill for nearly five months.

*   *   *

At this time we had a spy in close touch with Rommel's headquarters, who gave us accurate information of the fearful difficulties of Rommel's assertive but precarious position. We knew how narrow was the margin on which he hoped to maintain himself, and also the strong and strict injunctions of the German High Command that he was not to cast away his victories by asking too much of fortune.

Wavell, who had all our information, tried on his own initiative, even in the imminent advent of Crete, to claw down Rommel before the dreaded 15th Panzer Division arrived in full strength over the long road from Tripoli, and before Benghazi was effectively opened as a short cut for enemy supply. He wished to attack even before the tanks delivered by "Tiger" — "Tiger Cubs," as Wavell and I called them in our correspondence — could be brought into action. A small force under General Gott tried to do so, but the attempt failed, and by May 20 the opportunity of defeating Rommel before he could be reinforced had passed.

Despite preparations made in advance, the delays in unloading, refitting, and making desert-worthy the Tiger Cubs proved severe. The mechanical condition of many of the infantry tanks was found on arrival to be indifferent. Trouble soon descended. Rommel deployed the greater part of the 15th Panzer Division, and concentrated on the frontier between Capuzzo and Sidi Omar. He expected a serious attack to relieve Tobruk and was determined to recapture and hold Halfaya in order to make it more difficult. This famous pass was held by the 3rd Battalion Coldstream Guards, a regiment of field artillery, and two squadrons of tanks. The enemy advanced on May 26, and that evening captured a feature to the northward which gave good observation over the whole position held by the Coldstream. Next morning after heavy shelling a concerted onslaught by at least two battalions and sixty tanks placed us in great jeopardy. Reserves were too distant to be able to intervene, and it remained only to extricate the force without more ado. This was accomplished, but losses were severe; only two of our tanks remained effective. Rommel had gained his objective, and proceeded to install himself firmly at Halfaya. As he had hoped, his occupation of this position was to prove a considerable hindrance to us three weeks later.

*   *   *

Preparations for our main offensive, code-named "Battleaxe," continued actively; but there was a darker side. On May 31 Wavell reported the technical difficulties which he was having with the reforming of the 7th Armoured Division. The earliest date at which he would be able to launch "Battleaxe" was June 15. While he realised the dangers of postponement, with the risk of enemy air reinforcements and a heavy attack on Tobruk, he felt that as the forthcoming battle would be primarily a tank engagement he must give the Armoured Division every chance, and the extra days gained by waiting should "double the possibilities of success."

I now awaited in keen hope and fear our attack in the Desert, which might change in our favour the whole course of the campaign. In disquieting contrast with our own performances earlier in the year, the Germans had brought Benghazi rapidly into use, and the bulk of their forces was probably already being maintained to a large extent through that port. We now know that the Germans had succeeded in concentrating forward a large part of their own armour without our becoming aware of it. Actually they brought rather more than 200 tanks into action against our 180.

"Battleaxe" started early on June 15. At first things went reasonably well, but on the third day, June 17, everything went wrong, and it became clear that our stroke had failed. The withdrawal of the whole force was carried out in good order, protected by our fighter aircraft. The enemy did not press the pursuit, partly no doubt because his armour was heavily attacked by R.A.F. bombers. There was probably however another reason. As we now know, Rommel's orders were to act purely on the defensive and to build up resources for operations in the autumn. To have embroiled himself in a strong pursuit across the frontier, and suffered losses thereby, would have been in direct contravention of orders.

Although this action may seem small compared with the scale of the Mediterranean war in all its various campaigns, its failure was to me a most bitter blow. Success in the Desert would have meant the destruction of Rommel's audacious force. Tobruk would have been relieved, and the enemy's retreat might well have carried him back beyond Benghazi as fast as he had come. It was for this supreme object, as I judged it, that all the perils of "Tiger" had been dared. No news had reached me of the events of the 17th, and, knowing that the result must soon come in, I went down to Chartwell, which was all shut up, wishing to be alone. Here I got the reports of what had happened. I wandered about the valley disconsolately for some hours.

*   *   *

The reader who has followed the narrative will now be prepared in his mind for the decision which I took in the last ten days of June 1941. At home we had the feeling that Wavell was a tired man. It might well be said that we had ridden the willing horse to a standstill. The extraordinary convergence of five or six different theatres, with their ups and downs, especially downs, upon a single Commander-in-Chief constituted a strain to which few soldiers have been subjected. I was discontented with Wavell's provision for the defence of Crete, and especially that a few more tanks had not been sent. The Chiefs of Staff had overruled him in favour of the small but most fortunate plunge into Iraq which had resulted in the relief of Habbaniya and complete local success. Finally, there was "Battleaxe," which Wavell had undertaken in loyalty to the risks I had successfully run in sending out the Tiger Cubs. I was dissatisfied with the arrangements made by the Middle East Headquarters Staff for the reception of the Tiger Cubs, carried to his aid through the deadly Mediterranean at so much hazard and with so much luck. I admired the spirit with which he had fought this small battle, which might have been so important, and his extreme disregard of all personal risks in flying to and fro on the wide, confused field of fighting. But the operation seemed ill-concerted, especially from the failure to make a sortie from the Tobruk sally port as an indispensable preliminary and concomitant.

Above all this there hung the fact of the beating in of the desert flank by Rommel, which had undermined and overthrown all the Greek projects on which we had embarked, with all their sullen dangers and glittering prizes in what was for us the supreme sphere of the Balkan war. I am reminded of having commented: "Rommel has torn the new-won laurels from Wavell's brow and thrown them in the sand." This was not a true thought, but only a passing pang. Judgment upon all this can only be made in relation to the authentic documents written at the time, and no doubt also upon much other valuable evidence which the future will disclose. The fact remains that after "Battleaxe" I came to the conclusion that there should be a change.

General Auchinleck was now Commander-in-Chief in India. I had not altogether liked his attitude in the Norwegian campaign at Narvik. He had seemed to be inclined to play too much for safety and certainty, neither of which exists in war, and to be content to subordinate everything to the satisfaction of what he estimated as minimum requirements. However, I had been much impressed with his personal qualities, his presence, and high character. When after Narvik he had taken over the Southern Command I received from many quarters, official and private, testimony to the vigour and structure which he had given to that important region. His appointment as Commander-in-Chief in India had been generally acclaimed. We have seen how forthcoming he had been about sending the Indian forces to Basra, and the ardour with which he had addressed himself to the suppression of the revolt in Iraq. I had the conviction that in Auchinleck I should bring a new, fresh figure to bear the multiple strains of the Middle East, and that Wavell, on the other hand, would find in the great Indian command time to regain his strength before the new but impending challenges and opportunities arrived. I found that these views of mine encountered no resistance in our Ministerial and military circles in London. The reader must not forget that I never wielded autocratic powers, and always had to move with and focus political and professional opinion. On June 21 I telegraphed accordingly. Wavell received the decision with poise and dignity. He was at that time about to undertake a flight to Abyssinia which proved extremely dangerous. His biographer records that on reading my message he said, "The Prime Minister is quite right. There ought to be a new eye and a new hand in this theatre."

*   *   *

I had also for several months past been extremely distressed by the apparent inadequacy of the Cairo staff, and I increasingly realised the undue burdens of so many different kinds cast upon our struggling Commander-in-Chief. Wavell himself, together with the other Commanders-in-Chief, had as early as April 18 asked for some relief and assistance. His view was endorsed by his two professional colleagues. The Commanders-in-Chief had felt the convenience of having high political authority close at hand during Mr. Eden's visit. They were conscious of a vacuum after his departure.

My son Randolph, who had gone out with the commandos, now to some extent dispersed, was at this time in the Desert. He was a Member of Parliament and had considerable contacts. I did not hear much or often from him, but on June 7 I had received through the Foreign Office the following telegram which he had sent from Cairo with the knowledge and encouragement of our Ambassador, Sir Miles Lampson:

Do not see how we can start winning war out here until we have a competent civilian on the spot to provide day-to-day political and strategic direction. Why not send a member of the War Cabinet here to preside over whole war effort? Apart from small personal staff, he would need two outstanding men to co-ordinate supply and direct censorship, intelligence, and propaganda. Most thoughtful people here realise need for radical reform along these lines. No mere shunting of personnel will suffice, and the present time seems particularly ripe and favourable for a change of system. Please forgive me troubling you, but consider present situation deplorable and urgent action vital to any prospects of success.

It is the fact that this clinched matters in my mind. "I have been thinking," I replied to him a fortnight later, "a good deal for some time on the lines of your helpful and well-conceived telegram." And thereupon I took action.

I had brought Captain Oliver Lyttelton into the Government as President of the Board of Trade in October 1940. I had known him from his childhood. He served in the Grenadiers through the hardest fighting of the First World War, being wounded and decorated several times. After leaving the Army he had entered business and became the managing director of a large metal firm. Knowing his remarkable personal qualities, I did not hesitate to bring him into Parliament and high office. His administration had won respect from all parties in our National Government. I had not liked his proposals of 1941 for clothing coupons, but I found these were received with favour by the Cabinet and the House of Commons, and there is no doubt they were necessary at the time. He was an all-round man of action, and I now felt that he was in every way fitted for this new and novel post of a War Cabinet Minister resident in the Middle East. This would take another large slice of business off the shoulders of the military chiefs. I found this idea most readily acceptable to my colleagues of all parties. Accordingly he was appointed, with the prime duty "to relieve the High Command of all extraneous burdens, and to settle promptly on the spot in accordance with the policy of His Majesty's Government many questions affecting several departments or authorities which hitherto have required reference home."

All these new arrangements, with their consequential administerial reactions, fitted in with, and were appropriate to, the change in the command of the Middle East.

22

The Soviet Nemesis

NEMESIS PERSONIFIES "the Goddess of Retribution, who brings down all immoderate good fortune, checks the presumption that attends it . . . and is the punisher of extraordinary crimes." [1] We must now lay bare the error and vanity of cold-blooded calculation of the Soviet Government and enormous Communist machine, and their amazing ignorance about where they stood themselves. They had shown a total indifference to the fate of the Western Powers, although this meant the destruction of that "Second Front" for which they were soon to clamour. They seemed to have no inkling that Hitler had for more than six months resolved to destroy them. If their Intelligence Service informed them of the vast German deployment towards the East, which was now increasing every day, they omitted many needful steps to meet it. Thus they had allowed the whole of the Balkans to be overrun by Germany. They hated and despised the democracies of the West; but the four countries Turkey, Rumania, Bulgaria, and Yugoslavia, which were of vital interest to them and their own safety, could all have been combined by the Soviet Government in January with active British aid to form a Balkan front against Hitler. They let them all break into confusion, and all but Turkey were mopped up one by one. War is mainly a catalogue of blunders, but it may be doubted whether any mistake in history has equalled that of which Stalin and the Communist chiefs were guilty when they cast away all possibilities in the Balkans and supinely awaited, or were incapable of realising, the fearful onslaught which impended upon Russia. We had hitherto rated them as selfish calculators. In this period they were proved simpletons as well. The force, the mass, the bravery and endurance of Mother Russia had still to be thrown into the scales. But so far as strategy, policy, foresight,

[1] Quoted from the Oxford English Dictionary.

competence are arbiters Stalin and his commissars showed themselves at this moment the most completely outwitted bunglers of the Second World War.

*　　　　*　　　　*

Hitler's "Barbarossa" directive of December 18, 1940, had laid down the general grouping and primary tasks of the forces to be concentrated against Russia. At that date the total German strength on the Eastern Front was thirty-four divisions. To multiply that figure more than thrice was an immense process both of planning and preparation, and it fully occupied the early months of 1941. In January and February the Balkan adventure into which the Fuehrer allowed himself to be drawn caused a drain-away from the East to the South of five divisions, of which three were armoured. In May the German deployment in the East grew to eighty-seven divisions, and there were no less than twenty-five absorbed in the Balkans. Considering the magnitude and hazard of the invasion of Russia, it was improvident to disturb the concentration to the East by so serious a diversion. We shall now see how a delay of five weeks was imposed upon the supreme operation as the result of our resistance in the Balkans, and especially of the Yugoslav revolution. No one can measure exactly what consequences this had before winter set in upon the fortunes of the German-Russian campaign. It is reasonable to believe that Moscow was saved thereby. During May and the beginning of June many of the best-trained German divisions and all the armour were moved from the Balkans to the Eastern Front, and at the moment of their assault the Germans attacked with 120 divisions, seventeen of which were armoured and twelve motorised. Six Rumanian divisions were also included in their Southern Army Group. In general reserve a further 26 divisions were assembled or assembling; so that by early July the German High Command could count on at least 150 divisions, supported by the main striking power of their Air Force, about 2700 aircraft.

*　　　　*　　　　*

Up till the end of March I was not convinced that Hitler was resolved upon mortal war with Russia, nor how near it was. Our Intelligence reports revealed in much detail the extensive German troop movements towards and into the Balkan States which had marked the first three months of 1941. Our agents could move with a good deal of freedom in these quasi-neutral countries, and were able to keep us accurately posted about the heavy German forces gathering by rail and road to the southeast. But none of these necessarily involved the invasion of Russia, and all were readily explainable by German interests and policy in Rumania and Bulgaria, by her designs on Greece and arrangements with Yugoslavia and Hungary. Our information about the immense movement taking place through Germany towards the main Russian front, stretching from Rumania to the Baltic, was far more difficult to acquire. That Germany should at this stage, and before clearing the Balkan scene, open another major war with Russia seemed to me too good to be true.

There was no sign of lessening German strength opposite us across the Channel. The German air raids on Britain continued with intensity. The manner in which the German troop concentrations in Rumania and Bulgaria had been glozed over and apparently accepted by the Soviet Government, the evidence we had of large and invaluable supplies being sent to Germany from Russia, the obvious community of interest between the two countries in overrunning and dividing the British Empire in the East, all made it seem more likely that Hitler and Stalin would make a bargain at our expense rather than a war upon each other. This bargain we now know was within wide limits Stalin's aim.

These impressions were shared by our Joint Intelligence Committee. On April 7 they stated that there were a number of reports circulating in Europe of a German plan to attack Russia. Although Germany, they said, had considerable forces available in the East, and expected to fight Russia some time or other, it was unlikely that she would choose to make another major war front yet. Her main object in 1941 would, according to them, remain the defeat of the United Kingdom. As late as May 23 this committee from the three Services reported that rumours of impending attack on Russia had died down, and that there were reports that a new agreement between the two countries was impending.

Our Chiefs of Staff were ahead of their advisers; and more definite. "We have firm indications," they warned the Middle East Command on May 31, "that the Germans are now concentrating large army and air forces against Russia. Under this threat they will probably demand concessions most injurious to us. If the Russians refuse the Germans will march."

It was not till June 5 that the Joint Intelligence Committee reported that the scale of German military preparations in Eastern Europe seemed to indicate that an issue more vital than an economic agreement was at stake. It was possible that Germany desired to remove from her eastern frontier the potential threat of increasingly powerful Soviet forces. They considered it as yet impossible to say whether war or agreement would result.

I had not been content with this form of collective wisdom, and preferred to see the originals myself. I had arranged therefore, as far back as the summer of 1940, for Major Desmond Morton to make a daily selection of titbits, which I always read, thus forming my own opinion, sometimes at much earlier dates.

It was thus with relief and excitement that towards the end of March 1941 I read an Intelligence report from one of our most trusted sources of the movement and countermovement of German armour on the railway from Bucharest to Cracow. This showed that as soon as the Yugoslav Ministers made their submission in Vienna, three out of the five Panzer divisions which had moved through Rumania southward towards Greece and Yugoslavia had been sent northward to Cracow, and secondly that the whole of this transportation had been reversed after the Belgrade revolution and the three Panzer divisions sent back to Rumania. This shuffling and reversal of about sixty trains could not be concealed from our agents on the spot.

To me it illuminated the whole Eastern scene like a lightning flash. The suddent movement to Cracow of so much armour needed in the Balkan sphere could only mean Hitler's intention to invade Russia in May. This seemed to me henceforward certainly his major purpose. The fact that the Belgrade revolution had required their return to Rumania involved perhaps a delay from May to June. I cast about for some means of warning Stalin, and, by arousing him to his danger, establishing contacts with him like those I had made with President Roosevelt. I made the message short and cryptic, hoping that this very fact, and that it was the first message I had sent him since my formal telegram of June 25, 1940, commending Sir Stafford Cripps as Ambassador, would arrest his attention and make him ponder.

Prime Minister to Sir Stafford Cripps   3 Apr. 41

Following from me to M. Stalin, *provided it can be personally delivered by you:*

I have sure information from a trusted agent that when the Germans thought they had got Yugoslavia in the net — that is to say, after March 20 — they began to move three out of the five Panzer divisions from Rumania to Southern Poland. The moment they heard of the Serbian revolution this movement was countermanded. Your Excellency will readily appreciate the significance of these facts.

The British Ambassador did not reply till April 12, when he said that just before my telegram had been received he had himself addressed to Vyshinsky a long personal letter reviewing the succession of failures of the Soviet Government to counteract German encroachments in the Balkans, and urging in the strongest terms that the U.S.S.R. in her own interest must now decide on an immediate vigorous policy of co-operation with countries still opposing the Axis in that area. "Were I now," he said, "to convey through Molotov the Prime Minister's message, which expresses the same thesis in very much shorter and less emphatic form, I fear that the only effect would be probably to weaken impression already made by my letter to Vyshinsky. . . ."

I was vexed at this and at the delay which had occurred. This was the only message before the attack that I sent Stalin direct. Its brevity, the exceptional character of the communication, the fact that it came from the head of the Government and was to be delivered personally to the head of the Russian Government by the Ambassador, were all intended to give it special significance and arrest Stalin's attention. I was eventually told that Sir Stafford had handed it to Vyshinsky on April 19 and Vyshinsky had informed him in writing on April 23 that it had been conveyed to Stalin.

I cannot form any final judgment upon whether my message, if delivered with all the promptness and ceremony prescribed, would have altered the course of events. Nevertheless I still regret that my instructions were not carried out effectively. If I had had any direct contact with Stalin I might perhaps have prevented him from having so much of his air force destroyed on the ground.

*   *   *

We know now that Hitler's directive of December 18 had prescribed May 15 as the date for invading Russia, and that in his fury at the revolution in Belgrade this had been postponed for a month, and later till June 22. Until the middle of March the troop movements in the north on the main Russian front were not of a character to require special German measures of concealment. On March 13 however orders were issued by Berlin to terminate the work of the Russian commissions working in German territory and to send them home. The presence of Russians in this part of Germany could only be permitted up to March 25. During this time the 120 German divisions of the highest quality were assembling in their three army groups along the Russian front. The Southern Group, under Rundstedt, was, for the reasons explained, far from well found in armour. Its Panzer divisions had not only recently returned from Greece and Yugoslavia. Despite the postponement of the attack till June 22 they badly needed rest and overhaul after their mechanical wear and tear in the Balkans.

On April 13 Schulenburg came from Moscow to Berlin. Hitler received him on April 28, and treated his Ambassador to a tirade against

Russia. Schulenburg adhered to the theme which had governed all his reports. "I am convinced that Stalin is prepared to make even further concessions to us. It has already been indicated to our economic negotiators that (if we applied in due time) Russia could supply us with up to five million tons of grain a year." [2] Schulenburg returned to Moscow on April 30, profoundly disillusioned by his interview with Hitler. He had a clear impression that Hitler was bent on war. It seems that he had even tried to warn the Russian Ambassador in Berlin, Dekanosov, in this sense. And he fought persistently in the last hours for his policy of Russo-German understanding.

Weizsächer, the official head of the German Foreign Office, was a highly competent civil servant of the type to be found in the Government departments of many countries. He was not a politician with executive power, and would not, according to British custom, be held accountable for State policy. He was nevertheless condemned to seven years' penal servitude by decree of the courts set up by the conquerors. Although he is therefore classified as a war criminal, he certainly wrote good advice to his superiors, which we may be glad they did not take. He commented as follows upon this interview:

> I can summarise in one sentence my views on a German-Russian conflict. If every Russian city reduced to ashes were as valuable to us as a sunken British warship, I should advocate the German-Russian war for this summer; but I believe that we should be victors over Russia only in a military sense, and should, on the other hand, lose in an economic sense.
>
> It might perhaps be considered an alluring prospect to give the Communist system its death blow, and it might also be said that it was inherent in the logic of things to muster the Eurasian continent against Anglo-Saxondom and its following. But the sole decisive factor is whether this project will hasten the fall of England. . . .
>
> A German attack on Russia would only give the British new moral strength. It would be interpreted there as German uncertainty about the success of our fight against England. We should thereby not only be admitting that the war was going to last a long time yet, but we might actually prolong it in this way, instead of shortening it.

On May 7 Schulenburg hopefully reported that Stalin had taken over the chairmanship of the Council of People's Commissars in place of Molotov, and had thereby become head of the Government of the Soviet Union. ". . . I am convinced that Stalin will use his new

[2] *Nazi-Soviet Relations 1939-1941* (published in 1948 by the State Department in Washington), p. 332.

position in order to take part personally in the maintenance and development of good relations between the Soviets and Germany."

The German Naval Attaché, reporting from Moscow, expressed the same point in these words: "Stalin is the pivot of German-Soviet collaboration." Examples of Russian appeasement of Germany increased. On May 3 Russia had officially recognised the pro-German Government of Rashid Ali in Iraq. On May 7 the diplomatic representatives of Belgium and Norway were expelled from Russia. Even the Yugoslav Minister was flung out. At the beginning of June the Greek Legation was banished from Moscow. As General Thomas, the head of the economic section of the German War Ministry, later wrote in his paper on the war economy of the Reich: "The Russians executed their deliveries up to the eve of the attack, and in the last days the transport of rubber from the Far East was expedited by express trains."

*   *   *

We had not of course full information about the Moscow moods, but the German purpose seemed plain and comprehensible. On May 16 I had cabled to General Smuts: "It looks as if Hitler is massing against Russia. A ceaseless movement of troops, armoured forces, and aircraft northward from the Balkans and eastward from France and Germany is in progress." Stalin must have tried very hard to preserve his illusions about Hitler's policy. After another month of intense German troop movement and deployment Schulenburg could telegraph to the German Foreign Office on June 13:

> People's Commissar Molotov has just given me the following text of a Tass dispatch which will be broadcast tonight and published in the papers tomorrow:
> Even before the return of the English Ambassador Cripps to London, but especially since his return, there have been widespread rumours of an impending war between the U.S.S.R. and Germany in the English and foreign press. . . .
> Despite the obvious absurdity of these rumours, responsible circles in Moscow have thought it necessary to state that they are a clumsy propaganda manoeuvre of the forces arrayed against the Soviet Union and Germany, which are interested in a spread and intensification of the war.

Hitler had every right to be content with the success of his measures of deception and concealment, and with his victim's state of mind.

Molotov's final fatuity is worth recording. On June 22, at 1.17 A.M., Schulenburg telegraphed once more to the German Foreign Office:

Molotov summoned me to his office this evening at 9.30 P.M. After he had mentioned the alleged repeated border violations by German aircraft . . . Molotov stated as follows:

There were a number of indications that the German Government was dissatisfied with the Soviet Government. Rumours were even current that a war was impending between Germany and the Soviet Union. The Soviet Government was unable to understand the reasons for Germany's dissatisfaction. . . . He would appreciate it if I could tell him what had brought about the present situation in German-Soviet Russian relations.

I replied that I could not answer his question, as I lacked the pertinent information; that I would however transmit his communication to Berlin.

But the hour had now struck. At 4 A.M. on this same June 22, 1941, Ribbentrop delivered a formal declaration of war to the Russian Ambassador in Berlin. At daybreak Schulenburg presented himself to Molotov in the Kremlin. The latter listened in silence to the statement read by the German Ambassador, and then commented, "It is war. Your aircraft have just bombarded some ten open villages. *Do you believe that we deserved that?*" [3]

* * *

In the face of the Tass broadcast it had been vain for us to add to the various warnings which Eden had given to the Soviet Ambassador in London, or for me to make a renewed personal effort to arouse Stalin to his peril. Even more precise information had been constantly sent to the Soviet Government by the United States. Nothing that any of us could do pierced the purblind prejudice and fixed ideas which Stalin had raised between himself and the terrible truth. Although on German estimates 186 Russian divisions were massed behind the Soviet boundaries, of which 119 faced the German front, the Russian armies to a large extent were taken by surprise. The Germans found no signs of offensive preparations in the forward zone, and the Russian covering troops were swiftly overpowered. Something like the disaster which had befallen the Polish Air Force on September 1, 1939, was now to be repeated on a far larger scale on the Russian airfields, and many hundreds of Russian planes were caught at daybreak and destroyed before they could get into the air. Thus the ravings of hatred against Britain

[3] This was the last act of Count Schulenburg's diplomatic career. Late in 1943 his name appears in the secret circles of conspiracy against Hitler in Germany as possible Foreign Minister of a Government to succeed the Nazi régime, in view of his special qualifications to negotiate a separate peace with Stalin. He was arrested by the Nazis after the attempted assassination of Hitler in July 1944, and imprisoned in the Gestapo cells. On November 10 he was executed.

and the United States which the Soviet propaganda machine cast upon the midnight air were overwhelmed at dawn by the German cannonade. The wicked are not always clever, nor are dictators always right.

It is impossible to complete this account without referring to a terrible decision of policy adopted by Hitler towards his new foes, and enforced under all the pressure of the mortal struggle in vast barren· or ruined lands and winter horrors. Verbal orders were given by him at a conference on June 14, 1941, which to a large extent governed the conduct of the German Army towards the Russian troops and people, and led to many ruthless and barbarous deeds. According to the Nuremberg documents, General Halder testified:

> Prior to the attack on Russia the Fuehrer called a conference of all the commanders and persons connected with the Supreme Command on the question of the forthcoming attack on Russia. I cannot recall the exact date of this conference. . . . At this conference the Fuehrer stated that the methods used in the war against the Russians would have to be different from those used against the West. . . . He said that the struggle between Russia and Germany was a Russian struggle. He stated that since the Russians were not signatories of the Hague Convention the treatment of their prisoners of war did not have to follow the Articles of the Convention. . . . He [also] said that the so-called Commissars should not be considered prisoners of war.[4]

And according to Keitel:

> Hitler's main theme was that this was the decisive battle between the two ideologies, and that this fact made it impossible to use in this war [with Russia] methods, as we soldiers knew them, which were considered to be the only correct ones under international law.[5]

*   *   *

On the evening of Friday, June 20, I drove down to Chequers alone. I knew that the German onslaught upon Russia was a matter of days, or it might be hours. I had arranged to deliver a broadcast on Saturday night dealing with this event. It would of course have to be in guarded terms. Moreover, at this time the Soviet Government, at once haughty and purblind, regarded every warning we gave as a mere attempt of beaten men to drag others into ruin. As the result of my reflections in

---

[4] *Nuremburg Documents*, Part VI, pp. 310 ff.
[5] *Ibid.*, Part XI, p. 16.

the car I put off the broadcast till Sunday night, when I thought all would be clear. Thus Saturday passed with its usual toil.

When I awoke on the morning of Sunday, the 22nd, the news was brought to me of Hitler's invasion of Russia. This changed conviction into certainty. I had not the slightest doubt where our duty and our policy lay. Nor indeed what to say. There only remained the task of composing it. I asked that notice should immediately be given that I would broadcast at nine o'clock that night. Presently General Dill, who had hastened down from London, came into my bedroom with detailed news. The Germans had invaded Russia on an enormous front, had surprised a large portion of the Soviet Air Force grounded on the airfields, and seemed to be driving forward with great rapidity and violence. The Chief of the Imperial General Staff added, "I suppose they will be rounded up in hordes."

I spent the day composing my statement. There was not time to consult the War Cabinet, nor was it necessary. I knew that we all felt the same on this issue. Mr. Eden, Lord Beaverbrook, and Sir Stafford Cripps — he had left Moscow on the 10th — were also with me during the day. In the course of my broadcast I said:

"The Nazi regime is indistinguishable from the worst features of Communism. It is devoid of all theme and principle except appetite and racial domination. It excels all forms of human wickedness in the efficiency of its cruelty and ferocious aggression. No one has been a more consistent opponent of Communism than I have for the last twenty-five years. I will unsay no word that I have spoken about it. But all this fades away before the spectacle which is now unfolding. The past, with its crimes, its follies, and its tragedies, flashes away. I see the Russian soldiers standing on the threshold of their native land, guarding the fields which their fathers have tilled from time immemorial. I see them guarding their homes where mothers and wives pray — ah, yes, for there are times when all pray — for the safety of their loved ones, the return of the breadwinner, of their champion, of their protector. I see the ten thousand villages of Russia where the means of existence is wrung so hardly from the soil, but where there are still primordial human joys, where maidens laugh and children play. I see advancing upon all this in hideous onslaught the Nazi war machine, with its clanking, heel-clicking, dandified Prussian officers, its crafty expert agents fresh from the cowing and tying down of a dozen countries. I see also the dull, drilled, docile, brutish masses of the Hun soldiery plodding on like a swarm of crawling locusts. I see the German bombers and fighters in the sky, still smarting from many a British whipping, delighted to find what they believe is an easier and a safer prey.

"Behind all this glare, behind all this storm, I see that small group of villainous men who plan, organise, and launch this cataract of horrors upon mankind. . . .

"I have to declare the decision of His Majesty's Government — and I feel sure it is a decision in which the great Dominions will in due course concur — for we must speak out now at once, without a day's delay. I have to make the declaration, but can you doubt what our policy will be? We have but one aim and one single, irrevocable purpose. We are resolved to destroy Hitler and every vestige of the Nazi regime. From this nothing will turn us — nothing. We will never parley, we will never negotiate with Hitler or any of his gang. We shall fight him by land, we shall fight him by sea, we shall fight him in the air, until, with God's help, we have rid the earth of his shadow and liberated its peoples from his yoke. Any man or state who fights on against Nazidom will have our aid. Any man or state who marches with Hitler is our foe. . . . That is our policy and that is our declaration. It follows therefore that we shall give whatever help we can to Russia and the Russian people. We shall appeal to all our friends and allies in every part of the world to take the same course and pursue it, as we shall, faithfully and steadfastly to the end. . . .

"This is no class war, but a war in which the whole British Empire and Commonwealth of Nations is engaged, without distinction of race, creed, or party. It is not for me to speak of the action of the United States, but this I will say: if Hitler imagines that his attack on Soviet Russia will cause the slightest divergence of aims or slackening of effort in the great democracies who are resolved upon his doom, he is woefully mistaken. On the contrary, we shall be fortified and encouraged in our efforts to rescue mankind from his tyranny. We shall be strengthened and not weakened in determination and in resources.

"This is no time to moralise on the follies of countries and Governments which have allowed themselves to be struck down one by one, when by united action they could have saved themselves and saved the world from this castastrophe. But when I spoke a few minutes ago of Hitler's blood-lust and the hateful appetites which have impelled or lured him on his Russian adventure I said there was one deeper motive behind his outrage. He wishes to destroy the Russian power because he hopes that if he succeeds in this he will be able to bring back the main strength of his Army and Air Force from the East and hurl it upon this Island, which he knows he must conquer or suffer the penalty of his crimes. His invasion of Russia is no more than a prelude to an attempted invasion of the British Isles. He hopes, no doubt, that all this may be accomplished before the winter comes, and that he can overwhelm Great Britain before the Fleet and air-power of the United States may intervene. He hopes that he may once again repeat, upon a greater scale than ever before, that process of destroying his enemies one by one by which he has so long thrived and prospered, and that then the scene will be clear for the final act, without which all his conquests would be in vain — namely, the subjugation of the Western Hemisphere to his will and to his system.

"The Russian danger is therefore our danger, and the danger of the United States, just as the cause of any Russian fighting for his hearth and home is the cause of free men and free peoples in every quarter of the globe. Let us learn the lessons already taught by such cruel experience. Let us redouble our exertions, and strike with united strength while life and power remain."

BOOK III

The Grand Alliance

Sunday, December 7, 1941, and Onward

No American will think it wrong of me if I pro-claim that to have the United States at our side was to me the greatest joy.

THE
GERMAN
ATTACK
ON
RUSSIA

Period June – Sept 1941
Period Oct. – Dec.

0   100   200   300   400   500 Miles

White Sea

Archangel

FINLAND

Lake Onega

Lake Ladoga

Helsingfors

Reval

Lake Peipus

Leningrad

Baltic Sea

Riga

ARMY GRP "C"

E. PRUSSIA

Kalinin

Vitebsk

Moscow

Smolensk

Warsaw   ARMY GROUP "B"

Minsk

Tula

Bryansk

Pripet Marshes

Orel

Kursk

Brest Litovsk

Lublin

Voronezh

Lemberg   ARMY

GROUP   "A"

Konotop

Kiev

Stalingrad

Kharkov

Kremenchug

R. Volga

Carpathian Mts.

R. Dniester

R. Pruth

UKRAINE

R. Dnieper

R. Donetz

R. Don

Rostov

RUMANIA

Odessa

Sea of Azov

BULGARIA

Crimea

Sebastopol

Novorissisk

Caucasus Mts.

BLACK   SEA

Batum

T U R K E Y

I

Our Soviet Ally

THE ENTRY OF RUSSIA into the war was welcome but not immediately helpful to us. The German armies were so strong that it seemed that for many months they could maintain the invasion threat against England while at the same time plunging into Russia. Almost all responsible military opinion held that the Russian armies would soon be defeated and largely destroyed. The fact that the Soviet Air Force was allowed by its Government to be surprised on its landing grounds and that the Russian military preparations were far from being complete gave them a bad start. Frightful injuries were sustained by the Russian armies. In spite of heroic resistance, competent despotic war direction, total disregard of human life, and the opening of a ruthless guerrilla warfare in the rear of the German advance, a general retirement took place on the whole twelve-hundred-mile Russian front south of Leningrad for about four or five hundred miles. The strength of the Soviet Government, the fortitude of the Russian people, their immeasurable reserves of manpower, the vast size of their country, the rigours of the Russian winter, were the factors which ultimately ruined Hitler's armies. But none of these made themselves apparent in 1941. President Roosevelt was considered very bold when he proclaimed in September that the Russian front would hold and that Moscow would not be taken. The glorious strength and patriotism of the Russian people vindicated this opinion.

Even in August 1942, after my visit to Moscow and the conferences there, General Brooke, who had accompanied me, adhered to the opinion that the Caucasus Mountains would be traversed and the basin of the Caspian dominated by German forces, and we prepared accordingly on the largest possible scale for a defensive campaign in Syria and Persia. Throughout I took a more sanguine view than my military advisers of the Russian powers of resistance. I rested with confidence upon Stalin's assurance, given to me at Moscow, that he would hold the line of the Caucasus and that the Germans would not reach the Caspian in any strength. But we were vouchsafed so little information about Soviet resources and intentions that all opinions either way were hardly more than guesses.

It is true that the Russian entry into the war diverted the German air attack from Great Britain and diminished the threat of invasion. It gave us important relief in the Mediterranean. On the other hand, it imposed upon us most heavy sacrifices and drains. At last we were beginning to be well equipped. At last our munitions factories were pouring out their supplies of every kind. Our armies in Egypt and Libya were in heavy action and clamouring for the latest weapons, above all tanks and aeroplanes. The British armies at home were eagerly awaiting the long-promised modern equipment which in all its ever-widening complications was flowing at last towards them. At this moment we were compelled to make very large diversions of our weapons and vital supplies of all kinds, including rubber and oil. On us fell the burden of organising the convoys of British and still more of United States supplies and carrying them to Murmansk and Archangel through all the dangers and rigours of the Arctic passage. All the American supplies were a deduction from what had in fact been, or was to be, successfully ferried across the Atlantic for ourselves. In order to make this immense diversion and to forgo the growing flood of American aid without crippling our campaign in the Western Desert, we had to cramp all preparations which prudence urged for the defence of the Malay Peninsula and our Eastern Empire and possessions against the ever-growing menace of Japan.

Without in the slightest degree challenging the conclusion which history will affirm that the Russian resistance broke the power of the German armies and inflicted mortal injury upon the life energies of the German nation, it is right to make it clear that for more than a year after Russia was involved in the war she presented herself to our minds as a burden and not as a help. None the less we rejoiced to have this mighty nation in the battle with us, and we all felt that even if the Soviet armies were driven back to the Ural Mountains Russia would still exert an immense and, if she persevered in the war, an ultimately decisive force.

*   *   *

Up to the moment when the Soviet Government was set upon by Hitler they seemed to care for no one but themselves. Afterwards this mood naturally became more marked. Hitherto they had watched with stony composure the destruction of the front in France in 1940, and our vain efforts in 1941 to create a front in the Balkans. They had given important economic aid to Nazi Germany and had helped them in many minor ways. Now, having been deceived and taken by surprise, they were themselves under the flaming German sword. Their first impulse and lasting policy was to demand all possible succour from Great Britain and her Empire, the possible partition of which between Stalin and Hitler had for the last eight months beguiled Soviet minds from the progress of German concentration in the East. They did not hesitate to appeal in urgent and strident terms to harassed and struggling Britain to send them the munitions of which her armies were so short. They urged the United States to divert to them the largest quantities of the supplies on which we were counting, and, above all, even in the summer of 1941 they clamoured for British landings in Europe, regardless of risk and cost, to establish a Second Front. The British Communists, who had hitherto done their worst, which was not much, in our factories, and had denounced "the capitalist and imperialist war," turned about again overnight and began to scrawl the slogan "Second Front Now" upon the walls and hoardings.

We did not allow these somewhat sorry and ignominious facts to disturb our thought, and fixed our gaze upon the heroic sacrifices of the Russian people under the calamities which their Government had brought upon them, and their passionate defence of their native soil. This, while the struggle lasted, made amends for all.

The Russians never understood in the smallest degree the nature of the amphibious operation necessary to disembark and maintain a great army upon a well-defended hostile coast. Even the Americans were at this time largely unaware of the difficulties. Not only sea but air superiority at the invasion point was indispensable. Moreover, there was a third vital factor. A vast armada of specially constructed landing craft, above all tank-landing craft in numerous varieties, was the foundation of any successful heavily opposed landing. For the creation of this armada, as has been and will be seen, I had long done my best. It could not be ready even on a minor scale before the summer of 1943, and its power, as is now widely recognised, could not be developed on a sufficient scale till 1944. At the period we have now reached, in the summer of 1941, we had no mastery of the enemy air over Europe, except in the Pas de Calais, where the strongest German fortifications existed. The landing craft were only a-building. We had not even got an army in Britain as large, as well trained, as well equipped as the one we should have to meet on French soil. Yet Niagaras of folly and misstatement still pour out on this question of the Second Front. There was certainly no hope of convincing the Soviet Government at this or any other time. Stalin even suggested to me on one occasion later on that if the British were afraid he would be willing to send round three or four Russian Army Corps to do the job. It was not in my power, through lack of shipping and other physical facts, to take him at his word.

There was no response from the Soviet Government to my broadcast to Russia and the world on the day of the German attack, except that parts of it were printed in *Pravda* and other Russian Government organs, and that we were asked to receive a Russian Military Mission. The silence on the top level was oppressive, and I thought it my duty to break the ice. I quite understood that they might feel shy, considering all that had passed since the outbreak of the war between the Soviets and the Western Allies, and remembering what had happened twenty years before between me and the Bolshevik Revolutionary Government. On July 7 I therefore addressed myself to Stalin, and expressed our intention to bring all aid in our power to the Russian people. On the 10th I tried again. Official communications passed between the two Foreign Offices, but it was not until the 19th that I received the first direct communication from Stalin.

After thanking me for my two telegrams, he said:

> Perhaps it is not out of place to mention that the position of the Soviet forces at the front remains tense. . . . It seems to me therefore that the military situation of the Soviet Union, as well as of Great Britain, would be considerably improved if there could be established a front against Hitler in the West — Northern France, and in the North — the Arctic.
>
> A front in Northern France could not only divert Hitler's forces from the East, but at the same time would make it impossible for Hitler to invade Great Britain. The establishment of the front just mentioned would be popular with the British Army, as well as with the whole population of Southern England.
>
> I fully realise the difficulties involved in the establishment of such a front. I believe however that in spite of the difficulties it should be formed, not only in the interests of our common cause, but also in the interests of Great Britain herself. This is the most propitious moment for the establishment of such a front, because now Hitler's forces are diverted to the East and he has not yet had the chance to consolidate the position occupied by him in the East.
>
> It is still easier to establish a front in the North. Here, on the part of Great Britain, would be necessary only naval and air operations, without the landing of troops or artillery. The Soviet military, naval, and air forces would take part in such an operation.

> We would welcome it if Great Britain could transfer to this theatre of war something like one light division or more of the Norwegian volunteers, who could be used in Northern Norway to organise rebellion against the Germans.

Thus the Russian pressure for the establishment of a Second Front was initiated at the very beginning of our correspondence, and this theme was to recur throughout our subsequent relations with monotonous disregard, except in the Far North, for physical facts. This, my first telegram from Stalin, contained the only sign of compunction I ever perceived in the Soviet attitude. In this he volunteered a defence of the Soviet change of side and of his agreement with Hitler before the outbreak of the war, and dwelt, as I have already done, on the Russians' strategic need to hold a German deployment as far as possible to the west in Poland in order to gain time for the fullest development of Russian far-drawn military strength. I have never underrated this argument, and could well afford to reply in comprehending terms upon it.

From the first moment I did my utmost to help with munitions and supplies, both by consenting to severe diversions from the United States and by direct British sacrifices. Early in September the equivalent of two Hurricane squadrons were dispatched in H.M.S. *Argus* to Murmansk, to assist in the defence of the naval base and co-operate with Russian forces in that area. By September 11 the squadrons were in action, and they fought valiantly for three months. I was well aware that in the early days of our alliance there was little we could do, and I tried to fill the void by civilities, and to build up by frequent personal telegrams the same kind of happy relations which I had developed with the President. In this long Moscow series I received many rebuffs and only rarely a kind word. In many cases the telegrams were left unanswered altogether or for many days.

The Soviet Government had the impression that they were conferring a great favour on us by fighting in their own country for their own lives. The more they fought the heavier our debt became. This was not a balanced view. Two or three times in this long correspondence I had to protest in blunt language, but especially against the ill-usage of our sailors, who carried at so much peril the supplies to Murmansk and Archangel. Almost invariably however I bore hectoring and reproaches with "a patient shrug; for sufferance is the badge" of all who have to deal with the Kremlin. Moreover, I made constant allowances for the pressures under which Stalin and his dauntless Russian nation lay.

*   *   *

It will not be possible in this account to do more than place before the reader the salient features of the new colossal struggle of armies and populations which now began. In the first month the Germans bit and tore their way three hundred miles into Russia, but at the end of July there arose a fundamental clash of opinion between Hitler and Brauchitsch, the Commander-in-Chief. Brauchitsch held that Timoshenko's Army Group, which lay in front of Moscow, constituted the main Russian strength and must first be defeated. This was orthodox doctrine. Thereafter, Brauchitsch contended, Moscow, the main military, political, and industrial nerve centre of all Russia, should be taken. Hitler forcefully disagreed. He wished to gain territory and destroy Russian armies on the broadest front. In the north he demanded the capture of Leningrad, and in the south of the industrial Donetz Basin, the Crimea, and the entry to Russia's Caucasian oil supplies. Meanwhile Moscow could wait.

After vehement discussion Hitler overruled his Army chiefs. The Northern Army Group, reinforced from the centre, was ordered to press operations against Leningrad. The Central Army Group was relegated to the defensive. They were directed to send a Panzer group southward to take in flank the Russians who were being pursued across the Dnieper by Rundstedt. In this action the Germans prospered. By early September a vast pocket of Russian forces was forming around Kiev, and over half a million men were killed or captured in the desperate fighting which lasted all that month. In the north no such success could be claimed. Leningrad was encircled but not taken. Hitler's decision had not been right. He now turned his mind and will-power back to the centre. The besiegers of Leningrad were ordered to detach mobile forces and part of their supporting air force to reinforce a renewed drive on Moscow. The Panzer group which had been sent south to von Rundstedt came back again to join in the assault. At the end of September the stage was reset for the formerly discarded central thrust, while the southern armies drove on eastward to the lower Don, whence the Caucasus would lie open to them.

But by now there was another side to the tale. Despite their fearful losses Russian resistance remained tough and unbending. Their soldiers fought to the death, and their armies gained in experience and skill. Partisans rose up behind the German fronts and harassed the communications in a merciless warfare. The captured Russian railway system was proving inadequate; the roads were breaking up under the heavy traffic, and movement off the roads after rain was often impossible. Transport vehicles were showing many signs of wear. Barely two months remained before the dreaded Russian winter. Could Moscow be taken in that time? And if it were, would that be enough? Here then was the fateful question. Though Hitler was still elated by the victory at Kiev, the German generals might well feel that their early misgivings were justified. There had been four weeks of delay on what had now become the decisive front. The task of "annihilating the forces of the enemy in White Russia" which had been given to the Central Army Group was still not done.

But as the autumn drew on and the supreme crisis on the Russian front impended the Soviet demands upon us became more insistent.

*   *   *

Lord Beaverbrook returned from the United States having stimulated the already powerful forces making for a stupendous increase in production. He now became the champion in the War Cabinet of Aid to Russia. In this he rendered valuable service. When we remember the pressures that lay upon us to prepare the battle in the Libyan desert, and the deep anxieties about Japan which brooded over all our affairs in Malaya and the Far East, and that everything sent to Russia was subtracted from British vital needs, it was necessary that the Russian claims should be so vehemently championed at the summit of our war thought. I tried to keep the main proportion evenly presented in my own mind, and shared my stresses with my colleagues. We endured the unpleasant process of exposing our own vital security and projects to failure for the sake of our new ally — surly, snarly, grasping, and so lately indifferent to our survival.

I felt that when Beaverbrook and Averell Harriman got back from Washington and we could survey all the prospects of munitions and supplies they should go to Moscow and offer all we could spare and dare. Prolonged and painful discussions took place. The Service departments felt it was like flaying off pieces of their skin. However, we gathered together the utmost in our power, and consented to very large American diversions of all we longed for ourselves in order to make an effective contribution to the resistance of the Soviets. I brought the proposal to send Lord Beaverbrook to Moscow before my colleagues on August 28. The Cabinet were very willing that he should present the case to Stalin, and the President felt himself well represented by Harriman.

As a preliminary to this mission I outlined the position in general terms in a letter to Stalin, and on the evening of September 4, M. Maisky called to see me to deliver his reply. This was the first personal message since July. After thanking us for offering him another two hundred fighter aircraft, he came down to brass tacks.

> . . . The relative stabilisation at the front which we succeeded in achieving about three week ago [he cabled] has broken down during the last week, owing to transfer to Eastern Front of thirty to thirty-

four fresh German infantry divisions and of an enormous quantity of tanks and aircraft, as well as a large increase in activities of the twenty Finnish and twenty-six Rumanian divisions. Germans consider danger in the West a bluff, and are transferring all their forces to the East with impunity, being convinced that no second front exists in the West, and that none will exist. Germans consider it quite possible to smash their enemies singly: first Russia, then the English.

As a result we have lost more than one-half of the Ukraine, and in addition the enemy is at the gates of Leningrad. . . .

I think there is only one means of egress from this situation — to establish in the present year a second front somewhere in the Balkans or France, capable of drawing away from the Eastern Front 30 to 40 divisions, and at the same time of ensuring to the Soviet Union 30,000 tons of aluminium by the beginning of October next and a *monthly* minimum of aid amounting to 400 aircraft and 500 tanks (of small or medium size). . . .

The Soviet Ambassador, who was accompanied by Mr. Eden, stayed and talked with me for an hour and a half. He emphasised in' bitter terms how for the last eleven weeks Russia had been bearing the brunt of the German onslaught virtually alone. The Russian armies were now enduring a weight of attack never equalled before. He said that he did not wish to use dramatic language, but this might be a turning point in history. If Soviet Russia were defeated how could we win the war? M. Maisky emphasised the extreme gravity of the crisis on the Russian front in poignant terms which commanded my sympathy. But when presently I sensed an underlying air of menace in his appeal I was angered. I said to the Ambassador, whom I had known for many years, "Remember that only four months ago we in this island did not know whether you were not coming in against us on the German side. Indeed, we thought it quite likely that you would. Even then we felt sure we should win in the end. We never thought our survival was dependent on your action either way. Whatever happens, and whatever you do, you of all people have no right to make reproaches to us." As I warmed to the topic the Ambassador exclaimed, "More calm, please, my dear Mr. Churchill," but thereafter his tone perceptibly changed.

The discussion went over the ground already covered in the interchange of telegrams. The Ambassador pleaded for an immediate landing on the coast of France or the Low Countries. I explained the military reasons which rendered this impossible, and that it could be no relief to Russia. I said that I had spent five hours that day examining with our experts the means for greatly increasing the capacity of the

Trans-Persian railway. I spoke of the Beaverbrook-Harriman Mission and of our resolve to give all the supplies we could spare or carry. Finally Mr. Eden and I told him that we should be ready for our part to make it plain to the Finns that we would declare war upon them if they advanced into Russia beyond their 1918 frontiers. M. Maisky could not of course abandon his appeal for an immediate second front, and it was useless to argue further.

I at once consulted the Cabinet upon the issues raised in this conversation and in Stalin's message, and that evening sent a reply, of which the following paragraphs are pertinent:

> Although [I wrote] we should shrink from no exertion, there is in fact no possibility of any British action in the West, except air action, which would draw the German forces from the East before the winter sets in. There is no chance whatever of a second front being formed in the Balkans without the help of Turkey. I will, if your Excellency desires, gives all the reasons which have led our Chiefs of Staff to these conclusions. They have already been discussed with your Ambassador in conference today with the Foreign Secretary and the Chiefs of Staff. Action, however well meant, leading only to costly fiascos would be no help to anyone but Hitler. . . .
>
> We are ready to make joint plans with you now. Whether British armies will be strong enough to invade the mainland of Europe during 1942 must depend on unforeseeable events. It may be possible however to assist you in the extreme North when there is more darkness. We are hoping to raise our armies in the Middle East to a strength of three-quarters of a million before the end of the present year, and thereafter to a million by the summer of 1942. Once the German-Italian forces in Libya have been destroyed all these forces will be available to come into line on your southern flank, and it is hoped to encourage Turkey to maintain at the least a faithful neutrality. Meanwhile we shall continue to batter Germany from the air with increasing severity and to keep the seas open and ourselves alive. . . .

I thought the whole matter so important that I sent simultaneously the following telegram to the President while the impression was fresh in my mind:

> The Soviet Ambassador . . . used language of vague import about the gravity of the occasion and the turning-point character which would attach to our reply. Although nothing in his language war-

ranted the assumption, we could not exclude the impression that they might be thinking of separate terms. . . . I feel that the moment may be decisive. We can but do our best.

On September 15 I received another telegram from Stalin:

> I have no doubt that the British Government desires to see the Soviet Union victorious and is looking for ways and means to attain this end. If, as they think, the establishment of a second front in the West is at present impossible, perhaps another method could be found to render the Soviet Union an active military help?
>
> It seems to me that Great Britain could without risk land in Archangel twenty-five to thirty divisions, or transport them across Iran [Persia] to the southern regions of the U.S.S.R. In this way there could be established military collaboration between the Soviet and British troops on the territory of the U.S.S.R. A similar situation existed during the last war in France. The arrangement mentioned would constitute a great help. It would be a serious blow against the Hitler aggression. . . .

It is almost incredible that the head of the Russian Government with all the advice of their military experts could have committed himself to such absurdities. It seemed hopeless to argue with a man thinking in terms of utter unreality, and I sent the best answer I could.

*   *   *

Meanwhile the Beaverbrook-Harriman talks in London were completed, and on September 22 the Anglo-American Supply Mission set off in the cruiser *London* from Scapa Flow through the Arctic Sea to Archangel, and thence by air to Moscow. Much depended on them. Their reception was bleak and discussions not at all friendly. It might almost have been thought that the plight in which the Soviets now found themselves was our fault. The Soviet generals and officials gave no information of any kind to their British and American colleagues. They did not even inform them of the basis on which Russian needs of our precious war materials had been estimated. The Mission was given no formal entertainment until almost the last night, when they were invited to dinner at the Kremlin. It must not be thought that such an occasion among men preoccupied with the gravest affairs may not be helpful to the progress of business. On the contrary, many of the private interchanges which occur bring about that atmosphere where agreements can be reached. But there was little of this mood now, and it might almost have been we who had come to ask for favours.

One incident preserved by General Ismay in an apocryphal and somewhat lively form may be allowed to lighten the narrative. His orderly, a Royal Marine, was shown the sights of Moscow by one of the Intourist guides. "This," said the Russian, "is the Eden Hotel, formerly Ribbentrop Hotel. Here is Churchill Street, formerly Hitler Street. Here is the Beaverbrook railway station, formerly Goering railway station. Will you have a cigarette, comrade?" The Marine replied, "Thank you, comrade, formerly bastard!" This tale, though jocular, illustrates none the less the strange atmosphere of these meetings.

In the end a friendly agreement was reached. A protocol was signed setting out the supplies which Great Britain and the United States could make available to Russia within the period October 1941 to June 1942. This involved much derangement of our military plans, already hampered by the tormenting shortage of munitions. All fell upon us, because we not only gave our own production, but had to forgo most important munitions which the Americans would otherwise have sent to us. Neither the Americans nor ourselves made any promise about the transportation of these supplies across the difficult and perilous ocean and Arctic routes. In view of the insulting reproaches which Stalin uttered when we suggested that the convoys should not sail till the ice had receded, it should be noted that all we guaranteed was that the supplies would "be made available at British and United States centres of production." The preamble of the protocol ended with the words, "Great Britain and the United States will give aid to the transportation of these materials to the Soviet Union, and will help with the delivery."

"The effect of this agreement," Lord Beaverbrook telegraphed to me, "has been an immense strengthening of the morale of Moscow. The maintenance of this morale will depend on delivery. . . .

"I do not regard the military situation here as safe for the winter months. I do think that morale might make it safe."

Although General Ismay was fully empowered and qualified to discuss and explain the military situation in all its variants to the Soviet leaders, Beaverbrook and Harriman decided not to complicate their task by issues on which there could be no agreement. This aspect was not therefore dealt with in Moscow. Informally the Russians continued to demand the immediate establishment of the Second Front, and seemed quite impervious to any arguments showing its impossibility. Their agony is their excuse. Our Ambassador had to bear the brunt.

It was already late autumn. On October 2 the Central Army Group of von Bock renewed its advance on Moscow, with its two armies moving direct on the capital from the southwest and a Panzer group swinging wide on either flank. Orel on Ocober 8 and a week later

Kalinin on the Moscow-Leningrad road were taken. With his flanks thus endangered and under strong pressure from the central German advance, Marshal Timoshenko withdrew his forces to a line forty miles west of Moscow, where he again stood to fight. The Russian position at this moment was grave in the extreme. The Soviet Government, the Diplomatic Corps, and all industry that could be removed were evacuated from the city over five hundred miles farther east to Kuibyshev. On October 19 Stalin proclaimed a state of siege in the capital and issued an Order of the Day: "Moscow will be defended to the last." His commands were faithfully obeyed. Although Guderian's armoured group from Orel advanced as far as Tula, although Moscow was now three parts surrounded and there was some air bombardment, the end of October brought a marked stiffening in Russian resistance and a definite check to the German advance.

*　　　*　　　*

My wife felt very deeply that our inability to give Russia any military help disturbed and distressed the nation increasingly as the months went by and the German armies surged across the steppes. I told her that a Second Front was out of the question and that all that could be done for a long time would be the sending of supplies of all kinds on a large scale. Mr. Eden and I encouraged her to explore the possibility of obtaining funds by voluntary subscription for medical aid. This had already been begun by the British Red Cross and the Order of St. John, and my wife was invited by the joint organisation to head the appeal for "Aid to Russia." At the end of October, under their auspices, she issued her first appeal. A generous response was at once forthcoming. For the next four years she devoted herself to this task with enthusiasm and responsibility. In all nearly eight million pounds were collected by the contributions of rich and poor alike. Many wealthy people made munificent donations, but the bulk of the money came from the weekly subscriptions of the mass of the nation. Thus through the powerful organisation of the Red Cross and St. John and in spite of heavy losses in Arctic convoys medical and surgical supplies and all kinds of comforts and special appliances found their way in unbroken flow through the icy and deadly seas to the valiant Russian armies and people.

2

My Meeting with Roosevelt

IN THE MEANTIME a great deal had happened in the English-speaking world. In the middle of July Mr. Harry Hopkins arrived in Britain on his second mission from the President. The first topic which he opened to me was the new situation created by Hitler's invasion of Russia and its reaction upon all the Lend-Lease supplies we were counting on from the United States. Secondly, an American general, after being given the fullest facilities for inspection, had made a report throwing doubt upon our ability to withstand an invasion. This had caused the President anxiety. Thirdly, and in consequence, the President's misgivings about the wisdom of our trying to defend Egypt and the Middle East had been deepened. Might we not lose all through trying to do too much? Finally, there was the question of arranging a meeting between me and Roosevelt somehow, somewhere, soon.

This time Hopkins was not alone. There were in London a number of high United States officers of the Army and Navy, ostensibly concerned with Lend-Lease, and in particular Admiral Ghormley, who was working daily with the Admiralty on the Atlantic problem and the American share in its solution. I held a meeting with Hopkins's circle and the Chiefs of Staff on the night of July 24 at No. 10. Hopkins brought with him, besides Admiral Ghormley, Major General Chaney, who was called a "special observer," and Brigadier General Lee, the American Military Attaché. Averell Harriman, who had just returned from his tour in Egypt, in which by my directions he had been shown everything, completed the party.

Hopkins said that the "men in the United States who held the principal places and took decisions on defence matters" were of opinion that the Middle East was an indefensible position for the British Empire, and that great sacrifices were being made to maintain it. In their view the Battle of the Atlantic would be the final decisive battle of the war, and everything should be concentrated on it. The President, he said, was more inclined to support the struggle in the Middle East, because the enemy must be fought wherever he was found. General Chaney then placed the four problems of the British Empire in the following order: the defence of the United Kingdom and the Atlantic sea lanes; the defence of Singapore and the sea lanes to Australia and New Zealand; the defence of the ocean routes in general; and, fourth, the defence of the Middle East. All were important, but he placed them in that order. General Lee agreed with General Chaney. Admiral Ghormley was anxious about the supply line to the Middle East if American munitions were to go there in great volume. Might this not weaken the Atlantic battle?

I then asked the British Chiefs of Staff to express their views. The First Sea Lord explained why he felt even more confident of destroying an invading army this year than last. The Chief of the Air Staff showed how much stronger was the Royal Air Force compared with the German than in the previous September, and spoke of our newly increased power to batter the invasion ports. The Chief of the Imperial General Staff also spoke in a reassuring sense, and said that the Army was immeasurably stronger now than in the previous September. I interposed to explain the special measures we had taken for the defence of aerodromes after the lessons of Crete. I invited our visitors to visit any airfield in which they were interested. "The enemy may use gas, but if so it will be to his own disadvantage, since we have arranged for immediate retaliation and would have admirable concentrated targets in any lodgments he might make on the coast. Gas warfare would also be carried home to his own country." I then asked Dill to speak about the Middle East, and he gave a powerful exposition of some of the reasons which made it necessary for us to stay there.

My feeling at the end of our discussion was that our American friends were convinced by our statements and impressed by the solidarity among us.

Nevertheless the confidence which we felt about Home Defence did not extend to the Far East should Japan make war upon us. These anxieties also disturbed Sir John Dill. I retained the impression that Singapore had priority in his mind over Cairo. This was indeed a tragic issue, like having to choose whether your son or your daughter should be killed. For my part I did not believe that anything that might happen in Malaya could amount to a fifth part of the loss of Egypt, the Suez Canal, and the Middle East. I would not tolerate the idea of abandoning the struggle for Egypt, and was resigned to pay whatever forfeits were exacted in Malaya. This view also was shared by my colleagues.

* * *

One afternoon Harry Hopkins came into the garden of Downing Street and we sat together in the sunshine. Presently he said that the President would like very much to have a meeting with me in some lonely bay or other. I replied at once that I was sure the Cabinet would give me leave. Thus all was soon arranged. Placentia Bay, in Newfoundland, was chosen, the date of August 9 was fixed, and our latest battleship, the *Prince of Wales,* was placed under orders accordingly. I had the keenest desire to meet Mr. Roosevelt, with whom I had now corresponded with increasing intimacy for nearly two years. Moreover, a conference between us would proclaim the ever closer association of Britain and the United States, would cause our enemies concern, make Japan ponder, and cheer our friends. There was also much business to be settled about American intervention in the Atlantic, aid to Russia, our own supplies, and above all the increasing menace of Japan.

I took with me Sir Alexander Cadogan, of the Foreign Office, Lord Cherwell, Colonels Hollis and Jacob, of the Defence Office, and my personal staff. In addition there were a number of high officers of the technical and administrative branches and the Plans Division. The President said he would bring the chiefs of the United States fighting services with him, and Mr. Sumner Welles of the State Department. The utmost secrecy was necessary because of the large numbers of U-boats then in the North Atlantic, so the President, who was ostensibly on a holiday cruise, transhipped at sea to the cruiser *Augusta,* and left his yacht behind him as a blind. Meanwhile Harry Hopkins, though far from well, obtained Roosevelt's authority to fly to Moscow, a long, tiring, and dangerous journey, by Norway, Sweden, and Finland, in order to obtain directly from Stalin the fullest knowledge of the Soviet position and needs. He was to join the *Prince of Wales* at Scapa Flow.

The long special train which carried our whole company, including a large ciphering staff, picked me up at the station near Chequers. We boarded the *Prince of Wales* from a destroyer at Scapa. Before darkness fell on August 4 the *Prince of Wales* with her escort of destroyers steamed out into the broad waters of the Atlantic. I found Harry Hopkins much exhausted by his long air journeys and exacting conferences in Moscow. Indeed, he had arrived at Scapa two days before in such a condition that the Admiral had put him to bed at once and kept him there. Nevertheless he was as gay as ever, gathered strength slowly during the voyage, and told me all about his mission.

The spacious quarters over the propellers, which are most comfortable in harbour, become almost uninhabitable through vibration in heavy weather at sea, so I moved to the Admiral's sea-cabin on the bridge for working and sleeping. I took a great liking to our captain, Leach, a charming and lovable man and all that a British sailor should be. Alas! within four months he and many of his comrades and his splendid ship were sunk for ever beneath the waves. On the second day the seas were so heavy that we had to choose between slowing down and dropping our destroyer escort. Admiral Pound, First Sea Lord, gave the decision. Thenceforward we went on at high speed alone. There were several U-boats reported, which we made zigzags and wide diversions to avoid. Absolute wireless silence was sought. We could receive messages, but for a while we could not speak except at intervals. Thus there was a lull in my daily routine and a strange sense of leisure which I had not known since the war began. For the first time for many months I could read a book for pleasure. Oliver Lyttelton, Minister of State in Cairo, had given me *Captain Hornblower, R.N.*,[1] which I found vastly entertaining. When a chance came I sent him the message, "I find *Hornblower* admirable." This caused perturbation in the Middle East Headquarters, where it was imagined that "Hornblower" was the code word for some special operation of which they had not been told.

We arrived at our rendezvous at 9 A.M. on Saturday, August 9, and as soon as the customary naval courtesies had been exchanged I went aboard the *Augusta* and greeted President Roosevelt, who received me with all honours. He stood supported by the arm of his son Elliott while the National Anthems were played, and then gave me the warmest of welcomes. I gave him a letter from the King and presented the members of my party. Conversations were then begun between the President and myself, Mr. Summer Welles and Sir Alexander Cadogan, and the Staff officers on both sides, which proceeded more or less continuously for the remaining days of our visit, sometimes man to man and sometimes in larger conferences.

On Sunday morning, August 10, Mr. Roosevelt came aboard H.M.S. *Prince of Wales* and, with his Staff officers and several hundred representatives of all ranks of the United States Navy and Marines, attended divine service on the quarterdeck. This service was felt by us all to be a deeply moving expression of the unity of faith of our two peoples, and none who took part in it will forget the spectacle presented that sunlit morning on the crowded quarterdeck — the symbolism of the Union Jack and the Stars and Stripes draped side by side on the pulpit; the American and British chaplains sharing in the reading of the

[1] A novel by C. S. Forester.

prayers; the highest naval, military, and air officers of Britain and the United States grouped in one body behind the President and me; the close-packed ranks of British and American sailors, completely intermingled, sharing the same books and joining fervently together in the prayers and hymns familiar to both.

I chose the hymns myself — "For Those in Peril on the Sea" and "Onward, Christian Soldiers." We ended with "O God, Our Help in Ages Past." Every word seemed to stir the heart. It was a great hour to live. Nearly half those who sang were soon to die.

* * *

President Roosevelt told me at one of our first conversations that he thought it would be well if we could draw up a joint declaration laying down certain broad principles which should guide our policies along the same road. Wishing to follow up this most helpful suggestion, I gave him on this same Sunday a tentative outline of such a declaration, and after much discussion between ourselves and telegraphic debate with the War Cabinet in London, we produced the following document:

JOINT DECLARATION BY THE PRESIDENT AND
THE PRIME MINISTER
August 12, 1941

The President of the United States of America and the Prime Minister, Mr. Churchill, representing His Majesty's Government in the United Kingdom, being met together, deem it right to make known certain common principles in the national policies of their respective countries on which they base their hopes for a better future for the world.

First, their countries seek no aggrandisement, territorial or other.

Second, they desire to see no territorial changes that do not accord with the freely expressed wishes of the peoples concerned.

Third, they respect the right of all peoples to choose the form of government under which they will live; and they wish to see sovereign rights and self-government restored to those who have been forcibly deprived of them.

Fourth, they will endeavour, with due respect to their existing obligations, to further the enjoyment by all States, great or small, victor or vanquished, of access, on equal terms, to the trade and to the raw materials of the world which are needed for their economic prosperity.

Fifth, they desire to bring about the fullest collaboration between all nations in the economic field, with the object of securing for all improved labour standards, economic advancement, and social security.

Sixth, *after the final destruction of the Nazi tyranny* [2] they hope to see established a peace which will afford to all nations the means of dwelling in safety within their own boundaries, and which will afford assurance that all the men in all the lands may live out their lives in freedom from fear and want.

Seventh, such a peace should enable all men to traverse the high seas and oceans without hindrance.

Eighth, they believe that all the nations of the world, for realistic as well as spiritual reasons, must come to the abandonment of the use of force. Since no future peace can be maintained if land, sea, or air armaments continue to be employed by nations which threaten, or may threaten, aggression outside of their frontiers, they believe, pending the establishment of a wider and more permanent system of general security, that the disarmament of such nations is essential. They will likewise aid and encourage all other practicable measures which will lighten for peace-loving peoples the crushing burden of armaments.

The profound and far-reaching importance of what came to be called the "Atlantic Charter" was apparent. The fact alone of the United States, still technically neutral, joining with a belligerent Power in making such a declaration was astonishing. The inclusion in it of a reference to "the final destruction of the Nazi tyranny" (this was based on a phrase appearing in my original draft) amounted to a challenge which in ordinary times would have implied warlike action. Finally, not the least striking feature was the realism of the last paragraph, where there was a plain and bold intimation that after the war the United States would join with us in policing the world until the establishment of a better order.

Continuous conferences also took place between the naval and military chiefs and a wide measure of agreement was reached between them. The menace from the Far East was much in our minds. For several months the British and American Governments had been acting towards Japan in close accord. At the end of July the Japanese had completed their military occupation of Indo-China. By this naked act of aggression their forces were poised to strike at the British in Malaya, at the Americans in the Philippines, and at the Dutch in the East Indies. On July 24 the President asked the Japanese Government that, as a prelude to a general settlement, Indo-China should be neutralised and the Japanese troops withdrawn. To add point to these

[2] My subsequent italics.—W.S.C.

proposals an executive order was issued freezing all Japanese assets in the United States. This brought all trade to a standstill. The British Government took simultaneous action, and two days later the Dutch followed. The adherence of the Dutch meant that Japan was deprived at a stroke of her vital oil supplies.

*   *   *

The return voyage to Iceland was uneventful, although at one point it became necessary to alter course owing to the reported presence of U-boats nearby. Our escort included two United States destroyers, in one of which was Ensign Franklin D. Roosevelt, Jr., the President's son. On the 15th we met a combined homeward-bound convoy of seventy-three ships, all in good order and perfect station after a fortunate passage across the Atlantic. It was a heartening sight, and the merchant ships too were glad to look at the *Prince of Wales*.

We reached the island on Saturday morning, August 16, and anchored at Hvals Fiord, from which we travelled to Reykjavik in a destroyer. On arrival at the port I received a remarkably warm and vociferous welcome from a large crowd, whose friendly greetings were repeated whenever our presence was recognised during our stay, culminating in scenes of great enthusiasm on our departure in the afternoon, to the accompaniment of such cheers and handclapping as have, I was assured, seldom been heard in the streets of Reykjavik.

After a short visit to the Althingishus, to pay respects to the Regent and the members of the Icelandic Cabinet, I proceeded to a joint review of the British and American forces. There was a long march past in threes, during which the tune "United States Marines" bit so deeply into my memory that I could not get it out of my head. I found time to see the new airfields we were making, and also to visit the wonderful hot springs and the glasshouses they are made to serve. I thought immediately that they should also be used to heat Reykjavik and tried to further this plan even during the war. I am glad that it has now been carried out. I took the salute with the President's son standing beside me, and the parade provided another remarkable demonstration of Anglo-American solidarity.

On return to Hvals Fiord I visited the *Ramillies,* and addressed representatives of the crews of the British and American ships in the anchorage, including the destroyers *Hecla* and *Churchill.* As darkness fell after this long and very tiring ordeal we sailed for Scapa where we arrived without further incident early on the 18th, and I reached London on the following day.

3

Persia and the Desert

THE NEED TO PASS munitions and supplies of all kinds to the Soviet Government, and the extreme difficulties of the Arctic route, together with future strategic possibilities, made it eminently desirable to open the fullest communication with Russia through Persia. I was not without some anxiety about embarking on yet another new campaign in the Middle East, but the arguments were compulsive. The Persian oilfields were a prime war factor, and if Russia were defeated we would have to be ready to occupy them ourselves. And then there was the threat to India. The suppression of the revolt in Iraq and the Anglo-French occupation of Syria, achieved as they were by narrow margins, had blotted out Hitler's Oriental plan, but if the Russians foundered he might try again. An active and numerous German mission had installed itself in Teheran, and German prestige stood high. On the eve of my voyage to Placentia I had set up a special committee to co-ordinate the planning of an operation against Persia, and during my absence at sea they reported to me by telegram the results of their work, which had meanwhile been approved by the War Cabinet. It was clear that the Persians would not expel the German agents and residents from their country, and that we should have to resort to force. On August 13 Mr. Eden received M. Maisky at the Foreign Office, and the terms of our respective Notes to Teheran were agreed. A joint Anglo-Soviet Note of August 17 met with an unsatisfactory reply, and the date for the entry of British and Russian forces into Persia was fixed for the 25th.

In four days it was all over. The Abadan refinery was captured by an infantry brigade, which embarked at Basra and landed at dawn on August 25. The majority of the Persian forces were surprised but escaped in lorries. Some street fighting took place and a few Persian naval craft were seized. At the same time we captured the port of Khurramshahr from the landward side, and a force was sent north towards Ahwaz. As our troops were approaching Ahwaz news of the Shah's "Cease fire" order was received, and the Persian general ordered his troops back to barracks. In the north the oilfields were easily secured. Our casualties were 22 killed and 42 wounded.

All arrangements with the Russians were smoothly and swiftly agreed. The principal conditions imposed on the Persian Government were the cessation of all resistance, the ejection of Germans, neutrality in the war, and the Allied use of Persian communications for the transit of war supplies to Russia. The further occupation of Persia was peacefully accomplished. British and Russian forces met in amity, and Teheran was jointly occupied on September 17, the Shah having abdicated on the previous day in favour of his gifted twenty-two-year-old son. On September 20 the new Shah, under Allied advice, restored the Constitutional Monarchy, and his father shortly afterwards went into comfortable exile and died at Johannesburg in July 1944. Most of our forces withdrew from the country, leaving only detachments to guard the communications, and Teheran was evacuated by both British and Russian troops on October 18. Thereafter our forces, under General Quinan, were engaged in preparing defences against the possible incursion of German armies from Turkey or the Caucasus, and in making administrative preparations for the large reinforcements which would arrive if that incursion seemed imminent.

The creation of a major supply route to Russia through the Persian Gulf became our prime objective. With a friendly Government in Teheran ports were enlarged, river communications developed, roads built, and railways reconstructed. Starting in September 1941, this enterprise, begun and developed by the British Army, and presently to be adopted and expanded by the United States, enabled us to send to Russia, over a period of four and a half years, five million tons of supplies. Thus ended a brief and fruitful exercise of overwhelming force against a weak and ancient state. Britain and Russia were fighting for their lives. *Inter arma silent leges.* We may be glad that in our victory the independence of Persia has been preserved.

* * *

We must now return to the dominant theatre of the Mediterranean. General Auchinleck had assumed formal command of the Middle East on July 5, and I started my relations with our new Commander-in-Chief in high hopes, but an exchange of telegrams soon made it clear that there were serious divergences of views and values between us. He proposed to reinforce Cyprus as soon as possible by one division, he appreciated the need for regaining Cyrenaica, but he could not be confident that Tobruk could be held after September. He said that the features and armament of the new American tanks introduced modifications in tactical handling, and time must be allowed for these lessons to be studied. He agreed that by the end of July he would have about five hundred cruiser, Infantry, and American tanks. For any operation however fifty per cent reserves of tanks were required, thus permitting twenty-five per cent in the workshops and twenty-five per cent for immediate replacement of battle casualties. This was an almost prohibitive condition. Generals only enjoy such comforts in Heaven. And those who demand them do not always get there. Auchinleck stressed the importance of time both for individual and collective training, and the team spirit, which was essential for efficiency. He thought that the North (*i.e.*, a German attack through Turkey, Syria, and Palestine) might become the decisive front rather than the Desert.

All this caused me sharp disapopintment. The General's early decisions were also perplexing. By long persistence I had at last succeeded in having the 50th British Division brought to Egypt. I was sensitive to the hostile propaganda which asserted that it was the British policy to fight with any other troops but our own and thus avoid the shedding of United Kingdom blood. British casualties in the Middle East, including Greece and Crete, had in fact been greater than those of all our other forces put together, but the nomenclature which was customary gave false impression of the facts. The Indian divisions, of which one-third of the infantry and the whole of the artillery were British, were not described as British-Indian divisions. The armoured divisions, which had borne the brunt of the fighting, were entirely British, but this did not appear in their names. The fact that "British" troops were rarely mentioned in any reports of the fighting gave colour to the enemy's taunts, and provoked unfavourable comment not only in the United States but in Australia. I had looked forward to the arrival of the 50th Division as an effective means of countering these disparaging currents. General Auchinleck's decision to pick this as the division to send to Cyprus certainly seemed unfortunate, and lent substance to the reproaches to which we were unjustly subjected. The Chiefs of Staff at home were equally astonished on military grounds that so strange a use should be made of this magnificent body of men.

A far more serious resolve by General Auchinleck was to delay all action against Rommel in the Western Desert, at first for three and eventually for more than four and a half months. The vindication of Wavell's action of June 15, "Battleaxe," is found in the fact that although we were somewhat worsted and withdrew to our original position the Germans were utterly unable to advance for the whole of this prolonged period. Their communications, threatened by Tobruk, were insufficient to bring them the necessary reinforcements of armour or even of artillery ammunition to enable Rommel to do more than hold on by his will power and prestige. The feeding of his force imposed so heavy a strain upon him that its size could only grow gradually. In these circumstances he should have been engaged continuously by the British Army, which had ample road, rail, and sea communications, and was being continually strengthened at a much greater rate both in men and material.

A third misconception seemed to me to be a disproportionate concern for our northern flank. This indeed required the utmost vigilance and justified many defensive preparations and the construction of strong fortified lines in Palestine and Syria. The situation in this quarter however soon became vastly better than in June. Syria was conquered. The Iraq rebellion had been suppressed. All the key points in the desert were held by our troops. Above all, the struggle between Germany and Russia gave new confidence to Turkey. While this hung in the balance there was no chance of a German demand for the passage of her armies through Turkish territory. Persia was being brought into the Allied camp by British and Russian action. This would carry us beyond the winter. In the meanwhile the general situation favoured decisive action in the Western Desert.

Instead, I could not help feeling a stiffness in General Auchinleck's attitude, which would not be helpful to the interests we all served. Books written since the war have shown how subordinate but influential portions of the Cairo Operations Staff had deplored the decision to send the Army to Greece. They did not know how fully and willingly General Wavell had accepted this policy, still less how searchingly the War Cabinet and Chiefs of Staff had put the issue to him, almost inviting a negative. Wavell, it was suggested, had been led astray by the politicians, and the whole chain of disasters had followed on his compliance with their wishes. Now as a reward for his good nature he had been removed after all his victories in the moment of defeat. I cannot doubt that in these circles of the Staff there was a strong feeling that the new Commander should not let himself be pressed into hazardous adventures, but should take his time and work on certainties. Such a mood might well have been imparted to General Auchinleck. It was clear that not much progress would be made by correspondence, and in July I invited him to come to London.

His brief visit was from many points of view helpful. He placed himself in harmonious relations with members of the War Cabinet, with the Chiefs of Staff, and with the War Office. He spent a long week end with me at Chequers. As we got to know better this distinguished officer, upon whose qualities our fortunes were now so largely to depend, and as he became acquainted with the high circle of the British war machine and saw how easily and smoothly it worked, mutual confidence grew. On the other hand, we could not induce him to depart from his resolve to have a prolonged delay in order to prepare a set piece offensive on November 1. This was to be called "Crusader," and would be the largest operation we had yet launched. He certainly shook my military advisers with all the detailed argument he produced. I was myself unconvinced. But General Auchinleck's unquestioned abilities, his powers of exposition, his high, dignified, and commanding personality, gave me the feeling that he might after all be right, and that even if wrong he was still the best man. I therefore yielded to the November date for the offensive, and turned my energies to making it a success. We were all very sorry that we could not persuade him to entrust the battle, when it should come, to General Maitland Wilson. He preferred instead General Alan Cunningham, whose reputation stood high on the morrow of the Abyssinian victories. We had to make the best of it, and that is never worth doing by halves. Thus we shared his responsibility by endorsing his decisions. I must nevertheless record my conviction that General Auchinleck's four and a half months' delay in engaging the enemy in the Desert was alike a mistake and a misfortune.

We now have a very full knowledge of what the German High Command thought of Rommel's situation. They greatly admired his audacity and the incredible successes which had crowned it, but none the less they deemed him in great peril. They strictly forbade him to run any further risks until he could be strongly reinforced. Perhaps, with his prestige, he might bluff it out, in the precarious position in which he stood, until they could bring him the utmost aid in their power. His line of communications trailed back a thousand miles to Tripoli. Benghazi was a valuable short cut for a part at any rate of his supplies and fresh troops, but a toll of increasing severity had to be paid on the sea transport to both these bases. The British forces, already largely superior in numbers, were growing daily. The German tank superiority existed only in quality and organisation. They were weaker in the air. They were very short of artillery ammunition, and feared greatly to have to fire it off. Tobruk seemed a deadly threat in Rommel's rear, from which at any moment a sortie might be made, cutting his communications. However, while we remained motionless they could be thankful for every day that passed.

Both sides used the summer to reinforce their armies. For us the replenishment of Malta was vital. The loss of Crete deprived Admiral Cunningham's fleet of a fuelling base near enough to bring our protecting sea power into action. The possibilities of a sea-borne assault on Malta from Italy or Sicily grew, though, as we now know, it was not until 1942 that Hitler and Mussolini approved such a plan. Enemy air bases both in Crete and Cyrenaica menaced the convoy route from Alexandria to Malta so seriously that we had to depend entirely on the West for the passage of supplies. In this task Admiral Somerville, with Force H from Gibraltar, rendered distinguished service. The route the Admiralty had judged the more dangerous became the only one open. Fortunately at this time the demands of his Russian invasion compelled Hitler to withdraw his air force from Sicily, which gave a respite to Malta and restored to us the mastery in the air over the Malta Channel. This not only helped the approach of convoys from the West, but enabled us to strike harder at the transports and supply ships reinforcing Rommel.

Two considerable convoys were fought through successfully. The passage of each was a heavy naval operation. In October over 60 per cent of Rommel's supplies were sunk in passage. But my anxieties were not allayed, and I urged even greater efforts upon the Admiralty. I desired specially that a new surface force should be based upon Malta. The policy was accepted, though time was needed to bring it about. In October a striking force known as "Force K," comprising the cruisers *Aurora* and *Penelope* and the destroyers *Lance* and *Lively,* was formed at Malta. All these measures played their part in the struggle which was now to begin.

*   *   *

Descriptions of modern battles are apt to lose the sense of drama because they are spread over wide spaces and often take weeks to decide, whereas on the famous fields of history the fate of nations and empires was decided on a few square miles of ground in a few hours. The conflicts of fast-moving armoured and motorised forces in the desert present this contrast with the past in an extreme form.

Tanks had replaced the cavalry of former wars with a vastly more powerful and far-ranging weapon, and in many aspects their manoeuvres resembled naval warfare, with seas of sand instead of salt water. The fighting quality of the armoured column, like that of a cruiser squadron, rather than the position where they met the enemy, or the part of the horizon on which he appeared, was the decisive feature. Tank divisions or brigades, and still more smaller units, could form fronts in any direction so swiftly that the perils of being out-flanked or taken in rear or cut off had a greatly lessened significance. On the other hand, all depended from moment to moment upon fuel and ammunition, and the supply of both was far more complicated for armoured forces than for the self-contained ships and squadrons at sea. The principles on which the art of war is founded expressed themselves therefore in novel terms, and every encounter taught lessons of its own.

The magnitude of the war effort involved in these desert struggles must not be underrated. Although only about ninety or a hundred thousand fighting troops were engaged in each of the armies, these needed masses of men and material two or three times as large to sustain them in their trial of strength. The fierce clash of Sidi Rezegh, which marked the opening of General Auchinleck's offensive, when viewed as a whole, presents many of the most vivid features of war. The personal interventions of the two Commanders-in-Chief were as dominant and decisive and the stakes on both sides were as high as in the olden times.

Auchinleck's task was first to recapture Cyrenaica, destroying in the process the enemy's armour, and, secondly, if all went well, to capture Tripolitania. For these purposes General Cunningham was given command of the newly named Eighth Army, consisting of the XIIIth and XXXth Corps, and comprising, with the Tobruk garrison, about six divisions, with three brigades in reserve, and 724 tanks. The Western Desert Air Force totalled 1072 serviceable modern combat aircraft, in addition to ten squadrons operating from Malta. Seventy miles behind Rommel's front lay the garrison of Tobruk, comprising five brigade groups and an armoured brigade. This fortress was his constant preoccupation, and had hitherto prevented by its strategic threat any advance upon Egypt. To eliminate Tobruk was the settled

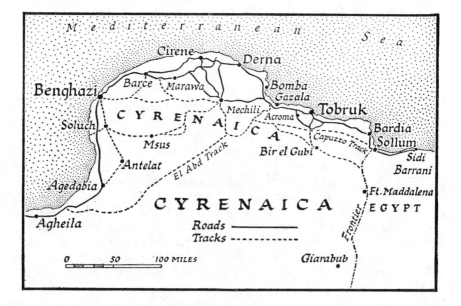

purpose of the German High Command, and all preparations possible had been made to begin the assault upon it on November 23. Rommel's army comprised the formidable Afrika Korps, consisting of the 15th and 21st Panzer Divisions and the 90th Light Division, and seven Italian divisions, of which one was armoured. The enemy had 558 tanks. Of the medium and heavy, two-thirds were German and carried heavier guns than the 2-pounders of our tanks. The enemy were moreover markedly superior in anti-tank weapons. The Axis Air Force consisted of 120 German and about 200 Italian serviceable aircraft at the moment of attack.

Early on November 18, in heavy rain, the Eighth Army leapt forward, and for three days all went well. Part of XXXth Corps' British 7th Armoured Division took Sidi Rezegh, but it was then attacked by the Afrika Korps, whose armour had been kept more concentrated. During the whole of the 21st and 22nd a savage struggle raged, mainly around and upon the airfield. Into this arena virtually all the armour on both sides was drawn, and surged to and fro in violent struggles under the fire of rival batteries. The stronger armament of the German tanks and the larger numbers they brought to the points of collision gave them the advantage. In spite of the heroic and brilliant leadership of Brigadier Jock Campbell the Germans prevailed, and we suffered more heavily than they in tanks. On the night of the 22nd the Germans recaptured Sidi Rezegh. Our force lost two-thirds of its armour, and was ordered to withdraw about twenty miles in order to reorganise. This was a heavy setback.

Meanwhile on November 21, the enemy armour being committed to battle, General Cunningham ordered the XIIIth Corps to advance. They captured the headquarters of the Afrika Korps, and on the 23rd nearly regained Sidi Rezegh, from which their comrades of the 7th Armoured Division had just been driven. On November 24 Freyberg concentrated the bulk of his New Zealanders five miles to the east of the airfield. A sortie from Tobruk had been launched, and was fighting hard against German infantry, but had not broken through. The New Zealand Division stood before Sidi Rezegh after a triumphant march The enemy frontier garrisons had been cut off, but their armour had won its battle against the XXXth Corps. Very heavy blows and severe losses had been exchanged, and the battle hung in the balance.

*   *   *

There was now a dramatic episode which recalls "Jeb" Stuart's ride round McClellan in 1862 on the Yorktown Peninsula in the American Civil War. It was however executed with an armoured force which was an army in itself, and whose destruction would have doomed the rest of the Axis army. Rommel resolved to seize the tactical initiative and to force his way eastward to the frontier with his armour in the hope of creating so much chaos and causing so much alarm as to prevail upon our command to give up the struggle and withdraw. He may well have had in his mind the fortune which had rewarded his armoured incursion in the preceding desert battle of June 15 and led to our retreat at the crucial moment. How nearly he succeeded this time will be apparent as the story proceeds.

He collected the greater part of the Afrika Korps, still the most formidable body in the field, and narrowly missing the headquarters of the XXXth Corps and two great dumps of supplies, without which we could not have continued the fight, he reached the frontier. Here he split his force into columns, some of which turned north and south, and others drove on twenty miles into Egyptian territory. He wrought havoc in our rearward areas and captured many prisoners. His columns however made no impression on the 4th Indian Division, and were pursued by hastily organised detachments. Above all our air force, which had now gained a high degree of mastery in the air above the contending armies, harried him all the time and all the way. Rommel's columns, virtually unsupported by their own Air, suffered the pangs our troops had known and endured when it was Germany who dominated the battle skies. On the 26th all the enemy's armour turned northward and sought haven in and near Bardia. Next day they hurried off to the west, back to Sidi Rezegh, whither they were urgently summoned. Rommel's daring stroke had failed, but, as will now be seen, only one man — the opposing Commander-in-Chief — stopped him.

The heavy blows we had received and the impression of disorder behind our front, caused by Rommel's raid, had led General Cunningham to represent to the Commander-in-Chief that a continuation of our offensive might result in the annihilation of our tank force, and so endanger the safety of Egypt. This would mean acknowledged defeat and failure of the whole operation. At this decisive moment General Auchinleck intervened personally. At Cunningham's request he flew with Air Marshal Tedder to the Desert Headquarters on November 23, and, with full knowledge of all the dangers, ordered General Cunningham "to continue to press the offensive against the enemy." By his personal action Auchinleck thus saved the battle and proved his outstanding qualities as a commander in the field.

On his return to Cairo on the 25th he decided to replace General Cunningham temporarily by General Ritchie, his Deputy Chief of Staff, "because I have reluctantly concluded that Cunningham, admirable as he has been up to date, has now begun to think defensively, mainly because of our large tank losses." The Minister of State, Oliver Lyttelton, explained and strongly supported the Commander-in-Chief's decision. To him I at once telegraphed our approval.

Here I shall leave this incident, so painful to the gallant officer concerned, to his brother the Naval Commander-in-Chief, and to General Auchinleck, who was a personal friend of both. I particularly admired General Auchinleck's conduct in rising superior to all personal considerations and to all temptations to compromise or delay action.

*   *   *

Meanwhile, Freyberg and his New Zealanders, supported by the 1st Army Tank Brigade, pressed hard upon Sidi Rezegh. After two days of severe fighting they recaptured it. Simultaneously the garrison of Tobruk resumed its sortie and on the night of the 26th joined hands with the relieving force. Some units entered beleaguered Tobruk. This brought Rommel back from Bardia. He fought his way to Sidi Rezegh, attacked in flank by the reorganised 7th Armoured Division, now mustering 120 tanks. He recaptured Sidi Rezegh, and drove back the New Zealand Brigade with crippling loss. Most of them were withdrawn southeastward to the frontier, where the heroic division reformed after losing more than three thousand men. The Tobruk garrison, again isolated, held on by a bold decision to all the ground gained.

General Ritchie now regrouped his army and Rommel made a final thrust to rescue his frontier garrisons. It was repulsed. The general retreat of the Axis army to the Gazala line then began.

On December 1 Auchinleck went himself to the Advanced Headquarters, and remained for ten days with General Ritchie. He did not assume the command himself, but closely supervised his subordinate. This did not seem to me the best arrangement for either of them. However, the power of the Eighth Army was now predominant, and on December 10 the Commander-in-Chief could tell me: "Enemy is apparently in full retreat towards the west . . . I think it now permissible to claim that the siege of Tobruk has been raised. We are pursuing vigorously in fullest co-operation with the Royal Air Force." We now know from German records that the enemy losses in the battle were about 33,000 men and 300 tanks. The comparable British and Imperial Army losses in the same period were about half, together with 278 tanks. Nine-tenths of this loss occurred in the first month of the offensive. Here then we reached a moment of relief, and indeed of rejoicing, about the Desert war.

But at this crucial moment our naval power in the Eastern Mediterranean was virtually destroyed by a series of disasters. Our interval of immunity and advantage came to its end. The U-boats arrived upon the scene. On November 12, while returning to Gibraltar after flying more aircraft into Malta, the *Ark Royal* had been struck by a torpedo from a German U-boat. All attempts to save the ship failed, and this famous veteran, which had played such a distinguished part in so many of our affairs, sank when only twenty-five miles from Gibraltar. A fortnight later the *Barham* was struck by three torpedoes and capsized in as many minutes with the loss of over 500 men. More was to follow. On the night of December 18 an Italian submarine approached Alexandria and launched three "human torpedoes," each controlled by two men. They penetrated the harbour while the boom gate was open for the passage of ships. They fixed time-bombs, which detonated early next morning under the battleships *Queen Elizabeth* and *Valiant*. Both ships were heavily injured and became a useless burden for months. We were successful in concealing the damage to the battle fleet for some time, but "Force K" was also stricken. On the very day of the Alexandria disaster news reached Malta of an important enemy convoy heading for Tripoli. Three cruisers and four destroyers at once went out to catch them. Approaching Tripoli our ships ran into a new minefield. Two of the cruisers were damaged, but were able to steam away. The third, drifting in the minefield, struck two more mines and sank. Only one man of her crew of over 700 survived — and he as a prisoner of war after four days on a raft, on which his captain, R. C. O'Connor, and thirteen others perished. All that remained of the British Eastern Mediterranean Fleet was a few destroyers and three cruisers of Admiral Vian's squadron.

On December 5 Hitler, realising at last Rommel's mortal peril, ordered the transfer of a whole air corps from Russia to Sicily and North Africa. A new air offensive against Malta was launched under General Kesselring's direction. The attacks on the island reached a new peak, and Malta could do no more than struggle for life. By the end of the year it was the Luftwaffe who held the mastery over the sea routes to Tripoli, and thus made possible the refit of Rommel's armies after their defeat. Seldom has the interaction of sea, air, and land warfare been so strikingly illustrated as in the events of these few months.

But now all paled under the stroke of world events.

4

Pearl Harbour!

I T WAS SUNDAY EVENING, December 7, 1941. Winant and Averell Harriman were alone with me at the table at Chequers. I turned on my small wireless set shortly after the nine o'clock news had started. There were a number of items about the fighting on the Russian front and on the British front in Libya, at the end of which some few sentences were spoken regarding an attack by the Japanese on American shipping at Hawaii, and also Japanese attacks on British vessels in the Dutch East Indies. There followed a statement that after the news Mr. Somebody would make a commentary, and that the Brains Trust programme would then begin, or something like this. I did not personally sustain any direct impression, but Averell said there was something about the Japanese attacking the Americans, and, in spite of being tired and resting, we all sat up. By now the butler, Sawyers, who had heard what had passed, came into the room, saying, "It's quite true. We heard it ourselves outside. The Japanese have attacked the Americans." There was a silence. At the Mansion House luncheon on November 11 I had said that if Japan attacked the United States a British declaration of war would follow "within the hour." I got up from the table and walked through the hall to the office, which was always at work. I asked for a call to the President. The Ambassador followed me out, and, imagining I was about to take some irrevocable step, said, "Don't you think you'd better get confirmation first?"

In two or three minutes Mr. Roosevelt came through. "Mr. President, what's this about Japan?" "It's quite true," he replied. "They have attacked us at Pearl Harbour. We are all in the same boat now." I put Winant onto the line and some interchanges took place, the Ambassador at first saying, "Good," "Good" — and then, apparently graver, "Ah!" I got on again and said, "This certainly simplifies things. God be with

505

you," or words to that effect. We then went back into the hall and tried to adjust our thoughts to the supreme world event which had occurred, which was of so startling a nature as to make even those who were near the centre gasp. My two American friends took the shock with admirable fortitude. We had no idea that any serious losses had been inflicted on the United States Navy. They did not wail or lament that their country was at war. They wasted no words in reproach or sorrow. In fact, one might almost have thought they had been delivered from a long pain.

*   *   *

Parliament would not have met till Tuesday, and the Members were scattered about the island, with all the existing difficulties of communication. I set the office to work to ring up the Speaker, the Whips, and others concerned, to call both Houses together next day. I rang the Foreign Office to prepare to implement without a moment's delay a declaration of war upon Japan, about which there were some formalities, in time for the meeting of the House, and to make sure all members of the War Cabinet were called up and informed, and also the Chiefs of Staff and the Service Ministers, who, I rightly assumed, had had the news.

No American will think it wrong of me if I proclaim that to have the United States at our side was to me the greatest joy. I could not foretell the course of events. I do not pretend to have measured accurately the martial might of Japan, but now at this very moment I knew the United States was in the war, up to the neck and in to the death. So we had won after all! Yes, after Dunkirk; after the fall of France; after the horrible episode of Oran; after the threat of invasion, when, apart from the Air and the Navy, we were an almost unarmed people; after the deadly struggle of the U-boat war — the first Battle of the Atlantic, gained by a hand's-breadth; after seventeen months of lonely fighting and nineteen months of my responsibility in dire stress. We had won the war. England would live; Britain would live; the Commonwealth of Nations and the Empire would live. How long the war would last or in what fashion it would end no man could tell, nor did I at this moment care. Once again in our long island history we should emerge, however mauled or mutilated, safe and victorious. We should not be wiped out. Our history would not come to an end. We might not even have to die as individuals. Hitler's fate was sealed. Mussolini's fate was sealed. As for the Japanese, they would be ground to powder. All the rest was merely the proper application of overwhelming force. The British Empire, the Soviet Union, and now the United States, bound together with every scrap of their life and strength, were, according to my lights, twice or even thrice the force of their antagonists. No doubt it would take a long time. I expected terrible forfeits in the East; but all this would be merely a passing phase. United we could subdue everybody else in the world. Many disasters, immeasurable cost and tribulation lay ahead, but there was no more doubt about the end.

Silly people, and there were many, not only in enemy countries, might discount the force of the United States. Some said they were soft, others that they would never be united. They would fool around at a distance. They would never come to grips. They would never stand blood-letting. Their democracy and system of recurrent elections would paralyse their war effort. They would be just a vague blur on the horizon to friend or foe. Now we should see the weakness of this numerous but remote, wealthy, and talkative people. But I had studied the American Civil War, fought out to the last desperate inch. American blood flowed in my veins. I thought of a remark which Edward Grey had made to me more than thirty years before — that the United States is like "a gigantic boiler. Once the fire is lighted under it there is no limit to the power it can generate." Being saturated and satiated with emotion and sensation, I went to bed and slept the sleep of the saved and thankful.

*   *   *

As soon as I woke I decided to go over at once to see Roosevelt. I put the matter to the Cabinet when we met at noon. On obtaining their approval I wrote to the King and his Majesty gave his Assent.

The War Cabinet authorised the immediate declaration of war upon Japan, for which all formal arrangements had been made. As Eden had already started on a journey to Moscow and I was in charge of the Foreign Office I sent the following letter to the Japanese Ambassador:

Foreign Office, December 8th

Sir,

On the evening of December 7th His Majesty's Government in the United Kingdom learned that Japanese forces without previous warning either in the form of a declaration of war or of an ultimatum with a conditional declaration of war had attempted a landing on the coast of Malaya and bombed Singapore and Hong Kong.

In view of these wanton acts of unprovoked aggression committed in flagrant violation of International Law and particularly of Article 1 of the Third Hague Convention relative to the opening of hostilities, to which both Japan and the United Kingdom are parties, His Majesty's Ambassador at Tokyo has been instructed to inform the Imperial Japanese Government in the name of His

Majesty's Government in the United Kingdom that a state of war exists between our two countries.

I have the honour to be, with high consideration,

Sir,

Your obedient servant,

WINSTON S. CHURCHILL

Some people did not like this ceremonial style. But after all when you have to kill a man it costs nothing to be polite.

Parliament met at 3 P.M., and in spite of the shortness of notice the House was full. Under the British Constitution the Crown declares war on the advice of Ministers, and Parliament is confronted with the fact. We were therefore able to be better than our word to the United States, and actually declared war upon Japan before Congress could act. The Royal Netherlands Government had also made their declaration. Both Houses voted unanimously in favour of the decision.

*   *   *

We were not told for some time any details of what had happened at Pearl Harbour, but the story has now been exhaustively recorded. Until early in 1941 the Japanese naval plan for war against the United States was for their main fleet to give battle in the waters near the Philippines when the Americans, as might be expected, fought their way across the Pacific to relieve their garrison in this outpost. The idea of a surprise attack on Pearl Harbour originated in the brain of Admiral Yamamoto, the Japanese Commander-in-Chief. Preparation for this treacherous blow before any declaration of war went forward with the utmost secrecy, and by November 22 the striking force of six carriers, with supporting battleships and cruisers, was concentrated in an unfrequented anchorage in the Kurile Islands, north of Japan proper. Already the date of the attack had been fixed for Sunday, December 7, and on November 26 (East longitude date) the force sailed under the command of Admiral Nagumo. Keeping far to the northward of Hawaii, amidst the fog and gales of these northern latitudes, Nagumo approached his goal undetected. Before sunrise on the fateful day the attack was launched from a position about 275 miles to the north of Pearl Harbour. Three hundred and sixty aircraft took part, comprising bombers of all types, escorted by fighters. At 7.55 A.M. the first bomb fell. Ninety-four ships of the United States Navy were present in the harbour. Among them the eight battleships of the Pacific Fleet were the prime targets. The carriers, with strong cruiser forces, were fortunately absent on missions elsewhere. By 10 A.M. the battle was over and the enemy withdrew. Behind them lay a shattered fleet hidden in a pall of fire and smoke, and the vengeance of the United States. The battleship *Arizona* had blown up, the *Oklahoma* had capsized, the *West Virginia* and *California* had sunk at their moorings, and every other battleship, except the *Pennsylvania,* which was in dry dock, had been heavily damaged. Over two thousand Americans had lost their lives, and nearly two thousand others were wounded. The mastery of the Pacific had passed into Japanese hands, and the strategic balance of the world was for the time being fundamentally changed.

In the Philippines, where General MacArthur commanded, our American Allies had yet another set of misfortunes. A warning indicating a grave turn in diplomatic relations had been received on November 20. Admiral Hart, commanding the modest United States Asiatic Fleet, had already been in consultation with the adjacent British and Dutch naval authorities, and, in accordance with his war plan, had begun to disperse his forces to the southward, where he intended to assemble a striking force in Dutch waters in conjunction with his prospective allies. He had at his disposal only one heavy and two light cruisers, besides a dozen old destroyers and various auxiliary vessels. His strength lay almost entirely in his submarines, of which he had twenty-eight. At 3 A.M. on December 8 Admiral Hart intercepted a message giving the staggering news of the attack on Pearl Harbour. He at once warned all concerned that hostilities had begun, without waiting for confirmation from Washington. At dawn the Japanese dive bombers struck, and throughout the ensuing days the air attacks continued on an ever-increasing scale. On the 10th the naval base at Cavite was completely destroyed by fire, and on the same day the Japanese made their first landing in the north of Luzon. Disasters mounted swiftly. Most of the American air forces were destroyed in battle or on the ground, and by December 20 the remnants had been withdrawn to Port Darwin, in Australia. Admiral Hart's ships had begun their southward dispersal some days before, and only the submarines remained to dispute the sea with the enemy. On December 21 the main Japanese invasion force landed in Lingayen Gulf, threatening Manila itself, and thereafter the march of events was not unlike that which was already in progress in Malaya; but the defence was more prolonged. Thus the long-nurtured plans of Japan exploded in a blaze of triumph.

Both Hitler and his staff were astonished. Jodl tells at his trial how Hitler "came in the middle of the night to my chart room [in East Prussia] in order to transmit this news to Field Marshal Keitel and myself. He was completely surprised." On the morning of December 8 however he gave the German Navy orders to attack American ships wherever found. This was three days before the official declaration of war by Germany on the United States.

*   *   *

I convened a meeting, mostly Admiralty, in the Cabinet War Room at ten o'clock on the night of the 9th to review the naval position. We were about a dozen. We tried to measure the consequences of this fundamental change in our war position against Japan. We had lost the command of every ocean except the Atlantic. Australia and New Zealand and all the vital islands in their sphere were open to attack. We had only one key weapon in our hands. The *Prince of Wales* and the *Repulse* had arrived at Singapore. They had been sent to these waters to exercise that kind of vague menace which capital ships of the highest quality whose whereabouts is unknown can impose upon all hostile naval calculations. How should we use them now? Obviously they must go to sea and vanish among the innumerable islands. There was general agreement on that.

I thought myself they should go across the Pacific to join what was left of the American Fleet. It would be a proud gesture at this moment, and would knit the English-speaking world together. We had already cordially agreed to the American Navy Department withdrawing their capital ships from the Atlantic. Thus in a few months there might be a fleet in being on the west coast of America capable of fighting a decisive sea battle if need be. The existence of such a fleet and of such a fact would be the best possible shield to our brothers in Australasia. We were all much attracted by this line of thought. But as the hour was late we decided to sleep on it, and settle the next morning what to do with the *Prince of Wales* and the *Repulse*.

Within a couple of hours they were at the bottom of the sea.

I was opening my boxes on the 10th when the telephone at my bedside rang. It was the First Sea Lord. His voice sounded odd. He gave a sort of cough and gulp, and at first I could not hear quite clearly. "Prime Minister, I have to report to you that the *Prince of Wales* and the *Repulse* have both been sunk by the Japanese — we think by aircraft. Tom Phillips is drowned." "Are you sure it's true?" "There is no doubt at all." So I put the telephone down. I was thankful to be alone. In all the war I never received a more direct shock. The reader of these pages will realise how many efforts, hopes, and plans foundered with these two ships. As I turned over and twisted in bed the full horror of the news sank in upon me. There were no British or American capital ships in the Indian Ocean or the Pacific except the American survivors of Pearl Harbour, who were hastening back to California. Over all this vast expanse of waters Japan was supreme, and we everywhere were weak and naked.

I went down to the House of Commons as soon as they met at eleven that morning to tell them myself what had happened, and next day I made them a full statement upon the new situation. There was much anxiety and not a little discontent with the long-drawn battle in Libya, which evidently hung in the balance. I did not at all conceal the prospect that very severe punishment awaited us at the hands of Japan. On the other hand, the Russian victories had revealed the fatal error of Hitler's Eastern campaign, and winter was still to assert its power. The U-boat war was at the moment under control, and our losses greatly reduced. Finally, four-fifths of the world were now fighting on our side. Ultimate victory was certain. In this sense I spoke.

I used the coldest form of factual narration, avoiding all promises of early success. The House was very silent, and seemed to hold its judgment in suspense. I did not seek or expect more.

5

A Voyage amid World War

MANY SERIOUS REASONS required my presence in London at this moment when so much was molten. I never had any doubt that a complete understanding between Britain and the United States outweighed all else, and that I must go to Washington at once with the strongest team of expert advisers who could be spared. It was thought too risky for us to go by air at this season in an unfavourable direction. Accordingly we travelled on the 12th to the Clyde. The *Prince of Wales* was no more. The *King George V* was watching the *Tirpitz*. The newborn *Duke of York* could carry us, and work herself up to full efficiency at the same time. The principals of our party were Lord Beaverbrook, a member of the War Cabinet; Admiral Pound, First Sea Lord; Air Marshal Portal, Chief of the Air Staff; and Field Marshal Dill, who had now been succeeded by General Brooke as Chief of the Imperial General Staff. I was anxious that Brooke should remain in London in order to grip the tremendous problems that awaited him. In his place I invited Dill, who was still in the centre of our affairs, trusted and respected by all, to come with me to Washington. Here a new sphere was to open to him.

With me also came Lord Moran, who had during 1941 become my constant medical adviser. This was his first voyage with me, but afterwards he came on all the journeys. To his unfailing care I probably owe my life. Although I could not persuade him to take my advice when he was ill, nor could he always count on my implicit obedience to all his instructions, we became devoted friends. Moreover, we both survived.

It was hoped to make the passage at an average of twenty knots in seven days, having regard to zigzags and detours to avoid the plotted U-boats. The Admiralty turned us down the Irish Channel into the Bay of Biscay. The weather was disagreeable. There was a heavy gale and a rough sea. The sky was covered with patchy clouds. We had to cross the out-and-home U-boat stream from the western French ports to their Atlantic hunting grounds. There were so many of them about that our captain was ordered by the Admiralty not to leave our flotilla behind us; but the flotilla could not make more than six knots in the heavy seas, and we paddled along at this pace round the South of Ireland for forty-eight hours. We passed within four hundred miles of Brest, and I could not help remembering how the *Prince of Wales* and the *Repulse* had been destroyed by shore-based torpedo aircraft attack the week before. The clouds had prevented all but an occasional plane of our air escort from joining us, but when I went on the bridge I saw a lot of unwelcome blue sky appearing. However, nothing happened, so all was well. The great ship with her attendant destroyers plodded on. But we became impatient with her slow speed. On the second night we approached the U-boat stream. Admiral Pound, who took the decision, said that we were more likely to ram a U-boat than to be torpedoed by one ourselves. The night was pitch-black. So we cast off our destroyers and ran through alone at the best speed possible in the continuing rough weather. We were battened down and great seas beat upon the decks. Lord Beaverbrook complained that he might as well have travelled in a submarine.

Our very large deciphering staff could of course receive by wireless a great deal of business. To a limited extent we could reply. When fresh escorts joined us from the Azores they could take in by daylight Morse signals from us in code, and then, dropping off a hundred miles or so, could transmit them without revealing our position. Still, there was a sense of radio claustrophobia — and we were in the midst of world war.

The fighting proceeded in all the theatres. Hong Kong had been attacked by Japan at nearly the same moment of time as Pearl Harbour. I had no illusions about its fate under the overwhelming impact of Japanese power. Twelve months earlier I had deprecated strengthening our garrison. Their loss was certain and they should have been reduced to a symbolical scale, but I had allowed myself to be drawn from this position and reinforcements had been sent. From the outset they were faced with a task beyond their powers. For a week they held out. Every man who could bear arms took part in a desperate resistance. Their tenacity was matched by the fortitude of the British civilian population. On Christmas Day the limit of endurance was reached and capitulation became inevitable. Another set of disasters loomed upon us in Malaya. The Japanese landings on the peninsula were accompanied by damaging raids on our airfields which badly crippled our already weak air forces and soon made the northerly aerodromes unusable. By the end of the month our troops, several times heavily engaged, were in action a full hundred and fifty miles from the position they had first held, and the Japanese had landed at least three full divisions, including their Imperial Guard. The quality of the enemy planes, speedily deployed on captured airfields, exceeded all expectations. We had been thrown on to the defensive and our losses were severe.

*   *   *

Everyone in our party worked incessantly while the *Duke of York* plodded westward, and all our thoughts were focused on the new and vast problems we had to solve. We looked forward with eagerness, but also with some anxiety, to our first direct contact as allies with the President and his political and military advisers. We knew before we left that the outrage of Pearl Harbour had stirred the people of the United States to their depths. The official reports and the Press summaries we had received gave the impression that the whole fury of the nation would be turned upon Japan. We feared lest the true proportion of the war as a whole might not be understood. We were conscious of a serious danger that the United States might pursue the war against Japan in the Pacific and leave us to fight Germany and Italy in Europe, Africa, and in the Middle East.

The first Battle of the Atlantic against the U-boats had turned markedly in our favour. We did not doubt our power to keep open our ocean paths. We felt sure we could defeat Hitler if he tried to invade the island. We were encouraged by the strength of the Russian resistance. We were unduly hopeful about our Libyan campaign. But all our future plans depended upon a vast flow of American supplies of all kinds, such as were now streaming across the Atlantic. Especially we counted on planes and tanks, as well as on the stupendous American merchant-ship construction. Hitherto, as a nonbelligerent, the President had been able and willing to divert large supplies of equipment from the American armed forces, since these were not engaged. This process was bound to be restricted now that the United States was at war with Germany, Italy, and above all Japan. Home needs would surely come first? Already, after Russia had been attacked, we had rightly sacrificed to aid the Soviet armies a large portion of the equipment and supplies now at last arriving from our factories. The United States had diverted to Russia even larger quantities of supplies than we otherwise would have received ourselves.

We had fully approved of all this on account of the splendid resistance which Russia was offering to the Nazi invader.

It had been none the less hard to delay the equipment of our own forces, and especially to withhold vitally needed weapons from our army, fiercely engaged in Libya. We must presume that "America first" would become the dominant principle with our Ally. We feared that there would be a long interval before American forces came into action on a great scale, and that during this period of preparation we should necessarily be greatly straitened. This would happen at a time when we ourselves had to face a new and terrible antagonist in Malaya, the Indian Ocean, Burma, and India. Evidently the partition of supplies would require profound attention and would be fraught with many difficulties and delicate aspects. Already we had been notified that all the schedules of deliveries under Lend-Lease had been stopped pending readjustment. Happily the output of the British munitions and aircraft factories was now acquiring scope and momentum, and would soon be very large indeed. But a long array of "bottlenecks" and possible denials of key items, which would affect the whole range of our production, loomed before our eyes as our battleship drove on through the incessant gales. Beaverbrook was, as usual in times of trouble, optimistic. He declared that the resources of the United States had so far not even been scratched; that they were immeasurable, and that once the whole force of the American people was diverted to the struggle results would be achieved far beyond anything that had been projected or imagined. Moreover, he thought the Americans did not yet realise their strength in the production field. All the present statistics would be surpassed and swept away by the American effort. There would be enough for all. In this his judgment was right.

All these considerations paled before the main strategic issue. Should we be able to persuade the President and the American Service chiefs that the defeat of Japan would not spell the defeat of Hitler, but that the defeat of Hitler made the finishing off of Japan merely a matter of time and trouble? Many long hours did we spend revolving this grave issue. The two Chiefs of Staff and General Dill with Hollis and his officers prepared several papers dealing with the whole subject and emphasising the view that the war was all one. As will be seen, these labours and fears both proved needless.

*   *   *

The eight days' voyage, with its enforced reduction of current business, with no Cabinet meetings to attend or people to receive, enabled me to pass in review the whole war as I saw and felt it in the light of

its sudden vast expansion. I recall Napoleon's remark about the value
of being able to focus objects in the mind for a long time without being
tired — *"fixer les objets longtemps sans être fatigué."* As usual I tried
to do this by setting forth my thought in typescript by dictation. In
order to prepare myself for meeting the President and for the American
discussions and to make sure that I carried with me the two Chiefs of
Staff, Pound and Portal, and General Dill, and that the facts could be
checked in good time by General Hollis and the Secretariat, I produced
three papers on the future course of the war, as I conceived it should be
steered. Each paper took four or five hours, spread over two or three
days. As I had the whole picture in my mind it all came forth easily,
but very slowly. In fact, it could have been written out two or three
times in longhand in the same period. As each document was completed
after being checked I sent it to my professional colleagues as an expression of my personal convictions. They were at the same time preparing
papers of their own for the combined Staff conferences. I was glad to
find that although my theme was more general and theirs more technical
there was our usual harmony on principles and values. No differences
were expressed which led to argument, and very few of the facts required correction. Thus, though nobody was committed in a precise or
rigid fashion, we all arrived with a body of doctrine of a constructive
character on which we were broadly united.

 The first paper assembled the reasons why our main objective for the
campaign of 1942 in the European theatre should be the occupation
of the whole coastline of Africa and of the Levant from Dakar to the
Turkish frontier by British and American forces. The second dealt with
the measures which should be taken to regain the command of the
Pacific, and specified May 1942 as the month when this could be
achieved. It dwelt particularly upon the need to multiply aircraft
carriers by improvising them in large numbers. The third declared as
the ultimate objective the liberation of Europe by the landing of large
Anglo-American armies wherever was thought best in the German-
conquered territory, and fixed the year 1943 as the date for this supreme
stroke.

 So many tales have been published of my rooted aversion from
large-scale operations on the Continent that it is important that the
truth should be emphasised. I always considered that a decisive assault
upon the German-occupied countries on the largest possible scale was
the only way in which the war could be won, and that the summer of
1943 should be chosen as the target date. The scale of the operation
contemplated by me was already before the end of 1941 set at forty
armoured divisions and a million other troops as essential for the opening phase. When I notice the number of books which have been writ-

ten on a false assumption of my attitude on this issue, I feel bound to direct the attention of the reader to the authentic and responsible documents written at the time, of which other instances will be given as the account proceeds.[1]

I gave these three papers to the President before Christmas. I explained that while they were my own personal views, they did not supersede any formal communications between the Staffs. I couched them in the form of memoranda for the British Chiefs of Staff Committee. Moreover, I told him they were not written expressly for his eye, but that I thought it important that he should know what was in my mind and what I wanted to have done and, so far as Great Britain was concerned, would try to bring to action. He read them immediately after receiving them, and the next day asked whether he might keep copies of them. To this I gladly assented.

I felt indeed that the President was thinking very much along the same lines as I was about action in French North Africa. We were now allies, and must act in common and on a greater scale. I was confident that he and I would find a large measure of agreement and that the ground had been well prepared. I was therefore in a hopeful mood, and, as will be seen, I eventually obtained the President's agreement to an expedition to North Africa ("Operation Torch"), which constituted our first great joint amphibious offensive.

While however it is vital to plan the future, and sometimes possible to forecast it in certain respects, no one can help the timetable of such mighty events being deranged by the actions and counterstrokes of the enemy. All the objectives in these memoranda were achieved by the British and United States forces in the order there set forth. My hopes that General Auchinleck would clear Libya in February 1942 were disappointed. He underwent a series of grievous reverses which will presently be described. Hitler, perhaps encouraged by this success, determined upon a large-scale effort to fight for Tunis, and presently moved about a hundred thousand fresh troops thither through Italy and across the Mediterranean. The British and American Armies therefore became involved in a larger and longer campaign in North Africa than I had contemplated. A delay of four months was for this reason enforced upon the timetable. The Anglo-American Allies did not obtain control of the whole North African shore from Tunis to Egypt until May 1943. The supreme plan of crossing the Channel to liberate France, for which I had earnestly hoped and worked, could not therefore be undertaken that summer, and was perforce postponed for one whole year, till the summer of 1944.

[1] The three papers here referred to may be studied in Chapter 34 of *The Grand Alliance*.

Subsequent reflection and the full knowledge we now possess have convinced me that we were fortunate in our disappointment. The year's delay in the expedition saved us from what would at that date have been at the best an enterprise of extreme hazard, with the probability of a world-shaking disaster. If Hitler had been wise he would have cut his losses in North Africa and would have met us in France with double the strength he had in 1944, before the newly raised American armies and staffs had reached their full professional maturity and excellence, and long before the enormous armadas of landing craft and the floating harbours (Mulberries) had been specially constructed. I am sure now that even if "Operation Torch" had ended as I hoped in 1942, or even if it had never been tried, the attempt to cross the Channel in 1943 would have led to a bloody defeat of the first magnitude, with measureless reactions upon the result of the war. I became increasingly conscious of this during the whole of 1943, and therefore accepted as inevitable the postponement of "Overlord," while fully understanding the vexation and anger of our Soviet Ally.

*   *   *

It had been intended that we should steam up the Potomac and motor to the White House, but we were all impatient after nearly ten days at sea to end our journey. We therefore arranged to fly from Hampton Roads, and landed after dark on December 22 at the Washington airport. There was the President waiting in his car. I clasped his strong hand with comfort and pleasure. We soon reached the White House, which was to be in every sense our home for the next three weeks. Here we were welcomed by Mrs. Roosevelt, who thought of everything that could make our stay agreeable.

I must confess that my mind was so occupied with the whirl of events and the personal tasks I had to perform that my memory till refreshed had preserved but a vague impression of these days. The outstanding feature was of course my contacts with the President. We saw each other for several hours every day, and lunched always together, with Harry Hopkins as a third. We talked of nothing but business, and reached a great measure of agreement on many points, both large and small. Dinner was a more social occasion, but equally intimate and friendly. The President punctiliously made the preliminary cocktails himself, and I wheeled him in his chair from the drawing room to the lift as a mark of respect, and thinking also of Sir Walter Raleigh spreading his cloak before Queen Elizabeth. I formed a very strong affection, which grew with our years of comradeship, for this formidable politician who had imposed his will for nearly ten years upon the American scene, and whose heart seemed to respond to many of the impulses that stirred my own. As we both, by need or habit, were forced to do much of our work in bed, he visited me in my room whenever he felt inclined, and encouraged me to do the same to him. Hopkins was just across the passage from my bedroom, and next door to him my travelling map room was soon installed. The President was much interested in this institution, which Captain Pim had perfected. He liked to come and study attentively the large maps of all the theatres of war which soon covered the walls, and on which the movement of fleets and armies was so accurately and swiftly recorded. It was not long before he established a map room of his own of the highest efficiency.

The days passed, counted in hours. Quite soon I realised that immediately after Christmas I must address the Congress of the United States, and a few days later the Canadian Parliament in Ottawa. These great occasions imposed heavy demands on my life and strength, and were additional to all the daily consultations and mass of current business. In fact, I do not know how I got through it all.

Simple festivities marked our Christmas. The traditional Christmas tree was set up in the White House garden, and the President and I made brief speeches from the balcony to enormous crowds gathered in the gloom. He and I went to church together on Christmas Day, and I found peace in the simple service and enjoyed singing the well-known hymns, and one, "O little town of Bethlehem," I had never heard before. Certainly there was much to fortify the faith of all who believe in the moral governance of the universe.

*   *   *

It was with heart stirrings that I fulfilled the invitation to address the Congress of the United States. The occasion was important for what I was sure was the all-conquering alliance of the English-speaking peoples. I had never addressed a foreign Parliament before. Yet to me, who could trace unbroken male descent on my mother's side through five generations from a lieutenant who served in George Washington's army, it was possible to feel a blood-right to speak to the representatives of the great Republic in our common cause. It certainly was odd that it should all work out this way; and once again I had the feeling, for mentioning which I may be pardoned, of being used, however unworthy, in some appointed plan.

I spent a good part of Christmas Day preparing my speech. The President wished me good luck when on December 26 I set out in the charge of the leaders of the Senate and House of Representatives from the White House to the Capitol. There seemed to be great crowds along the broad approaches, but the security precautions, which in the United States go far beyond British custom, kept them a long way off, and two or three motorcars filled with armed plain-clothes policemen clustered around as escort. On getting out I wished to walk up to the cheering masses in a strong mood of brotherhood, but this was not allowed. Inside the scene was impressive and formidable, and the great semicircular hall, visible to me through a grille of microphones, was thronged.

I must confess that I felt quite at home, and more sure of myself than I had sometimes been in the House of Commons. What I said was received with the utmost kindness and attention. I got my laughter and applause just where I expected them. The loudest response was when, speaking of the Japanese outrage, I asked, "What sort of people do they think we are?" The sense of the might and will power of the American nation streamed up to me from the august assembly. Who could doubt that all would be well? Afterwards the leaders came along with me close up to the crowds which surrounded the building, so that I could give them an intimate greeting; and then the Secret Service men and their cars closed round and took me back to the White House, where the President, who had listened in, told me I had done quite well.

*   *   *

I travelled by the night train of December 28–29 to Ottawa, to stay with Lord Athlone, the Governor-General. On the 29th I attended a meeting of the Canadian War Cabinet. Thereafter Mr. Mackenzie King, the Prime Minister, introduced me to the leaders of the Conservative Opposition, and left me with them. These gentlemen were unsurpassed in loyalty and resolution, but at the same time they were rueful not to have the honour of waging the war themselves, and at having to listen to so many of the sentiments which they had championed all their lives expressed by their Liberal opponents.

On the 30th I spoke to the Canadian Parliament. The preparation of my two transatlantic speeches, transmitted all over the world, amid all the flow of executive work, which never stopped, was an extremely hard exertion. Delivery is no serious burden to a hard-bitten politician, but choosing what to say and what not to say in such an electric atmosphere is anxious and harassing. I did my best. The most successful point in the Canadian speech was about the Vichy Government, with whom Canada was still in relations.

It was their duty [in 1940] and it was also their interest to go to North Africa, where they would have been at the head of the French Empire. In Africa, with our aid, they would have had overwhelming sea-power. They would have had the recognition of the United States, and the use of all the gold they had lodged beyond the seas. If they had done this Italy might have been driven out of the war before the end of 1940, and France would have held her place as a nation in the councils of the Allies and at the conference table of the victors. But their generals misled them. When I warned them that Britain would fight on alone whatever they did, their generals told their Prime Minister and his divided Cabinet, "In three weeks England will have her neck wrung like a chicken." Some chicken! Some neck!

This went very well. I quoted, to introduce a retrospect, Sir Harry Lauder's song of the last war which began:

If we all look back on the history of the past
We can just tell where we are.

The words "that grand old comedian" were on my notes. On the way down I thought of the word "minstrel." What an improvement! I rejoice to know that he was listening and was delighted at the reference. I am so glad I found the right word for one who, by his inspiring songs and valiant life, rendered measureless service to the Scottish race and to the British Empire.

I was lucky in the timing of these speeches in Washington and Ottawa. They came at the moment when we could all rejoice at the creation of the Grand Alliance, with its overwhelming potential force, and before the cataract of ruin fell upon us from the long, marvellously prepared assault of Japan. Even while I spoke in confident tones I could feel in anticipation the lashes which were soon to score our naked flesh. Fearful forfeits had to be paid not only by Britain and Holland but by the United States, in the Pacific and Indian Oceans, and in all the Asiatic lands and islands they lap with their waves. An indefinite period of military disaster lay certainly before us. Many dark and weary months of defeat and loss must be endured before the light would come again. When I returned in the train to Washington on New Year's Eve I was asked to go into the carriage filled with many leading pressmen of the United States. It was with no illusions that I wished them all a glorious New Year. "Here's to 1942. Here's to a year of toil — a year of struggle and peril, and a long step forward towards victory. May we all come through safe and with honour!"

6

Anglo-American Accords

THE FIRST MAJOR design which was presented to me by Mr. Roosevelt after my arrival from England had been the drawing up of a solemn Declaration to be signed by all the nations at war with Germany and Italy, or with Japan. The President and I, repeating our methods in framing the Atlantic Charter, prepared drafts of the Declaration and blended them together. In principle, in sentiment, and indeed in language, we were in full accord. At home the War Cabinet was at once surprised and thrilled by the scale on which the Grand Alliance was planned. There was much rapid correspondence, and some difficult points arose about what Governments and authorities should sign the Declaration, and also on the order of precedence. We gladly accorded the first place to the United States, and on my return to the White House all was ready for the signature of the United Nations Pact. Many telegrams had passed between Washington, London, and Moscow, but now all was settled. The President had exerted his most fervent efforts to persuade Litvinov, the Soviet Ambassador, newly restored to favour by the turn of events, to accept the phrase "religious freedom." He was invited to luncheon with us in the President's room on purpose. After his hard experiences in his own country he had to be careful. Later on the President had a long talk with him alone about his soul and the dangers of hell fire. The accounts which Mr. Roosevelt gave us on several occasions of what he said to the Russian were impressive. Indeed, on one occasion I promised Mr. Roosevelt to recommend him for the position of Archbishop of Canterbury if he should lose the next Presidential election. I did not however make any official recommendation to the Cabinet or the Crown upon this point, and as he won the election in 1944 it did not arise. Litvinov reported the issue about "religious freedom" in evident fear and trembling to Stalin, who accepted it as a matter of course. The War Cabinet also got their point in about "social security," with which, as the author of the first Unemployment Insurance Act, I cordially concurred. After a spate of telegrams had flowed about the world for a week agreement was reached throughout the Grand Alliance.

The title of "United Nations" was substituted by the President for that of "Associated Powers." I thought this a great improvement. I showed my friend the lines from Byron's *Childe Harold*:

> *Here, where the sword United Nations drew,*
> *Our countrymen were warring on that day!*
> *And this is much — and all — which will not pass away.*

The President was wheeled in to me on the morning of January 1. I got out of my bath, and agreed to the draft. The Declaration could not by itself win battles, but it set forth who we were and what we were fighting for. Later that day Roosevelt, I, Litvinov, and Soong, representing China, signed this majestic document in the President's study. It was left to the State Department to collect the signatures of the remaining twenty-two nations. The final text must be recorded here.

A Joint Declaration by the United States of America, the United Kingdom of Great Britain and Northern Ireland, the Union of Soviet Socialist Republics, China, Australia, Belgium, Canada, Costa Rica, Cuba, Czechoslovakia, the Dominican Republic, El Salvador, Greece, Guatemala, Haiti, Honduras, India, Luxemburg, the Netherlands, New Zealand, Nicaragua, Norway, Panama, Poland, South Africa, and Yugoslavia.

The Governments signatory hereto,

Having subscribed to a common programme of purposes and principles embodied in the Joint Declaration of the President of the United States of America and the Prime Minister of the United Kingdom of Great Britain and Northern Ireland, dated August 14, 1941, known as the Atlantic Charter,

Being convinced that complete victory over their enemies is essential to defend life, liberty, independence, and religious freedom, and to preserve human rights and justice in their own lands as well as in other lands, and that they are now engaged in a common struggle against savage and brutal forces seeking to subjugate the world, DECLARE:

(1) Each Government pledges itself to employ its full resources, military or economic, against those members of the Tripartite Pact and its adherents with which such Government is at war.

(2) Each Government pledges itself to co-operate with the Governments signatory hereto, and not to make a separate armistice or peace with the enemies.

The foregoing declaration may be adhered to by other nations which are, or which may be, rendering material assistance and contributions in the struggle for victory over Hitlerism.

*   *   *

It may well be thought by future historians that the most valuable and lasting result of our first Washington conference — "Arcadia," as it was code-named — was the setting up of the now famous "Combined Chiefs of Staff Committee." Its headquarters were in Washington, but since the British Chiefs of Staff had to live close to their own Government they were represented by high officers stationed there permanently. These representatives were in daily, indeed hourly, touch with London, and were thus able to state and explain the views of the British Chiefs of Staff to their U.S. colleagues on any and every war problem at any time of the day or night. The frequent conferences that were held in various parts of the world — Casablanca, Washington, Quebec, Teheran, Cairo, Malta, and the Crimea — brought the principals themselves together for sometimes as much as a fortnight. Of the two hundred formal meetings held by the Combined Chiefs of Staff Committee during the war no fewer than eighty-nine were at these conferences; and it was at these full-dress meetings that the majority of the most important decisions were taken.

The usual procedure was that in the early morning each Chiefs of Staff Committee met among themselves. Later in the day the two teams met and became one; and often they would have a further combined meeting in the evening. They considered the whole conduct of the war, and submitted agreed recommendations to the President and me. Our own direct discussions had of course gone on meanwhile by talks or telegrams, and we were in intimate contact with our own staff. The proposals of the professional advisers were then considered in plenary meetings, and orders given accordingly to all commanders in the field. However sharp the conflict of views at the Combined Chiefs of Staff meeting, however frank and even heated the argument, sincere loyalty to the common cause prevailed over national or personal interests. Decisions once reached and approved by the heads of Governments were pursued by all with perfect loyalty, especially by those whose original opinions had been overruled. There was never a failure to reach effective agreement for action, or to send clear instructions to the commanders in every theatre. Every executive officer knew that the orders he received bore with them the combined conception and expert authority of both Governments. There never was a more serviceable war machinery established among allies, and I rejoice that in fact if not in form it continues to this day.

The Russians were not represented on the Combined Chiefs of Staff Committee. They had a far-distant, single, independent front, and there was neither need nor means of staff integration. It was sufficient that we should know the general sweep and timing of their movements and that they should know ours. In these matters we kept in as close touch with them as they permitted. I shall in due course describe the personal visits which I paid to Moscow. And at Teheran, Yalta, and Potsdam the Chiefs of Staff of all three nations met round the table.

*   *   *

I have described how Field Marshal Dill, though no longer Chief of the Imperial General Staff, had come with us in the *Duke of York*. He had played his full part in all the discussions, not only afloat, but even more when we met the American leaders. I at once perceived that his prestige and influence with them was upon the highest level. No British officer we sent across the Atlantic during the war ever acquired American esteem and confidence in an equal degree. His personality, discretion, and tact gained him almost at once the confidence of the President. At the same time he established a true comradeship and private friendship with General Marshall.

Immense expansions were ordered in the production sphere. In all these Beaverbrook was a potent impulse. The official American history of their industrial mobilisation for war [1] bears generous testimony to this. Donald Nelson, the Executive Director of American War Production, had already made gigantic plans. "But," says the American account, "the need for boldness had been dramatically impressed upon Nelson by Lord Beaverbrook . . ." What happened is best portrayed by Mr. Nelson's own words:

> Lord Beaverbrook emphasised the fact that we must set our production sights much higher than for the year 1942, in order to cope with a resourceful and determined enemy. He pointed out that we had as yet no experience in the losses of material incidental to a war of the kind we are now fighting. . . . The ferment Lord Beaverbrook was instilling in the mind of Nelson he was also imparting to the President. In a note to the President Lord Beaverbrook set the expected 1942 production of the United States, the United Kingdom, and Canada against British, Russian, and American requirements. The comparison exposed tremendous deficits in 1942

[1] *History of the War Production Board, 1940–1945.*

planned production. For tanks these deficits were 10,500; for air-craft 26,730; for artillery 22,600; and for rifles 1,600,000. Pro-duction targets had to be increased, wrote Lord Beaverbrook, and he pinned his faith on their realisation in "the immense possibilities of American industry." . . . The outcome was a set of production objectives whose magnitude exceeded even those Nelson had pro-posed. The President was convinced that the concept of our indus-trial capacity must be completely overhauled. . . . He directed the fulfilment of a munitions schedule calling for 45,000 combat aircraft, 45,000 tanks, 20,000 anti-aircraft guns, 14,900 anti-tank guns, and 500,000 machine guns in 1942.

These remarkable figures were achieved or surpassed by the end of 1943. In shipping, for example, the new tonnage built in the U.S.A. was as follows:

1942	..	..	..	5,339,000 tons
1943	..	..	..	12,384,000 tons

*   *   *

Continued concentration of mind upon the war as a whole, my con-stant discussions with the President and his principal advisers and with my own, my two speeches and my journey to Canada, together with the heavy flow of urgent business requiring decision and all the tele-grams interchanged with my colleagues at home, made this period in Washington not only intense and laborious but even exhausting. My American friends thought I was looking tired and ought to have a rest. Accordingly Mr. Stettinius very kindly placed his small villa in a seaside solitude near Palm Beach at my disposal, and on January 4 I flew down there, and found time to deal with several difficult questions which pursued me. The Italian "human torpedo" attack in Alexandria harbour which had disabled the *Queen Elizabeth* and *Valiant* has al-ready been described. This misfortune, following upon all our other naval losses at this moment, was most untimely and disturbing. I saw its gravity at once. The Mediterranean battle fleet was for the time being nonexistent, and our naval power to guard Egypt from direct overseas invasion in abeyance. It seemed necessary in the emergency to send whatever torpedo planes could be gathered from the south coast of England. This had, as will presently be seen, an unpleasant sequel.

I was also much disturbed by the reports which Mr. Eden had brought back with him from Moscow of Soviet territorial ambitions, especially in the Baltic States. These were the conquests of Peter the Great, and had been for two hundred years under the Czars. Since the Russian revolution they had been the outpost of Europe against Bolshevism. They were what are now called "social democracies," but very lively and truculent. Hitler had cast them away like pawns in his deal with the Soviets before the outbreak of war in 1939. There had been a severe Russian and Communist purge. All the dominant personalities and elements had been liquidated in one way or another. The life of these strong peoples was henceforward underground. Presently, as we shall see, Hitler came back with a Nazi counterpurge. Finally, in the general victory the Soviets had control again. Thus the deadly comb ran back and forth, and back again, through Esthonia, Latvia, and Lithuania. There was no doubt however where the right lay. The Baltic States should be sovereign independent peoples.

* * *

I set out by train to return to Washington on the night of the 9th, and reached the White House on the 11th. There I found that great progress had been made by the Combined Chiefs of Staff, and that it was mostly in harmony with my views. The President convened a meeting on January 12, when there was complete agreement upon the broad principles and targets of the war. The differences were confined to priorities and emphasis, and all was ruled by that harsh and despotic factor, shipping. "The President," says the British record, "set great store on organising . . . a combined United States–British expedition to North Africa. A tentative timetable had been worked out for putting 90,000 United States and 90,000 British troops, together with a considerable air force, into North Africa." On "Grand Strategy" the Staffs agreed that *"only the minimum of forces necessary for the safeguarding of vital interests in other theatres should be diverted from operations against Germany."* No one had more to do with obtaining this cardinal decision than General Marshall.

On the 14th I took leave of Mr. Roosevelt. He seemed concerned about the dangers of the voyage. Our presence in Washington had been for many days public to the world, and the charts showed more than twenty U-boats on our homeward courses. We flew in beautiful weather from Norfolk to Bermuda, where the *Duke of York,* with escorting destroyers, awaited us inside the coral reefs. I travelled in an enormous Boeing flying boat, which made a most favourable impression upon me. During the three hours' trip I made friends with the chief pilot, Captain Kelly Rogers, who seemed a man of high quality and experience. I took the controls for a bit, to feel this ponderous machine of thirty or more tons in the air. I got more and more attached to the flying boat. Presently I asked the captain, "What about flying from Bermuda to England? Can she carry enough petrol?" Under his stolid exterior he became visibly excited. "Of course we can do it. The present weather forecast would give a forty-mile-an-hour wind behind us. We could do it in twenty hours." I asked how far it was, and he said, "About three thousand five hundred miles." At this I became thoughtful.

However, when we landed I opened the matter to Portal and Pound. Formidable events were happening in Malaya; we ought all to be back at the earliest moment. The Chief of the Air Staff said at once that he thought the risk wholly unjustifiable, and he could not take the responsibility for it. The First Sea Lord supported his colleagues. There was the *Duke of York*, with her destroyers, all ready for us, offering comfort and certainty. I said, "What about the U-boats you have been pointing out to me?" The Admiral made a disdainful gesture about them, which showed his real opinion of such a menace to a properly escorted and fast battleship. It occurred to me that both these officers thought my plan was to fly myself and leave them to come back in the *Duke of York*, so I said, "Of course there would be room for all of us." They both visibly changed countenance at this. After a considerable pause Portal said that the matter might be looked into, and that he would discuss it at length with the captain of the flying boat and go into weather prospects with the meteorological authorities. I left it at that.

Two hours later they both returned, and Portal said that he thought it might be done. The aircraft could certainly accomplish the task under reasonable conditions; the weather outlook was exceptionally favourable on account of the strong following wind. No doubt it was very important to get home quickly. Pound said he had formed a very high opinion of the aircraft skipper, who certainly had unrivalled experience. Of course there was a risk, but on the other hand there were the U-boats to consider. So we settled to go unless the weather deteriorated. The starting time was 2 P.M. the next day. It was thought necessary to reduce our baggage to a few boxes of vital papers. Dill was to remain behind in Washington as my personal military representative with the President. Our party would consist only of myself, the two Chiefs of Staff, and Max Beaverbrook, Charles Moran, and Hollis. All the rest would go by the *Duke of York*.

I woke up unconscionably early next morning with the conviction that I should certainly not go to sleep. I must confess that I felt rather frightened. I thought of the ocean spaces, and that we should never be within a thousand miles of land until we approached the British Isles. I thought perhaps I had done a rash thing, that there were too many eggs in one basket. I had always regarded an Atlantic flight with awe. But the die was cast. Still, I must admit that if at breakfast, or even before luncheon, they had come to me to report that the weather had changed and we must go by sea I should have easily reconciled myself to a voyage in the splendid ship which had come all this way to fetch us.

It was, as the captain had predicted, quite a job to get off the water. Indeed, I thought that we should hardly clear the low hills which closed the harbour. There was really no danger; we were in sure hands. The flying boat lifted ponderously a quarter of a mile from the reef, and we had several hundred feet of height to spare. There is no doubt about the comfort of these great flying boats. The motion was smooth, the vibration not unpleasant, and we passed an agreeable afternoon and had a merry dinner. These boats have two storeys, and one walks up a regular staircase to the control room. Darkness had fallen, and all the reports were good. We were now flying through dense mist at about seven thousand feet. One could see the leading edge of the wings, with their great flaming exhausts pouring back over the wing surfaces. In these machines at this time a large rubber tube which expanded and contracted at intervals was used to prevent icing. The captain explained to me how it worked, and we saw from time to time the ice splintering off as it expanded. I went to bed and slept soundly for several hours.

*   *   *

I woke just before the dawn, and went forward to the controls. The daylight grew. Beneath us was an almost unbroken floor of clouds.

After sitting for an hour or so in the co-pilot's seat I sensed a feeling of anxiety around me. We were supposed to be approaching England from the southwest and we ought already to have passed the Scilly Islands, but they had not been seen through any of the gaps in the cloud floor. As we had flown for more than ten hours through mist and had had only one sight of a star in that time, we might well be slightly off our course. Wireless communication was of course limited by the normal wartime rules. It was evident from the discussions which were going on that we did not know where we were. Presently Portal, who had been studying the position, had a word with the captain, and then said to me, "We are going to turn north at once." This was done, and after another half-hour in and out of the clouds we sighted England, and soon arrived over Plymouth, where, avoiding the balloons, which were all shining, we landed comfortably.

As I left the aircraft the captain remarked, "I never felt so much relieved in my life as when I landed you safely in the harbour." I did not appreciate the significance of his remark at the moment. Later on I learnt that if we had held on our course for another five or six minutes before turning northward we should have been over the German batteries in Brest. We had slanted too much to the southward during the night. Moreover, the decisive correction which had been made brought us in, not from the southwest, but from just east of south — that is to say, from the enemy's direction rather than from that from which we were expected. This had the result, as I was told some weeks later, that we were reported as a hostile bomber coming in from Brest, and six Hurricanes from Fighter Command were ordered out to shoot us down. However, they failed in their mission.

To President Roosevelt I cabled, "We got here with a good hop from Bermuda and a thirty-mile wind."

7

The Fall of Singapore

I WAS EXPECTED to make a full statement to Parliament about
my mission to Washington and all that had happened in the five
weeks I had been away. Two facts stood out in my mind. The first was
that the Grand Alliance was bound to win the war in the long run. The
second was that a vast, measureless array of disasters approached us in
the onslaught of Japan. Everyone could see with intense relief that our
life as a nation and Empire was no longer at stake. On the other hand,
the fact that the sense of mortal danger was largely removed set every
critic, friendly or malevolent, free to point out the many errors which
had been made. Moreover, many felt it their duty to improve our
methods of conducting the war and thus shorten the fearful tale. I was
myself profoundly disturbed by the defeats which had already fallen
upon us, and no one knew better than I that these were but the beginnings of the deluge. The demeanour of the Australian Government, the
well-informed and airily detached criticism of the newspapers, the
shrewd and constant girding of twenty or thirty able Members of
Parliament, the atmosphere of the lobbies, gave me the sense of an
embarrassed, unhappy, baffled public opinion, albeit superficial, swelling and mounting about me on every side.

On the other hand, I was well aware of the strength of my position.
I could count on the goodwill of the people for the share I had had in
their survival in 1940. I did not underrate the broad, deep tide of national fidelity that bore me forward. The War Cabinet and the Chiefs
of Staff showed me the highest loyalty. I was sure of myself. I made it
clear, as occasion required, to those about me that I would not consent
to the slightest curtailment of my personal authority and responsibility.
The press was full of suggestions that I should remain Prime Minister
and make the speeches but cede the actual control of the war to some-

one else. I resolved to yield nothing to any quarter, to take the prime and direct personal responsibility upon myself, and to demand a vote of confidence from the House of Commons. I also remembered that wise French saying, *"On ne règne sur les âmes que par le calme."*

It was necessary above all to warn the House and the country of the misfortunes which impended upon us. There is no worse mistake in public leadership than to hold out false hopes soon to be swept away. The British people can face peril or misfortune with fortitude and buoyancy, but they bitterly resent being deceived or finding that those responsible for their affairs are themselves dwelling in a fool's paradise. I felt it vital, not only to my own position but to the whole conduct of the war, to discount future calamities by describing the immediate outlook in the darkest terms. It was also possible to do so at this juncture without prejudicing the military situation or disturbing that underlying confidence in ultimate victory which all were now entitled to feel. In spite of the shocks and stresses which each day brought, I did not grudge the twelve or fourteen hours of concentrated thought which ten thousand words of original composition on a vast, many-sided subject demanded, and while the flames of adverse war in the Desert licked my feet I succeeded in preparing my statement and appreciation of our case.

*   *   *

Even before I left the White House, my hopes of a victory, in which Rommel would be destroyed, had faded. Rommel had escaped. The results of Auchinleck's successes at Sidi Rezegh and at Gazala had not been decisive. The revival of the enemy air power in the Mediterranean during December and January and the virtual disappearance for several months of our sea command was to deprive him of the fruits of the victory for which he had struggled so hard and waited too long. The prestige which he had given us in the making of all our plans for the Anglo-American descent on French North Africa was definitely weakened, and this operation was obviously set back for months.

Worse was now to come. Space forbids a detailed account of the military disaster which, for the second time, at this same fatal corner and one year later, was now to ruin the whole British campaign in the Desert for 1942. Suffice it to say that on January 21, from his position at Agheila, Rommel launched a reconnaissance in force, consisting of three columns each of about a thousand motorised infantry supported by tanks. These rapidly found their way through the gaps between our contact troops, who had no armour working with them, and were ordered to withdraw. He again proved himself a master of desert tactics, and, outwitting our commanders, regained the greater part of Cyrenaica. A retreat of nearly three hundred miles ruined our hopes and lost us Benghazi and all the stores General Auchinleck had been gathering for his hoped-for offensive in the middle of February. General Ritchie reassembled his crippled forces in the neighbourhood of Gazala and Tobruk. Here pursuers and pursued gasped and glared at each other until the end of May, when Rommel was able to strike again.

*   *   *

On January 27 the debate began, and I laid our case before the House. I could see they were in a querulous temper, because when I had asked as soon as I got home that my forthcoming statement might be electrically recorded so that it could be used for broadcasting to the Empire and the United States, objection was taken on various grounds which had no relation to the needs of the hour. I therefore withdrew my request, although it would not have been denied in any other Parliament in the world. It was in such an atmosphere that I rose to speak.

I gave them some account of the Desert battle, but the House did not of course appreciate the significance of Rommel's successful counterstroke, for they could be given no inkling of the larger plans that would be opened by a swift British conquest of Tripolitania. The loss of Benghazi and Agedabia, which had already become public, seemed to be a part of the sudden ebbs and flows of desert warfare. Moreover, I had at the time no precise information as to what had happened, and why.

I presently came to the larger issue of our nakedness in the Far East:

> There never has been a moment, there never could have been a moment, when Great Britain or the British Empire, single-handed, could fight Germany and Italy, could wage the Battle of Britain, the Battle of the Atlantic, and the Battle of the Middle East, and at the same time stand thoroughly prepared in Burma, the Malay Peninsula, and generally in the Far East, against the impact of a vast military empire like Japan, with more than seventy mobile divisions, the third navy in the world, a great air force, and the thrust of eighty or ninety millions of hardy, warlike Asiatics. If we had started to scatter our forces over these immense areas in the Far East we should have been ruined. If we had moved large armies of troops urgently needed on the war fronts to regions which were not at war and might never be at war, we should have been altogether wrong. We should have cast away the chance, which has now become something more than a chance, of all of us emerg-

ing safely from the terrible plight in which we have been
plunged. . . .

The decision was taken to make our contribution to Russia, to
try to beat Rommel, and to form a stronger front from the Levant
to the Caspian. It followed from that decision that it was in our
power only to make a moderate and partial provision in the Far East
against the hypothetical danger of a Japanese onslaught. Sixty
thousand men, indeed, were concentrated at Singapore, but priority
in modern aircraft, in tanks, and in anti-aircraft and anti-tank artillery was accorded to the Nile Valley.

I had to burden the House for nearly two hours. They took what
they got without enthusiasm. But I had the impression that they were
not unconvinced by the argument. In view of what I saw coming
towards us I thought it well to end by putting things at their worst,
and making no promises while not excluding hope.

The debate then ran on for three days. But the tone was to me unexpectedly friendly. There was no doubt what the House would do.
My colleagues in the War Cabinet, headed by Mr. Atlee, sustained
the Government case with vigour and even fierceness. I had to wind up
on the 29th. At this time I feared that there would be no division. I
tried by taunts to urge our critics into the lobby against us without at
the same time offending the now thoroughly reconciled assembly. But
nothing that I dared say could spur any of the disaffected figures in the
Conservative, Labour, and Liberal Parties into voting. Luckily, when
the division was called the vote of confidence was challenged by the
Independent Labour Party who numbered three. Two were required
as tellers, and the result was four hundred and sixty-four to one. I was
grateful to James Maxton, the leader of the minority, for bringing the
matter to a head. Such a fuss had been made by the press that telegrams
of relief and congratulation flowed in from all over the Allied world.
The warmest were from my American friends at the White House. I
had sent congratulations to the President on his sixtieth birthday. "It is
fun," he cabled, "to be in the same decade with you." The naggers in
the press were not however without resource. They spun round with
the alacrity of squirrels. How unnecessary it had been to ask for a vote
of confidence! Who had ever dreamed of challenging the National
Government? These "shrill voices," as I called them, were but the un-
knowing heralds of approaching catastrophe.

*   *   *

I judged it impossible to hold an inquiry by Royal Commission
into the circumstances of the fall of Singapore while the war was

raging. We could not spare the men, the time, or the energy. Parliament accepted this view; but I certainly thought that in justice to the officers and men concerned there should be an inquiry into all the circumstances as soon as the fighting stopped. This however has not been instituted by the Government of the day.[1] Years have passed, and many of the witnesses are dead. It may well be that we shall never have a formal pronouncement by a competent court upon the worst disaster and largest capitulation in British history. In these pages I do not attempt to set myself up in the place of such a court or pronounce an opinion on the conduct of individuals. I have recorded elsewhere [2] the salient facts as I believe them. From these and from the documents written at the time the reader must form his own opinion.

It is at least arguable whether it would not have been better to concentrate all our strength on defending Singapore Island, merely containing the Japanese advance down the Malay Peninsula with light mobile forces. The decision of the commanders on the spot, which I approved, was to fight the battle for Singapore in Johore, but to delay the enemy's approach thereto as much as possible. The defence of the mainland consisted of a continuous retreat, with heavy rearguard actions and stubborn props. The fighting reflects high credit on the troops and commanders engaged. It drew in to itself however nearly all the reinforcements piecemeal as they arrived. Every advantage lay with the enemy. There had been minute prewar study of the ground and conditions. Careful large-scale plans and secret infiltration of agents, including even hidden reserves of bicycles for Japanese cyclists, had been made. Superior strength and large reserves, some of which were not needed, had been assembled. All the Japanese divisions were adept in jungle warfare.

The Japanese mastery of the air, arising, as has been described, from our bitter needs elsewhere, and for which the local commanders were in no way responsible, was another deadly fact. In the result the main fighting strength of such an army as we had assigned to the defence of Singapore, and almost all the reinforcements sent after the Japanese declaration of war, were used up in gallant fighting on the peninsula, and when these had crossed the causeway to what should have been their supreme battleground their punch was gone. Here they rejoined the local garrison and the masses of base details which swelled our numbers though not our strength. The army which could fight the decisive struggle for Singapore and had been provided for that supreme objective in this theatre was dissipated before the Japanese attack began. It might be a hundred thousand men; but it was an army no more.

* * *

[1] Written in 1951. [2] *The Hinge of Fate:* Chapter VI.

It soon became clear that General Wavell, now Supreme Allied Commander of these eastern regions, had already doubts of our ability to maintain a prolonged defence of Singapore. I had counted much upon the island and fortress standing a siege requiring heavy artillery to be landed, transported, and mounted by the Japanese. Before I left Washington I still contemplated a resistance of at least two months. I watched with misgivings but without effective intervention the consumption of our forces in their retreat through the Malay Peninsula. On the other hand, there was the gain of precious time.

But on January 16 Wavell telegraphed: "Until quite recently all plans were based on repulsing sea-borne attacks on [Singapore] island and holding land attack in Johore or farther north, and little or nothing was done to construct defences on north side of island to prevent crossing Johore Straits, though arrangements have been made to blow up the causeway. The fortress cannon of heaviest nature have all-round traverse, but their flat trajectory makes them unsuitable for counter-battery work. Could certainly not guarantee to dominate enemy siege batteries with them . . ."

It was with feelings of painful surprise that I read this message on the morning of the 19th. So there were no permanent fortifications covering the landward side of the naval base and of the city! Moreover, even more astounding, no measures worth speaking of had been taken by any of the commanders since the war began, and more especially since the Japanese had established themselves in Indo-China, to construct field defences. They had not even mentioned the fact that they did not exist.

All that I had seen or read of war had led me to the conviction that, having regard to modern fire power, a few weeks will suffice to create strong field defences, and also to limit and canalise the enemy's front of attack by minefields and other obstructions. Moreover, it had never entered into my head that no circle of detached forts of a permanent character protected the rear of the famous fortress. I cannot understand how it was I did not know this. But none of the officers on the spot and none of my professional advisers at home seem to have realised this awful need. At any rate, none of them pointed it out me — not even those who saw my telegrams based upon the false assumption that a regular siege would be required. I had read of Plevna in 1877, where, before the era of machine guns, defences had been improvised by the Turks in the actual teeth of the Russian assault; and I had examined Verdun in 1917, where a field army lying in and among detached forts had a year earlier made so glorious a record. I had put my faith in the enemy being compelled to use artillery on a very large scale in order to pulverise our strong points at Singapore, and in the almost prohibitive difficulties and long delays which would impede such an artillery concentration and the gathering of ammunition along Malayan communications. Now, suddenly, all this vanished away, and I saw before me the hideous spectacle of the almost naked island and of the wearied, if not exhausted, troops retreating upon it.

I do not write this in any way to excuse myself. I ought to have known. My advisers ought to have known and I ought to have been told, and I ought to have asked. The reason I had not asked about this matter, amid the thousands of questions I put, was that the possibility of Singapore having no landward defences no more entered into my mind than that of a battleship being launched without a bottom. I am aware of the various reasons that have been given for this failure: the preoccupation of the troops in training and in building defence works in Northern Malaya; the shortage of civilian labour; prewar financial limitations and centralised War Office control; the fact that the Army's role was to protect the naval base, situated on the north shore of the island, and that it was therefore their duty to fight in front of that shore and not along it. I do not consider these reasons valid. Defences should have been built.

My immediate reaction was to repair the neglect so far as time allowed, but when I awoke on the morning of the 21st the following most pessimistic telegram from General Wavell lay at the top of my box:

> Officer whom I had sent to Singapore for plans of defence of island has now returned. Schemes are now being prepared for defence of northern part of island. *Number of troops required to hold island effectively probably are as great as or greater than number required to defend Johore.*[3] I have ordered Percival [the Commander-in-Chief] to fight out the battle in Johore, but to work out plans to prolong resistance on island as long as possible should he lose Johore battle. I must warn you however that I doubt whether island can be held for long once Johore is lost. The fortress guns are sited for use against ships, and have mostly ammunition for that purpose only; many can only fire seawards.[4] Part of garrison has already been sent into Johore, and many troops remaining are doubtful value. I am sorry to give you depressing picture, but I do not want you to have false picture of the island fortress. Singapore defences were constructed entirely to meet seaward attack. I still hope Johore may be held till next convoy arrives.

[3] My italics.—W.S.C.
[4] This is inaccurate. The majority of the guns could fire landward also.

I pondered over this message for a long time. So far I had thought only of animating, and as far as possible compelling, the desperate defence of the island, the fortress, and the city, and this in any case was the attitude which should be maintained unless any decisive change of policy was ordered. But now I began to think more of Burma and of the reinforcements on the way to Singapore. These could be doomed or diverted. There was still ample time to turn their prows northward to Rangoon. I therefore prepared the following minute to the Chiefs of Staff, and gave it to General Ismay in time for their meeting at 11.30 A.M. on the 21st. I confess freely however that my mind was not made up. I leaned upon my friends and counsellors. We all suffered extremely at this time.

In view of this very bad telegram from General Wavell, we must reconsider the whole position at a Defence Committee meeting tonight.

We have already committed exactly the error which I feared . . . Forces which might have made a solid front in Johore, or at any rate along the Singapore waterfront, have been broken up piecemeal. No defensive line has been constructed on the landward side. No defence has been made by the Navy to the enemy's turning movements on the west coast of the peninsula. General Wavell has expressed the opinion that it will take more troops to defend Singapore Island than to win the battle in Johore. The battle in Johore is almost certainly lost.

His message gives little hope for prolonged defence. It is evident that such defence would be only at the cost of all the reinforcements now on the way. If General Wavell is doubtful whether more than a few weeks' delay can be obtained, the question arises whether we should not at once blow the docks and batteries and workshops to pieces and concentrate everything on the defence of Burma and keeping open the Burma Road.

2. It appears to me that this question should be squarely faced now and put bluntly to General Wavell. What is the value of Singapore [to the enemy] above the many harbours in the Southwest Pacific if all naval and military demolitions are thoroughly carried out? On the other hand, the loss of Burma would be very grievous. It would cut us off from the Chinese, whose troops have been the most successful of those yet engaged against the Japanese. We may, by muddling things and hesitating to take an ugly decision, lose *both* Singapore and the Burma Road. Obviously the decision depends upon how long the defence of Singapore Island can be maintained. If it is only for a few weeks, it is certainly not worth losing all our reinforcements and aircraft.

3. Moreover, one must consider that the fall of Singapore, accompanied as it will be by the fall of Corregidor, will be a tremendous shock to India, which only the arrival of powerful forces and successful action on the Burma front can sustain.

Pray let all this be considered this morning.

The Chiefs of Staff reached no definite conclusion, and when we met in the evening at the Defence Committee a similar hesitation to commit ourselves to so grave a step prevailed. The direct initial responsibility lay with General Wavell as Allied Supreme Commander. Personally I found the issue so difficult that I did not press my new view, which I should have done if I had been resolved. We could none of us foresee the collapse of the defence which was to occur in little more than three weeks. A day or two could at least be spared for further thought.

*   *   *

Sir Earle Page, the Australian representative, did not of course attend the Chiefs of Staff Committee, nor did I invite him to the Defence Committee. By some means or other he was shown a copy of my minute to the Chiefs of Staff. He immediately telegraphed to his Government, and on January 24 we received a message from the Australian Prime Minister, Mr. Curtin, of which the following passages are material:

> ... Page has reported that the Defence Committee has been considering the evacuation of Malaya and Singapore. After all the assurances we have been given, the evacuation of Singapore would be regarded here and elsewhere as an inexcusable betrayal. . . . We understood that it was to be made impregnable, and in any event it was to be capable of holding out for a prolonged period until the arrival of the main fleet.
>
> Even in an emergency diversion of reinforcements should be to the Netherlands East Indies and not Burma. Anything else would be deeply resented, and might force the Netherlands East Indies to make a separate peace.
>
> On the faith of the proposed flow of reinforcements, we have acted and carried out our part of the bargain. We expect you not to frustrate the whole purpose by evacuation. . . .

Every allowance must be made for the state of mind into which the Australian Government were thrown by the hideous efficiency of the Japanese war machine. The command of the Pacific was lost; their three best divisions were in Egypt and a fourth at Singapore. They realised that Singapore was in deadly peril, and they feared an actual invasion of Australia itself. All their great cities, containing more than half the whole population of the continent, were on the sea coast. A mass exodus into the interior and the organising of a guerrilla without arsenals or supplies stared them in the face. Help from the Mother Country was far away, and the power of the United States could only slowly be established in Australasian waters. I did not myself believe that the Japanese would invade Australia across three thousand miles of ocean, when they had so much alluring prey in their clutch in the Dutch East Indies and Malaya. The Australian Cabinet saw the scene in a different light, and deep forebodings pressed upon them all. Even in these straits they maintained their party divisions rigidly. The Labour Government majority was only two. They were opposed to compulsory service even for Home Defence. Although the Opposition was admitted to the War Council no National Government was formed.

Mr. Curtin's telegram was nevertheless both serious and unusual. The expression "inexcusable betrayal" was not in accordance with the truth or with military facts. A frightful disaster was approaching. Could we avoid it? How did the balance of loss and gain stand? At this time the destination of important forces still rested in our control. There is no "betrayal" in examining such issues with a realistic eye. Moreover, the Australian War Committee could not measure the whole situation. Otherwise they would not have urged the complete neglect of Burma, which was proved by events to be the only place we still had the means to save.

It is not true to say that Mr. Curtin's message decided the issue. If we had all been agreed upon the policy we should, as I had suggested, certainly have put the case "bluntly" to Wavell. I was conscious however of a hardening of opinion against the abandonment of this renowned key point in the Far East. The effect that would be produced all over the world, especially in the United States, of a British "scuttle" while the Americans fought on so stubbornly at Corregidor was terrible to imagine. There is no doubt what a purely military decision should have been. By general agreement or acquiescence however all efforts were made to reinforce Singapore and to sustain its defence. The 18th British Division, part of which had already landed, went forward on its way.

The value of these and other reinforcements however was less than their numbers suggest. They needed time to get on their tactical feet, and had to be thrown into the losing battle as soon as they were landed. Great hopes were pinned on the Hurricane fighters, of which a considerable quantity had been sent. Here at last were aircraft of quality to match the Japanese. They were assembled with all speed and took the air. For a few days indeed they did much damage, but the conditions were strange to the newly arrived pilots, and before long the Japanese superiority in numbers began increasingly to take its toll. They dwindled fast. The Japanese now had a full five divisions. They came rapidly down the coast and on January 27 General Percival decided to retire to Singapore Island. Every man and vehicle had in the final stage to pass over the Causeway thither. The greater part of one brigade was lost in the early stages, but on the morning of January 31 the rest of the force had crossed and the Causeway was blown up behind them.

At home we no longer nursed illusions about a protracted defence. The only question was how long. Those of the heavy guns of the coast defences which could fire northward were not of much use, with their limited ammunition, against the jungle-covered country in which the enemy was gathering. Only one squadron of fighter aircraft remained on the island, and only a single aerodrome was now usable. Losses and wastage had reduced the numbers of the garrison, now finally concentrated, from the 106,000 estimated by the War Office to about 85,000 men, including base and administrative units and various non-combatant corps. Of this total probably 70,000 were armed. The preparation of field defences and obstacles, though representing a good deal of local effort, bore no relation to the mortal needs which now arose. There were no permanent defences on the front about to be attacked. The spirit of the Army had been largely reduced by the long retreat and hard fighting on the peninsula. Behind all lay the city of Singapore, which at that time sheltered a population of perhaps a million of many races and a host of refugees.

*   *   *

On the morning of February 8, patrols reported that the enemy were massing in the plantations northwest of the island, and our positions were heavily shelled. At 10.45 P.M. the leading waves of assault were carried across the Johore Strait in armoured landing craft brought, as the result of long and careful planning, to the launching sites by road. There was very heavy fighting and many craft were sunk, but the Australians were thin on the ground and enemy parties got ashore at many points. Next evening a new and similar attack developed around the Causeway, and again the enemy succeeded in gaining a footing. February 11 was a day of confused fighting on the whole front. The Causeway had been breached towards the enemy's end, and they were able to repair it rapidly as soon as our covering troops withdrew. The Japanese Imperial Guards advanced across it that night. On the 13th the prepared scheme for evacuating to Java by sea some three thousand nominated individuals was put into effect. Those ordered to go included key men, technicians, surplus staff officers, nurses, and others whose services would be of special value for the prosecution of the war.

By now conditions in the city of Singapore were shocking. Civil labour had collapsed, failure of the water supply seemed imminent, and reserves of food and ammunition for the troops had been seriously depleted by the loss of depots now in enemy hands. By this time the programme of organised demolitions had been put in hand. The guns of the fixed defences and nearly all field and anti-aircraft guns were destroyed, together with secret equipment and documents. All aviation petrol and aircraft bombs were burnt or blown up. Some confusion arose concerning demolitions in the naval base. The orders were issued, the floating dock was sunk and the caisson and pumping machinery of the graving-dock destroyed, but much else in the full plan was left incomplete. On the 14th Wavell sent me the following message, which seemed conclusive:

> Have received telegram from Percival that enemy are close to town and that his troops are incapable of further counterattack. Have ordered him to continue inflict maximum damage to enemy by house-to-house fighting if necessary. Fear however that re-sistance not likely to be very prolonged.

*　　　　*　　　　*

The reader will recall my minute to the Chiefs of Staff of January 21 about abandoning the defence of Singapore and diverting reinforcements to Rangoon, and how I did not press this point of view. When all our hearts hardened on fighting it out at Singapore, the only chance of success, and indeed of gaining time, which was all we could hope for, was to give imperative orders to fight in desperation to the end. These orders were accepted and endorsed by General Wavell, who indeed put the utmost pressure on General Percival. It is always right that whatever may be the doubts at the summit of war direction the general on the spot should have no knowledge of them and should receive instructions which are simple and plain. But now when it was certain that all was lost at Singapore I was sure it would be wrong to enforce needless slaughter, and without hope of victory to inflict the horrors of street fighting on the vast city, with its teeming, helpless, and now panic-stricken population. I told General Brooke where I stood, and found that he also felt that we should put no more pressure from home upon General Wavell, and should authorise him to take the inevitable decision, for which we should share the responsibility.

Sunday, February 15, 1942, was the day of the capitulation. There were only a few days of military food reserves, gun ammunition was very short, there was practically no petrol left for vehicles. Worst of all, the water supply was expected to last only another twenty-four hours. General Percival was advised by his senior commanders that of the two alternatives, counterattack or surrender, the first was beyond the capacity of the exhausted troops. He decided upon capitulation. The Japanese demanded and received unconditional surrender. Hostilities closed at 8.30 P.M.

8

The U-Boat Paradise

DESPITE A SUBSTANTIAL RECONSTRUCTION of the Government, my own position had not seemed to be affected in all this period of political tension and change at home and disaster abroad. I was too much occupied with hourly business to have much time for brooding upon it. My personal authority even seemed to be enhanced by the uncertainties affecting several of my colleagues or would-be colleagues. I did not suffer from any desire to be relieved of my responsibilities. All I wanted was compliance with my wishes after reasonable discussion. Misfortunes only brought me and the Chiefs of Staff closer together, and this unity was felt through all the circles of the Government. There was no whisper of intrigue or dissidence, either in the War Cabinet or in the much larger number of Ministers of Cabinet rank. From outside however there was continuous pressure to change my method of conducting the war, with a view to obtaining better results than were now coming in. "We are all with the Prime Minister, but he has too much to do. He should be relieved of some of the burdens that fall upon him." This was the persistent view, and many theories were pressed. But I was entirely resolved to keep my full power of war-direction. This could only be exercised by combining the offices of Prime Minister and Minister of Defence. More difficulty and toil are often incurred in overcoming opposition and adjusting divergent and conflicting views than by having the right to give decisions oneself. It is most important that at the summit there should be one mind playing over the whole field, faithfully aided and corrected, but not divided in its integrity. I should not of course have remained Prime Minister for an hour if I had been deprived of the office of Minister of Defence. The fact that this was widely known repelled all challenges, even under the most unfavourable conditions, and many well-

meant suggestions of committees and other forms of impersonal machinery consequently fell to the ground. I must record my gratitude to
all who helped me to succeed.

But the year 1942 was to provide many rude shocks. For the first
six months everything went ill. In the Atlantic it proved the toughest
of the whole war. The U-boat fleet had grown to nearly two hundred
and fifty, of which Admiral Doenitz could report nearly a hundred
operational, with fifteen more a month. They ravaged American waters
almost uncontrolled. By the end of January thirty-one ships, of nearly
200,000 tons, had been sunk off the United States and Canadian coast.
Soon the attack spread southward off Hampton Roads and Cape Hat-
teras, and thence to the cost of Florida. The great sea highway teemed
with defenceless American and Allied shipping. Along it the precious
tanker fleet moved in unbroken procession to and from the oil ports of
Venezuela and the Gulf of Mexico, and here and in the Caribbean, amid
a wealth of targets, the U-boats chose to prey chiefly on the tankers.
Neutrals of all kinds were assailed. Week by week the scale of massacre grew. In February they destroyed seventy-one ships, of 384,000
tons, in the Atlantic, of which all but two were sunk in the American
zone. This was the highest rate of loss which we had so far suffered. It
was soon to be surpassed.

All this destruction, far exceeding anything known in this war, though
not reaching the catastrophic figures of the worst period of 1917, was
caused by no more than twelve to fifteen boats working in the area at
one time. The protection afforded by the United States Navy was for
several months hopelessly inadequate. It is surprising indeed that during two years of the advance of total war towards the American continent more provision had not been made against this deadly onslaught.
Under the President's policy of "all aid to Britain short of war" much
had been done for us. We had acquired fifty old destroyers and
ten American revenue cutters. In exchange we had given the invaluable
West Indian bases. But the vessels were now sadly missed by our Ally.
After Pearl Harbour the Pacific pressed heavily on the United States
Navy. Still, with all the information they had about our protective
measures, both before and during the struggle, it is remarkable that no
plans had been made for coastal convoys and for multiplying small craft.
Neither had the Coastal Air Defence been developed. The American
Army Air Force, which controlled almost all military shore-based aircraft, had no training in anti-submarine warfare, whereas the Navy,
equipped with float planes and amphibians, had not the means to carry
it out, and in these crucial months an effective American defence system
was only achieved with painful, halting steps.

Our disasters might have been far greater had the Germans sent their

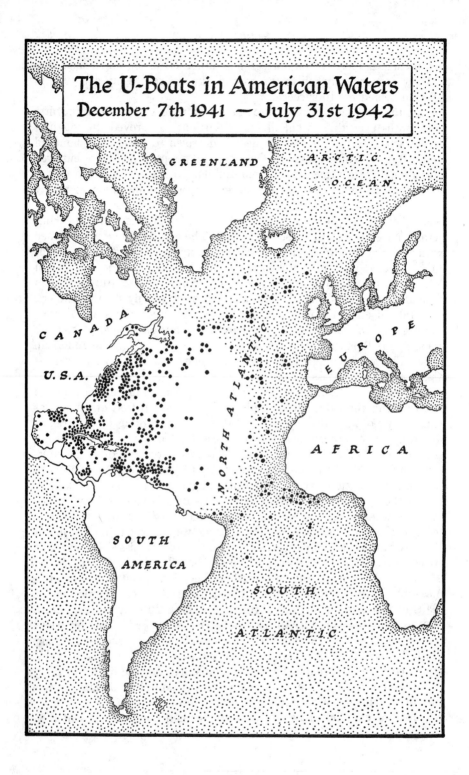

The U-Boats in American Waters
December 7th 1941 — July 31st 1942

GREENLAND

ARCTIC
OCEAN

CANADA

U.S.A.

NORTH ATLANTIC

EUROPE

AFRICA

SOUTH
AMERICA

SOUTH

ATLANTIC

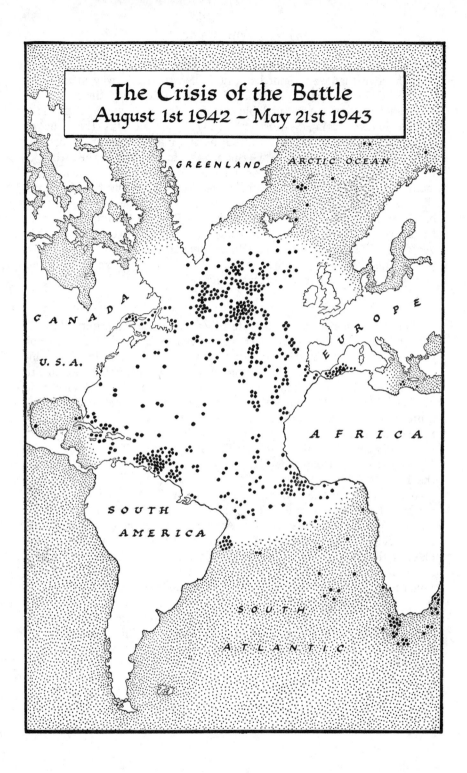

The Crisis of the Battle
August 1st 1942 ~ May 21st 1943

GREENLAND

ARCTIC OCEAN

CANADA

U.S.A.

EUROPE

AFRICA

SOUTH
AMERICA

SOUTH
ATLANTIC

heavy surface ships raiding into the Atlantic, but Hitler was obsessed with the idea that we intended to invade Northern Norway at an early date. With his powerful one-track mind he sacrificed a glittering chance and concentrated every available surface ship and many a precious U-boat in Norwegian waters. "Norway," he said, "is the zone of destiny in this war." It was indeed, as the reader is aware, most important, but at this time the German opportunity lay in the Atlantic. In vain the admirals argued for a naval offensive. Their Fuehrer remained adamant, and his strategic decision was strengthened by the shortage of oil fuel. Already in January he had sent the *Tirpitz,* his only battleship, but the strongest in the world, to Trondheim, and on the 12th he determined to recall to their home ports the battlecruisers *Scharnhorst* and *Gneisenau,* which had been blockaded in Brest for nearly a year. This led to an incident which caused so much commotion and outcry in England that it requires a digression.

*   *   *

The serious losses in the Mediterranean and the temporary disablement of our whole Eastern Fleet had forced us to send almost all our torpedo-carrying aircraft to protect Egypt against potential overseas invasion. But all possible preparations were made to watch Brest and to attack any sortie with bomb and torpedo by air and sea. Mines were also laid along the presumed route both in the Channel and near the Dutch coast. The Admiralty expected that the passage of the Dover Strait would be attempted by night; but the German admiral preferred to use darkness to elude our patrols when leaving Brest and run the Dover batteries in daylight. He sailed from Brest before midnight on the 11th.

The morning of the 12th was misty, and when the enemy ships were spotted, the radar of our patrolling aircraft broke down. Our shore radar also failed to detect them. At the time we thought this an unlucky accident. We have learnt since the war that General Martini, the chief of the German radar, had made a careful plan. The German jamming, hitherto somewhat ineffective, was invigorated by the addition of much new equipment, but in order that nothing should be suspicious it was brought into operation gradually, so that the jamming should appear only a little more vicious each day. Our operators therefore did not complain unduly, and nobody suspected anything unusual. By February 12 however the jamming had grown so strong that our sea-watching radar was in fact useless. It was not until 11.25 A.M. that the Admiralty received the news. By then the escaping cruisers and their powerful air and destroyer escort were within twenty miles of Boulogne. Soon after noon the Dover batteries opened fire with their heavy guns, and the first striking force of five motor torpedo boats immediately put to sea and attacked. Six torpedo-carrying Swordfish aircraft from Manston, in Kent, led by Lieutenant-Commander Esmonde (who had led the first attack on the *Bismarck*), set off without waiting for more than ten Spitfires in support. The Swordfish, fiercely attacked by enemy fighters, discharged their torpedoes against the enemy, but at a heavy cost. None returned, and only five survivors were rescued. Esmonde was awarded a posthumous V.C.

Successive waves of bombers and torpedo bombers assailed the enemy till nightfall. There was much bitter and confused fighting with the German fighters, in which we suffered more severe losses than the enemy with his superior numbers. When the German cruisers were off the Dutch coast at about 3.30 P.M. five destroyers from Harwich pressed home an attack, launching their torpedoes at about three thousand yards under tremendous fire. Unscathed either by the Dover batteries or the torpedoes, the squadron held its course, and by the morning of the 13th all their ships had reached home. The news astonished the British public, who could not understand what appeared to them, not unnaturally, to be a proof of the German mastery of the English Channel. Our Secret Service soon found out that both the *Scharnhorst* and the *Gneisenau* had fallen victims to our air-laid mines. It was six months before the *Scharnhorst* was capable of service, and the *Gneisenau* never appeared again in the war. This however could not be made public and national wrath was vehement.

To allay complaints an official inquiry was held, which reported the publishable facts. Viewed in the afterlight and in its larger aspects the episode was highly advantageous to us. "When I speak on the radio next Monday evening," cabled the President, "I shall say a word about these people who treat the episode in the Channel as a defeat. I am more and more convinced that the location of all the German ships in Germany makes our joint North Atlantic naval problem more simple." But it looked very bad at the time to everyone in the Grand Alliance outside our most secret circles.

*   *   *

Meanwhile havoc continued to reign along the Atlantic coast of the United States. A U-boat commander reported to Doenitz that ten times as many U-boats could find ample targets. Resting on the bottom during daylight, they used their high surface speed at night to select the richest prey. Nearly every torpedo they carried claimed its victim, and when torpedoes were expended the gun was almost as effective. The towns of the Atlantic shore, where for a while the waterfronts remained fully lighted, heard nightly the sounds of battle, saw the burning, sinking ships offshore, and rescued the survivors and wounded. There was bitter anger against the Administration, which was much embarrassed. It is however easier to infuriate Americans than to cow them.

In London we had marked these misfortunes with anxiety and grief. On February 10 we offered unasked twenty-four of our best-equipped anti-submarine trawlers and ten corvettes with their trained crews to the American Navy. These were welcomed by our Ally, and the first arrived in New York early in March. It was little enough, but the utmost we could spare. "'Twas all she gave — 'twas all she had to give." Coastal convoys could not begin until an organisation had been built up and escorts scraped together. The available fighting ships and aircraft were at first used only to patrol threatened areas. The enemy easily evaded them and hunted elsewhere. The main stress now fell between Charleston and New York, while single U-boats prowled over all the Caribbean and the Gulf of Mexico, with a freedom and insolence which were hard to bear. The sinkings were nearly half a million tons, mostly within three hundred miles of the American coast, and nearly half were tankers. Only two U-boats were sunk in American waters by American aircraft, and first kill off the American coast by a surface vessel was not made until April 14, by the United States destroyer *Roper*.

In Europe March closed with the brilliant and heroic exploit of St. Nazaire. This was the only place along all the Atlantic coast where the *Tirpitz* could be docked for repair if she were damaged. If the dock, one of the largest in the world, could be destroyed a sortie of the *Tirpitz* from Trondheim into the Atlantic would become far more dangerous and might not be deemed worth making. Our Commandos were eager for the fray, and here was a deed of glory intimately involved in high strategy. Led by Commander Ryder of the Royal Navy, with Colonel Newman of the Essex Regiment, an expedition of destroyers and light coastal craft sailed from Falmouth on the afternoon of March 26 carrying about two hundred and fifty Commando troops. They had four hundred miles to traverse through waters under constant enemy patrol, and five miles up the estuary of the Loire.

The goal was the destruction of the gates of the great lock. The *Campbeltown*, one of the fifty old American destroyers, carrying three tons of high explosive in her bows, drove into the lock gates, in the teeth of a close and murderous fire. Here, led by Lieutenant-Commander Beattie, she was scuttled, and the fuzes of her main demolition charges set to explode later. From her decks Major Copeland, with a landing party, leaped ashore to destroy the dock machinery. The Germans met them in overwhelming strength, and furious fighting began. All but five of the landing party were killed or captured. Commander Ryder's craft, although fired on from all sides, miraculously remained afloat during his break for the open sea with the remnants of his force, and got safely home. But the great explosion was still to come. Something had gone wrong with the fuze. It was not till the next day, when a large group of German officers and technicians were inspecting the wreck of the *Campbeltown,* jammed in the lock gates, that the ship blew up, with devastating force, killing hundreds of Germans and shattering the great lock for the rest of the war. The Germans treated the prisoners, four of whom received the Victoria Cross, with respect, but severe punishment was inflicted on the brave Frenchmen who on the spur of the moment rushed from every quarter to the aid of what they hoped was the vanguard of liberation.

*   *   *

On April 1 it at last became possible for the United States Navy to start a partial convoy system. At first this could be no more than daylight hops of about a hundred and twenty miles between protected anchorages by groups of vessels under escort, and all shipping was brought to a standstill at night. On any one day there ware upwards of a hundred and twenty ships requiring protection between Florida and New York. The consequent delays were misfortune in another form. It was not until May 14 that the first fully organised convoy sailed from Hampton Roads for Key West. Thereafter the system was quickly extended northward to New York and Halifax, and by the end of the month the chain along the east coast from Key West northward was at last complete. Relief was immediate, and the losses fell.

Admiral Doenitz forthwith changed his point of attack to the Caribbean and the Gulf of Mexico, where convoys were not yet working. Ranging farther, the U-boats also began to appear off the coast of Brazil and in the St. Lawrence river. It was not until the end of the year that a complete interlocking convoy system covering all these immense areas became fully effective. But June saw an improvement, and the last days of July may be taken as closing the terrible massacre along the American coast. In seven months the Allied losses in the Atlantic from U-boats alone amounted to over three million tons, which included 181 British ships of 1,130,000 tons. Less than one-tenth occurred in convoys. All this cost the enemy up to July no more than fourteen U-boats sunk throughout the Atlantic and Arctic Oceans, and of these kills only six were in North American waters.

Thereafter we regained the initiative. In July alone five U-boats were destroyed off the Atlantic coast, besides six more German and three Italian elsewhere. This total of fourteen for the month, half by convoy escorts, gave us encouragement. It was the best so far achieved; but the number of new boats coming into service each month still exceeded the rate of our kills. Moreover, whenever we began to win Admiral Doenitz shifted his U-boats. With the oceans to play in he could always gain a short period of immunity in a new area. In May a transatlantic convoy lost seven ships, about 700 miles west of Ireland. This was followed by an onslaught near Gibraltar and the reappearance of U-boats around Freetown. Once more Hitler came to our aid by insisting that a group of U-boats should be held ready to ward off an Allied attempt to occupy the Azores or Madeira. His thought in this direction was not altogether misplaced, but his demand coincided with the end of the halcyon days on the American coast.

The U-boat attack was our worst evil. It would have been wise for the Germans to stake all upon it. I remember hearing my father say, "In politics when you have got hold of a good thing, stick to it." This is also a strategic principle of importance. Just as Goering repeatedly shifted his air targets in the Battle of Britain in 1940, so now the U-boat warfare was to some extent weakened for the sake of competing attractions. Nevertheless it constituted a terrible event in a very bad time.

*   *   *

It will be well here to relate the course of events elsewhere and to record briefly the progress of the Atlantic battle up to the end of 1942.

In August the U-boats turned their attention to the area around Trinidad and the north coast of Brazil, where the ships carrying bauxite to the United States for the aircraft industry and the stream of outward-bound ships with supplies for the Middle East offered the most attractive targets. Others were at work near Freetown; some ranged as far south as the Cape of Good Hope, and a few even penetrated into the Indian Ocean. For a time the South Atlantic caused us anxiety. Here in September and October five large homeward-bound liners sailing independently were sunk, but all our troop transports outward bound for the Middle East in convoy came through unscathed. Among the big ships lost was the *Laconia*, of nearly 20,000 tons, carrying two thousand Italian prisoners of war to England. Many were drowned.

The main battle was by now once more joined along the great convoy routes in the North Atlantic. The U-boats had already learned to respect the power of the air, and in their new assault they worked almost entirely in the central section, beyond the reach of aircraft based on Iceland and Newfoundland. Two convoys were severely mauled in August, one of them losing eleven ships, and during this month U-boats sank 108 vessels, amounting to over half a million tons. In September and October the Germans reverted to the earlier practice of submerged attack by day. With the larger numbers now working in "wolf packs," and with our limited resources, serious losses in convoy could not be prevented, and we felt most acutely the lack of enough very long-range (V.L.R.) aircraft in the Coastal Command. Air cover still ranged no more than about six hundred miles from our shore bases, and only about four hundred from Newfoundland, leaving a large unguarded gap in the centre of the Atlantic Ocean where the surface escorts could gain no help from the air. Against this distressing background our airmen did their utmost.

Naval escorts could never range widely from the convoys and break up the heavy concentrations on the flanks. Thus, when the "wolf packs" struck they could saturate our defence. The only remedy was to surround each convoy with enough aircraft to find and force any nearby U-boats to dive, and thus provide an unmolested lane. Even this was not enough. We must seek out and attack them vigorously wherever we could find them, both by sea and air. The aircraft, the trained air crews, and the air weapons were still too few, but we now made a start by forming a "Support Group" of surface forces.

The idea had long been advocated, but the means were lacking. The first of these Support Groups, which later became a most potent factor in the U-boat war, consisted of two sloops, four of the new frigates now coming out of the builders' yards, and four destroyers. With highly trained and experienced crews and the latest weapons, working independently of the convoy escorts and untrammelled by other responsibilities, their task, in co-operation with the Air, was to seek, hunt, and destroy. In 1943 an aircraft would often guide a Support Group to its prey, the pursuit of one U-boat would disclose others, and a "pack" would be discovered.

Aircraft which would accompany the convoys were also provided. By the end of 1942 six "escort carriers" were in service. Eventually many were built in America, besides others in Great Britain, and the first of them, the *Avenger*, sailed with a North Russian convoy in September. They made their first effective appearance with the North African convoys in late October. Equipped with naval Swordfish aircraft, they met the need — namely, all-round reconnaissance in depth, independent of land bases, and in intimate collaboration with the surface escorts. Thus by the utmost exertions and ingenuity we began to win; but the power of the enemy was also growing and we had many setbacks.

Between January and October 1942 the number of U-boats had more than doubled. One hundred ninety-six were operational, and our North Atlantic convoys were subjected to fiercer and larger packs than ever before. All our escorts had to be cut to the bone for the sake of our main operations in Africa, and in November our losses at sea were the heaviest of the whole war, including 117 ships, of over 700,000 tons, by U-boats alone, and another 100,000 tons from other causes.

So menacing were the conditions in the outer waters that on November 4 I personally convened a new Anti-U-boat Committee. Its power to take far-reaching decisions played no small part in the conflict. In a great effort to lengthen the range of our radar-carrying Liberator aircraft, we decided to withdraw them from action until the necessary improvements were made. The President at my request sent all suitable American aircraft, fitted with the latest type of radar, to work from the United Kingdom. We were presently able to resume operations in the Bay of Biscay in greater strength and with far better equipment. All this was to bring its reward in 1943.

9

American Naval Victories[1]

The Coral Sea and Midway Island

STIRRING EVENTS affecting the whole course of the war now occurred in the Pacific Ocean. By the end of March the first phase of the Japanese war plan had achieved a success so complete that it surprised even its authors. Japan was master of Hong Kong, Siam, Malaya, and nearly the whole of the immense island region forming the Dutch East Indies. Japanese troops were plunging deeply into Burma. In the Philippines the Americans still fought on at Corregidor, but without hope of relief.

Japanese exultation was at its zenith. Pride in their martial triumphs and confidence in their leadership was strengthened by the conviction that the Western Powers had not the will to fight to the death. Already the Imperial armies stood on the frontiers so carefully chosen in their pre-war plans as the prudent limit of their advance. Within this enormous area, comprising measureless resources and riches, they could consolidate their conquests and develop their newly won power. Their long-prepared scheme had prescribed a pause at this stage to draw breath, to resist an American counterattack or to organise a further advance. But now in the flush of victory it seemed to the Japanese leaders that the fulfillment of their destiny had come. They must not be unworthy of it. These ideas arose not only from the natural temptations to which dazzling success exposes mortals, but from serious military reasoning. Whether it was wiser to organise their new perimeter thoroughly or by surging forward to gain greater depth for its defence seemed to them a balanced strategic problem.

After deliberation in Tokyo the more ambitious course was adopted. It was decided to extend the grasp outward to include the Western Aleutians, Midway Island, Samoa, Fiji, New Caledonia, and Port

[1] See *Coral Sea, Midway, and Submarine Actions*, by Captain S. E. Morison, U.S. Navy.

Moresby in Southern New Guinea.[2] This expansion would threaten Pearl Harbour, still the main American base. It would also, if maintained, sever direct communication between the United States and Australia. It would provide Japan with suitable bases from which to launch further attacks.

The Japanese High Command had shown the utmost skill and daring in making and executing their plans. They started however upon a foundation which did not measure world forces in true proportion. They never comprehended the latent might of the United States. They thought still, at this stage, that Hitler's Germany would triumph in Europe. They felt in their veins the surge of leading Asia forward to measureless conquests and their own glory. Thus they were drawn into a gamble, which even if it had won would only have lengthened their predominance by perhaps a year, and, as they lost, cut it down by an equal period. In the actual result they exchanged a fairly strong and gripped advantage for a wide and loose domain, which it was beyond their power to hold; and, being beaten in this outer area, they found themselves without the forces to make a coherent defence of their inner and vital zone.

Nevertheless at this moment in the world struggle no one could be sure that Germany would not break Russia, or drive her beyond the Urals, and then be able to come back and invade Britain; or as an alternative spread through the Caucasus and Persia to join hands with the Japanese vanguards in India. To put things right for the Grand Alliance there was needed a decisive naval victory by the United States, carrying with it predominance in the Pacific, even if the full command of that ocean were not immediately established. This victory was not denied us. I had always believed that the command of the Pacific would be regained by the American Navy, with any help we could give from or in the Atlantic, by May. Such hopes were based only upon a computation of American and British new construction, already maturing, of battleships, aircraft carriers, and other vessels. We may now describe in a necessarily compressed form the brilliant and astonishing naval battle which asserted this majestic fact in an indisputable form.

*   *   *

At the end of April 1942 the Japanese High Command began their new policy of expansion. This was to include the capture of Port Moresby and the seizure of Tulagi, in the Southern Solomons, opposite the large island of Guadalcanal. The occupation of Port Moresby would complete the first stage of their domination of New Guinea and give added security to their advanced naval base at Rabaul, in New Britain.

---

[2] See map on facing page.

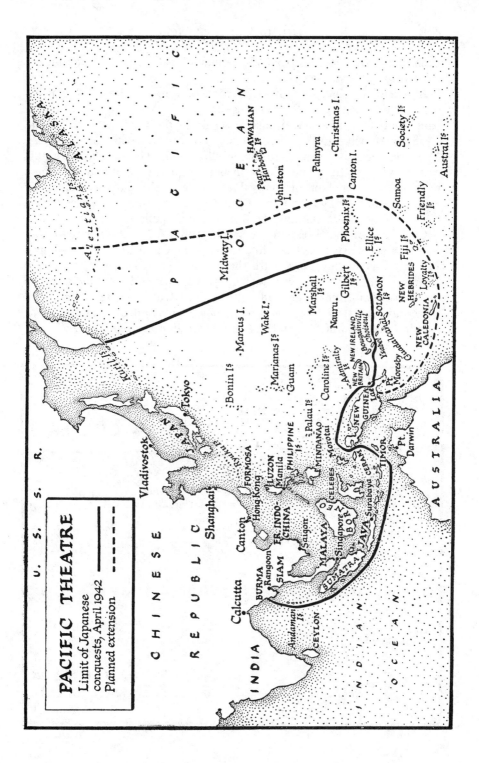

PACIFIC THEATRE

Limit of Japanese
conquests, April 1942 ———
Planned extension ------

From New Guinea and from the Solomons they could begin the envelopment of Australia.

American Intelligence quickly became aware of a Japanese concentration in these waters. Forces were observed to be assembling at Rabaul from their main naval base at Truk, in the Caroline Islands, and a southward drive was clearly imminent. It was even possible to forecast May 3 as the date when operations would begin. The American aircraft carriers were at this time widely dispersed on various missions. These included the launching of General Doolittle's bold and spectacular air attack against Tokyo itself on April 18. This even may indeed have been a factor in determining the new Japanese policy.

Conscious of the threat in the south, Admiral Nimitz at once began to assemble the strongest possible force in the Coral Sea. Rear Admiral Fletcher was already there, with the carrier *Yorktown* and three heavy cruisers. On May 1 he was joined by the carrier *Lexington* and two more cruisers from Pearl Harbour under Rear Admiral Fitch, and three days later by a squadron commanded by a British officer, Rear Admiral Crace, which comprised the Australian cruisers *Australia* and

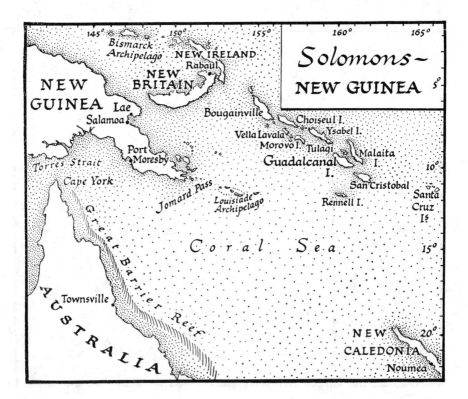

Hobart and the American cruiser *Chicago*. The only other carriers immediately available, the *Enterprise* and the *Hornet*, had been engaged in the Tokyo raid, and though they were sent south as rapidly as possible they could not join Admiral Fletcher until the middle of May. Before then the impending battle had been fought.

On May 3, while refuelling at sea about four hundred miles south of Guadalcanal, Admiral Fletcher learnt that the enemy had landed at Tulagi, apparently with the immediate object of establishing there a seaplane base from which to observe the eastern approaches to the Coral Sea. In view of the obvious threat to this outpost the small Australian garrison had been withdrawn two days previously. Fletcher at once set off to attack the island with only his own task group; Fitch's group were still fuelling. Early on the following morning aircraft from the *Yorktown* struck at Tulagi in strength. The enemy covering forces had however withdrawn and only a few destroyers and small craft remained. The results were therefore disappointing.

The next two days passed without important incident, but it was evident that a major clash could not be long delayed. Fletcher's three groups, having refuelled, were now all in company, standing to the northwestward towards New Guinea. He knew that the Port Moresby invasion force had left Rabaul, and would probably pass through the Jomard Passage, in the Louisiade Archipelago, on either the 7th or 8th. He knew also that three enemy carriers were in the neighbourhood, but not their positions. The Japanese striking force, comprising the carriers *Zuikaku* and *Shokaku*, with two heavy cruisers in support, had come south from Truk, keeping to the eastward of the Solomons, well out of range of air reconnaissance, and had entered the Coral Sea from the east on the evening of the 5th. On the 6th they were fast closing in on Fletcher, and at one time in the evening were only seventy miles away, but neither side was aware of the presence of the other. During the night the forces drew apart, and on the morning of the 7th Fletcher reached his position south of the Louisiades, whence he intended to strike at the invasion force.[3] He now detached Crace's group to go on ahead and cover the southern exit from the Jomard Passage, where the enemy might be expected that day. Crace was soon spotted, and in the afternoon was heavily attacked by successive waves of shore-based torpedo bombers, comparable in strength with those which had sunk the *Prince of Wales* and *Repulse*. By skilful handling and good fortune not a ship was hit, and he continued onward towards Port Moresby, until, hearing that the enemy had turned back, he withdrew to the southward.

Meanwhile the enemy carriers, of which Admiral Fletcher still had

[3] See map on facing page.

no precise news, remained his chief concern. At dawn he commenced a wide search, and at 8.15 A.M. he was rewarded by a report of two carriers and four cruisers north of the Louisiades. In fact the enemy sighted was not the carrier striking force, but the weak escort group covering the invasion transports, which included the light carrier *Shoho.* However, Fletcher struck with all his strength, and three hours later the *Shoho* was overwhelmed and sunk. This event deprived the invasion force of its air cover and made it turn back. Thus the transports intended for Port Moresby never entered the Jomard Passage, and remained north of the Louisiades until finally ordered to withdraw.

*   *   *

Fletcher's whereabouts were now disclosed to the enemy and he was in a serious plight. An enemy attack must be expected at any time, and his own striking forces would not be rearmed and ready for further action until the afternoon. Luckily for him the weather was thick and getting worse and the enemy had no radar. The Japanese carrier force was in fact well within striking distance to the eastward. They launched an attack during the afternoon, but in the squally, murky weather the planes missed their target. Returning empty-handed to their carriers, they passed close to Fletcher's force, and were detected on the radar screen. Fighters were sent out to intercept, and in a confused melee in the gathering darkness many Japanese planes were destroyed. Few of the twenty-seven bombers which had set out regained their ships to take part in the battle next day.

Both sides, knowing how close they were together, contemplated making a night attack with surface forces. Both judged it too risky. During the night they once more drew apart, and on the morning of the 8th the luck of the weather was reversed. It was now the Japanese who had the shelter of low cloud, while Fletcher's ships were bathed in brilliant sunshine. The game of hide-and-seek began again. At 8.38 a search plane from the *Lexington* at last located the enemy, and about the same time an intercepted signal made it plain that the enemy had also sighted the American carriers. A full-scale battle between two equal and well-balanced forces was at hand.

Before 9 A.M. an American striking force of eighty-two aircraft was being launched, and by 9.25 all were on their way. About the same time the enemy were launching a similar strike of sixty-nine. The American attack developed about 11 A.M., the Japanese some twenty minutes later. By 11.40 all was over. The American aircraft had trouble with low cloud round the target. When they found it one of the enemy carriers headed for the cover of a rain squall and the whole of the attack was thrown against the other, the *Shokaku.* Three bomb hits were scored and the ship was set on fire, but the damage was less than it seemed. Although put out of action for the time being, the *Shokaku* was able to get home for repair. The *Zuikaku* remained unscathed.

Meanwhile in clear weather the Japanese attack went in against the *Yorktown* and *Lexington*. By most skilful manoeuvring the *Yorktown* evaded nearly all attacks, but suffered many near misses. One bomb hit caused severe casualties and started a fire. This was soon mastered and the ship's fighting efficiency was little impaired. The less handy *Lexington* was not so fortunate, taking two torpedo hits and two or three bombs. The end of the action found her heavily on fire and listing to port, with three boiler rooms flooded. By gallant exertions the fires were brought under control, the list was corrected, and the ship was soon making twenty-five knots. The aircraft losses on both sides in this fierce encounter, the first in history between carriers, were assessed after the war: American, 33; Japanese, 43.

* * *

If events in the Coral Sea had ended here, the balance would clearly have been in the Americans' favour. They had sunk the light carrier *Shoho*, severely damaged the *Shokaku*, and turned back the invasion force intended for Port Moresby. Their own two carriers seemed to be in fair shape, and their only loss up to this point was a fleet tanker, and her attendant destroyer, which had been sunk the day before by the Japanese carriers. But a disaster now overtook them. An hour after the battle ended, the *Lexington* was heavily shaken by an internal explosion. Fires broke out below which spread and became uncontrollable. Valiant efforts to save the ship proved of no avail, and that evening she was abandoned without further loss of life and sunk by an American torpedo. Both sides now withdrew from the Coral Sea, and both claimed the victory. The Japanese propaganda, in strident terms, declared that not only both Admiral Fletcher's carriers, but also a battleship and a heavy cruiser, had been sunk. Their own actions after the battle were inconsistent with this belief. They postponed until July their advance towards Port Moresby, although the way was now open to them. By then the whole scene had changed, and the stroke was abandoned in favour of an overland advance from the bases they had already gained in New Guinea. These days marked the limit of the Japanese drive by sea towards Australia.

On the American side the conservation of their carrier forces was the prime necessity. Admiral Nimitz was well aware that greater events were looming farther north, which would require his whole strength. He was content to have arrested for the time being the Japanese move into the Coral Sea, and instantly recalled to Pearl Harbour all his carriers, including the *Enterprise* and *Hornet*, then hastening to join Fletcher. Wisely, too, the loss of the *Lexington* was concealed until after the Midway Island battle, as the Japanese were obviously uncertain about the true state of affairs and were groping for information.

This encounter had an effect out of proportion to its tactical importance. Strategically it was a welcome American victory, the first against Japan. Nothing like it had ever been seen before. It was the first battle at sea in which surface ships never exchanged a shot. It also carried the chances and hazards of war to a new pitch. The news blazed round the world with tonic effect, bringing immense relief and encouragement to Australia and New Zealand as well as to the United States. The tactical lessons learnt here at heavy cost were soon applied with outstanding success in the Battle of Midway Island, the opening moves of which were now about to begin.

*   *   *

The advance into the Coral Sea was only the opening phase in the more ambitious Japanese policy. Even while it was in progress Yamamoto, the Japanese Admiralissimo, was preparing to challenge American power in the Central Pacific by seizing Midway Island, with its airfield, from which Pearl Harbour itself, another thousand miles to the east, could be threatened and perhaps dominated. At the same time a diversionary force was to seize points of vantage in the Western Aleutians. By careful timing of his movements Yamamoto hoped to draw the American fleet north to counter the threat to the Aleutians and leave him free to throw his main strength against Midway Island. By the time the Americans could intervene here in force he hoped to have possession of the island and to be ready to meet the counterattack with overwhelming force. The importance to the United States of Midway, the outpost of Pearl Harbour, was such that these movements must inevitably bring about a major clash. Yamamoto felt confident that he could force a decisive battle on his own terms, and that with his great superiority, particularly in fast battleships, he would stand an excellent chance of annihilating his enemy. That was the broad plan which he imparted to his subordinate, Admiral Nagumo. All depended however on Admiral Nimitz falling into the trap, and equally on his having no countersurprise of his own.

But the American commander was vigilant and active. His Intelligence kept him well informed, even as to the date when the expected blow was to fall. Although the plan against Midway might be a blind to conceal a real stroke against the Aleutian chain of islands and an advance towards the American continent, Midway was incomparably the more likely and greater danger, and he never hesitated to deploy his strength in that direction. His chief anxiety was that his carriers must at best be weaker than Nagumo's experienced four, which had fought with outstanding success from Pearl Harbour to Ceylon. Two others of this group had been diverted to the Coral Sea, and one of them had been damaged; but Nimitz, on the other hand, had lost the *Lexington,* the *Yorktown* was crippled, the *Saratoga* had not yet rejoined him after making good battle damage, and the *Wasp* was still near the Mediterranean, where she had succoured Malta. Only the *Enterprise* and the *Hornet,* hurrying back from the South Pacific, and the *Yorktown,* if she could be repaired in time, could be made ready for the coming battle. Admiral Nimitz had no battleships nearer than San Francisco, and these were too slow to work with carriers; Yamamoto had eleven, three of them among the strongest and fastest in the world. The odds against the Americans were heavy, but Nimitz could now count on powerful shore-based air support from Midway Island itself.

✻   ✻   ✻

During the last week of May the main strength of the Japanese Navy began to move from their bases. The first to go was the Aleutian diversionary force, which was to attack Dutch Harbour on June 3 and draw the American fleet in that direction. Thereafter landing forces were to seize the islands of Attu, Kiska, and Adak, farther to the westward. Nagumo with his group of four carriers would strike at Midway the following day, and on June 5 the landing force would arrive and capture the island. No serious opposition was expected. Yamamoto with his battle fleet would meanwhile lie well back to the westward, outside the range of air search, ready to strike when the expected American counter-attack developed.

This was the second supreme moment for Pearl Harbour. The carriers *Enterprise* and *Hornet* arrived from the south on May 26. The *Yorktown* appeared next day, with damage calculated to take three months to repair, but by a decision worthy of the crisis within forty-eight hours she was made taut and fit for battle and was rearmed with a new air group. She sailed again on the 30th to join Admiral Spruance, who had left two days before with the other two carriers. Admiral Fletcher remained in tactical command of the combined force. At Midway the airfield was crammed with bombers, and the ground forces for the defence of the island were at the highest "Alert." Early information of the approach of the enemy was imperative, and continuous air search began on May 30. United States submarines kept their watch west and north of Midway. Four days passed in acute suspense. At 9 A.M. on June 3 a Catalina flying boat on patrol more than seven hundred miles west of Midway sighted a group of eleven enemy ships.

The bombing and torpedo attacks which followed were unsuccessful, except for a torpedo hit on a tanker, but the battle had begun, and all uncertainty about the enemy's intentions was dispelled. Admiral Fletcher through his Intelligence sources had good reason to believe that the enemy carriers would approach Midway from the northwest, and he was not put off by the reports received of the first sighting, which he correctly judged to be only a group of transports. He turned his carriers to reach his chosen position about two hundred miles north of Midway by dawn on the 4th, ready to pounce on Nagumo's flank if and when he appeared.

June 4 broke clear and bright, and at 5.34 A.M. a patrol from Midway at last broadcast the long-awaited signal reporting the approach of the Japanese aircraft carriers. Reports began to arrive thick and fast. Many planes were seen heading for Midway, and battleships were sighted supporting the carriers. At 6.30 A.M. the Japanese attack came in hard and strong. It met a fierce resistance, and probably one-third of the attackers never returned. Much damage was done and many casualties suffered, but the airfield remained serviceable. There had been time to launch a counterattack at Nagumo's fleet. His crushing superiority in fighters took heavy toll, and the results of this gallant stroke, on which great hopes were set, were disappointing. The distraction caused by their onslaught seems however to have clouded the judgment of the Japanese commander, who was also told by his airmen that a second strike at Midway would be necessary. He had retained on board a sufficient number of aircraft to deal with any American carriers which might appear, but he was not expecting them, and his search had been underpowered and at first fruitless. Now he decided to break up the formations which had been held in readiness for this purpose and to rearm them for another stroke at Midway. In any case it was necessary to clear his flight decks to recover the aircraft returning from the first attack. This decision exposed him to a deadly peril, and although Nagumo later heard of an American force, including one carrier, to the eastward, it was too late. He was condemned to receive the full weight of the American attack with his flight decks encumbered with useless bombers, refuelling and rearming.

*   *   *

Admirals Fletcher and Spruance by their earlier cool judgment were well placed to intervene at this crucial moment. They had intercepted the news streaming in during the early morning, and at 7 A.M. the *Enterprise* and *Hornet* began to launch a strike with all the planes they had, except for those needed for their own defence. The *Yorktown*, whose aircraft had been carrying out the morning search, was delayed while they were recovered, but her striking force was in the air soon after 9 A.M., by which time the first waves from the other two carriers were approaching their prey. The weather near the enemy was cloudy, and the dive bombers failed at first to find their target. The *Hornet's* group, unaware that the enemy had turned away, never found them and missed the battle. Owing to this mischance the first attacks were made by torpedo bombers alone from all three carriers, and, although pressed home with fierce courage, were unsuccessful in the face of the overwhelming opposition. Of forty-one torpedo bombers which attacked only six returned. Their devotion brought its reward. While all Japanese eyes and all available fighter strength were turned on them, the thirty-seven dive bombers from the *Enterprise* and *Yorktown* arrived on the scene. Almost unopposed, their bombs crashed into Nagumo's flagship, the *Akagi,* and her sister the *Kaga,* and about the same time another wave of seventeen bombers from the *Yorktown* struck the *Soryu.* In a few minutes the decks of all three ships were a shambles, littered with blazing and exploding aircraft. Tremendous fires broke out below, and it was soon clear that all three ships were doomed. Admiral Nagumo could but shift his flag to a cruiser and watch three-quarters of his fine command burn.

It was past noon by the time the Americans had recovered their aircraft. They had lost over sixty, but the prize they had gained was great. Of the enemy carriers only the *Hiryu* remained, and she at once resolved to strike a blow for the banner of the Rising Sun. As the American pilots were telling their tale on board the *Yorktown* after their return news came that an attack was approaching. The enemy, reported to be about forty strong, pressed it home with vigour, and although heavily mauled by fighters and gunfire they scored three bomb hits on the *Yorktown.* Severely damaged but with her fires under control, she carried on, until two hours later the *Hiryu* struck again, this time with torpedoes. This attack ultimately proved fatal. Although the ship remained afloat for two days she was sunk by a Japanese submarine.

The *Yorktown* was avenged even while she still floated. The *Hiryu* was marked at 2.45 P.M., and within the hour twenty-four dive bombers from the *Enterprise* were winging their way towards her. At 5 P.M. they struck, and in a few minutes she too was a flaming wreck, though she did not sink until the following morning. The last of Nagumo's four fleet carriers had been smashed, and with them were lost all their highly trained air crews. These could never be replaced. So ended the battle of June 4, rightly regarded as the turning-point of the war in the Pacific.

* * *

The victorious American commanders had other perils to face. The Japanese Admiralissimo with his formidable battle fleet might still assail Midway. The American air forces had suffered heavy losses, and there were no heavy ships capable of successfully engaging Yamamoto if he chose to continue his advance. Admiral Spruance, who now assumed command of the carrier group, decided against a pursuit to the westward, not knowing what strength the enemy might have, and having no heavy support for his own carriers. In this decision he was unquestionably right. The action of Admiral Yamamoto in not seeking to retrieve his fortunes is less easily understood. At first he resolved to press on, and ordered four of his most powerful cruisers to bombard Midway in the early hours of June 5. At the same time another powerful Japanese force was advancing to the northeastward, and had Spruance chosen to pursue the remnants of Nagumo's group he might have been caught in a disastrous night action. During the night however the Japanese commander abruptly changed his mind, and at 2.55 A.M. on June 5 he ordered a general retirement. His reasons are by no means clear, but it is evident that the unexpected and crushing defeat of his precious carriers had deeply affected him. One more disaster was to befall him. Two of the heavy cruisers proceeding to bombard Midway came into collision while avoiding attack by an American submarine. Both were severely damaged, and were left behind when the general retirement began. On June 6 these cripples were attacked by Spruance's airmen, who then sank one and left the other apparently in a sinking condition. This much-battered ship, the *Mogami*, eventually succeeded in making her way home.

After seizing the small islands of Attu and Kiska, in the western group of the Aleutians, the Japanese withdrew as silently as they had come.

*   *   *

Reflection on Japanese leadership at this time is instructive. Twice within a month their sea and air forces had been deployed in battle with aggressive skill and determination. Each time when their Air Force had been roughly handled they had abandoned their goal, even though on each occasion it seemed to be within their grasp. The men of Midway, Admirals Yamamoto, Nagumo, and Kondo, were those who planned and carried out the bold and tremendous operations which in four months destroyed the Allied Fleets in the Far East and drove the British Eastern Fleet out of the Indian Ocean. Yamamoto withdrew at Midway because, as the entire course of the war had shown, a fleet without air cover and several thousand miles from its base could not risk remaining within range of a force accompanied by carriers with air groups largely intact. He ordered the transport force to retire because it would have been tantamount to suicide to assault, without air support, an island defended by air forces and physically so small that surprise was impossible.

The rigidity of the Japanese planning and the tendency to abandon the object when their plans did not go according to schedule is thought to have been largely due to the cumbersome and imprecise nature of their language, which rendered it extremely difficult to improvise by means of signalled communications.

One other lesson stands out. The American Intelligence system succeeded in penetrating the enemy's most closely guarded secrets well in advance of events. Thus Admiral Nimitz, albeit the weaker, was twice able to concentrate all the forces he had in sufficient strength at the right time and place. When the hour struck this proved decisive. The importance of secrecy and the consequences of leakage of information in war are here proclaimed.

* * *

This memorable American victory was of cardinal importance, not only to the United States, but to the whole Allied cause. The moral effect was tremendous and instantaneous. At one stroke the dominant position of Japan in the Pacific was reversed. The glaring ascendancy of the enemy, which had frustrated our combined endeavours throughout the Far East for six months, was gone for ever. From this moment all our thoughts turned with sober confidence to the offensive. No longer did we think in terms of where the Japanese might strike the next blow, but where we could best strike at him to win back the vast territories that he had overrun in his headlong rush. The road would be long and hard, and massive preparations were still needed to win victory in the East, but the issue was not in doubt; nor need the demands from the Pacific bear too heavily on the great effort the United States was preparing to exert in Europe.

* * *

The annals of war at sea present no more intense, heart-shaking shock than these two battles, in which the qualities of the United States Navy and Air Force and of the American race shone forth in splendour. The novel and hitherto utterly unmeasured conditions which air warfare had created made the speed of action and the twists of fortune more intense than has ever been witnessed before. But the bravery and self-devotion of the American airmen and sailors and the nerve and skill of their leaders was the foundation of all. As the Japanese Fleet withdrew to their far-off home ports their commanders knew not only that their aircraft-carrier strength was irretrievably broken, but that they were confronted with a will power and passion in the foe they had challenged equal to the highest traditions of their Samurai ancestors, and backed with a development of power, numbers, and science to which no limit could be set.

IO

"Second Front Now!"

O N A P R I L 8 Hopkins and General Marshall arrived in London. They brought with them a comprehensive memorandum prepared by the United States Joint Staff and approved by the President. Its importance justifies the publication of its text in full.

OPERATIONS IN WESTERN EUROPE

April 1942

Western Europe is favoured as the theatre in which to stage the first major offensive by the United States and Great Britain. Only there could their combined land and air resources be fully developed and the maximum support given to Russia.

The decision to launch this offensive must be made *at once,* because of the immense preparations necessary in many directions. Until it can be launched the enemy in the West must be pinned down and kept in uncertainty by ruses and raids; which latter would also gain useful information and provide valuable training.

The combined invasion forces should consist of forty-eight divisions (including nine armoured), of which the British share is eighteen divisions (including three armoured). The supporting air forces required amount to 5800 combat aircraft, 2550 of them British.

Speed is the essence of the problem. The principal limiting factors are shortages of landing craft for the assault and of shipping to transport the necessary forces from America to the U.K. *Without affecting essential commitments in other theatres, these forces can be brought over by April 1, 1943, but only if sixty per cent of the lift is carried by non-U.S. ships.*[1] If the movement is dependent

[1] My italics.—W.S.C.

only on U.S. shipping the date of the assault must be postponed to the late summer of 1943.

About 7000 landing craft will be needed, and current construction programmes must be greatly accelerated to achieve this figure. Concurrently, preparatory work to receive and operate the large U.S. land and air contingents must be speeded up.

The assault should take place on selected beaches between Havre and Boulogne, and be carried out by a first wave of at least six divisions, supplemented by air-borne troops. It would have to be nourished at the rate of at least 100,000 men a week. As soon as the beachheads are secure armoured forces would move rapidly to seize the line of the Oise-St. Quentin. Thereafter the next objective would be Antwerp.

Since invasion on this scale cannot be mounted before April 1, 1943, at earliest, a plan must be prepared, and kept up to date, for immediate action by such forces as may be available from time to time. This may have to be put into effect as an emergency measure either (a) to take advantage of a sudden German disintegration, or (b) "as a sacrifice" to avert an imminent collapse of Russian resistance. In any such event local air superiority is essential. On the other hand, during the autumn of 1942 probably not more than five divisions could be dispatched and maintained. In this period the chief burden would fall on the U.K. For example, on September 15 the U.S. could find two and a half divisions of the five needed, but only 700 combat aircraft; so that the contribution required from the U.K. might amount to 5000 aircraft.

*   *   *

Hopkins, much exhausted by his journey, fell ill for two or three days, but Marshall started talks with our Chiefs of Staff at once. It was not possible to arrange the formal conference with the Defence Committee till Tuesday the 14th. Meanwhile I talked the whole position over with the Chiefs of Staff as well as with my colleagues. We were all relieved by the evident strong American intention to intervene in Europe, and to give the main priority to the defeat of Hitler. This had always been the foundation of our strategic thought. On the other hand, neither we nor our professional advisers could devise any practical plan for crossing the Channel with a large Anglo-American army and landing in France before the late summer of 1943. This, as is recorded earlier, had always been my aim and timetable. There was also before us the new American idea of a preliminary emergency landing on a much smaller but still substantial scale in the autumn of 1942. We were most willing to study this, and also any other plan of diversion, for the sake of Russia and also for the general waging of the war.

On the night of the 14th the Defence Committee met our American friends at 10 Downing Street. The discussion was crucial and the conclusion was unanimous. We all agreed that there should be a cross-Channel operation in 1943. It was now named, though not by me, "Round-up."

But in planning this gigantic enterprise it was not possible for us to lay aside all other duties. Our first Imperial obligation was to defend India from the Japanese invasion, by which it seemed it was already menaced. Moreover, this task bore a decisive relation to the whole war. To leave four hundred millions of His Majesty's Indian subjects, to whom we were bound in honour, to be ravaged and overrun, as China had been, by the Japanese would have been a deed of shame. But also to allow the Germans and Japanese to join hands in India or the Middle East involved a measureless disaster to the Allied cause. It ranked in my mind almost as the equal of the retirement of Soviet Russia behind the Urals, or even of their making a separate peace with Germany. At this date I did not deem either of these contingencies likely. I had faith in the power of the Russian armies and nation fighting in defence of their native soil. Our Indian Empire however, with all its glories, might fall an easy prey. I had to place this point of view before the American envoys. Without active British aid India might be conquered in a few months. Hitler's subjugation of Soviet Russia would be a much longer, and to him more costly, task. Before it was accomplished the Anglo-American command of the air would have been established beyond challenge. Even if all else failed this would be finally decisive.

I was in complete accord with what Hopkins had called "a frontal assault upon the enemy in Northern France in 1943." But what was to be done in the interval? The main armies could not simply be preparing all that time. Here there was a wide diversity of opinion. General Marshall had advanced the proposal that we should attempt to seize Brest or Cherbourg, preferably the latter, or even both, during the early autumn of 1942. The operation would have to be almost entirely British. The Navy, the air, two-thirds of the troops, and such landing craft as were available must be provided by us. Only two or three American divisions could be found. These, it must be remembered, were very newly raised. It takes at least two years and a very strong professional cadre to form first-class troops. The enterprise was therefore one on which British Staff opinion would naturally prevail. Clearly there must be an intensive technical study of the problem.

Nevertheless I by no means rejected the idea at the outset; but there were other alternatives which lay in my mind. The first was the descent on French North Africa (Morocco, Algeria, and Tunisia), which for the present was known as "Gymnast," and which ultimately emerged in the great operation "Torch." I had a second alternative plan for which I always hankered and which I thought could be undertaken as well as the invasion of French North Africa. This was "Jupiter" — namely, the liberation of Northern Norway. Here was direct aid to Russia. Here was the only method of direct combined military action with Russian troops, ships, and air. Here was the means, by securing the northern tip of Europe, of opening the broadest flood of supplies to Russia. Here was an enterprise which, as it had to be fought in Arctic regions, involved neither large numbers of men nor heavy expenditure of supplies and munitions. The Germans had got those vital strategic points by the North Cape very cheaply. They might also be regained at a small cost compared with the scale which the war had now attained. My own choice was for "Torch," and if I could have had my full way I should have tried "Jupiter" also in 1942.

The attempt to form a bridgehead at Cherbourg seemed to me more difficult, less attractive, less immediately helpful or ultimately fruitful. It would be better to lay our right claw on French North Africa, tear with our left at the North Cape, and wait a year without risking our teeth upon the German fortified front across the Channel.

Those were my views then, and I have never repented of them. I was however very ready to give "Sledgehammer," as the Cherbourg assault was called, a fair run with other suggestions before the Planning Committees. I was almost certain the more it was looked at the less it would be liked. If it had been in my power to give orders I would have settled upon "Torch" and "Jupiter," properly synchronised for the autumn, and would have let "Sledgehammer" leak out as a feint through rumour and ostentatious preparation. But I had to work by influence and diplomacy in order to secure agreed harmonious action with our cherished Ally, without whose aid nothing but ruin faced the world. I did not therefore open any of the alternatives at our meeting on the 14th.

On the supreme issue we welcomed with relief and joy the decisive proposal of the United States to carry out a mass invasion of Germany as soon as possible, using England as the springboard. We might so easily, as will be seen, have been confronted with American plans to assign the major priority to helping China and crushing Japan. But from the very start of our alliance after Pearl Harbour the President and General Marshall, rising superior to powerful tides of public opinion, saw in Hitler the prime and major foe. Personally I longed to see British and American armies shoulder to shoulder in Europe. But I had little doubt myself that study of details — landing craft and all that — and also reflection on the main strategy of the war, would rule out "Sledgehammer." In the upshot no military authority — Army, Navy, or Air — on either side of the Atlantic was found capable of preparing such a plan, or, so far as I was informed, ready to take the responsibility for executing it. United wishes and goodwill cannot overcome brute facts.

To sum up: I pursued always the theme set forth in my memorandum given to the President in December 1941, namely, (1) that British and American liberating armies should land in Europe in 1943. And how could they land in full strength otherwise than from Southern England? Nothing must be done which would prevent this and anything that would promote it. (2) In the meantime, with the Russians fighting on a gigantic scale from hour to hour against the main striking force of the German Army, we could not stand idle. We must engage the enemy. This resolve also lay at the root of the President's thought. What then should be done in the year or fifteen months that must elapse before a heavy cross-Channel thrust could be made? Evidently the occupation of French North Africa was in itself possible and sound, and fitted into the general strategic scheme.

I hoped that this could be combined with a descent upon Norway, and I still believe both might have been simultaneously possible. But in these tense discussions of unmeasurable things it is a great danger to lose simplicity and singleness of purpose. Though I hoped for both "Torch" and "Jupiter," I never had any intention of letting "Jupiter" queer the pitch of "Torch." The difficulties of focusing and combining in one vehement thrust all the efforts of two mighty countries were such that no ambiguity could be allowed to darken counsel. (3) The only way therefore to fill the gap, before large masses of British and United States troops could be brought in contact with the Germans in Europe in 1943, was by the forcible Anglo-American occupation of French North Africa in conjunction with the British advance westward across the desert towards Tripoli and Tunis.

Eventually, when all other plans and arguments had worn themselves out and perished by the way, this became the united decision of the Western Allies.

*   *   *

In May we had other visitors. Molotov arrived to negotiate an Anglo-Russian Alliance and to learn our views upon the opening of a Second Front. The Alliance was concluded and the Second Front was discussed in detail. Our Russian guests had expressed the wish to be lodged in the country outside London during their stay, and I therefore placed Chequers at their disposal. I remained meanwhile at the Storey's Gate Annexe. However, I went down for two nights to Chequers. Here

I had the advantage of having long private talks with Molotov and Ambassador Maisky, who was the best of interpreters, translating quickly and easily, and possessing a wide knowledge of affairs. With the aid of good maps I tried to explain what we were doing, and the limitations and peculiar characteristics in the war capacity of an island Power. I also went at length into the technique of amphibious operations, and described the perils and difficulties of maintaining our lifeline across the Atlantic in the face of U-boat attack. I think Molotov was impressed with all this, and realised that our problem was utterly different from that of a vast land Power. At any rate, we got closer together than at any other time.

The inveterate suspicion with which the Russians regarded foreigners was shown by some remarkable incidents during Molotov's stay at Chequers. On arrival they had asked at once for keys to all the bedrooms. These were provided with some difficulty, and thereafter our guests always kept their doors locked. When the staff at Chequers succeeded in getting in to make the beds they were disturbed to find pistols under the pillows. The three chief members of the mission were attended not only by their own police officers, but by two women who looked after their clothes and tidied their rooms. When the Soviet envoys were absent in London these women kept constant guard over their masters' rooms, only coming down one at a time for their meals. We may claim, however, that presently they thawed a little, and even chatted in broken French and signs with the household staff.

Extraordinary precautions were taken for Molotov's personal safety. His room had been thoroughly searched by his police officers, every cupboard and piece of furniture and the walls and floors being meticulously examined by practised eyes. The bed was the object of particular attention; the mattresses were all prodded in case of infernal machines, and the sheets and blankets were rearranged by the Russians so as to leave an opening in the middle of the bed out of which the occupant could spring at a moment's notice, instead of being tucked in. At night a revolver was laid out beside his dressing gown and his dispatch case. It is always right, especially in time of war, to take precautions against danger, but every effort should be made to measure its reality. The simplest test is to ask oneself whether the other side have any interest in killing the person concerned. For myself, when I visited Moscow I put complete trust in Russian hospitality.

*   *   *

Molotov flew on to Washington and came back full of the plans for a cross-Channel operation in 1942. We ourselves were still actively studying this in conjunction with the American Staff, and nothing but difficulties had as yet emerged. There could be no harm in a public statement, which might make the Germans apprehensive and consequently hold as many of their troops in the West as possible. We therefore agreed to the issue of a communiqué, which was published on June 11, containing the following sentence: "In the course of the conversations full understanding was reached with regard to the urgent tasks of creating a Second Front in Europe in 1942."

I felt it above all important that in this effort to mislead the enemy we should not mislead our Ally. At the time of drafting the communiqué therefore I handed Molotov personally in the Cabinet Room and in the presence of some of my colleagues an *aide-mémoire* which made it clear that while we were trying our best to make plans we were not committed to action and that we could give no promise. When subsequent reproaches were made by the Soviet Government, and when Stalin himself raised the point personally with me, we always produced the *aide-mémoire* and pointed to the words *"We can therefore give no promise."*

AIDE–MÉMOIRE

We are making preparations for a landing on the Continent in August or September 1942. As already explained, the main limiting factor to the size of the landing force is the availability of special landing craft. Clearly however it would not further either the Russian cause or that of the Allies as a whole if, for the sake of action at any price, we embarked on some operation which ended in disaster and gave the enemy an opportunity for glorification at our discomfiture. It is impossible to say in advance whether the situation will be such as to make this operation feasible when the time comes. *We can therefore give no promise in the matter,* but provided that it appears sound and sensible we shall not hesitate to put our plans into effect.

*   *   *

During the weeks which followed, professional opinion marched forward. I gave all my thought to the problem of "Sledgehammer," and called for constant reports. Its difficulties soon became obvious. The storm of Cherbourg by a sea-landed army in the face of German opposition, probably in superior numbers and with strong fortifications, was a hazardous operation. If it succeeded, the Allies would be penned up in Cherbourg and the tip of the Cotentin peninsula, and would have to maintain themselves in this confined bomb and shell trap for nearly a year under ceaseless bombardment and assault. They could be supplied only by the port of Cherbourg, which would have to be defended all the winter and spring against potentially continuous and occasionally overwhelming air attack. The drain which such a task would impose must be a first charge upon all our resources of shipping and air power. It would bleed all other operations. If we succeeded we should have to debouch in the summer from the narrow waist of the Cotentin peninsula, after storming a succession of German fortified lines defended by whatever troops the Germans might care to bring. Even so there was only one railroad along which our army could advance, and this would certainly have been destroyed. Moreover, it was not apparent how this unpromising enterprise would help Russia. The Germans had left twenty-five mobile divisions in France. We could not have more than nine ready by August for "Sledgehammer," and of these seven must be British. There would therefore be no need for the recall of German divisions from the Russian front.

As these facts and many more presented themselves in an ugly way to the military staffs a certain lack of conviction and ardour manifested itself, not only among the British but among our American comrades. The ceaseless Staff discussions continued during the summer. "Sledge-hammer" was knocked out by general assent. On the other hand, I did not receive much positive support for "Jupiter" — Northern Norway. We were all agreed upon the major cross-Channel invasion in 1943. The question arose irresistibly, what to do in the interval? It was impossible for the United States and Britain to stand idle all that time without fighting, except in the desert. The President was determined that Americans should fight Germans on the largest possible scale *during* 1942. Where then could this be achieved? Where else but in French North Africa, upon which the President had always smiled? Out of many plans the fittest might survive.

I was content to wait for the answer.

II

My Second Visit to Washington. Tobruk

ALTHOUGH GENERAL AUCHINLECK had not felt himself strong enough to seize the initiative in the Desert, he awaited with some confidence the enemy's attack. General Ritchie, commanding the Eighth Army, had under his chief's supervision prepared an elaborate defensive position stretching from Gazala to Bir Hacheim, forty-five miles due south. composed of fortified points called "boxes," held in strength by brigades or larger forces, the whole being covered by an immense spread of mine fields. Behind this the whole of our armour and the XXXth Corps were held in reserve.

All the Desert battles, except Alamein, began by swift, wide turning movements of armour on the Desert Flank. Rommel started by moonlight on the night of May 26–27, and swept forward with all his armour, intending to engage and destroy our own, and hoping as we now know, to seize Tobruk on the second day of his attack. This he failed to accomplish, and on June 10, after much bitter and gallant fighting, General Auchinleck sent us an estimate of the casualties on both sides. The figures of tanks, guns, and aircraft were satisfactory, and also precise. But I was naturally struck by the following statement: "Our own losses in personnel are estimated very approximately at 10,000, of whom some 8000 may be prisoners, but the casualties of the 5th Indian Division not yet accurately known." This extraordinary disproportion between killed and wounded on the one hand and prisoners on the other revealed that something must have happened of an unpleasant character. It showed also that the Cairo headquarters were in important respects unable to measure the event. I did not dwell on this in my reply.

Throughout June 12 and 13 a fierce battle was fought for possession of the ridges that lie between El Adem and "Knightsbridge." This was the culmination of the tank battle; at its close the enemy were masters

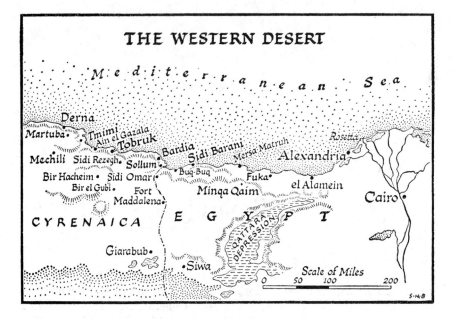

THE WESTERN DESERT

of the field, and our own armour gravely reduced. "Knightsbridge," the focus of communications in that neighbourhood, had to be evacuated, after a stubborn defence, and by the 14th it became clear that the battle had taken a heavy adverse turn. Mr. Casey, the Minister of State, sent me a telegram which emphasised the Service messages, and contained the following passage:

> As to Auchinleck himself, I have all possible confidence in him as regards his leadership and the way he is conducting the battle with the forces that are available to him. My only wish is that he could be at two places at once, both here at the centre of the web and forward directing the Eighth Army battle in person. I have even thought at times in recent days that it would be a good thing for him to go forward and take charge of the battle, leaving his Chief of Staff here temporarily in charge, but he does not think so and I do not want to press him on it. It is Auchinleck's battle, and decisions as to leadership subordinate to himself are for him to make.

Mr. Casey's remark about the advantages of Auchinleck's taking personal command of the Desert battle confirmed my own feelings which I had expressed to the General a month before. The Commander-in-Chief of the Middle East was embarrassed and hampered by his too extensive responsibilities. He thought of the battle, on which all in his work depended, only as a part of his task. There was always the danger from the north, to which he felt it his duty to attach an importance to which we at home, in a better position to judge, no longer subscribed.

The arrangement which he had made was a compromise. He left the fighting of the decisive battle to General Ritchie, who had so recently ceased to be his Deputy Chief of Staff. At the same time he kept this officer under strict supervision, sending him continuous instructions. It was only after the disaster had occurred that he was induced, largely by the urgings of the Minister of State, to do what he should have done from the beginning and take over the direct command of the battle himself. It is to this that I ascribe his personal failure, some of the blame for which undoubtedly falls on me and my colleagues for the unduly wide responsibilities assigned a year before to the Middle East Command. Still, we had done our best to free him from these undue burdens by precise, up-to-date, and superseding advice, which he had not accepted. Personally I believe that if he had taken command from the outset and, as was fully in his power, left a deputy in Cairo to keep an eye on the north and discharge the mass of varied business belonging to the rest of the immense theatre over which he presided he might well have won the battle, and certainly when late in the day he took command he saved what was left of it.

The reader will presently see how these impressions bit so deeply into me that in my directive to General Alexander of August 10 I made his main duty clear beyond a doubt. One lives and learns.

*   *   *

Immediately Tobruk glared upon us, and, as in the previous year, we had no doubt that it should be held at all costs. Now also, after a month's needless delay, General Auchinleck ordered up the New Zealand Division from Syria, but not in time for it to take part in the battle for Tobruk. We were not satisfied with his orders to General Ritchie, which did not positively require him to defend the fortress. To make sure I sent the following telegram:

> We are glad to have your assurance that you have no intention of giving up Tobruk. War Cabinet interpret [your telegram] to mean that, if the need arises, General Ritchie would leave as many troops in Tobruk as are necessary to hold the place for certain.

The reply left no doubt, and on this we rested with confidence based upon the experience of the previous year. Moreover, our position, as General Auchinleck had pointed out, appeared on paper much better than in 1941. We had an army deployed on a fortified front, in close proximity to Tobruk, with a newly constructed direct broad-gauge railway sustaining it. We were no longer formed to a flank with our communications largely dependent on the sea, but according to the orthodox principles of war, running back at right angles from the centre of our front to our main base. In these circumstances, though grieved by what had happened, I still felt, from a survey of all the forces on both sides, and of Rommel's immense difficulties of supply, that all would be well. With the New Zealand Division now not far away, and with powerful reinforcements approaching by sea, I did not myself feel that the continuance of hard fighting in the greatest possible strength on both sides would be to our detriment in the long run. I did not therefore cancel the plans I had made for a second visit to Washington, where business of the highest importance to the general strategy of the war had to be transacted. In this I was supported by my colleagues.

* * *

The main object of my journey was to reach a final decision on the operations for 1942–43. The American authorities in general, and Mr. Stimson and General Marshall in particular, were anxious that some plan should be decided upon at once, which would enable the United States to engage the Germans in force on land and in the air in 1942. Failing this, there was the danger that the American Chiefs of Staff would seriously consider a radical revision of the strategy of "Germany first." Another matter lay heavy on my mind. It was the question of "Tube Alloys," which was our code word for what afterwards became the atomic bomb. Our research and experiments had now reached a point where definite agreements must be made with the United States, and it was felt this could only be achieved by personal discussions between me and the President. The fact that the War Cabinet decided that I should leave the country and London with the Chief of the Imperial General Staff and General Ismay at the height of the Desert battle measures the importance which we attached to a settlement of the grave strategic issues which were upon us.

On account of the urgency and crisis of our affairs in these very difficult days, I decided to go by air rather than by sea. This meant that we should be barely twenty-four hours cut off from the full stream of information. Efficient arrangements were made for the immediate transmission of messages from Egypt and for the rapid passage and decoding of all reports, and no harmful delays in taking decisions were expected or in fact occurred.

Although I now knew the risks we had run on our return voyage flight from Bermuda in January, my confidence in the chief pilot, Kelly Rogers, and his Boeing flying boat was such that I asked specially that he should take charge. We left Stranraer on the night of June 17, shortly before midnight. The weather was perfect and the moon full.

I sat for two hours or more in the co-pilot's seat admiring the shining sea, revolving my problems, and thinking of the anxious battle. I slept soundly in the "bridal suite" until in broad daylight we reached Gander. Here we could have refuelled, but this was not thought necessary, and after making our salutes to the airfield we pursued our voyage. As we were travelling with the sun the day seemed very long. We had two luncheons with a six-hour interval, and contemplated a late dinner after arrival.

For the last two hours we flew over the land, and it was about seven o'clock by American time when we approached Washington. As we gradually descended towards the Potomac River I noticed that the top of the Washington Monument, which is over five hundred and fifty feet high, was about our level, and I impressed upon Captain Kelly Rogers that it would be peculiarly unfortunate if we brought our story to an end by hitting this of all other objects in the world. He assured me that he would take special care to miss it. Thus we landed safely and smoothly on the Potomac after a journey of twenty-seven flying hours. Lord Halifax, General Marshall, and several high officers of the United States welcomed us. I repaired to the British Embassy for dinner. It was too late for me to fly on to Hyde Park that night. We read all the latest telegrams — there was nothing important — and dined agreeably in the open air. The British Embassy, standing on the high ground, is one of the coolest places in Washington, and compares very favourably in this respect with the White House.

Early the next morning, the 19th, I flew to Hyde Park. The President was on the local airfield, and saw us make the roughest bump landing I have experienced. He welcomed me with great cordiality, and, driving the car himself, took me to the majestic bluffs over the Hudson River on which Hyde Park, his family home, stands. The President drove me all over the estate showing me its splendid views. In this drive I had some thoughtful moments. Mr. Roosevelt's infirmity prevented him from using his feet on the brake, clutch, or accelerator. An ingenious arrangement enabled him to do everything with his arms, which were amazingly strong and muscular. He invited me to feel his biceps, saying that a famous prize fighter had envied them. This was reassuring; but I confess that when on several occasions the car poised and backed on the grass verges of the precipices over the Hudson I hoped the mechanical devices and brakes would show no defects. All the time we talked business, and though I was careful not to take his attention off the driving we made more progress than we might have done in formal conference.

The President was very glad to hear I had brought the C.I.G.S. with me. His field of interest was always brightened by recollections of his youth. It had happened that the President's father had entertained at

Hyde Park the father of General Brooke. Mr. Roosevelt therefore expressed keen interest to meet the son, who had reached such a high position. When they met two days later he received him with the utmost cordiality, and General Brooke's personality and charm created an almost immediate intimacy which greatly helped the course of business.

*   *   *

I told Harry Hopkins about the different points on which I wanted decisions, and he talked them over with the President, so that the ground was prepared and the President's mind armed upon each subject. Of these "Tube Alloys" was one of the most complex, and, as it proved, overwhelmingly the most important. I had my papers with me, but the discussion was postponed till the next day, the 20th, as the President needed more information from Washington. Our talk took place after luncheon, in a tiny little room which juts out on the ground floor. The room was dark and shaded from the sun. Mr. Roosevelt was ensconced at a desk almost as big as the apartment. Harry sat or stood in the background. My two American friends did not seem to mind the intense heat.

I told the President in general terms of the great progress we had made, and that our scientists were now definitely convinced that results might be reached before the end of the present war. He said his people were getting along too, but no one could tell whether anything practical would emerge till a full-scale experiment had been made. We both felt painfully the dangers of doing nothing. We knew what efforts the Germans were making to procure supplies of "heavy water" — a sinister term, eerie, unnatural, which began to creep into our secret papers. What if the enemy should get an atomic bomb before we did! However sceptical one might feel about the assertions of scientists, much disputed among themselves and expressed in jargon incomprehensible to laymen, we could not run the mortal risk of being outstripped in this awful sphere.

I strongly urged that we should at once pool all our information, work together on equal terms, and share the results, if any, equally between us. The question then arose as to where the research plant was to be set up. We were already aware of the enormous expense that must be incurred, with all the consequent grave diversion of resources and brain-power from other forms of war effort. Considering that Great Britain was under close bombing attack and constant enemy air reconnaissance, it seemed impossible to erect in the island the vast and conspicuous factories that were needed. We conceived ourselves at least as far advanced as our Ally, and there was of course the alternative of Canada, who had a vital contribution herself to make through the supplies of uranium she had actively gathered. It was a hard decision to spend several hundred million pounds sterling, not so much of money as of competing forms of precious war energy, upon a project the success of which no scientist on either side of the Atlantic could guarantee. Nevertheless, if the Americans had not been willing to undertake the venture we should certainly have gone forward on our own power in Canada, or, if the Canadian Government demurred, in some other part of the Empire. I was however very glad when Mr. Roosevelt said he thought the United States would have to do it. We therefore took this decision jointly, and settled a basis of agreement. I shall continue the story in a later chapter. But meanwhile I have no doubt that it was the progress we had made in Britain and the confidence of our scientists in ultimate success imparted to the President that led him to his grave and fateful decision.

*   *   *

Late on the night of the 20th the Presidential train bore us back to Washington, which we reached about eight o'clock the next morning. We were heavily escorted to the White House, and I was again accorded the very large air-conditioned room, in which I dwelt in comfort at about thirty degrees below the temperature of most of the rest of the building. I glanced at the newspapers, read telegrams for an hour, had my breakfast, looked up Harry across the passage, and then went to see the President in his study. General Ismay came with me. Presently a telegram was put into the President's hands. He passed it to me without a word. It said, "Tobruk has surrendered, with twenty-five thousand men taken prisoners." This was so surprising that I could not believe it. I therefore asked Ismay to inquire of London by telephone. In a few minutes he brought the following message, which had just arrived from Admiral Harwood at Alexandria.[1]

> Tobruk has fallen, and situation deteriorated so much that there is a possibility of heavy air attack on Alexandria in near future, and in view of approaching full moon period I am sending all Eastern Fleet units south of the Canal to await events. I hope to get H.M.S. *Queen Elizabeth* out of dock towards end of this week.[2]

This was one of the heaviest blows I can recall during the war. Not only were its military effects grievous, but it affected the reputation

[1] Admiral Harwood had succeeded Admiral Cunningham in the Mediterranean Command on May 31.
[2] Admiral Harwood made this decision because Alexandria could now be attacked by dive bombers with fighter cover.

of the British armies. At Singapore 85,000 men had surrendered to inferior numbers of Japanese. Now in Tobruk a garrison of 25,000
(actually 33,000) seasoned soldiers had laid down their arms to perhaps
one-half of their number. If this was typical of the morale of the
Desert army, no measure could be put upon the disasters which impended in Northeast Africa. I did not attempt to hide from the President the shock I had received. It was a bitter moment. Defeat is one
thing; disgrace is another. Nothing could exceed the sympathy and
chivalry of my two friends. There were no reproaches; not an unkind
word was spoken. "What can we do to help?" said Roosevelt. I replied
at once, "Give us as many Sherman tanks as you can spare, and ship
them to the Middle East as quickly as possible." The President sent
for General Marshall, who arrived in a few minutes, and told him of my
request. Marshall replied, "Mr. President, the Shermans are only just
coming into production. The first few hundred have been issued to
our own armoured divisions, who have hitherto had to be content with
obsolete equipment. It is a terrible thing to take the weapons out of a
soldier's hands. Nevertheless, if the British need is so great they must
have them; and we could let them have a hundred 105-mm. self-
propelled guns in addition."

To complete the story it must be stated that the Americans were
better than their word. Three hundred Sherman tanks with engines not
yet installed and a hundred self-propelled guns were put into six of
their fastest ships and sent off to the Suez Canal. The ship containing
the engines for all the tanks was sunk by a submarine off Bermuda.
Without a single word from us the President and Marshall put a
further supply of engines into another fast ship and dispatched it to
overtake the convoy. "A friend in need is a friend indeed."

*   *   *

On June 21, when we were alone together after lunch, Harry said to
me, "There are a couple of American officers the President would like
you to meet, as they are very highly thought of in the Army, by
Marshall, and by him." At five o'clock therefore Major Generals
Eisenhower and Clark were brought to my air-cooled room. I was
immediately impressed by these remarkable but hitherto unknown
men. They had both come from the President, whom they had just
seen for the first time. We talked almost entirely about the major
cross-Channel invasion in 1943, "Round-up," as it was then called, on
which their thoughts had evidently been concentrated. We had a most
agreeable discussion, lasting for over an hour. I felt sure that these
officers were intended to play a great part in it, and that was the
reason why they had been sent to make my acquaintance. Thus began

a friendship which across all the ups and downs of war I have preserved with deep satisfaction to this day.

Meanwhile the surrender of Tobruk reverberated round the world. On the 22nd Hopkins and I were at lunch with the President in his room. Presently Mr. Elmer Davis, the head of the Office of War Information, arrived with a bunch of New York newspapers, showing flaring headlines about "ANGER IN ENGLAND," "TOBRUK FALL MAY BRING CHANGE OF GOVERNMENT," "CHURCHILL TO BE CENSURED," etc. I had been invited by General Marshall to visit one of the American Army camps in South Carolina. We were to start by train with him and Mr. Stimson on the night of June 23. Mr. Davis asked me seriously whether, in view of the political situation at home, I thought it wise to carry out the programme, which of course had been elaborately arranged. Might it not be misinterpreted if I were inspecting troops in America when matters of such vital consequence were taking place both in Africa and London? I replied that I would certainly carry out the inspections as planned, and that I doubted whether I should be able to provoke twenty members into the Lobby against the Government on an issue of confidence. This was in fact about the number which the malcontents eventually obtained.

Accordingly I started by train next night for South Carolina, and arrived at Fort Jackson the next morning. The train drew up, not at a station, but in the open plain. It was a very hot day, and we got out of the train straight onto the parade ground, which recalled the plains of India in the hot weather. We went first to an awning and saw the American armour and infantry march past. Next we watched the parachute exercises. They were impressive and convincing. I had never seen a thousand men leap into the air at once. I was given a "walkie-talkie" to carry. This was the first time I had ever handled such a convenience. In the afternoon we saw the mass-produced American divisions doing field exercises with live ammunition. At the end I said to Ismay (to whom I am indebted for this account), "What do you think of it?" He replied, "To put these troops against German troops would be murder." Whereupon I said, "You're wrong. They are wonderful material and will learn very quickly." To my American hosts however I consistently pressed my view that it takes two years or more to make a soldier. Certainly two years later the troops we saw in Carolina bore themselves like veterans.

We flew back to Washington on the afternoon of the 24th, where I received various reports, and next evening I set out for Baltimore, where my flying boat lay. The President bade me farewell at the White House with all his grace and courtesy, and Harry Hopkins and Averell Harriman came to see me off. The narrow, closed-in gangway which led to the water was heavily guarded by armed American police. There seemed to be an air of excitement, and the officers looked serious. Before we took off I was told that one of the plain-clothes men on duty had been caught fingering a pistol and heard muttering that he would "do me in," with some other expressions of an unappreciative character. He had been pounced upon and arrested. Afterwards he turned out to be a lunatic. Crackpates are a special danger to public men, as they do not worry about the "get away."

We came down at Botwood the next morning in order to refuel, and took off again after a meal of fresh lobsters. Thereafter I ate at stomach-time — *i.e.,* with the usual interval between meals — and slept whenever possible. I sat in the co-pilot's seat as, after flying over Northern Ireland, we approached the Clyde at dawn, and landed safely. My train was waiting, with Peck, one of my personal secretaries, and a mass of boxes, and four or five days' newspapers. In an hour we were off to the South. It appeared that we had lost a by-election by a sweeping turnover at Maldon. This was one of the by-products of Tobruk.

This seemed to me to be a bad time. I went to bed, browsed about in the files for a while, and then slept for four or five hours till we reached London. What a blessing is the gift of sleep! The War Cabinet were on the platform to greet me on arrival, and I was soon at work in the Cabinet Room.

12

The Vote of Censure

THE CHATTER AND CRITICISMS of the press, where the
sharpest pens were busy and many shrill voices raised, found its
counterpart in the activities of a few score of Members in the House
of Commons, and a fairly glum attitude on the part of our immense
majority. A party Government might well have been overturned at this
juncture, if not by a vote, by the kind of intensity of opinion which led
Mr. Chamberlain to relinquish power in May 1940. But the National
Coalition Government, fortified by a reconstruction in February, was
massive and overwhelming in its strength and unity. All its principal
Ministers stood together around me, with never a thought that was not
loyal and robust. I seemed to have maintained the confidence of all
those who watched with full knowledge the unfolding story and shared
the responsibilities. No one faltered. There was not a whisper of intrigue. We were a strong, unbreakable circle, and capable of withstanding any external political attack and of persevering in the common cause
through every disappointment.

We had had a long succession of misfortunes and defeats — Malaya,
Singapore, Burma; Auchinleck's lost battle in the Desert; Tobruk,
unexplained, and, it seemed, inexplicable; the rapid retreat of the
Desert army and the loss of all our conquests in Libya and Cyrenaica;
four hundred miles of retrogression towards the Egyptian frontier;
over fifty thousand of our men casualties or prisoners. We had lost
vast masses of artillery, ammunition, vehicles, and stores of all kinds.
We were back again at Mersa Matruh, at the old positions of two years
before, but this time with Rommel and his Germans triumphant, pressing forward in our captured lorries fed with our oil supplies, in many
cases firing our own ammunition. Only a few more marches, one
more success, and Mussolini and Rommel would enter Cairo, or its

ruins, together. All hung in the balance, and after the surprising re-verses we had sustained, and in face of the unknown factors at work, who would predict how the scales would turn?

The Parliamentary situation required prompt definition. It seemed however rather difficult to demand another Vote of Confidence from the House so soon after that which had preceded the collapse of Singapore. It was therefore very convenient when on June 25 the discontented Members decided among themselves to place a Vote of Censure on the Order Paper. It read as follows:

> That this House, while paying tribute to the heroism and endurance of the Armed Forces of the Crown in circumstances of exceptional difficulty, has no confidence in the central direction of the war.

It stood in the name of Sir John Wardlaw-Milne, an influential member of the Conservative Party. He was chairman of the powerful all-party Finance Committee, whose reports of cases of administrative waste and inefficiency I had always studied with close attention. The Committee had a great deal of information at their disposal and many contacts with the outer circle of our war machine. When it was also announced that the motion would be seconded by Admiral of the Fleet Sir Roger Keyes, and supported by the former Secretary of State for War, Mr. Hore-Belisha, it was at once evident that a serious challenge had been made. Indeed, in some newspapers and in the lobbies the talk ran of an approaching political crisis which would be decisive.

I said at once that we would give full opportunity for public debate, and fixed July 1 for the occasion. There was one announcement I felt it necessary to make, and I telegraphed to Auchinleck: "When I speak in the Vote of Censure debate on Thursday, about 4 P.M., I deem it necessary to announce that you have taken the command in supersession of Ritchie as from June 25."

The battle crisis in Egypt grew steadily worse, and it was widely believed that Cairo and Alexandria would soon fall to Rommel's flaming sword. Mussolini indeed made preparations to fly to Rommel's headquarters with the idea of taking part in the triumphal entry to one or both of these cities. It seemed that we should reach a climax on the Parliamentary and Desert fronts at the same moment. When it was realised by our critics that they would be faced by our united National Government some of their ardour evaporated, and the mover of the motion offered to withdraw it if the critical situation in Egypt rendered public discussion untimely. We had however no intention of letting them escape so easily. Considering that for nearly three weeks the whole world, friend or foe, had been watching with anxiety the mounting political and military tension, it was impossible not to bring matters to a head.

*   *   *

The debate was opened by Sir John Wardlaw-Milne in an able speech in which he posed the main issue. This motion was "not an attack upon officers in the field. It is a definite attack upon the central direction here in London, and I hope to show that the causes of our failure lie here far more than in Libya or elsewhere. The first vital mistake that we made in the war was to combine the offices of Prime Minister and Minister of Defence." He dilated upon the "enormous duties" cast upon the holder of the two offices. "We must have a strong, full-time leader as the chief of the Chiefs of Staff Committee. I want a strong and independent man appointing his generals and his admirals and so on. I want a strong man in charge of all three branches of the Armed Forces of the Crown . . . strong enough to demand all the weapons which are necessary for victory . . . to see that his generals and admirals and air marshals are allowed to do their work in their own way and are not interfered with unduly from above. Above all, I want a man who, if he does not get what he wants, will immediately resign. . . . We have suffered both from the want of the closest examination by the Prime Minister of what is going on here at home, and also by the want of that direction which we should get from the Minister of Defence, or other officer, whatever his title might be, in charge of the Armed Forces. . . . It is surely clear to any civilian that the series of disasters of the past few months, and indeed of the past two years, is due to fundamental defects in the central administration of the war."

All this was making its point, but Sir John then made a digression. "It would be a very desirable move — if His Majesty the King and His Royal Highness would agree — if His Royal Highness the Duke of Gloucester were to be appointed Commander-in-Chief of the British Army — without of course administrative duties." This proved injurious to his case, as it was deemed a proposal to involve the Royal Family in grievous controversial responsibilities. Also the appointment of a Supreme War Commander with almost unlimited powers and his association with a Royal Duke seemed to have some flavour of dictatorship about it. From this moment the long and detailed indictment seemed to lose some of its pith. Sir John concluded, "The House should make it plain that we require one man to give his whole time to the winning of the war, in complete charge of all the Armed Forces of the Crown, and when we have got him let the House strengthen him to carry out the task with power and independence."

The motion was seconded by Sir Roger Keyes. The Admiral, who had been pained by his removal from the position of Director of Combined Operations, and still more by the fact that I had not always been able to take his advice while he was there, was hampered in his attack by his long personal friendship with me. He concentrated his criticisms mainly upon my expert advisers — meaning of course the Chiefs of Staff. "It is hard that three times in the Prime Minister's career he should have been thwarted — in Gallipoli, in Norway, and in the Mediterranean   in carrying out strategical strokes which might have altered the whole course of two wars, each time because his constitutional naval adviser declined to share the responsibility with him if it entailed any risk." The inconsistency between this argument and that of the mover did not pass unnoticed. One of the members of the Independent Labour Party, Mr. Stephen, interrupted to point out that the mover had proposed "a Vote of Censure on the ground that the Prime Minister has interfered unduly in the direction of the war; whereas the seconder seems to be seconding because the Prime Minister has not sufficiently interfered in the direction of the war." This point was apparent to the House.

"We look to the Prime Minister," said Admiral Keyes, "to put his house in order, and to rally the country once again for its immense task." Here another Socialist made a pertinent intervention. "The motion is directed against the central direction of the war. If the motion is carried the Prime Minister has to go; but the honourable and gallant Member is appealing to us to keep the Prime Minister there." "It would be," said Sir Roger, "a deplorable disaster if the Prime Minister had to go." Thus the debate was ruptured from its start.

Nevertheless, as it continued the critics increasingly took the lead. The new Minister of Production, Captain Oliver Lyttelton, who dealt with the complaints made against our equipment, had a stormy passage in the full, detailed account which he gave of this aspect. Strong Conservative support was given to the Government from their back benches, Mr. Boothby in particular making a powerful and helpful speech. Lord Winterton, the Father of the House, revived the force of the attack, and concentrated it upon me. "Who is the Minister of the Government who practically controlled the Narvik operation? It is the present Prime Minister, who was then First Lord of the Admiralty. . . . No one dares put the blame, where it should be put constitutionally, on the Prime Minister. . . . If whenever we have disasters we get the same answer, that whatever happens you must not blame the Prime Minister, we are getting very close to the intellectual and moral position of the German people — 'The Fuehrer is always right.' . . . During the thirty-seven years in which I have been in this House I have never seen such attempts to absolve a Prime Minister from Ministerial responsibility as are going on at present. . . . We never had anything in the last war comparable with this series of disasters. Now, see what this Government get off with — because 'the Fuehrer is always right.' We all agree that the Prime Minister was the Captain-General of our courage and constancy in 1940. But a lot has happened since 1940. If this series of disasters goes on, the right honourable gentleman, by one of the greatest acts of self-abnegation which any man could carry out, should go to his colleagues — and there is more than one suitable man for Prime Minister on the Treasury Bench now — and suggest that one of them should form a Government, and that the right honourable gentleman himself would take office under him. He might do so, perhaps, as Foreign Secretary, because his management of our relations with Russia and with the United States has been perfect."

It was not possible for me to listen to more than half the speeches of the animated debate, which lasted till nearly three in the morning. I had of course to be shaping my rejoinder for the next day; but my thoughts were centred on the battle which seemed to hang in the balance in Egypt.

* * *

The debate, which had talked itself out in the small hours of its first day, was resumed with renewed vigour on July 2. Certainly there was no denial of free speech or lack of it. One member even went so far as to say, " We have in this country five or six generals, members of other nations, Czechs, Poles, and French, all of them trained in the use of these German weapons and this German technique. I know it is hurtful to our pride, but would it not be possible to put some of those men temporarily in charge in the field, until we can produce trained men of our own? Is there anything wrong in sending out these men, of equal rank with General Ritchie? Why should we not put them in the field in charge of our troops? They know how to fight this war; our people do not, and I say that it is far better to win battles and save British soldiers' lives under the leadership of other members of the United Nations than to lose them under our own inefficient officers. The Prime Minister must realise that in this country there is a taunt on everyone's lips that if Rommel had been in the British Army he would still have been a sergeant.[1] Is that not so? It is a taunt right through the Army. There is a man in the British Army — and this shows how we are using our trained men — who flung one hundred and fifty thousand men across the Ebro in Spain: Michael

---

[1] This of course showed complete ignorance of Rommel's long and distinguished professional career in both wars.

Dunbar. He is at present a sergeant in an armoured brigade in this country. He was Chief of Staff in Spain; he won the battle of the Ebro, and he is a sergeant in the British Army. The fact of the matter is that the British Army is ridden by class prejudice. You have got to change it, and you will have to change it. If the House of Commons has not the guts to make the Government change it, events will. Although the House may not take any notice of me today, you will be doing it next week. Remember my words next Monday and Tuesday. It is events which are criticising the Government. All that we are doing is giving them a voice, inadequately perhaps, but we are trying to do it."

The main case against the Government was summed up by Mr. Hore-Belisha, the former Secretary of State for War. He concluded, "We may lose Egypt or we may not lose Egypt — I pray God we may not — but when the Prime Minister, who said that we would hold Singapore, that we would hold Crete, that we had smashed the German army in Libia . . . when I read that he had said that we are going to hold Egypt, my anxieties became greater. . . . How can one place reliance in judgments that have so repeatedly turned out to be misguided? That is what the House of Commons has to decide. Think what is at stake. In a hundred days we lost our Empire in the Far East. What will happen in the next hundred days? Let every Member vote according to his conscience."

I followed this powerful speech in winding up the debate. The House was crammed. Naturally I made every point which occurred to me. Mr. Hore-Belisha had dwelt upon the failures of the British tanks and the inferiority of our equipment in armour. He was not in a very strong position to do this on account of the pre-war record of the War Office. I was able to turn the tables upon him:

"The idea of the tank was a British conception. The use of armoured forces as they are now being used was largely French, as General de Gaulle's book shows. It was left to the Germans to convert those ideas to their own use. For three or four years before the war they were busily at work with their usual thoroughness upon the design and manufacture of tanks, and also upon the study and practice of armoured warfare. One would have thought that even if the Secretary of State for War of those days could not get the money for large-scale manufacture he would at any rate have had full-size working models made and tested out exhaustively, and the factories chosen and the jigs and gauges supplied, so that he could go into mass production of tanks and anti-tank weapons when the war began.

"When what I may call the Belisha period ended we were left with some two hundred and fifty armoured vehicles, very few of which carried even a two-pounder gun. Most of these were captured or destroyed in France.

"I willingly accept, indeed I am bound to accept, what the noble Lord [Earl Winterton] has called the 'constitutional responsibility' for everything that has happened, and I consider that I discharged that responsibility by not interfering with the technical handling of armies in contact with the enemy. But before the battle began I urged General Auchinleck to take the command himself, because I was sure nothing was going to happen in the vast area of the Middle East in the next month or two comparable in importance to the fighting of this battle in the Western Desert, and I thought he was the man to handle the business. He gave me various good reasons for not doing so, and General Ritchie fought the battle. As I told the House on Tuesday, General Auchinleck on June 25 superseded General Ritchie and assumed command himself. We at once approved his decision, but I must frankly confess that the matter was not one on which we could form any final judgment, so far as the superseded officer is concerned. I cannot pretend to form a judgment upon what has happened in this battle. I like commanders on land and sea and in the air to feel that between them and all forms of public criticism the Government stands like a strong bulkhead. They ought to have a fair chance, and more than one chance. Men may make mistakes and learn from their mistakes. Men may have bad luck, and their luck may change. But anyhow you will not get generals to run risks unless they feel they have behind them a strong Government. They will not run risks unless they feel that they need not look over their shoulders or worry about what is happening at home, unless they feel they can concentrate their gaze upon the enemy. And you will not, I may add, get a Government to run risks unless they feel that they have got behind them a loyal, solid majority. Look at the things we are being asked to do now, and imagine the kind of attack which would be made on us if we tried to do them and failed. In wartime if you desire service you must give loyalty. . . .

"I wish to speak a few words 'of great truth and respect' — as they say in the diplomatic documents — and I hope I may be granted the fullest liberty of debate. This Parliament has a peculiar responsibility. It presided over the beginning of the evils which have come on the world. I owe much to the House, and it is my hope that it may see the end of them in triumph. This it can do only if, in the long period which may yet have to be travelled, the House affords a solid foundation to the responsible Executive Government, placed in power by its own choice. The House must be a steady stabilising factor in the State, and not an instrument by which the disaffected sections of the press can attempt to promote one crisis after another. If democracy and Parliamentary institutions are to triumph in this war it is absolutely necessary that Governments resting upon them shall be able to act and dare, that the servants of the Crown shall not be harassed by nagging and snarling, that enemy propaganda shall not be fed needlessly out of our own hands, and our reputation disparaged and undermined throughout the world. On the contrary, the will of the whole House should be made manifest upon important occasions. It is important that not only those who speak, but those who watch and listen and judge, should also count as a factor in world affairs. After all, we are still fighting for our lives, and for causes dearer than life itself. We have no right to assume that victory is certain; it will be certain only if we do not fail in our duty. . . . Sober and constructive criticism, or criticism in Secret Session, has its high virtue; but the duty of the House of Commons is to sustain the Government or to change the Government. If it cannot change it it should sustain it. There is no working middle course in wartime. . . . Only the hostile speeches are reported abroad, and much play is made with them by our enemy.

". . . The mover of this Vote of Censure has proposed that I should be stripped of my responsibilities for defence in order that some military figure or some other unnamed personage should assume the general conduct of the war, that he should have complete control of the Armed Forces of the Crown, that he should be the Chief of the Chiefs of Staff, that he should nominate or dismiss the generals or the admirals, that he should always be ready to resign — that is to say, to match himself against his political colleagues, if colleagues they could be considered — if he did not get all he wanted, that he should have under him a Royal Duke as Commander-in-Chief of the Army, and finally, I presume, though this was not mentioned, that this unnamed personage should find an appendage in the Prime Minister to make the necessary explanations, excuses, and apologies to Parliament when things go wrong, as they often do and often will. That is at any rate a policy. It is a system very different from the Parliamentary system under which we live. It might easily amount to or be converted into a dictatorship. I wish to make it perfectly clear that as far I am concerned I shall take no part in such a system."

Sir John J. Wardlaw-Milne here injected, "I hope my right honourable friend has not forgotten the original sentence, which was 'subject to the War Cabinet'?"

I continued: " 'Subject to the War Cabinet,' against which this all-powerful potentate is not to hesitate to resign on every occasion if he cannot get his way. It is a plan, but it is not a plan in which I should personally be interested to take part, and I do not think that it is one which would commend itself to this House.

"The setting down of this Vote of Censure by Members of all parties is a considerable event. Do not, I beg of you, let the House underrate the gravity of what has been done. It has been trumpeted all round the world to our disparagement, and when every nation, friend and foe, is waiting to see what is the true resolve and conviction of the House of Commons, it must go forward to the end. All over the world, throughout the United States, as I can testify, in Russia, far away in China, and throughout every subjugated country, all our friends are waiting to know whether there is a strong, solid Government in Britain and whether its national leadership is to be challenged or not. Every vote counts. If those who have assailed us are reduced to contemptible proportions and their Vote of Censure on the National Government is converted to a vote of censure upon its authors, make no mistake, a cheer will go up from every friend of Britain and every faithful servant of our cause, and the knell of disappointment will ring in the ears of the tyrants we are striving to overthrow."

The House divided, and Sir John Wardlaw-Milne's motion of "No Confidence" was defeated by 475 votes to 25.

My American friends awaited the issue with real anxiety. They were delighted by the result, and I woke to receive their congratulations.

*   *   *

A curious historical point had been made in the debate by Mr. Walter Elliot when he recalled Macaulay's account of Mr. Pitt's Administration. "Pitt was at the head of a nation engaged in a life-and-death struggle. . . . But the fact is that after eight years of war, after a vast expenditure of life and . . . wealth, the English Army under Pitt was the laughing-stock of all Europe. They could not boast of a single brilliant exploit. It had never shown itself on the Continent but to be beaten, chased, forced to re-embark." However, Macaulay proceeded to record that Pitt was always sustained by the House of Commons. "Thus through a long and calamitous period every disaster that happened without the walls of Parliament was regularly followed by triumph within them. At length he had no longer an Opposition to encounter, and in the eventful year 1799 the largest majority that could be mustered to vote against the Government was twenty-five." "It is odd," said Mr. Elliot, "how history is in some ways repeated." He could not know before the division how true this was. I too was astonished that the figure of twenty-five was almost exactly the one I had named to the President and Harry Hopkins when I was with them at the White House on the day of the Tobruk news.

13

The Eighth Army at Bay

THE CAPTURE OF TOBRUK without a long siege revolutionised the Axis plans. Hitherto it had been intended that after it was taken Rommel should stand on the Egyptian frontier and Malta should be seized by air-borne and sea-borne forces. As late as June 21 Mussolini reiterated these orders. The day after Tobruk fell Rommel reported that he proposed to destroy the small British forces left on the frontier, and thus open the way to Egypt. The condition and morale of his forces, the large captures of munitions and supplies, and the weakness of the British position prompted pursuit "into the heart of Egypt." He requested approval. A letter also arrived from Hitler pressing Rommel's proposals upon Mussolini.

> Destiny has offered us a chance which will never occur twice in the same theatre of war. . . . The English Eighth Army has been practically destroyed. In Tobruk the port installations are almost intact. You now possess, Duce, an auxiliary base whose significance is all the greater because the English themselves have built from there a railway leading almost into Egypt. If at this moment the remains of this British Army are not pursued to the last breath of each man, the same thing will happen as when the British were deprived of success when they nearly reached Tripoli and suddenly stopped in order to send forces to Greece. . . .
>
> The goddess of Battles visits warriors only once. He who does not grasp her at such a moment never reaches her again.[1]

The Duce needed no persuasion. Elated at the prospect of conquering Egypt, he postponed the assault on Malta till the beginning of

[1] Quoted in Cavallero, *Commando Supremo*, p. 277.

September, and Rommel — now a Field Marshal, rather to Italian surprise — was authorised to occupy the relatively narrow passage between Alamein and the Qattara Depression as the starting point for
future operations whose final objective was the Suez Canal. Kesselring
held a different view. Believing that the Axis position in the Desert
would never be secure until Malta was captured, he was alarmed at the
change of plan. He pointed out to Rommel the dangers of this "fool-
hardy enterprise."

Hitler himself had not been confident of success against Malta, as he
mistrusted the ability of the Italian troops who would have formed the
major part of the expedition. The attack might well have failed. Nevertheless it now seems certain that the shattering and grievous loss of
Tobruk spared the island from the supreme trial. This is a consolation
of which no good soldier, whether involved or not, should avail himself. The burden falls upon the High Command rather than on the
Generals concerned, and still less upon the troops.

*   *   *

Rommel swiftly organised his pursuit, and on June 24 crossed the
frontier to Egypt, opposed only by our light mobile columns, and the
stubborn and magnificent fighter squadrons of the Royal Air Force, who
really covered the retreat of the Eighth Army to Mersa Matruh. Their
position here was not strong. About the town itself there was an organised defensive system, but south of it were only some lines of unconnected mine fields inadequately guarded. As in the case of the rejected frontier position, the Matruh line, if it were to be successfully
held, needed a powerful armoured force to guard its southern flank. The
7th Armoured Division, though now rebuilt to nearly a hundred tanks,
was not yet capable of such a task.

General Auchinleck himself came forward to Matruh on June 25, and
decided to take over direct operational command of the Army from
General Ritchie. He should have done this when I asked him to in May.
He quickly concluded that it was not possible to make a final stand at
Matruh. Arrangements were already in hand for the preparation and
occupation of the Alamein position, a hundred and twenty miles farther
back. Dispositions were made to halt the enemy, if only for a time, and
the New Zealand Division, which had arrived at Matruh from Syria on
June 21, were at length moved on the 26th into action on the ridge about
Minqa Qaim. That evening the enemy broke through the front of the
29th Indian Infantry Brigade, where the mine field was incomplete.
The next morning they streamed through the gap, and then, passing behind the New Zealanders, encircled and attacked them from three sides.
Desperate fighting continued all day, and at the end it seemed that the

division was doomed. General Freyberg had been severely wounded. But he had a worthy successor. Brigadier Inglis was determined to break out. Shortly after midnight the 4th New Zealand Brigade moved due east across country with all its battalions deployed and bayonets fixed. For a thousand yards no enemy were met. Then firing broke out. The whole brigade charged in line. The Germans were taken completely by surprise, and were routed in hand-to-hand fighting under the moon. The rest of the New Zealand Division struck south by circuitous routes. This is how Rommel has described the episode:

> The wild flare-up which ensued involved my own battle head-quarters. . . . The exchanges of fire between my forces and the New Zealanders reached an extraordinary pitch of intensity. Soon my headquarters were surrounded by burning vehicles, making them the target for continuous enemy fire at close range. I had enough of this after a while, and ordered the troops with the staff to move back southeastwards. The confusion reigning on that night can scarcely be imagined.[2]

Thus the New Zealanders broke clear, and the whole division was reunited in a high state of discipline and ardour near the Alamein position, eighty miles away. So little were they disorganised that they were used forthwith to stiffen its defences.

The remainder of the Eighth Army were also brought back to safety, though with difficulty. The troops were amazed rather than depressed, but with the advantage of short communications, and with Alexandria only forty miles away, reorganisation did not take long. Auchinleck, once in direct command, seemed a different man from the thoughtful strategist with one eye on the decisive battle and the other on the vague and remote dangers in Syria and Persia. He sought at once to regain the tactical initiative. As early as July 2 he made the first of a series of counterattacks which continued until the middle of the month. These challenged Rommel's precarious ascendancy. I sent my encouragement, on the morrow of the Vote of Censure debate, which had been an accompaniment to the cannonade.

In fact Rommel's communications were strained to the utmost limit and his troops exhausted. Only a dozen German tanks were still fit for action, and the superiority of the British Air Force, especially in fighters, was again becoming dominant. Rommel reported on July 4 that he was suspending his attacks and going over to the defensive for a while in order to regroup and replenish his forces. He was still confident however of taking Egypt, and his opinion was shared by Mussolini and by

---

[2] *Rommel*, by Desmond Young, p. 269.

Hitler. The Fuehrer indeed, without reference either to the Italians or to his own naval command, postponed the attack on Malta until the conquest of Egypt was complete.

Auchinleck's counterattacks pressed Rommel very hard for the first fortnight of July. He then took up the challenge, and from July 15 to July 20 renewed his attempts to break the British line. On the 21st he had to report that he was checked: "The crisis still exists." On the 26th he was contemplating withdrawal to the frontier. He complained that he had received little in the way of replenishments; he was short of men, tanks, and artillery; the British Air Force was extremely active. And so the battle swayed back and forth until the end of the month, by which time both sides had fought themselves to a standstill. The Eighth Army under Auchinleck had weathered the storm, and in its stubborn stand had taken seven thousand prisoners. Egypt was still safe.

It was at this juncture, when I was politically at my weakest and without a gleam of military success, that I had to procure from the United States the decision which, for good or ill, dominated the next two years of the war. This was the abandonment of all plans for crossing the Channel in 1942 and the occupation of French North Africa in the autumn or winter by a large Anglo-American expedition.

*   *   *

I had made a careful study of the President's mind and its reactions for some time past, and I was sure that he was powerfully attracted by the North African plan. This had always been my aim, as was set forth in my papers of December 1941. Everyone in our British circle was by now convinced that a Channel crossing in 1942 would fail, and no military man on either side of the ocean was prepared to recommend such a plan or to take responsibility for it. I stated the case with whatever force I could command and in the plainest terms in an important telegram to the President, which was dated July 8:

> No responsible British general, admiral, or air marshal is prepared to recommend "Sledgehammer" [3] as a practicable operation in 1942. The Chiefs of Staff have reported, "The conditions which would make 'Sledgehammer' a sound, sensible enterprise are very unlikely to occur." They are now sending their paper to your Chiefs of Staff.

[3] The following shortly explain the code-names occurring in this chapter:
GYMNAST: The landing in Northwest Africa, later called "Torch."
JUPITER: Operations in Northern Norway.
ROUNDUP: The invasion of Europe, afterwards called "Overlord."
SLEDGEHAMMER: The attack on Brest or Cherbourg in 1942.

2. The taking up of the shipping is being proceeded with by us for camouflage purposes, though it involves a loss in British imports of perhaps 250,000 tons. But far more serious is the fact that, according to Mountbatten, if we interrupt the training of the troops we should, apart from the loss of landing craft, etc., delay [our main invasion of France] for at least two or three months, even if the enterprise were unsuccessful and the troops had to be withdrawn after a short stay.

3. In the event of a lodgment being effected and maintained it would have to be nourished, and the bomber effort on Germany would have to be greatly curtailed. All our energies would be involved in defending the bridgehead. The possibility of mounting a large-scale operation in 1943 would be marred, if not ruined. All our resources would be absorbed piecemeal on the very narrow front which alone is open. It may therefore be said that premature action in 1942, while probably ending in disaster, would decisively injure the prospect of well-organised large-scale action in 1943.

4. I am sure myself that French North Africa ["Gymnast"] is by far the best chance for effecting relief to the Russian front in 1942. This has all along been in harmony with your ideas. In fact, it is your commanding idea. Here is the true Second Front of 1942. I have consulted the Cabinet and Defence Committee, and we all agree. Here is the safest and most fruitful stroke that can be delivered this autumn.

5. We of course can aid in every way, by transfer of either American or British landing forces from the United Kingdom to "Gymnast," and with landing craft, shipping, etc. You can, if you choose, put the punch in partly from here and the rest direct across the Atlantic.

6. It must be clearly understood that we cannot count upon an invitation or a guarantee from Vichy. But any resistance would not be comparable to that which would be offered by the German Army in the Pas de Calais. Indeed, it might be only token resistance. The stronger you are, the less resistance there would be and the more to overcome it. This is a political more than a military issue. It seems to me that we ought not to throw away the sole great strategic stroke open to us in the Western theatre during this cardinal year.

7. Besides the above we are studying very hard the possibility of an operation in Northern Norway, or, if this should prove impracticable, elsewhere in Norway. The difficulties are great owing to the danger of shore-based aircraft attack upon our ships. We are having frightful difficulties about the Russian convoys. All the more

is it necessary to try to clear the way and maintain the contact with Russia.

But before the final decision for action could be obtained there was a pause. Strong tensions grew in the supreme American war direction. General Marshall was divided from Admiral King as between Europe and the Pacific. Neither was inclined to the North African venture. In this deadlock the President's liking for North Africa grew steadily stronger. Field Marshal Dill's qualities had won him the confidence of all the rival schools of thought, and his tact preserved their goodwill. The President was conscious of the strength of the arguments against "Sledgehammer." If he placed it in the forefront of his communications to us, it was to convince General Marshall that it would have every chance. But if no one would touch it, what then? There was the wave of American Staff opinion which argued, "If nothing can be done this year in Europe let us concentrate on Japan, and thus bring the United States Army and Navy thought together and unite General Marshall with Admiral King."

The President withstood and brushed aside this fatal trend of thought. He was convinced that the United States Army must fight against the Germans in 1942. Where then could this be but in French North Africa? "This was," says Mr. Stimson, "his secret war baby." The movement of the force of the argument and of the President's mind to this conclusion was remorseless.

On Saturday, July 18, General Marshall, Admiral King, and Harry Hopkins landed at Prestwick, and travelled by train to London. Here they went into immediate conference with the American Service Chiefs now established in the capital, Eisenhower, Clark, Stark, and Spaatz. The debate on "Sledgehammer" was renewed. Opinion among the American leaders was still strongly in favour of pressing on exclusively with this operation. Only the President himself seemed to have been impressed by my arguments. He had drafted for the delegation the most massive and masterly document on war policy that I ever saw from his hand.[4]

MEMORANDUM FOR HON. HARRY L. HOPKINS, GENERAL MARSHALL, AND ADMIRAL KING

Subject: Instructions for London Conference, July 1942

16 July 42

1. You will proceed immediately to London as my personal representatives for the purpose of consultation with appropriate British authorities on the conduct of the war.

[4] Robert Sherwood, *Roosevelt and Hopkins*, pp. 603–5.

2. The military and naval strategic changes have been so great since Mr. Churchill's visit to Washington that it became necessary to reach immediate agreement on joint operational plans between the British and ourselves along two lines:

(a) Definite plans for the balance of 1942.

(b) Tentative plans for the year 1943, which of course will be subject to change in the light of occurrences in 1942, but which should be initiated at this time in all cases involving preparation in 1942 for operations in 1943.

3. (a) The common aim of the United Nations must be the defeat of the Axis Powers. There cannot be compromise on this point.

(b) We should concentrate our efforts and avoid dispersion.

(c) Absolute co-ordinated use of British and American forces is essential.

(d) All available U.S. and British forces should be brought into action as quickly as they can be profitably used.

(e) It is of the highest importance that U.S. ground troops be brought into action against the enemy in 1942.

4. British and American material promises to Russia must be carried out in good faith. If the Persian route of delivery is used preference must be given to combat material. This aid must continue as long as delivery is possible, and Russia must be encouraged to continue resistance. Only complete collapse, which seems unthinkable, should alter this determination on our part.

5. In regard to 1942, you will carefully investigate the possibility of executing "Sledgehammer." Such an operation would definitely sustain Russia this year. "Sledgehammer" is of such grave importance that every reason calls for accomplishment of it. You should strongly urge immediate all-out preparations for it, that it be pushed with utmost vigour, and that it be executed whether or not Russian collapse becomes imminent. In the event Russian collapse becomes probable, "Sledgehammer" becomes not merely advisable but imperative. The principal objective of "Sledgehammer" is the positive diversion of German air forces from the Russian front.

6. Only if you are completely convinced that "Sledgehammer" is impossible of execution with reasonable chance of serving its intended purpose inform me.

7. *If "Sledgehammer" is finally and definitely out of the picture I want you to consider the world situation as it exists at that time, and determine upon another place for U.S. troops to fight in 1942.*[5]

[5] My italics.—W.S.C.

It is my present view of the world picture that:

(a) If Russia contains a large German force against her, "Round-up" [the invasion of Europe] becomes possible in 1943, and plans for "Round-up" should be immediately considered and preparations made for it.

(b) If Russia collapses and German air and ground forces are released "Round-up" may be impossible of fulfilment in 1943.

8. The Middle East should be held as strongly as possible whether Russia collapses or not. I want you to take into consideration the effect of losing the Middle East. Such loss means in series:

(1) Loss of Egypt and the Suez Canal.
(2) Loss of Syria.
(3) Loss of Mosul oil wells.
(4) Loss of the Persian Gulf through attacks from the north and west, together with access to all Persian Gulf oil.
(5) Joining hands between Germany and Japan and the probable loss of the Indian Ocean.
(6) The very important probability of German occupation of Tunis, Algiers, Morocco, Dakar, and the cutting of the ferry route through Freetown and Liberia.
(7) Serious danger to all shipping in the South Atlantic, and serious danger to Brazil and the whole of the east coast of South America. I include in the above possibilities the use by the Germans of Spain, Portugal, and their territories.
(8) You will determine the best methods of holding the Middle East. These methods include definitely either or both of the following:
(a) Sending aid and ground forces to the Persian Gulf, to Syria, and to Egypt.
(b) *A new operation in Morocco and Algeria intended to drive in against the back door of Rommel's armies. The attitude of French colonial troops is still in doubt.*[6]

9. I am opposed to an American all-out effort in the Pacific against Japan with the view to her defeat as quickly as possible. It is of the utmost importance that we appreciate that defeat of Japan

---

[6] My italics.—W.S.C.

does not defeat Germany and that American concentration against Japan this year or in 1943 increases the chance of complete German domination of Europe and Africa. On the other hand, it is obvious that defeat of Germany or the holding of Germany in 1942 or in 1943 means probable eventual defeat of Germany in the European and African theatre and in the Near East. *Defeat of Germany means the defeat of Japan, probably without firing a shot or losing a life.*[7]

10. Please remember three cardinal principles — speed of decision on plans, unity of plans, attack combined with defence but not defence alone. This affects the immediate objective of U.S. ground forces fighting against Germans in 1942.

11. I hope for total agreement within one week of your arrival.

FRANKLIN D. ROOSEVELT
Commander-in-Chief

But in spite of this last injunction, General Marshall told me on the afternoon of July 22 that he and his colleagues had reached a deadlock in their talks with the British Chiefs of Staff, and they would have to report to the President for instructions.

I replied that I fully shared the ardent desire of the President and his Service advisers "to engage the enemy in the greatest possible strength at the earliest possible moment," but that I felt sure that, with the limited forces at our disposal, we should not be justified in attempting "Sledgehammer" in 1942. I pointed to the number of ugly possibilities looming in front of us. There might, for example, be a collapse in Russia, or the Germans might move into the Caucasus, or they might beat General Auchinleck and occupy the Nile Delta and the Suez Canal, or again they might establish themselves in North Africa and West Africa and thereby put an almost prohibitive strain on our shipping. Nevertheless, disagreement between Great Britain and America would have far greater consequences than all the above possibilities. It was therefore agreed that the American Chiefs of Staff should report to the President that the British were not prepared to go ahead with "Sledgehammer" and ask for instructions.

Mr. Roosevelt replied at once that he was not surprised at the disappointing outcome of the London talks. He agreed that it was no use continuing to press for "Sledgehammer" in the face of British opposition, and instructed his delegation to reach a decision with us on some operation which would involve American land forces being brought into action against the enemy in 1942. Thus "Sledgehammer" fell by the

[7] My italics.—W.S.C.

wayside and "Gymnast" came into its own. Marshall and King, though naturally disappointed, bowed to the decision of their Commander-in-Chief, and the greatest goodwill between us all again prevailed.

I now hastened to rechristen my favourite. "Gymnast," and its variants, vanished from our code names. On July 24 in an instruction from me to the Chiefs of Staff "Torch" became the new and master term. On July 25 the President cabled to Hopkins that plans for landings in North Africa to take place "not later than October 30" should go ahead at once. That evening our friends set off on their journey back to Washington.

All was therefore agreed and settled in accordance with my long-conceived ideas and those of my colleagues, military and political. This was a great joy to me, especially as it came in what seemed to be the darkest hour. At every point except one the plans I cherished were adopted. "Jupiter" alone (the Norway enterprise) I could not carry, although its merits were not disputed. I did not give up this plan yet, but in the end I failed to establish it. For months past I had sought "No 'Sledgehammer,' " but instead the North African invasion *and* "Jupiter." "Jupiter" fell by the way. But I had enough to be thankful for.

"The President," cabled Field Marshal Dill from Washington, "has gone to Hyde Park for short rest, but before going he issued orders for full steam ahead 'Torch' at the earliest possible moment. He has asked Combined Chiefs of Staff to tell him on August 4 earliest date when landing could take place. Risk of whittling to Pacific may still exist, but President entirely sound on this point.

"In the American mind 'Round-up' in 1943 is excluded by acceptance of 'Torch.' We need not argue about that. A one-track mind on 'Torch' is what we want at present . . . May what you are at have the success which courage and imagination deserve."

This message reached me at midnight on August 1, 1942, on the Lyneham Airfield, where I was about to set forth upon a journey of which the next chapter will offer both explanation and account.

14

My Journey to Cairo:
Changes in Command

THE DOUBTS I HAD about the High Command in the Middle
East were fed continually by the reports which I received from
many quarters. It became urgently necessary for me to go there and
settle the decisive questions on the spot. It was at first accepted that
this journey would be by Gibraltar and Takoradi and thence across
Central Africa to Cairo, involving five or even six days' flying. However,
at this juncture there arrived in England a young American pilot, Captain Vanderkloot, who had just flown from the United States in the
aeroplane "Commando," a Liberator plane from which the bomb racks
had been removed and some sort of passenger accommodation substituted. This machine was certainly capable of flying along the route
prescribed with good margins in hand at all stages. Portal, the Chief
of the Air Staff, saw this pilot and cross-examined him about "Commando." Vanderkloot, who had already flown about a million miles,
asked why it was necessary to fly all round by Takoradi, Kano, Fort
Lamy, El Obeid, etc. He said he could make one bound from Gibraltar
to Cairo, flying from Gibraltar eastward in the afternoon, turning sharply
south across Spanish or Vichy territory as dusk fell, and then proceeding eastward till he struck the Nile about Assiout, when a turn to the
northward would bring us in another hour or so to the Cairo landing
ground northwest of the Pyramids. This altered the whole picture. I
could be in Cairo in two days. Portal was convinced.

We were all anxious about the reaction of the Soviet Government to
the unpleasant though inevitable news that there would be no crossing
of the Channel in 1942. It happened that on the night of July 28 I had
the honour of entertaining the King to dinner with the War Cabinet
in the propped-up garden room at Number 10, which we used for dining.
I obtained His Majesty's approval privately for my journey, and im-

mediately he had gone brought the Ministers, who were in a good frame of mind, into the Cabinet Room and clinched matters. It was settled that I should go to Cairo in any case, and should propose to Stalin that I should go on to see him. On the 30th I therefore telegraphed to him as follows:

> We are making preliminary arrangements for another effort to run a large convoy through to Archangel in the first week of September.
>
> 2. I am willing, if you invite me, to come myself to meet you in Astrakhan, the Caucasus, or similar convenient meeting place. We could then survey the war together and take decisions hand-in-hand. I could then tell you plans we have made with President Roosevelt for offensive action in 1942. I would bring the Chief of the Imperial General Staff with me.
>
> 3. I am starting for Cairo forthwith. I have serious business there, as you may imagine. From there I will, if you desire it, fix a convenient date for our meeting, which might, so far as I am concerned, be between August 10 and 13, all being well.
>
> 4. The War Cabinet have endorsed my proposals.

The reply came next day.

> On behalf of the Soviet Government I invite you to the U.S.S.R. to meet the members of the Government. . . . I think the most suitable meeting place would be Moscow, as neither I nor the members of the Government and the leading men of the General Staff could leave the capital at the moment of such an intense struggle against the Germans. The presence of the Chief of the Imperial General Staff would be extremely desirable.
>
> The date of the meeting please fix yourself in accordance with the time necessary for completion of your business in Cairo. You may be sure beforehand that any date will suit me.
>
> Let me express my gratitude for your consent to send the next convoy with war materials for the U.S.S.R. at the beginning of September. In spite of the extreme difficulty of diverting aircraft from the battlefront we will take all possible measures to increase the aerial protection of the convoy.

Thus all was arranged, and we started after midnight on Sunday August 2, from Lyneham in the bomber "Commando." This was a very different kind of travel from the comforts of the Boeing flying boats. The bomber was at this time unheated, and razor-edged draughts cut in through many chinks. There were no beds, but two shelves in the after cabin enabled me and Lord Moran to lie down. There were plenty of blankets for all. We flew low over the South of England in order to be recognised by our batteries, who had been warned, but who were also under "Alert" conditions. As we got out to sea I left the cockpit and retired to rest, fortified by a good sleeping cachet.

We reached Gibraltar uneventfully on the morning of August 3, spent the day looking round the fortress, and started at 6 P.M. for Cairo, a hop of two thousand miles or more, as the detours necessary to avoid the hostile aircraft around the Desert battle were considerable. Vander-kloot, in order to have more petrol in hand, did not continue down the Mediterranean till darkness fell, but flew straight across the Spanish zone and the Vichy quasi-hostile territory. Therefore, as we had an armed escort till nightfall of four Beaufighters we in fact openly violated the neutrality of both these regions. No one molested us in the air, and we did not come within cannon shot of any important town. All the same I was glad when darkness cast her shroud over the harsh landscape and we could retire to such sleeping accommodation as "Commando" could offer. It would have been very tiresome to make a forced landing on neutral territory, and even descent in the desert, though preferable, would have raised problems of its own. However, all "Commando's" four engines purred happily, and I slept sound as we sailed through the starlit night.

It was my practice on these journeys to sit in the co-pilot's seat before sunrise, and when I reached it on this morning of August 4 there in the pale, glimmering dawn the endless winding silver ribbon of the Nile stretched joyously before us. Often had I seen the day break on the Nile. In war and peace I had traversed by land or water almost its whole length, except the "Dongola Loop," from Lake Victoria to the sea. Never had the glint of daylight on its waters been so welcome to me.

Now for a short spell I became "the man on the spot." Instead of sitting at home waiting for the news from the front I could send it myself. This was exhilarating.

*   *   *

The following issues had to be settled in Cairo. Had General Auchinleck or his staff lost the confidence of the Desert Army? If so, should he be relieved, and who could succeed him? In dealing with a commander of the highest character and quality, of proved ability and resolution, such decisions are painful. In order to fortify my own judgment I had urged General Smuts to come from South Africa to the scene and he was already at the Embassy when I arrived. We spent the morning together, and I told him all our troubles and the choices that were open. In the afternoon I had a long talk with Auchinleck, who explained the military position very clearly. After luncheon next day General Wavell arrived from India, and at six o'clock I held a meeting about the Middle East, attended by all the authorities — Smuts, Casey, who had succeeded Lyttelton as Minister of State in the Middle East, General Brooke, the C.I.G.S., Wavell, Auchinleck, Admiral Harwood, and Tedder for the Air. We did a lot of business with a very great measure of agreement. But all the time my mind kept turning to the prime question of the command.

It is not possible to deal with changes of this character without reviewing the alternatives. In this part of the problem the Chief of the Imperial General Staff, whose duty it was to appraise the quality of our generals, was my adviser. I first offered the Middle East Command to him. General Brooke would of course have greatly liked this high operational appointment, and I knew that no man would fill it better. He thought it over, and had a long talk the next morning with General Smuts. Finally he replied that he had been C.I.G.S. for only eight months, he believed he had my full confidence, and the Staff machine was working very smoothly. Another change at this moment might cause a temporary dislocation at this critical time. It may well be also that out of motives of delicacy he did not wish to be responsible for advising General Auchinleck's supersession and then taking the post himself. His reputation stood too high for such imputations; but I had now to look elsewhere.

Alexander and Montgomery had both fought with him in the battle which enabled us to get back to Dunkirk in May 1940. We both greatly admired Alexander's magnificent conduct in the hopeless campaign to which he had been committed in Burma. Montgomery's reputation stood high. If it were decided to relieve Auchinleck we had no doubt that Alexander must be ordered to carry the load in the Middle East. But the feelings of the Eighth Army must not be overlooked. Might it not be taken as a reproach upon them and all their commanders of every grade if two men were sent from England to supersede all those who had fought in the desert? Here General Gott, one of the Corps Commanders, seemed in every way to meet the need. The troops were devoted to him and he had not earned the title "Strafer" by nothing. But then there was the view which Brooke reported to me, that he was very tired and needed a rest. It was at this moment too early to take decisions. I had travelled all this way to have the chance of seeing and hearing what was possible in the short time which might be claimed and spared.

* * *

The hospitality of our Ambassador, Sir Miles Lampson, was princely. I slept in his air-cooled bedroom and worked in his air-cooled study. It was intensely hot, and those were the only two rooms in the house where the temperature was comfortable. In these otherwise agreeable surroundings we dwelt for more than a week, sensing the atmosphere, hearing opinions, and visiting the front or the large camps to the east of Cairo in the Kassassin area, where our powerful reinforcements were now steadily arriving.

On August 5 I visited the Alamein positions. I drove with General Auchinleck in his car to the extreme right flank of the line west of El Ruweisat. Thence we proceeded along the front to his headquarters behind the Ruweisat Ridge, where were were given breakfast in a wire-netted cube, full of flies and important military personages. I had asked for various officers to be brought, but above all General "Strafer" Gott. It was said that he was worn down with his hard service. This was what I wanted to find out. Having made the acquaintance of the various Corps and Divisional Commanders who were present, I therefore asked that General Gott should drive with me to the airfield, which was my next stop. Objection was raised by one of Auchinleck's staff officers that this would take him an hour out of his way; but I insisted he should come with me. And here was my first and last meeting with Gott. As we rumbled and jolted over the rough tracks I looked into his clear blue eyes and questioned him about himself. Was he tired, and had he any views to give? Gott said that no doubt he was tired, and he would like nothing better than three months' leave in England, which he had not seen for several years, but he declared himself quite capable of further immediate efforts and of taking any responsibilities confided to him. We parted at the airfield at two o'clock on this afternoon of August 5. By the same hour two days later he had been killed by the enemy in almost the very air spaces through which I now flew.

At the airfield I was handed over to Air Vice-Marshal Coningham, who, under Tedder, commanded all the air power which had worked with the Army, and without whose activity the immense retreat of five hundred miles could never have been accomplished without even greater disasters than we had suffered. We flew in a quarter of an hour to his headquarters, where luncheon was provided, and where all the leading Air officers, from Group Captains upward, were gathered. I was conscious of an air of nervousness in my hosts from the moment of my arrival. The food had all been ordered from Shepheard's Hotel. A special car was bringing down the dainties of Cairo. But it had gone astray. Frantic efforts were being made to find it. At last it arrived.

This turned out to be a gay occasion in the midst of care — a real oasis in a very large desert. It was not difficult to perceive how critical the Air was of the Army, and how both Air and Army were astonished at the reverse which had befallen our superior forces. In the evening I flew back to Cairo and reported my general impressions to Mr. Attlee.

All the next day, the 6th, I spent with Brooke and Smuts, and in drafting the necessary telegrams to the Cabinet. The questions that had now to be settled not only affected the high personalities but also the entire structure of command in this vast theatre. I had always felt that the name "Middle East" for Egypt, the Levant, Syria, and Turkey was ill chosen. This was the Near East. Persia and Iraq were the Middle East; India, Burma, and Malaya the East; and China and Japan the Far East. But, far more important than changing names, I felt it necessary to divide the existing Middle East Command, which was far too diverse and expansive. Now was the time to effect this change in organisation. At 8.15 P.M. I therefore telegraphed to Mr. Attlee as follows:

> . . . I have come to the conclusion that a drastic and immediate change is needed in the High Command.
>
> 2. I therefore propose that the Middle East Command shall be reorganised into two separate Commands, namely:
>> (a) "Near East Command," comprising Egypt, Palestine, and Syria, with its centre in Cairo, and
>> (b) "Middle East Command," comprising Persia and Iraq, with its centre in Basra or Baghdad.
>
> The Eighth and Ninth Armies fall within the first and the Tenth Army in the second of these Commands.
>
> 3. General Auchinleck to be offered the post of C.-in-C. the new Middle East Command. . . .
>
> 4. General Alexander to be Commander-in-Chief the Near East.
>
> 5. General Montgomery to succeed Alexander in "Torch." I regret the need of moving Alexander from "Torch," but Montgomery is in every way qualified to succeed [him in that].
>
> 6. General Gott to command the Eighth Army under Alexander.
>
> . . . The above constitute the major simultaneous changes which the gravity and urgency of the situation here require. I shall be grateful to my War Cabinet colleages if they will approve them. Smuts and C.I.G.S. wish me to say they are in full agreement that amid many difficulties and alternatives this is the right course to pursue. The Minister of State is also in full agreement. I have no doubt the changes will impart a new and vigorous impulse to the Army and restore confidence in the Command, which I regret does not exist at the present time. Here I must emphasise the need of a new start and vehement action to animate the whole of this vast

but baffled and somewhat unhinged organisation. The War Cabinet will not fail to realise that a victory over Rommel in August or September may have a decisive effect upon the attitude of the French in North Africa when "Torch" begins.

The War Cabinet accepted my view about drastic and immediate changes in the High Command. They warmly approved the selection of General Alexander, and said that he would leave England at once. They did not however like the idea of reorganising the Middle East Command into two separate Commands. It seemed to them that the reasons which led to the setting up of the Unified Command were now stronger than they had been when the decision to do so was taken in December 1941. They agreed that Montgomery should take Alexander's place in "Torch," and had summoned him to London at once. Finally, they were content to leave it to me to settle the other appointments.

The next morning I sent a further explanation of my proposals. The War Cabinet replied that I had not entirely removed their misgivings, but that as I was on the spot with Smuts and the C.I.G.S., who both agreed with the proposal, they were prepared to authorise the action proposed. They strongly represented however that the continuance of the title of Commander-in-Chief Middle East if General Auchinleck were appointed to command in Persia and Iraq would lead to confusion and misrepresentation. I saw this was right and accepted their advice.

*   *   *

I spent all August 7 visiting the 51st Highland Division, who had just landed. As I went up the stairs after dinner at the Embassy I met Colonel, now Sir Ian, Jacob. "This is bad about Gott," he said, "What has happened?" "He was shot down this afternoon flying into Cairo." I certainly felt grief and impoverishment at the loss of this splendid soldier, to whom I had resolved to confide the most direct fighting task in the impending battle. All my plans were dislocated. The removal of Auchinleck from the Supreme Command was to have been balanced by the appointment to the Eighth Army of Gott, with all his Desert experience and prestige, and the whole covered by Alexander's assumption of the Middle East. What was to happen now? There could be no doubt who his successor should be, and I telegraphed to Mr. Atlee: "C.I.G.S. decisively recommends Montgomery for Eighth Army. Smuts and I feel this post must be filled at once. Pray send him by special plane at earliest moment. Advise me when he will arrive."

It appeared that the War Cabinet had already assembled at 11.15 P.M. on August 7 to deal with my telegrams of that day, which had just been decoded. Discussion was still proceeding upon them when a secretary came in with my new messages, stating that Gott was dead, and secondly asking that General Montgomery should be sent out at once. I have been told this was an acute moment for our friends in Downing Street. However, as I have several times observed, they had been through much and took it doggedly. They sat till nearly dawn, agreed in all essentials to what I had proposed, and gave the necessary orders about Montgomery.

*   *   *

When sending my message to the Cabinet telling them of Gott's death I had asked that General Eisenhower should not be told that we had proposed to give him Montgomery in place of Alexander. But this was too late: he had been told already. The further change of plan involved a consequent dislocation of a vexatious kind in the preparation of "Torch." Alexander had been chosen to command the British First Army in that great enterprise. He had already started to work with General Eisenhower. They were getting on splendidly together, as they always did. Now Alexander had been taken from him for the Middle East. Ismay was sent to convey the news and my apologies to Eisenhower for this break in continuity and disturbance of contacts which the hard necessity of war compelled. Ismay dilated upon Montgomery's brilliant qualities as a commander in the field. Montgomery arrived at Eisenhower's headquarters almost immediately, and all the civilities of a meeting of this kind between the commanders of armies of different nations woven into a single enterprise had been discharged. The very next morning, the 8th, Eisenhower had to be informed that Montgomery must fly that day to Cairo to command the Eighth Army. This task also fell to Ismay. Eisenhower was a broad-minded man, practical, serviceable, dealing with events as they came in cool selflessness. He naturally however felt disconcerted by the two changes in two days in this vital post in the vast operation confided to him. He was now to welcome a third British Commander. Can we wonder that he asked Ismay, "Are the British really taking 'Torch' seriously?" Nevertheless the death of Gott was a war fact which a good soldier understood. General Anderson was appointed to fill the vacancy, and Montgomery started for the airfield with Ismay, who thus had an hour or more to give him the background of these sudden changes.

A story —alas, not authenticated — has been told of this conversation. Montgomery spoke of the trials and hazards of a soldier's career. He gave his whole life to his profession, and lived long years of study and self-restraint. Presently fortune smiled, there came a gleam of success, he gained advancement, opportunity presented itself, he had a great command. He won a victory, he became world-famous, his name was on every lip. Then the luck changed. At one stroke all his life's work flashed away, perhaps through no fault of his own, and he was flung into the endless catalogue of military failures. "But," expostulated Ismay, "you ought not to take it so badly as all that. A very fine army is gathering in the Middle East. It may well be that you are not going to disaster." "What!" cried Montgomery, sitting up in the car. "What do you mean? I was talking about Rommel!"

*   *   *

I now had to inform General Auchinleck that he was to be relieved of his command, and, having learned from past experience that that kind of unpleasant thing is better done by writing than orally, I sent Colonel Jacob by air to his headquarters with the following letter:

CAIRO

August 8, 1942

Dear General Auchinleck,

On June 23 you raised in your telegram to the C.I.G.S. the question of your being relieved in this Command, and you mentioned the name of General Alexander as a possible successor. At that time of crisis for the Army His Majesty's Government did not wish to avail themselves of your high-minded offer. At the same time you had taken over the effective command of the battle, as I had long desired and had suggested to you in my telegram of May 20. You stemmed the adverse tide, and at the present time the front is stabilised.

2. The War Cabinet have now decided, for the reasons which you yourself had used, that the moment has come for a change. It is proposed to detach Iraq and Persia from the present Middle Eastern theatre. Alexander will be appointed to command the Middle East, Montgomery to command the Eighth Army, and I offer you the command of Iraq and Persia, including the Tenth Army, with headquarters at Basra or Baghdad. It is true that this sphere is today smaller than the Middle East, but it may in a few months become the scene of decisive operations, and reinforcements for the Tenth Army are already on the way. In this theatre, of which you have special experience, you will preserve your associations with India. I hope therefore that you will comply with my wish and directions with the same disinterested public spirit that you have shown on all occasions. Alexander will arrive almost immediately, and I hope that early next week, subject of course to the movements of the enemy, it may be possible to effect

the transfer of responsibility on the Western battlefront with the utmost smoothness and efficiency.

3. I shall be very glad to see you at any convenient time if you should so desire.

Believe me,
Yours sincerely,
WINSTON S. CHURCHILL

P.S. Colonel Jacob, who bears this letter, is also charged by me to express my sympathy in the sudden loss of General Gott.

In the evening Jacob returned. Auchinleck had received this stroke with soldierly dignity. He was unwilling to accept the new command, and would come to see me the next day. Jacob's diary records:

> The Prime Minister was asleep. He awoke at six o'clock, and I had to recount to him as best I could what had passed between me and General Auchinleck. C.I.G.S. joined us. . . . The Prime Minister's mind is entirely fixed on the defeat of Rommel, and on getting General Alexander into complete charge of the operations in the Western Desert. He does not understand how a man can remain in Cairo while great events are occurring in the Desert and leave the conduct of them to someone else. He strode up and down declaiming on this point, and he means to have his way. "Rommel, Rommel, Rommel, Rommel!" he cried. "What else matters but beating him?"

General Auchinleck reached Cairo just after midday, and we had an hour's conversation, which was at once bleak and impeccable.

General Alexander came to see me that evening, and final arrangements for the changes in command were drafted. I reported these accomplishments to London in a telegram of which the following passage is crucial:

> . . . I have given General Alexander the following directive, which is most agreeable to him, and in which C.I.G.S. concurs:
>
> "1. Your prime and main duty will be to take or destroy at the earliest opportunity the German-Italian Army commanded by Field Marshal Rommel, together with all its supplies and establishments in Egypt and Libya.
>
> "2. You will discharge or cause to be discharged such other duties as pertain to your Command, without prejudice to the task

described in paragraph 1, which must be considered paramount in His Majesty's interests."

It may no doubt be possible in a later phase of the war to alter the emphasis of this directive, but I am sure that simplicity of task and singleness of aim are imperative now.

Alexander's reply, sent six months later, will be recorded in due course.

15

Moscow:

The First Meeting

LATE ON THE NIGHT of August 10, after a dinner of notables at the genial Cairo Embassy, we started for Moscow. My party, which filled three planes, now included the C.I.G.S., General Wavell, who spoke Russian, Air Marshal Tedder, and Sir Alexander Cadogan. Averell Harriman had lately arrived from America on my special request to the President. He and I travelled together. By dawn we were approaching the mountains of Kurdistan. The weather was good and Vanderkloot in high spirits. As we drew near to these serrated uplands I asked him at what height he intended to fly them. He said nine thousand feet would do. However, looking at the map I found several peaks of eleven and twelve thousand feet, and there seemed one big one of eighteen or twenty thousand, though that was farther off. So long as you are not suddenly encompassed by clouds, you can wind your way through mountains with safety. Still, I asked for twelve thousand feet, and we began sucking our oxygen tubes. As we descended about 8.30 A.M. on the Teheran airfield and were already close to the ground I noticed the altimeter registered four thousand five hundred feet, and ignorantly remarked, "You had better get that adjusted before we take off again." But Vanderkloot said, "The Teheran airfield is over four thousand feet above sea level."

Sir Reader Bullard, His Majesty's Minister in Teheran, met me on arrival. He was a tough Briton, with long experience of Persia and no illusions.

We were too late to leap the northern range of the Elburz Mountains before dark, and I found myself graciously bidden to lunch with the Shah in a palace with a lovely swimming pool amid great trees on an abrupt spur of the mountains. The mighty peak I had noticed in the morning gleamed brilliant pink and orange. In the afternoon in the

garden of the British Legation there was a long conference with Averell
Harriman and various high British and American railway authorities,
and it was decided that the United States should take over the whole
Trans-Persian railway from the Gulf to the Caspian. This railway,
newly completed by a British firm, was a remarkable engineering
achievement. There were 390 major bridges on its track through the
mountain gorges. Harriman said the President was willing to undertake the entire responsibility for working it to full capacity, and could
provide locomotives, rolling stock, and skilled men in military units to
an extent impossible for us. I therefore agreed to this transfer, subject
to stipulations about priority for our essential military requirements.
On account of the heat and noise of Teheran, where every Persian seems
to have a motorcar and blows his horn continually, I slept amid tall trees
at the summer residence of the British Legation about a thousand feet
above the city.

At six-thirty next morning, Wednesday, August 12, we started, gaining height as we flew through the great valley which led to Tabriz, and
then turned northward to Enzeli, on the Caspian. We passed this
second range of mountains at about eleven thousand feet, avoiding both
clouds and peaks. Two Russian officers were now in the plane, and
the Soviet Government assumed responsibility for our course and safe
arrival. The snow-clad giant gleamed to the eastward. I noticed that
we were flying alone, and a wireless message explained that our second
plane, with the C.I.G.S., Wavell, Cadogan, and others, had had to
turn back over Teheran because of engine trouble. In two hours the
waters of the Caspian Sea shone ahead. Beneath was Enzeli. I had
never seen the Caspian, but I remembered how a quarter of a century
before I had, as Secretary of State for War, inherited a fleet upon it
which for nearly a year ruled its pale, placid waters. We now came
down to a height where oxygen was no longer needed. On the western
shore, which we could dimly see, lay Baku and its oilfields. The German armies were now so near the Caspian that our course was set for
Kuibyshev, keeping well away from Stalingrad and the battle area.
This took us near the delta of the Volga. As far as the eye could reach
spread vast expanses of Russia, brown and flat and with hardly a sign
of human habitation. Here and there sharp rectilineal patches of
ploughed land revealed an occasional State farm. For a long way the
mighty Volga gleamed in curves and stretches as it flowed between its
wide, dark margins of marsh. Sometimes a road, straight as a ruler,
ran from one wide horizon to the other. After an hour or so of this I
clambered back along the bomb bay to the cabin and slept.

I pondered on my mission to this sullen, sinister Bolshevik State I
had once tried so hard to strangle at its birth, and which, until Hitler

appeared, I had regarded as the mortal foe of civilised freedom. What was it my duty to say to them now? General Wavell, who had literary inclinations, summed it all up in a poem. There were several verses, and the last line of each was, "No Second Front in nineteen forty-two." It was like carrying a large lump of ice to the North Pole. Still, I was sure it was my duty to tell them the facts personally and have it all out face to face with Stalin, rather than trust to telegrams and intermediaries. At least it showed that one cared for their fortunes and understood what their struggle meant to the general war. We had always hated their wicked regime, and, till the German flail beat upon them, they would have watched us being swept out of existence with indifference and gleefully divided with Hitler our Empire in the East.

The weather being clear, the wind favourable, and my need to get to Moscow urgent, it was arranged to cut the corner of Kuibyshev and go on straight to the capital. I fear a splendid banquet and welcome in true Russian hospitality was thus left on one side. At about five o'clock the spires and domes of Moscow came in sight. We circled around the city by carefully prescribed courses along which all the batteries had been warned, and landed on the airfield, which I was to revisit during the struggle.

Here was Molotov at the head of a concourse of Russian generals and the entire Diplomatic Corps, with the very large outfit of photographers and reporters customary on these occasions. A strong guard of honour, faultless in attire and military punctilio, was inspected, and marched past after the band had played the national anthems of the three Great Powers whose unity spelt Hitler's doom. I was taken to the microphone and made a short speech. Averell Harriman spoke on behalf of the United States. He was to stay at the American Embassy. Molotov drove me in his car to my appointed residence, eight miles out of Moscow, "State Villa No. 7." While going through the streets of Moscow, which seemed very empty, I lowered the window for a little more air, and to my surprise felt that the glass was over two inches thick. This surpassed all records in my experience. "The Minister says it is more prudent," said Interpreter Pavlov. In a little more than half an hour we reached the villa.

*   *   *

Everything was prepared with totalitarian lavishness. There was placed at my disposal, as aide-de-camp, an enormous, splendid-looking officer (I believe of a princely family under the Czarist regime), who also acted as our host and was a model of courtesy and attention. A number of veteran servants in white jackets and beaming smiles waited on every wish or movement of the guests. A long table in the dining room and various sideboards were laden with every delicacy and stimulant that supreme power can command. I was conducted through a spacious reception room to a bedroom and bathroom of almost equal size. Blazing, almost dazzling, electric lights displayed the spotless cleanliness. The hot and cold water gushed. I longed for a hot bath after the length and the heat of the journey. All was instantly prepared. I noticed that the basins were not fed by separate hot and cold water taps and that they had no plugs. Hot and cold turned on at once through a single spout, mingled to exactly the temperature one desired. Moreover, one did not wash one's hands in the basin, but under the flowing current of the taps. In a modest way I have adopted this system at home. If there is no scarcity of water it is far the best.

After all necessary immersions and ablutions we were regaled in the dining room with every form of choice food and liquor, including of course caviare and vodka, but with many other dishes and wines from France and Germany far beyond our mood or consuming powers. Besides, we had but little time before starting for Moscow. I had told Molotov that I should be ready to see Stalin that night, and he proposed seven o'clock.

I reached the Kremlin, and met for the first time the great Revolutionary Chief and profound Russian statesman and warrior with whom for the next three years I was to be in intimate, rigorous, but always exciting, and at times even genial, association. Our conference lasted nearly four hours. As our second aeroplane had not arrived with Brooke, Wavell, and Cadogan, there were present only Stalin, Molotov, Voroshilov, myself, Harriman, and our Ambassador, with interpreters. I have based this account upon the record which we kept, subject to my own memory, and to the telegrams I sent home at the time.

The first two hours were bleak and sombre. I began at once with the question of the Second Front, saying that I wished to speak frankly and would like to invite complete frankness from Stalin. I would not have come to Moscow unless he had felt sure that he would be able to discuss realities. When Molotov had come to London I had told him that we were trying to make plans for a diversion in France. I had also made it clear to Molotov that I could make no promises about 1942, and had given Molotov a memorandum to this effect. Since then an exhaustive Anglo-American examination of the problem had been carried out. The British and American Governments did not feel themselves able to undertake a major operation in September, which was the latest month in which the weather was to be counted upon. But, as Stalin knew, they were preparing for a very great operation in 1943. For this purpose a million American troops were now scheduled to reach the United Kingdom at their point of assembly in the spring of

1943, making an expeditionary force of twenty-seven divisions, to which the British Government were prepared to add twenty-one divisions. Nearly half of this force would be armoured. So far only two and a half American divisions had reached the United Kingdom, but the big transportation would take place in October, November, and December.

I told Stalin that I was well aware that this plan offered no help to Russia in 1942, but thought it possible that when the 1943 plan was ready it might well be that the Germans would have a stronger army in the West than they now had. At this point Stalin's face crumpled up into a frown, but he did not interrupt. I then said I had good reasons against an attack on the French coast in 1942. We had only enough landing craft for an assault landing on a fortified coast — enough to throw ashore six divisions and maintain them. If it were successful, more divisions might be sent, but the limiting factor was landing craft, which were now being built in very large numbers in the United Kingdom, and especially in the United States. For one division which could be carried this year it would be possible next year to carry eight or ten times as many.

Stalin, who had begun to look very glum, seemed unconvinced by my argument, and asked if it was impossible to attack any part of the French coast. I showed him a map which indicated the difficulties of making an air umbrella anywhere except actually across the Straits. He did not seem to understand, and asked some questions about the range of fighter planes. Could they not, for instance, come and go all the time? I explained that they could indeed come and go, but at this range they would have no time to fight, and I added that an air umbrella to be of any use had to be kept open. He then said that there was not a single German division in France of any value, a statement which I contested. There were in France twenty-five German divisions, nine of which were of the first line. He shook his head. I said that I had brought the Chief of the Imperial General Staff and General Sir Archibald Wavell with me in order that such points might be examined in detail with the Russian General Staff. There was a point beyond which statesmen could not carry discussions of this kind.

Stalin, whose glumness had by now much increased, said that, as he understood it, we were unable to create a second front with any large force and unwilling even to land six divisions. I said that this was so. We could land six divisions, but the landing of them would be more harmful than helpful, for it would greatly injure the big operation planned for next year. War was war but not folly, and it would be folly to invite a disaster which would help nobody. I said I feared the news I brought was not good news. If by throwing in 150,000 to 200,000 men we could render him aid by drawing away from the Russian front appreciable German forces, we would not shrink from this course on the grounds of loss. But if it drew no men away and spoiled the prospects for 1943 it would be a great error.

Stalin, who had become restless, said that his view about war was different. A man who was not prepared to take risks could not win a war. Why were we so afraid of the Germans? He could not understand. His experience showed that troops must be blooded in battle. If you did not blood your troops you had no idea what their value was. I inquired whether he had ever asked himself why Hitler did not come to England in 1940, when he was at the height of his power and we had only twenty-thousand trained troops, two hundred guns, and fifty tanks. He did not come. The fact was that Hitler was afraid of the operation. It was not so easy to cross the Channel. Stalin replied that this was no analogy. The landing of Hitler in England would have been resisted by the people, whereas in the case of a British landing in France the people would be on the side of the British. I pointed out that it was all the more important therefore not to expose the people of France by a withdrawal to the vengeance of Hitler and to waste them when they would be needed in the big operation in 1943.

There was an oppressive silence. Stalin at length said that if we could not make a landing in France this year he was not entitled to demand it or to insist upon it, but he was bound to say that he did not agree with my arguments.

*   *   *

I then unfolded a map of Southern Europe, the Mediterranean, and North Africa. What was a "Second Front"? Was it only a landing on a fortified coast opposite England? Or could it take the form of some other great enterprise which might be useful to the common cause? I thought it better to bring him southward by steps. If, for instance, we could hold the enemy in the Pas de Calais by our concentrations in Britain, and at the same time attack elsewhere — for instance, in the Loire, the Gironde, or alternatively the Scheldt — this was full of promise. There indeed was a general picture of next year's big operation. Stalin feared that it was not practicable. I said that it would indeed be difficult to land a million men, but that we should have to persevere and try.

We then passed on to the bombing of Germany, which gave general satisfaction. Stalin emphasised the importance of striking at the morale of the German population. He said he attached the greatest importance to bombing, and that he knew our raids were having a tremendous effect in Germany.

After this interlude, which relieved the tension, Stalin observed that from our long talk it seemed that all we were going to do was no "Sledgehammer," no "Round-up," and pay our way by bombing Germany. I decided to get the worst over first and to create a suitable background for the project I had come to unfold. I did not therefore try at once to relieve the gloom. Indeed, I asked specially that there should be the plainest speaking between friends and comrades in peril. However, courtesy and dignity prevailed.

*   *   *

The moment had now come to bring "Torch" into action. I said that I wanted to revert to the question of a Second Front in 1942, which was what I had come for. I did not think France was the only place for such an operation. There were other places, and we and the Americans had decided upon another plan, which I was authorised by the American President to impart to Stalin secretly. I would now proceed to do so. I emphasised the vital need of secrecy. At this Stalin sat up and grinned and said that he hoped that nothing about it would appear in the British Press.

I then explained precisely Operation "Torch." As I told the whole story Stalin became intensely interested. His first question was what would happen in Spain and Vichy France. A little later on he remarked that the operation was militarily right, but he had political doubts about the effect on France. He asked particularly about the timing, and I said not later than October 30, but the President and all of us were trying to pull it forward to October 7. This seemed a great relief to the Russians.

I then described the military advantages of freeing the Mediterranean, whence still another front could be opened. In September we must win in Egypt, and in October in North Africa, all the time holding the enemy in Northern France. If we could end the year in possession of North Africa we could threaten the belly of Hitler's Europe, and this operation should be considered in conjunction with the 1943 operation. That was what we and the Americans had decided to do.

To illustrate my point I had meanwhile drawn a picture of a crocodile, and explained to Stalin with the help of this picture how it was our intention to attack the soft belly of the crocodile as we attacked his hard snout. And Stalin, whose interest was now at a high pitch, said, "May God prosper this undertaking."

I emphasised that we wanted to take the strain off the Russians. If we attempted that in Northern France we should meet with a rebuff. If we tried in North Africa we had a good chance of victory, and then we could help in Europe. If we could gain North Africa Hitler would have to bring his Air Force back, or otherwise we would destroy his allies, even, for instance, Italy, and make a landing. The operation would have an important influence on Turkey and on the whole of Southern Europe, and all I was afraid of was that we might be forestalled. If North Africa were won this year we could make a deadly attack upon Hitler next year. This marked the turning point in our conversation.

Stalin then began to present various political difficulties. Would not an Anglo-American seizure of "Torch" regions be misunderstood in France? What were we doing about de Gaulle? I said that at this stage we did not wish him to intervene in the operation. The Vichy French were likely to fire on de Gaullists but unlikely to fire on Americans. Harriman backed this very strongly by referring to reports, on which the President relied, by American agents all over "Torch" territories, and also to Admiral Leahy's opinion.

*   *   *

At this point Stalin seemed suddenly to grasp the strategic advantages of "Torch." He recounted four main reasons for it: first, it would hit Rommel in the back; second, it would overawe Spain; third, it would produce fighting between Germans and Frenchmen in France; and, fourth, it would expose Italy to the whole brunt of the war.

I was deeply impressed with this remarkable statement. It showed the Russian Dictator's swift and complete mastery of a problem hitherto novel to him. Very few people alive could have comprehended in so few minutes the reasons which we had all so long been wrestling with for months. He saw it all in a flash.

I mentioned a fifth reason, namely, the shortening of the sea route through the Mediterranean. Stalin was concerned to know whether we were able to pass through the Straits of Gibraltar. I said it would be all right. I also told him about the change in the command in Egypt, and of our determination to fight a decisive battle there in late August or September. Finally, it was clear that they all liked "Torch," though Molotov asked whether it could not be in September.

I then added, "France is down and we want to cheer her up." France had understood Madagascar and Syria. The arrival of the Americans would send the French nation over to our side. It would intimidate Franco. The Germans might well say at once to the French, "Give us your Fleet and Toulon." This would stir anew the antagonisms between Vichy and Hitler.

I then opened the prospect of our placing an Anglo-American Air Force on the southern flank of the Russian armies in order to defend the Caspian and the Caucasian mountains and generally to fight in this theatre. I did not however go into details, as of course we had to win our battle in Egypt first, and I had not the President's plans for the American contribution. If Stalin liked the idea we would set to work in detail upon it. He replied that they would be most grateful for this aid, but that the details of location, etc., would require study. I was very keen on this project, because it would bring about more hard fighting between the Anglo-American air power and the Germans, all of which aided the gaining of mastery in the air under more fertile conditions than looking for trouble over the Pas de Calais.

We then gathered round a large globe, and I explained to Stalin the immense advantages of clearing the enemy out of the Mediterranean. I told Stalin I should be available should he wish to see me again. He replied that the Russian custom was that the visitor should state his wishes and that he was ready to receive me at any time. He now knew the worst, and yet we parted in an atmosphere of goodwill.

The meeting had now lasted nearly four hours. It took half an hour or more to reach State Villa No. 7. Tired as I was, I dictated my telegram to the War Cabinet and the President after midnight, and then, with the feeling that at least the ice was broken and a human contact established, I slept soundly and long.

16

MOSCOW:
A Relationship Established

LATE THE NEXT MORNING I awoke in my luxurious quarters. It was Thursday, August 13 — to me always "Blenheim Day." I had arranged to visit Molotov in the Kremlin at noon in order to explain to him more clearly and fully the character of the various operations we had in mind. I pointed out how injurious to the common cause it would be if owing to recriminations about dropping "Sledgehammer" we were forced to argue publicly against such enterprises. I also explained in more detail the political setting of "Torch." He listened affably, but contributed nothing. I proposed to him that I should see Stalin at 10 P.M. that night, and later in the day got word that eleven o'clock would be more convenient, and as the subjects to be dealt with would be the same as those of the night before, would I wish to bring Harriman? I said "Yes," and also Cadogan, Brooke, Wavell, and Tedder, who had meanwhile arrived safely from Teheran in a Russian plane. They might have had a very dangerous fire in their Liberator.

Before leaving this urbane, rigid diplomatist's room I turned to him and said, "Stalin will make a great mistake to treat us roughly when we have come so far." For the first time Molotov unbent. "Stalin," he said, "is a very wise man. You may be sure that, however he argues, he understands all. I will tell him what you say."

I returned in time for luncheon to State Villa Number Seven. Out of doors the weather was beautiful. It was just like what we love most in England — when we get it. I thought we would explore the domain. State Villa Number Seven was a fine large, brand-new country house standing in its own extensive lawns and gardens in a fir wood of about twenty acres. There were agreeable walks, and it was pleasant in the beautiful August weather to lie on the grass or pine needles. There were several fountains, and a large glass tank filled with many kinds of goldfish, who were all so tame that they would eat out of your hand. I made a point of feeding them every day. Around the whole was a stockade, perhaps fifteen feet high, guarded on both sides by police and soldiers in considerable numbers. About a hundred yards from the house was an air-raid shelter. At the first opportunity we were conducted over it. It was of the latest and most luxurious type. Lifts at either end took you down eighty or ninety feet into the ground. Here were eight or ten large rooms inside a concrete box of massive thickness. The rooms were divided from each other by heavy sliding doors. The lights were brilliant. The furniture was stylish, sumptuous and brightly coloured. I was more attracted by the goldfish.

*　　　*　　　*

We all repaired to the Kremlin at 11 P.M., and were received only by Stalin and Molotov, with their interpreter. Then began a most unpleasant discussion. I said he must understand we had made up our minds upon the course to be pursued and that reproaches were vain. We argued for about two hours, during which he said a great many disagreeable things, especially about our being too much afraid of fighting the Germans, and if we tried it like the Russians we should find it not so bad; that we had broken our promise about "Sledgehammer"; that we had failed in delivering the supplies promised to Russia and only sent remnants after we had taken all we needed for ourselves. Apparently these complaints were addressed as much to the United States as to Britain.

I repulsed all his contentions squarely, but without taunts of any kind. I suppose he was not used to being contradicted repeatedly, but he did not become at all angry, or even animated. He reiterated his view that it should be possible for the British and Americans to land six or eight divisions on the Cherbourg peninsula, since they had domination of the air. He felt that if the British Army had been fighting the Germans as much as the Russian Army it would not be so frightened of them. The Russians, and indeed the R.A.F., had shown that it was possible to beat the Germans. The British infantry could do the same provided they acted at the same time as the Russians.

I interposed that I pardoned the remarks which Stalin had made on account of the bravery of the Russian Army. The proposal for a landing in Cherbourg overlooked the existence of the Channel. Finally Stalin said we could carry it no further. He must accept our decision. He then abruptly invited us to dinner at eight o'clock the next night.

Accepting the invitation, I said I would leave by plane at dawn the following morning — i.e., the 15th. Joe seemed somewhat concerned at this, and asked could I not stay longer. I said certainly, if there was any good to be done, and that I would wait one more day anyhow. I then exclaimed that there was no ring of comradeship in his attitude. I had travelled far to establish good working relations. We had done our utmost to help Russia, and would continue to do so. We had been left entirely alone for a year against Germany and Italy. Now that the three great nations were allied, victory was certain, provided we did not fall apart, and so forth. I was somewhat animated in this passage, and before it could be translated he made the remark that he liked the tone of my utterance. Thereafter the talk began again in a somewhat less tense atmosphere.

He plunged into a long discussion of two Russian trench mortars firing rockets, which he declared were devastating in their effects, and which he offered to demonstrate to our experts if they could wait. He said he would let us have all information about them, but should there not be something in return? Should there not be an agreement to exchange information about inventions? I said that we would give them everything without any bargaining, except only those devices which, if carried in aeroplanes over the enemy lines and shot down, would make our bombing of Germany more difficult. He accepted this. He also agreed that his military authorities should meet our generals, and this was arranged for three o'clock in the afternoon. I said they would re-quire at least four hours to go fully into the various technical questions involved in "Sledgehammer," "Round-up," and "Torch." He observed at one moment that "Torch" was "militarily correct," but that the political side required more delicacy — i.e., more careful handling. From time to time he returned to "Sledgehammer," grumbling about it. When he said our promise had not been kept I replied, "I repudiate that statement. Every promise has been kept," and I pointed to the *aide-mémoire* I gave Molotov.[1] He made a sort of apology, saying that he was expressing his sincere and honest opinions, that there was no mis-trust between us, but only a difference of view.

Finally I asked about the Caucasus. Was he going to defend the mountain chain, and with how many divisions? At this he sent for a relief model, and, with apparent frankness and evident knowledge, ex-plained the strength of this barrier, for which he said twenty-five divi-sions were available. He pointed to the various passes and said they would be defended. I asked were they fortified, and he said, "Yes, certainly." The Russian front line, which the enemy had not yet reached, was north of the main range. He said they would have to hold out for two months, when the snow would make the mountains im-passable. He declared himself quite confident of their ability to do this, and also recounted in detail the strength of the Black Sea Fleet, which was gathered at Batum.

[1] See page 575.

All this part of the talk was easier, but when Harriman asked about the plans for bringing American aircraft across Siberia, to which the Russians had only recently consented after long American pressing, he replied, curtly, "Wars are not won with plans." Harriman backed me up throughout, and we neither of us yielded an inch nor spoke a bitter word.

Stalin made his salute and held out his hand to me on leaving, and I took it.

*   *   *

I reported to the War Cabinet on August 14:

> We asked ourselves what was the explanation of this perform-ance and transformation from the good ground we had reached the night before. I think the most probable is that his Council of Commissars did not take the news I brought as well as he did. They may have more power than we suppose, and less knowledge. Perhaps he was putting himself on the record for future purposes and for their benefit, and also letting off steam for his own. Cado-gan says a similar hardening up followed the opening of the Eden interview at Christmas, and Harriman says that this technique was also used at the beginning of the Beaverbrook mission.
>
> It is my considered opinion that in his heart, so far as he has one, Stalin knows we are right, and that six divisions on "Sledge-hammer" would do him no good this year. Moreover, I am cer-tain that his surefooted and quick military judgment makes him a strong supporter of "Torch." I think it not impossible that he will make amends. In that hope I persevere. Anyhow, I am sure it was better to have it out this way than any other. There was never at any time the slightest suggestion of their not fighting on, and I think myself that Stalin has good confidence that he will win. . . .

That evening we attended an official dinner at the Kremlin, where about forty people, including several of the military commanders, mem-bers of the Politburo, and other high officials, were present. Stalin and Molotov did the honours in cordial fashion. These dinners were lengthy, and from the beginning many toasts were proposed and responded to in very short speeches. Silly tales have been told of how these Soviet dinners became drinking bouts. There is no truth whatever in this. The Marshal and his colleagues invariably drank their toasts from tiny glasses, taking only a sip on each occasion. I had been well brought up.

During the dinner Stalin talked to me in lively fashion through the interpreter Pavlov. "Some years ago," he said, "we had a visit from

Mr. George Bernard Shaw and Lady Astor." Lady Astor suggested that Mr. Lloyd George should be invited to visit Moscow, to which Stalin had replied, "Why should we ask him? He was the head of the intervention." On this Lady Astor said, "That is not true. It was Churchill who misled him." "Anyhow," said Stalin, "Lloyd George was head of the Government and belonged to the Left. He was responsible, and we like a downright enemy better than a pretending friend." "Well, Churchill is finished finally," said Lady Astor. "I am not so sure," Stalin had answered. "If a great crisis comes the English people might turn to the old war horse." At this point I interrupted saying, "There is much in what she said. I was very active in the intervention, and I do not wish you to think otherwise." He smiled amicably, so I said, "Have you forgiven me?" "Premier Stalin, he say," said Interpreter Pavlov, "all that is in the past, and the past belongs to God."

In the course of one of my later talks with Stalin I said, "Lord Beaverbrook has told me that when he was on his mission to Moscow in October 1941 you asked him, 'What did Churchill mean by saying in Parliament that he had given me warnings of the impending German attack?' I was of course," said I, "referring to the telegram I sent you in April '41," and I produced the telegram which Sir Stafford Cripps had tardily delivered. When it was read and translated to him Stalin shrugged his shoulders. "I remember it. I did not need any warnings. I knew war would come, but I thought I might gain another six months or so." In the common cause I refrained from asking what would have happened to us all if we had gone down forever while he was giving Hitler so much valuable material, time, and aid.

As soon as I could I gave a more formal account of the banquet to Mr. Attlee and the President.

> The dinner passed off in a very friendly atmosphere and the usual Russian ceremonies. Wavell made an excellent speech in Russian. I proposed Stalin's health, and Alexander Cadogan proposed death and damnation to the Nazis. Though I sat on Stalin's right I got no opportunity of talking about serious things. Stalin and I were photographed together, also with Harriman. Stalin made quite a long speech proposing the "Intelligence Service," in the course of which he made a curious reference to the Dardanelles in 1915, saying that the British had won and the Germans and Turks were already retreating, but we did not know because the intelligence was faulty. This picture, though inaccurate, was evidently meant to be complimentary to me.
>
> 2. I left about 1.30 A.M., as I was afraid we should be drawn into a lengthy film and was fatigued. When I said good-bye to Stalin he

said that any differences that existed were only of method. I said we would try to remove even those differences by deeds. After a cordial handshake I then took my departure, and got some way down the crowded room, but he hurried after me and accompanied me an immense distance through corridors and staircases to the front door, where we again shook hands.

3. Perhaps in my account to you of the Thursday night meeting I took too gloomy a view. I feel I must make full allowance for the really grievous disappointment which they feel here that we can do nothing more to help them in their immense struggle. In the upshot they have swallowed this bitter pill. Everything for us now turns on hastening "Torch" and defeating Rommel.

I had been offended by many things which had been said at our conferences. I made every allowance for the strain under which the Soviet leaders lay, with their vast front flaming and bleeding along nearly two thousand miles, and the Germans but fifty miles from Moscow and advancing towards the Caspian Sea. The technical military discussions had not gone well. Our generals had asked all sorts of questions to which their Soviet colleagues were not authorised to give answers. The only Soviet demand was for "A Second Front NOW." In the end Brooke was rather blunt, and the military conference came to a somewhat abrupt conclusion.

We were to start at dawn on the 16th. On the evening before I went at seven o'clock to say good-bye to Stalin. We had a useful and important talk. I asked particularly whether he would be able to hold the Caucasus mountain passes, and also prevent the Germans reaching the Caspian, taking the oilfields round Baku, with all that meant, and then driving southwards through Turkey or Persia. He spread out the map, and then said with quiet confidence. "We shall stop them. They will not cross the mountains." He added, "There are rumours that the Turks will attack us in Turkestan. If they do I shall be able to deal with them as well." I said there was no danger of this. The Turks meant to keep out, and would certainly not quarrel with England.

Our hour's conversation drew to its close, and I got up to say good-bye. Stalin seemed suddenly embarrassed, and said in a more cordial tone than he had yet used with me, "You are leaving at daybreak. Why should we not go to my house and have some drinks?" I said that I was in principle always in favour of such a policy. So he led the way through many passages and rooms till we came out into a still roadway within the Kremlin, and in a couple of hundred yards gained the apartment where he lived. He showed me his own rooms, which were of moderate size, simple, dignified, and four in number — a dining room, working room, bedroom, and a large bathroom. Presently there appeared, first a very aged housekeeper and later a handsome red-haired girl, who kissed her father dutifully. He looked at me with a twinkle in his eye, as if, so I thought, to convey, "You see, even we Bolsheviks have family life." Stalin's daughter started laying the table, and in a short time the housekeeper appeared with a few dishes. Meanwhile Stalin had been uncorking various bottles, which began to make an imposing array. Then he said, "Why should we not have Molotov? He is worrying about the communiqué. We could settle it here. There is one thing about Molotov — he can drink." I then realised that there was to be a dinner. I had planned to dine at State Villa Number Seven, where General Anders, the Polish commander, was awaiting me, but I told my new and excellent interpreter, Major Birse, to telephone that I should not be back till after midnight. Presently Molotov arrived. We sat down, and, with the two interpreters, were five in number. Major Birse had lived twenty years in Moscow, and got on very well with the Marshal, with whom he for some time kept up a running conversation, in which I could not share.

We actually sat at this table from 8.30 P.M. till 2.30 the next morning, which, with my previous interview, made a total of more than seven hours. The dinner was evidently improvised on the spur of the moment, but gradually more and more food arrived. We pecked and picked, as seemed to be the Russian fashion, at a long succession of choice dishes, and sipped a variety of excellent wines. Molotov assumed his most affable manner, and Stalin, to make things go, chaffed him unmercifully.

Presently we talked about the convoys to Russia. This led him to make a rough and rude remark about the almost total destruction of an Arctic convoy in June.

"Mr. Stalin asks," said Pavlov, with some hesitation, "has the British Navy no sense of glory?" I answered, "You must take it from me that what was done was right. I really do know a lot about the Navy and sea war." "Meaning," said Stalin, "that I know nothing." "Russia is a land animal," I said; "the British are sea animals." He fell silent and recovered his good humour. I turned the talk on to Molotov. "Was the Marshal aware that his Foreign Secretary on his recent visit to Washington had said he was determined to pay a visit to New York entirely by himself, and that the delay in his return was not due to any defect in the aeroplane, but because he was off on his own?"

Although almost anything can be said in fun at a Russian dinner, Molotov looked rather serious at this. But Stalin's face lit with merriment as he said, "It was not to New York he went. He went to Chicago, where the other gangsters live."

Relations having thus been entirely restored, the talk ran on. I opened the question of a British landing in Norway, with Russian support, and explained how, if we could take the North Cape in the winter and destroy the Germans there, the path of the convoys would henceforward be open. This idea was always, as has been seen, one of my favourite plans. Stalin seemed much attracted by it, and, after talking of ways and means, we agreed we must do it if possible.

*   *   *

It was now past midnight, and Cadogan had not appeared with the draft of the communiqué.

"Tell me," I asked, "have the stresses of this war been as bad to you personally as carrying through the policy of the Collective Farms?"

This subject immediately roused the Marshal.

"Oh, no," he said, "the Collective Farm policy was a terrible struggle."

"I thought you would have found it bad," said I, "because you were not dealing with a few score thousands of aristocrats or big landowners, but with millions of small men."

"Ten millions," he said holding up his hands. "It was fearful. Four years it lasted. It was absolutely necessary for Russia, if we were to avoid periodic famines, to plough the land with tractors. We must mechanise our agriculture. When we gave tractors to the peasants they were all spoiled in a few months. Only Collective Farms with workshops could handle tractors. We took the greatest trouble to explain it to the peasants. It was no use arguing with them. After you have said all you can to a peasant he says he must go home and consult his wife, and he must consult his herder." This last was a new expression to me in this connection.

"After he has talked it over with them he always answers that he does not want the Collective Farm and he would rather do without the tractors."

"These were what you call Kulaks?"

"Yes," he said, but he did not repeat the word. After a pause, "It was all very bad and difficult — but necessary."

"What happened?" I asked.

"Oh, well," he said, "many of them agreed to come in with us. Some of them were given land of their own to cultivate in the province of Tomsk or the province of Irkutsk or farther north, but the great bulk were very unpopular and were wiped out by their labourers."

There was a considerable pause. Then, "Not only have we vastly increased the food supply, but we have improved the quality of the grain beyond all measure. All kinds of grain used to be grown. Now no one

is allowed to sow any but the standard Soviet grain from one end of
our country to the other. If they do they are severely dealt with.
This means another large increase in the food supply."

I record as they come back to me these memories, and the strong
impression I sustained at the moment of millions of men and women
being blotted out or displaced for ever. A generation would no doubt
come to whom their miseries were unknown, but it would be sure of
having more to eat and bless Stalin's name. I did not repeat Burke's
dictum, "If I cannot have reform without injustice, I will not have
reform." With the World War going on all round us it seemed vain to
moralise aloud.

About 1 A.M. Cadogan arrived with the draft communiqué, and we
set to work to put it into final form. A considerable sucking pig was
brought to the table. Hitherto Stalin had only tasted the dishes, but
now it was half past one in the morning and around his usual dinner
hour. He invited Cadogan to join him in the conflict, and when my
friend excused himself our host fell upon the victim singlehanded.
After this had been achieved he went abruptly into the next room to
receive the reports from all sectors of the front, which were delivered to
him from 2 A.M. onward. It was about twenty minutes before he returned, and by that time we had the communiqué agreed. Finally, at
2.30 A.M. I said must go. I had half an hour to drive to the villa, and as
long to drive back to the airport. I had a splitting headache, which for
me was very unusual. I still had General Anders to see. I begged Molotov not to come and see me off at dawn, for he was clearly tired out.
He looked at me reproachfully, as if to say, "Do you really think I
would fail to be there?"

We took off at 5.30 A.M. I was very glad to sleep in the plane, and I
remember nothing of the landscape or journey till we reached the foot
of the Caspian and began to climb over the Elburz Mountains. At
Teheran I did not go to the Legation, but to the cool, quiet glades of
the summer residence, high above the city. Here a great press of telegrams awaited me. I had planned a conference for the next day at
Baghdad with most of our high authorities in Persia and Iraq, but I did
not feel I could face the heat of Baghdad in the August noonday, and
it was quite easy to change the venue to Cairo. I dined with the Legation party that night in the agreeable woodland, and was content to
forget all cares till morning.

17

Strain and Suspense

O N T H E 17th O F A U G U S T I received news of the attack on
Dieppe, plans for which had been started in April after the brilliant and audacious raid on St. Nazaire. On May 13 the outline plan
was approved by the Chiefs of Staff Committee as a basis for detailed
planning by the Force Commanders. More than ten thousand men were
to be employed by the three Services. This was of course the most considerable enterprise of its kind which we had attempted against the
occupied French coastline. From available intelligence it appeared that
Dieppe was held only by German low-category troops amounting to
one battalion, with supporting units making no more than 1400 men
in all. The assault was originally fixed for July 4, and the troops embarked at ports in the Isle of Wight. The weather was unfavourable
and the date was postponed till July 8. Four German aircraft made
an attack upon the shipping which had been concentrated. The weather
continued bad and the troops disembarked. It was now decided to cancel the operation altogether. General Montgomery, who, as Commander-
in-Chief of Southeastern Command, had hitherto supervised the plans,
was strongly of opinion that it should not be remounted, as the troops
concerned had all been briefed and were now dispersed ashore.

However, I thought it most important that a large-scale operation
should take place this summer, and military opinion seemed unanimous
that until an operation on that scale was undertaken no responsible
general would take the responsibility of planning for the main invasion.

In discussion with Admiral Mountbatten it became clear that time
did not permit a new large-scale operation to be mounted during the
summer, but that Dieppe could be remounted (the code name was
"Jubilee") within a month, provided extraordinary steps were taken
to ensure secrecy.

For this reason no records were kept, but after the Canadian authorities and the Chiefs of Staff had given their approval, I personally went through the plans with the C.I.G.S., Admiral Mountbatten, and the Naval Force Commander, Captain J. Hughes-Hallett. It was clear that no substantial change between "Jubilee" and "Rutter" was suggested, beyond substituting Commandos to silence the flank coastal batteries in place of air-borne troops. This was now possible as two more infantry landing ships had become available to carry the Commandos, and the chances of weather conditions causing "Jubilee" once more to be abandoned were considerably reduced by omitting the air-borne drop. In spite of an accidental encounter between the landing craft carrying one of the Commandos and a German coastal convoy, one of the batteries was completely destroyed and the other prevented from seriously interfering with the operation; so that this change in no way affected the outcome of the operation.

Our postwar examination of their records shows that the Germans did not receive, through leakages of information, any special warning of our intention to attack. However, their general estimate of the threat to the Dieppe sector led to an intensification of defence measures along the whole front. Special precautions were ordered for periods like that between August 10 and August 19, when moon and tide were favourable for landings. The division responsible for the defence of the Dieppe sector had been reinforced during July and August, and was at full strength and on routine alert at the moment of the raid. The Canadian Army in Britain had long been eager and impatient for action, and the main part of the landing force was provided by them. The story is vividly told by the official historian of the Canadian Army [1] and in other official publications, and need not be repeated here. Although the utmost gallantry and devotion were shown by all the troops and by the British Commandos and by the landing craft and their escorts, and many splendid deeds were done, the results were disappointing and our casualties were very heavy. In the Canadian 2nd Division eighteen per cent of the five thousand men embarked lost their lives and nearly two thousand of them were taken prisoner.

Looking back, the casualties of this memorable action may seem out of proportion to the results. It would be wrong to judge the episode solely by such a standard. Dieppe occupies a place of its own in the story of the war, and the grim casualty figures must not class it as a failure. It was a costly but not unfruitful reconnaissance in force. Tactically it was a mine of experience. It shed revealing light on many shortcomings in our outlook. It taught us to build in good time various new types of craft and appliances for later use. We learnt again the

[1] Colonel C. P. Stacey, *The Canadian Army, 1939–45.*

value of powerful support by heavy naval guns in an opposed landing, and our bombardment technique, both marine and aerial, was thereafter improved. Above all it was shown that individual skill and gallantry without thorough organisation and combined training would not prevail, and that teamwork was the secret of success. This could only be provided by trained and organised amphibious formations. All these lessons were taken to heart.

Strategically the raid served to make the Germans more conscious of danger along the whole coast of Occupied France. This helped to hold troops and resources in the West, which did something to take the weight off Russia. Honour to the brave who fell. Their sacrifice was not in vain.

*   *   *

On August 19 I paid another visit to the Desert Front. I drove with Alexander in his car out from Cairo past the Pyramids, about one hundred thirty miles through the desert to the sea at Abusir. I was cheered by all he told me. As the shadows lengthened we reached Montgomery's headquarters at Burg el Arab. Here the afterwards famous caravan was drawn up amid the sand dunes by the sparkling waves. The General gave me his own wagon, divided between office and bedroom. After our long drive we all had a delicious bathe. "All the armies are bathing now at this hour all along the coast," said Montgomery as we stood in our towels. He waved his arm to the westward. Three hundred yards away about a thousand of our men were disporting themselves on the beach. Although I knew the answer, I asked, "Why do the War Office go to the expense of sending out white bathing drawers for the troops? Surely this economy should be made." They were in fact tanned and burnt to the darkest brown everywhere except where they wore their short pants.

How fashions change! When I marched to Omdurman forty-four years before, the theory was that the African sun must at all costs be kept away from the skin. The rules were strict. Special spine pads were buttoned onto the back of all our khaki coats. It was a military offence to appear without a pith helmet. We were advised to wear thick underclothing, following Arab custom enjoined by a thousand years of experience. Yet now halfway through the twentieth century many of the white soldiers went about their daily toil hatless and naked except for the equal of a loin cloth. Apparently it did them no harm. Though the process of changing from white to bronze took several weeks and gradual application, sunstroke and heatstroke were rare. I wonder how the doctors explain all this.

After we had dressed for dinner — my zip suit hardly takes a minute to put on — we gathered in Montgomery's map wagon. There he gave us a masterly exposition of the situation, showing that in a few days he had firmly gripped the whole problem. He accurately predicted Rommel's next attack, and explained his plans to meet it. All of which proved true and sound. He then described his plans for taking the offensive himself. He must however have six weeks to get the Eighth Army into order. He would re-form the divisions as integral tactical units. We must wait till the new divisions had taken their place at the front and until the Sherman tanks were broken in. Then there would be three Army Corps, each under an experienced officer, whom he and Alexander knew well. Above all the artillery would be used as had never been possible before in the Desert. He spoke of the end of September. I was disappointed at the date, but even this was dependent upon Rommel. Our information showed that a blow from him was imminent. I was myself already fully informed, and was well content that he should try a wide turning movement round our Desert Flank in order to reach Cairo, and that a manoeuvre battle should be fought on his communications.

At this time I thought much of Napoleon's defeat in 1814. He too poised to strike at the communications, but the Allies marched straight on into an almost open Paris. I thought it of the highest importance that Cairo should be defended by every able-bodied man in uniform not required for the Eighth Army. Thus alone would the field army have full manoeuvring freedom and be able to take risks in letting its flank be turned before striking. It was with great pleasure that I found we were all in agreement. Although I was always impatient for offensive action on our part at the earliest moment, I welcomed the prospect of Rommel breaking his teeth upon us before our main attack was launched. But should we have time to organise the defence of Cairo? Many signs pointed to the audacious commander who faced us only a dozen miles away striking his supreme blow before the end of August. Any day indeed, my friends said, he might make his bid for continued mastery. A fortnight or three weeks' delay would be all to our good.

                    *   *   *

On August 20 we sallied forth early to see the prospective battlefield and the gallant troops who were to hold it. I was taken to the key point southeast of the Ruweisat Ridge. Here, amid the hard, rolling curves and creases of the desert, lay the mass of our armour, camouflaged, concealed, and dispersed, yet tactically concentrated. Here I met the young Brigadier Roberts, who at that time commanded the whole of our armoured force in this vital position. All our best tanks were under him. Montgomery explained to me the disposition of our artillery of all natures. Every crevice of the desert was packed with camouflaged concealed batteries. Three or four hundred guns would fire at the German armour before we hurled in our own.

Although of course no gatherings of troops could be allowed under the enemy's continuous air reconnaissance, I saw a great many soldiers that day, who greeted me with grins and cheers. I inspected my own regiment, the 4th Hussars, or as many of them as they dared to bring together — perhaps fifty or sixty — near the field cemetery, in which a number of their comrades had been newly buried. All this was moving, but with it all there grew a sense of the reviving ardour of the Army. Everybody said what a change there was since Montgomery had taken command. I could feel the truth of this with joy and comfort.

We were to lunch with Bernard Freyberg. My mind went back to a similar visit I had paid him in Flanders, at his battle post in the valley of the Scarpe, a quarter of a century before, when he already commanded a brigade. Then he had blithely offered to take me for a walk along his outposts. But knowing him and knowing the line as I did I declined. Now it was the other way round. I certainly hoped to see at least a forward observation post of these splendid New Zealanders, who were in contact about five miles away. Alexander's attitude showed he would not forbid but rather accompany the excursion. But Bernard Freyberg flatly refused to take the responsibility, and this was not a matter about which orders are usually given, even by the highest authority.

Instead we went into his sweltering mess tent, and were offered a luncheon, far more magnificent than the one I had eaten on the Scarpe. This was an August noonday in the Desert. The set piece of the meal was a scalding broth of tinned New Zealand oysters, to which I could do no more than was civil. Presently Montgomery, who had left us some time before, drove up. Freyberg went out to salute him, and told him his place had been kept and that he was expected to luncheon. But "Monty," as he was already called, had, it appeared, made it a rule not to accept hospitality from any of his subordinate commanders. So he sat outside in his car eating an austere sandwich and drinking his lemonade with all formalities. Napoleon also might have stood aloof in the interests of discipline. *Dur aux grands* was one of his maxims. But he would certainly have had an excellent roast chicken, served him from his own *fourgon*. Marlborough would have entered and quaffed the good wine with his officers — Cromwell, I think, too. The technique varies, and the results seem to have been good in all these cases.

We spent all the afternoon among the Army, and it was past seven when we got back to the caravan and the pleasant waves of its beach. I was so uplifted by all I had seen that I was not at all tired and sat up late talking. Before Montgomery went to bed at ten o'clock, in accordance with his routine, he asked me to write something in his personal diary. I did so now and on several other occasions during the long war. Here is what I wrote this time:

"May the anniversary of Blenheim, which marks the opening of the new Command, bring to the Commander-in-Chief of the Eighth Army and his troops the fame and fortune they will surely deserve."

On August 22 I visited the Tura caves, near Cairo, where vital repair work was being done. Out of these caves the stones of the Pyramids had been cut some time before. They came in very handy now. Everything looked very smart and efficient on the spot, and an immense amount of work was being done day and night by masses of skilled men. But I had my tables of facts and figures and remained dissatisfied. The scale was far too small. The original fault lay with the Pharaohs for not having built more and larger Pyramids. Other responsibilities were more difficult to assign. We spent the rest of the day flying from one airfield to another, inspecting the installations and addressing the ground staffs. At one point two or three thousand airmen were assembled. I also visited, brigade by brigade, the Highland Division, just landed. It was late when we got back to the Embassy.

During these last days of my visit all my thought rested upon the impending battle. At any moment Rommel might attack with a devastating surge of armour. He could come in by the Pyramids with hardly a check except a single canal till he reached the Nile, which flowed serenely by at the bottom of the Residency lawn. Lady Lampson's baby son smiled from his pram amid the palm trees. I looked out across the river at the flat expanses beyond. All was calm and peaceful, but I suggested to the mother that it was very hot and sultry in Cairo and could not be good for children. "Why not send the baby away to be braced by the cool breezes of the Lebanon?" But she did not take my advice, and none can say she did not judge the military situation rightly.

In the fullest accord with General Alexander and the C.I.G.S., I set on foot a series of extreme measures for the defence of Cairo and the waterlines running northward to the sea. Rifle pits and machine-gun posts were constructed, bridges mined and their approaches wired, and inundations loosed over the whole wide front. All the office population of Cairo, numbering thousands of staff officers and uniformed clerks, were armed with rifles and ordered to take their stations, if need be, along the fortified water line. The 51st Highland Division was not yet regarded as "desert-worthy," but these magnificent troops were now ordered to man the new Nile front. The position was one of great strength because of the comparatively few causeways which cross the canalised flooded or floodable area of the Delta. It seemed quite practicable to arrest an armoured rush along the causeways. The defence of Cairo would normally have belonged to the British general who commanded the Egyptian Army, all of whose forces were also arrayed. I thought it better however to place the responsibility, should an emergency occur, upon General Maitland-Wilson — "Jumbo" — who had been appointed to the Persia-Iraq Command, but whose headquarters during these critical weeks were forming in Cairo. To him I issued a directive to inform himself fully of the whole defence plan, and to take responsibility from the moment when General Alexander told him that Cairo was in danger.

I had now to go home on the eve of battle and return to far wider but by no means less decisive affairs. I had already obtained the Cabinet's approval of the directive to be given to General Alexander. He was the supreme authority with whom I now dealt in the Middle East. Montgomery and the Eighth Army were under him. So also, if it became necessary, was Maitland-Wilson and the defence of Cairo. "Alex," as I had long called him, had already moved himself and his personal headquarters into the Desert by the Pyramids. Cool, gay, comprehending all, he inspired quiet, deep confidence in every quarter.

* * *

We sailed off from the Desert airfield at 7.05 P.M. on August 23, and I slept the sleep of the just till long after daylight. When I clambered along the bomb bay to the cockpit of the "Commando" we were already approaching Gibraltar. I must say it looked very dangerous. All was swathed in morning mist. One could not see a hundred yards ahead, and we were not flying more than thirty feet above the sea. I asked Vanderkloot if it was all right, and said I hoped he would not hit the Rock of Gibraltar. His answers were not particularly reassuring, but he felt sufficiently sure of his course not to go up high and stand out to sea, which personally I should have been glad to see him do. We held on for another four or five minutes. Then suddenly we flew into clear air, and up towered the great precipice of Gibraltar, gleaming on the isthmus and strip of neutral ground which joins it to Spain and the mountain called the Queen of Spain's Chair. After three or four hours' flying in mist Vanderkloot had been exact. We passed the grim rock face a few hundred yards away without having to alter our course, and made a perfect landing. I still think it would have been better to go aloft and circle round for an hour or two. We had the petrol and were not pressed for time. But it was a fine performance. We spent the morning with the Governor, and flew home in the afternoon, taking a wide sweep across the Bay of Biscay when darkness fell.

* * *

When I set out on my missions to Cairo and Moscow the commander for "Torch" had not been chosen. I had suggested on July 31 that if General Marshall were designated for the Supreme Command of the cross-Channel operation in 1943 General Eisenhower should act as his deputy and forerunner in London and work at "Torch," which he would himself command, with General Alexander as his second. Opinion moved forward on these lines, and before I started from Cairo for Moscow the President had sent me his agreement. Much however remained to be decided about the final shaping of our plans, and on the day following my return to London, Generals Eisenhower and Clark came to dine with me to discuss the state of the operation.

I was at this time in very close and agreeable contact with these American officers. From the moment they arrived in June I had arranged a weekly luncheon at Number 10 on Tuesdays. These meetings seemed to be a success. I was nearly always alone with them, and we talked all our affairs over, back and forth, as if we were all of one country. We also had a number of informal conferences in our downstairs dining-room, beginning at about ten o'clock at night and sometimes running late. Several times the American generals came for a night or a week end to Chequers. Nothing but shop was ever talked on any of these occasions. I am sure these close relationships were necessary for the conduct of the war, and I could not have grasped the whole position without them.

On September 22 at a Chiefs of Staff meeting at which I presided and Eisenhower was present, the final decision was taken. The date of "Torch" was fixed for November 8.

*   *   *

In the midst of all this, Rommel made his determined but, as it proved, his last thrust towards Cairo. Until this was over, my thoughts lay in the Desert and the trial of strength impending there. I had full confidence in our new commanders, and was sure that our numerical superiority in troops, armour, and air power was higher than it had ever been before. But after the unpleasant surprises of the past two years it was difficult to banish anxiety. As I had been so lately over the very ground where the battle was to be fought, and had the picture of the creased and curving rocky desert, with its hidden batteries and tanks and our Army crouched for a counterspring, so vividly in my mind's eye, the whole scene was fiercely lighted. Another reverse would not only be disastrous in itself, but would damage British prestige and influence in the discussions we were having with our American Allies. On the other hand, if Rommel were repulsed growing confidence and the feeling that the tide was about to turn in our favour would help carry all our other affairs to agreement.

General Alexander had promised to send me the word "Zip" (which I took from the clothes I so often wore) when it actually began. "What do you now think," I asked him on August 28, "of the probabilities of 'Zip' coming this moon? Military Intelligence opinion now does not regard it as imminent. All good wishes." " 'Zip' now equal money every day," he replied, "from now onwards. Odds against [it are] increasing till September 2, when it can be considered unlikely." On the 30th I received the monosyllabic signal "Zip," and telegraphed to Roosevelt and Stalin: "Rommel has begun the attack for which we have been preparing. An important battle may now be fought."

Rommel's plan, correctly deduced by Montgomery, was to pass his armour through the weakly defended mine belt in the southern part of the British front and then swing north to roll up our position from flank and rear. The critical ground for the success of this manoeuvre was the Alam Halfa ridge, and Montgomery's dispositions were made principally to ensure that this did not fall into enemy hands.

During the night of August 30 the two armoured divisions of the German Afrika Korps penetrated the mine belt, and next morning moved to the Ragil Depression. Our 7th Armoured Division, withdrawing steadily before them, took station on the eastward flank. To the north of the German armour two Italian armoured divisions and one motorised also attempted to cross the mine field. They had little success. It was deeper than they had expected, and they found themselves under severe harassing fire from the enfilading artillery of the New Zealand Division. The German 90th Light Division however successfully penetrated, to form a hinge for the armour's northern swing. At the other end of the line simultaneous holding attacks were made on the 5th Indian and 9th Australian Divisions; these were repulsed after some stiff fighting. From the Ragil Depression the German-Italian armour had the option of striking north against the Alam Halfa ridge or northeast towards Hammam. Montgomery hoped that they would not take the latter course. He preferred to fight on his chosen battleground, the ridge. A map which showed easy going for tanks in that direction and bad going farther east had been planted upon Rommel. General von Thoma, captured two months later, stated that this false information had its intended effect. Certainly the battle now took the precise form that Montgomery desired.

On the evening of the 31st a northward thrust was repulsed and the enemy's armoured mass went into laager for a night, uncomfortably spent under continuous artillery fire and violent air bombardment. Next morning they advanced against the centre of the British line, where the 10th Armoured Division were now concentrated to meet them. The sand was much heavier than they had been led to believe and the resistance far stronger than they had hoped. The attack, though renewed in the afternoon, failed. Rommel was now deeply committed. The Italians had foundered. He had no hope of reinforcing his forward armour and the heavy going had consumed much of his scanty fuel. He had probably heard also of the sinking of three more tankers in the Mediterranean. So on September 2 his armour took up a defensive posture and awaited attack.

Montgomery did not accept the invitation, and Rommel had no alternative but to withdraw. On the 3rd the movement began, harassed in flank by the 7th British Armoured Division, which took a heavy toll of unarmoured transport vehicles. That night the British counterattack began, not on the enemy armour, but on the 90th Light and the Trieste Motorised Divisions. If these could be broken, then the gaps in the mine field might be blocked before the German armour could return through them. The New Zealand Division made strong attacks, but they were fiercely resisted and the Afrika Korps escaped. Montgomery now stopped the pursuit. He planned to seize the initiative when the time was ripe, but not yet. He was content to have repulsed Rommel's final thrust for Egypt with such heavy loss. At relatively little cost to themselves the Eighth Army and the Desert Air Force had inflicted a heavy stroke upon the enemy and caused another crisis in his supply. From documents captured later we know that Rommel was in dire straits and of his insistent demands for help. We know too that he was a wearied, ailing man at the time. The consequences of Alam Halfa, as the engagement was called, were effective two months later.

Although our two greater operations at both ends of the Mediterranean were now settled and all preparations for them were moving forward, the period of waiting was one of suppressed but extreme tension. The inner circle who knew were anxious about what would happen. All those who did not know were disquieted that nothing was happening.

I had now been twenty-eight months at the head of affairs, during which we had sustained an almost unbroken series of military defeats. We had survived the collapse of France and the air attack on Britain. We had not been invaded. We still held Egypt. We were alive and at bay; but that was all. On the other hand, what a cataract of disasters had fallen upon us! The fiasco of Dakar, the loss of all our Desert conquests from the Italians, the tragedy of Greece, the loss of Crete, the unrelieved reverses of the Japanese war, the loss of Hong Kong, the overrunning of the Dutch East Indies, the catastrophe of Singapore, the Japanese conquest of Burma, Auchinleck's defeat in the Desert, the surrender of Tobruk, the failure, as it was judged, at Dieppe — all these were galling links in a chain of misfortune and frustration to which no parallel could be found in our history. The fact that we were no longer alone, but instead had the two most mighty nations in the world in alliance fighting desperately at our side, gave indeed assurances of ultimate victory. But this, by removing the sense of mortal peril, only made criticism more free. Was it strange that the whole character and system of the war direction, for which I was responsible, should have been brought into question and challenge?

It is indeed remarkable that I was not in this bleak lull dismissed from power, or confronted with demands for changes in my methods, which it was known I should never accept. I should then have vanished from the scene with a load of calamity on my shoulders, and the harvest, at last to be reaped, would have been ascribed to my belated disappearance. For indeed the whole aspect of the war was about to be transformed. Henceforward increasing success, marred hardly by a mishap, was to be our lot. Although the struggle would be long and hard, requiring the most strenuous effort from all, we had reached the top of the pass, and our road to victory was not only sure and certain, but accompanied by constant cheering events. I was not denied the right to share in this new phase of the war because of the unity and strength of the War Cabinet, the confidence which I preserved of my political and professional colleagues, the steadfast loyalty of Parliament, and the persisting goodwill of the nation. All this shows how much luck there is in human affairs, and how little we should worry about anything except doing our best.

In the meantime I found some relief in examining the proposals which the Foreign Office were elaborating, in consultation with the State Department in Washington, on the future of world government after the war. The Foreign Secretary circulated to the War Cabinet in October an important document on this subject entitled "The Four-Power Plan," under which the supreme direction would have come from a council composed of Great Britain, the United States, Russia, and China. I am glad that I found strength to put my own opinions on record in the following minute to the Foreign Secretary, which was dated 21 October 1942:

> In spite of the pressure of events, I will endeavour to write a reply. It sounds very simple to pick out these four Big Powers. We cannot however tell what sort of a Russia and what kind of Russian demands we shall have to face. A little later on it may be possible. As to China, I cannot regard the Chungking Government as representing a great world Power. Certainly there would be a faggot vote on the side of the United States in any attempt to liquidate the British overseas Empire.

2. I must admit that my thoughts rest primarily in Europe — the revival of the glory of Europe, the parent continent of the modern nations and of civilisation. It would be a measureless disaster if Russian barbarism overlaid the culture and independence of the ancient States of Europe. Hard as it is to say now, I trust that the European family may act unitedly as one under a Council of Europe. I look forward to a United States of Europe in which the barriers between the nations will be greatly minimised and unrestricted travel will be possible. I hope to see the economy of Europe studied as a whole. I hope to see a Council consisting of perhaps ten units, including the former Great Powers, with several confederations — Scandinavian, Danubian, Balkan, etc. — which would possess an international police and be charged with keeping Prussia disarmed. Of course we shall have to work with the Americans in many ways, and in the greatest ways, but Europe is our prime care, and we certainly do not wish to be shut up with the Russians and the Chinese when Swedes, Norwegians, Danes, Dutch, Belgians, Frenchmen, Spaniards, Poles, Czechs, and Turks will have their burning questions, their desire for our aid, and their very great power of making their voices heard. It would be easy to dilate upon these themes. Unhappily the war has prior claims on your attention and on mine.

Thus we approached the great military climax upon which all was to be staked.

18

The Battle of Alamein

IN THE WEEKS WHICH FOLLOWED the changes in command, planning preparations and training went forward ceaselessly in Cairo and at the front. The Eighth Army was strengthened to an extent never before possible. The 51st and 44th Divisions had arrived from home and become "desert-worthy." Our strength in armour rose to seven brigades of over a thousand tanks, nearly half of them Grants and Shermans from the United States; we now had a two-to-one superiority in numbers and at least a balance of quality. A powerful and highly trained artillery was for the first time in the Western Desert massed to support the impending attack.

The Air Force in the Middle East was subordinated to the military conceptions and requirements of the Commander-in-Chief. However, under Air Marshal Tedder there was no need for hard-and-fast precedents. The relations between the Air Command and the new generals were in every way agreeable. The Western Desert Air Force, under Air Marshal Coningham, had now attained a fighting strength of 550 aircraft. There were two other groups, in addition to the aircraft based on Malta, numbering 650 planes, whose task it was to harry enemy ports and supply routes across both the Mediterranean and the Desert. Together with a hundred United States fighters and medium bombers, our total strength amounted to about 1200 serviceable aircraft.

Alexander told us in various telegrams that about October 24 had been chosen for "Lightfoot," as the operation was to be called. "Since there is no open flank," he said, "the battle must be so stage-managed that a hole is blown in the enemy's front." Through this the Xth Corps, comprising the main armour, which was to be the spearhead of our attack, would advance in daylight. This corps would not have all its weapons and equipment before October 1. It would then require nearly a month's training for its role. "In my view it is essential that the initial break-in attack should be launched in the full-moon period. This will be a major operation, which will take some time, and an adequate gap in the enemy's lines must be made if our armoured forces are to have a whole day in which to make their operation decisive. . . ."

The weeks passed, and the date drew near. The Air Force had already begun their battle, attacking enemy troops, airfields, and communications. Special attention was paid to their convoys. In September thirty per cent of Axis shipping supplying North Africa was sunk, largely by air action. In October the figure rose to forty per cent. The loss of petrol was sixty-six per cent. In the four autumn months over 200,000 tons of Axis shipping was destroyed. This was a severe injury to Rommel's army. At last the word came. General Alexander telegraphed: "Zip!"

In the full moon of October 23 nearly a thousand guns opened upon the enemy batteries for twenty minutes, and then turned onto their infantry positions. Under this concentration of fire, deepened by bombing from the air, the XXXth (General Leese) and XIIIth Corps (General Horrocks) advanced. Attacking on a front of four divisions, the whole XXXth Corps sought to cut two corridors through the enemy's fortifications. Behind them the two armoured divisions of the Xth Corps (General Lumsden) followed to exploit success. Strong advances were made under heavy fire, and by dawn deep inroads had been made. The engineers had cleared the mines behind the leading troops. But the mine-field system had not been pierced in its depth, and there was no early prospect of our armour breaking through. Farther south the 1st South African Division fought their way forward to protect the southern flank of the bulge, and the 4th Indian Division launched raids from the Ruweisat Ridge, while the 7th Armoured and 44th Divisions of the XIIIth Corps broke into the enemy defences opposite to them. This achieved its object of inducing the enemy to retain his two armoured divisions for three days behind this part of the front while the main battle developed in the north.

So far however no hole had been blown in the enemy's deep system of mine fields and defences. In the small hours of the 25th, Montgomery held a conference of his senior commanders, at which he ordered the armour to press forward again before dawn in accordance with his original instructions. During the day more ground was indeed gained, after hard fighting; but the feature known as Kidney Ridge became the focus of an intense struggle with the enemy's 15th Panzer and Ariete armoured divisions, which made a series of violent counterattacks. On the front of the XIIIth Corps our attack was pressed no farther, in order to keep the 7th Armoured Division intact for the climax.

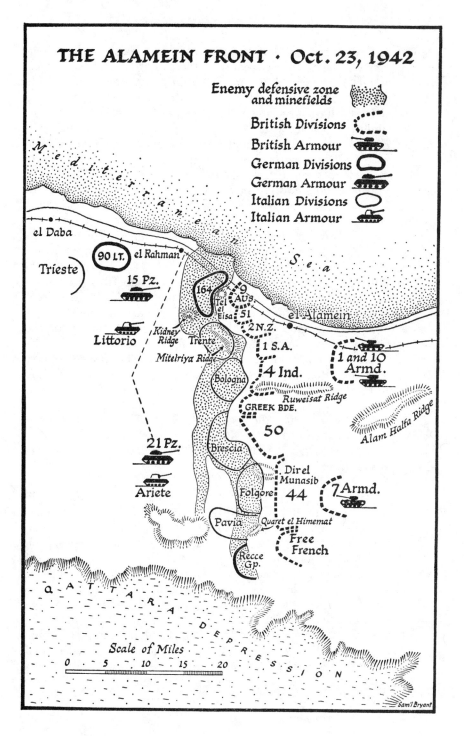

THE ALAMEIN FRONT · Oct. 23, 1942

Enemy defensive zone and minefields

British Divisions

British Armour

German Divisions

German Armour

Italian Divisions

Italian Armour

Mediterranean Sea

el Daba

Trieste

90 LT. el Rahman

15 Pz.

164 9 AUS

Tel el Eisa 51

2 N.Z.

el Alamein

Littorio

Kidney Ridge Trente

Mitelriya Ridge

1 S.A.

4 Ind.

1 and 10 Armd.

Bologna

Ruweisat Ridge

GREEK BDE.

Alam Halfa Ridge

50

21 Pz.

Brescia

Dir el Munasib

Ariete

Folgore 44 7 Armd.

Pavia Quaret el Himemat

Free French

Recce Gp.

QATTARA DEPRESSION

Scale of Miles

0   5   10   15   20

Sam'l Bryant

There had been serious derangements in the enemy's command. Rommel had gone to hospital in Germany at the end of September, and his place was taken by General Stumme. Within twenty-four hours of the start of the battle Stumme died of a heart attack. Rommel, at Hitler's request, left hospital and resumed his command late on the 25th.

Hard fighting continued on October 26 all along the deep bulge so far forced into the enemy line, and especially again at Kidney Ridge. The enemy Air Force, which had been quiescent on the previous two days, now made its definite challenge to our air superiority. There were many combats, ending mostly in our favour. The efforts of the XIIIth Corps had delayed but could not prevent the movement of the German armour to what they now knew was the decisive sector of their front. This movement however was severely smitten by our Air Force.

At this moment a new and fruitful thrust was made by the 9th Australian Division, under General Morshead. They struck northwards from the bulge towards the sea. Montgomery was prompt to exploit this notable success. He held back the New Zealanders from their westward drive and ordered the Australians to continue their advance towards the north. This threatened the retreat of part of the German infantry division on the northern flank. At the same time he now felt that the momentum of his main attack was beginning to falter in the midst of the mine fields and strongly posted anti-tank guns. He therefore regathered his forces and reserves for a renewed and revived assault.

All through the 27th and the 28th a fierce conflict raged for Kidney Ridge against the repeated attacks of the 15th and 21st Panzer Divisions, now arrived from the southern sector. General Alexander has described the struggle in these words: [1]

> On October 27 came a big armoured counterattack in the old style. Five times they attacked with all available tanks, both German and Italian, but gained no ground and suffered heavy and, worse still, disproportionate casualties, for our tanks, fighting on the defensive, suffered but lightly. On October 28 [the enemy] came again, [after] prolonged and careful reconnaissance all the morning, to find the weak spots and locate our anti-tank guns, followed by a smashing concentrated attack in the afternoon with the setting sun behind him. The reconnaissance was less successful than in the old days, since both our tanks and anti-tank guns could engage him with longer range. When the enemy attempted to concentrate for the final attack the R.A.F. once more intervened on a devastating scale. In two and a half hours bomber sorties dropped

[1] In a telegram dated November 9, sent to me after the battle.

eighty tons of bombs in his concentration area, measuring three miles by two, and the enemy's attack was defeated before he could even complete his forming up. This was the last occasion on which the enemy attempted to take the initiative.

In these days of October 26 and 28 three enemy tankers of vital importance were sunk by air attack, thus rewarding the long series of air operations which were an integral part of the land battle.

*   *   *

Montgomery now made his plans and dispositions for the decisive break-through ("Operation Supercharge"). He took out of the line the 2nd New Zealand and the 1st British Armoured Divisions, the latter being in special need of reorganisation after its notable share in the repulse of the German armour at Kidney Ridge. The British 7th Armoured and 51st Divisions and a brigade of the 44th were brought together and the whole welded into a new reserve. The break-through was to be led by the New Zealanders, the 151st and 152nd British Infantry Brigades, and the 9th British Armoured Brigade.

The magnificent forward drive of the Australians, achieved by ceaseless bitter fighting, had swung the whole battle in our favour. At 1 A.M. on November 2 "Supercharge" began. Under a barrage of three hundred guns the British brigades attached to the New Zealand Division broke through the defended zone, and the 9th Armoured drove on ahead. They found however that a new line of defence strong in anti-tank weapons was facing them along the Rahman track. In a long engagement the brigade suffered severely, but the corridor behind was held open and the 1st British Armoured Division moved forward through it. Then came the last clash of armour in the battle. All the remaining enemy tanks attacked our salient on each flank, and were repulsed. Here was the final decision; but even next day, the 3rd, when our air reports indicated that the enemy's retirement had begun, his covering rearguard on the Rahman track still held the main body of our armour at bay. An order came from Hitler forbidding any retreat, but the issue was no longer in German hands. Only one more hole had to be punched. Very early on November 4, five miles south of Tel el Aggagir, the 5th Indian Brigade launched a quickly mounted attack which was completely successful. The battle was now won, and the way finally cleared for our armour to pursue across the open desert.

Rommel was in full retreat, but there was transport and petrol for only a part of his force, and the Germans, though they had fought valiantly, gave themselves priority in vehicles. Many thousands of men from six Italian divisions were left stranded in the desert, with little

food or water, and no future but to be rounded up into prison camps.
The battlefield was strewn with masses of destroyed or useless tanks,
guns, and vehicles. According to their own records, the German armoured divisions, which had started the battle with two hundred forty
serviceable tanks, on November 5 mustered only thirty-eight. The
German Air Force had given up the hopeless task of combating our
superior Air, which now operated almost unhindered, attacking with all
its resources the great columns of men and vehicles struggling westward.
Rommel has himself paid notable tribute to the great part played by the
Royal Air Force.[2] His army had been decisively beaten; his lieutenant,
General von Thoma, was in our hands, with nine Italian generals.

There seemed good hopes of turning the enemy's disaster into annihilation. The New Zealand Division was directed on Fuka, but when
they reached it on November 5 the enemy had already passed. There
was still a chance that they might be cut off at Mersa Matruh, upon
which the 1st and 7th British Armoured Divisions had been thrust. By
nightfall on the 6th they were nearing their objective, while the enemy
were still trying to escape from the closing trap. But then rain came
and forward petrol was scarce. Throughout the 7th our pursuit was
halted. The twenty-four-hour respite prevented complete encirclement.
Nevertheless four German divisions and eight Italian divisions had
ceased to exist as fighting formations. Thirty thousand prisoners were
taken, with enormous masses of material of all kinds. Rommel has left
on record his opinion of the part played by our gunners in his defeat:
"The British artillery demonstrated once again its well-known excellence. Especially noteworthy was its great mobility and speed of reaction to the requirements of the assault troops."[3]

*   *   *

The Battle of Alamein differed from all previous fighting in the
Desert. The front was limited, heavily fortified, and held in strength.
There was no flank to turn. A break-through must be made by whoever
was the stronger and wished to take the offensive. In this way we are
led back to the battles of the First World War on the Western Front.
We see repeated here in Egypt the same kind of trial of strength as was
presented at Cambrai at the end of 1917, and in many of the battles of
1918, namely, short and good communications for the assailants, the use
of artillery in its heaviest concentration, the "drum-fire barrage," and
the forward inrush of tanks.

In all this General Montgomery and his chief, Alexander, were deeply
versed by experience, study, and thought. Montgomery was a great

[2] Desmond Young, *Rommel*, p. 258.
[3] Desmond Young, *op. cit.*, p. 279.

artillerist. He believed, as Bernard Shaw said of Napoleon, that cannons kill men. Always we shall see him trying to bring three or four hundred guns into action under one concerted command, instead of the skirmishing of batteries which was the inevitable accompaniment of swoops of armour in wide desert spaces. Of course everything was on a far smaller scale than in France and Flanders. We lost more than 13,500 men at Alamein in twelve days, but nearly 60,000 on the first day of the Somme. On the other hand, the fire power of the defensive had fearfully increased since the previous war, and in those days it was always considered that a concentration of two or three to one was required, not only in artillery but men, to pierce and break a carefully fortified line. We had nothing like this superiority at Alamein. The enemy's front consisted not only of successive lines of strong points and machine-gun posts, but of a whole deep area of such a defensive system. And in front of all there lay the tremendous shield of minefields of a quality and density never known before. For these reasons the Battle of Alamein will ever make a glorious page in British military annals.

There is another reason why it will survive. It marked in fact the turning of "the Hinge of Fate." It may almost be said, "Before Alamein we never had a victory. After Alamein we never had a defeat."

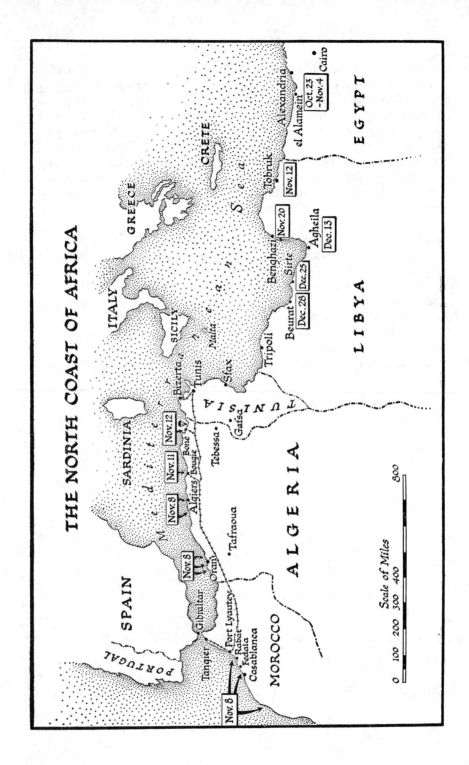

THE NORTH COAST OF AFRICA

SPAIN

PORTUGAL

Gibraltar

Tangier
Port Lyautey
Rabat
Fedala
Casablanca

Nov. 8

MOROCCO

Tafraoua

Oran
Nov. 8

ALGERIA

Algiers
Bougie
Bone
Nov. 8
Nov. 11
Nov. 12

SARDINIA

Bizerta
Tunis
Sfax

Tebessa
Gafsa

TUNISIA

Tripoli

SICILY

ITALY

Malta

Mediterranean Sea

GREECE

CRETE

LIBYA

Sirte
Dec. 25
Benghazi
Nov. 20
Beurat
Dec. 28
Agheila
Dec. 13

Tobruk
Nov. 12

el Alamein
Oct. 23
-Nov. 4

Alexandria

EGYPT

Cairo

Scale of Miles

0   100 200 300 400   800

19

The Torch Is Lit

PRESIDENT ROOSEVELT'S PREJUDICES against General de Gaulle, the contacts he possessed through Admiral Leahy with Vichy, and our memories of the leakage about Dakar two years before led to a decision to withhold all information about "Torch" from the Free French. I did not contest these resolves. I was none the less conscious of our British relationships with de Gaulle, and of the gravity of the affront which he would have to suffer by being deliberately excluded from all share in the design. I planned to tell him just before the blow fell. As some means of softening this slight to him and his Movement, I arranged to confide the trusteeship of Madagascar to his hands. All the facts before us in the months of preparation and everything we have learnt since justify the view that bringing de Gaulle into the business would have been deeply injurious to French reactions in North Africa.

But the need to find some outstanding French figure was obvious, and to British and American eyes none seemed more appropriate than General Giraud, the fighting General of high rank whose dramatic, audacious escape from his prison in Germany was a famous tale. I have mentioned my meeting with Giraud at Metz in 1937,[1] when I visited the Maginot Line, of which he commanded the principal sector. He told me about his adventures in the First World War as an escaped prisoner behind the German lines. As fellow escapees this gave us something in common. Now he had as an Army Commander repeated his youthful exploits in an even more sensational fashion. The Americans entered into secret parleys with the General, and plans were made to bring him from the Riviera to Gibraltar at the decisive moment. Many hopes were based on "King-pin," as he was called in our code,

[1] See page 190.

and not without danger from the sea Giraud and his two sons were safely transported.

*   *   *

Meanwhile our great armadas were approaching the scene. Most of the convoys which sailed from British ports had to cross the Bay of Biscay, traversing all the U-boat routes. Heavy escorts were needed, and we had somehow to conceal not only the concentration of shipping which from the beginning of October began to crowd the Clyde and other western ports, but also the actual sailing of the convoys. We were completely successful. The Germans were led by their own Intelligence to believe that Dakar was again our aim. By the end of the month about forty German and Italian U-boats were stationed to the south and east of the Azores. They mauled severely a large convoy homeward bound for Sierra Leone, and sank thirteen ships. In the circumstances this could be borne. The first of the "Torch" convoys left the Clyde on October 22. By the 26th all the fast troopships were under way and American forces were sailing for Casablanca direct from the United States. The whole expedition of about six hundred and fifty ships was now launched upon the enterprise. They traversed the Bay of Biscay or the Atlantic unseen by the U-boats or by the Luftwaffe.

All our resources were at full strain. Far to the north our cruisers watched the Denmark Strait and the exits from the North Sea to guard against intervention by enemy surface ships. Others covered the American approach near the Azores, and Anglo-American bombers attacked the U-boat bases along the French Atlantic seaboard. The leading ships began to enter the Mediterranean on the night of November 5th–6th still undetected. It was not until the 7th, when the Algiers convoy was less than twenty-four hours from its destination, that it was sighted, and even then only one ship was attacked.

On November 5 Eisenhower by a hazardous flight reached Gibraltar. I had placed the fortress within his command as the temporary headquarters of the leader of this first large-scale American and British enterprise. Here the great concentration of aircraft for "Torch" was made. The whole isthmus was crowded with machines, and fourteen squadrons of fighters were assembled for zero hour. All this activity necessarily took place in full view of German observers, and we could only hope they would think it was for the reinforcement of Malta. We did all we could make them think so. Apparently they did.

General Eisenhower, in his memoirs, has given a vivid account of his anxious experiences during the night of November 7–8, and all through the next few days. He was always very good at bearing stresses of this kind. The immensity of the stake that was being played, the uncertainty of the weather, by which all might be wrecked, the fragmentary news which arrived, the extraordinary complications of the French attitude, the danger from Spain — all, apart from the actual fighting, must have made this a very hard trial to the Commander, whose responsibilities were enormous and direct.

Upon all this there descended General Giraud. He had come with the idea that he would be appointed Supreme Commander in North Africa, and that the American and British armies, of whose strength he had no prior knowledge, would be placed under his authority. He himself strongly urged a landing in France instead of or in addition to Africa, and for some time seemed to imagine that this picture possessed reality. Argument, protracted over forty-eight hours, proceeded between him and General Eisenhower before this brave Frenchman could be convinced of the proportion of affairs. We had all counted overmuch upon "King-pin," but no one was to be more undeceived than he about the influence he had with the French governors, generals, and indeed the Officer Corps, in North Africa.

* * *

A curious but in the upshot highly fortunate complication now occurred. Admiral Darlan, having completed his tour of inspection in North Africa, returned to France. His son was stricken by infantile paralysis and taken into hospital at Algiers. The news of his dangerous condition led the Admiral to fly back on November 5. He thus happened to be in Algiers on the eve of the Anglo-American descent. This was an odd and formidable coincidence. Mr. Robert Murphy, the American political representative in North Africa, hoped he would depart before the assault struck the shores. But Darlan, absorbed in his son's illness, tarried for a day, staying in the villa of a French official, Admiral Fénard.

Our leading hope in Algiers in recent weeks had been General Juin, the French Military Commander. His relations with Mr. Murphy had been intimate, although the actual date had not been imparted to him. A little after midnight on the 7th Murphy visited Juin to tell him that the hour had struck. A mighty Anglo-American army, sustained by overwhelming naval and air forces, was approaching, and would begin landing in Africa in a few hours. General Juin, although deeply engaged and loyal to the enterprise, was staggered by the news. He had conceived himself to possess full command of the situation in Algiers. But he knew that Darlan's presence completely overrode his authority. At his disposal were a few hundred ardent young Frenchmen. He knew only too well that all control of the military and political government had passed from his hands into those of the Minister-Admiral. Now he would certainly not be obeyed. Why, he asked, had he not been told earlier of zero hour? The reasons were obvious, and the fact would have made no difference to his authority. Darlan was on the spot and Darlan was master of all Vichy-French loyalties. Murphy and Juin decided to ask Darlan by telephone to come to them at once. Before two in the morning Darlan, roused from slumber by the urgent message from General Juin, came. On being told of the imminent stroke he turned purple and said, "I have known for a long time that the British were stupid, but I always believed that the Americans were more intelligent. I begin to believe that you make as many mistakes as they do."

Darlan, whose aversion to Britain was notorious, had for a long time been committed to the Axis. In May 1941 he had agreed to grant facilities to the Germans both at Dakar and for the passage of supplies to Rommel's armies through Tunisia. At the time this treacherous move had been stopped by General Weygand, who commanded in North Africa, and who succeeded in persuading Pétain to refuse this German demand. Hitler, at that time fully preoccupied with the impending Russian campaign, did not press the matter, despite contrary advice from his naval staff. In November of the same year Weygand, deemed unreliable by the Germans, was relieved of his command. Although nothing more was heard of the Axis plans to use Dakar against us, the Tunisian ports were later opened to Axis shipping and played a part in feeding Rommel's armies during the summer of 1942. Now circumstances had changed, and with them Darlan's attitude, but whatever thoughts he might have nourished of aiding an Anglo-American occupation of North Africa he was still bound to Pétain in form and in fact. He knew that if he went over to the Allies he would become personally responsible for the invasion by Germany of Unoccupied France. The most he could be prevailed upon to do therefore was to ask Pétain by telegram for liberty of action. In the hideous plight in which he had become involved by the remorseless chain of events this was his only course.

Soon after 1 A.M. on November 8, British and American landings began at many points east and west of Algiers under the direction of Rear Admiral Burrough, R.N. Most careful preparations had been made for guiding the landing craft to the chosen beaches. In the west, leading units of the British 11th Brigade were completely successful, but farther east the ships and craft carrying the Americans were driven some miles from their planned positions by an unexpected tidal set, and in the darkness there was some confusion and delay. Fortunately we gained surprise and opposition along the coast was nowhere serious. Mastery was soon complete. An aircraft of the Fleet Air Arm, observing friendly signals from the ground, landed at Blida Airfield, and with the cooperation of the Local French commander held it until Allied troops arrived from the beaches.

The most severe fighting was in the port of Algiers itself. Here the British destroyers *Broke* and *Malcolm* tried to force an entrance and land American Rangers on the mole so as to take over the harbour, occupy the batteries, and prevent the scuttling of ships. This brought them under the point-blank fire of the defending batteries, and ended in disaster. The *Malcolm* was soon crippled, but the *Broke* entered the harbour at the fourth attempt and landed her troops. Later she was heavily damaged while withdrawing, and eventually sank. Many of the troops were trapped ashore and had to surrender.

At 5 P.M. Darlan sent a telegram to his chief, saying, "American troops having entered into the city in spite of our delaying action, I have authorised General Juin, the Commander-in-Chief, to negotiate the surrender of the city of Algiers only." The surrender of Algiers took effect from 7 P.M. From that moment Admiral Darlan was in American power, and General Juin resumed control of his command under Allied direction.

At Oran there was stronger opposition. Regular French units who had fought the British in Syria, and men under naval command with bitter recollections of our attack on the French fleet in 1940, battled with a United States "Task Force." An American parachute battalion which set out from England to seize the airfields became scattered over Spain in stormy weather. The leading elements pressed on, but their navigation was faulty and they descended some miles from their target.

Two small British warships tried to land a party of American troops in Oran harbour. Their object, as at Algiers, was to prevent the French from sabotaging the installations or scuttling the ships and turn it into an Allied base at the earliest moment. Led by Captain F. T. Peters, R.N., the *Walney* and the *Hartland,* both ex-American coastguard cutters transferred to us under Lend-Lease, encountered murderous fire at point-blank range, and were destroyed, with most of those on board. Captain Peters miraculously survived, only to meet his death a few days later in an aircraft disaster while returning to England. He was posthumously awarded the Victoria Cross and the American Distinguished Service Cross. By dawn French destroyers and submarines were active in Oran Bay, but were either sunk or dispersed. Coastal batteries were bombarded and bombed by British naval forces, including the *Rodney*. Fighting continued until the morning of the 10th, when the Americans launched their final attack on the city. By noon the French capitulated.

The "Western Task Force" reached the Moroccan coast before dawn on November 8. The main assault was near Casablanca, with flanking attacks to the north and south. The weather was fair but hazy, and the surf on the beaches less severe than had been feared. Later it got worse, but by then a firm foothold had been gained. For a time there was severe fighting. At sea a fierce action took place. In Casablanca lay the

unfinished new battleship *Jean Bart,* incapable of movement but able
to use her four 15-inch guns. She was soon engaged in a duel with the
American battleship *Massachusetts,* while the French flotilla, supported
by the cruiser *Primauguet,* sailed out to oppose the landing. They met
the whole of the American fleet. Seven French ships and three submarines were destroyed, with a thousand casualties. The *Jean Bart* was
gutted by fire and beached, and it was not until the morning of November 11 that Noguès, the French Resident-General, under Darlan's orders,
surrendered. "I have lost," he reported, "all our fighting ships and aircraft after three days of violent combat." Captain Mercier, of the
Primauguet, longed for the Allied victory, but he died on her bridge in
the execution of his orders. We may all be thankful if our lives have not
been rent by such dire problems and conflicting loyalties.

*　　　*　　　*

Fragmentary news of all this began to come in to General Eisenhower's headquarters at Gibraltar, and he was now faced with a grave
political situation. He had agreed with Giraud to put him in command
of such French forces as might rally to the Allied cause. Now there had
suddenly and accidentally appeared in the centre of the scene a man who
could in fact decide whether any of them would come over in an orderly
fashion. The hope that they would rally to Giraud had not yet been
put to the test, and first reactions were not encouraging. On the morning of November 9 therefore General Giraud, and a little later General
Clark, acting as General Eisenhower's personal deputy, flew to Algiers.
The reception of Giraud by the leading French commanders was icy.
The local Resistance organisation, fostered by American and British
agents, had already collapsed. Clark's first conference with Darlan produced no agreement. It was obvious that no one of importance would
accept Giraud as Supreme French Commander. Next morning General
Clark arranged a second meeting with the Admiral. He told Eisenhower
by radio that a deal with Darlan was the only solution. There was no
time to engage in telegraphic discussions with London and Washington. Giraud was not present. Darlan hesitated on the ground of
lack of instructions from Vichy. Clark gave him half an hour to
make up his mind. The Admiral at length agreed to order a general
"Cease fire" throughout North Africa. "In the name of the Marshal"
he assumed complete authority throughout the French North African
territories, and ordered all officials to remain on duty.

*　　　*　　　*

In Tunisia Darlan ordered the French Resident-General, Admiral
Esteva, to join the Allies. Esteva was a faithful servant of Vichy. He

followed the cataract of events with mounting confusion and alarm. As he was closer to the enemy in Sicily and on his eastern frontier, his position was worse than that of either Darlan or Noguès. His high subordinates matched him in equal indecision. Already on November 9 units of the German Air Force occupied the important airfield at El Aouina. On the same day German and Italian troops arrived. Depressed and wavering, Esteva clung to a formal allegiance to Vichy, while the Axis forces in Tripolitania were coming from the east, and the Allies hastened from the west. The French General Barré, at first baffled by a problem the like of which, gentle reader, you have not yet been asked to solve, finally moved the bulk of the French garrison westwards and placed himself under the orders of General Giraud. At Bizerta however three torpedo boats and nine submarines surrendered to the Axis.

In Alexandria, where the French naval squadron had been immobilised since 1940, parleys took place without effect. Admiral Godefroy, its commander, persisted in his loyalty to Vichy and refused to recognise the authority of Darlan. In his view, until the Allies had conquered Tunisia they could not claim that it was in their power to liberate France. Thus his ships continued in idleness until in the fullness of time we conquered Tunis. At Dakar the Vichy Governor-General Boisson accepted Darlan's order to cease resistance on November 23, but the units of the French Navy there refused to join the Allies. Only after the completion of our conquest of all North Africa did the battleship *Richelieu* and her three cruisers rally to our cause.

*   *   *

The Anglo-American descent in North Africa brought an immediate sequel in France. As early as December 1940 the Germans had drawn up detailed plans for the occupation of the free zone of France. These were now put into force. Hitler's main object was to capture the principal units of the French Fleet, which lay at Toulon. General Eisenhower was just as anxious to lay his hands on this great prize, but while he was negotiating with Darlan, and Darlan was sending messages to Vichy, the Germans were marching rapidly towards the Mediterranean coast and occupying the whole of France. This simplified the Admiral's position. He could now maintain, and his word would be accepted by local officials and commanders, that Marshal Pétain was no longer a free agent. The German move also struck Darlan's vital nerve. As in 1940, the fate of the French Fleet was again in the balance. He was the only man who could save it. He acted decisively. On the afternoon of November 11 he telegraphed to

Metropolitan France that the Toulon fleet was to put to sea if in danger of imminent capture by the Germans.

Admiral Auphan, the Minister of Marine at Vichy, wished to stand by Darlan, but he was powerless in the face of Laval and of the attitude of the French commanders at Toulon. Admiral de Laborde, the Commander of the French Mediterranean Fleet, was fanatically anti-British. On hearing the news of the landings he wished to put to sea and attack the Allied convoys. He rejected Darlan's appeals to come over, and when the Germans arrived at the perimeter of the French naval base an agreement was made whereby a free zone round the harbour was to be garrisoned by French troops. Attempts were made to put the port in a serious state of defence. But on November 18 the Germans demanded the withdrawal of all French troops from the zone, and the following day Auphan resigned.

The Germans now planned a *coup de main* against the Fleet. The operation took place on November 27. The courage and resource of a few officers, including Laborde, who rallied at last, made possible the wholesale scuttling of the Fleet. One battleship, two battle cruisers, seven cruisers, twenty-nine destroyers and torpedo boats, and sixteen submarines were among the seventy-three ships which sank in the port.

*   *   *

Less than a month afterwards, Admiral Darlan was murdered. On the afternoon of December 24 he drove down from his villa to his offices in the Palais d'Eté. At the door of his bureau he was shot down by a young man of twenty named Bonnier de la Chapelle. The Admiral died within the hour on the operating table of a nearby hospital. The youthful assassin had under much persuasion worked himself into an exalted state of mind as the saviour of France from wicked leadership. He was tried by court-martial under Giraud's orders, and, much to his surprise, was executed by a firing squad shortly after dawn on the 26th.

Few men have paid more heavily for errors of judgment and failure of character than Admiral Darlan. He was a professional figure, and a strong personality. His life's work had been to re-create the French Navy, and he had raised it to a position it had never held since the days of the French kings. He commanded the allegiance not only of the Naval Officer Corps but of the whole Naval Service. In accordance with his repeated promises, he ought in 1940 to have ordered the fleets to Britain, to the United States, the African ports, anywhere out of German power. He was under no treaty of obligation to do so except assurances which he had voluntarily given. But this was his resolve until on that deadly June 20, 1940, he accepted from Marshal Pétain's hands the office of Minister of Marine. Then, perhaps influenced by motives of a departmental character, he gave his allegiance to Marshal Pétain's Government. Ceasing to be a sailor and becoming a politician, he exchanged a sphere in which he had profound knowledge for one where his chief guide was his anti-British prejudices, dating, as I have mentioned, from the Battle of Trafalgar, where his great-grandfather had fallen.

In this new situation he showed himself a man of force and decision who did not wholly comprehend the moral significance of much that he did. Ambition stimulated his errors. His vision as an Admiral had not gone beyond his Navy, nor as a Minister beyond immediate local or personal advantages. For a year and a half he had been a great power in shattered France. At the time when we descended upon North Africa he was the undoubted heir of the aged Marshal. Now suddenly a cataract of amazing events fell upon him.

We have recounted the stresses which he underwent. All French North and West Africa looked to him. The invasion of Vichy France by Hitler gave him the power, and it may be the right, to make a new decision. He brought to the Anglo-American Allies exactly what they needed, namely, a French voice which all French officers and officials in this vast theatre, now plunged in the war, would obey. He struck his final blow for us, and it is not for those who benefited enormously from his accession to our side to revile his memory. A stern, impartial judge may say that he should have refused all parley with the Allies he had injured, and defied them to do their worst with him. We may all be glad he took the opposite course. It cost him his life, but there was not much left in life for him. It seemed obvious at the time that he was wrong in not sailing the French Fleet to Allied or neutral ports in June 1940; but he was right in this second fearful decision. Probably his sharpest pang was his failure to bring over the Toulon fleet. Always he declared it should never fall into German hands. In this undertaking before history he did not fail. Let him rest in peace, and let us all be thankful we have never had to face the trials under which he broke.

20

The Casablanca Conference

AMERICAN MILITARY OPINION, not only in the highest circles, was convinced that the decision for "Torch" ruled out all prospect of a major crossing of the Channel into Occupied France in 1943. I had not yet brought myself to accept this view. I still hoped that French North Africa, including the Tunisian tip, might fall into our hands after a few months' fighting. In this case the main invasion of Occupied France from England would still be possible in July or August 1943. I was therefore most anxious that the strongest build-up of American power in Britain which our shipping would allow should proceed at the same time as "Torch." This idea of being able to use our left as well as our right hand, and the fact that the enemy must prepare himself against blows from either, seemed wholly in accordance with the highest economy of war. Events would decide whether we should thrust across the Channel or follow our luck in the Mediterranean, or do both. It seemed imperative, in the interests of the war as a whole and especially of aiding Russia, that the Anglo-American armies should enter Europe either from the west or from the east in the coming year.

There was however a danger that we might do neither. Even if our campaign in Algeria and Tunisia prospered swiftly, we might have to content ourselves with capturing Sardinia or Sicily or both, and put off crossing the Channel till 1944. This would mean a wasted year for the Western Allies, with results which might be fatal, not indeed to our survival, but to a decisive victory. We could not go on losing five or six hundred thousand tons of shipping a month indefinitely. A stalemate was Germany's last hope.

Before we knew what was going to happen at Alamein or to "Torch," and while the terrific struggle in the Caucasus seemed undecided, the

British Chiefs of Staff were weighing all these issues. The Planners under them were also busy. Their reports were in my opinion unduly negative, and from both sides of the Atlantic we were reaching a sort of combined deadlock. The British Staffs favoured the Mediterranean and an attack upon Sardinia and Sicily, with Italy as the goal. The United States experts had given up all hopes of crossing the Channel in 1943, but were most anxious not to be entangled in the Mediterranean in such a way as to prevent their great design in 1944. "It would seem," I wrote in November, "that the sum of all American fears is to be multiplied by the sum of all British fears, faithfully contributed by each Service."

It will no doubt be said that the course of events proved that I took too sanguine a view about the prospects in North Africa, and the United States Staffs were right in believing that the decision for "Torch" which we had taken in July closed the possibility of crossing the Channel in 1943. Certainly that was what happened. No one could foresee at this time that Hitler would make his immense effort to reinforce the Tunisian tip by sending thither by air and sea, in spite of heavy losses, nearly a hundred thousand of his best troops. This was on his part a grave strategic error. It certainly delayed for several months our victory in Africa. If he had held the forces which were captured or destroyed there in May he might either have reinforced his retreating front against Russia, or have gathered the strength in Normandy which would have deterred us, even if we were so resolved, from trying to land in 1943. Hardly anyone now disputes the wisdom of the decision to wait till 1944. My conscience is clear that I did not deceive or mislead Stalin. I tried my best. On the other hand, provided we invaded the mainland of Europe from the Mediterranean in the coming campaign and that the Anglo-American armies were in full contact with the enemy. I was not ill-content with the decision which Fate and facts were to impose.

Indeed there now came a definite check and setback in North Africa. Although we had the initiative and the advantage of surprise our build-up was inevitably slow. Shipping imposed its harsh limits. Unloading was hampered by air attacks on Algiers and Bone. Road transport was lacking. The single-line coastal railway, five hundred miles long, was in poor condition, with hundreds of bridges and culverts, any one of which might be sabotaged. With the arrival of German troops in large numbers by air in Tunis a high-class, stubborn, and violent resistance began. The French forces who had now joined our cause were over a hundred thousand strong. The majority were native troops of good quality, but as yet ill equipped and unorganised. General Eisenhower thrust forward every American unit on which

he could lay his hands. We put in all we could. On November 28 a
British infantry brigade, with part of the United States 1st Armoured
Division, nearly reached Djedeida, only twelve miles from Tunis.
This was the climax of the winter fighting.

Now came the rainy season. It poured. Our improvised airfields
became quagmires. The German Air Force, though not yet strong in
numbers, worked from good all-weather airfields. On December 1
they counterattacked, frustrating the advance we had planned, and
in a few days we were forced back to Medjez. Supplies could only
reach the forward troops by sea on a small scale. Indeed, it was
barely possible to nourish them, far less to make any accumulations.
It was not till the night of December 22 that a renewed attack could
be launched. This met with some initial success, but at dawn began
three days of torrential rain. Our airfields became useless and vehicles
could only move along the indifferent roads.

At a conference on Christmas Eve, General Eisenhower decided to
give up the plan for the immediate capture of Tunis and until campaigning could begin again, to guard his forward airfields on the
general line already gained. Although the Germans suffered important
losses at sea, their strength in Tunisia continually grew. By the end
of December their numbers approached fifty thousand.

The Eighth Army had meanwhile covered immense distances. Rommel succeeded in withdrawing his shattered forces from Alamein. His
rearguards were heavily pressed, but an attempt to head him off south
of Benghazi failed. He paused at Agheila, while Montgomery, after
his long advance, contended with the same difficulties of transport
and supply on which his predecessors had foundered. On December 13
Rommel was dislodged and nearly cut off by a wide turning movement
of the 2nd New Zealand Division. He suffered severely, and the
Desert Air Force took heavy toll of his transport on the coast road.
Montgomery could follow at first only with light forces. The Eighth
Army had advanced twelve hundred miles since Alamein. After occupying Sirte and its landing grounds on Christmas Day our troops closed
with Rommel's next main position near Buerat at the end of the year.

*   *   *

The Chiefs of Staff Committee meanwhile produced two papers for
the War Cabinet summarising their considered views upon future
strategy. In reaching their conclusions they emphasised a serious
divergence of view between themselves and their American colleagues.
It was one of emphasis and priority rather than of principle. The
British Chiefs of Staff thought the best policy was to follow up "Torch"
vigorously, accompanied by as large a preparation for crossing the

Channel in 1943 as possible, while the American Chiefs of Staff favoured putting our main European effort into crossing the Channel and standing fast in North Africa. Here was a crucial issue. It could only be resolved by the President and myself, and after considerable debate we decided to meet and settle it at Casablanca.

I flew there on January 12, 1943. My journey was a little anxious. In order to heat the "Commando" they had established a petrol engine inside which generated fumes and raised various heating points to very high temperatures. I was woken up at two in the morning, when we were over the Atlantic five hundred miles from anywhere, by one of these heating points burning my toes, and it looked to me as if it might soon get red-hot and light the blankets. I therefore climbed out of my bunk and woke up Peter Portal, who was sitting in the well beneath, asleep in his chair, and drew his attention to this very hot point. We looked around the cabin and found two others, which seemed equally on the verge of becoming red-hot. We then went down into the bomb alley (it was a converted bomber), and found two men industriously keeping alive this petrol heater. From every point of view I thought this was most dangerous. The hot points might start a conflagration, and the atmosphere of petrol would make an explosion imminent. Portal took the same view. I decided that it was better to freeze than to burn, and I ordered all heating to be turned off, and we went back to rest shivering in the ice-cold winter air about eight thousand feet up, at which we had to fly to be above the clouds. I am bound to say this struck me as rather an unpleasant moment.

When we got to Casablanca we found beautiful arrangements made. There was a large hotel in the suburb of Anfa with ample accomodation for all the British and American Staffs and big conference rooms. Round this hotel were dotted a number of extremely comfortable villas which were reserved for the President, for me, for General Giraud, and also for General de Gaulle, should he come. The whole enclave was wired in and closely guarded by American troops. I and the Staff were there two days before the President arrived. I had some nice walks with Pound and the other Chiefs of Staff on the rocks and the beach. Wonderful waves rolling in, enormous clouds of foam, made one marvel that anybody could have got ashore at the landing. There was not one calm day. Waves fifteen feet high were roaring up terrible rocks. No wonder so many landing craft and ships' boats were turned over with all their men. My son Randolph had come across from the Tunisian front. There was plenty to think about, and the two days passed swiftly by. Meanwhile the Staffs consulted together for long hours every day.

The President arrived in the afternoon of the 14th. We had a most

warm and friendly meeting, and it gave me intense pleasure to see my
great colleague here on conquered or liberated territory which he and
I had secured in spite of the advice given him by all his military experts. The next day General Eisenhower arrived, after a very hazardous flight. He was most anxious to know what line the Combined
Chiefs of Staff would take, and to keep in touch with them. Their
plane of command was altogether above his. A day or two later
Alexander came in, and reported to me and the President about the
progress of the Eighth Army. He made a most favourable impression
upon the President, who was greatly attracted by him, and also by
his news, which was that the Eighth Army would take Tripoli in the
near future. He explained how Montgomery, who had two strong
Army Corps, had dismounted one and taken all the vehicles to bring
the other on alone, and that this would be strong enough to drive
Rommel right back through Tripoli to the Mareth frontier line, which
was a very serious obstacle. Everyone was much cheered by this news,
and the easy, smiling grace of Alexander won all hearts. His unspoken
confidence was contagious.

After ten days' work on the main issues, the Combined Chiefs of
Staff reached agreement. Both the President and I kept in daily touch
with their work and agreed between ourselves about it. It was settled
that we should concentrate all upon taking Tunis, both with the Desert
Army and with all forces that could be found by the British, and from
Eisenhower's army, and that Alexander should be Eisenhower's Deputy and virtually in charge of all the operations. On the other immediate step, namely, whether we should attack Sicily or Sardinia, agreement was also reached. The differences did not run along national
lines, but were principally between the Chiefs of Staff and the Joint
Planners. I was myself sure that Sicily should be the next objective,
and the Combined Chiefs of Staff took the same view. The Joint Planners, on the other hand, together with Lord Mountbatten, felt that we
should attack Sardinia rather than Sicily, because they thought it
could be done three months earlier; and Mountbatten pressed this view
on Hopkins and others. I remained obdurate, and, with the Combined
Chiefs of Staff solid behind me, insisted on Sicily. The Joint Planners,
respectful but persistent, then said that this could not be done until
August 30. At this stage I personally went through all the figures with
them, and thereafter the President and I gave orders that D-Day was
to be during the favourable July moon period, or, if possible, the favourable June moon period. In the event the air-borne troops went in
on the night of July 9, and the landings started on the morning of
July 10.

*   *   *

The question of de Gaulle had meanwhile been raised. Darlan's murder, however criminal, had relieved the Allies of their embarrassment at working with him. His authority had passed smoothly to the organisation created in agreement with the Americans during the months of November and December. Giraud filled the gap. The path was cleared for the French forces now rallied in North and Northwest Africa to unite with the Free French Movement round de Gaulle, and comprising all Frenchmen throughout the world outside German control. I was now most anxious for de Gaulle to come, and the President agreed generally with this view. I asked Mr. Roosevelt also to telegraph inviting him. The General was very haughty and refused several times. I then got Eden to put the utmost pressure upon him, even to the point of saying that if he would not come we should insist on his being replaced by someone else at the head of the French Liberation Committee in London. At last on January 22 he arrived. He was taken to his villa, which was next to Giraud's. He would not call upon Giraud, and it was some hours before he could be prevailed upon to meet him. I had a very stony interview with de Gaulle, making it clear that if he continued to be an obstacle we should not hesitate to break with him finally. He was very formal, and stalked out of the villa and down the little garden with his head high in the air. Eventually he was prevailed upon to have a talk with Giraud, which lasted for two or three hours and must have been extremely pleasant to both of them. In the afternoon he went to see the President, and to my relief they got on unexpectedly well. The President was attracted by "the spiritual look in his eyes"; but very little could be done to bring them into accord.

In these pages various severe statements, based on events of the moment, are set down about General de Gaulle, and certainly I had continuous difficulties and many sharp antagonisms with him. There was however a dominant element in our relationship. I could not regard him as representing captive and prostrate France, nor indeed the France that had a right to decide freely the future for herself. I knew he was no friend of England. But I always recognised in him the spirit and conception which, across the pages of history, the word "France" would ever proclaim. I understood and admired, while I resented, his arrogant demeanour. Here he was — a refugee, an exile from his country under sentence of death, in a position entirely dependent upon the goodwill of the British Government, and also now of the United States. The Germans had conquered his country. He had no real foothold anywhere. Never mind; he defied all. Always, even when he was behaving worst, he seemed to express the personality of France — a great nation, with all its pride, authority, and ambition. It was said in mockery that he thought himself the living representative of Joan of Arc, whom one of his ancestors is supposed to have served as a faithful adherent. This did not seem to me as absurd as it looked. Clemenceau, with whom it was said he also compared himself, was a far wiser and more experienced statesman. But they both gave the impression of being unconquerable Frenchmen.

*   *   *

One other matter requires mention. In a report to the War Cabinet, I made the following suggestion:

> . . . We propose to draw up a statement of the work of the conference for communication to the Press at the proper time. I should be glad to know what the War Cabinet would think of our including in this statement a declaration of the firm intention of the United States and the British Empire to continue the war relentlessly until we have brought about the "unconditional surrender" of Germany and Japan. The omission of Italy would be to encourage a break-up there. The President liked this idea, and it would stimulate our friends in every country . . .

The reader should note this telegram, as the use by the President at the subsequent meeting with the press of the words "unconditional surrender" raised issues which will recur in this story and certainly be long debated. There is a school of thought, both in England and America, which argues that the phrase prolonged the war and played into the dictators' hands by driving their peoples and armies to desperation. I do not myself agree with this, for reasons which the course of this narrative will show. Nevertheless, as my own memory has proved defective on some points, it will be well to state the facts as my archives reveal them.

The records of the War Cabinet show that this was brought before them at their afternoon meeting on January 20. The discussion seems to have turned, not upon the principle of "unconditional surrender," but on making an exception in favour of Italy. Accordingly on January 21 Mr. Attlee and Mr. Eden sent us the following message:

> The Cabinet were unanimously of opinion that balance of advantage lay against excluding Italy, because of misgivings which would inevitably be caused in Turkey, in the Balkans, and elsewhere. Nor are we convinced that effect on Italians would be good. Knowledge of all rough stuff coming to them is surely more likely to have desired effect on Italian morale.

There can therefore be no doubt that the phrase "unconditional surrender" in the proposed joint statement that was being drafted was mentioned by me to the War Cabinet, and not disapproved in any way by them. On the contrary, their only wish was that Italy should not be omitted from its scope. I do not remember nor have I any record of anything that passed between me and the President on the subject after I received the Cabinet message, and it is quite possible that in the pressure of business, especially the discussions about the relations of Giraud and de Gaulle and interviews with them, the matter was not further referred to between us. Meanwhile the official joint statement was being prepared by our advisers and by the Chiefs of Staff. This was a careful and formally worded document, which both the President and I considered and approved. It seems probable that as I did not like applying unconditional surrender to Italy I did not raise the point again with the President, and we had certainly both agreed to the communiqué we had settled with our advisers. There is no mention in it of "unconditional surrender." It was submitted to the War Cabinet, who approved it in this form.

It was with some feeling of surprise that I heard the President say at the Press Conference on January 24 that we would enforce "unconditional surrender" upon all our enemies. It was natural to suppose that the agreed communiqué had superseded anything said in conversation. General Ismay, who knew exactly how my mind was working from day to day, and was also present at all the discussions of the Chiefs of Staff when the communiqué was prepared, was also surprised. In my speech which followed the President's I of course supported him and concurred in what he had said. Any divergence between us, even by omission, would on such an occasion and at such a time have been damaging or even dangerous to our war effort. I certainly take my share of the responsibility, together with the British War Cabinet.

The President's account to Hopkins seems however conclusive:

> We had so much trouble getting those two French generals together that I thought to myself that this was as difficult as arranging the meeting of Grant and Lee — and then suddenly the Press Conference was on, and Winston and I had had no time to prepare for it, and the thought popped into my mind that they had called Grant "Old Unconditional Surrender," and the next thing I knew I had said it.[1]

I do not feel that this frank statement is in any way weakened by the fact that the phrase occurs in the notes from which he spoke.

[1] Robert Sherwood, *Roosevelt and Hopkins*, p. 696.

Memories of the war may be vivid and true, but should never be trusted without verification, especially where the sequence of events is concerned. I certainly made several erroneous statements about the "unconditional surrender" incident, because I said what I thought and believed at the moment without looking up the records. Mine was not the only memory at fault, for Mr. Ernest Bevin in the House of Commons on July 21, 1949, gave a lurid account of the difficulties he had had to encounter in rebuilding Germany after the war through the policy of "unconditional surrender," on which he said neither he nor the War Cabinet had ever been consulted at the time. I replied on the spur of the moment, with equal inaccuracy and good faith, that the first time I heard the words was from the lips of the President at the Casablanca Press Conference. It was only when I got home and searched my archives that I found the facts as they have been set out here. I am reminded of the professor who in his declining hours was asked by his devoted pupils for his final counsel. He replied, "Verify your quotations."

*   *   *

The use of the expression "unconditional surrender," although widely hailed at the time, has since been described by various authorities as one of the great mistakes of Anglo-American war policy. It requires to be dealt with at this point. It is said that it prolonged the struggle and made recovery afterwards more difficult. I do not believe that this is true. Indeed, my principal reason for opposing, as I always did, an alternative statement on peace terms, which was so often urged, was that a statement of the actual conditions on which the three great Allies would have insisted, and would have been forced by public opinion to insist, would have been far more repulsive to any German peace movement than the general expression "unconditional surrender." I remember several attempts being made to draft peace conditions which would satisfy the wrath of the conquerors against Germany. They looked so terrible when set forth on paper, and so far exceeded what was in fact done, that their publication would only have stimulated German resistance. They had in fact only to be written out to be withdrawn.

In several public utterances I made clear what the President and I had in mind.

"The term 'unconditional surrender,'" I said in the House of Commons on February 22, 1944, "does not mean that the German people will be enslaved or destroyed. It means however that the Allies will not be bound to them at the moment of surrender by any pact or obligation. . . . Unconditional surrender means that the victors have a free hand. It does not mean that they are entitled to behave in a barbarous manner, nor that they wish to blot out Germany from among the nations of Europe. If we are bound, we are bound by our own consciences to civilisation. We are not to be bound to the Germans as the result of a bargain struck. That is the meaning of 'unconditional surrender.' "

It cannot be contended that in the closing years of the war there was any misconception in Germany.[2]

*　　　　*　　　　*

We were now to wind up our affairs. Our last formal and plenary meeting with the Chiefs of Staff took place on January 23, when they presented to us their final report on "The Conduct of the War in 1943." It may be epitomised as follows:

> The defeat of the U-boat must remain a first charge on the resources of the United Nations. The Soviet forces must be sustained by the greatest volume of supplies that can be transported to Russia.
>
> Operations in the European theatre will be conducted with the object of defeating Germany in 1943 with the maximum forces that can be brought to bear upon her by the United Nations.
>
> The main lines of offensive action will be:
>
> *In the Mediterranean*
> (*a*) The occupation of Sicily, with the object of:
> > (i) Making the Mediterranean line of communications more secure.
> > (ii) Diverting German pressure from the Russian front.
> > (iii) Intensifying the pressure on Italy.
>
> (*b*) To creat a situation in which Turkey can be enlisted as an active ally.
>
> . . . Operations in the Pacific and Far East shall continue, with the object of maintaining pressure on Japan, and for the full-scale offensive against Japan as soon as Germany is defeated. These operations must be kept within such limits as will not, in the opinion of the Joint Chiefs of Staff, jeopardise the capacity of the United Nations to take advantage of any favourable opportunity for the decisive defeat of Germany in 1943. . . .

Finally, on the morning of the 24th, we came to the Press Conference, where de Gaulle and Giraud were made to sit in a row of

[2] But a full account of this issue may be found in *The Hinge of Fate*, Chapter 38.

chairs, alternating with the President and me, and we forced them to shake hands in public before all the reporters and photographers. They did so, and the pictures of this event cannot be viewed even in the setting of these tragic times without a laugh. The fact that the President and I were at Casablanca had been a well-kept secret. When the press reporters saw us both they could scarcely believe their eyes, or, when they were told we had been there for nearly a fortnight, their ears.

After the compulsory, or "shotgun," marriage (as it is called in the United States) of the bride and bridegroom, about whom such pains had been taken, the President made his speech to the reporters, and I supported him.

*   *   *

The President prepared to depart. But I said to him, "You cannot come all this way to North Africa without seeing Marrakesh. Let us spend two days there. I must be with you when you see the sunset on the snows of the Atlas Mountains." I worked on Harry Hopkins also in this sense. It happened there was a most delightful villa, of which I knew nothing, at Marrakesh which the American Vice-Consul, Mr. Kenneth Pendar, had been lent by an American lady, Mrs. Taylor. This villa would accommodate the President and me, and there was plenty of outside room for our entourages. So it was decided that we should all go to Marrakesh. Roosevelt and I drove together the one hundred and fifty miles across the desert — already it seemed to me to be beginning to get greener — and reached the famous oasis. My description of Marrakesh was "the Paris of the Sahara," where all the caravans had come from Central Africa for centuries to be heavily taxed en route by the tribes in the mountains and afterwards swindled in the Marrakesh markets, receiving the return, which they greatly valued, of the gay life of the city, including fortunetellers, snake charmers, masses of food and drink, and on the whole the largest and most elaborately organised brothels in the African continent. All these institutions were of long and ancient repute.

It was agreed between us that I should provide the luncheon, and the President and I drove together all the way, five hours, and talked a great deal of shop, but also touched on lighter matters. Many thousand American troops were posted along the road to protect us from any danger, and aeroplanes circled ceaselessly overhead. In the evening we arrived at the villa, where we were very hospitably and suitably entertained by Mr. Pendar. I took the President up the tower of the villa. He was carried in a chair, and sat enjoying a wonderful sunset on the snows of the Atlas. We had a very jolly dinner, about fifteen or six-

teen, and we all sang songs. I sang, and the President joined in the
choruses, and at one moment was about to try a solo. However, someone interrupted and I never heard this.

My illustrious colleague was to depart just after dawn on the 25th
for his long flight by Lagos and Dakar and so across to Brazil and then
up to Washington. We had parted the night before, but he came round
in the morning on the way to the aeroplane to say another good-bye.
I was in bed, but would not hear of letting him go to the airfield alone,
so I jumped up and put on my zip, and nothing else except slippers, and
in this informal garb I drove with him to the airfield, and went on the
plane and saw him comfortably settled down, greatly admiring his
courage under all his physical disabilities and feeling very anxious about
the hazards he had to undertake. These aeroplane journeys had to be
taken as a matter of course during the war. None the less I always regarded them as dangerous excursions. However, all was well. I then
returned to the Villa Taylor, where I spent another two days in correspondence with the War Cabinet about my future movements, and
painting from the tower the only picture I ever attempted during the
war.

21

Turkey, Stalingrad, and Tunis

THE STRATEGIC SCENE in the Mediterranean had been transformed by the Allied occupation of North Africa, and with the acquisition of a solid base on its southern shores a forward movement against the enemy became possible. The President and I had long sought to open a new route to Russia and to strike at Germany's southern flank. Turkey was the key to all such plans. To bring Turkey into the war on our side had for many months been our aim. It now acquired new hope and urgency.

Stalin was in full agreement with Mr. Roosevelt and myself, and I now wished to clinch the matter by a personal meeting with President Inönü on Turkish soil. There was also much business to be done in Cairo, and I hoped on the way home to visit the Eighth Army in Tripoli, if it were taken, and also to call at Algiers. There were many things I could settle on the spot, and more which I needed to see with my own eyes. On January 20 therefore I telegraphed from Casablanca to the Deputy Prime Minister and the Foreign Secretary that I proposed to fly from Marrakesh to Cairo, stay there for two or three days and then get into direct touch with the Turks.

The War Cabinet thought a direct approach to Turkey was premature and urged my return direct to London to give an account to Parliament of my meeting with Mr. Roosevelt, but after some telegraphic debate they acquiesced in my plan. Accordingly on the afternoon of the 26th we sailed off in the "Commando," and after having an extremely good dinner, provided by Mr. Pendar at the Taylor Villa, I slept soundly till once again I went to the co-pilot's seat and sat by Captain Vanderkloot, and we saw together for the second time dawn gleam upon the waters of the Nile. This time we had not to go so far to the south, because the victory of Alamein had swept our foes fifteen hundred miles farther to the west. We arrived at the airfield, ten miles from the Pyramids, and were welcomed by the Ambassador, Lord Killearn, and received by the Cairo Command. We then repaired to the Embassy. Here I was joined by Sir Alexander Cadogan, Permanent Under-Secretary of State at the Foreign Office, sent from England by the Cabinet at my desire. We were all able to contrast the situation with what it had been in August 1942 with feelings of relief and satisfaction.

Messages now reached me to say that the Turkish President, Ismet Inönü, was delighted at the idea of the proposed meeting, and arrangements were made for it to take place at Adana, on the coast near the Turkish-Syrian border, on January 30. I went in the "Commando" to meet the Turks. It is only a four-hour flight across the Mediterranean, most of it in sight of Palestine and Syria, and I had with me in another plane Cadogan and Generals Brooke, Alexander, Wilson, and other officers. We landed not without some difficulty on the small Turkish airfield, and we had hardly completed the salutations and ceremonials before a very long enamelled caterpillar began to crawl out of the mountain defiles, containing the President, the entire Turkish Government, and Marshal Chakmak. They received us with the utmost cordiality and enthusiasm. Several saloon carriages had been put on the train for our accommodation, there being none other in the neighbourhood. We spent two nights in the train, having long daily discussions with the Turks and very agreeable talks at meals with President Inönü.

The general discussion turned largely on to two questions, the structure of the postwar world, and the arrangements for an international organisation, and the future relations of Turkey and Russia. I give only a few examples of the remarks which, according to the record, I made to the Turkish leaders. I said that I had seen Molotov and Stalin, and my impression was that both desired a peaceful and friendly association with the United Kingdom and the United States. In the economic sphere both Western Powers had much to give to Russia, and they could help in the reparation of Russia's losses. I could not see twenty years ahead, but we had nevertheless made a treaty for twenty years. I thought Russia would concentrate on reconstruction for the next ten years. There would probably be changes: Communism had already been modified. I thought we should live in good relations with Russia, and if Great Britain and the United States acted together and maintained a strong air force they should be able to ensure a period of stability. Russia might even gain by this. She possessed vast undeveloped areas — for instance, in Siberia.

The Turkish Prime Minister observed that I had expressed the view that Russia might become imperialistic. This made it necessary for Turkey to be very prudent. I replied that there would be an international organisation to secure peace and security, which would be stronger than the League of Nations. I added that I was not afraid of Communism. Mr. Saracoglu remarked that he was looking for something more real. All Europe was full of Slavs and Communists. All the defeated countries would become Bolshevik and Slav if Germany was beaten. I said that things did not always turn out as bad as was expected; but if they did so it was better that Turkey should be strong and closely associated with the United Kingdom and the United States. If Russia, without any cause, were to attack Turkey the whole international organisation of which I had spoken would be applied on behalf of Turkey, and the guarantees after the present war would be much more severe, not only where Turkey was concerned, but in the case of all Europe. I would not be a friend of Russia if she imitated Germany. If she did so we should arrange the best possible combination against her, and I would not hesitate to say so to Stalin.

During these general political discussions military conversations were conducted by the C.I.G.S. and our and our other high commanders. The two main points to be considered were the provision of equipment for the Turkish forces, prior and subsequent to any political move by Turkey, and the preparation of plans for their reinforcement by British units in the event of their coming into the war. The results of these talks were embodied in a military agreement.

My parleys with Turkey were intended to prepare the way for her entry into the war in the autumn of 1943. That this did not take place after the collapse of Italy and with the further Russian advances against Germany north of the Black Sea was due to unfortunate events in the Aegean later in the year, which will be described in their proper place.

*   *   *

I flew back from Adana to Cairo, stopping at Cyprus on the way, and then on to Tripoli. It had been taken punctually by the Eighth Army on January 23. The port was found severely damaged. The entrance had been completely blocked by sunken ships, and the approaches lavishly sown with mines. This had been foreseen, and the first supply ship entered the harbour on February 2. A week later two thousand tons a day were being handled. Although the Eighth Army had still great distances to travel, its maintenance during the fifteen-hundred-mile advance from Alamein, crowned by the rapid opening up of Tripoli, was an administrative feat for which credit lay with General Lindsell in Cairo and General Robertson with the Eighth Army. At the end of the month the Eighth Army was joined by General

Leclerc, who had led a mixed force of Free French about 2500 strong fifteen hundred hundred miles across the Desert from French Equatorial Africa. Leclerc placed himself unreservedly under Montgomery's orders. He and his troops were to play a valuable part in the rest of the Tunisian campaign.

The Eighth Army crossed the frontier into Tunisia on February 4, thus completing the conquest of the Italian Empire by Great Britain. In accordance with the decisions taken at the Casablanca Conference, this Army now came under General Eisenhower, with General Alexander as his deputy in executive command of land operations. The reader may remember the directive I had given Alexander on leaving Cairo six months earlier.[1] He now sent me the following reply:

> Sir,
> The orders you gave me on August [10], 1942, have been fulfilled. His Majesty's enemies, together with their impedimenta, have been completely eliminated from Egypt, Cyrenaica, Libya, and Tripolitania. I now await your further instructions.

After two long and vivid days I set off with my party to visit Eisenhower and all the others at Algiers. Here the tension was acute. The murder of Darlan still imposed many precautions on all prominent figures. The Cabinet continued to show concern about my safety, and evidently wanted me home as soon as possible. This at least was complimentary. On Sunday night, February 7, 1943, we took off, and flew directly and safely home. This was my last flight in "Commando," which later perished with all hands, though with a different pilot and crew.

* * *

My first task on getting home was to make a full statement to the House of Commons on the Casablanca Conference, my tour of the Mediterranean, and on the general position. It took me more than two hours on February 11 to make my speech. But I was more tired by my journeying than I had realised at the time, and I must have caught a chill. A few days later a cold and sore throat obliged me to lie up. In the evening of the 16th, when I was alone with Mrs. Churchill, my temperature suddenly rose, and Lord Moran, who had been watching me, took a decided view and told me that I had inflammation of the base of a lung. His diagnosis led him to prescribe the drug called M and B. The next day elaborate photographs were taken and confirmed the diagnosis, and Dr. Geoffrey Marshall of Guy's Hospital was called

[1] See page 615.

in consultation. All my work had come to me hour by hour at the Annexe, and I had maintained my usual output though feeling far from well. But now I became aware of a marked reduction in the number of papers which reached me. When I protested the doctors, supported by my wife, argued that I ought to quit my work entirely. I would not agree to this. What should I have done all day? They then said I had pneumonia, to which I replied, "Well, surely you can deal with that. Don't you believe in your new drug?" Doctor Marshall said he called pneumonia "the old man's friend." "Why?" I asked, "Because it takes them off so quietly." I made a suitable reply, but we reached an agreement on the following lines. I was only to have the most important and interesting papers sent me, and to read a novel. I chose *Moll Flanders*, about which I had heard excellent accounts, but had not found time to test them. On this basis I passed the next week in fever and discomfort, and I sometimes felt very ill. There is a blank in my flow of minutes from the 19th to the 25th. Soon the President, General Smuts, and other friends who had heard about my illness sent repeated telegrams urging me to obey the doctor's orders, and I kept faithfully to my agreement. When I finished *Moll Flanders* I gave it to Doctor Marshall to cheer him up. The treatment was successful.

*　　　　*　　　　*

Stalin at this time sent me a film of the Stalingrad victory, with all its desperate fighting wonderfully portrayed, and this is the point at which to tell, all too briefly, the tale of the magnificent and decisive struggle of the Russian armies.

The German drive to the Caucasus had culminated and foundered during the summer and autumn of 1942. At first all had gone very much according to plan, though not quite so swiftly as had been hoped. The Southern Army group cleared the Russians from within the bend of the Lower Don. It was then divided into Army Group A, under List, and Army Group B, under Bock, and on July 23 Hitler had given them their tasks. Army Group A was to capture the entire eastern shore of the Black Sea and the adjacent oilfields, and Army Group B, having established a defensive flank along the River Don, was to advance on Stalingrad, "smash the enemy forces being assembled there, and occupy the city." The troops in front of Moscow would conduct holding operations, and Leningrad in the north would be captured in early September.

General von Kleist's First Panzer Army of fifteen divisions led the onrush to the Caucasus. Once across the Don they made much headway against little opposition. They reached the Maikop oilfields on August 9, to find them thoroughly destroyed. They failed to reach the

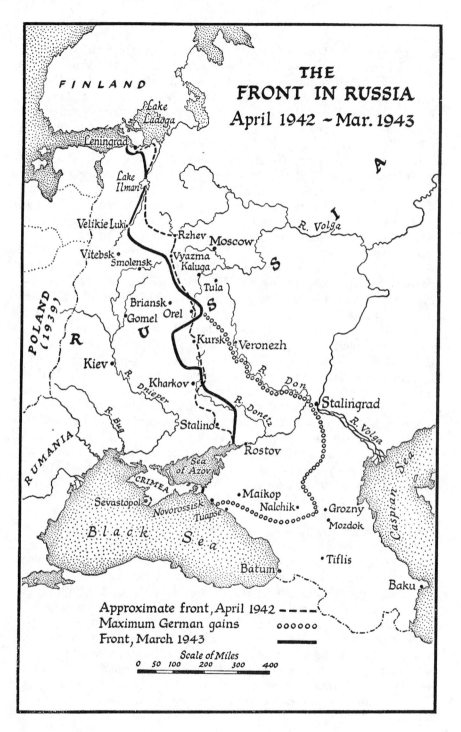

FINLAND

Lake
Ladoga

Leningrad

Lake
Ilman

Velikie Luki

Vitebsk

Smolensk

Rzhev   Moscow

Vyazma
Kaluga

Tula

Briansk

Gomel   Orel

Kursk

Veronezh

Kiev

Kharkov

Stalino

Rostov

Sea
of Azov

CRIMEA

Sevastopol

Novorossisk

Tuapse

Maikop

Nalchik

Grozny

Mozdok

Black   Sea

Batum

Tiflis

Baku

POLAND
(1939)

RUMANIA

R. Bug

R. Dnieper

R. Donetz

R. Don

Stalingrad

R. Volga

R. Volga

Caspian Sea

THE
FRONT IN RUSSIA
April 1942 ~ Mar. 1943

Approximate front, April 1942 – – – –
Maximum German gains   oooooo
Front, March 1943   ▬▬▬▬

Scale of Miles
0   50 100   200   300   400

Grozny oilfields. Those of Baku, the greatest of them all, were still three hundred miles away, and Hitler's orders to seize the whole of the Black Sea littoral could not be carried out. Reinforced by fresh troops sent down by railway along the western shore of the Caspian, the Russians everywhere held firm. Kleist, weakened by diversions for the Stalingrad effort, struggled on till November amid the Caucasian foothills. Winter then descended. His bolt was shot.

On the front of Army Group B worse than failure befell. The lure of Stalingrad fascinated Hitler; its very name was a challenge. The city was a considerable centre of industry and a strong point on the defensive flank protecting his main thrust to the Caucasus. It became a magnet drawing to itself the supreme effort of the German Army and Air Force. Resistance grew daily stiffer. It was not till September 15 that, after heavy fighting between the Don and the Volga, the outskirts of Stalingrad were reached. The battering-ram attacks of the next month made some progress at the cost of terrible slaughter. Nothing could overcome the Russians, fighting with passionate devotion amid the ruins of their city.

The German generals, long uneasy, had now good cause for anxiety. After three months of fighting, the main objectives of the campaign, the Caucasus, Stalingrad, and Leningrad, were still in Russian hands. Casualties had been very heavy and replacements insufficient. Hitler, instead of sending fresh contingents forward to replace losses, was forming them into new and untrained divisions. In military opinion it was high time to call a halt, but "The Carpet-eater" would not listen. At the end of September, Halder, Hitler's Chief of Staff, finally resisted his master, and was dismissed. Hitler scourged his armies on.

By mid-October the German position had markedly worsened. Army Group B was stretched over a front of seven hundred miles. General Paulus's Sixth Army had expended its effort, and now lay exhausted with its flanks thinly protected by allies of dubious quality. Winter was near, when the Russians would surely make their counterstroke. If the Don front could not be held the safety of the armies on the Caucasus front would be undermined. But Hitler would not countenance any suggestion of withdrawal. On November 19 the Russians delivered their long and valiantly prepared encircling assault, striking both north and south of Stalingrad upon the weakly defended German flanks. Four days later the Russian pincers met and the Sixth Army was trapped between the Don and the Volga. Paulus proposed to break out. Hitler ordered him to hold his ground. As the days passed, the Army was compressed into an ever-lessening space. On December 12, in bitter weather, the Germans made a desperate effort to break through the Russian cordon and relieve their besieged comrades. They failed.

Thereafter, though Paulus and his army held out for seven more terrible weeks, their doom was certain.

Great efforts were made to supply him from the air, but little got through, and at the expense of heavy losses in aircraft. The cold was intense; food and ammunition were scarce, and an outbreak of typhus added to the miseries of his men. On January 8 he rejected an ultimatum to surrender, and next day the last phase began with violent Russian attacks from the west. The Germans fought strongly, so that only five miles were gained in as many days. But at last they began to crack, and by the 17th the Russians were within ten miles of Stalingrad itself. Paulus threw into the fight every man who could bear arms, but it was no use. On January 22 the Russians surged forward again, until the Germans were thrown back on the outskirts of the city they had tried in vain to take. Here the remains of a once-great army were pinned in an oblong only four miles deep by eight long. Under intense artillery fire and air bombardment the survivors defended themselves in violent street fighting, but their plight was hopeless, and as the Russians pressed forward, exhausted units began to surrender wholesale. Paulus and his staff were captured, and on February 2 Marshal Voronov reported that all resistance had ceased and ninety thousand prisoners had been taken. These were the survivors of twenty-one German and one Rumanian divisions. Thus ended Hitler's prodigious effort to conquer Russia by force and destroy Communism by an equally odious form of totalitarian tyranny.

The spring of 1943 marked the turning point of the war on the Eastern Front. Even before Stalingrad the mounting Russian tide had swept the enemy back all along the line. The German army of the Caucasus was skilfully withdrawn, but the Russians pressed the enemy from the Don and back beyond the Donetz River, the starting line of Hitler's offensive of the previous summer. Farther north again the Germans lost ground, until they were more than two hundred and fifty miles from Moscow. The investment of Leningrad was broken. The Germans and their satellites suffered immense losses in men and material. The ground gained in the past year was taken from them. They were no longer superior to the Russians on land. In the air they had now to reckon with the growing power of the British and American Air Forces, operating both from Britain and in Africa.

*   *   *

Victory, however, made Stalin no more genial. If he could have come to Casablanca the three Allies might have worked out a common plan face to face. But this was not to be, and discussions were pursued by telegram. We told him of our military decisions, and on my return

THE GRAND ALLIANCE

home, with the President's authority, I had sent him an additional explanation of our plans, namely, to liberate Tunisia in April, capture Sicily, and push our preparations to the limit for crossing the Channel in August or September.

> . . . It is evident [he replied promptly] that, contrary to your previous calculations, the end of operations in Tunis is expected in April instead of February. I hardly need to tell you how disappointing is such a delay. . . . It is [also] evident from your message that the establishment of the Second Front, in particular in France, is envisaged only in August–September. It seems to me that the present position demands the greatest possible speeding up of the action contemplated — *i.e.*, of the opening of the Second Front in the West at a considerably earlier date than indicated. In order not to give the enemy any respite it is extremely important to deliver the blow from the West in the spring or in the early summer and not to postpone it until the second half of the year. . . .

And a month later (March 15):

> Fully realising the importance of Sicily, I must however point out that it cannot replace the Second Front in France. . . . I deem it my duty to warn you in the strongest possible manner how dangerous would be from the view-point of our common cause further delay in the opening of the Second Front in France. This is the reason why the uncertainty of your statements concerning the contemplated Anglo-American offensive across the Channel arouses grave anxiety in me, about which I feel I cannot be silent.

*　　　*　　　*

It was evident that the most effective aid which we could offer the Russians was the speedy clearing of the Axis forces from North Africa and the stepping up of the air war against Germany, but although the pace of our advance from the east had surpassed expectations, the Allied situation had for some time remained anxious. Malta was indeed re-victualled and rearmed and had again sprung into full activity. From our new bases in Algeria and Cyrenaica our naval and air forces ranged widely, protecting Allied shipping and taking heavy toll of enemy supplies and reinforcements. Besides blockading Tunis, where German air forces were still strong, we reached out to the ports on the Italian mainland. Palermo, Naples, and Spezia all felt the lash as our strength mounted, and R.A.F. bombers from home took over the at-

tack on Northern Italy. The Italian Fleet made no attempt to interfere. Apart from the presence of the British Fleet, the lack of oil was
serious. There were days when there was not one ton of fuel in all
Sicily for the escort vessels covering supplies to Tunis.

But all this could not disguise the fact that after the failure to conquer
Tunisia in December our initial blow was spent. Refusing to recognise
that he could not safeguard by sea or air even the short passage from
Sicily, Hitler ordered the creation of a new army to meet the impend-'
ing Allied attacks from both east and west. Rommel, promoted to
command all the Axis troops, concentrated two German armoured
divisions east of Faid to throw back the opposing U.S. Corps and
prevent them from coming down on his flank and rear while he was
engaged against the hard pressure of the Eighth Army. The attack
began on February 14. It had been mistakenly expected that the main
blow would come through Fondouk and not Faid. Consequently the
1st U.S. Armoured Division, under General Anderson's orders, was
much dispersed. On the 17th Kasserine, Feriana, and Sbeitla were in
German hands. Rommel then struck northward. A fierce fight ensued,
but by noon on the 22nd he began a general withdrawal in good order
and eventually our original line was re-established. But Rommel was
not yet finished. Four days later he began a series of strong attacks on
the front of the British Vth Corps. South of Medjez the enemy were
repulsed without significant gains; to the north they won several
miles, leaving the town itself in an awkward salient. Near the coast our
troops were forced back twenty miles, but they then held firm.

In the last week of February General Alexander took command of
the whole front. At the same time, in accordance with the Casablanca
agreement, Air Marshal Tedder assumed control of the Allied Air
Forces. The battle in Tunisia was now at its height. On March 6
Rommel made four major attacks on the advancing Eighth Army, using
all three of the German Panzers. Every one of them was beaten off
with heavy loss. This was probably Rommel's sharpest rebuff in all his
African exploits. Moreover, it was his last action there. Shortly afterwards he was invalided to Germany, and von Arnim succeeded him.

The Eighth Army then moved forward to close with the enemy's
main position, the Mareth Line. This was a highly organised twenty-
mile-long defense system constructed by the French before the war to
prevent Italian incursion into Tunisia. Now Italians were manning it
against the British! A fortnight was needed to prepare a deliberate
assault against such strongly held defences. The blow was struck during the third week in March, the enemy were outflanked, and on
April 7, after bitter and complicated fighting, a patrol of the 4th Indian Division met one from the U.S. IInd Corps. The American

greeting, "Hello, Limey," [2] although not understood, was accepted with the utmost cordiality. The two armies which had started nearly two thousand miles apart were now at last joined together. On the 18th a great enemy air convoy a hundred strong was set upon by our Spitfires and American Warhawks off Cape Bon. The convoy was scattered in confusion; over fifty were brought down. Next day South African Kittyhawks destroyed fifteen out of eighteen; and finally on April 22 a further thirty, including many laden with petrol, went flaming into the sea. This virtually ended Hitler's obstinate attempt, which Germany could ill afford. No more transport aircraft dared to fly by day. Their achievement had been great. In the four months December to March they had ferried more than 40,000 men and 14,000 tons of supplies to Africa.

On May 6 Alexander launched his culminating attack. The Allied Air Forces put forth a supreme effort, with 2500 sorties in the day. The Axis had been gradually worn down, and at this crisis could only make sixty sorties in reply. The climax was at hand. The relentless blockade by sea and air was fully established. Enemy movement over the sea was at a standstill, their air effort ended. The British IXth Corps made a clean break in the enemy front. The two armoured divisions passed through the infantry and reached Massicault, halfway to Tunis. Next day, May 7, they pressed on, the 7th Armoured Division entered Tunis, and then swerved north to join hands with the United States force. Resistance on the main American front had cracked at the same time, and their 9th Infantry Division reached Bizerta. Three German divisions were thus trapped between the Allied troops, and surrendered on May 9.

The 6th Armoured Division, followed by the 4th British and with the 1st Armoured on their right, drove east, through and beyond Tunis. They were held up by a hastily organised resistance at a defile by the sea a few miles east of the city, but their tanks splashed through along the beach, and at nightfall on May 10 reached Hammamet, on the east coast. Behind them the 4th Division swept round the Cape Bon peninsula, meeting no opposition. All the remaining enemy were caught in the net to the south.

> . . . I expect all organised resistance to collapse within the next forty-eight hours [cabled General Alexander on May 11] and final liquidation of whole Axis forces in the next two or three days. I calculate that prisoners up to date exceed 100,000, but this is not yet confirmed, and they are still coming in. Yesterday I saw a

---

[2] A name for British sailors in vogue in the United States Navy, arising from the use of lime juice on British ships in bygone days to prevent scurvy.

horse-drawn gig laden with Germans driving themselves to the prisoners' cage. As they passed we could not help laughing, and they laughed too. The whole affair was more like Derby Day . . .

Admiral Cunningham had made full preparation for the final collapse, and he ordered all available naval forces to patrol the straits and stop an Axis "Dunkirk" evacuation. The appropriate code name of this operation was "Retribution." On the 8th he signalled, "Sink, burn, and destroy. Let nothing pass." But only a few barges tried to escape, and nearly all were captured or sunk. On the 12th the encircling ring was closed. The enemy laid down their arms.

At 2.15 P.M. on May 13 Alexander signalled to me: "Sir: It is my duty to report that the Tunisian campaign is over. All enemy resistance has ceased. We are masters of the North African shores."

No one could doubt the magnitude of the victory of Tunis. It held its own with Stalingrad. Nearly a quarter of a million prisoners were taken. Very heavy loss of life had been inflicted on the enemy. One-third of their supply ships had been sunk. Africa was clear of our foes. One continent had been redeemed. In London there was, for the first time in the war, a real lifting of spirits. Parliament received the Ministers with regard and enthusiasm, and recorded its thanks in the warmest terms to the commanders. I had asked that the bells of all the churches should be rung. I was sorry not to hear their chimes, but I had more important work to do on the other side of the Atlantic.

22

Italy the Goal

THE REASONS WHICH LED ME to hasten to Washington, once the decision in Africa was certain, were serious. What should we do with our victory? Were its fruits to be gathered only in the Tunisian tip, or should we drive Italy out of the war and bring Turkey in on our side? These were fateful questions, which could only be answered by a personal conference with the President. Second only to these were the plans for action in the Indian theatre. I was conscious of serious divergences beneath the surface which, if not adjusted, would lead to grave difficulties and feeble action during the rest of the year. I was resolved to have a conference on the highest possible level.

The doctors did not want me to fly at the great height required in a bomber, and it was therefore decided to go by sea. We left London on the night of May 4, and went aboard the *Queen Mary* in the Clyde on the following day. The ship had been admirably fitted up to meet all our needs. The whole delegation was accommodated on the main deck, which was sealed off from the rest of the ship. Offices, conference rooms, and of course the Map Room, stood ready for immediate use. From the moment we got on board our work went forward ceaselessly. The conference, which I had christened "Trident," was to last at least a fortnight, and was intended to cover every aspect of the war. Our party had therefore to be a large one. The "regulars" were in full force: the Chiefs of Staff, with a goodly number of Staff Officers; Lord Leathers, with senior officials of the Ministry of War Transport; and Ismay, with members of my Defence Office. The Commanders-in-Chief in India, Field Marshal Wavell, Admiral Somerville, and Air Chief Marshal Peirse, were also with us. I had summoned them because I was sure that our American friends would be very anxious that we should do everything possible — and even impossible — in the way of immediate operations from India. The conference must hear at first hand the views of the men who would have to do whatever task was chosen.

There was much to be settled among ourselves before we reached Washington, and now we were all under one deck. The Joint Planning and Intelligence Staffs were in almost continuous session. The Chiefs of Staff met daily, and sometimes twice a day. I adhered to my usual practice of giving them my thoughts each morning in the shape of minutes and directives, and I generally had a discussion with them each afternoon or evening. These processes of probing, sifting, and arguing continued throughout the voyage, and grave decisions were reached in measured steps.

*   *   *

We had to think about all the theatres at once. Upon the operations in Europe, following the victory in Africa, we were in complete agreement. It had been decided at Casablanca to attack Sicily, and all preparations were far advanced. The British Chiefs of Staff were convinced that an attack upon the mainland of Italy should follow, or even overlap, the capture of Sicily. They proposed the seizure of a bridgehead on the toe of Italy, to be followed by a further assault on the heel as prelude to an advance on Bari and Naples. A paper setting out these views and the arguments which led up to them was prepared on board ship and handed to the American Chiefs of Staff as a basis for discussion on our arrival in Washington.

We anticipated more difficulties in reaching agreement with our American friends over the second great sphere of British military action, namely, the operations from India. Many plans had been set forth on paper, but we had little to show in fact. The President and his circle still cherished exaggerated ideas of the military power which China could exert if given sufficient arms and equipment. They also feared unduly the imminence of a Chinese collapse if support were not forthcoming. I disliked thoroughly the idea of reconquering Burma by an advance along the miserable communications in Assam. I hated jungles — which go to the winner anyway — and thought in terms of air power, sea power, amphibious operations, and key points. It was however an essential to all our great business that our friends should not feel we had been slack in pulling our weight and be convinced that we were ready to make the utmost exertions to meet their wishes. What happened in Burma will be recounted later.

On May 11 we arrived off Staten Island. Harry Hopkins was there to meet us, and we immediately entrained for Washington. The President was on the platform to greet me, and whisked me off to my old rooms at the White House. The next afternoon, May 12, at 2.30

P.M., we all met in his oval study to survey and lay out our work at the conference.

Mr. Roosevelt asked me to open the discussion. According to the record, the essence of my thought was as follows:

". . . We should never forget that there were 185 German divisions on the Russian front. We had destroyed the German army in Africa, but soon we should not be in contact with them anywhere. The Russian effort was prodigious, and placed us in their debt. The best way of taking the weight off the Russian front in 1943 would be to get, or knock, Italy out of the war, thus forcing the Germans to send a large number of troops to hold down the Balkans. . . . We had a large army and the Metropolitan Fighter Air Force in Great Britain. We had our finest and most experienced troops in the Mediterranean. The British alone had thirteen divisions in Northwest Africa. Supposing that Sicily was completed by the end of August, what should these troops do between that time and the date [in 1944], seven or eight months later, when the cross-Channel operation might first be mounted? They could not possibly stand idle, and so long a period of apparent inaction would have a serious effect on Russia, who was bearing such a disproportionate weight."

Mr. Roosevelt agreed that to relieve Russia we must engage the Germans. But he questioned the occupation of Italy, which would release German troops to fight elsewhere. He thought that the best way of forcing Germany to fight would be to launch an operation across the Channel.

I replied that as we were now agreed we could not do this till 1944, it seemed imperative to use our great armies to attack Italy. I did not think that an occupation of the whole peninsula would be necessary. If Italy collapsed the United Nations would occupy the ports and airfields needed for further operations into the Balkans and Southern Europe. An Italian Government could control the country, subject to Allied supervision. All these grave issues were now to be thrashed out by our combined Staffs and their experts.

At first the differences seemed insuperable and it looked like a hopeless breach. During this period leakages from high American officers were made to Democratic and Republican senators, leading to a debate in the Senate. By patience and perserverance our difficulties were gradually overcome. The fact that the President and I were living side by side seeing each other at all hours, that we were known to be in close agreement, and that the President intended to decide himself on the ultimate issues — all this, together with the priceless work of Hopkins, exercised throughout a mollifying and also a dominating influence on the course of Staff discussions. After a serious crisis of opinions, side by side with the most agreeable personal relations between the professional men, an almost complete agreement was reached about invading Sicily.

But although so much had gone well, I was extremely concerned that no definite recommendations had been made by the Combined Staffs to follow up the conquest of Sicily by the invasion of Italy. I knew that the American Staff's mind had been turned to Sardinia. They thought that this should be the sole remaining objective for the mighty forces which were gathered in the Mediterranean during the whole of the rest of 1943. On every ground, military and political, I deplored this prospect. The Russians were fighting every day on their enormous front, and their blood flowed in a torrent. Were we then to keep over a million and a half fine troops, and all their terrific air and naval power, idle for nearly a year?

The President had not seemed ready to press his advisers to become more precise about invading Italy, but as this was the main purpose for which I had crossed the Atlantic I could not let the matter rest. Hopkins said to me privately, "If you wish to carry your point you will have to stay here another week, and even then there is no certainty." I was deeply distressed at this, and on May 25 appealed personally to the President to let General Marshall come to Algiers with me. I explained to the conference that I should feel awkward in discussing these matters with General Eisenhower without the presence of a United States representative on the highest level. If decisions were taken it might subsequently be thought that I had exerted an undue influence. I was accordingly very gratified to hear that General Marshall would accompany me, and I was sure that it would now be possible to arrange for a report to be sent back to the Combined Chiefs of Staff for their consideration.

Early the next day, General Marshall, the C.I.G.S., Ismay, and the rest of my party took off from the Potomac River in a flying boat. We had some very agreeable talks during the long flight, and took advantage of our leisure to clear away some accumulation of papers. As we approached Gibraltar we looked around for our escort. There was no escort. Everyone's attention was attracted by an unknown aircraft, which we thought at first was taking an interest in us. As it came no closer we concluded it was a Spaniard; but they all seemed quite concerned about it till it disappeared. On alighting, at about 5 P.M., we were met by the Governor. It was too late to continue our journey to Algiers that night, and he conveyed us to the Convent, where he resides, the nuns having been removed two centuries ago.

We did not leave Gibraltar for Algiers until the following afternoon. There was therefore an opportunity to show General Marshall the Rock, and we all made a few hours' pilgrimage, and inspected the new distillery which assured the fortress a permanent supply of fresh water, and various important guns, some hospitals, and a large number of troops. I finally went below to see the Governor's special pet, the new Rock gallery, cut deep in the rock, with its battery of eight quick-firing guns commanding the isthmus and the neutral ground between Britain and Spain. An immense amount of work had been put into this, and it certainly seemed, as we walked along it, that whatever perils Gibraltar might have to fear attack from the mainland was no longer one of them. The Governor's pride in his achievement was shared by his British visitors. It was not until we said good-bye upon the flying boat that General Marshall somewhat hesitatingly observed, "I admired your gallery, but we had one like it at Corregidor. The Japanese fired their artillery at the rock several hundred feet above it, and in two or three days blocked it off with an immense bank of rubble." I was grateful to him for his warning, but the Governor seemed thunderstruck. All the smiles vanished from his face.

We flew off in the early afternoon with a dozen Beaufighters circling far above us, and in the evening light reached the Algiers airfield, where Generals Eisenhower and Bedell Smith, Admiral Andrew Cunningham, General Alexander, and other friends were waiting for us. I motored straight to Admiral Cunningham's villa, next door to General Eisenhower, which he placed at my disposal.

*   *   *

I have no more pleasant memories of the war than the eight days in Algiers and Tunis. I telegraphed to Eden to come out and join me so as to make sure we saw eye to eye on the meeting we had arranged between Giraud and de Gaulle, and all our other business.

I was determined to obtain before leaving Africa the decision to invade Italy should Sicily be taken. Brooke and I imparted our views to General Alexander, Admiral Andrew Cunningham, and Air Marshal Tedder, and later to Montgomery. All these leading figures in the recent battles were inclined to action on the greatest scale, and saw in the conquest of Italy the natural fruition of our whole series of victories from Alamein onward. We had however to procure the agreement of our great Ally. Eisenhower was very reserved. He listened to all our arguments, and I am sure agreed with their purpose. But Marshall remained up till almost the last moment silent or cryptic.

The circumstances of our meeting were favourable to the British. We had three times as many troops, four times as many warships, and almost as many aeroplanes available for actual operations as the Americans. We had since Alamein, not to speak of the earlier years, lost in the Mediterranean eight times as many men and three times as many ships as our Allies. But what ensured for these potent facts the fairest and most attentive consideration with the American leaders was that notwithstanding our immense preponderance of strength we had continued to accept General Eisenhower's Supreme Command and to preserve for the whole campaign the character of a United States operation. The American chiefs do not like to be outdone in generosity. No people respond more spontaneously to fair play. If you treat Americans well they always want to treat you better. Nevertheless I consider that the argument which convinced the Americans was on its merits overwhelming.

We held our first meeting at General Eisenhower's villa in Algiers at five o'clock on May 29. General Eisenhower, as our host, presided, and had with him Marshall and Bedell Smith, as his two principals. I sat opposite to him, with Brooke, Alexander, Cunningham, Tedder, Ismay, and some others. Marshall said the United States Chiefs of Staff felt that no decision could be made about invading Italy until the result of the attack on Sicily and the situation in Russia were known. The logical approach would be to set up two forces, each with its own staff, in separate places. One force would train for an operation against Sardinia and Corsica, and the other for an operation on the mainland of Italy. When the situation was sufficiently clear to enable a choice to be made, the necessary air forces, landing craft, etc., would be made over to the force charged with implementing the selected plan. Ike said at once that if Sicily was polished off easily he would be willing to go straight to Italy. General Alexander agreed.

The G.I.G.S. then made his general statement. A hard struggle between the Russians and the Germans was imminent, and we should do all in our power to help. We should make the Germans disperse their strength. They were already widely stretched, and could not reduce their forces either in Russia or in France. The place where they could most conveniently do this was Italy. If the foot of Italy were found to be packed with troops we should try elsewhere. If Italy were knocked out of the war Germany would have to replace the twenty-six Italian divisions in the Balkans, and reinforce the Brenner Pass, the Riviera, and the Spanish and Italian frontiers. This dispersal was just what we needed for crossing the Channel, and we should do everything in our power to increase it.

Eisenhower then declared that the discussion had seemed to simplify his problem. If Sicily were to succeed, say within a week, he would at once cross the Straits of Messina and establish a bridgehead. I expressed a personal veiw that Sicily would be finished by August 15. If so, and if the strain had not been too heavy, we should at once go for the toe of Italy, provided that not too many German divisions had been moved there. The Balkans represented a greater danger to Germany than the loss of Italy, as Turkey might react to our advantage.

Brooke thereupon set out our whole Mediterranean strength. Deducting seven divisions to be sent home for the cross-Channel operation and two to cover British commitments to Turkey, there would be twenty-seven Allied divisions available in the Mediterranean area. With such forces in our hands it would be bad indeed if nothing happened between August or September and the following May.

*   *   *

Although much lay in the balance, I was well satisfied with this opening discussion. The desire of all the leaders to go forward on the boldest lines was clear, and I felt myself that the reservations made on account of the unknowable would be settled by events in accordance with my hopes.

We met again on the afternoon of May 31. Mr. Eden arrived in time to be present. I tried to clinch matters, and said my heart lay in an invasion of Southern Italy, but the fortunes of battle might necessitate a different course. At any rate, the alternative between Southern Italy and Sardinia involved the difference between a glorious campaign and a mere convenience. General Marshall was in no way hostile to these ideas, but did not wish for a clear-cut decision to be taken at this moment. It would be better to decide what to do after we had started the attack on Sicily. He felt it would be necessary to know something of the German reactions in order to determine whether there would be real resistance in Southern Italy; whether the Germans would withdraw to the Po, and, for example, whether they could organise and handle the Italians with any finesse; what preparations had been made in Sardinia, Corsica, or in the Balkans; what readjustments they would make on the Russian front. There were two or three different ways in which Italy might fall; a great deal could happen between now and July. He, General Eisenhower, and the Combined Chiefs of Staff were fully, aware of my feelings about invading Italy, but their only desire was to select the "Post-Sicily" alternative, which would give the best results.

I said that I very passionately wanted to see Italy out of the way and Rome in our possession. I could not endure to see a great army standing idle when it might be engaged in striking Italy out of the war. Parliament and the people would become impatient if the Army were not active, and I was willing to take almost desperate steps in order to prevent such a calamity.

*   *   *

An incident now occurred which, as it relates to matters which have become the subject of misunderstandings and controversy after the war, must be related. Mr. Eden, at my request, commented on the Turkish situation, and said that knocking Italy out of the war would go a long way towards bringing the Turks in. They would become much more friendly "when our troops had reached the Balkan area." Eden and I were in full agreement on war policy, but I feared that the turn of his pharse might mislead our American friends. The record states, "The Prime Minister intervened to observe emphatically that he was not advocating sending an army into the Balkans now or in the near future." Mr. Eden agreed that it would not be necessary to put an army into the Balkans, since the Turks would begin to show favourable reactions as soon as we were able to constitute an immediate threat to the Balkans.

Before we separated I asked General Alexander to give his view. He did so in an extremely impressive speech. Securing a bridgehead on the Italian mainland should be part of the plan. It would be impossible for us to win a great victory unless we could exploit it by moving ahead, preferably up into Italy. All this however would be clarified as the Sicily operation moved along. It was not impossible, although it seemed unlikely, that the toe of Italy would be so strongly held as to require a complete restaging of our operations, and we should be ready to keep moving, with no stop at all, once the attack on Sicily started. Modern warfare allowed us to forge ahead very rapidly, with radio controlling troops at a great distance and with air providing protection and support over a wide area. The going might become more difficult as we moved up the Italian mainland, but this was no argument against going as far as we could on the momentum of the Sicily drive. In war the incredible often occurred. A few months before it would have been impossible for him to believe what had actually happened to Rommel and his Afrika Korps. A few weeks since he would have found it difficult to believe that three hundred thousand Germans would collapse in a week. The enemy air forces had been swept out of the skies so completely that we could have a parade, if we chose, of all our North Africa forces on one field in Tunisia without any danger from enemy aircraft.

He was at once supported by Admiral Cunningham, who said that if all went well in Sicily we should go directly across the Straits. Eisenhower concluded the meeting by expressing appreciation of the journey which General Marshall and I had made to clarify for him what the Combined Chiefs of Staff had done. He understood it was his responsibility to get information regarding the early phases of the invasion of Sicily and forward them to the Combined Chiefs of Staff in time for them to decide on the plan which would follow, without a break or a stop. He would send not only information but also strong recommendations, based upon the conditions of the moment. He hoped that his three top commanders (Alexander, Cunningham, and Tedder) would have an opportunity to comment more formally on these matters, although he agreed completely with what they had said thus far.

* * *

On the two following days we travelled by plane and car to some beautiful places rendered historic by the battles of a month before. General Marshall went on a brief American tour of his own, and then travelled with General Alexander and myself, meeting all the commanders and seeing stirring sights of troops. The sense of victory was in the air. The whole of North Africa was cleared of the enemy. A quarter of a million prisoners were cooped in our cages. Everyone was very proud and delighted. There is no doubt that people like winning very much, I addressed many thousand soldiers at Carthage in the ruins of an immense amphitheatre. Certainly the hour and setting lent themselves to oratory. I have no idea what I said, but the whole audience clapped and cheered as doubtless their predecessors of two thousand years ago had done as they watched gladiatorial combats.

*   *   *

I felt that great advances had been made in our discussions and that everybody wanted to go for Italy. I therefore, in summing up at our last meeting on June 3, stated the conclusions in a most moderate form and paid my tribute to General Eisenhower.

Eden and I flew home together by Gibraltar. As my presence in North Africa had been fully reported, the Germans were exceptionally vigilant, and this led to a tragedy which much distressed me. The regular commercial aircraft was about to start from the Lisbon airfield when a thick-set man smoking a cigar walked up and was thought to be a passenger on it. The German agents therefore signalled that I was on board. Although these passenger planes had plied unmolested for many months between Portugal and England, a German war plane was instantly ordered out, and the defenceless aircraft was ruthlessly shot down. Thirteen passengers perished, and among them the well-known British actor Leslie Howard, whose grace and gifts are still preserved for us by the records of the many delightful films in which he took part. The brutality of the Germans was only matched by the stupidity of their agents. It is difficult to understand how anyone could imagine that with all the resources of Great Britain at my disposal I should have booked a passage in an unarmed and unescorted plane from Lisbon and flown home in broad daylight. We of course made a wide loop out by night from Gibraltar into the ocean, and arrived home without incident. It was a painful shock to me to learn what had happened to others in the inscrutable workings of Fate.

BOOK IV

Triumph
and Tragedy

1943–1945

*The overwhelming victory of the Grand Alliance
has failed so far to bring general peace to our
anxious world.*

☆

I

The Capture of Sicily and the Fall of Mussolini

W E H A V E N O W reached the turning point of the Second World War. The entry of the United States into the struggle after the Japanese assault on Pearl Harbour made it certain that the cause of Freedom would not be cast away. The aggressors, both in Europe and Asia, had been driven to the defensive. Stalingrad in February 1943 marked the turn of the tide in Russia. By May all German and Italian forces in the African continent had been killed or captured. The American victories in the Coral Sea and at Midway Island a year before had stopped Japanese expansions in the Pacific Ocean. Australia and New Zealand were freed from the threat of invasion, and the leaders of Japan were already conscious that their onslaught had passed its zenith. Hitler had still to pay the full penalty of his fatal error in trying to conquer Russia by invasion. He had still to squander the immense remaining strength of Germany in many theatres not vital to the main result. Soon the German nation was to be alone in Europe, surrounded by an infuriated world in arms.

But between survival and victory there are many stages. Over two years of intense and bloody fighting lay before us all. Henceforward however the danger was not Destruction, but Stalemate. The Americans' armies had to mature and their vast construction of shipping to become effective before the full power of the Great Republic could be hurled into the struggle, and the Western Allies could never strike home at Hitler's Europe, and thus bring the war to a decisive end, unless another major favourable change came to pass. Anglo-American "maritime power," a modern term expressing the combined strength of naval and air forces properly woven together, became supreme on and under the surface of the seas and oceans during 1943. Without this no am-

phibious operations on the enormous scale required to liberate Europe
would have been possible. Soviet Russia would have been left to face
Hitler's whole remaining strength while most of Europe lay in his
grip.

*   *   *

The singlehanded British struggle against the U-boats, the magnetic
mines, and the surface raiders in the first two and a half years of the
war has already been described. The long-awaited supreme event of
the American alliance which arose from the Japanese attack on Pearl
Harbour seemed at first to have increased our perils at sea. In 1940 and
1941 we lost four million tons of merchant shipping a year. In 1942,
after the United States was our Ally, this figure nearly doubled, and
the U-boats sank ships faster than the Allies could build them. During
1943, thanks to the immense shipbuilding programme of the United
States, the new tonnage at last surpassed losses at sea from all causes,
and the second quarter saw, for the first time, U-boat losses exceed their
rate of replacement. The time was presently to come when more U-boats
would be sunk in the Atlantic than merchant ships. But before this
lay a long and bitter conflict.

The Battle of the Atlantic was the dominating factor all through the
war. Never for one moment could we forget that everything happening
elsewhere, on land, at sea, or in the air, depended ultimately on its
outcome, and amid all other cares we viewed its changing fortunes day
by day with hope or apprehension. The tale of hard and unremitting
toil, often under conditions of acute discomfort and frustration and
always in the presence of unseen danger, is lighted by incident and
drama. But for the individual sailor or airman there were few moments
of exhilarating action to break the monotony of an endless succession
of anxious, uneventful days. Vigilance could never be relaxed. Dire
crisis might at any moment flash upon the scene with brilliant fortune
or glare with mortal tragedy. Many gallant actions and incredible feats
of endurance are recorded, but the deeds of those who perished will
never be known. Our merchant seamen displayed their highest qualities,
and the brotherhood of the sea was never more strikingly shown than in
their determination to defeat the U-boat.

In April 1943 we could see the balance turn. The U-boat packs were
kept underwater and harried continually, while the air and surface
escort of the convoys coped with the attackers. We were now strong
enough to form independent flotilla groups to act like cavalry divisions,
apart from all escort duties. This I had long desired to see. Two hundred and thirty-five U-boats, the greatest number the Germans ever
achieved, were in action. But their crews were beginning to waver.

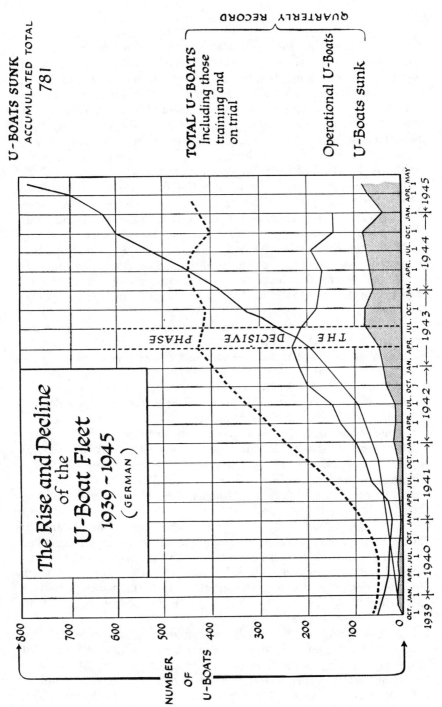

The Rise and Decline of the U-Boat Fleet
1939 – 1945
(GERMAN)

U-BOATS SUNK
ACCUMULATED TOTAL
781

QUARTERLY RECORD

TOTAL U-BOATS
Including those training and on trial

Operational U-Boats

U-Boats sunk

THE DECISIVE PHASE

NUMBER OF U-BOATS

800
700
600
500
400
300
200
100
0

1939 ← OCT. | JAN. APR. JUL. OCT. ← 1940 → | JAN. APR. JUL. OCT. ← 1941 → | JAN. APR. JUL. OCT. ← 1942 → | JAN. APR. JUL. OCT. ← 1943 → | JAN. APR. JUL. OCT. ← 1944 → | JAN. APR. MAY ← 1945

They could never feel safe. Their attacks, even when conditions were favourable, were no longer pressed home, and our Atlantic losses fell by nearly 300,000 tons. In May alone forty U-boats perished in the ocean. The German Admiralty watched their charts with strained attention, and at the end of the month Admiral Doenitz recalled the remnants of his fleet to rest or to fight in less hazardous waters. By June the sinkings fell to the lowest figure since the United States had entered the war. The convoys came through intact, the supply line was safe, the decisive battle had been fought and won.

Our armies could now be launched across the sea against the underbelly of Hitler's Europe. The extirpation of Axis power in North Africa opened to our convoys the direct route to Egypt, India, and Australia, protected from Gibraltar to Suez by sea and air forces working from the newly won bases along the route. The long haul round the Cape, which had cost us so dear in time, effort, and tonnage, would soon be ended. The saving of an average of forty-five days for each convoy to the Middle East increased magnificently at one stroke the fertility of our shipping.

As the defeat of the U-boats affected all subsequent events we must here carry the story forward. For a time they dispersed over the remote wastes of the South Atlantic and Indian Oceans, where our defences were relatively weak but we presented fewer targets. Our air offensive in the Bay of Biscay continued to gather strength. In July thirty-seven U-boats were sunk, mostly by air attack, and of these nearly half perished in the Bay. In the last three months of the year fifty-three were destroyed, while we lost only forty-seven merchant ships.

Throughout a stormy autumn the U-boats struggled vainly, and with small results, to retrieve their ascendancy in the North Atlantic. Although in the face of the harsh facts Admiral Doenitz was forced to recoil, he continued to maintain as many U-boats at sea as ever. But their attack was blunted and they seldom tried to cut through our defences. He did not however despair. "The enemy," he declared in January 1944, "has succeeded in gaining the advantage in defence. The day will come when I shall offer Churchill a first-rate submarine war. The submarine weapon has not been broken by the setbacks of 1943. On the contrary, it has become stronger. In 1944, which will be a successful but a hard year, we shall smash Britain's supply [line] with a new submarine weapon."

This confidence was not wholly unfounded. A gigantic effort was being made in Germany to develop a new type of U-boat which could move more quickly underwater and travel much farther. At the same time many of the older boats were withdrawn so that they could be fitted with the "Schnorkel" and work in British coastal waters. This

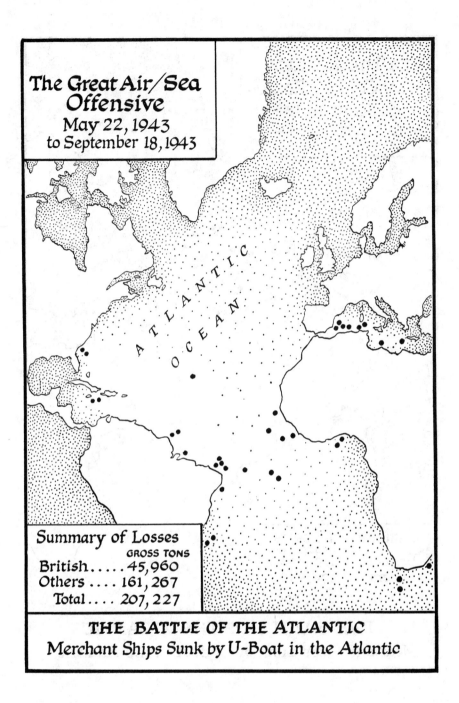

The Great Air/Sea
Offensive
May 22, 1943
to September 18, 1943

A T L A N T I C
O C E A N

Summary of Losses
GROSS TONS
British..... 45,960
Others 161,267
Total 207,227

THE BATTLE OF THE ATLANTIC
Merchant Ships Sunk by U-Boat in the Atlantic

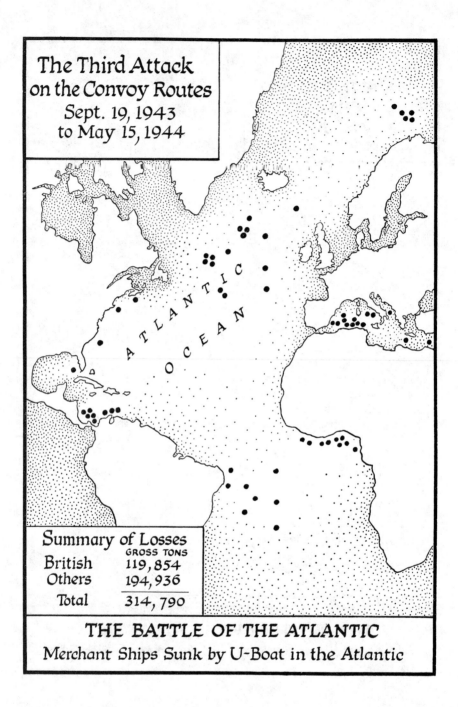

The Third Attack
on the Convoy Routes
Sept. 19, 1943
to May 15, 1944

ATLANTIC OCEAN

Summary of Losses

	GROSS TONS
British	119,854
Others	194,936
Total	314,790

THE BATTLE OF THE ATLANTIC
Merchant Ships Sunk by U-Boat in the Atlantic

new device enabled them to recharge their batteries while submerged with only a small tube for the intake of air remaining above the surface. Their chances of eluding detection from the air were thus improved, and it soon became evident that the Schnorkel-fitted boats were intended to dispute the passage of the English Channel whenever the Allied invasion was launched. What happened will be narrated in due course. It is high time to return to the Mediterranean scene and the month of July 1943.

*   *   *

General Eisenhower considered that Sicily should only be attacked if our purpose was to clear the Mediterranean sea-route. If our real aim was to invade and defeat Italy he thought that our proper initial objectives were Sardinia and Corsica, "since these islands lie on the flank of the long Italian boot and would force a very much greater dispersion of enemy strength in Italy than the mere occupation of Sicily, which lies off the mountainous toe of the peninsula." [1] This was no doubt a military opinion of high authority, although one I could not share. But political forces play their part, and the capture of Sicily and the direct invasion of Italy were to bring about results of a far more swift and far-reaching character.

"Husky," as the capture of Sicily was called in our code names, was an undertaking of the first magnitude. Although eclipsed by events in Normandy, its importance and its difficulties should not be underrated. The landing was based on the experience gained in North Africa, and those who planned "Overlord" learned much from "Husky." In the initial assault nearly 3000 ships and landing craft took part, carrying between them 160,000 men, 14,000 vehicles, 600 tanks, and 1800 guns. These forces had to be collected, trained, equipped, and eventually embarked, with all the vast impedimenta of amphibious warfare, at widely dispersed bases in the Mediterranean, in Great Britain, and in the United States. In spite of anxieties all went forward smoothly and proved a remarkable example of joint Staff work. For reasons of policy we had hitherto yielded the command and direction of the campaign in North Africa to the United States. But now we had entered upon a new stage — the invasion of Sicily, and what should follow from it. It was agreed that action against Italy should be decided in the light of the fighting in Sicily. As the Americans became more attracted to this larger adventure, instead of being content for the rest of the year with Sardinia, and while the prospects of another joint campaign unfolded, I felt it necessary that the British should be at least equal partners with our Allies. The proportions of the armies

[1] Dwight D. Eisenhower, *Crusade in Europe*, p. 159.

available in July were: British, eight divisions; United States, six. Air, the United States, 55 per cent; British, 45 per cent. Naval, 80 per cent British. Besides all this there remained the considerable British armies in the Middle East and in the Eastern Mediterranean, including Libya, which were independently commanded by General Maitland Wilson, from the British headquarters at Cairo. It did not seem too much in these circumstances that we should have at least an equal share of the High Command. And this was willingly conceded by our loyal comrades. We were moreover given the direct conduct of the fighting. Alexander was to command the Fifteenth Army Group, consisting of the Seventh United States Army under General Patton and the Eighth British Army under Montgomery. Air Chief Marshal Tedder commanded the Allied Air Force, and Admiral Cunningham the Allied naval forces. The whole was under the overall command of General Eisenhower.

Intense air attack upon the island began on July 3 with the bombing of airfields both there and in Sardinia, which made many unusable. The enemy fighters were thrown onto the defensive, and their long-range bombers forced to withdraw their bases to the Italian mainland. Four of the five train ferries operating across the Straits of Messina were sunk. By the time our convoys were approaching the island air superiority was firmly established, and Axis warships and aircraft made no serious effort to interfere with the sea-borne assault. By our cover plans the enemy were kept in doubt until the last moment where our stroke would fall. Our naval movements and military preparations in Egypt suggested an expedition to Greece. Since the fall of Tunis they had sent more planes to the Mediterranean, but the additional squadrons had gone, not to Sicily, but to the Eastern Mediterranean, Northwest Italy, and Sardinia. July 10 was the appointed day. On the morning of July 9 the great armadas from east and west were converging south of Malta, and it was time for all to steam for the beaches of Sicily. On my way to Chequers, where I was to await the result, I spent an hour in the Admiralty War Room. The map covered an entire wall, and showed the enormous convoys, escorts, and supporting detachments moving towards their assault beaches. This was the greatest amphibious operation so far attempted in history. But all depended on the weather.

*   *   *

The morning was fine, but by noon a fresh and unseasonable northwest wind sprang up. During the afternoon the wind increased, and by evening there was a heavy swell, which would make landings hazardous, particularly on the western beaches in the American sector. The landing-craft convoys plunging northward from Malta and from many African ports between Bizerta and Benghazi were having a rough voyage.

Arrangements had been made for postponing the landing in case of necessity, but a decision would have to be taken not later than midday. Watching anxiously from the Admiralty, the First Sea Lord inquired by signal about the weather conditions. Admiral Cunningham replied at 8 P.M., "Weather not favourable, but operation proceeding." "It was," he says, "manifestly too late for postponement, but considerable anxiety was felt, particularly for the small-craft convoys making up against the sea." They were indeed much delayed and became scattered. Many ships arrived late, but fortunately no great harm resulted. "The wind," says Cunningham, "mercifully eased during the night, and by the morning of the 10th had ceased, leaving only a tiresome swell and surf on the western beaches."

The bad weather helped to give us surprise. Admiral Cunningham continues: "The very efficient cover plan and deceptive routing of convoys played their part. In addition the vigilance of the enemy was undoubtedly relaxed owing to the unfavourable phase of the moon. Finally came this wind, dangerously close at the time to making some, if not all, of the landings impracticable. These apparently unfavourable factors had actually the effect of making the weary Italians, who had been alert for many nights, turn thankfully in their beds, saying, 'Tonight at any rate they can't come.' BUT THEY CAME."

The air-borne forces met hard fortune. More than one-third of the gliders carrying our 1st Air Landing Brigade were cast off too early by their American towing aircraft and many of the men they carried were drowned. The rest were scattered over southeastern Sicily, and only twelve gliders arrived at the important bridge which was their aim. Out of eight officers and sixty-five men who seized and held it until help came twelve hours later only nineteen survived. This was a forlorn feat of arms. On the American front the air landings were also too widely dispersed, but many small parties creating damage and confusion inland worried the Italian coastal divisions.

The sea-borne landings, under continuous fighter protection, were everywhere highly successful. Twelve airfields were soon in our hands, and by July 18 there were only twenty-five servicable German aircraft in the island. Eleven hundred planes, more than half of them German, were left behind destroyed or damaged. The enemy, once they had recovered from the initial surprise, had fought stubbornly. The difficulties of the ground were great. The roads were narrow, and cross-country movement was often impossible except for men on foot. On the Eighth Army front the towering mass of Mount Etna blocked the way, and enabled the enemy to watch our moves. As they lay on the low ground of the Catania plain, malaria ran riot among our men. Nevertheless, once we were safely ashore and our air forces were operating from captured airfields the issue was never in doubt. Contrary to our earlier hopes, the bulk of the Germans successfully withdrew across the Straits of Messina, but after thirty-eight days of fighting General Alexander telegraphed: "By 10.00 A.M. this morning, August 17, 1943, the last German soldier was flung out of Sicily and the whole island is now in our hands."

Our next strategic move was still in suspense. Should we cross the Straits of Messina and seize the toe of Italy, should we seize the heel at Taranto, or should we land higher up the west coast, in the Gulf of Salerno, and capture Naples? Or, again, must we restrict ourselves to the occupation of Sardinia? The progress now achieved clarified the scene. On July 19 a strong force of American bombers had attacked the railway yards and airport at Rome. Havoc was wrought and the shock was severe. The speedy collapse of Italy became probable. The Americans, however, held that none of the operations elsewhere, especially "Overlord," should be prejudiced by more vigorous action in the Mediterranean. This reservation was to cause keen anxiety during the landing at Salerno. While somewhat sharp discussions were in progress the scene had been completely transformed by the fall of Mussolini.

The Duce had now to bear the brunt of the military disasters into which he had, after so many years of rule, led his country. He had exercised almost absolute control and could not cast the burden on the Monarchy, Parliamentary institutions, the Fascist Party, or the General Staff. All fell on him. Now that the feeling that the war was lost spread throughout well-informed circles in Italy the blame fell upon the man who had so imperiously thrust the nation on to the wrong and the losing side. These convictions formed and spread widely during the early months of 1943. The lonely dictator sat at the summit of power, while military defeat and Italian slaughter in Russia, Tunis, and Sicily were the evident prelude to direct invasion.

In vain he made changes among the politicians and generals. In February General Ambrosio had succeeded Cavallero as Chief of the Italian General Staff. Ambrosio, together with the Duke of Acquarone, the Minister of Court, were personal advisers of the King and had the confidence of the Royal circle. For months they had been hoping to overthrow the Duce and put an end to the Fascist regime. But Mussolini still dwelt in the European scene as if he were a principal factor. He was affronted when his new military chief proposed the immediate withdrawal of the Italian divisions from the Balkans. He regarded these forces as the counterpoise to German predominance in Europe. He did not realise that defeats abroad and internal demoralisation had robbed him of his status as Hitler's ally. He cherished the illusion of power and consequence when the reality had gone. Thus he resisted Ambrosio's formidable request. So durable however was the impression of his authority and the fear of his personal action in extremity that there was prolonged hesitation throughout all the forces of Italian society about how to oust him. Who would "bell the cat"? Thus the spring had passed with invasion by a mighty foe, possessing superior power by land, sea, and air, drawing ever nearer.

The climax had now come. Since February the taciturn, cautious-minded, constitutional King had been in contact with Marshal Badoglio, who had been dismissed after the Greek disasters in 1940. He found in him at length a figure to whom he might entrust the conduct of the state. A definite plan was made. It was resolved that Mussolini should be arrested on July 26, and General Ambrosio agreed to find the agents and create the situation for this stroke. The General was aided un-wittingly by elements in the Fascist Old Guard, who sought a new revival of the party, by which, in many cases, they would not be the losers. They saw in the summoning of the highest party organ, the Fascist Grand Council, which had not met since 1939, the means of confronting the Duce with an ultimatum. On July 13 they called on Mussolini and induced him to convene a formal session of the Council of July 24. The two movements appear to have been separate and inde-pendent, but their close coincidence in date is significant.

On this same July 19, accompanied by General Ambrosio, he left by air to meet Hitler at a villa at Feltre, near Rimini. "There was a most beautiful cool and shady park," writes Mussolini in his memoirs, "and a labyrinthine building which some people found almost uncanny. It was like a crossword puzzle frozen into a house." All preparations had been made to entertain the Fuehrer for at least two days, but he left the same afternoon. "The meeting," says Mussolini, "was, as usual, cordial, but the entourage and the attitude of the higher Air Force offi-cers and of the troops was chilly." [2]

The Fuehrer held forth lengthily upon the need for a supreme effort. The new secret weapons, he said, would be ready for use against England by the winter. Italy must be defended, "so that Sicily may become for the enemy what Stalingrad was for us." [3] The Italians must produce both manpower and the organisation. Germany could not provide the reinforcements and equipment asked for by Italy owing to the pressure on the Russian front.

Ambrosio urged his chief to tell Hitler plainly that Italy could not continue in the war. It is not clear what advantage would have come

[2] Mussolini, *Memoirs*, 1942–43 (English edition), p. 50.
[3] Rizzoli, *Hitler e Mussolini: Lettere e Documenti*, p. 173.

from this, but the fact that Mussolini seemed almost dumbstruck finally decided Ambrosio and the other Italian generals present that no further leadership could be expected from him.

In the midst of Hitler's discourse on the situation an agitated Italian official entered the room with the news, "At this moment Rome is undergoing a violent enemy air bombardment." Apart from a promise of further German reinforcements for Sicily, Mussolini returned to Rome without anything to show. As he approached he flew into a huge black cloud of smoke rising from hundreds of wagons on fire in the Littorio railway station. He had an audience of the King, whom he found "frowning and nervous." "A tense situation," said the King. "We cannot go on much longer. Sicily has gone west now. The Germans will double-cross us. The discipline of the troops has broken down. . . ." Mussolini answered, according to the records, that he hoped to disengage Italy from the Axis alliance by September 15. The date shows how far he was out of contact with reality.

The chief actor in the final drama now appeared on the scene. Dino Grandi, veteran Fascist, former Foreign Minister and Ambassador to Britain, a man of strong personal determination, who had hated the Italian declaration of war upon Britain, but had hitherto submitted to the force of events, arrived in Rome to take the lead at the meeting of the Grand Council. He called on his old leader on July 22, and told him brutally that he intended to propose the formation of a National Government and the restoration to the King of the supreme command of the armed forces.

*   *   *

At 5 P.M. on the 24th the Grand Council met. Care appears to have been taken by the Chief of Police that they should not be disturbed by violence. Mussolini's musketeers, his personal bodyguard, were relieved of their duty to guard the Palazzo Venezia, which was also filled with armed police. The Duce unfolded his case, and the Council who were all dressed in their black Fascist uniform, took up the discussion. Mussolini ended: "War is always a party war — a war of the party which desires it; it is always one man's war — the war of the man who declared it. If today this is called Mussolini's war, the war in 1859 could have been called Cavour's war. This is the moment to tighten the reins and assume the necessary responsibility. I shall have no difficulty in replacing men, in turning the screw, in bringing forces to bear not yet engaged, in the name of our country, whose territorial integrity is today being violated."

Grandi then moved a resolution calling upon the Crown to assume more power and upon the King to emerge from obscurity and assume his responsibilities. He delivered what Mussolini describes as "a violent philippic," "the speech of a man who was at last giving vent to a long-cherished rancour." The contacts between members of the Grand Council and the Court became evident. Mussolini's son-in-law, Ciano, supported Grandi. Everyone present was now conscious that a political convulsion impended. The debate continued till midnight, when Scorza, secretary of the Fascist Party, proposed adjourning till next day. But Grandi leaped to his feet, shouting, "No, I am against the proposal. We have started this business and we must finish it this very night!" It was after two o'clock in the morning when the voting took place. "The position of each member of the Grand Council," writes Mussolini, "could be discerned even before voting. There was a group of traitors who had already negotiated with the Crown, a group of accomplices, and a group of uninformed who probably did not realise the seriousness of the vote, but they voted just the same." Nineteen replied "Yes" to Grandi's motion and seven "No." Two abstained. Mussolini rose. "You have provoked a crisis of the regime. So much the worse. The session is closed." The party secretary was about to give the salute to the Duce when Mussolini checked him with a gesture, saying, "No, you are excused." They all went away in silence. None slept at home.

Meanwhile the arrest of Mussolini was being quietly arranged. The Duke of Acquarone, the Court Minister, sent instructions to Ambrosio, whose deputies and trusted agents in the police and the Carabinieri acted forthwith. The key telephone exchanges, the police headquarters and the offices of the Ministry of the Interior were quietly and unobtrusively taken over. A small force of military police was posted out of sight near the Royal villa.

Mussolini spent the morning of Sunday, July 25, in his office, and visited some quarters in Rome which had suffered by bombing. He asked to see the King, and was granted an audience at five o'clock. "I thought the King would withdraw his delegation of authority of June 1, 1940, concerning the command of the armed forces, a command which I had for some time past been thinking of relinquishing. I entered the villa therefore with a mind completely free from any forebodings, in a state which, looking back on it, might really be called utterly unsuspecting." On reaching the Royal abode he noticed that there were everywhere reinforcements of Carabinieri. The King, in Marshal's uniform, stood in the doorway. The two men entered the drawing room. The King said, "My dear Duce, it's no longer any good. Italy has gone to bits. Army morale is at rock bottom. The soldiers don't want to fight any more. . . . The Grand Council's vote is terrific — nineteen votes for Grandi's motion, and among them four holders of the Order of the Annunciation! . . . At this moment you are the most hated man in

Italy. You can no longer count on more than one friend. You have one friend left, and I am he. That is why I tell you that you need have no fears for your personal safety, for which I will ensure protection. I have been thinking that the man for the job now is Marshal Badoglio."

Mussolini replied, "You are taking an extremely grave decision. A crisis at this moment would mean making the people think that peace was in sight, once the man who declared war had been dismissed. The blow to the Army's morale would be serious. The crisis would be considered as a triumph for the Churchill-Stalin setup, especially for Stalin. I realise the people's hatred. I had no difficulty in recognising it last night in the midst of the Grand Council. One can't govern for such a long time and impose so many sacrifices without provoking resentments. In any case, I wish good luck to the man who takes the situation in hand." The King accompanied Mussolini to the door. "His face," says Mussolini, "was livid and he looked smaller than ever, almost dwarfish. He shook my hand and went in again. I descended the few steps and went towards my car. Suddenly a Carabinieri captain stopped me and said, 'His Majesty has charged me with the protection of your person.' I was continuing towards my car when the captain said to me, pointing to a motor ambulance standing nearby, 'No. We must get in there.' I got into the ambulance, together with my secretary. A lieutenant, three Carabinieri, and two police agents in plain clothes got in as well as the captain, and placed themselves by the door armed with machine guns. When the door closed the ambulance drove off at top speed. I still thought that all this was being done, as the King had said, in order to protect my person."

Later that afternoon Badoglio was charged by the King to form a new Cabinet of Service chiefs and civil servants, and in the evening the Marshal broadcast the news to the world. Two days later the Duce was taken on Marshal Badoglio's order to be interned on the island of Ponza.

Thus ended Mussolini's twenty-one year's dictatorship in Italy, during which he had raised the Italian people from the Bolshevism into which they might have sunk in 1919 to a position in Europe such as Italy had never held before. A new impulse had been given to the national life. The Italian Empire in North Africa was built. Many important public works in Italy were completed. In 1935 the Duce had by his will power overcome the League of Nations — "Fifty nations led by one" — and was able to complete his conquest of Abyssinia. His regime was far too costly for the Italian people to bear, but there is no doubt that it appealed during its period of success to very great numbers of Italians. He was, as I had addressed him at the time of the fall of France, "the Italian lawgiver." The alternative to his rule might well have been a Communist Italy, which would have brought perils and misfortunes of a different character both upon the Italian people and Europe. His fatal mistake was the declaration of war on France and Great Britain following Hitler's victories in June 1940. Had he not done this he could well have maintained Italy in a balancing position, courted and rewarded by both sides and deriving an unusual wealth and prosperity from the struggles of other countries. Even when the issue of the war became certain Mussolini would have been welcomed by the Allies. He had much to give to shorten its course. He could have timed his movement to declare war on Hitler with art and care. Instead he took the wrong turning. He never understood the strength of Britain, nor the long-enduring qualities of Island resistance and sea-power. Thus he marched to ruin. His great roads will remain a monument to his personal power and long reign.

*   *   *

At this time Hitler made a crowning error in strategy and war direction. The impending defection of Italy, the victorious advance of Russia, and the evident preparations for a cross-Channel attack by Britain and the United States should have led him to concentrate and develop the most powerful German army as a central reserve. In this way only could he use the high qualities of the German command and fighting troops, and at the same time take full advantage of the central position which he occupied, with its interior lines and remarkable communications. As General von Thoma said while a prisoner of war in our charge, "Our only chance is to create a situation where we can use the Army." Hitler, as I have pointed out earlier in this account, had in fact made a spider's web and forgotten the spider. He tried to hold everything he had won. Enormous forces were squandered in the Balkans and in Italy which could play no part in the main decisions. A central reserve of thirty or forty divisions of the highest quality and mobility would have enabled him to strike at any one of his opponents advancing upon him and fight a major battle with good prospects of success. He could, for instance, have met the British and Americans at the fortieth or fiftieth day after their landing in Normandy a year later with fresh and greatly superior forces. There was no need to consume his strength in Italy and the Balkans, and the fact that he was induced to do so must be taken as the waste of his last opportunity.

Knowing that these choices were open to him, I wished also to have the options of pressing right-handed in Italy or left-handed across the Channel, or both. The wrong dispositions which he made enabled us to undertake the main direct assault under conditions which offered good prospects and achieved success.

Hitler had returned from the Feltre meeting convinced that Italy could only be kept in the war by purges in the Fascist Party and increasing pressure by the Germans on the Fascist leaders. Mussolini's sixtieth birthday fell on July 29, and Goering was chosen to pay him an official visit on this occasion. But during the course of July 25 alarming reports from Rome began to come in to Hitler's headquarters. By the evening it was clear that Mussolini had resigned or had been removed, and that Badoglio had been nominated by the King as his successor. It was finally decided that any major operation against the new Italian Government would require withdrawals of more divisions than could be spared from the Eastern Front in the event of the expected Russian offensive. Plans were made to rescue Mussolini, to occupy Rome, and to support Italian Fascism wherever possible. If Badoglio signed an armistice with the Allies, further plans were drawn up for seizing the Italian Fleet and occupying key positions throughout Italy, and for overawing Italian garrisons in the Balkans and in the Aegean.

"We must act," Hitler told his advisers on July 26. "Otherwise the Anglo-Saxons will steal a march on us by occupying the airports. The Fascist Party is at present only stunned, and will recover behind our lines. The Fascist Party is the only one that has the will to fight on our side. We must therefore restore it. All reasons advocating further delays are wrong; thereby we run the danger of losing Italy to the Anglo-Saxons. These are matters which a soldier cannot comprehend. Only a man with political insight can see his way clear."

2

Synthetic Harbours

PROSPECTS OF VICTORY in Sicily, the Italian situation, and the progress of the war made me feel the need in July for a new meeting with the President and for another Anglo-American Conference. It was Roosevelt who suggested that Quebec should be the scene. Mr. Mackenzie King welcomed the proposal, and nothing could have been more agreeable to us. No more fitting or splendid setting for a meeting of those who guided the war policy of the Western world could have been chosen at this cardinal moment than the ancient citadel of Quebec, at the gateway of Canada, overlooking the mighty St. Lawrence River. The President, while gladly accepting Canadian hospitality, did not feel it possible that Canada should be formally a member of the Conference, as he apprehended similar demands by Brazil and other American partners in the United Nations. We also had to think of the claims of Australia and the other Dominions. This delicate question was solved and surmounted by the broadminded outlook of the Canadian Prime Minister and Government. I for my part was determined that we and the United States should have the Conference to ourselves, in view of all the vital business we had in common. A triple meeting of the heads of the three major Powers was a main object of the future; now it must be for Britain and the United States alone. We assigned to it the name "Quadrant."

I left London for the Clyde, where the *Queen Mary* awaited us, on the night of August 4, in a train which carried the very heavy staffs we needed. We were, I suppose, over two hundred, besides about fifty Royal Marine orderlies. The scope of the Conference comprised not only the Mediterranean campaign, now at its first climax, but even more preparations for the cross-Channel design of 1944, the whole conduct of the war in the Indian theatre, and our share in the struggle against Japan. For the Channel crossing we took with us three officers sent by Lieutenant General F. E. Morgan, Chief of Staff to the Supreme Allied Commander, yet to be finally chosen, who with his combined Anglo-American staff had completed our joint outline plan. As the whole of our affairs in the Indian and Far Eastern theatres were under examination I brought with me General Wavell's Director of Military Operations, who had flown specially from India.

I also took with me a young Brigadier named Wingate, who had already made his mark as a leader of irregulars in Abyssinia, and had greatly distinguished himself in the jungle fighting in Burma. These new brilliant exploits won him in some circles of the Army in which he served the title of "the Clive of Burma." I had heard much of all this, and knew also how the Zionists had sought him as a future Commander-in-Chief of any Israelite army that might be formed. I had him summoned home in order that I might have a look at him before I left for Quebec. I was about to dine alone on the night of August 4 at Downing Street when the news that he had arrived by air and was actually in the house was brought to me. I immediately asked him to join me at dinner. We had not talked for half an hour before I felt myself in the presence of a man of the highest quality. He plunged at once into his theme of how the Japanese could be mastered in jungle warfare by long-range penetration groups landed by air behind the enemy lines. This interested me greatly. I wished to hear much more about it, and also to let him tell his tale to the Chiefs of Staff.

I decided at once to take him with me on the voyage. I told him our train would leave at ten. It was then nearly nine. Wingate had arrived just as he was after three days' flight from the actual front, and with no clothes but what he stood up in. He was of course quite ready to go, but expressed regret that he would not be able to see his wife, who was in Scotland and had not even heard of his arrival. However, the resources of my Private Office were equal to the occasion. Mrs. Wingate was aroused at her home by the police and taken to Edinburgh in order to join our train on its way through and to go with us to Quebec. She had no idea of what it was all about until, in the early hours of the morning, she actually met her husband on a platform at Waverley Station. They had a very happy voyage together.

As I knew how much the President liked meeting young, heroic figures, I had also invited Wing Commander Guy Gibson, fresh from leading the attack which had destroyed the Möhne and Eder Dams. These supplied the industries of the Ruhr, and fed a wide area of fields, rivers, and canals. A special type of mine had been invented for their destruction, but it had to be dropped at night from a height of no more than sixty feet. After months of continuous and concentrated practice sixteen

Lancasters of No 617 Squadron of the Royal Air Force attacked on the night of May 16. Half were lost, but Gibson had stayed to the end, circling under fierce fire over the target to direct his squadron. He now wore a remarkable set of decorations — the Victoria Cross, a Distinguished Service Order and bar, and a Distinguished Flying Cross and bar — but no other ribbons. This was unique.

My wife came with me, and my daughter Mary, now a subaltern in an anti-aircraft battery, was my aide-de-camp. We sailed on August 5, this time for Halifax, in Nova Scotia, instead of New York.

*   *   *

The *Queen Mary* drove on through the waves, and we lived in the utmost comfort on board her, with a diet of prewar times. As usual on these voyages, we worked all day long. Our large cipher staff, with attendant cruisers to dispatch outgoing messages, kept us in touch with events from hour to hour. Each day I studied with the Chiefs of Staff the various aspects of the problems we were to discuss with our American friends. The most important of these was of course "Overlord."

One morning on our voyage, at my request, Brigadier K. G. McLean, with two other officers from General Morgan's staff, came to me as I lay in my bed in the spacious cabin, and, after they had set up a large-scale map, explained in a tense and cogent tale the plan which had been prepared for the cross-Channel descent upon France. The reader is perhaps familiar with all the arguments of 1941 and 1942 upon this burning question in all its variants, but this was the first time that I had heard the whole coherent plan presented in precise detail both of numbers and tonnage as the result of prolonged study by officers of both nations.

The choice narrowed to the Pas de Calais or Normandy. The former gave us the best air cover, but here the defences were the most formidable, and although it promised a shorter sea voyage this advantage was only apparent. While Dover and Folkstone are much closer to Calais and Boulogne than the Isle of Wight is to Normandy, their harbours were far too small to support an invasion. Most of our ships would have had to sail from ports along the whole south coast of England and from the Thames estuary, and so cross a lot of salt water in any case. General Morgan and his advisers recommended the Normandy coast, which from the first had been advocated by Mountbatten. There can be no doubt now that this decision was sound. Normandy gave us the greatest hope. The defences were not so strong as in the Pas de Calais. The seas and beaches were on the whole suitable, and were to some extent sheltered from the westerly gales by the Cotentin peninsula. The hinterland favoured the rapid deployment of large forces, and was sufficiently remote from the main strength of the enemy. The port of Cherbourg could be isolated and captured early in the operation. Brest could be outflanked and taken later.

All the coast between Havre and Cherbourg was of course defended with concrete forts and pillboxes, but as there was no harbour capable of sustaining a large army in this fifty-mile half-moon of sandy beaches it was thought that the Germans would not assemble large forces in immediate support of the sea front. Their High Command had no doubt said to themselves, "This is a good sector for raids up to ten or twenty thousand men, but unless Cherbourg is taken in working order no army in any way equal to the task of an invasion can be landed or supplied. It is a coast for a raid, but not for wider operations." If only there were harbours which could nourish great armies, here was the front on which to strike.

*   *   *

Of course, as the reader will have seen, I was well abreast of all the thought about landing craft and tank landing craft. I had also long been a partisan of piers with their heads floating out in the sea. Much work had since been done on them, following a minute which in the course of our discussions I had issued to Lord Louis Mountbatten, the Chief of Combined Operations, as long ago as May 30, 1942.

> They *must* float up and down with the tide. The anchor problem must be mastered. The ships must have a side flap cut in them, and a drawbridge long enough to overreach the moorings of the piers. Let me have the best solution worked out. Don't argue the matter. The difficulties will argue for themselves.

Thought later moved to the creation of a large area of sheltered water protected by a breakwater based on blockships brought to the scene by their own power and then sunk in a prearranged position. This idea originated with Commodore J. Hughes-Hallett in June 1943, while he was serving as Naval Chief of Staff in General Morgan's organisation. Imagination, contrivance, and experiment had been ceaseless, and now in August 1943 there was a complete project for making two full-scale temporary harbours which could be towed over and brought into action within a few days of the original landing. These synthetic harbours were called "Mulberries," a code name which certainly did not reveal their character or purpose.

The whole project was majestic. On the beaches themselves would be the great piers, with their seaward ends afloat and sheltered. At these piers coasters and landing craft would be able to discharge at all states of the tide. To protect them against the wanton winds and waves breakwaters would be spread in a great arc to seaward, enclosing a large area of sheltered water. Thus sheltered, deep-draught ships could lie at anchor and discharge, and all types of landing craft could ply freely to and from the beaches. These breakwaters would be composed of sunken concrete structures and blockships. I have described the similar structures which I thought might in the First World War have been used to create artificial harbours in the Heligoland Bight. Now they were to form a principal part of the great plan.

*   *   *

Further discussions on succeeding days led into more technical detail. The Channel tides have a play of more than twenty feet, with corresponding scours along the beaches. The weather is always uncertain, and winds and gales may whip up in a few hours irresistible forces against frail human structures. The fools or knaves who had chalked "Second Front Now" on our walls for the past two years had not had their minds burdened by such problems. I had long pondered upon them.

I was now convinced of the enormous advantages of attacking the Havre-Cherbourg sector, provided these unexpected harbours could be brought into being from the first and thus render possible the landing and sustained advance of armies of a million rising to two million men, with all their immense modern equipment and impedimenta. This would mean being able to unload at least twelve thousand tons a day.

Three dominating assumptions were made both by the framers of the plan and the British Chiefs of Staff. With these I was in entire agreement, and, as will be seen later, they were approved by the Americans and accepted by the Russians.

1. That there must be a substantial reduction in the strength of the German fighter aircraft in Northwest Europe before the assault took place.

2. That there should be not more than twelve mobile German divisions in Northern France at the time the operation was launched, and that it must not be possible for the Germans to build up more than fifteen divisions in the succeeding two months.

3. That the problem of beach maintenance of large forces in the tidal waters of the English Channel over a prolonged period must be overcome. To ensure this it was essential that we should be able to construct at least two effective synthetic harbours.

I was very well satisfied with the prospect of having the whole of this story presented to the President with my full support. At least it would

convince the American authorities that we were not insincere about
"Overlord" and had not grudged thought or time in preparation. I
arranged to assemble in Quebec the best experts in such matters from
London and Washington. Together they could pool resources and find
the best answers to the many technical problems.

*   *   *

I also had many discussions with the Chiefs of Staff on our affairs in
the Indian and Far Eastern theatres. We had none too good a tale to tell.
A division had advanced at the end of 1942 down the Arakan coast of
Burma to recapture the port of Akyab. Though strengthened until a
complete corps was engaged, the operation had failed, and our troops
were forced back over the Indian frontier.

Although there was much to be said in explanation, I felt that the
whole question of the British High Command against Japan must come
under review. New methods and new men were needed. I had long felt
that it was a bad arrangement for the Commander-in-Chief of India to
command the operations in Burma in addition to his other far-reaching
responsibilities. It seemed to me that the vigorous prosecution of large-
scale operations against the Japanese in Southeast Asia necessitated the
creation of a separate Supreme Allied Command. The Chiefs of Staff
were in complete agreement, and prepared a memorandum on these lines
for discussion with their American colleagues in Quebec. There remained the question of the commander of this new theatre, and we were
in no doubt that he should be British. Of the various names that were
put forward, I was sure in my own mind that Admiral Mountbatten had
superior qualifications for this great command, and I determined to make
this proposal to the President at the first opportunity. The appointment
of an officer of the substantive rank of Captain R.N. to the Supreme
Command of one of the main theatres of the war was an unusual step;
but, having carefully prepared the ground beforehand, I was not surprised when the President cordially agreed.

It is astonishing how quickly a voyage can pass if one has enough to
do to occupy every waking minute. I had looked forward to an interval
of rest and a change from the perpetual clatter of the war. But as we
approached our destination the holiday seemed to be over before it had
begun.

Halifax was reached on August 9. The great ship drew in to the
landing jetty and we went straight to our train. In spite of all precautions about secrecy, large crowds were assembled. As my wife and I sat
in our saloon at the end of the train the people gathered round and
gave us welcome. Before we started I made them sing "The Maple
Leaf" and "O Canada!" I feared they did not know "Rule, Britannia,"

though I am sure they would have enjoyed it if we had had a band. After about twenty minutes of hand-shakings, photographs, and autographs we left for Quebec. On August 17 the President and Harry Hopkins arrived and Eden and Brendan Bracken flew in from England. As the delegations gathered, news of Italian peace moves came out to us, and it was under the impression of Italy's approaching surrender that our talks were held.

The first plenary session was held on August 19. Highest strategic priority "as a prerequisite to 'Overlord' " was given to the combined bomber offensive against Germany. The lengthy discussions upon Operation "Overlord" were then summarised in the light of the combined planning in London by General Morgan. The Chiefs of Staff now reported as follows:

OPERATION "OVERLORD"

This operation will be the primary United States–British ground and air effort against the Axis in Europe. (Target date, May 1, 1944.) . . .

As between Operation "Overlord" and operations in the Mediterranean, where there is a shortage of resources available resources will be distributed and employed with the main object of ensuring the success of "Overlord." Operations in the Mediterranean theatre will be carried out with the forces allotted at "Trident" [the previous Conference at Washington in May], except in so far as these may be varied by decision of the Combined Chiefs of Staff. . . .

These paragraphs produced some discussion at our meeting. I pointed out that the success of "Overlord" depended on certain conditions being fulfilled in regard to relative strength. I emphasised that I strongly favoured "Overlord" in 1944, though I had not been in favour of trying to attack Brest or Cherbourg in 1942 or 1943. The objections which I had to the cross-Channel operation were however now removed. I thought that every effort should be made to add at least twenty-five per cent to the first assault. This would mean finding more landing craft. There were still nine months to go, and much could be done in that time. The beaches selected were good, and it would be better if at the same time a landing were to be made on the inside beaches of the Cotentin peninsula. "Above all," I said, "the initial lodgment must be strong."

As the United States had the African command, it had been earlier agreed between the President and me that the commander of "Overlord" should be British, and I proposed for this purpose, with the President's agreement, General Brooke, the Chief of the Imperial General

Staff, who, it may be remembered, had commanded a corps in the decisive battle on the road to Dunkirk, with both Alexander and Montgomery as his subordinates. I had informed General Brooke of this intention early in 1943. This operation was to begin with equal British and American forces, and as it was to be based on Great Britain it seemed right to make such an arrangement. However, as the year advanced and the immense plan of the invasion began to take shape I became increasingly impressed with the very great preponderance of American troops that would be employed after the original landing with equal numbers had been successful, and now at Quebec I myself took the initiative of proposing to the President that an American commander should be appointed for the expedition to France. He was gratified at this suggestion, and I dare say his mind had been moving that way. We therefore agreed that an American officer should command "Overlord" and that the Mediterranean should be entrusted to a British commander, the actual date of the change being dependent upon the progress of the war. I informed General Brooke, who had my entire confidence, of this change, and of the reasons for it. He bore the great disappointment with soldierly dignity.

*   *   *

As for the Far East, the main dispute between the British and American Chiefs of Staff was on the issue that Britain demanded a full and fair place in the war against Japan from the moment when Germany was beaten. She demanded a share of the airfields, a share of the bases for the Royal Navy, and a proper assignment of duties to whatever divisions she could transport to the Far East after the Hitler business was finished. My friends on the Chiefs of Staff Committee had been pressed by me to fight this point to the utmost limit, because at this stage in the war what I most feared was that American critics would say, "England, having taken all she could from us to help her beat Hitler, stands out of the war against Japan and will leave us in the lurch." However, at the Quebec Conference this impression was effectively removed. No decision was reached on the actual operations to be undertaken, though it was decided that the main effort should be put into offensive operations with the object of "establishing land communications with China and improving and securing the air route." In the "overall strategic concept" of the Japan war plans were to be made to bring about the defeat of Japan within twelve months after the collapse of Germany.

Finally there was the Mediterranean scene. On August 10 Eisenhower held a meeting of his commanders to select from a variety of proposals the means by which the campaign should be carried into Italy. He had to take special account of the enemy dispositions at that time. Eight of the sixteen German divisions in Italy were in the north under Rommel, two were near Rome, and six were farther south under Kesselring. These might be reinforced from another twenty which had been withdrawn from the Russian front to refit in France. Nothing we might gather for a long time could equal such strength, but the British and Americans had command of sea and air, and also the initiative. The assault upon which all minds were now set was a daring enterprise. It was hoped to gain the ports of Naples and Taranto, whose combined facilities were proportioned to the scale of the armies we must use. The early capture of airfields was a prime aim. Those near Rome were as yet beyond our reach, but there was an important group at Foggia adaptable for heavy bombers, and our tactical air forces sought others in the heel of Italy and Montecorvino, near Salerno.

General Eisenhower decided to begin the assault in early September by an attack across the Straits of Messina, with subsidiary descents on the Calabrian coast. This would be the prelude to the capture of Naples (Operation "Avalanche") by a British and an American army corps landing on the good beaches in the Gulf of Salerno. This was at the extreme range of fighter cover from the captured Sicilian airfields. As soon as possible after the landings the Allied forces would drive north to capture Naples.

The Combined Chiefs of Staff advised the President and me to accept this plan, and to authorise the seizure of Sardinia and Corsica in second priority. We did so with alacrity; indeed, it was exactly what I had hoped and striven for. Later it was proposed to land an air-borne division to capture the airfields south of Rome. This also we accepted. The circumstances in which this feature was cancelled will be recounted in due course.

3

The Invasion of Italy

THE QUEBEC CONFERENCE ended on August 24, and
our notable colleagues departed and dispersed. They flew off in
every direction like the fragments of a shell. After all the study and
argument there was a general desire for a few days' rest. One of my
Canadian friends, Colonel Clarke, who had been attached to me by the
Dominion Government during the proceedings, owned a ranch about
seventy-five miles away amid the mountains and pine forests from
which the newspapers get their pulp to guide us on life's journey. Here
lay the Lake of the Snows, an enormous dammed-up expanse of water
reported to be full of the largest trout. Brooke and Portal were ardent
and expert anglers, and a plan had been made among other plans at the
Conference for them to see what they could do. I promised to join them
later if I could, but I had undertaken to deliver a broadcast on the 31st,
and this hung overhead like a vulture in the sky. I remained for a few
days in the Citadel, pacing the ramparts for an hour each afternoon,
and brooding over the glorious panorama of the St. Lawrence and all
the tales of Wolfe and Quebec. I had promised to drive through the
city, and I had a lovely welcome from all its people. I attended a meeting of the Canadian Cabinet, and told them all that they did not already
know about the Conference and the war. I had the honour to be sworn
a Privy Counsellor of the Dominion Cabinet. This compliment was paid
me at the instance of my old friend of forty years' standing and trusted
colleague, Mr. Mackenzie King.

There was so much to say and not to say in the broadcast that I could
not think of anything, so my mind turned constantly to the Lake of the
Snows, of which glittering reports had already come in from those who
were there. I thought I might combine fishing by day with preparing
the broadcast after dark. I resolved to take Colonel Clarke at his word,

and set out with my wife by car. I had noticed that Admiral Pound had not gone with the other two Chiefs of Staff to the lake, and I suggested that he should come with us now. His Staff officer said that he had a lot of cleaning up to do after the Conference. I had been surprised by the subdued part he had taken in the far-ranging naval discussions, but when he said he could not come fishing I had a fear that all was not well. We had worked together in the closest comradeship from the first days of the war. I knew his worth and courage. I also knew that at home he would get up at four or five in the morning for a few hours' fishing before returning to the Admiralty whenever he saw the slightest chance. However, he kept to his quarters and I did not see him before starting.

We had a wonderful all-day drive up the river valley, and after sleeping at a rest house on the way my wife and I reached the spacious log cabin on the lake. Brooke and Portal were leaving the next day. It was just as well. They had caught a hundred fish apiece each day, and had only to continue at this rate to lower the level of the lake appreciably. My wife and I sallied forth in separate boats for several hours, and though we are neither of us experts we certainly caught a lot of fine fish. We were sometimes given rods with three separate hooks, and I once caught three fish at the same time. I do not know whether this was fair. We did not run at all short of fresh trout at the excellent meals. The President had wanted to come himself, but other duties claimed him. I sent the biggest fish I caught to him at Hyde Park. The broadcast made progress, but original composition is more exhausting than either arguing or fishing.

We returned to Quebec for the night of the 29th. I attended another meeting of the Canadian Cabinet, and at the right time on the 31st, before leaving for Washington, I spoke to the Canadian people and to the Allied world. A quotation is pertinent to this account:

> The contribution which Canada has made to the combined effort of the British Commonwealth and Empire in these tremendous times has deeply touched the heart of the Mother Country and of all the other members of our widespread family of States and races.
>
> From the darkest days the Canadian Army, growing stronger year by year, has played an indispensable part in guarding our British homeland from invasion. Now it is fighting with distinction in wider and ever-widening fields. The Empire Air Training Organisation, which has been a wonderful success, has found its seat in Canada, and has welcomed the flower of the manhood of Great Britain, of Australia, and New Zealand to her spacious flying fields and to comradeship with her own gallant sons.
>
> Canada has become in the course of this war an important sea-

faring nation, building many scores of warships and merchant ships, some of them thousands of miles from salt water, and sending them forth manned by hardy Canadian seamen to guard the Atlantic convoys and our vital lifeline across the ocean. The munitions industries of Canada have played a most important part in our war economy. Last, but not least, Canada has relieved Great Britain of what would otherwise have been a debt for these munitions of no less than two thousand million dollars.

All this of course was dictated by no law. It came from no treaty or formal obligation. It sprang in perfect freedom from sentiment and tradition and a generous resolve to serve the future of mankind. I am glad to pay my tribute on behalf of the people of Great Britain to the great Dominion, and to pay it from Canadian soil. I only wish indeed that my other duties, which are exacting, allowed me to travel still farther afield and tell Australians, New Zealanders, and South Africans to their faces how we feel towards them for all they have done, and are resolved to do. . . .

Next day I reached the White House. The President and I sat talking after dinner in his study, and Admiral Pound came to see us upon a naval point. The President asked him several questions about the general aspects of the war, and I was pained to see that my trusted naval friend had lost the outstanding matter-of-fact precision which characterised him. Both the President and I were sure he was very ill. Next morning Pound came to see me in my big bed–sitting room and said abruptly, "Prime Minister, I have come to resign. I have had a stroke and my right side is largely paralysed. I thought it would pass off, but it gets worse every day and I am no longer fit for duty." I at once accepted the First Sea Lord's resignation, and expressed my profound sympathy for his breakdown in health. I told him he was relieved at that moment from all responsibility, and urged him to rest for a few days and then come home with me in the *Renown*. He was completely master of himself, and his whole manner was instinct with dignity. As soon as he left the room I cabled to the Admiralty placing Vice-Admiral Syfret in responsible charge from that moment pending the appointment of a new First Sea Lord.

*   *   *

Throughout the talks at Quebec events had been marching forward in Italy. The President and I had directed during these critical days the course of secret armistice negotiations with the Badoglio Government, and had also been following anxiously and closely the military arrangement for a landing on Italian soil. I deliberately prolonged my stay in the United States in order to be in close contact with our American friends at this crucial moment in Italian affairs. On the day of my arrival in Washington the first definite and official news was received that Badoglio had agreed to capitulate to the Allies, and on September 3 in an olive grove near Syracuse General Castellano signed the military terms of the surrender of Italy. Before dawn on the same day the British Eighth Army crossed the Straits of Messina to enter the Italian mainland.

It now remained to co-ordinate the terms of the Italian surrender with our military strategy. The American General Taylor, of the 82nd Air-borne Division, was sent to Rome on September 7. His secret mission was to arrange with the Italian General Staff for the airfields around the capital to be seized during the night of the 9th. But the situation had radically changed since General Castellano had asked for Allied protection. The Germans had powerful forces at hand, and appeared to be in possession of the airfields. The Italian Army was demoralised and short of ammunition. Divided counsels seethed round Badoglio. Taylor demanded to see him. Everything hung in the balance. The Italian leaders feared that any announcement of the surrender, which had already been signed, would lead to the immediate German occupation of Rome and the end of the Badoglio Government. At two o'clock on the morning of September 8 General Taylor saw Badoglio, who, since the airfields were lost, begged for delay in broadcasting the armistice terms. He had in fact already telegraphed to Algiers that the security of the Rome airfields could not be guaranteed. The air descent was therefore cancelled.

Eisenhower now had to make a quick decision. The attack on Salerno was due to be launched within less than twenty-four hours. He refused Badoglio's request and at 6 P.M. broadcast the announcement of the armistice, followed by the text of the declaration, which Marshal Badoglio himself announced about an hour later from Rome. The surrender of Italy had been completed.

During the night of September 8–9 German forces began the encirclement of Rome. Badoglio and the Royal Family installed themselves in a state of siege in the building of the Ministry of War. There were hasty discussions in an atmosphere of mounting tension and panic. In the small hours a convoy of five vehicles passed through the eastern gates of Rome on the road to the Adriatic port of Pescara. Here two corvettes took on board the party, which contained the Italian Royal Family, together with Badoglio and his Government and senior officials. They reached Brindisi in the early morning of September 10, when the essential services of an anti-Fascist Italian Government were rapidly set up on territory occupied by Allied forces.

After the departure of the fugitives the veteran Marshal Caviglia, the victor of Vittorio Veneto in the First World War, arrived in Rome to take upon himself the responsibility of negotiating with the German forces closing in round the city. Scattered fighting was already taking place at the gates. Certain regular units of the Italian Army and Partisan bands of Roman citizens engaged the Germans on the outskirts. On September 11 opposition ceased with the signature of a military truce, and the Nazi divisions were free to move through the city.

Meanwhile, after dark on September 8, in accordance with Allied instructions, the main body of the Italian Fleet left Genoa and Spezia on a daring voyage of surrender to Malta, unprotected either by Allied or Italian aircraft. Next morning when steaming down the west coast of Sardinia it was attacked by German aircraft from bases in France. The flagship *Roma* was hit, and blew up with heavy loss of life, including the Commander-in-Chief, Admiral Bergamini. The battleship *Italia* was also damaged. Leaving some light craft to rescue survivors, the rest of the Fleet continued its painful journey. On the 10th they were met at sea by British forces, including the *Warspite* and *Valiant,* which had so often sought them under different circumstances, and were escorted to Malta. A squadron from Taranto, including two battleships, had also sailed on the 9th, and on the morning of the 11th Admiral Cunningham informed the Admiralty that "the Italian battle fleet now lies at anchor under the guns of the fortress of Malta."

*   *   *

On the whole, therefore, things had so far gone very smoothly for the Allies. After crossing the Straits of Messina the Eighth Army had encountered practically no opposition. Reggio was speedily taken, and the advance began along the narrow and hilly roads of Calabria. "The Germans," cabled Alexander on September 6, "are fighting their rearguard action more by demolitions than by fire. . . . While in Reggio this morning there was not a warning sound to be heard or a hostile plane to be seen. On the contrary, on this lovely summer day naval craft of all types were plying backwards and forwards between Sicily and the mainland, carrying men, stores, and munitions. In its lively setting it was more like a regatta in peace-time than a serious operation of war." There was little fighting, but the advance was severely delayed by the physical difficulties of the country, demolitions carried out by the enemy, and his small but skilfully handled rearguards.

But on the night of the 8th Alexander sent me his "Zip" message. It had been planned that I and those of our party who had not already flown to England should go home by sea, and the *Renown* awaited us at Halifax. I broke the train journey to say good-bye to the President,

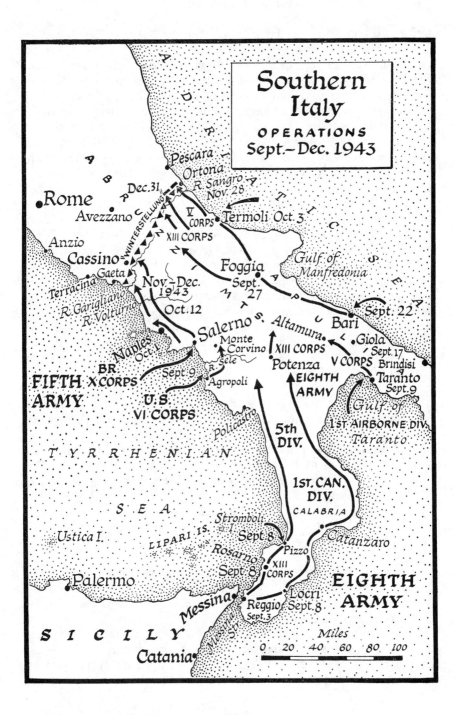

Southern Italy
OPERATIONS
Sept.–Dec. 1943

and was thus with him at Hyde Park when the Battle of Salerno began. I resumed my train journey on the night of the 12th, to reach Halifax on the morning of the 14th. The various reports which reached me on the journey, as well as the newspapers, made me deeply anxious. Evidently a most critical and protracted struggle was in progress. My concern was all the greater because I had always strongly pressed for this sea-borne landing, and felt a special responsibility for its success. Surprise, violence, and speed are the essence of all amphibious landings. After the first twenty-four hours the advantage of sea power in striking where you will may well have vanished. Where there were ten men there are soon ten thousand. My mind travelled back over the years. I thought of General Stopford waiting nearly three days on the beach at Suvla Bay in 1915 while Mustafa Kemal marched two Turkish divisions from the lines at Bulair to the hitherto undefended battlefield. I had had a more recent experience when General Auchinleck had remained at his headquarters in Cairo surveying orthodoxly from the summit and centre the wide and varied sphere of his command, while the battle, on which everything turned, was being decided against him in the Desert. I had the greatest confidence in Alexander, but all the same I passed a painful day while our train rumbled forward through the pleasant lands of Nova Scotia. At length I wrote out the following message for Alexander, feeling sure he would not resent it. It was not sent till after I had sailed:

> I hope you are watching above all the Battle of "Avalanche," which dominates everything. None of the commanders engaged has fought a large-scale battle before. The Battle of Suvla Bay was lost because Ian Hamilton was advised by his Chief of Staff to remain at a remote central point where he would know everything. Had he been on the spot he could have saved the show. At this distance and with time lags I cannot pretend to judge, but I feel it my duty to set before you this experience of mine from the past.
>
> 2. *Nothing* should be denied which will nourish the decisive battle for Naples. . . .

His answer was prompt and comforting. He was already at Salerno. "Many thanks," he replied, "for your offer of help. Everything possible is being done to make 'Avalanche' a success. Its fate will be decided in the next few days."

I was also relieved to learn that Admiral Cunningham had not hesitated to hazard his battleships close inshore in support of the Army. On the 14th he sent up the *Warspite* and *Valiant*, which had just arrived at Malta conducting to surrender the main body of the Italian

Fleet. Next day they were in action, and their accurate air-directed bombardment with heavy guns impressed both friend and foe and greatly contributed to the defeat of the enemy. Unhappily, on the afternoon of the 16th the *Warspite* was disabled by a new type of glider bomb, about which we had heard something, and were to learn more.

It was a relief to board the *Renown*. The splendid ship lay alongside the quay. Admiral Pound was already on board, having come through direct from Washington. He bore himself as erect as ever, and no one looking at him would have dreamed that he was stricken. I invited him to join us at my table on the homeward voyage, but he said he would prefer to take his meals in his cabin with his staff officer. He died on October 21, Trafalgar Day. He had been a true comrade to me, both at the Admiralty and on the Chiefs of Staff Committee. He was succeeded as First Sea Lord by Admiral Sir Andrew Cunningham.

* * *

While we zigzagged our way across the ocean a remarkable stroke was made upon Taranto, for which not only Alexander but Admiral Cunningham, on whom fell the brunt of execution, deserve the highest credit for well-run risks. The first-class port was capable of serving a whole army. The Italian surrender seemed to Alexander to justify daring. There were no transport aircraft to lift the British 1st Airborne Division, nor any ordinary shipping to carry it by sea. Six thousand of these picked men were embarked on British warships, and on September 9, the day of the landing on Salerno beaches, the Royal Navy steamed boldly into Taranto harbour and deposited the troops ashore, unopposed. One of our cruisers which struck a mine and sank was our only naval loss.[1]

All the time the Battle of Salerno went on. The telegrams flowed in. Alexander was kind enough to keep me fully informed, and his vivid messages can be read in their relation to the whole event. For three critical days the issue hung in the balance, but after bitter fighting, in which we suffered moments of grave hazard, the Germans failed to throw us back into the sea. Kesselring realised he could not succeed. Pivoting his right on the high ground above Salerno, he began to swing his whole line back. The Eighth Army, spurred on by Montgomery, joined hands with the hard-pressed Fifth. The British Xth Corps, with the United States VIth Corps on their right, drove back the enemy's rearguards around Vesuvius, marched past the ruins of Pompeii and Herculaneum, and entered Naples on October 1. We had won.

[1] I have in my home the Union Jack, the gift of General Alexander, that was hoisted at Taranto, and was one of the first Allied flags to be flown in Europe since our expulsion from France.

4

Deadlock in the Mediterranean [1]

A FEW DAYS AFTER my return from Halifax I had sent General Eisenhower a telegram which should be borne in mind in reading my account of the autumn and winter. The second paragraph sought to establish the proportion of effort, especially where bottlenecks were concerned, which should be devoted to our various enterprises. These proportions should not be overlooked by those who wish to understand the controversies with which this chapter deals. War presents the problem of the correct employment of available means, and cannot be epitomised as "One thing at a time."

> As I have been pressing for action in several directions, I feel I ought to place before you the priorities which I assign in my own mind to these several desirable objectives.
>
> 2. Four-fifths of our effort should be the build-up of Italy. One-tenth should be our making sure of Corsica (which will soon finish) and in the Adriatic. The remaining tenth should be concentrated on Rhodes. This of course applies to the limiting factors only. These I presume, are mainly landing craft and assault shipping, with light naval craft.
>
> 3. I send this as a rough guide to my thought only because I do not want you to feel I am pressing for everything in all directions without understanding how grim are your limitations.

Eisenhower replied next day:

> We are examining resources carefully to give Mid-East necessary support in this project, and feel sure that we can meet minimum requirements of Mid-East.

[1] See map on page 443. "Crete and the Aegean."

When Montgomery can get the bulk of his forces forward to support the right of the Fifth Army things will begin to move more rapidly on the Naples front. As is always the case following the early stages of a combined operation, we have been badly stretched both tactically and administratively. We are working hard to improve the situation and you will have good news before long.

This answer did not refer as specifically as I had hoped to what I deemed the all-important part of my message, namely, the small proportion of troops required for subsidiary enterprises, and of these there were many.

The surrender of Italy gave us the chance of gaining important prizes in the Aegean at very small cost and effort. The Italian garrisons obeyed the orders of the King and Marshal Badoglio, and would come over to our side if we could reach them before they were overawed and disarmed by the Germans in the islands. These were much inferior in numbers, but it is probable that for some time past they had been suspicious of their allies' fidelity and had their plans laid. Rhodes, Leros, and Cos were island fortresses which had long been for us strategic objectives of a high order in the secondary sphere, and their occupation had been specifically approved by the Combined Chiefs of Staff in their final summary of the Quebec decisions on September 10. Rhodes was the key to the group, because it had good airfields from which we could defend any other islands we might occupy and complete our naval control of these waters. Moreover, the British air forces in Egypt and Cyrenaica could guard Egypt just as well, or even better, if some of them moved forward to Rhodes. It seemed to me a rebuff to fortune not to pick up these treasures. The command of the Aegean by air and by sea was within our reach. The effect of this might be decisive upon Turkey, at that time deeply moved by the Italian collapse. If we could use the Aegean and the Dardanelles the naval short cut to Russia was established. There would be no more need for the perilous and costly Arctic convoys, or the long and wearisome supply line through the Persian Gulf.

General Wilson was eager for action, and plans and preparations for the capture of Rhodes had been perfected in the Middle East Command over several months. In August the 8th Indian Division had been trained and rehearsed in the operation, and was made ready to sail on September 1. But the American pressure to disperse our trained assault shipping from the Mediterranean, either westward for the preparations for a still remote "Overlord" or to the Indian theatre, was very strong. Agreements made before the Italian collapse and appropriate to a totally different situation were rigorously invoked, at least at the secondary level, and on August 26, in pursuance of a minor decision at the Washington Conference in the previous May, the Combined Chiefs of Staff ordered the dispatch to the Far East, for an operation against the coast of Burma, of the shipping that could have transported the division to Rhodes. Thus Wilson's well-conceived plans for rapid action in the Dodecanese were harshly upset. He had sent with great promptitude small parties by sea and air to a number of other islands, but once Rhodes was denied to us our gains throughout the Aegean became precarious. Only a powerful use of air forces could give us what we needed. It would have taken very little of their time had there been accord. General Eisenhower and his staff seemed unaware of what lay at our fingertips, although we had voluntarily placed all our considerable resources entirely in their hands.

*   *   *

We now know how deeply the Germans were alarmed at the deadly threat which they expected us to develop on their southeastern flank. At a conference at the Fuehrer's headquarters on September 24 both the Army and the Navy representatives strongly urged the evacuation of Crete and other islands in the Aegean while there was still time. They pointed out that these advanced bases had been seized for offensive operations in the Eastern Mediterranean, but that now the situation was entirely changed. They stressed the need to avoid the loss of troops and material which would be of decisive importance for the defence of the continent. Hitler overruled them. He insisted that he could not order evacuation, particularly of Crete and the Dodecanese, because of the political repercussions which would follow. He said, "The attitude of our allies in the southeast and Turkey's attitude is determined solely by their confidence in our strength. Abandonment of the islands would create a most unfavourable impression." In this decision to fight for the Aegean islands he was justified by events. He gained large profits in a subsidiary theatre at small cost to the main strategic position. In the Balkans he was wrong. In the Aegean he was right.

For a time our affairs prospered in the outlying small islands. By the end of September Cos, Leros, and Samos were occupied by a battalion each, and detachments occupied a number of other islands. Italian garrisons, where encountered, were friendly enough, but their vaunted coast and anti-aircraft defences were found to be in poor shape, and the transport of our own heavier weapons and vehicles was hardly possible with the shipping at our disposal.

Apart from Rhodes, the island of Cos was strategically the most important. It alone had an airfield from which our fighter aircraft could operate. This was rapidly brought into use and twenty-four Bofors guns landed for its defence. Naturally it became the objective of the first enemy counterattack, and at dawn on October 3 German parachutists descended on the central airfield and overwhelmed the solitary company defending it. The rest of the battalion, in the north of the island, was cut off by an enemy landing from the sea which the Navy, by an unlucky event, had been unable to intercept. The island fell.

On September 22 Wilson reported his minimum and modest needs for a new attempt on Rhodes. Using the 10th Indian Division and part of an armoured brigade, he required only naval escorts and bombarding forces, three L.S.T.s (Landing Ship, Tanks), a few motor transport ships, a hospital ship, and enough transport aircraft to lift one parachute battalion. I was greatly troubled at our inability to support these operations, and cabled for help to General Eisenhower. The small aids needed seemed very little to ask from our American friends. The concessions which they had made to my unceasing pressure during the last three months had been rewarded by astounding success. The landing craft for a single division, a few days' assistance from the main Allied Air Force, and Rhodes would be ours. The Germans, who had now regripped the situation, had moved many of their planes to the Aegean to frustrate the very purpose which I had in mind. On October 7 I also laid the issue before the President in its full scope, but was pained to receive a telegram which practically amounted to a refusal, and left me, already committed, with his and the American Chiefs of Staff's approval, to face the impending blow. The negative forces which hitherto had been so narrowly overcome had indeed resumed their control. This is what Mr. Roosevelt said:

> I do not want to force on Eisenhower diversions which limit the prospects for the early successful development of the Italian operations to a secure line north of Rome.
>
> I am opposed to any diversion which will in Eisenhower's opinion jeopardise the security of his current situation in Italy, the build-up of which is exceedingly slow, considering the well-known characteristics of his opponent, who enjoys a marked superiority in ground troops and Panzer divisions.
>
> It is my opinion that no diversion of forces or equipment should prejudice "Overlord" as planned. The American Chiefs of Staff agree. I am transmitting a copy of this message to Eisenhower.

I noticed in particular the sentence "It is my opinion that no diversion of forces or equipment should prejudice 'Overlord' as planned." To pretend that the delay of six weeks in the return of nine landing craft for "Overlord" out of over five hundred involved, which would in any

case have had six months in hand, would comprise the main operation
of May 1944 was to reject all sense of proportion. On October 8 I
therefore made a further earnest appeal. Looking back upon the far-
reaching favourable results which had followed from my journey with
General Marshall to Algiers in June, from which the whole of our good
fortune had sprung, I thought I might ask for the same procedure, and
I made all preparations to fly at once to Tunis, where the Commanders-
in-Chief were now assembling in conference.

But Mr. Roosevelt's reply quenched my last hopes. He thought my
attendance would be inappropriate. I accordingly cancelled my proposed flight. At the critical moment of the conference, information was
received that Hitler had decided to reinforce his army in Italy and fight
a main battle south of Rome. This tipped the scales against the small
reinforcement required for the attack on Rhodes.

Although I could understand how, in the altered situation, the opinion of the generals engaged in our Italian campaign had been affected,
I remained — and remain — in my heart unconvinced that the capture
of Rhodes could not have been fitted in. Nevertheless, with one of
the sharpest pangs I suffered in the war I submitted. If one has to
submit it is wasteful not to do so with the best grace possible. When
so many grave issues were pending I could not risk any jar in my personal relations with the President. I therefore took advantage of the
news from Italy to accept what I thought, and think, to have been an
improvident decision.

Nothing was gained by all the overcaution. The capture of Rome
proved to be eight months distant. Twenty times the quantity of shipping that would have helped to take Rhodes in a fortnight was employed throughout the autumn and winter to move the Anglo-American
heavy-bomber bases from Africa to Italy. Rhodes remained a thorn
in our side. Turkey, witnessing the extraordinary inertia of the Allies
near her shores, became much less forthcoming, and denied us her airfields.

*   *   *

The American Staff had enforced their view; the price had now to be
paid by the British. Although we strove to maintain our position in
Leros the fate of our small force there was virtually sealed. The garrison was brought up to the strength of a brigade — three fine battalions
of British infantry who had undergone the whole siege and famine of
Malta [2] and were still regaining their physical weight and strength.
The Admiralty did their best, and General Eisenhower dispatched two

[2] 4th Battalion, the Buffs, 2nd Battalion, Royal Irish Fusiliers, 1st Battalion, King's
Own.

groups of long-range fighters to the Middle East as a temporary measure. There they soon made their presence felt. But on October 11 they were withdrawn. Thereafter the enemy had air mastery, and it was only by night that our ships could operate without crippling loss. Early on November 12, German troops came ashore, and in the afternoon six hundred parachutists cut the defence in two. In the last stages the garrison of Samos, the 2nd Royal West Kents, had been dispatched to Leros, but all was over. They fell themselves a prey. With little air support of their own and heavily attacked by enemy aircraft, the battalions fought on till the evening of the 16th, when, exhausted, they could fight no more. Thus this fine brigade of troops fell into enemy power. All our hopes in the Aegean were for the time being ended. We tried at once to evacuate the small garrisons in Samos and other islands, and to rescue survivors from Leros. Over a thousand British and Greek troops were brought off, as well as many friendly Italians and German prisoners, but our naval losses were again severe. Six destroyers and two submarines were sunk by aircraft or mine and four cruisers and four destroyers damaged. These trials were shared by the Greek Navy, which played a gallant part throughout.

I have recounted the painful episodes of Rhodes and Leros in some detail. They constitute, happily on a small scale, the most acute difference I ever had with General Eisenhower. For many months, in the face of endless resistances, I had cleared the way for his successful campaign in Italy. Instead of only gaining Sardinia, we had established a large group of armies on the Italian mainland. Corsica was a bonus in our hands. We had drawn an important part of the German reserves away from other theatres. The Italian people and Government had come over to our side. Italy had declared war on Germany. Their Fleet was added to our own. Mussolini was a fugitive. The liberation of Rome seemed not far distant. Nineteen German divisions, abandoned by their Italian comrades, lay scattered through the Balkans, in which we had not used a thousand officers and men. The date for "Overlord" had not been decisively affected.

I had been instrumental in finding from the British and Imperial forces in Egypt four first-class divisions over and above those which had been deemed possible. Not only had we aided General Eisenhower's Anglo-American Staff upon their victorious career, but we had furnished them with substantial unexpected resources, without which disaster might well have occurred. I was grieved that the small requests I had made for strategic purposes almost as high as those already achieved should have been so obdurately resisted and rejected. Of course, when you are winning a war almost everything that happens can be claimed to be right and wise. It would however have been easy, but for pedantic denials in the minor sphere, to have added the control of the Aegean, and very likely the accession of Turkey, to all the fruits of the Italian campaign.

*   *   *

At the same time, on Kesselring's advice, Hitler changed his mind about his Italian strategy. Till then he had meant to withdraw his forces behind Rome and hold only Northern Italy. Now he ordered them to fight as far south as possible. The line selected, the so-called "Winterstellung," ran behind the river Sangro, on the Adriatic side, across the mountainous spine of Italy, to the mouth of the Garigliano on the west. The natural features of the country, its steep mountains and swift rivers, made this position, several miles in depth, immensely strong. After a year of almost continuous retreat in Africa, Sicily, and Italy the German troops were glad to turn about and fight. They now had nineteen divisions in Italy, and the Allies the equivalent of thirteen. Large reinforcements and much consolidation were required to hold our rapid and brilliant conquests. All this put a strain on our shipping. The first probing efforts at the German line met with little success. Our men had been fighting hard for two months, the weather was shocking, and the troops needed rest and re-grouping. Bridgeheads were thrown across the river, but the main enemy defences lay on high ground beyond. Bad weather, with rain, mud, and swollen rivers, postponed the Eighth Army attack until November 28, but then it made good progress. After a week of heavy fighting we were established ten miles beyond the Sangro. But the enemy still held firm, and more reinforcements came to them from Northern Italy. Some more ground was gained during December, but no vital objectives were taken, and winter weather brought active operations to a close. The U.S. Fifth Army (which included the British Xth Corps), under General Clark, struggled on up the road towards Cassino, and attacked the foremost defences of the German main positions. The enemy were strongly posted on mountains overlooking the road on either side. The formidable Monte Cassino massif to the west was attacked and finally cleared after a tough struggle. But it was not till the beginning of the New Year that the Fifth Army was fully aligned along the river Garigliano and its tributary, the Rapido, where it faced the heights of Cassino and the famous monastery.

Thus the position in Italy was changed greatly to our disadvantage. The Germans were strongly reinforced and ordered to resist instead of to withdraw. The Allies, on the contrary, were sending eight of their best divisions from Italy and the Mediterranean back to England for the cross-Channel attack in 1944. The four extra divisions I was gathering or had sent did not repair the loss. A deadlock supervened, and was not relieved during eight months of severe fighting.

Nevertheless, in spite of these disappointments, the Italian campaign had attracted to itself twenty good German divisions. I had called it the Third Front. If the garrisons kept in the Balkans for fear of attack there are added, nearly forty divisions were retained facing the Allies in the Mediterranean. Our Second Front, Northwest Europe, had not yet flared into battle, but its existence was real. About thirty divisions was the least number ever opposite it, and this rose to sixty as the invasion loomed closer. Our strategic bombing from Britain forced the enemy to divert great numbers of men and masses of material to defend their homeland. These were not negligible contributions to the Russians on what they had every right to call the First Front.

*   *   *

I must end this chapter with a summary.

In this period in the war all the great strategic combinations of the Western Powers were restricted and distorted by the shortage of tank-landing craft for the transport, not so much of tanks, but of vehicles of all kinds. The letters "L.S.T." (Landing Ship, Tanks) are burnt in upon the minds of all those who dealt with military affairs in this period. We had invaded Italy in strong force. We had an army there which, if not supported, might be entirely cast away, giving Hitler the greatest triumph he had had since the fall of France. On the other hand, there could be no question of our not making the "Overlord" attack in 1944. The utmost I asked for was an easement, if necessary, of two months — *i.e.,* from some time in May 1944 to some time in July. This would meet the problem of the landing craft. Instead of their having to return to England in the late autumn of 1943 before the winter gales, they could go in the early spring of 1944. If however the May date were insisted upon pedantically, and interpreted as May 1, the peril to the Allied Army in Italy seemed beyond remedy. If some of the landing craft earmarked for "Overlord" were allowed to stay in the Mediterranean over the winter there would be no difficulty in making a success of the Italian campaign. There were masses of troops not in action in the Mediterranean: three or four French divisions, two or three American divisions, at least four (including the Poles) British or British-controlled divisions. The one thing that stood between these and effective operation in Italy was the L.S.T.s, and the main thing that stood between us and the L.S.T.s was the insistence upon an early date for their return to Britain.

The reader of this narrative must not be misled into thinking (*a*) that I wanted to abandon "Overlord," (*b*) that I wanted to deprive

"Overlord" of vital forces, or (c) that I contemplated a campaign by armies operating in the Balkan peninsula. These are legends. Never had such a wish entered my mind. Give me an easement of six weeks or two months from May 1 in the date of "Overlord" and I could for several months use the landing craft in the Mediterranean in order to bring really effective forces to bear in Italy, and thus not only take Rome but draw off German divisions from either or both the Russian and Normandy fronts. All these matters had been discussed in Washington without regard to the limited character of the issues with which my argument was concerned.

As we shall see presently, in the end everything that I asked for was done. The landing craft not only were made available for upkeep in the Mediterranean; they were even allowed a further latitude for the sake of the Anzio operation in January. This in no way prevented the successful launching of "Overlord" on June 6 with adequate forces. What happened however was that the long fight about trying to get these small easements and to prevent the scrapping of one vast front in order to conform to a rigid date upon the other led to prolonged, unsatisfactory operations in Italy.

5

Arctic Convoys

T HE YEAR 1942 had closed in Arctic waters with a spirited action by British destroyers escorting a convoy to North Russia, which had led to a crisis in the German High Command and the dismissal of Admiral Raeder from control of naval affairs. Between January and March, in the remaining months of almost perpetual darkness, two more convoys of forty-two ships, and six ships sailing independently, set out on this hazardous voyage. Forty arrived. During the same period thirty-six ships were safely brought back from Russian ports and five were lost. The return of daylight made it easier for the enemy to attack the convoys. What was left of the German Fleet, including the *Tirpitz*, was now concentrated in Norwegian waters, and presented a formidable and continuing threat along a large part of the route. The Atlantic battle with the U-boats was moving to a violent crisis. The strain on our destroyers was more than we could bear. The March convoy had to be postponed, and in April the Admiralty proposed, and I agreed, that supplies to Russia by this route should be suspended till the autumn darkness.

*   *   *

The decision was taken with deep regret because of the tremendous battles on the Russian front which distinguished the campaign of 1943. After the spring thaw both sides gathered themselves for a momentous struggle. The Russians, both on land and in the air, had now the upper hand, and the Germans can have had few hopes of ultimate victory. They gained no advantages to make up for their heavy losses, and the new "Tiger" tanks, on which they had counted for success, were mauled by the Russian artillery. Their Army had already been depleted by its previous campaigns in Russia and diluted by inclusion of its second-rate allies. Now, when the Russian blows began to fall, it was unable to parry them. Three immense battles, of Kursk, Orel, and Kharkov, all within a space of two months, marked the ruin of the German army on the Eastern Front. Everywhere they were outfought and overwhelmed. The Russian plan, vast though it was, never outran their resources. It was not only on land that the Russians proved their new superiority. In the air about twenty-five hundred German aircraft were opposed by at least twice as many Russian planes, whose efficiency had been much improved. The German Air Force at this period of the war was at the peak of its strength, numbering about six thousand aircraft in all. That less than half could be spared to support this crucial campaign is proof enough of the value to Russia of our operations in the Mediterranean and of the growing Allied bomber effort based on Britain. In fighter aircraft especially the Germans felt the pinch. Although inferior on the Eastern Front, yet in September they had to weaken it still more in order to defend themselves in the West, where by the winter nearly three-quarters of the total German fighter strength was deployed. The swift and overlapping Russian blows gave the Germans no opportunity to make the best use of their air resources. Air units were frequently moved from one battle area to another in order to meet a fresh crisis, and wherever they went, leaving a gap behind them, they found the Russian planes in overmastering strength.

In September the Germans were in retreat along the whole of their southern front, from opposite Moscow to the Black Sea. The Russians swung forward in full pursuit. At the northern hinge a Russian thrust took Smolensk on September 25. No doubt the Germans hoped to stand on the Dnieper, the next great river line, but by early October the Russians were across it north of Kiev, and to the south at Pereyaslav and Kremenchug. Farther south again Dniepropetrovsk was taken on October 25. Only near the mouth of the river were the Germans still on the western bank of the Dnieper; all the rest had gone. The retreat of the strong German garrison in the Crimea was cut off. Kiev, outflanked on either side, fell on November 6, with many prisoners. By December, after a three months' pursuit, the German armies in Central and South Russia had been thrust back more than two hundred miles, and, failing to hold the Dnieper River line, lay open and vulnerable to a winter campaign in which, as they knew from bitter experience, their opponents excelled. Such was the grand Russian story of 1943.

*　　　　　*　　　　　*

It was natural that the Soviet Government should look reproachfully at the suspension of the convoys, for which their armies hungered. On

Operations in Russia
July – Dec. 1943

Front July 1, 1943	——————
Front Dec. 31, 1943	- - - - - -
German attacks	(arrow)
Russian attacks	(arrow)
Russian pursuits	- ▶ - ▶ - ▶
International frontiers	-·-·-·-

FINLAND

Lake Ladoga

Baltic Sea

Leningrad

ESTONIA

LATVIA

Lake Ilmen

Velikie Luki

Moscow

Viazma

Vitebsk

Smolensk

POLAND (1939)

Mogilev

Bryansk

Minsk

Orel

Voronezh

Pripet Marshes

Gomel

Kursk

R. Don

Bielgorod

Korosten

Kharkov

Kiev

Pereyaslav

U K R A I N E

Jitomir

Kremenchug

R. Dnieper

Dniepropetrovsk

R. Donetz

R. Bug

Krivoi Rog

Nikopol Taganrog

Rostov

Perekop

Sea of Azov

Odessa

RUMANIA

Kerch

| 0 | 50 | 100 | 200 |

MILES

Black

CRIMEA

Sea

Sevastopol

the evening of September 21, Molotov sent for our Ambassador in Moscow and asked for the sailings to be resumed. He pointed out that the Italian Fleet had been eliminated and the U-boats had abandoned the North Atlantic for the southern route. The Persian railway could not carry enough. For three months the Soviet Union had been undertaking a wide and most strenuous offensive, yet in 1943 they had received less than a third of the previous year's supplies. The Soviet Government therefore "insisted" upon the urgent resumption of the convoys, and expected His Majesty's Government to take all necessary measures within the next few days.

When we met in London on the night of the 29th to discuss all this an agreeable new fact was before us. The *Tirpitz* had been disabled by the audacious and heroic attack of our midget submarines. Of six craft which took part, two penetrated all the elaborate defences. Their commanding officers, Lieutenant Cameron, R.N.R., and Lieutenant Place, R.N., rescued by the Germans, survived as prisoners of war, and received the Victoria Cross. Later air reconnaissance showed that the battleship was heavily damaged and would require refit in a dockyard before she could again be ready for action. The *Lützow* had already gone to the Baltic. Thus we had an easement, probably of some months, in the Arctic waters.

But Mr. Eden had serious complaints about the Russian treatment of our men, and I accordingly sent the following telegram to Stalin:

> . . . It is a very great pleasure to me to tell you that we are planning to sail a series of four convoys to North Russia in November, December, January, and February, each of which will consist of approximately thirty-five ships, British and American. . . .

To avoid new charges of breach of faith from the Soviet, if our efforts to help them proved vain, I inserted a safeguarding paragraph:

> However, I must put it on record that this is no contract or bargain, but rather a declaration of our solemn and earnest resolve. On this basis I have ordered the necessary measures to be taken for the sending of these four convoys of thirty-five ships.

I then proceeded with our list of grievances about the treatment of our men in North Russia.

> . . . The present numbers of naval personnel are below what is necessary, even for our present requirements, owing to men having to be sent home without relief. Your civil authorities have refused us all visas for men to go to North Russia, even to relieve

those who are seriously overdue for relief. M. Molotov has pressed His Majesty's Government to agree that the number of British Service personnel in North Russia should not exceed that of the Soviet Service personnel and trade delegation in this country. We have been unable to accept this proposal, since their work is quite dissimilar and the number of men needed for war operations cannot be determined in such an unpractical way. . . .

I must therefore ask you to agree to the immediate grant of visas for the additional personnel now required, and for your assurance that you will not in future withhold visas when we find it necessary to ask for them in connection with the assistance that we are giving you in North Russia. I emphasise that of about one hundred and seventy naval personnel at present in the North over one hundred and fifty should have been relieved some months ago, but Soviet visas have been withheld. The state of health of these men, who are unaccustomed to the climatic and other conditions, makes it very necessary to relieve them without further delay. . . .

I must also ask your help in remedying the conditions under which our Service personnel and seamen at present find themselves in North Russia. These men are of course engaged in operations against the enemy in our joint interest, and chiefly to bring Allied supplies to your country. They are, I am sure you will admit, in a wholly different position from ordinary individuals proceeding to Russian territory. Yet they are subjected by your authorities to the following restrictions, which seem to me inappropriate for men sent by an ally to carry out operations of the greatest interest to the Soviet Union:

(a) No one may land from one of H.M. ships or from a British merchant ship except by a Soviet boat in the presence of a Soviet official and after examination of documents on each occasion.

(b) No one from a British warship is allowed to proceed alongside a British merchantman without the Soviet authorities being informed beforehand. This even applies to the British admiral in charge.

(c) British officers and men are required to obtain special passes before they can go from ship to shore or between two British shore stations. These passes are often much delayed, with consequent dislocation of the work in hand.

(d) No stores, luggage, or mail for this operational force may be landed except in the presence of a Soviet official, and numerous formalities are required for the shipment of all stores and mail.

(e) Private Service mail is subjected to censorship, although
for an operational force of this kind censorship should, in
our view, be left in the hands of British Service authorities.

The imposition of these restrictions makes an impression upon
officers and men alike which is bad for Anglo-Soviet relations, and
would be deeply injurious if Parliament got to hear of it. The
cumulative effect of these formalities has been most hampering to
the efficient performance of the men's duties, and on more than one
occasion to urgent and important operations. No such restrictions
are placed upon Soviet personnel here. . . . I trust indeed, M.
Stalin, that you will find it possible to have these difficulties
smoothed out in a friendly spirit, so that we may help each other,
and the common cause, to the utmost of our strength.

These were modest requests considering the efforts we were now to
make. I did not receive Stalin's answer for nearly a fortnight. This was
the reply I got:

I received your message of October 1 informing me of the intention to send four convoys to the Soviet Union by the northern
route in November, December, January, and February. However,
this communication loses its value by your statement that this intention to send northern convoys to the U.S.S.R. is neither an
obligation nor an agreement, but only a statement, which, as it
may be understood, is one the British side can at any moment re-
nounce regardless of any influence it may have on the Soviet
armies at the front. I must say that I cannot agree with such a
posing of the question. Supplies from the British Government to
the U.S.S.R., armaments and other military goods, cannot be considered otherwise than as an obligation, which, by special agreement between our countries, the British Government undertook in
respect of the U.S.S.R., which bears on its shoulders, already for
the third year, the enormous burden of struggle with the common
enemy of the Allies — Hitlerite Germany. . . . As experience has
shown, delivery of armaments and military supplies to the
U.S.S.R. through Persian ports cannot compensate in any way for
those supplies which were not delivered by the northern route .
. . . It is impossible to consider this posing of the question to be
other than a refusal of the British Government to fulfil the obligations it undertook, and as a kind of threat addressed to the
U.S.S.R.
Concerning your mention of controversial points allegedly con-

tained in the statement of M. Molotov, I have to say that I do not find any foundation for such a remark. . . . I do not see the necessity for increasing the number of British servicemen in the North of the U.S.S.R., since the great majority of British servicemen who are already there are not adequately employed, and for many months have been doomed to idleness, as has already been pointed out several times by the Soviet side. . . . There are also regrettable facts of the inadmissible behaviour of individual British servicemen who attempted, in several cases, to recruit, by bribery, certain Soviet citizens for Intelligence purposes. Such instances, offensive to Soviet citizens, naturally gave rise to incidents which led to undesirable complications.

Concerning your mention of formalities and certain restrictions existing in northern ports, it is necessary to have in view that such formalities and restrictions are unavoidable in zones near and at the front, if one does not forget the war situation which exists in the U.S.S.R. . . . Nevertheless the Soviet authorities granted many privileges in this respect to the British servicemen and seamen, about which the British Embassy was informed as long ago as last March. Thus your mention of many formalities and restrictions is based on inaccurate information.

Concerning the question of censorship and prosecution of British servicemen, I have no objection if the censorship of private mail for British personnel in northern ports would be made by the British authorities themselves, on condition of reciprocity. . . .

"I have now," I commented to the President, "received a telegram from U.J. which I think you will feel is not exactly all one might hope for from a gentleman for whose sake we are to make an inconvenient, extreme, and costly exertion. . . . I think, or at least I hope, this message came from the machine rather than from Stalin, as it took twelve days to prepare. The Soviet machine is quite convinced it can get everything by bullying, and I am sure it is a matter of some importance to show that this is not necessarily always true."

*　　　*　　　*

On the 18th I asked the Soviet Ambassador to come to see me. As this was the first occasion on which I had met M. Gousev, who had succeeded Maisky, he gave me the greetings of Marshal Stalin and Molotov, and I told him of the good reputation he had made for himself with us in Canada. After these compliments we had a short discussion about the Second Front. I spoke to him earnestly about the great desire we had to work with Russia and to be friends with her, how we saw that she should have a great place in the world after the war, that we should welcome this, and that we would do our best also to make good relations between her and the United States.

I then turned to Stalin's telegram about the convoys. I said very briefly that I did not think this message would help the situation, that it had caused me a good deal of pain, that I feared any reply which I could send would only make things worse, that the Foreign Secretary was in Moscow and I had left it to him to settle the matter on the spot and that therefore I did not wish to receive the message. I then handed back to the Ambassador an envelope. Gousev opened the envelope to see what was inside it, and, recognising the message, said he had been instructed to deliver it to me. I then said, "I am not prepared to receive it," and got up to indicate in a friendly manner that our conversation was at an end. I moved to the door and opened it. We had a little talk in the doorway about his coming to luncheon in the near future and discussing with Mrs. Churchill some questions connected with her Russian fund, which I told him had now reached four million pounds. I did not give M. Gousev a chance of recurring to the question of the convoys or of trying to hand me back the envelope, and bowed him out.

The War Cabinet endorsed my refusal to receive Stalin's telegram. It was certainly an unusual diplomatic incident, and, as I learnt later, it impressed the Soviet Government. In fact, Molotov referred to it several times in conversation. Even before it could be reported to Moscow there were misgivings in Soviet circles. On October 19 Mr. Eden, who had arrived there for a long-planned conference of the Foreign Secretaries of the three major Allies, telegraphed that Molotov had called on him at the Embassy and said that his Government greatly valued the convoys and had sadly missed them. The northern route was the shortest and quickest way of getting supplies to the front, where the Russians were going through a difficult time. The German winter defence line had to be broken. Molotov promised to speak to Stalin about it all and arrange a meeting.

The important discussion took place on the 21st. Meanwhile, in order to strengthen Eden's hands, and at his suggestion, I suspended the sailing of the British destroyers, which was the first move in the resumption of the convoys. Eventually it was arranged that the convoys should be resumed. The first started in November, and a second followed it in December. Between them they comprised seventy-two ships. All arrived safely, and at the same time return convoys of empty ships were successfully brought out.

*　　　*　　　*

The December outward-bound convoy was to bring about a gratifying naval engagement. The disablement of the *Tirpitz* had left the *Scharnhorst* the only heavy enemy ship in Northern Norway. She sallied forth from Alten Fiord with five destroyers on the evening of Christmas Day, 1943, to attack the convoy about fifty miles south of Bear Island. The reinforced convoy escort comprised fourteen destroyers, with a covering force of three cruisers. The Commander-in-Chief, Admiral Fraser, lay to the southwestward in his flagship, the *Duke of York*, with the cruiser *Jamaica* and four destroyers.

Twice the *Scharnhorst* tried to strike at the convoy. Each time she was intercepted and engaged by the escort cruisers and destroyers, and after indecisive fighting, in which both the *Scharnhorst* and the British cruiser *Norfolk* were hit, the Germans broke off the action and withdrew to the southward, shadowed and reported by our cruisers. The German destroyers were never seen and took no part. Meanwhile the Commander-in-Chief was approaching at his utmost speed through heavy seas. At 4.17 P.M., when the last of the Arctic twilight had long since gone, the *Duke of York* detected the enemy by radar at about twenty-three miles. The *Scharnhorst* remained unaware of her approaching doom, until, at 4.50 P.M., the *Duke of York* opened fire at 12,000 yards with the aid of star shell. At the same time Admiral Fraser sent his four destroyers in to attack when opportunity offered. One of these, the *Stord,* was manned by the Royal Norwegian Navy. The *Scharnhorst* was surprised, and turned away to the eastward. In a running fight she suffered several hits, but was able with her superior speed gradually to draw ahead. However, by 6.20 P.M. it became apparent that her speed was beginning to fall and our destroyers were able to close in on either flank. At about 7 P.M. they all pressed home their attacks. Four torpedoes struck. Only one destroyer was hit.

The *Scharnhorst* turned to drive off the destroyers, and thus the *Duke of York* was able to close rapidly to about ten thousand yards and reopen fire with crushing effect. In half an hour the unequal battle between a battleship and a wounded battle cruiser was over, and the *Duke of York* left the cruisers and destroyers to complete the task. The *Scharnhorst* soon sank, and of her company of 1970 officers and men, including Rear Admiral Bey, we could only save thirty-six men.

Although the fate of the crippled *Tirpitz* was delayed for nearly a year, the sinking of the *Scharnhorst* not only removed the worst menace to our Arctic convoys but gave new freedom to our Home Fleet. We no longer had to be prepared at our average moment against German heavy ships breaking out into the Atlantic at their selected moment. This was an important relief. When in April 1944 there were signs that the *Tirpitz* had been repaired sufficiently to move for refit to a Baltic port, aircraft from the carriers *Victorious* and *Furious* attacked her with heavy bombs, and she was once more immobilised. The Royal Air Force now took up the attack from a base in North Russia. They succeeded in causing further damage, which led to the *Tirpitz* being removed to Tromsö Fiord, which was two hundred miles nearer to Britain, and within the extreme range of our home-based heavy bombers. The Germans had now abandoned hope of getting the ship home for repair and had written her off as a seagoing fighting unit. On November 12 twenty-nine specially fitted Lancasters of the Royal Air Force, including those of 617 Squadron, famous for the Möhne Dam exploit, struck the decisive blow, with bombs of twelve thousand pounds weight. They had to fly over two thousand miles from their bases in Scotland, but the weather was clear and three bombs hit their target. The *Tirpitz* capsized at her moorings, more than half of her crew of 1900 men being killed, at the cost of one bomber, whose crew survived.

All British heavy ships were now free to move to the Far East.

*   *   *

In the whole of the war ninety-one merchant ships were lost on the Arctic route, amounting to 7.8 per cent of the loaded vessels outward bound and 3.8 per cent of those returning. Only fifty-five of these were in escorted convoys. Of about four million tons of cargo dispatched from America and the United Kingdom, an eighth was lost. In this arduous work the Merchant Navy lost 829 lives, while the Royal Navy paid a still heavier price. Two cruisers and seventeen other warships were sunk and 1840 officers and men died.

The forty convoys to Russia carried the huge total of £428,000,000 worth of material, including 5000 tanks and over 7000 aircraft from Britain alone. Thus we redeemed our promise, despite the many hard words of the Soviet leaders and their harsh attitude towards our rescuing sailors.

6

Teheran: The Opening

I HAD hardly got home after my visits to the Citadel, the White House, and Hyde Park during the Quebec Conference in August and September 1943, when I turned to the theme of a meeting of the three heads of Governments which logically followed the Anglo-American conversations. In principle there was general agreement that this was urgent and imperative, but no one who did not live through it can measure the worries and complications which attended the fixing of the time, place, and conditions of this, the first conference of what were then called the Big Three.

Several serious aspects of the impending conference absorbed my mind. The selection of a Supreme Commander for "Overlord," our cross-Channel entry into Europe in 1944, was urgent. This of course affected in the most direct manner the military conduct of the war, and raised a number of personal issues of importance and delicacy. At the Quebec Conference I had agreed with the President that "Overlord" should fall to an American officer, and had so informed General Brooke, to whom I had previously offered the task. I understood from Mr. Roosevelt that he would choose General Marshall, and this was entirely satisfactory to us. However, in the interval between Quebec and our meeting in Cairo I became conscious that the President had not finally made up his mind about Marshall. None of the other arrangements could of course be made before the main decision had been taken. Meanwhile rumour became rife in the American press, and there was the prospect of Parliamentary reactions in London.

I also thought it most important that the British and American Staffs, and above them the President and I, should reach a general agreement on the policy of "Overlord" and its impingement on the Mediterranean. The whole armed strength overseas of our two countries was involved, and the British forces were to be equal at the out-set of "Overlord," twice as strong as the Americans in Italy, and three times as numerous in the rest of the Mediterranean. Surely we ought to reach some solid understanding before inviting the Soviet repre-sentatives, either political or military, to join us. The President ap-peared to favour the idea, but not the timing. There was emerging a strong current of opinion in American Government circles which seemed to wish to win Russian confidence even at the expense of co-ordinating the Anglo-American war effort. I on the other hand felt it of the utmost importance that we should meet the Russians with a clear and united view both on the outstanding problems of "Overlord" and upon the question of the High Commands. I wished the proceedings to take three stages: first, a broad Anglo-American agreement at Cairo; secondly, a Supreme Conference between the three heads of the Govern-ments of the three major Powers at Teheran; and, thirdly, on returning to Cairo, the discussion of what was purely Anglo-American business about the war in the Indian theatre and the Indian Ocean, which was certainly urgent. I did not want the short time we had at our disposal to be absorbed in what were after all comparatively minor matters, when the decision involving the course of the whole war demanded at least provisional settlement. Mr. Roosevelt agreed to come to Cairo first, but he wanted Molotov to come as well and also the Chinese. Nothing, however, would induce Stalin to compromise his relations with the Japanese by entering a four-Power conference with their three enemies. All question of Soviet representatives coming to Cairo was thus nega-tived. This was in itself a great relief. It was obtained however at a serious inconvenience and subsequent cost.

*   *   *

On the afternoon of November 12 I sailed in the *Renown* from Plymouth with my personal staff on a journey which was to keep me from England for more than two months. After calling at Algiers and Malta, we reached Alexandria on the morning of the 21st. I flew at once to the desert landing ground near the Pyramids. Here Mr. Casey had placed at my disposal the agreeable villa he was using. We lay in a broad expanse of Kasserine woods, thickly dotted with the luxurious abodes and gardens of the cosmopolitan Cairo magnates. Generalissimo Chiang Kai-Shek and Madame had already been ensconced half a mile away. The President was to occupy the spacious villa of the American Ambassador Kirk, about three miles down the road to Cairo. I went to the desert airfield to welcome him when he arrived in the "Sacred Cow" next morning, and we drove to his villa together.

The Staffs congregated rapidly. The headquarters of the Conference and the venue of all the British and American Chiefs of Staff was at the Mena House Hotel, opposite the Pyramids, and I was but half a mile away. The whole place bristled with troops and anti-aircraft guns, and the strictest cordons guarded all approaches. Everyone set to work at once at their various levels upon the immense mass of business which had to be decided or adjusted.

What we had apprehended from Chiang Kai-shek's presence now in fact occurred. The talks of the British and American Staffs were sadly distracted by the Chinese story, which was lengthy, complicated, and minor. Moreover, as will be seen, the President, who took an exaggerated view of the Indian-Chinese sphere, was soon closeted in long conferences with the Generalissimo. All hope of persuading Chiang and his wife to go and see the Pyramids and enjoy themselves till we returned from Teheran fell to the ground, with the result that Chinese business occupied first instead of last place at Cairo. The President, in spite of my arguments, gave the Chinese the promise of a considerable amphibious operation across the Bay of Bengal within the next few months. This would have cramped "Overlord" for landing- and tank-landing craft, which had now become the bottleneck, far more than any of my Turkey and Aegean projects. It would also have hampered grievously the immense operations we were carrying out in Italy. On November 29 I wrote to the Chiefs of Staff: "The Prime Minister wishes to put on record the fact that he specifically refused the Generalissimo's request that we should undertake an amphibious operation simultaneously with the land operations in Burma." It was not until we returned from Teheran to Cairo that I at length prevailed upon the President to retract his promise. Even so, many complications arose. Of this more anon.

I of course took occasion to visit the Generalissimo at his villa, where he and his wife were suitably installed. This was the first time I had met Chiang Kai-shek. I was impressed by his calm, reserved, and efficient personality. At this moment he stood at the height of his power and fame. To American eyes he was one of the dominant forces in the world. He was the champion of "the New Asia." He was certainly a steadfast defender of China against Japanese invasion. He was a strong anti-Communist. The accepted belief in American circles was that he would be the head of the great Fourth Power in the world after the victory had been won. All these views and values have since been cast aside by many of those who held them. I, who did not in those day share the excessive estimates of Chiang Kai-shek's power or of the future helpfulness of China, may record the fact that the Generalissimo is still serving the same causes which at this time had gained him such wide renown. He has however since been beaten by the Communists in his own country, which is a very bad thing to be. I had a very pleasant conversation with Madame Chiang Kai-shek, and found her a most remarkable and charming personality. The President had us all photographed together at one of our meetings at his villa, and although both the Generalissimo and his wife are now regarded as wicked and corrupt reactionaries by many of their former admirers I am glad to keep this as a souvenir.

*   *   *

On November 24 a meeting of our Combined Chiefs of Staff was held by the President, without the presence of the Chinese delegation, to discuss operations in Europe and the Mediterranean. We sought to survey the relations of the two theatres and to exchange our views before going on to Teheran. The President opened upon the effect on "Overlord" of any possible action we could take in the meantime in the Mediterranean, including the problem of Turkey's entry into the war.

When I spoke I said "Overlord" remained top of the bill, but this operation should not be such a tyrant as to rule out every other activity in the Mediterranean; for example, a little flexibility in the employment of landing craft ought to be conceded. General Alexander had asked that the date of their leaving for "Overlord" should be deferred from mid-December to mid-January. Eighty additional L.S.T.s had been ordered to be built in Britain and Canada. We should try to do even better than this. The points which were at issue between the American and British Staffs would probably be found to affect no more than a tenth of our common resources, apart from the Pacific. Surely some degree of elasticity could be arranged. Nevertheless I wished to remove any idea that we had weakened, cooled, or were trying to get out of "Overlord." We were in it up to the hilt. To sum up, I said that the programme I advocated was to try to take Rome in January and Rhodes in February; to renew supplies to the Yugoslavs, settle the Command arrangements, and to open the Aegean, subject to the outcome of an approach to Turkey; all preparations for "Overlord" to go ahead full steam within the framework of the foregoing policy for the Mediterranean.

Mr. Eden now joined us from England, whither he had flown after his discussions in Moscow. His arrival was a great help to me. On the way back from the Moscow Conference he and General Ismay had met the Turkish Foreign Minister and other Turks. Mr. Eden pointed out that we had urgent need of air bases in the southwest of Anatolia. He explained that our military situation at Leros and Samos was precarious, owing to German air superiority. Both places had since been lost. Mr.

Eden also dwelt on the advantages that would be derived from Turkey's entry into the war. In the first place, it would oblige the Bulgarians to concentrate their forces on the frontier, and thus compel the Germans to replace Bulgarian troops in Greece and Yugoslavia to the extent of some ten divisions. Secondly, it would be possible to attack the one target which might be decisive — the oil wells at Ploesti. Thirdly, Turkish chrome would be cut off from Germany. Finally, there was the moral advantage. Turkey's entry into the war might well hasten the process of disintegration in Germany and among her satellites. By all this argument the Turkish delegation were unmoved. They said, in effect, that the granting of bases in Anatolia would amount to intervention in the war, and that if they intervened in the war there was nothing to prevent a German retaliation on Constantinople, Angora, and Smyrna. They refused to be comforted by the assurances that we would give them sufficient fighters to deal with any air attack that the Germans could launch and that the Germans were so stretched everywhere that they had no divisions available to attack Turkey. The only result of the discussions was that the Turkish delegation promised to report to their Government. Considering what had been happening under their eyes in the Aegean, the Turks can hardly be blamed for their caution.

Finally, there was the issue of the High Commands. Neither the President nor any of his immediate circle referred to the matter in any way on the occasions, formal and informal but always friendly, when we came into contact. I therefore rested under the impression that General Marshall would command "Overlord," that General Eisenhower would succeed him in Washington, and that it would fall to me, representing His Majesty's Government, to choose the Mediterranean commander, who at that time I had no doubt would be Alexander, already waging the war in Italy. Here the issue rested till we returned to Cairo.

*   *   *

There have been many misleading accounts of the line I took, with the full agreement of the British Chiefs of Staff, at the Triple Conference at Teheran. It has become a legend in America that I strove to prevent the cross-Channel enterprise called "Overlord," and that I tried vainly to lure the Allies into some massive invasion of the Balkans, or a large-scale campaign in the Eastern Mediterranean, which would effectively kill it. Much of this nonsense has already in previous chapters been exposed and refuted, but it may be worth while to set forth what it was I actually sought, and what, in a very large measure, I got.

"Overlord," now planned in great detail, should be launched in May

or June, or at the lastest in the opening days of July 1944. The troops
and all the ships to carry them still had first priority. Secondly, the
great Anglo-American army in action in Italy must be nourished to
achieve the capture of Rome and advance to secure the airfields north
of the capital, from which the air attack on Southern Germany became
possible. After these were gained there should be no advance in Italy
beyond the Pisa–Rimini line — *i.e.,* we should not extend our front into
the broader part of the Italian peninsula. These operations, if resisted
by the enemy, would attract and hold very large German forces, would
give the Italians the chance to "work their passage," and keep the flame
of war burning continually upon the hostile front.

I was not opposed at this time to a landing in the South of France,
along the Riviera, with Marseilles and Toulon as objectives, and thereafter an Anglo-American advance northward up the Rhone Valley in
aid of the main invasion across the Channel. Alternatively, I preferred
a right-handed movement from the north of Italy, using the Istrian
peninsula and the Ljubljana Gap, towards Vienna. I was delighted
when the President suggested this, and tried, as will be seen, to engage
him in it. If the Germans resisted we should attract many of their divisions from the Russian or Channel fronts. If we were not resisted we
should liberate at little cost enormous and invaluable regions. I was
sure we should be resisted, and thus help "Overlord" in a decisive
manner.

My third request was that the Eastern Mediterranean, with all the
prizes that it afforded, should not be neglected, provided no strength
which could be applied across the Channel should be absorbed. In all
this I adhered to the proportions which I had mentioned to General
Eisenhower two months earlier — namely, four-fifths in Italy, one-
tenth in Corsica and the Adriatic, and one-tenth in the Eastern Mediterranean. From this I never varied — not an inch in a year.

We were all agreed, British, Russians, and Americans, upon the first
two, involving nine-tenths of our available strength. All I had to plead
was the effective use of one-tenth of our strength in the Eastern Mediterranean. Simpletons will argue, "Would it not have been much better
to centre all upon the decisive operation and dismiss all other opportunities as wasteful diversions?" But this ignores the governing facts. All
the available shipping in the Western Hemisphere was already committed to the last ton to the preparation of "Overlord" and the maintenance of our front in Italy. Even if more shipping had been found it
could not have been used, because the programmes of disembarkation
filled to the utmost limit all the ports and camps involved. As for the
Eastern Mediterranean, nothing was needed that could be applied
elsewhere. The Air Force massed for the defence of Egypt could equally

well or better discharge its duty if used from a forward frontier. All the troops, two or three divisions at the outside, were already in that theatre, and there were no ships, except local vessels, to carry them to the larger scenes. To get the active, vigorous use of these forces, who otherwise would be mere lookers-on, might inflict grave injury upon the enemy. If Rhodes were taken the whole Aegean could be dominated by our Air Force and direct sea contact established with Turkey. If, on the other hand, Turkey could be persuaded to enter the war, or to strain her neutrality by lending us the airfields we had built for her, we could equally dominate the Aegean and the capture of Rhodes would not be necessary. Either way it would work.

And of course the prize was Turkey. If we could gain Turkey it would be possible without the subtraction of a single man, ship, or aircraft from the main and decisive battles to dominate the Black Sea with submarines and light naval forces, and to give a right hand to Russia and carry supplies to her armies by a route far less costly, far more swift, and far more abundant than either the Arctic or the Persian Gulf.

This was the triple theme which I pressed upon the President and Stalin on every occasion, not hesitating to repeat the arguments remorselessly. I could have gained Stalin, but the President was oppressed by the prejudices of his military advisers and drifted to and fro in the argument, with the result that the whole of these subsidiary but gleaming opportunities were cast aside unused. Our American friends were comforted in their obstinacy by the reflection that "at any rate we have stopped Churchill entangling us in the Balkans." No such idea had ever crossed my mind. I regard the failure to use otherwise unemployable forces to bring Turkey into the war and dominate the Aegean as an error in war direction which cannot be excused by the fact that in spite of it victory was won.

*   *   *

The first plenary meeting was held at the Soviet Embassy on Sunday, November 28, at 4 P.M. The conference room was spacious and handsome, and we seated ourselves at a large, round table. I had with me Eden, Dill, the three Chiefs of Staff, and Ismay. The President had Harry Hopkins, Admiral Leahy, Admiral King, and two other officers. General Marshall and General Arnold were not present: "they had misunderstood the time of the meeting," says Hopkins's biographer, "and had gone off on a sight-seeing tour round Teheran."[1] I had my admirable interpreter of the previous year, Major Birse. Pavlov again performed this service for the Soviets, and Mr. Bohlen, a new figure,

---

[1] Robert E. Sherwood, *Roosevelt and Hopkins*, p. 778.

for the United States. Molotov and Marshal Voroshilov alone accompanied Stalin. He and I sat almost opposite one another. The discussion on this first day came to a crucial point. The record says:

> Marshal Stalin addressed the following questions to the Prime Minister:
>
> *Question:* "Am I right in thinking that the invasion of France is to be undertaken by thirty-five divisions?"
>
> *Answer:* "Yes. Particularly strong divisions."
>
> *Question:* "Is it intended that this operation should be carried out by the forces now in Italy?"
>
> *Answer:* "No. Seven divisions have already been, or are in process of being, withdrawn from Italy and North Africa to take part in 'Overlord.' These seven divisions are required to make up the thirty-five divisions mentioned in your first question. After they have been withdrawn about twenty-two divisions will be left in the Mediterranean for Italy or other objectives. Some of these could be used either for an operation against Southern France or for moving from the head of the Adriatic towards the Danube. Both these operations will be timed in conformity with 'Overlord.' Meanwhile it should not be difficult to spare two or three divisions to take the islands in the Aegean."

The formal conferences were interspersed with what may be thought to be even more important talks between Roosevelt, Stalin, and myself at luncheons and dinners. Here there were very few things that could not be said and received in good humour. On this night the President was our host for dinner. We were a party of ten on eleven, including the interpreters, and conversation soon became general and serious.

After dinner, when we were strolling about the room, I led Stalin to a sofa and suggested that we talk for a little on what was to happen after the war was won. He assented, and we sat down. Eden joined us. "Let us," said the Marshal, "first consider the worst that might happen." He thought that Germany had every possibility of recovering from this war, and might start on a new one within a comparatively short time. He feared the revival of German nationalism. After Versailles peace had seemed assured, but Germany had recovered very quickly. We must therefore establish a strong body to prevent Germany starting a new war. He was convinced that she would recover. When I asked, "How soon?" he replied, "Within fifteen to twenty years." I said that the world must be made safe for at least fifty years. If it was only for fifteen to twenty years then we should have betrayed our soldiers.

Stalin thought we should consider restraints on Germany's manufacturing capacity. The Germans were an able people, very industrious
and resourceful, and they would recover quickly. I replied that there
would have to be certain measures of control. I would forbid them all
aviation, civil and military, and I would forbid the General Staff
system. "Would you," asked Stalin, "also forbid the existence of watch-
makers' and furniture factories for making parts of shells? The Germans produced toy rifles which were used for teaching hundreds of
thousands of men how to shoot."

"Nothing," I said, "is final. The world rolls on. We have now learnt
something. Our duty is to make the world safe for at least fifty years by
German disarmament, by preventing rearmament, by supervision of
German factories, by forbidding all aviation, and by territorial changes
of a far-reaching character. It all comes back to the question whether
Great Britain, the United States, and the U.S.S.R. can keep a close
friendship and supervise Germany in their mutual interest. We ought
not to be afraid to give orders as soon as we see any danger."

"There was control after the last war," said Stalin, "but it failed."

"We were inexperienced then," I replied. "The last war was not to
the same extent a national war, and Russia was not a party at the Peace
Conference. It will be different this time." I had a feeling that Prussia
should be isolated and reduced, that Bavaria, Austria, and Hungary
might form a broad, peaceful, unaggressive confederation. I thought
Prussia should be dealt with more sternly than the other parts of the
Reich, which might thus be influenced against throwing in their lot
with her. It must be remembered that these were wartime moods.

"All very good, but insufficient," was Stalin's comment.

Russia, I continued, would have her army, Great Britain and the
United States their navies and air forces. In addition, all three Powers
would have their other resources. All would be strongly armed, and
must not assume any obligation to disarm. "We are the trustees for
the peace of the world. If we fail there will be perhaps a hundreds years
of chaos. If we are strong we can carry out our trusteeship. There is
more," I went on, "than merely keeping the peace. The three Powers
should guide the future of the world. I do not want to enforce any
system on other nations. I ask for freedom and for the right of all
nations to develop as they like. We three must remain friends in order
to ensure happy homes in all countries."

Stalin asked again what was to happen to Germany.

I replied that I was not against the toilers in Germany, but only
against the leaders and against dangerous combinations. He said that
there were many toilers in the German divisions who fought under
orders. When he asked German prisoners who came from the labouring

classes (such is the record, but he probably meant "Communist Party") why they fought for Hitler they replied that they were carrying out orders. He shot such prisoners.

* * *

I then suggested we should discuss the Polish question. He agreed and invited me to begin. I said that we had declared war on account of Poland. Poland was therefore important to us. Nothing was more important than the security of the Russian western frontier. But I had given no pledges about frontiers. I wanted heart-to-heart talks with the Russians about this. When Marshal Stalin felt like telling us what he thought about it the matter could be discussed and we could reach some agreement, and the Marshal should tell me what was necessary for the defence of the western frontiers of Russia. After this war in Europe, which might end in 1944, the Soviet Union would be overwhelmingly strong and Russia would have a great responsibility in any decision she took with regard to Poland. Personally I thought Poland might move westward, like soldiers taking two steps "left close." If Poland trod on some German toes that could not be helped, but there must be a strong Poland. Poland was an instrument needed in the orchestra of Europe.

Stalin said the Polish people had their culture and their language, which must exist. They could not be extirpated.

"Are we to try," I asked, "to draw frontier lines?"

"Yes."

"I have no power from Parliament, nor, I believe, has the President, to define any frontier lines. But we might now, in Teheran, see if the three heads of Governments, working in agreement, could form some sort of policy which we could recommend to the Poles and advise them to accept."

Stalin asked whether this was possible without Polish participation. I said "Yes," and that when this was all informally agreed between ourselves we could go to the Poles later. Mr. Eden here remarked that he had been much struck by Stalin's statement that afternoon that the Poles could go as far west as the Oder. He saw hope in that and was much encouraged. Stalin asked whether we thought he was going to swallow Poland up. Eden said he did not know how much the Russians were going to eat. How much would they leave undigested? Stalin said the Russians did not want anything belonging to other people, although they might have a bite at Germany. Eden said that what Poland lost in the east she might gain in the west. Stalin replied that possibly she might, but he did not know. I then demonstrated with the help of three matches my idea of Poland moving westward. This pleased Stalin, and on this note our group parted for the moment.

* * *

The morning of the 29th was occupied by a conference of the British, Soviet, and American military chiefs. As I knew that Stalin and Roosevelt had already had a private conversation, and were in fact staying at the same Embassy, I suggested that the President and I might lunch together before the second plenary meeting that afternoon. Roosevelt however declined, and sent Harriman to me to explain that he did not want Stalin to know that he and I were meeting privately. I was surprised at this, for I thought we all three should treat each other with equal confidence. The President after luncheon had a further interview with Stalin and Molotov, at which many important matters were discussed, including particularly Mr. Roosevelt's plan for the government of the postwar world. This should be carried out by the "Four Policemen," namely, the U.S.S.R., the United States, Great Britain, and China. Stalin did not react favourably to this. He said the "Four Policemen" would not be welcomed by the small nations of Europe. He did not believe that China would be very powerful when the war ended, and even if she were, European states would resent having China as an enforcement authority for themselves. In this the Soviet leader certainly showed himself more prescient and possessed of a truer sense of values than the President. When Stalin proposed as an alternative that there should be one committee for Europe and another for the Far East — the European committee to consist of Britain, Russia, the United States, and possibly one other European nation — the President replied that this was somewhat similar to my idea of regional committees, one for Europe, one for the Far East, and one for the Americas. He does not seem to have made it clear that I also contemplated a Supreme United Nations Council, of which the three regional committees would be the components. As I was not informed till much later of what had taken place I was not able to correct this erroneous presentation.

Before our second plenary session began at four o'clock I presented, by the King's command, the Sword of Honour which His Majesty had had specially designed and wrought to commemorate the glorious defence of Stalingrad. The large outer hall was filled with Russian officers and soldiers. When, after a few sentences of explanation, I handed the splendid weapon to Stalin he raised it in a most impressive gesture to his lips and kissed the scabbard. He then passed it to Voroshilov, who dropped it. It was carried from the room in great solemnity escorted by a Russian guard of honour. As this procession moved away I saw the President sitting at the side of the room, obviously stirred by the ceremony. We then moved to the conference chamber and took our seats again at the round table, this time with all the Chiefs of Staff, who now reported the result of their morning's labours.

In the discussions which followed I reminded Stalin of the three conditions on which the success of "Overlord" depended. First, there must be a satisfactory reduction in the strength of the German fighter force in Northwest Europe between now and the assault. Secondly, German reserves in France and the Low Countries must not be more on the day of the assault than about twelve full-strength first-quality mobile divisions. Thirdly, it must not be possible for the Germans to transfer from other fronts more than fifteen first-quality divisions during the first sixty days of the operation. To obtain these conditions we should have to hold as many Germans as possible in Italy and Yugoslavia. If Turkey entered the war, this would be an added help, but not an essential condition. The Germans now in Italy had for the most part come from France. If we slackened off our pressure in Italy they would go back again. We must continue to engage the enemy on the only front where at present we could fight them. If we engaged them as fiercely as possible during the winter months in the Mediterranean this would make the best possible contribution towards creating the conditions needed for a successful "Overlord."

Stalin asked what would happen if there were thirteen or fourteen mobile divisions in France and more than fifteen available from other fronts. Would this rule out "Overlord"?

I said, "No, certainly not."

Before we separated, Stalin looked at me across the table and said, "I wish to pose a very direct question to the Prime Minister about 'Overlord.' Do the Prime Minister and the British Staff really believe in 'Overlord'?" I replied, "Provided the conditions previously stated for 'Overlord' are established when the time comes, it will be our stern duty to hurl across the Channel against the Germans every sinew of our strength." On this we separated.

7

Teheran: Crux and Conclusions

N OVEMBER 30 was for me a crowded and memorable day. It was my sixty-ninth birthday, and was passed almost entirely in transacting some of the most important business with which I have ever been concerned. The fact that the President was in private contact with Marshal Stalin and dwelling at the Soviet Embassy, and that he had avoided ever seeing me alone since we left Cairo, in spite of our hitherto intimate relations and the way in which our vital affairs were interwoven, led me to seek a direct personal interview with Stalin. I felt that the Russian leader was not deriving a true impression of the British attitude. The false idea was forming in his mind that, to put it shortly, "Churchill and the British Staffs mean to stop 'Overlord' if they can, because they want to invade the Balkans instead." It was my duty to remove this double misconception.

The exact date of "Overlord" depended upon the movements of a comparatively small number of landing craft. These landing craft were not required for any operation in the Balkans. The President had committed us to an operation against the Japanese in the Bay of Bengal. If this were cancelled there would be enough landing craft for all I wanted, namely, the amphibious power to land against opposition two divisions at a time on the coasts of Italy or Southern France, and also to carry out "Overlord" as planned in May. I had agreed with the President that May should be the month, and he had, for his part, given up the specific date of May 1. This would give me the time I needed. If I could persuade the President to obtain relief from his promise to Chiang Kai-shek and drop the Bay of Bengal plan, which had never been mentioned in our Teheran conferences, there would be enough landing craft both for the Mediterranean and for a punctual "Overlord." In the event the great landings began on June 6, but this

date was decided much later on, not by any requirements of mine, but
by the moon and the weather. I also succeeded when we returned to
Cairo, as will be seen, in persuading the President to abandon the
enterprise in the Bay of Bengal. I therefore consider that I got what
I deemed imperative. But this was far from certain at Teheran on this
November morning. I was determined that Stalin should know the
main facts. I did not feel entitled to tell him that the President and I
had agreed upon May for "Overlord." I knew that Roosevelt wanted to
tell him this himself at our luncheon which was to follow my conversation with the Marshal.

The following is founded upon the record made by Major Birse, my
trusted interpreter, of my private talk with Stalin.

*   *   *

I began by reminding the Marshal that I was half American and had
a great affection for the American people. What I was going to say was
not to be understood as disparaging to the Americans, and I would be
perfectly loyal towards them, but there were things which it was better
to say outright between two persons.

We had a preponderance of troops over the Americans in the Mediterranean. There were two or three times more British troops than American there. That was why I was anxious that the armies in the Mediterranean should not be hamstrung if it could be avoided. I wanted to
use them all the time. In Italy there were some thirteen to fourteen
divisions, of which nine or ten were British. There were two armies,
the Fifth Anglo-American Army, and the Eighth Army, which was
entirely British. The choice had been represented as keeping to the
date of "Overlord" or pressing on with the operations in the Mediterranean. But that was not the whole story. The Americans wanted me
to undertake an amphibious operation in the Bay of Bengal against the
Japanese in March. I was not keen about it. If we had the landing
craft needed for the Bay of Bengal in the Mediterranean we should have
enough to do all we wanted there and still be able to keep an early
date for "Overlord." It was not a choice between the Mediterranean
and the date of "Overlord," but between the Bay of Bengal and the
date of "Overlord." However, the Americans had pinned us down to a
date for "Overlord" and operations in the Mediterranean had suffered
in the last two months. Our army in Italy was somewhat disheartened
by the removal of seven divisions. We had sent home three divisions,
and the Americans were sending four of theirs, all in preparation for
"Overlord." That was why we had not been able to take full advantage
of the Italian collapse. But it also proved the earnestness of our preparations for "Overlord." Stalin said that was good.

I then turned to the question of landing craft, and explained once again how and why they were the bottleneck. We had plenty of troops in the Mediterranean, even after the removal of the seven divisions, and there would be an adequate invading British and American army in the United Kingdom. All turned on landing craft. When Stalin had made his momentous announcement two days before about Russia's coming into the war against Japan after Hitler's surrender I had immediately suggested to the Americans that they might find more landing craft for the operations we had been asked to carry out in the Indian Ocean, or that they might send some landing craft from the Pacific to help the first lift of "Overlord." In that case there might be enough for all. But the Americans were very touchy about the Pacific. I had pointed out to them that Japan would be beaten much sooner if Russia joined in the war against her, and that they could therefore afford to give us more help.

The issue between myself and the Americans was in fact a very narrow one. It was not that I was in any way lukewarm about "Overlord." I wanted to get what I needed for the Mediterranean and at the same time keep to the date for "Overlord." The details had to be hammered out between the Staffs, and I had hoped that this might be done in Cairo. Unfortunately Chiang Kai-shek had been there and Chinese questions had taken up nearly all the time. But I was sure that in the end enough landing craft would be found for all.

Now about "Overlord." The British would have ready by the date fixed in May or June nearly sixteen divisions, with their corps troops, landing-craft troops, anti-aircraft, and services, a total of slightly over half a million men. These would consist of some of our best troops, including battle-trained men from the Mediterranean. In addition the British would have all that was needed from the Royal Navy to handle transportation and to protect the Army, and there would be the Metropolitan Air Forces of about four thousand first-line British aircraft in continuous action. The American import of troops was now beginning. Up till now they had sent mainly air troops and stores for the Army, but in the next four or five months I thought one hundred and fifty thousand men or more would come every month, making a total of seven to eight thousand men by May. The defeat of the submarines in the Atlantic had made this movement possible. I was in favour of launching the operation in the South of France about the same time as "Overlord" or at whatever moment was found correct. We should be holding enemy troops in Italy, and of the twenty-two or twenty-three divisions in the Mediterranean as many as possible would go to the South of France and the rest would remain in Italy.

A great battle was impending in Italy. General Alexander had about half a million men under him. There were thirteen or fourteen Allied divisions against nine to ten German. The weather had been bad and bridges had been swept away, but in December we intended to push on, with General Montgomery leading the Eighth Army. An amphibious landing would be made near the Tiber. At the same time the Fifth Army would be fiercely engaged holding the enemy. It might turn into a miniature Stalingrad. We did not intend to push into the wide part of Italy, but to hold the narrow leg.

Stalin said he must warn me that the Red Army was depending on the success of our invasion of Northern France. If there were no operations in May 1944 then the Red Army would think that there would be no operations at all that year. The weather would be bad and there would be transport difficulties. If the operation did not take place he did not want the Red Army to be disappointed. Disappointment could only create bad feeling. If there was no big change in the European war in 1944 it would be very difficult for the Russians to carry on. They were war-weary. He feared that a feeling of isolation might develop in his troops. That was why he had tried to find out whether "Overlord" would be undertaken on time as promised. If not, he would have to take steps to prevent bad feeling in the Red Army. It was most important.

I said "Overlord" would certainly take place, provided the enemy did not bring into France larger forces than the Americans and British could gather there. If the Germans had thirty to forty divisions in France I did not think the force we were going to put across the Channel would be able to hold on. I was not afraid of going on shore, but of what would happen on the thirtieth, fortieth, or fiftieth day. However, if the Red Army engaged the enemy and we held them in Italy, and possibly the Turks came into the war, then I thought we could win.

Stalin said that the first steps of "Overlord" would have a good effect on the Red Army, and if he knew that it was going to take place in May or June he could already prepare blows against Germany. The spring was the best time. March and April were months of slackness, during which he could concentrate troops and material, and in May and June he could attack. Germany would have no troops for France. The transfer of German divisions to the East was continuing. The Germans were afraid of their eastern front, because it had no Channel which had to be crossed and there was no France to be entered. The Germans were afraid of the Red Army advance. The Red Army would advance if it saw that help was coming from the Allies. He asked when "Overlord" would begin.

I said that I could not disclose the date for "Overlord" without the

President's agreement, but the answer would be given at lunch time, and I thought he would be satisfied.

*   *   *

After a short interval Stalin and I separately proceeded to the President's quarters for the luncheon of "Three Only" (with our interpreters) to which he had invited us. Roosevelt then told him that we were both agreed that "Overlord" should be launched during the month of May. Stalin was evidently greatly pleased and relieved by this solemn and direct engagement which we both made. The conversation turned on lighter subjects, and the only part of which I have a record was the question of Russia's outlet upon the seas and oceans. I had always thought it was a wrong thing, capable of breeding disastrous quarrels, that a mighty land mass like the Russian Empire, with its population of nearly two hundred millions, should be denied during the winter months all effective access to the broad waters.

After a brief interval the third plenary session began as before in the Russian Embassy at four o'clock. There was a full attendance and we numbered nearly thirty. General Brooke then announced that, after sitting in combined session, the United States and British Chiefs of Staff had recommended us to launch "Overlord" in May, "in conjunction with a supporting operation against the South of France, on the largest scale permitted by the landing craft available at that time."

Stalin said he understood the importance of the decision and the difficulties inherent in carrying it out. The danger period for "Overlord" would be at the time of deployment from the landings. At this point the Germans might transfer troops from the East in order to create the maximum difficulties for "Overlord." In order to prevent any movement from the east of any considerable German forces he undertook to organise a large-scale Russian offensive in May.[1] I asked if there would be any difficulty in the three Staffs concerting cover plans. Stalin explained that the Russians had made considerable use of deception by means of dummy tanks, aircraft, and airfields. Radio deception had also proved effective. He was entirely agreeable to the Staffs collaborating with the object of devising joint cover and deception schemes. "In wartime," I said, "truth is so precious that she should always be attended by a bodyguard of lies." Stalin and his comrades greatly appreciated this remark when it was translated, and upon this note our formal conference ended gaily.

*   *   *

[1] The main Russian attack began on June 23.

Hitherto we had assembled for our conferences or meals in the Soviet Embassy. I had claimed however that I should be the host at the third dinner, which should be held in the British Legation. This could not well be disputed. Great Britain and I myself both came first alphabetically, and in seniority I was four or five years older than Roosevelt or Stalin. We were by centuries the longest established of the three Governments; I might have added, but did not, that we had been the longest in the war; and, finally, November 30 was my birthday. These arguments, particularly the last one, were conclusive, and all preparations were made by our Minister for a dinner of nearly forty persons, including not only the political and military chiefs but some of their higher staffs. The Soviet Political Police, the N.K.V.D., insisted on searching the British Legation from top to bottom, looking behind every door and under every cushion, before Stalin appeared; and about fifty armed Russian policemen, under their own General, posted themselves near all the doors and windows. The American Security men were also much in evidence. Everything however passed off agreeably. Stalin, arriving under heavy guard, was in the best of tempers, and the President, from his wheeled chair, beamed on us all in pleasure and goodwill.

This was a memorable occasion in my life. On my right sat the President of the United States, on my left the master of Russia. Together we controlled a large preponderance of the naval and three-quarters of all the air forces in the world, and could direct armies of nearly twenty millions of men, engaged in the most terrible of wars that had yet occurred in human history. I could not help rejoicing at the long way we had come on the road to victory since the summer of 1940, when we had been alone, and, apart from the Navy and the Air, practically unarmed, against the triumphant and unbroken might of Germany and Italy, with almost all Europe and its resources in their grasp. Mr. Roosevelt gave me for a birthday present a beautiful Persian porcelain vase, which, although it was broken into fragments on the homeward journey, has been marvellously reconstructed and is one of my treasures.

During dinner I had a most pleasant conversation with both my august guests. Stalin repeated the question he had posed at the Conference, "Who will command 'Overlord'?" I said that the President had not yet finally made up his mind, but that I was almost certain it would be General Marshall, who sat opposite us at no great distance, and that was how it had stood hitherto. He was evidently very pleased at this. He then spoke about General Brooke. He thought that he did not like the Russians. He had been very abrupt and rough with them at our first Moscow meeting in August 1942. I reassured him, remark-

TEHERAN: CRUX AND CONCLUSIONS

ing that military men were apt to be blunt and hard-cut when dealing with war problems with their professional colleagues. Stalin said that he liked them all the better for that. He gazed at Brooke intently across the room.

When the time came I proposed the health of our illustrious guests, and the President proposed my health and wished me many happy returns of the day. He was followed by Stalin, who spoke in a similar strain.

Many informal toasts were then proposed, according to the Russian custom, which is certainly very well suited to banquets of this kind. Hopkins made a speech couched in a happy vein, in the course of which he said that he had made "a very long and thorough study of the British Constitution, which is unwritten, and of the War Cabinet, whose authority and composition are not specifically defined." As the result of this study, he said, "I have learnt that the provisions of the British Constitution and the powers of the War Cabinet are just whatever Winston Churchill wants them to be at any given moment." This caused general laughter. The reader of this tale will know how little foundation there was in this jocular assertion. It is true that I received a measure of loyal support in the direction of the war from Parliament and my Cabinet colleagues which may well be unprecedented, and that there were very few large issues upon which I was overruled; but it was with some pride that I reminded my two great comrades on more than one occasion that I was the only one of our trinity who could at any moment be dismissed from power by the vote of a House of Commons freely elected on universal franchise, or could be controlled from day to day by the opinion of a War Cabinet representing all parties in the State. The President's term of office was fixed, and his powers not only as President but as Commander-in-Chief were almost absolute under the American Constitution. Stalin appeared to be, and at this moment certainly was, all-powerful in Russia. They could order; I had to convince and persuade. I was glad that this should be so. The process was laborious, but I had no reason to complain of the way it worked.

As the dinner proceeded there were many speeches, and most of the principal figures, including Molotov and General Marshall, made their contribution. But the speech which stands out in my memory came from General Brooke. I quote the account he was good enough to write for me.

Halfway through the dinner [he says] the President very kindly proposed my health, referring to the time when my father had visited his father at Hyde Park. Just as he was finishing, and I was thinking what an easy time I should have replying to such kind words, Stalin got up and said he would finish the toast. He then proceeded to imply that I had failed to show real feelings of friendship towards the Red Army, that I was lacking in a true appreciation of its fine qualities, and that he hoped in future I should be able to show greater comradeship towards the soldiers of the Red Army!

I was very much surprised by these accusations, as I could not think what they were based on. I had however seen enough of Stalin by then to know that if I sat down under these insults I should lose any respect he might ever have had for me, and that he would continue such attacks in the future.

I therefore rose to thank the President most profusely for his very kind expressions, and then turned to Stalin in approximately the following words:

"Now, Marshal, may I deal with your toast. I am surprised that you should have found it necessary to raise accusations against me that are entirely unfounded. You will remember that this morning while we were discussing cover plans Mr. Churchill said that 'in war truth must have an escort of lies.' You will also remember that you yourself told us that in all your great offensives your real intentions were always kept concealed from the outer world. You told us that all your dummy tanks and dummy aeroplanes were always massed on those fronts that were of an immediate interest, while your true intentions were covered by a cloak of complete secrecy.

"Well, Marshal, you have been misled by dummy tanks and dummy aeroplanes, and you have failed to observe those feelings of true friendship which I have for the Red Army, nor have you seen the feelings of genuine comradeship which I bear towards all its members."

As this was translated by Pavlov, sentence by sentence, to Stalin I watched his expression carefully. It was inscrutable. But at the end he turned to me and said with evident relish, "I like that man. He rings true. I must have a talk with him afterwards."

At length we moved into the antechamber, and here everyone moved about in changing groups. I felt that there was a greater sense of solidarity and good-comradeship than we had ever reached before in the Grand Alliance. I had not invited Randolph and Sarah to the dinner, though they came in while my birthday toast was being proposed, but now Stalin singled them out and greeted them most warmly, and of course the President knew them well.

As I moved around I saw Stalin in a small circle face to face with "Brookie," as I called him. The General's account continues:

> As we walked out of the room the Prime Minister told me that he had felt somewhat nervous as to what I should say next when I had referred to "truth" and "lies." He comforted me however by telling me that my reply to the toast had had the right effect on Stalin. I therefore decided to return to the attack in the anteroom. I went up to Stalin and told him how surprised I was, and grieved, that he should have found it necessary to raise such accusations against me in his toast. He replied at once through Pavlov, "The best friendships are those founded on misunderstandings," and he shook me warmly by the hand.

It seemed to me that all the clouds had passed away, and in fact Stalin's confidence in my friend was established on a foundation of respect and goodwill which was never shaken while we all worked together.

It must have been after two in the morning when we finally separated. The Marshal resigned himself to his escort and departed, and the President was conveyed to his quarters in the Soviet Embassy. I went to bed tired out but content, feeling sure that nothing but good had been done. It certainly was a happy birthday for me.

*   *   *

On December 1 our long and hard discussions at Teheran reached their end. The military conclusions governed in the main the future of the war. The cross-Channel invasion was fixed for May, subject naturally to tides and the moon. It was to be aided by a renewed major Russian offensive. At first sight I liked the proposed descent upon the French southern shore by part of the Allied Armies in Italy. The project had not been examined in detail, but the fact that both the Americans and the Russians favoured it made it easier to secure the landing craft necessary for the success of our Italian campaign and the capture of Rome, without which it would have been a failure. I was of course more attracted by the President's alternative suggestion of a right-handed move from Italy by Istria and Trieste, with ultimate designs for reaching Vienna through the Ljubljana Gap. All this lay five or six months ahead. There would be plenty of time to make a final choice as the general war shaped itself, if only the life of our armies in Italy was not paralysed by depriving them of their modest requirements in landing craft. Many amphibious or semi-amphibious schemes were open. I expected that the sea-borne operations in the Bay of Bengal would be abandoned, and this, as the next chapter will show, proved correct.

I was glad to feel that several important options were still preserved. Our strong efforts were to be renewed to bring Turkey into the war, with all that might accompany this in the Aegean, and follow from it in the Black Sea. In this we were to be disappointed. Surveying the whole military scene, as we separated in an atmosphere of friendship and unity of immediate purpose, I personally was well content.

The political aspects were at once more remote and speculative. Obviously they depended upon the results of the great battles yet to be fought, and after that upon the mood of each of the Allies when victory was gained. It would not have been right at Teheran for the Western democracies to found their plans upon suspicions of the Russian attitude in the hour of triumph and when all her dangers were removed. Stalin's promise to enter the war against Japan as soon as Hitler was overthrown and his armies defeated was of the highest importance. The hope of the future lay in the most speedy ending of the war and the establishment of a World Instrument to prevent another war, founded upon the combined strength of the three Great Powers, whose leaders had joined hands in friendship around the table.

We had procured a mitigation for Finland, which on the whole is operative today. The frontiers of the new Poland had been broadly outlined both in the East and in the West. The "Curzon Line," subject to interpretation in the East, and the line of the Oder, in the West, seemed to afford a true and lasting home for the Polish nation after all its sufferings. At the time the question between the Eastern and Western Neisse, which flow together to form the Oder River, had not arisen. When in July 1945 it arose in a violent form and under totally different conditions at the Potsdam Conference I at once declared that Great Britain adhered only to the eastern tributary. And this is still my position.

The supreme question of the treatment to be accorded to Germany by the victors could at this milestone only be the subject of "a preliminary survey of a vast political problem," and, as Stalin described it, "certainly very preliminary." It must be remembered that we were in the midst of a fearful struggle with the mighty Nazi Power. All the hazards of war lay around us, and all its passions of comradeship among Allies, of retribution upon the common foe, dominated our minds. The President's tentative projects for the partition of Germany into five self-governing states and two territories, of vital consequence, under the United Nations, were of course far more acceptable to Stalin than the proposal which I made for the isolation of Prussia and the constitution of a Danubian Confederation, or of a South Germany and also a Danubian Confederation. This was only my personal view. But I do not at all repent having put it forward in the circumstances which lay about us at Teheran.

We all deeply feared the might of a united Germany. Prussia had a great history of her own. It would be possible, I thought, to make a stern but honourable peace with her, and at the same time to recreate in modern forms what had been in general outline the Austro-Hungarian Empire, of which it has been well said, "If it did not exist it would have to be invented." Here would be a great area in which not only peace but friendship might reign at a far earlier date than in any other solution. Thus a United Europe might be formed in which all the victors and vanquished might find a sure foundation for the life and freedom of all their tormented millions.

I do not feel any break in the continuity of my thought in this immense sphere. But vast and disastrous changes have fallen upon us in the realm of fact. The Polish frontiers exist only in name, and Poland lies quivering in the Russian-Communist grip. Germany has indeed been partitioned, but only by a hideous division into zones of military occupation. About this tragedy it can only be said IT CANNOT LAST.

8

Carthage and Marrakesh

ON DECEMBER 2 I got back to Cairo from Teheran, and was once more installed in the villa near the Pyramids. The President arrived the same evening, and we resumed our intimate discussions on the whole scene of the war and on the results of our talks with Stalin. Meanwhile the Combined Chiefs of Staff, who had refreshed themselves by a visit to Jerusalem on their way back from Teheran, were to carry forward their discussions on all their great business the next day. Admiral Mountbatten had returned to India, whence he had submitted the revised plan he had been instructed to make for an amphibious attack on the Andaman Islands (Operation "Buccaneer"). This would absorb the vitally needed landing craft already sent to him from the Mediterranean. I wished to make a final attempt to win the Americans to the alternative enterprise against Rhodes.

The next evening I dined again with the President. Eden was with me. We remained at the table until after midnight, still discussing our points of difference. I shared the views of our Chiefs of Staff, who were much worried by the promise which the President had made to Generalissimo Chiang Kai-shek before Teheran to launch an early attack across the Bay of Bengal. This would have swept away my hopes and plans of taking Rhodes, on which I believed the entry of Turkey into the war largely depended. But Mr. Roosevelt's heart was set upon it. When our Chiefs of Staff raised it in the military conferences the United States Staffs simply declined to discuss the matter. The President, they said, had taken his decision and they had no choice but to obey.

On the afternoon of December 4 we held our first plenary meeting since Teheran, but made little headway. The President began by saying that he must leave on December 6, and that all reports should be ready for the final agreement of both parties by the evening of Sunday,

December 5. Apart from the question of the entry of Turkey into the war, the only outstanding point seemed to be the comparatively small one of the use to be made of a score of landing craft and their equipment. It was unthinkable that one could be beaten by a petty item like that, and he felt bound to say that the detail *must* be disposed of.

I said that I did not wish to leave the Conference in any doubt that the British delegation viewed our early dispersal with great apprehension. There were still many questions of first-class importance to be settled. Two decisive events had taken place in the last few days. In the first place, Stalin had voluntarily proclaimed that the Soviet would declare war on Japan the moment Germany was defeated. This would give us better bases than we could ever find in China, and made it all the more important that we should concentrate on making "Overlord" a success. It would be necessary for the Staffs to examine how this new fact would affect operations in the Pacific and Southeast Asia.

The second event of first-class importance was the decision to cross the Channel during May. I myself would have preferred a July date, but I was determined nevertheless to do all in my power to make a May date a complete success. It was a task transcending all others. A million Americans were to be thrown in eventually, and five or six hundred thousand British. Terrific battles were to be expected, on a scale far greater than anything that we had experienced before. In order to give "Overlord" the greatest chance, it was thought necessary that the descent on the Riviera (code-named "Anvil") should be as strong as possible. It seemed to me that the crisis for the invading armies would come at about the thirtieth day, and it was essential that every possible step should be taken by action elsewhere to prevent the Germans from concentrating a superior force against our beachheads. As soon as the "Overlord" and "Anvil" forces got into the same zone they would come under the same commander.

The President, summing up the discussion, asked whether he was correct in thinking that there was general agreement on the following points:

(*a*) Nothing should be done to hinder "Overlord."
(*b*) Nothing should be done to hinder "Anvil."
(*c*) By hook or by crook we should scrape up sufficient landing craft to operate in the Eastern Mediterranean if Turkey came into the war.
(*d*) Admiral Mountbatten should be told to go ahead and do his best [in the Bay of Bengal] with what had already been allocated to him.

On this last point I suggested that it might be necessary to withdraw resources from Mountbatten in order to strengthen "Overlord" and "Anvil." The President said that he could not agree. We had a moral obligation to do something for China, and he would not be prepared to forgo the amphibious operation except for some very good and readily apparent reason. I replied that this "very good reason" might be provided by our supreme adventure in France. At present the "Overlord" assault was only on a three-division basis, whereas we had put nine divisions ashore in Sicily on the first day. The main operation was at present on a very narrow margin.

Reverting to the Riviera attack, I expressed the view that it should be planned on the basis of an assault force of at least two divisions. This would provide enough landing craft to do the outflanking operations in Italy, and also, if Turkey came into the war soon, to capture Rhodes. I then pointed out that operations in Southeast Asia must be judged in their relation to the predominating importance of "Overlord." I said that I was surprised at the demands for taking the Andamans which had reached me from Admiral Mountbatten. In the face of Stalin's promise that Russia would come into the war operations in the Southeast Asia Command had lost a good deal of their value, while, on the other hand, their cost had been put up to a prohibitive extent.

The discussion continued on whether or not to persist in the Andamans project. The President resisted the British wish to drop it. No conclusion was reached, except that the Chiefs of Staff were directed to go into details.

*   *   *

On December 5 we met again, and the report of the Combined Staffs on operations in the European theatre was read out by the President and agreed. Everything was now narrowed down to the Far Eastern operation. Rhodes had receded in the picture, and I concentrated on getting the landing craft for "Anvil" and the Mediterranean. A new factor had presented itself. The estimates of the Southeast Asia Command of the force needed to storm the Andamans had been startling. The President said that fourteen thousand should be sufficient. Anyhow, the fifty thousand men proposed certainly broke the back of the Andamans expedition so far as this meeting was concerned. It was agreed for the moment that Mountbatten should be asked what amphibious operations he could undertake on a smaller scale, on the assumption that most of the landing craft and assault shipping were withdrawn from Southeast Asia during the next few weeks. Thus we parted, leaving Mr. Roosevelt much distressed.

Before anything further could be done the deadlock in Cairo was broken. In the afternoon the President, in consultation with his advisers, decided to abandon the Andaman Islands plan. He sent me a laconic private message: " 'Buccaneer' is off." General Ismay reminds me that when I told him the welcome news cryptically on the telephone that the President had changed his mind and was so informing Chiang Kai-shek I said, "He is a better man that ruleth his spirit than he that taketh a city." We all met together at 7.30 P.M. the next evening to go over the final report of the Conference. The Southern France assault operation was formally approved, and the President read out his signal to Generalissimo Chiang Kai-shek, informing him of the decision to abandon the Andamans plan.

* * *

One of the main purposes of our Cairo meeting had been to resume talks with the Turkish leaders. I had telegraphed President Inönü on December 1 from Teheran suggesting that he should join the President and myself in Cairo. It was arranged that Vyshinsky should also be present. These conversations arose out of the meeting between Mr. Eden and the Turkish Foreign Minister in Cairo at the beginning of November on the former's journey home from Moscow. The Turks now came again to Cairo on December 4, and the following evening I entertained the Turkish President to dinner. My guest displayed great caution, and in subsequent meetings showed to what extent his advisers were still impressed by the German military machine. I pressed the case hard. With Italy out of the war the advantages of Turkey's entry were manifestly increased and her risks lessened.

The Turks soon departed to report to their Parliament. It was agreed that in the meantime British specialists should be assembled to implement the first stages of establishing an Allied force in Turkey. And there the matter rested. By the time Christmas arrived I was becoming resigned to Turkish neutrality.

In all our many talks at Cairo the President never referred to the vital and urgent issue of the command of "Overlord," and I was under the impression that our original arrangement and agreement held good. But on the day before his departure from Cairo he told me his final decision. We were driving in his motorcar from Cairo to the Pyramids. He then said, almost casually, that he could not spare General Marshall, whose great influence at the head of military affairs and of the war direction, under the President, was invaluable, and indispensable to the successful conduct of the war. He therefore proposed to nominate Eisenhower to "Overlord," and asked me for my opinion. I said it was for him to decide, but that we had also the warmest regard for General Eisenhower, and would trust our fortunes to his direction with hearty goodwill.

Up to this time I had thought Eisenhower was to go to Washington as Military Chief of Staff, while Marsall commanded "Overlord." Eisenhower had heard of this too, and was very unhappy at the prospect of leaving the Mediterranean for Washington. Now it was all settled: Eisenhower for "Overlord," Marshall to stay at Washington, and a British commander for the Mediterranean.

The full story of the President's long delay and hesitations and of his final decision is referred to by Mr. Hopkins's biographer, who says that Roosevelt made the decision on Sunday, December 5, "against the almost impassioned advice of Hopkins and Stimson, against the known preference of both Stalin and Churchill, against his own proclaimed inclination." Then Mr. Sherwood quotes the following extract from a note which he had from General Marshall after the war. "If I recall," said Marshall, "the President stated, in completing our conversation, 'I feel I could not sleep at night with you out of the country.'" There can be little doubt that the President felt that the command only of "Overlord" was not sufficient to justify General Marshall's departure from Washington.[1]

At last our labours were finished. I gave a dinner at the villa to the Combined Chiefs of Staff, Mr. Eden, Mr. Casey, and one or two others. I remember being struck by the optimism which prevailed in high Service circles. The idea was mooted that Hitler would not be strong enough to face the spring campaign, and might collapse even before "Overlord" was launched in the summer. I was so much impressed by the current of opinion that I asked everybody to give his view in succession round the table. All the professional authorities were inclined to think that the German collapse was imminent. The three politicians present took the opposite view. Of course, on these vast matters on which so many lives depend there is always a great deal of guesswork. So much is unknown and immeasurable. Who can tell how weak the enemy may be behind his flaming fronts and brazen mask? At what moment will his will power break? At what moment will he be beaten down?

The President had found no time for sightseeing, but I could not bear his leaving without seeing the Sphinx. One day after tea I said, "You must come now." We motored there forthwith, and examined this wonder of the world from every angle. Roosevelt and I gazed at her for some minutes in silence as the evening shadows fell. She told us nothing and maintained her inscrutable smile. There was no use waiting

---

[1] Robert E. Sherwood, *Roosevelt and Hopkins*, pp. 802–3. There had at one time been a proposal that General Marshall should command *both* theatres, i.e. the Mediterranean as well as "Overlord." This had fallen to the ground.

longer. On December 7 I bade farewell to my great friend when he flew off from the airfield beyond the Pyramids.

*   *   *

I had not been at all well during this journey and conterence, and as it drew to its close I became conscious of being very tired. For instance, I noticed that I no longer dried myself after my bath, but lay on the bed wrapped in my towel till I dried naturally. A little after midnight on December 11 I and my personal party left in our aircraft for Tunis. I had planned to spend one night there at General Eisenhower's villa, and to fly next day to Alexander's and then Montgomery's headquarters in Italy, where the weather was reported to be absolutely vile and all advances were fitful.

Morning saw us over the Tunis airfields. We were directed by a signal not to land where we had been told, and were shifted to another field some forty miles away. We all got out, and they began to unload the luggage. It would be an hour before motorcars could come, and then a long drive. As I sat on my official boxes near the machines I certainly did feel completely worn out. Now however came a telephone message from General Eisenhower, who was waiting at the first airfield, that we had been wrongly transferred and that landing was quite possible there. So we scrambled back into our plane, and in ten minutes were with him, quite close to his villa. Ike, always the soul of hospitality, had waited two hours with imperturbable good humour. I got into his car, and after we had driven for a little while I said, "I am afraid I shall have to stay with you longer than I had planned. I am completely at the end of my tether, and I cannot go on to the front until I have recovered some strength." All that day I slept, and the next day came fever and symptoms at the base of my lung which were adjudged to portend pneumonia. So here I was at this pregnant moment on the broad of my back amid the ruins of ancient Carthage.

When the X-ray photographs showed that there was a shadow on one of my lungs I found that everything had been diagnosed and foreseen by Lord Moran. Dr. Bedford and other high medical authorities in the Mediterranean and excellent nurses arrived from all quarters as if by magic. The admirable M and B, from which I did not suffer any inconvenience, was used at the earliest moment, and after a week's fever the intruders were repulsed. Although Moran records that he judged that the issue was at one time in doubt, I did not share his view. I did not feel so ill in this attack as I had the previous February. The M and B, which I also called Moran and Bedford, did the work most effectively. There is no doubt that pneumonia is a very different illness from what it was before this marvellous drug was discovered. I did not at any time relinquish my part in the direction of affairs, and there was not the slightest delay in giving the decisions which were required from me.

My immediate task, as British Minister of Defence responsible to the War Cabinet, was to propose a British Supreme Commander for the Mediterranean. This post we confided to General Wilson, it being also settled that General Alexander should command the whole campaign in Italy, as he had done under General Eisenhower. It was also arranged that General Devers, of the U.S. Army, should become General Wilson's Deputy in the Mediterranean, and Air Chief Marshal Tedder General Eisenhower's Deputy in "Overlord," and that Montgomery should actually command the whole cross-Channel invasion force until such time as the Supreme Commander could transfer his headquarters to France and assume the direct operational control. All this was carried out with the utmost smoothness in perfect agreement by the President and by me, with Cabinet approval, and worked in good comradeship and friendship by all concerned.

But the days passed in much discomfort. Fever flickered in and out. I lived on my theme of the war, and it was like being transported out of oneself. The doctors tried to keep the work away from my bedside, but I defied them. They all kept on saying, "Don't work, don't worry," to such an extent that I decided to read a novel. I had long ago read Jane Austen's *Sense and Sensibility,* and now I thought I would have *Pride and Prejudice.* Sarah read it to me beautifully from the foot of the bed. I had always thought it would be better than its rival. What calm lives they had, those people! No worries about the French Revolution, or the crashing struggle of the Napoleonic wars. Only manners controlling natural passion so far as they could, together with cultured explanations of any mischances. All this seemed to go very well with M and B.

One morning Sarah was absent from her chair at the foot of my bed, and I was about to ask for my box of telegrams in the prohibited hours when in she walked with her mother. I had no idea that my wife was flying out from England to join me. She had hurried to the airport to fly in a two-engined Dakota. The weather was bad, but Lord Beaverbrook was vigilant. He got to the airport first, and stopped her flight until a four-engined plane could be procured. (I always think it better to have four engines when flying long distances across the sea.) Now she had arrived after a very rough journey in an unheated plane in midwinter. Jock Colville had escorted her, and was a welcome addition to my hard-pressed personal staff, through whom so much business was being directed. "My love to Clemmie," cabled the President. "I feel relieved that she is with you as your superior officer."

*   *   *

As I lay prostrate I felt we were at one of the climaxes of the war. The mounting of "Overlord" was the greatest event and duty in the world. But must we sabotage everything we could have in Italy, where the main strength overseas of our country was involved? Were we to leave it a stagnant pool from which we had drawn every fish we wanted? As I saw the problem, the campaign in Italy, in which a million or more of our British, British-controlled, and Allied troops were engaged, was the faithful and indispensable comrade and counterpart of the main cross-Channel operation. Here the American clear-cut, logical, large-scale, mass-production style of thought was formidable. In life people have first to be taught "Concentrate on essentials." This is no doubt the first step out of confusion and fatuity; but it is only the first step. The second stage in war is a general harmony of war effort by making everything fit together, and every scrap of fighting strength play its full part all the time. I was sure that a vigorous campaign in Italy during the first half of 1944 would be the greatest help to the supreme operation of crossing the Channel, on which all minds were set and all engagements made. But every item which any Staff officer could claim as "essential" or "vital," to use these hard-worked words, had to be argued out as if it carried with it the success or failure of our main purpose. Twenty or a dozen vehicle landing craft had to be fought for as if the major issue turned upon them.

The case seemed to me brutally simple. All the ships we had would be used to carry to England everything the United States could produce in arms and men. Surely the enormous forces we could not possibly move by sea from the Italian theatre should play their part. Either they would gain Italy easily and immediately bite upon the German inner front, or they would draw large German forces from the front which we were to attack across the Channel in the last days of May, or the early days of June, as the moon and the tides prescribed.

The deadlock to which our armies in Italy had been brought by the stubborn German resistance on the eighty-mile front from sea to sea had already led General Eisenhower to contemplate an amphibious flanking attack. He had planned to land with one division south of the Tiber and make a dart for Rome, in conjunction with an attack by the main armies. The arrest of these armies and the distance of the landing point from them made everyone feel that more than one division was required. I had of course always been a partisan of the "end run," as the Americans call it, or "cat-claw," which was my term. I had never succeeded in getting this manoeuvre open to sea power included in any of our Desert advances. In Sicily however General Patton had twice used the command of the sea flank as he advanced along the northern coast of the island with great effect.

There was a great deal of professional support. Eisenhower was already committed in principle, though his new appointment to the command of "Overlord" now gave him a different sense of values and a new horizon. Alexander, Deputy Supreme Commander and commanding the armies in Italy, though the operation right and necessary; Bedell Smith was ardent and helpful in every direction. This was also true of Admiral John Cunningham, who held all the naval cards, and of Air Marshal Tedder. I had therefore a powerful array of Mediterranean authorities. Moreover, I felt sure the British Chiefs of Staff would like the plan, and that with their agreement I could obtain the approval of the War Cabinet. When you cannot give orders hard and lengthy toils must be faced.

*　　　　*　　　　*

I began my effort on December 19, when the C.I.G.S. arrived at Carthage to see me on his way home from Montgomery's headquarters in Italy. We had hoped to go there together, but my illness had prevented me. We had a full discussion, and I found that General Brooke had by a separate route of thought arrived at the same conclusion as I had. We agreed on the policy, and also that while I should deal with the commanders on the spot he would do his best to overcome all difficulties at home. He then left by air for London. The Chiefs of Staff had evidently been thinking on the same lines, and, after hearing his account, telegraphed on the 22nd: "We are in full agreement with you that the present stagnation cannot be allowed to continue. . . . The solution, as you say, clearly lies in making use of our amphibious power to strike round on the enemy's flank and open up the way for a rapid advance on Rome. . . . We think the aim should be to provide a lift for at least two divisions. . . ." After explaining that the new plan would involve giving up both the capture of Rhodes and also a minor amphibious operation on the Arakan coast of Burma, they ended: "If you approve the above line of thought we propose to take the matter up with the Combined Chiefs of Staff with a view to action being taken on these lines at once."

This led to a hard scrutiny of our resources. Some landing craft for the cancelled operation against the Andamans were on their way to the Mediterranean across the Indian Ocean. Others were due to return home for "Overlord." All were in extreme demand.

The whole morning of Christmas Day I held a conference at Carthage. Eisenhower, Alexander, Bedell Smith, General Wilson, Tedder, Admiral John Cunningham, and other high officers were present. The only one not there was General Mark Clark, of the Fifth Army. This was an oversight which I regret, as it was to his army that the operation was eventually entrusted and he ought to have had the background in his mind. We were all agreed that nothing less than a two-division lift would suffice. At this time I contemplated an assault by two British divisions from the Eighth Army, in which Montgomery was about to be succeeded by General Leese. I thought the amphibious operation involved potential mortal risks to the landed forces, and I preferred to run them with British troops, because it was to Britain that I was responsible. Moreover, the striking force would then have been homogeneous instead of half and half.

Everything turned on landing craft, which held for some weeks all our strategy in the tightest ligature. What with the rigid date prescribed for "Overlord" and the movement, repair, and refitting of less than a hundred of these small vessels, all plans were in a strait jacket. We escaped, though mauled, from this predicament. But I must also admit that I was so much occupied in fighting for the principle that I did not succeed in getting, and indeed did not dare to demand, the necessary weight and volume for the "cat-claw." Actually there were enough L.S.T.s for the operation as planned, and in my opinion, if the extravagant demands of the military machine had been reduced, we could, without prejudice to any other pledge or commitment, have flung ashore south of the Tiber a still larger force with full mobility. However, the issue was fought out in terms of routine Army requirements and the exact dates when L.S.T.s could be free for "Overlord," making of course all allowances for their return home in winter Biscay weather, and with the time-margins for their refits stated at their maximum. If I had asked for a three-division lift I should not have got anything. How often in life must one be content with what one can get! Still, it would be better to do it right.

At the close of our discussion I sent the following to the President, and a similar telegram home. I was careful to state the root fact bluntly.

. . . Having kept fifty-six L.S.T.s in the Mediterranean so long, it would seem irrational to remove them for the very week when they can render decisive service. What, also, could be more dangerous than to let the Italian battle stagnate and fester on for another three months? We cannot afford to go forward leaving a vast half-finished job behind us. It therefore seemed to those present that every effort should be made to bring off Anzio on a two-division basis around January 20, and orders have been issued to General Alexander to prepare accordingly. If this opportunity is not grasped we must expect the ruin of the Mediterranean campaign of 1944. I earnestly hope therefore that you may agree to the three weeks' delay in return of the fifty-six landing craft, and that all the authorities will be instructed to make sure that the May "Overlord" is not prejudiced thereby. . . .

* * *

Lord Moran thought it possible for me to leave Carthage after Christmas, but insisted that I must have three weeks' convalescence somewhere. And where could be better than the lovely villa at Marrakesh, where the President and I had stayed after Casablanca a year before? All plans had been made during the past few days. I was to be the guest of the United States Army at Marrakesh. It was also thought that I had been long enough at Carthage to be located. Small vessels had ceaselessly to patrol the bay in front of the villa in case some U-boat turned up for a surprise raid. There might also be a long-range air attack. I had my own protection in a battalion of the Coldstream Guards. I was too ill, or too busy, to be consulted about all this, but I saw in my beloved Marrakesh a haven where I could regain my strength.

Outside the villa a magnificent guard of the Coldstream was drawn up. I had not realised how much I had been weakened by my illness. I found it quite a difficulty to walk along the ranks and climb into the motorcar. The flight at six thousand feet had been planned on the weather forecast that the skies would be clear. However, as we sailed on and the uplands of Tunisia began to rise about us I saw a lot of large fleecy and presently blackish clouds gathering around, and after a couple of hours we were more often in mist than in sunlight. I have always had a great objection to what are called "stuffed clouds" — i.e., clouds with mountains inside them — and flying an intricate route through the various valleys before us in order to keep under six thousand feet seemed to me an unfair proposition for the others in the plane. I therefore sent for the pilot and told him to fly at least two thousand feet above the highest mountain within a hundred miles of his route. Lord Moran agreed. Oxygen was brought by a skilled administrator, specially provided for the journey. We sailed up into the blue. I got along all right, and we made a perfect landing at about four o'clock on the Marrakesh airfield. Our second plane, which had adhered strictly to its instructions, had a very severe and dangerous flight through the various gorges and passes, many of which were traversed with only fleeting glimpses of the towering mountains. At this low height the weather was by no means good. The plane arrived safely an hour behind us with one of its doors blown off and nearly everybody very sick. I was sorry indeed they should have been put to so much discomfort and risk on my account. They could have flown it all out comfortably under blue skies at twelve or even eleven thousand feet.

*   *   *

Nothing could exceed the comfort, and even luxury, of my new abode, or the kindness of everyone concerned. But one thing rose above all others in my mind — what answer would the President give to my telegram? When I thought of the dull, dead-weight resistance, taking no account of timing and proportion, that I had encountered about all Mediterranean projects I awaited the answer with deep anxiety. What I asked for was a hazardous enterprise on the Italian coast, and a possible delay of three weeks from May 1 — four if the moon phase was to be observed — in the date of the Channel crossing. I had gained the agreement of the commanders on the spot. The British Chiefs of Staff had always agreed in principle, and were now satisfied in detail. But what would the Americans say to a four weeks' postponement of "Overlord"? However, when one is thoroughly tired out the blessing of sleep is not usually denied.

It was with joy, not, I confess, unmingled with surprise, that on December 28 I received a telegram from Mr. Roosevelt agreeing to delay the departure of the fifty-six L.S.T.s "on the basis that 'Overlord' remains the paramount operation and will be carried out on the date agreed to at Cairo and Teheran." "I thank God," I replied, "for this fine decision, which engages us once again in wholehearted unity upon a great enterprise. . . ."

Great efforts had indeed been made by the Staffs at home, and especially by the Admiralty, to accomplish the "cat-claw," and I hastened to congratulate them. The President's telegram was a marvel. I was sure that I owed it not only to his goodwill but to Marshall's balance of mind, to Eisenhower's loyalty to the show he was about to quit, and Bedell Smith's active, knowledgeable, fact-armed diplomacy. On the same day Alexander sent us his plan. After conferring with General Mark Clark and General Brian Robertson, he had decided to use an American and a British division. Armour, paratroops, and Commandos would be on a fifty-fifty basis, and the whole would be under an American corps commander. The attack would go in about January 20. Ten days beforehand he would launch a big offensive against Cassino to draw off the German reserves. The forward plunge of the main armies would follow. I was well content. So far so good.

I determined to be at home before the shock of Anzio occurred. On January 14 therefore we all flew in beautiful weather to Gibraltar, where the *King George V* awaited me. On the 15th she made her way out of Algeciras Bay wide into the Atlantic, and thence to Plymouth. After a restful voyage we were welcomed by the War Cabinet and Chiefs of Staff, who really seemed quite glad to see me back. I had been more than two months away from England, and they had been through a lot of worry on account both of my illness and my activities. It was indeed a homecoming, and I felt deeply grateful to all these trusty friends and fellow workers.

9

Marshal Tito: The Greek Torment

T H E   R E A D E R   M U S T   N O W go back to a fierce and sombre tale, which the main narrative has outstripped. Yugoslavia since Hitler's invasion and conquest in April 1941 had been the scene of fearful events. The spirited boy King took refuge in England with such of Prince Paul's ministers and other members of the Government as had defied the German assault. In the mountains there began again the fierce guerrilla with which the Serbs had resisted the Turks for centuries. General Mihailović was its first and foremost champion, and round him rallied the surviving *élite* of Yugoslavia. In the vortex of world affairs their struggle was hardly noticeable. It belongs to the "unestimated sum of human pain." Mihailović suffered as a guerrilla leader from the fact that many of his followers were well-known people with relations and friends in Serbia, and property and recognisable connections elsewhere. The Germans pursued a policy of murderous blackmail. They retaliated for guerrilla activities by shooting batches of four or five hundred selected people in Belgrade. Under this pressure Mihailović drifted gradually into a posture where some of his commanders made accommodations with the German and Italian troops to be left alone in certain mountain areas in return for doing little or nothing against the enemy. Those who have triumphantly withstood such strains may brand his name, but history, more discriminating, should not erase it from the scroll of Serbian patriots. By the autumn of 1941 Serbian resistance to the German terror had become only a shadow. The national struggle could only be sustained by the innate valour of the common people. This however was not lacking.

A wild and furious war for existence against the Germans broke into flame among the partisans. Among those Tito stood forth, pre-eminent and soon dominant. Tito, as he called himself, was a Soviet-trained

Communist who, until Russia was invaded by Hitler, and after Yugoslavia had been assailed, had fomented political strikes along the Dalmatian coast, in accordance with the general Comintern policy. But once he united in his breast and brain his Communist doctrine with his burning ardour for his native land in her extreme torment he became a leader, with adherents who had little to lose but their lives, who were ready to die, and if to die to kill. This confronted the Germans with a problem which could not be solved by the mass executions of notables or persons of substance. They found themselves confronted by desperate men who had to be hunted down in their lairs. The partisans under Tito wrested weapons from German hands. They grew rapidly in numbers. No reprisals, however bloody, upon hostages or villages deterred them. For them it was death or freedom. Soon they began to inflict heavy injury upon the Germans and became masters of wide regions.

It was inevitable that the partisan movement should also come into savage quarrels with their fellow countrymen who were resisting half-heartedly or making bargain for immunity with the common foe. The partisans deliberately violated any agreements made with the enemy by the Cetniks — as the followers of General Mihailović were called. The Germans then shot Cetnik hostages, and in revenge Cetniks gave the Germans information about the partisans. All this happened sporadically and uncontrollably in these wild mountain regions. It was a tragedy within a tragedy.

$$*\qquad*\qquad*$$

I had followed these events amid other preoccupations so far as was possible. Except for a trickle of supplies dropped from aircraft, we were not able to help. Our headquarters in the Middle East was responsible for all operations in this theatre, and maintained a system of agents and liaison officers with the followers of Mihailović. When in the summer of 1943 we broke into Sicily and Italy, the Balkans, and especially Yugoslavia, never left my thoughts. Up to this point our missions had only gone to the bands under Mihailović, who represented the official resistance to the Germans and the Yugoslav Government in Cairo. In May 1943 we took a new departure. It was decided to send small parties of British officers and N.C.O.s to establish contact with the Yugoslav Partisans, in spite of the fact that cruel strife was proceeding between them and the Cetniks, and that Tito was waging war as a Communist not only against the German invaders but against the Serbian monarchy and Mihailović. At the end of that month Captain Deakin, an Oxford don who had helped me for five years before the war in my literary work, was dropped by parachute to set up a mission with Tito. Other British missions followed, and by June much evidence had accumulated. The Chiefs of Staff reported on June 6: "It is clear from information available to the War Office that the Cetniks are hopelessly compromised in their relations with the Axis in Herzegovina and Montenegro. During the recent fighting in the latter area it has been the well-organised Partisans rather than the Cetniks who have been holding down the Axis forces."

Towards the end of the month my attention was drawn to the question of obtaining the best results from local resistance to the Axis in Yugoslavia. Having called for full information, I presided at a Chiefs of Staff conference at Downing Street on June 23. In the course of the discussion I emphasised the very great value of giving all possible support to the Yugoslav anti-Axis movement, which was containing about thirty-three Axis divisions in that area. This matter was of such importance that I directed that the small number of additional aircraft required to increase our aid must be provided, if necessary at the expense of the bombing of Germany and of the U-boat war.

Before leaving for Quebec I decided to pave the way for further action in the Balkans by appointing a senior officer to lead a larger mission to the Partisans in the field, and with authority to make direct recommendations to me about our future action towards them. Mr. Fitzroy Maclean was a member of Parliament, a man of daring character, and with Foreign Office training. This mission landed in Yugoslavia by parachute in September 1943, to find the situation revolutionised. The news of the Italian surrender had reached Yugoslavia only with the official broadcast announcements. But, in spite of complete absence of any warning by us, Tito took quick and fruitful action. Within a few weeks six Italian divisions had been disarmed by the Partisan forces, and another two went over to fight with them against the Germans. With Italian equipment the Yugoslavs were now able to arm eighty thousand more men, and to occupy for the moment most of the Adriatic coastline. There was a good chance of strengthening our general position in the Adriatic in relation to the Italian front. The Yugoslav Partisan army, totalling two hundred thousand men, although fighting primarily as guerrillas, was engaged in widespread action against the Germans, who continued their violent reprisals with increasing fury.

*   *   *

One effect of this increased activity in Yugoslavia was to exacerbate the conflict between Tito and Mihailović. Tito's growing military strength raised in an increasingly acute form the ultimate position of the Yugoslav monarchy and the exiled Government. Till the end of the war sincere and prolonged efforts were made both in London and within

Yugoslavia to reach a working compromise between both sides. I had hoped that the Russians would use their good offices in this matter. When Mr. Eden went to Moscow in October 1943 the subject was placed on the Conference agenda. He made a frank and fair statement of our attitude in the hope of securing a common Allied policy towards Yugoslavia, but the Russians displayed no wish either to pool information or to discuss a plan of action.

Even after many weeks I saw little prospect of any working arrangement between the hostile factions in Yugoslavia. "The fighting," I telegraphed to Roosevelt, "is of the most cruel and bloody character, with merciless reprisals and executions of hostages by the Huns. . . . We hope soon to compose the Greek quarrels, but the differences between Tito's Partisans and Mihailović's Serbs are very deep-seated."

My gloomy forecast proved true. At the end of November Tito summoned a political congress of his movement at Jajce, in Bosnia, and not only set up a Provisional Government, "with sole authority to represent the Yugoslav nation," but also formally deprived the Royal Yugoslav Government in Cairo of all its rights. The King was forbidden to return to the country until after the liberation. The Partisans had established themselves without question as the leading elements of resistance in Yugoslavia, particularly since the Italian surrender. But it was important that no irrevocable political decisions about the future regime in Yugoslavia should be made in the atmosphere of occupation, civil war, and *émigré* politics. The tragic figure of Mihailović had become the major obstacle. We had to maintain close military contact with the Partisans, and therefore to persuade the King to dismiss Mihailović from his post as Minister of War. Early in December we withdrew official support from Mihailović and recalled the British missions operating in his territory.

Yugoslav affairs were considered at the Teheran Conference against this background. Although it was decided by the three Allied Powers to give the maximum support to the Partisans, the role of Yugoslavia in the war was dismissed by Stalin as of minor importance, and the Russians even disputed our figures of the number of Axis divisions in the Balkans. The Soviet Government however agreed to send a Russian mission to Tito as a result of Mr. Eden's initiative. They also wished to keep contact with Mihailović.

On my return from Teheran to Cairo I saw King Peter, and told him about the strength and significance of the Partisan movement and that it might be necessary for him to dismiss Mihailović from his Cabinet. The only hope which the King possessed of returning to his country would be, with our mediation, to reach some provisional arrangement with Tito without delay and before the Partisans further extended their hold upon the country. The Russians too professed their willingness to work for some kind of compromise and I received almost unanimous advice as to what course to pursue in this disagreeable situation. Officers who had served with Tito and the commanders of missions to Mihailović presented similar pictures. The British Ambassador to the Royal Yugoslav Government, Mr. Stevenson, telegraphed to the Foreign Office: "Our policy must be based on three new factors: The Partisans will be the rulers of Yugoslavia. They are of such value to us militarily that we must back them to the full, subordinating political considerations to military. It is extremely doubtful whether we can any longer regard the monarchy as a unifying element in Yugoslavia."

By January 1944 I had been convinced by the arguments of men I knew and trusted that Milhailović was a millstone tied round the neck of the King, and he had no chance till he got rid of him. The Foreign Secretary agreed, and I wrote to Tito in this sense. But for two months longer the political wrangle over Yugoslav affairs continued in *émigré* circles in London. Each day lost diminished the chances of a balanced arrangement, and it was not until nearly the end of May that Mihailović was dismissed, and a moderate politician, Dr. Subašić, was asked to form a new Administration. Neither was I able to bring Tito and Subašić together until I met them in Naples in August, where, as will be related in due course, I did what I could to assuage the torments both of Yugoslavia and of her southernmost neighbour, Greece, to whose affairs and fortunes we must now turn.

*   *   *

After the withdrawal of the Allies in April 1941 Greece, like Yugoslavia, was occupied by the Axis Powers. The collapse of the Army and the retirement of the King and his Government into exile revived the bitter controversies of Greek politics. Both in the homeland and in Greek circles abroad there was hard criticism of the monarchy, which had sanctioned the dictatorship of General Metaxas, and thereby directly associated itself with the regime which had been defeated. There was much famine in the first winter, partially relieved by Red Cross shipments. The country was exhausted by the fighting and the Army was destroyed, but at the time of the surrender weapons were hidden in the mountains, and in sporadic fashion, and on a minor scale, resistance to the enemy was planned. In the towns of Central Greece starvation provided plenty of recruits. In April 1942 the body calling itself the National Liberation Front (known by its initials in Greek as E.A.M.), which had come into being in the previous autumn, announced the formation of the People's Liberation Army (E.L.A.S.). Small fighting groups were recruited during the following year, while in Epirus and the mountains of the northwest remnants of the Greek Army and local mountaineers gathered round the person of Colonel Napoleon Zervas. The E.A.M.-E.L.A.S. organisation was dominated by a hard core of Communist leaders. The adherents of Zervas, originally Republican in sympathy, became as time passed exclusively anti-Communist. Around these two centres Greek resistance to the Germans gathered. Neither of them had any sympathy or direct contact with the Royalist Government in London.

On the eve of Alamein we decided to attack the German supply lines leading down through Greece to the Piraeus, the port of Athens and an important base on the German route to North Africa. The first British Military Mission, under Lieutenant Colonel Myers, was accordingly dropped by parachute and made contact with the guerrillas. A viaduct on the main Athens railway line was destroyed and Greek agents made brilliant and daring sabotage against Axis shipping in the Piraeus. During the following summer the British missions were strengthened, and special efforts were made to convince the enemy that we would land in Greece on a large scale after our victory in Tunis. Anglo-Greek parties blew another railway bridge on the main Athens line, and other operations were so successful that two German divisions were moved into Greece which might have been used in Sicily. But this was the last direct military contribution which the Greek guerrillas made to the war.

The three divergent elements, E.L.A.S., numbering twenty thousand men, and predominantly under Communist control, the Zervas bands, known as E.D.E.S., totalling five thousand, and the Royalist politicians, grouped in Cairo or in London round King George II, all now thought that the Allies would probably win, and the struggle among them for political power began in earnest, to the advantage of the common foe. When the Italians capitulated in September 1943 E.L.A.S. was able to acquire most of their equipment, including the weapons of an entire division, and thus gained military supremacy. In October E.L.A.S. forces attacked E.D.E.S. (Zervas), and the British Headquarters in Cairo suspended all shipments of arms to the former.

Every effort was made by our missions on the spot to limit and end the civil war which sprawled across the ruined and occupied country, and in February 1944, British officers succeeded in establishing an uneasy truce between the two factions. But the Soviet armies were now on the borders of Rumania, and as the chances of a German evacuation of the Balkans increased, and with them the possibilities of a return of the Royal Government, with British support, the E.A.M. leaders decided on a Communist *coup d'état*.

A Political Committee of National Liberation was set up in the mountains, and the news broadcast to the world. This was a direct challenge to the future authority of the Royal Government, and the signal for trouble in the Greek armed forces in the Middle East and in Greek Government circles abroad. On March 31 a group of officers from the Army, Navy, and Air Force called on the Premier, M. Tsouderos, in Cairo to demand his resignation. The 1st Brigade of the Greek Army, which I was hoping could take part in the Italian campaign, mutinied against its officers. Five ships of the Royal Hellenic Navy declared for a republic, and on April 8 a Greek destroyer refused to put to sea unless a Government was formed which would include representatives of E.A.M.

I was at this time in charge of the Foreign Office, owing to Mr. Eden's absence. I thus had all the threads directly in my hands and with my support and personal encouragement, General Paget, who commanded the British forces in Egypt, surrounded the Brigade, which numbered forty-five hundred men and over fifty guns, all deployed in defensive positions against us. On the evening of the 23rd the ships were boarded by loyal Greek sailors and with about fifty casualities the mutineers were collected and sent ashore. Next day the Brigade surrendered and laid down its arms, and was evacuated to a prisoner-of-war cage, where the ringleaders were arrested. There were no Greek casualties, but one British officer was killed. The naval mutineers had surrendered unconditionally twenty-four hours earlier.

Meanwhile the King had arrived in Cairo, and on April 12 issued a proclamation that a representative Government composed largely of Greeks from within Greece would be formed. Steps were taken in secret to bring out representatives from metropolitan Greece, including M. Papandreou, the leader of the Greek Social Democratic Party, and on the 26th he took office. In May a conference of all parties, including leaders from the Greek mountains, met at a mountain resort in the Lebanon. Here it was agreed, after a fierce debate lasting three days, to set up an Administration in Cairo in which all groups would be represented under the Premiership of Papandreou, while in the mountains of Greece a united military organisation would continue to struggle against the Germans. The difficulties and struggles which lay before us all in this nerve-centre of Europe and the world will be recounted in their proper place. We may now leave this scene for others not less convulsive but larger.

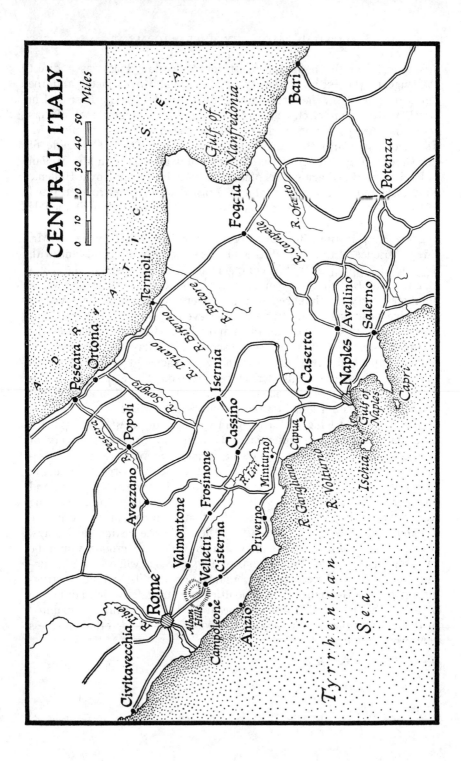

CENTRAL ITALY

Miles
0 10 20 30 40 50

10

The Anzio Stroke

RETROSPECT IS NOW INEVITABLE if the Italian setting is to be understood. After the surrender in September 1943, the organisation of resistance to the Germans fell by default into the hands of an underground Committee of Liberation in Rome, and linked with the mounting activity of Partisan bands which now began activities throughout the peninsula. The members of this committee were politicians driven from power by Mussolini in the early 1920's or representatives of groups hostile to Fascist rule. Over all hung the menace of a recrudescence of the hard core of Fascism in the hour of defeat. The Germans certainly did their best to promote it.

Mussolini had been interned on the island of Ponza, and later at La Maddalena, off the coast of Sardinia. Fearing a German *coup de main*, Badoglio had at the end of August moved his former master to a small mountain resort high in the Abruzzi, in Central Italy. In the haste of the flight from Rome no precise instructions were given to the police agents and Carabinieri guarding the fallen Dictator. On the morning of Sunday, September 12, ninety German parachutists landed by glider near the hotel where Mussolini was confined. He was removed, without casualties, in a light German aircraft, and carried to yet another meeting in Munich with Hitler.

During the succeeding days the two men debated how to extend the life of Italian Fascism in those parts of Italy still occupied by the German troops. On the 15th the Duce announced that he had reassumed the leadership of Fascism and that a new Republican-Fascist Party, purged and uplifted from traitorous elements, would rebuild a faithful Government in the North. For a moment it seemed that the old system, now dressed up in a pseudo-revolutionary garb, might flare again into life. The results disappointed the Germans, but there was to be no

795

turning back. Mussolini's halfhearted "Hundred Days" began. At the end of September he set up his headquarters on the shores of Lake Garda. This pitiful shadow Government is known as the "Republic of Salo." Here the squalid tragedy was played out. The dictator and lawgiver of Italy for more than twenty years dwelt with his mistress in the hands of his German masters, ruled by their will, and cut off from the outside world by carefully chosen German guards and doctors.

The Italian surrender caught their armies in the Balkans completely unawares, and many troops were trapped in desperate positions between local guerrilla forces and the vengeful Germans. There were savage reprisals. The Italian garrison of Corfu, over seven thousand strong, were almost annihilated by their former allies. The Italian troops of the island of Cephalonia held out until September 22. Many of the survivors were shot, and the rest deported. Some of the garrisons of the Aegean islands managed to escape in small parties to Egypt. In Albania, on the Dalmatian coast, and inside Yugoslavia a number of detachments joined the partisans. More often they were taken off to forced labour and their officers executed. In Montenegro the greater part of two Italian divisions were formed by Tito into the "Garibaldi Divisions," which suffered heavy losses by the end of the war. In the Balkans and Aegean the Italian armies lost nearly forty thousand men after the announcement of the armistice on September 8, not including those who died in deportation camps.

Italy herself was plunged into the horrors of civil war. Officers and men of the Italian Army stationed in the German-occupied North and patriots from the towns and countryside began to form partisan units and to operate against the Germans and against their compatriots who still adhered to the Duce. Contacts were made with the Allied armies south of Rome and with the Badoglio Government. In these months the network of Italian resistance to the German occupation was created in a cruel atmosphere of civil strife, assassinations, and executions. The insurgent movement in Central and Northern Italy here as elsewhere in occupied Europe convulsed all classes of the people.

Not the least of their achievements was the succour and support given to our prisoners of war trapped by the armistice in camps in Northern Italy. Out of about eighty thousand of these men, conspicuously clothed in battle dress, and in the main with little knowledge of the language or geography of the country, at least ten thousand, mostly helped by the local population with civilian clothes, were guided to safety, thanks to the risks taken by members of the Italian Resistance and the simple people of the countryside.

The bitterness and confusion were heightened in the New Year. Mussolini's phantom republic came under mounting pressure from the Germans. The governing circles around Badoglio in the South were assailed by intrigues in Italy and despised by public opinion in Britain and the United States. Mussolini was the first to react. When he arrived in Munich after his escape he found there his daughter Edda and her husband, Count Ciano. These two had fled from Rome at the time of the surrender, and although Ciano had voted against his father-in-law at the fateful meeting of the Grand Council, he hoped, thanks to the influence of his wife, for a reconciliation. During these days in Munich this in fact happened. This aroused the indignation of Hitler, who had already placed the Ciano family under house arrest on their arrival. The reluctance of the Duce to punish the traitors to Fascism, and particularly Ciano, was perhaps the main reason why Hitler formed such a low opinion of his colleague at this critical time.

It was not until the declining strength of the "Republic of Salo" had fallen far, and the impatience of its German masters had sharpened, that Mussolini agreed to let loose a wave of calculated vengeance. All those leaders of the old Fascist regime who had voted against him in July and who could be caught in German-occupied Italy were brought to trial at the end of 1943, in the medieval fortress at Verona. Among them was Ciano. Without exception they received the death sentence. In spite of the entreaties and threats of Edda, the Duce could not relent. In January 1944 the group, which included not only Ciano but also the seventy-eight-year-old Marshal de Bono, a colleague in the march on Rome, were taken out to die a traitor's death — to be shot in the back tied to a chair. They all died bravely.

The end of Ciano was in keeping with all the elements of Renaissance tragedy. Mussolini's submission to Hitler's vengeful demands brought him only shame, and the miserable neo-Fascist republic dragged on by Lake Garda — a relic of the Broken Axis.

*   *   *

We had meanwhile spent the first weeks of January in intensive preparations for Operation "Shingle," as Anzio was called in our codes, and in preliminary operations by the Fifth Army to draw the enemy's attention and reserves away from the beachhead. Fighting was bitter, for the Germans clearly meant to prevent us from breaking into the Gustav Line, which, with Cassino as its central feature, was the rearmost position of their deep defensive zone. In these rocky mountains a great fortified system had been created, with lavish use of concrete and steel. From their observation posts on the heights the enemy could direct their guns on all movements in the valleys below. Our troops made great efforts which, though gaining little ground, had the desired effect on the enemy. It distracted their attention from the approaching threat to their vulnerable seaward flank and caused them to bring up three good divisions from reserve to restore the situation.

By the afternoon of the 21st the convoys for Anzio were well out to sea, covered by our aircraft. The weather was well suited to a concealed approach. Our heavy attacks on enemy airfields, and especially at Perugia, the German air reconnaissance base, kept many of their aircraft grounded, and it was with tense, but I trust suppressed, excitement that I awaited the outcome of this considerable stroke. Presently I learned that the VIth Corps, consisting of the 3rd United States and 1st British Divisions under the American General Lucas, had landed on the Anzio beaches at 2 A.M. on the 22nd. There was very little opposition and practically no casualties. By midnight thirty-six thousand men and over three thousand vehicles were ashore. "We appear," signalled Alexander, who was on the spot, "to have got almost complete surprise. I have stressed the importance of strong-hitting mobile patrols being boldly pushed out to gain contact with the enemy, but so far have not received reports of their activities." I was in full agreement with this, and replied: "Thank you for all your messages. Am very glad you are pegging out claims rather than digging in beachheads."

But now came disaster, and the ruin in its prime purpose of the enterprise. General Lucas confined himself to occupying his beachhead and having equipment and vehicles brought ashore. General Penney, commanding the British 1st Division, was anxious to push inland. His reserve brigade was however held back with the corps. Minor probing attacks towards Cisterna and Campoleone occupied the 22nd and 23rd. No general attempt to advance was made by the commander of the expedition. By the evening of the 23rd the whole of the two divisions and their attached troops, including two British Commandos, the United States Rangers, and parachutists, had been landed, with masses of impedimenta. The defences of the beachhead growing, but the opportunity for which great exertions had been made was gone.

Kesselring reacted quickly to his critical situation. The bulk of his reserves were already committed against us on the Cassino front, but he pulled in whatever units were available, and in forty-eight hours the equivalent of about two divisions was assembled to resist our further advance. On the 27th, serious news arrived. The Guards Brigade had gone forward, but they were still about a mile and a half short of Campoleone, and the Americans were still south of Cisterna. Alexander said that neither he nor General Clark was satisfied with the speed of the advance, and that Clark was going to the beachhead at once. I replied:

> I am glad to learn that Clark is going to visit the beachhead. It would be unpleasant if your troops were sealed off there and the main army could not advance up from the south.

This however was exactly what was going to happen.

*   *   *

Meanwhile our attacks on the Cassino positions continued. The threat to his flank did not weaken Kesselring's determination to withstand our assaults. The German resolve was made crystal-clear by an order from Hitler captured on the 24th:

> The Gustav Line must be held at all costs for the sake of the political consequences which would follow a completely successful defence. The Fuehrer expects the bitterest struggle for every yard.

He was certainly obeyed. At first we made good progress. We crossed the river Rapido above Cassino town and attacked southwards against Monastery Hill; but the Germans had reinforced and held on fanatically, and by early February our strength was expended. A New Zealand Corps of three divisions was brought over from the Adriatic, and on the 15th our second major attack began with the bombing of the monastery itself. The height on which the monastery stood surveyed the junction of the rivers Rapido and Liri and was the pivot of the whole German defence. It had already proved itself a formidable, strongly defended obstacle. Its steep sides, swept by fire, were crowned by the famous building, which several times in previous wars had been pillaged, destroyed, and rebuilt. There is controversy about whether it should have been destroyed once again. The monastery did not contain German troops, but the enemy fortifications were hardly separate from the building itself. It dominated the whole battlefield, and naturally General Freyberg, the Corps Commander concerned, wished to have it heavily bombarded from the air before he launched the infantry attack. The Army Commander, General Mark Clark, unwillingly sought and obtained permission from General Alexander, who accepted the responsibility. On February 15 therefore, after the monks had been given full warning, over 450 tons of bombs were dropped and heavy damage was done. The great outer walls and gateway still stood. The result was not good. The Germans had now every excuse for making whatever use they could of the rubble of the ruins, and this gave them even better opportunities for defence than when the building was intact.

It fell to the 4th Indian Division, which had recently relieved the Americans on the ridges north of the monastery, to make the attack. On two successive nights they tried in vain to seize a knoll that lay between their position and Monastery Hill. On the night of February 18 a third attempt was made. The fighting was desperate, and all our men who reached the knoll were killed. Later that night a brigade by-passed the knoll and moved directly at the monastery, only to encounter a concealed ravine heavily mined and covered by enemy machine guns at shortest range. Here they lost heavily and were stopped. While this fierce conflict was raging on the heights above them the New Zealand Division succeeded in crossing the river Rapido; but they were counter-attacked by tanks before their bridgehead was secure and forced back again. The direct attack on Cassino had failed.

*   *   *

We must now return to the beachhead. By January 30 the 1st U.S. Armoured Division had landed at Anzio and the 45th U.S. Division was on its way. All this had to be done over the difficult beaches or through the tiny fishing port. "The situation as it now stands," signalled Admiral John Cunningham, "bears little relation to the light-ning thrust by two or three divisions envisaged at Marrakesh, but you may rest assured that no effort will be spared by the Navies to provide the sinews of victory." This promise, as will be seen, was amply re-deemed.

On the same day the VIth Corps made its first attack in strength. Some ground was gained, but on February 3 the enemy launched a counterstroke which drove in the salient of the 1st British Division and was clearly only a prelude to harder things to come. In the words of General Wilson's report, "the perimeter was sealed off and our forces therein are not capable of advancing." Though General Lucas had achieved surprise he had failed to take advantage of it. All this was a great disappointment at home and in the United States. I did not of course know what orders had been given to General Lucas, but it is a root principle to push out and join issue with the enemy, and it would seem that his judgment was against it from the beginning. As I said at the time, I had hoped that we were hurling a wildcat onto the shore, but all we had got was a stranded whale. We were apparently still stronger than the Germans in fighting power. The ease with which they moved their pieces about on the board and the rapidity with which they adjusted the perilous gaps they had to make on their southern front was most impressive. It all seemed to give us very adverse data for "Overlord."

The expected major effort to drive us back into the sea opened on the 16th when the enemy employed over four divisions, supported by four hundred and fifty guns, in a direct thrust southward from Campoleone. Hitler's special order of the day was read out to the troops before the attack. He demanded that our beachhead "abscess" be eliminated in three days. The attack fell at an awkward moment, as the 45th U.S. and 56th British Divisions, transferred from the Cassino front, were just relieving our gallant 1st Division, who soon found themselves in full action again. A deep, dangerous wedge was driven into our line, which was forced back here to the original beachhead. All hung in the balance. No further retreat was possible. Even a short advance would have given the enemy the power to use not merely their long-range guns in harassing fire upon the landing stages and shipping, but to put down a proper field artillery barrage upon all intakes or departures. I had no illusions about the issue. It was life or death.

But fortune, hitherto baffling, rewarded the desperate valour of the British and American armies. Before Hitler's stipulated three days the German attack was stopped. Then their own salient was counterattacked in flank and cut out under fire from all our artillery and bombardment by every aircraft we could fly. The fighting was intense, losses on both sides were heavy, but the deadly battle was won.

One more attempt was made by Hitler — for his was the will power at work — at the end of February. The 3rd U.S. Division, on the eastern flank, was attacked by three German divisions. These were weakened and shaken by their previous failure. The Americans held stubbornly, and the attack was broken in a day, when the Germans had suffered more than twenty-five hundred casualties. On March 1, Kesselring accepted his failure. He had frustrated the Anzio expedition. He could not destroy it.

At the beginning of March the weather brought about a deadlock. Napoleon's fifth element — mud — bogged down both sides. We could not break the main front at Cassino, and the Germans had equally failed to drive us into the sea at Anzio. In numbers there was little to choose between the two combatants. By now we had twenty divisions in Italy, but both Americans and French had had very heavy losses. The enemy had eighteen or nineteen divisions south of Rome, and five more in Northern Italy, but they too were tired and worn.

There could be no hope now of a break-out from the Anzio beachhead and no prospect of an early link-up between our two separated forces until the Cassino front was broken. The prime need therefore was to make the beachhead really firm, to relieve and reinforce the troops, and to pack in stores to withstand a virtual siege and nourish a subsequent sortie. Time was short, since many of the landing craft must soon leave for "Overlord." Their move had so far been rightly postponed, but no further delay was possible. The Navies put all their strength into the effort, with admirable results. The previous average daily tonnage landed had been three thousand; in the first ten days of March this was more than doubled.

＊　　　＊　　　＊

But although Anzio was now no longer an anxiety the campaign in Italy as a whole had dragged. We had hoped that by this time the Germans would have been driven north of Rome and that a substantial part of our armies would have been set free for a strong landing on the French Riviera coast to help the main cross-Channel invasion. This operation, "Anvil," had been agreed in principle at Teheran. It was soon to become a cause of contention between ourselves and our American Allies. The campaign in Italy had obviously to be carried forward a long way before this issue arose, and the immediate need was to break the deadlock on the Cassino front. Preparations for the third Battle of Cassino were begun soon after the February failure, but the bad weather delayed it until March 15.

This time Cassino town was the primary objective. After a heavy bombardment, in which nearly one thousand tons of bombs and twelve hundred shells were expended, our infantry advanced. "It seemed to me inconceivable," said Alexander, "that any troops should be left alive after eight hours of such terrific hammering." But they were. The 1st German Parachute Division, probably the toughest fighters in all their Army, fought it out amid the heaps of rubble with the New Zealanders and Indians. By nightfall the greater part of the town was in our hands, while the 4th Indian Division, coming down from the north, made equally good progress and next day were two-thirds of the way up Monastery Hill. Then the battle swung against us. Our tanks could not cross the large craters made by the bombardment and follow up the infantry assault. Nearly two days passed before they could help. The enemy filtered in reinforcements. The weather broke in storm and rain. The struggle in the ruins of Cassino town continued until the 23rd, with hard fighting in attacks and counterattacks. The New Zealanders and the Indians could do no more. We had however established a firm bridgehead over the river Rapido, which, with a deep bulge made across the lower Garigliano in January, was of great value when the final, successful battle came. Here and at the Anzio bridgehead we had pinned down in Central Italy nearly twenty good German divisions. Many of them might have gone to France.

Such is the story of the struggle of Anzio; a story of high opportunity and shattered hopes, of skilful inception on our part and swift recovery by the enemy, of valour shared by both. We now know that early in January the German High Command had intended to transfer five of their best divisions from Italy to Northwest Europe. Kesselring protested that in such an event he could no longer carry out his orders to fight south of Rome and he would have to withdraw. Just as the argument was at its height the Anzio landing took place. The High Command dropped the idea, and instead of the Italian front contributing forces to Northwest Europe the reverse took place. We knew nothing of all these changes of plan at the time, but it proves that the aggressive action of our armies in Italy, and specifically the Anzio stroke, made its full contribution towards the success of "Overlord." We shall see later on the part it played in the liberation of Rome.

II

"Overlord"

THOUGHT ARISING FROM factual experience may be a
bridle or a spur. The reader will be aware that while I was always
willing to join with the United States in a direct assault across the Channel on the German sea front in France, I was not convinced that this
was the only way of winning the war, and I knew that it would be a
very heavy and hazardous adventure. The fearful price we had had to
pay in human life and blood for the great offensives of the First World
War was graven in my mind. It still seemed to me, after a quarter of a
century, that fortifications of concrete and steel armed with modern fire
power, and fully manned by trained, resolute men, could only be overcome by surprise in time or place by turning their flanks, or by some
new and mechanical device like the tank. Superiority of bombardment,
terrific as it may be, was no final answer. The defenders could easily
have ready other lines behind their first, and the intervening ground
which the artillery could conquer would become impassable crater fields.
These were the fruits of knowledge which the French and British had
bought so dearly from 1915 to 1917.

Since then new factors had appeared, but they did not all tell the
same way. The fire power of the defence had vastly increased. The
development of mine fields both on land and in the sea was enormous.
On the other hand, we, the attackers, held air supremacy, and could
land large numbers of paratroops behind the enemy's front, and above
all block and paralyse the communications by which he could bring
reinforcements for a counterattack.

Throughout the summer months of 1943 General Morgan and his
Allied Inter-Service Staff had laboured at the plan. In a previous chap-
ter I have described how it was presented to me during my voyage to
Quebec for the "Quadrant" Conference. There the scheme was generally

approved, but Eisenhower and Montgomery disagreed with one important feature. They wanted an assault in greater strength and on a wider front, so as to gain quickly a good-sized bridgehead in which to build up their forces for the break-out. Also it was important to capture the docks at Cherbourg earlier than had been planned. They wanted a first assault by five divisions instead of three. Of course this was perfectly right. General Morgan himself had advocated an extension of the initial landing, but had not been given enough resources. But where were the extra landing craft to come from? Southeast Asia had already been stripped. There were sufficient in the Mediterranean to carry two divisions, but these were needed for "Anvil," the sea-borne assault on Southern France which was to take place at the same time as "Overlord" and draw German troops away from the North. If "Anvil" were to be reduced it would be too weak to be helpful. It was not until March that General Eisenhower, in conference with the British Chiefs of Staff, made his final decision. The American Chiefs of Staff had agreed that he should speak for them. Having recently come from the Mediterranean, he knew all about "Anvil," and now as Supreme Commander of "Overlord" he could best judge the needs of both. It was agreed to take the ships of one division from "Anvil" and to use them for "Overlord." The ships for a second division could be found by postponing "Overlord" till the June moon period. The output of new landing craft in that month would fill the gap.

*   *   *

Once the size of the expedition had been determined, it was possible to go ahead with intensive training. Not the least of our difficulties was to find enough room. A broad partition was arranged between British and American forces, whereby the British occupied the southeastern and the Americans the southwestern parts of England. The inhabitants of coastal areas accepted all the inconveniences in good part. One British division with its naval counterpart did all its earlier training in the Moray Firth area in Scotland. The winter prepared them for the rough-and-tumble of D-Day.

The theory and practice of amphibious operations had long been established by the Combined Operations Staff, under Admiral Mountbatten, who had been succeeded by General Laycock. It had now to be taught to all concerned, in addition to the thorough general training needed for modern warfare. This of course had long been going on in Britain and America in exercises great and small with live ammunition. Many officers and men entered into battle for the first time, but all bore themselves like seasoned troops.

Lessons from previous large-scale exercises, and of course from our hard experience at Dieppe, were applied in final rehearsals by all three Services, which culminated in early May. All this activity did not pass unnoticed by the enemy. We did not object, and special pains were taken that they should be remarked by watchers in the Pas de Calais, where we wanted the Germans to believe we were coming. Constant air reconnaissance kept us informed of what was going on across the Channel. And of course there were other ways of finding out. Many trips were made by parties in small craft to resolve some doubtful point, to take soundings inshore, to examine new obstacles, or to test the slope and nature of a beach. All this had to be done in darkness, with silent approach, stealthy reconnaissance, and timely withdrawal.

An intricate decision was the choice of D-Day and "H-Hour," the moment at which the leading assault craft should hit the beach. From this many other timings had to be worked backwards. It was agreed to approach the enemy coast by moonlight, because this would help both our ships and our air-borne troops. A short period of daylight before H-Hour was also needed to give order to the deployment of the small craft and accuracy to the covering bombardment. But if the interval between first light and H-Hour was too long the enemy would have more time to recover from their surprise and fire on our troops in the act of landing.

Then there were the tides. If we landed at high tide the underwater obstacles would obstruct our approach; if at low tide the troops would have far to go across the exposed beaches. Many other factors had to be considered, and it was finally decided to land about three hours before high water. But this was not all. The tides varied by forty minutes between the eastern and western beaches, and there was a submerged reef in one of the British sectors. Each sector had to have a different H-Hour, which varied from one place to another by as much as eighty-five minutes.

Only on three days in each lunar month were all the desired conditions fulfilled. The first three-day period after May 31, General Eisenhower's target date, was June 5, 6, and 7. Thus was June 5 chosen. If the weather were not propitious on any of those three days the whole operation would have to be postponed at least a fortnight — indeed, a whole month if we waited for the moon.

*   *   *

Of course we had not only to plan for what we were really going to do. The enemy were bound to know that a great invasion was being prepared; we had to conceal the place and time of attack and make him think we were landing somewhere else and at a different moment. This alone involved an immense amount of thought and action. Coastal areas were banned to visitors; censorship was tightened; letters after a certain date were held back from delivery; foreign embassies were forbidden to send cipher telegrams, and even their diplomatic bags were delayed. Our major deception was to pretend that we were coming across the Straits of Dover. It would not be proper even now to describe all the methods employed to mislead the enemy, but the obvious ones of simulated concentrations of troops in Kent and Sussex, of fleets of dummy ships collected in the Cinque Ports, of landing exercises on the nearby beaches, of increased wireless activity, were all used. More reconnaissances were made at or over the places we were *not* going to than at the places we were. The final result was admirable. The German High Command firmly believed the evidence we put at their disposal. Rundstedt, the Commander-in-Chief on the Western Front, was convinced that the Pas de Calais was our objective.

*   *   *

The concentration of the assaulting forces — 176,000 men, 20,000 vehicles, and many thousand tons of stores, all to be shipped in the first two days — was in itself an enormous task. From their normal stations all over Britain the troops were brought to the southern counties. The three air-borne divisions which were to drop on Normandy before the sea assault were assembled close to the airfields whence they would set out. From their concentration areas in rear troops were brought forward for embarkation in assigned priority to camps in marshalling areas near the coast. At the marshalling camps they were divided up into detachments corresponding to the ship- or boat-loads in which they would be embarked. Here every man received his orders. Once briefed, none were permitted to leave camp. The camps themselves were situated near to the embarkation points. These were ports or "hards" — *i.e.,* stretches of beach concreted to allow of easy embarkation on smaller craft. Here they were to be met by the naval ships.

It seemed most improbable that all this movement by sea and land would escape the attentions of the enemy. There were many tempting targets for their Air, and full precautions were taken. Nearly seven thousand guns and rockets and over a thousand balloons protected the great masses of men and vehicles. But there was no sign of the Luftwaffe. How different things were four years before! The Home Guard, who had so patiently waited for a worth-while job all those years, now found it. Not only were they manning sections of anti-aircraft and coast defences, but they also took over many routine and security duties, thus releasing other soldiers for battle. All Southern England thus became a vast military camp, filled with men trained, instructed, and eager to come to grips with the Germans across the water.

On Monday, May 15, three weeks before D-Day, we held a final conference in London at Montgomery's headquarters in St. Paul's School.
The King, Field Marshal Smuts, the British Chiefs of Staff, the commanders of the expedition, and many of their principal Staff officers were
present. On the stage was a map of the Normandy beaches and the
immediate hinterland, set at a slope so that the audience could see it
clearly, and so constructed that the high officers explaining the plan of
operation could walk about on it and point out the landmarks. General
Eisenhower opened the proceedings, and the forenoon session closed
with an address by His Majesty. Montgomery made an impressive
speech. He was followed by several Naval, Army, and Air commanders,
and also by the principal administrative officer, who dwelt upon the
elaborate preparations that had been made for the administration of the
force when it got ashore.

Events now began to move swiftly and smoothly to the climax.
There was still no sign that the enemy had penetrated our secrets. We
observed some reinforcement of light naval forces at Cherbourg and
Havre, and there was more minelaying activity in the Channel, but in
general he remained quiescent, awaiting a definite lead about our intentions. On May 28 subordinate commanders were informed that
D-Day would be June 5. From this moment all personnel committed
to the assault were "sealed" in their ships or at their camps and assembly points ashore. All mail was impounded and private messages
of all kinds forbidden except in case of personal emergency.

The weather now began to cause anxiety. A fine spell was giving
way to unsettled conditions, and henceforward a commanders' meeting
was held twice daily to study the weather reports. At their first meeting
poor conditions were predicted for D-Day, with low clouds. This was
of prime importance to the air forces, affecting both the bombing and
the air-borne landings. On June 2 the first warships sailed from the
Clyde, as well as two midget submarines from Portsmouth, whose duty
was to mark the assault areas. June 3 brought little encouragement.
A rising westerly wind was whipping up a moderate sea; there was
heavy cloud and a lowering cloud base. Predictions for June 5 were
gloomy.

That afternoon I drove down to Portsmouth with Mr. Bevin and
Field Marshal Smuts and saw a large number of troops embarking for
Normandy. We visited the headquarters ship of the 50th Division, and
then cruised down the Solent in a launch, boarding one ship after
another. On the way back we stopped at General Eisenhower's camp
and wished him luck. We got back to the train in time for a very late
dinner. While it was in progress Ismay was called to the telephone by
Bedell Smith, who told him that the weather was getting worse and that

the operation would probably have to be postponed for twenty-four hours. General Eisenhower would wait until the early hours of June 4 before making a definite decision. Meanwhile units of the great armada would continue to put to sea according to programme.

Ismay came back and reported the bleak news. Those who had seen the array in the Solent felt that the movement was now as impossible to stop as an avalanche. We were haunted by the knowledge that if the bad weather continued and the postponement had to be prolonged beyond June 7 we could not again get the necessary combination of moon and tide for at least another fortnight. Meanwhile the troops had all been briefed. They clearly could not be kept on board these tiny ships indefinitely. How was a leakage to be prevented?

But the anxiety that everyone felt was in no way apparent at the dinner table in the train. Field Marshal Smuts was at his most entertaining pitch. He told the story of the Boer surrender at Vereeniging in 1902 — how he had impressed on his colleagues that it was no use fighting on and that they must throw themselves on the mercy of the British. He had been assailed as a coward and a defeatist by his own friends, and he had spent the most difficult hour of his life. In the end however he had won through, had gone to Vereeniging, and peace was made. He went on to speak about his experiences at the outbreak of the Second World War, when he had to cross the floor of the House and fight his own Prime Minister, who wished to remain neutral.

We went to bed at about half past one. Ismay told me that he would wait up to hear the result of the morning conference. As there was nothing I could do about it, I said that I was not to be woken to hear the result. At 4.15 A.M. Eisenhower again met his commanders, and heard from the weather experts the ominous report, sky overcast, cloud ceiling low, strong southwesterly wind, with rain and moderate sea. The forecast for the 5th was even worse. Reluctantly he ordered a postponement of the attack for twenty-four hours, and the whole vast array was put into reverse in accordance with a carefully prepared plan. All convoys at sea turned about and small craft sought shelter in convenient anchorage. Only one large convoy, comprising 138 small vessels, failed to receive the message, but this too was overtaken and turned round without arousing the suspicions of the enemy. It was a hard day for the thousands of men cooped up in landing craft all round the coast. The Americans who came from the West Country ports had the greatest distance to go and suffered most.

At about five o'clock that morning Bedell Smith again telephoned Ismay confirming the postponement, and Ismay went to bed. Half an hour later I woke up and sent for him. He told me the news. He says I made no comment.

The hours dragged slowly by until, at 9.15 P.M. on the evening of June 4, another fateful conference opened at Eisenhower's battle headquarters. Conditions were bad, typical of December rather than June, but the weather experts gave some promise of a temporary improvement on the morning of the 6th. After this they predicted a return of rough weather for an indefinite period. Faced with the desperate alternatives of accepting the immediate risks or of postponing the attack for at least a fortnight, General Eisenhower, with the advice of his commanders, boldly, and as it proved wisely, chose to go ahead with the operation, subject to final confirmation early on the following morning. At 4 A.M. on June 5 the die was irrevocably cast: the invasion would be launched on June 6.

In retrospect this decision rightly evokes admiration. It was amply justified by events, and was largely responsible for gaining us the precious advantage of surprise. We now know that the German meteorological officers informed their High Command that invasion on the 5th or 6th of June would not be possible owing to stormy weather, which might last for several days.

All day on June 5 the convoys bearing the spearhead of the invasion converged on the rendezvous south of the Isle of Wight. Thence, in an endless stream, led by the mine sweepers on a wide front and protected on all sides by the Allied Navies and Air Forces, the greatest armada that ever left our shores set out for the coast of France. The rough conditions at sea were a severe trial to troops on the eve of battle, particularly in the terrible discomfort of the smaller craft. Yet, in spite of all, the vast movement was carried through with almost the precision of a parade, and, although not wholly without loss, such casualties and delays as did occur, mostly to small craft in tow, had no appreciable effect on events.

Round all our coasts the network of defence was keyed to the highest pitch of activity. The Home Fleet was alert against any move by German surface ships, while air patrols watched the enemy coast from Norway to the Channel. Far out at sea, in the Western Approaches and in the Bay of Biscay, aircraft of Coastal Command, in great strength, supported by flotillas of destroyers, kept watch for enemy reactions. Our Intelligence told us that over fifty U-boats were concentrated in the French Biscay ports, ready to intervene when the moment came. While I sat in my chair in the Map Room of the Annexe the thrilling news of the capture of Rome arrived.

12

Rome and D-Day

DEADLOCK at Anzio and Cassino imposed a halt in the Allied advance in Italy which lasted for nearly two months. Our troops had to be rested and regrouped. Most of the Eighth Army had to be brought over from the Adriatic side and the two armies concentrated for the next assault. In the meantime General Wilson used all his air power to impede and injure the enemy, who, like us, were using the pause for reorganising and replenishing themselves for further battle.

The potent Allied Air joined in attacking enemy land communications in the hope that these could be kept cut and their troops forced to withdraw for lack of supplies. This operation, optimistically called "Strangle," aimed at blocking the three main railway lines from Northern Italy, the principal targets being bridges, viaducts, and other bottlenecks. They tried to starve the Germans out. The effort lasted more than six weeks, and did great damage. Railway movement was consistently stopped far north of Rome, but it failed to attain all we hoped. By working their coastal shipping to the utmost, transferring loads to motor transport, and making full use of the hours of darkness the enemy contrived to maintain themselves. But they could not build enough reserve stocks for protracted and heavy fighting, and in the severe land battles at the end of May they were much weakened. The junction of our separated armies and the capture of Rome took place more rapidly than we had forecast. The German Air Force suffered severely and in early May it could muster only a bare seven hundred planes against our thousand combat aircraft.

By then General Clark, of the Fifth Army, had over seven divisions, four of them French, on the front from the sea to the river Liri; thence the Eighth Army, now under General Leese, continued the line through Cassino into the mountains with the equivalent of nearly twelve. In all the Allies mustered over twenty-eight divisions, of which the equivalent of only three remained in the Adriatic sector.

Opposed to them were twenty-three German divisions, but our deception arrangements had puzzled Kesselring so well that they were widely spread. Between Cassino and the sea, where our main blows were to fall, there were only four, and reserves were scattered and at a distance. Our attack came unexpectedly. The Germans were carrying out reliefs opposite the British front, and one of their Army Commanders had planned to go on leave

The great offensive began at 11 P.M. on May 11, when the artillery of both our armies, two thousand guns, opened a violent fire, reinforced at dawn by the full weight of the Tactical Air Force. After much heavy fighting the enemy began to weaken. On the morning of May 18 Cassino town was finally cleared by the 4th British Division, and the Poles triumphantly hoisted their red and white standard over the ruins of the monastery. Kesselring had been sending down reinforcements as fast as he could muster them, but they were arriving piecemeal, only to be thrown into the battle to check the flood of the Allied advance. By the 25th the Germans were in full retreat and hotly pursued on the whole of the Eighth Army front.

Six divisions under the American General Truscott had been packed into the Anzio beachhead and burst forth with the simultaneous onslaught of the Eighth Army. After two days of stiff fighting they gained contact with the U.S. IInd Corps. At long last our forces were reunited, and we began to reap the harvest from our winter sowing. The enemy in the south were in full retreat, and the Allied Air did its utmost to impede movement and break up concentrations. But obstinate rearguards frequently checked our pursuing forces, and their retirement did not degenerate into a rout. The mountainous country stopped us using our great strength in armour, which otherwise could have been employed to much advantage.

But on the night of June 2 the German resistance broke and next day Truscott's Corps in the Alban Hills, with the British 1st and 5th Divisions on its left, pressed on towards Rome. The IInd American Corps led them by a short head. They found the bridges mostly intact, and at 7.15 P.M. on June 4 the head of their 88th Division entered the Piazza Venezia, in the heart of the capital. From many quarters came messages of warm congratulations. I even got a pat from the Bear.

At noon on D-Day, June 6, 1944, I asked the House of Commons to "take formal cognisance of the liberation of Rome by the Allied Armies under the command of General Alexander," the news of which had been released the previous night. There was intense excitement about the landings in France, which everyone knew were in progress at the moment. Nevertheless I devoted ten minutes to the campaign in Italy and in paying my tribute to the Allied Armies there. After thus keeping them on tenterhooks for a little I gave them an account of what had happened, so far as we were then informed. By the afternoon I felt justified in reporting to Stalin:

> Everything has started well. The mines, obstacles, and land batteries have been largely overcome. The air landings were very successful, and on a large scale. Infantry landings are proceeding rapidly, and many tanks and self-propelled guns are already ashore. Weather outlook moderate to good.

His answer was prompt, and contained welcome news of the highest importance. "I have received," he cabled, "your communication about the success of the beginning of the 'Overlord' operations. It gives joy to us all and hope of further successes. The summer offensive of the Soviet forces, organised in accordance with the agreement at the Teheran Conference, will begin towards the middle of June on one of the important sectors of the front. . . . At the end of June and during July offensive operations will become a general offensive of the Soviet forces."

I was actually sending him a fuller account of our progress when this telegram arrived. "I am well satisfied," I answered, "with the situation up to noon today [June 7]. Only at one American beach has there been serious difficulty, and that has now been cleared up. Twenty thousand air-borne troops are safely landed behind the flanks of the enemy's lines, and have made contact in each case with the American and British sea-borne forces. We got across with small losses. We had expected to lose about ten thousand men. . . ."

Stalin telegraphed again a few days later:

> As is evident, the landing, conceived on a grandiose scale, has succeeded completely. My colleagues and I cannot but admit that the history of warfare knows no other like undertaking from the point of view of its scale, its vast conception, and its masterly execution. As is well known, Napoleon in his time failed ignominiously in his plan to force the Channel. The hysterical Hitler, who boasted for two years that he would effect a forcing of the Channel, was unable to make up his mind even to hint at attempting to carry out his threat. Only our Allies have succeeded

in realising with honour the grandiose plan of the forcing of the Channel. History will record this deed as an achievement of the highest order.

The word "grandiose" is the translation from the Russian text which was given me. I think "majestic" was probably what Stalin meant. At any rate, harmony was complete.

*   *   *

On June 10 General Montgomery reported that he was sufficiently established ashore to receive a visit. I therefore set off in my train to Portsmouth, with Smuts, Brooke, General Marshall, and Admiral King. All three American Chiefs of Staff had flown to the United Kingdom on June 8 in case any vital military decision had to be taken at short notice. A British and an American destroyer awaited us. Smuts, Brooke, and I embarked in the former, and General Marshall and Admiral King, with their staffs, in the latter, and we crossed the Channel without incident to our respective fronts. Montgomery, smiling and confident, met me at the beach as we scrambled out of our landing craft. His army had already penetrated seven or eight miles inland. There was very little firing or activity. The weather was brilliant. We drove through our limited but fertile domain in Normandy. It was pleasant to see the prosperity of the countryside. The fields were full of lovely red and white cows basking or parading in the sunshine. The inhabitants seemed quite buoyant and well nourished and waved enthusiastically. Montgomery's headquarters, about five miles inland, were in a château with lawns and lakes around it. We lunched in a tent looking towards the enemy. The General was in the highest spirits. I asked him how far away was the actual front. He said about three miles. I asked him if he had a continuous line. He said, "No." "What is there then to prevent an incursion of German armour breaking up our luncheon?" He said he did not think they would come. The staff told me the château had been heavily bombed the night before, and certainly there were a good many craters around it. I told him he was taking too much of a risk if he made a habit of such proceedings. Anything can be done once or for a short time, but custom, repetition, prolongation, is always to be avoided when possible in war. He did in fact move two days later, though not till he and his staff had had another dose.

It continued fine, and apart from occasional air alarms and anti-aircraft fire there seemed to be no fighting. We made a considerable inspection of our limited bridgehead. I was particularly interested to see the local ports of Port-en-Bessin, Courseulles, and Ouistreham. We had not counted much on these little harbours in any of the plans we had made for the great descent. They proved a most valuable acquisition, and soon were discharging about two thousand tons a day. I dwelt on these agreeable facts as we drove or walked round our interesting but severely restricted conquest.

Smuts, Brooke, and I went home in the destroyer *Kelvin*. Admiral Vian, who now commanded all the flotillas and light craft protecting the Arromanches harbour, was on board. He proposed that we should go and watch the bombardment of the German position by the battleships and cruisers protecting the British left flank. Accordingly we passed between the two battleships, which were firing at twenty thousand yards, and through the cruiser squadron, firing at about fourteen thousand yards, and soon we were within seven or eight thousand yards of the shore, which was thickly wooded. The bombardment was leisurely and continuous, but there was no reply from the enemy. As we were about to turn I said to Vian, "Since we are so near, why shouldn't we have a plug at them ourselves before we go home?" He said "Certainly," and in a minute or two all our guns fired on the silent coast. We were of course well within the range of their artillery, and the moment we had fired, Vian made the destroyer turn about and depart at the highest speed. We were soon out of danger and passed through the cruiser and battleship lines. This is the only time I have ever been on board a naval vessel when she fired "in anger" — if it can be so called. I admired the Admiral's sporting spirit. Smuts too was delighted. I slept soundly on the four-hour voyage to Portsmouth. Altogether it had been a most interesting and enjoyable day.

* * *

Soon afterwards I wrote to the President about various questions, including the visit of de Gaulle to France, which I had arranged without consulting Roosevelt beforehand, and added:

> I had a jolly day on Monday on the beaches and inland. There is a great mass of shipping extended more than fifty miles along the coast. It is being increasingly protected against weather by the artificial harbours, nearly every element of which has been a success, and will soon have effective shelter against bad weather. The power of our air and of our anti-U-boat forces seems to ensure it a very great measure of protection. After doing much laborious duty we went and had a plug at the Hun from our destroyer, but although the range was six thousand yards he did not honour us with a reply.

Marshall and King came back in my train. They were greatly reassured by all they saw on the American side, and Marshall wrote out a charming telegram to Mountbatten, saying how many of these new craft had been produced under his organisation and what a help they had been. You used the word "stupendous" in one of your early telegrams to me. I must admit that what I saw could only be described by that word, and I think your officers would agree as well. . . . How I wish you were here!

13

Normandy to Paris

L ET US SURVEY the enemy's dispositions and plans as we now
know them. Marshal Rundstedt, with sixty divisions, was in command of the whole Atlantic Wall, from the Low Countries to the Bay
of Biscay, and from Marseilles along the southern French shore. Under
him Rommel held the coast from Holland to the Loire. His Fifteenth
Army with nineteen divisions held the sector about Calais and Boulogne,
and his Seventh Army had nine infantry and one Panzer division at
hand in Normandy. The ten Panzer divisions on the whole Western
Front were spreadeagled from Belgium to Bordeaux. How strange that
the Germans, now on the defensive, made the same mistake as the
French in 1940 and dispersed their most powerful weapon of counterattack!

It is indeed remarkable that the vast, long-planned assault fell on the
enemy as a surprise both in time and place. Early on June 5 Rommel
left his headquarters to visit Hitler at Berchtesgaden, and was in
Germany when the blow fell. There had been much argument about
which front the Allies would attack. Rundstedt had consistently believed that our main blow would be launched across the Straits of
Dover, as that was the shortest sea route and gave the best access to
the heart of Germany. Rommel for long agreed with him. Hitler and
his staff however appeared to have had reports indicating that Normandy would be the principal battleground.[1] Even after we had landed
uncertainties continued. Hitler lost a whole critical day in making up
his mind to release the two nearest Panzer divisions to reinforce the
front. The German Intelligence Service grossly overestimated the number of divisions and the amount of suitable shipping available in England.
On their showing there were ample resources for a second big landing,

[1] Blumentritt, *Von Rundstedt*, pp. 218, 219.

so Normandy might be only a preliminary and subsidiary one. It was not until the third week in July, six weeks after D-Day, that reserves from the Fifteenth Army were sent south from the Pas de Calais to join the battle. Our deception measures both before and after D-Day had aimed at creating this confused thinking. Their success was admirable and had far-reaching results on the battle.

But the enemy fought stubbornly and were not easily overcome. In the American sector the marshes near Carentan and at the mouth of the river Vire hampered our movements, and everywhere the country was suited to infantry defence. The *bocage* which covers much of Normandy consists of a multitude of small fields divided by banks with ditches and very high hedges. Artillery support was hindered by lack of good observation and it was extremely difficult to use tanks. It was infantry fighting all the way, with every little field a potential strong point. Nevertheless good progress was made, except for the failure to capture Caen.

This small but famous town was to be the scene of bitter struggles over many days. To us it was important, because there was good ground to the east for constructing air strips, and it was also the hinge

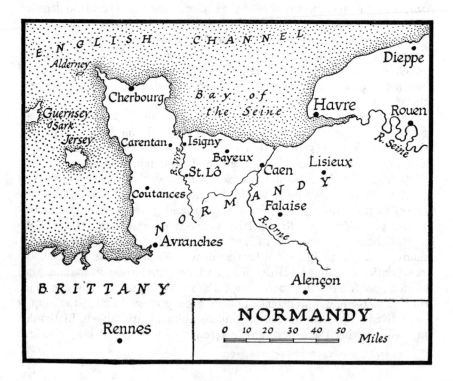

on which our whole plan turned, and on which Montgomery intended to make a great left wheel by the American forces. It was equally important for the Germans. If their lines were pierced the whole of their Seventh Army would be forced southeastward towards the Loire, opening a gap between it and the Fifteenth Army in the north. The way to Paris would then be open. Thus Caen became the scene of ceaseless attacks and the most stubborn defence, drawing towards it a great part of the German divisions, and especially their armour. This was a help as well as a hindrance.

The Germans, though the reserve divisions of their Fifteenth Army were still intact north of the Seine, had of course been reinforced from elsewhere, and by June 12 twelve divisions were in action, four of them Panzers. This was less than we expected. Our tremendous air offensive had destroyed every bridge across the Seine below Paris and the principal bridges across the Loire. Most of the reinforcing troops had to use the roads and railways running through the gap between Paris and Orléans, and endured continuous and damaging attacks by day and night from our air forces. Their divisions arrived piecemeal, short of equipment, and fatigued by long night marches, and were thrown into the line as they came. The German command had no chance to form a striking force behind the battle for a powerful well-concerted counteroffensive.

By June 11 the Allies had formed a continuous front and our fighters were operating from half a dozen forward air strips. The Americans thrust westward and northward, and after sharp fighting stood before the outer defences of Cherbourg on the 22nd. The enemy resisted stoutly till the 26th so as to carry out demolitions. These were so thorough that heavy loads could not be brought in through the port till the end of August.

*   *   *

Beyond the battlefield other events influenced the future. On the night of June 12–13 the first flying bombs fell on London. They were launched in Northern France from places remote from our landed armies. Their early conquest would bring relief to our civil population, once again under bombardment. Part of the Strategic Air Force renewed attacks on these sites, but there could of course be no question of distorting the land battle on this account. As I said in Parliament, the people at home could feel they were sharing the perils of their soldiers.

On June 17, at Margival, near Soissons, Hitler held a conference with Rundstedt and Rommel. His two generals pressed on him strongly the folly of bleeding the German Army to death in Normandy. They urged that before it was destroyed the Seventh Army should make an orderly withdrawal towards the Seine, where, together with the Fifteenth Army it could fight a defensive but mobile battle with at least some hope of success. But Hitler would not agree. Here, as in Russia and Italy, he demanded that no ground should be given up and all should fight where they stood. The generals were of course right.

We were meanwhile consolidating our strength. In the first six days 326,000 men, 54,000 vehicles, and 104,000 tons of stores were landed. An immense supply organisation came rapidly into being. By June 19 the two "Mulberry" harbours, one at Arromanches, the other ten miles farther west, in the American sector, were taking shape. The submarine pipelines ("Pluto") were to come into action later, but meanwhile Port-en-Bessin was being developed as the main supply port for petrol.[2] But then a four-day gale began which almost entirely prevented the landing of men and material, and did great damage to the newly sunk breakwaters. Many floating structures which were not designed for such conditions broke from their moorings and crashed into other breakwaters and the anchored shipping. The harbour in the American sector was ruined, and its serviceable parts were used to repair Arromanches. This gale, the like of which had not been known in June for forty years, was a severe misfortune. We were already behind our programme of unloading. The break-out was equally delayed, and on June 23 we stood only on the line we had prescribed for the 11th.

*   *   *

In the last week of June the British established a bridgehead south of Caen. Efforts to extend it southward and eastward were repelled and the southern sector was twice attacked by several Panzer divisions. In violent conflicts the Germans were severely defeated, with heavy losses from our air and powerful artillery.[3] It was now our turn to strike, and on July 8 a strong attack on Caen was launched from the north and northwest. Royal Air Force heavy bombers dropped more than 2000 tons on the German defences, and at dawn British infantry, hampered unavoidably by the bomb craters and the rubble of fallen buildings, made good progress. By the 10th all of Caen on our side of the river was gained, and by the middle of July thirty Allied divisions were ashore. Half were American and half British and Canadian. Against these the Germans had gathered twenty-seven divisions. But they had already suffered 160,000 casualties, and General Eisenhower estimated their fighting value as no higher than sixteen divisions.

[2] The "Pluto" project included first the laying of pipelines in the assault area through which seagoing tankers could discharge petrol direct to the shore. Submarine pipelines across the Channel were laid later from the Isle of Wight to Cherbourg and from Dungeness to Boulogne.

[3] These attacks were the result of Hitler's instructions at the Soissons Conference. On July 1 Keitel telephoned Rundstedt and asked, "What shall we do?" Rundstedt answered, "Make peace, you idiots. What else can you do?"

An important event now occurred. On July 17 Rommel was severely wounded. His car was attacked by our low-flying fighters, and he was carried to hospital in what was thought a dying condition. He made a wonderful recovery, in time to meet his death later on at Hitler's orders. In early July, Rundstedt was replaced in the overall command of the Western Front by von Kluge, a general who had won distinction in Russia, and on the 20th there took place a renewed, unsuccessful attempt on Hitler's life. According to the most trustworthy story, Colonel von Stauffenberg had placed under Hitler's table, at a staff meeting, a small case containing a time bomb. Hitler was spared from the full effect of the explosion by the heavy tabletop and its supporting crosspieces, and also by the light structure of the building itself, which allowed an instantaneous dispersal of the pressures. Several officers present were killed, but the Fuehrer, though badly shaken and wounded, arose exclaiming, "Who says I am not under the special protection of God?" All the fury of his nature was aroused by this plot, and the vengeance which he inflicted on all suspected of being in it makes a terrible tale.

*   *   *

Montgomery's general offensive, planned for July 18, now approached. The British Army attacked with three corps, preceded by an even greater bombardment by the Allied air. The Luftwaffe was totally prevented from interfering. Good progress was made to the east of Caen, until clouded skies began to hamper our planes and led to a week's delay in launching the break-out from the American sector. I thought this was an opportunity to visit Cherbourg and to spend a few days in the "Mulberry" harbour. On the 20th I flew direct in an American Army Dakota to their landing ground on the Cherbourg peninsula, and was taken all round the harbour by the United States commander. Here I saw for the first time a flying-bomb launching point. It was a very elaborate affair. I was shocked at the damage the Germans had done to the town, and shared the staff disappointment at the inevitable delay in getting the port to work. The basins of the harbour were thickly sown with contact mines. A handful of devoted British divers were at work day and night disconnecting these at their mortal peril. Warm tributes were paid to them by their American comrades. After a long and dangerous drive to the United States beachhead known as Utah Beach I went aboard a British motor torpedo boat, and thence had a rough passage to Arromanches. As one gets older seasickness retreats. I did not succumb, but slept soundly till we were in the calm waters of our synthetic lagoon. I went aboard the cruiser *Enterprise*, where I remained for three days, making myself thoroughly acquainted with the whole working of the harbour, on which all the armies now almost entirely depended, and at the same time transacting my London business.

The nights were very noisy, there being repeated raids by single aircraft, and more numerous alarms. By day I studied the whole process of the landing of supplies and troops, both at piers, in which I had so long been interested, and on the beaches. On one occasion six tank-landing craft came to the beach in line. When their prows grounded, their drawbridges fell forward and out came the tanks, three or four from each, and splashed ashore. In less than eight minutes by my stop watch the tanks stood in column of route on the highroad ready to move into action. This was an impressive performance, and typical of the rate of discharge which had now been achieved. I was fascinated to see the D.U.K.W.s, the American amphibious load carriers, swimming through the harbour, waddling ashore, and then hurrying up to the great dump where the lorries were waiting to take their supplies to the various units. Upon the wonderful efficiency of this system, now yielding results far greater than we had ever planned, depended the hopes of a speedy and victorious action.

On my last day at Arromanches I visited Montgomery's headquarters, a few miles inland. The Commander-in-Chief was in the best of spirits on the eve of his largest operation, which he explained to me in all detail. He took me into the ruins of Caen and across the river, and we also visited other parts of the British front. Then he placed at my disposal his captured Storch aeroplane, and the Air Commander himself piloted me all over the British positions. This aircraft could land at a pinch almost anywhere, and consequently one could fly at a few hundred feet from the ground, gaining a far better view and knowledge of the scene than by any other method. I also visited several of the air stations and said a few words to gatherings of officers and men. Finally, I went to the field hospital, where, though it was a quiet day, a trickle of casualties was coming in. One poor man was to have a serious operation, and was actually on the table about to take the anaesthetic. I was slipping away when he said he wanted me. He smiled wanly and kissed my hand. I was deeply moved, and very glad to learn later on that the operation had been entirely successful.

*   *   *

At this time the orders which had held the German Fifteenth Army behind the Seine were cancelled, and several fresh divisions were sent to reinforce the hard-pressed Seventh. Their transference, by rail or road, or across the Seine by the ferry system which had replaced the broken bridges, was greatly delayed and injured by our air forces. The long-withheld aid reached the field too late to turn the scale.

The hour of the great American break-out under General Omar

Bradley had come at last. On July 25 their VIIth Corps struck south-ward from St. Lô, and the next day the VIIIth Corps, on their right, joined the battle. The bombardment by the United States Air Force had been devastating, and the infantry assault prospered. Then the armour leaped through and swept on to the key point of Coutances. The German escape route down that coast of Normandy was cut, and the whole German defence west of the Vire was in jeopardy and chaos. The roads were jammed with retreating troops, and the Allied bombers and fighter bombers took a destructive toll of men and vehicles. The advance drove forward. Avranches was taken on July 31, and soon afterwards the sea corner, opening the way to the Brittany peninsula, was turned. The Canadians, under General Crerar, made a simultane-ous attack from Caen down the Falaise road. This was effectively opposed by four Panzer divisions. Montgomery, who still commanded the whole battle line, thereupon transferred the weight of the British attack to the other front, and gave orders to the British Second Army, under General Dempsey, for a new thrust from Caumont to Vire. Pre-ceded again by heavy air bombing, this started on July 30, and Vire was reached a few days later.

On August 7 I went again to Montgomery's headquarters by air and after he had given me a vivid account with his maps an American colonel arrived to take me to General Bradley. The route had been carefully planned to show me the frightful devastation of the towns and villages through which the United States troops had fought their way. All the buildings were pulverised by air bombing. We reached Bradley's head-quarters about four o'clock. The General welcomed me cordially, but I could feel there was great tension, as the battle was at its height and every few minutes messages arrived. I therefore cut my visit short and motored back to my aeroplane, which awaited me. I was about to go on board when, to my surprise, Eisenhower arrived. He had flown from London to his advanced headquarters, and, hearing of my movements, intercepted me. He had not yet taken over the actual command of the army in the field from Montgomery; but he supervised everything with a vigilant eye, and no one knew better than he how to stand close to a tremendous event without impairing the authority he had delegated to others.

* * *

The Third United States Army, under General Patton, had now been formed and was in action. He detached two armoured and three in-fantry divisions for the westward and southerly drive to clear the Brit-tany peninsula. The cut-off enemy at once retreated towards their for-tified ports. The French Resistance Movement, which here numbered thirty thousand men, played a notable part, and the peninsula was quickly overrun. By the end of the first week in August the Germans, amounting to forty-five thousand garrison troops and remnants of four divisions, had been pressed into defensive perimeters at St. Malo, Brest, Lorient, and St. Nazaire. Here they could be penned and left to wither, thus saving the unnecessary losses which immediate assaults would have required.

While Brittany was thus being cleared or cooped the rest of Patton's Army drove eastward in the "long hook" which was to carry them to the gap between the Loire and Paris and down the Seine towards Rouen. The town of Laval was entered on August 6, and Le Mans on the 9th. Few Germans were found in all this wide region, and the main difficulty was supplying the advancing Americans over long and ever-lengthening distances. Except for a limited air lift, everything had still to come from the beaches of the original landing and pass down the western side of Normandy through Avranches to reach the front. Avranches therefore became the bottleneck, and offered a tempting opportunity for a German attack striking westward from the neighbourhood of Falaise. The idea caught Hitler's fancy, and he gave orders for the maximum possible force to attack Mortain, burst its way through to Avranches, and thus cut Patton's communications. The German commanders were unanimous in condemning the project. Realising that the battle for Normandy was already lost, they wished to use four divisions which had just arrived from the Fifteenth Army in the north to carry out an orderly retreat to the Seine. They thought that to throw any fresh troops westward was merely to "stick out their necks," with the certain prospect of having them severed. Hitler insisted on having his way, and on August 7 five Panzer and two infantry divisions delivered a vehement attack on Mortain from the east.

The blow fell on a single U.S. division, but it held firm and three others came to its aid. After five days of severe fighting and concentrated bombing from the air the enemy were thrown back in confusion, and, as their generals had predicted, the whole salient from Falaise to Mortain was at the mercy of converging attacks from three sides. The Allied forces swept on to the crowded Germans within the long and narrow pocket, and with the artillery inflicted fearful slaughter. The Germans held stubbornly on to the jaws of the gap at Falaise and Argentan, and, giving priority to their armour, tried to extricate all that they could. But on August 17 command and control broke down and the scene became a shambles. The jaws closed on August 20, and although by then a considerable part of the enemy had been able to scramble eastward no fewer than eight German divisions were annihilated. What had been the Falaise pocket was their grave. Von Kluge reported to Hitler: "The enemy air superiority is terrific and smothers almost all our movements. Every movement of the enemy however is prepared and protected by his air forces. Losses in men and material are extraordinary. The morale of the troops has suffered very heavily under constant murderous enemy fire."

The Third United States Army, besides clearing the Brittany peninsula and contributing with their "short hook" to the culminating victory at Falaise, thrust three corps eastward and northeastward from Le Mans. On August 17 they reached Orléans, Chartres, and Dreux. Thence they drove northwestward to meet the British advancing on Rouen. Our Second Army had experienced some delay. They had to reorganise after the Falaise battle, and the enemy found means to improvise rearguard positions. However, the pursuit was pressed hotly, and all the Germans south of the Seine were soon seeking desperately to retreat across it, under destructive air attacks. None of the bridges destroyed by previous air bombardments had been repaired, but there were a few pontoon bridges and a fairly adequate ferry service. Very few vehicles could be saved. South of Rouen immense quantities of transport were abandoned. Such troops as escaped were in no condition to resist on the farther bank of the river.

*   *   *

Eisenhower, now in supreme command, was determined to avoid a battle for Paris. Stalingrad and Warsaw had proved the horrors of frontal assaults and patriotic risings, and he resolved to encircle the capital and force the garrison to surrender or flee. By August 20 the time for action had come. Patton had crossed the Seine near Mantes, and his right flank had reached Fountainebleau. The French Underground had revolted. The police were on strike. The Prefecture was in Patriot hands. An officer of the Resistance reached Patton's headquarters with vital reports, and on the Wednesday morning these were delivered to Eisenhower at Le Mans.

Attached to Patton was the French 2nd Armoured Division, under General Leclerc, which had landed in Normandy on August 1, and played an honourable part in the advance. De Gaulle arrived the same day, and was assured by the Allied Supreme Commander that when the time came — and as had been long agreed — Leclerc's troops would be the first in Paris. That evening, news of street fighting in the capital decided Eisenhower to act, and Leclerc was told to march. The operation orders, dated August 23, began with the words "Mission (1) s'emparer de Paris . . ."

On August 24 the main thrust, led by Colonel Billotte, son of the commander of the First French Army Group, who was killed in May 1940, moved up from Orléans. That night a vanguard of tanks reached the Porte d'Orléans, and entered the square in front of the Hôtel de Ville. Early next morning Billotte's armoured columns held both

banks of the Seine opposite the Cité. By the afternoon the headquarters of the German commander, General von Choltitz, in the Hôtel
Meurice, were surrounded. Von Choltitz was taken before Leclerc.
This was the end of the road from Dunkirk to Lake Chad and home
again. In a low voice Leclerc spoke his thoughts aloud: "Maintenant,
ça y est," and then in German he introduced himself to the vanquished.
After a brief and brusque discussion the capitulation of the garrison
was signed, and one by one their remaining strong points were occupied by the Resistance and the regular troops.

The city was given over to a rapturous demonstration. German
prisoners were spat at, collaborators dragged through the streets, and
the liberating troops fêted. On this scene of long-delayed triumph
there arrived General de Gaulle. At the Hôtel de Ville, in company
with the main figures of the Resistance and Generals Leclerc and Juin,
he appeared for the first time as the leader of Free France before the
jubilant population. There was a spontaneous burst of wild enthusiasm. In the afternoon of August 26, de Gaulle made his formal entry
on foot down the Champs Elysées to the Place de la Concorde, and
then in a file of cars to Notre Dame. There was some firing from inside and outside the cathedral by hidden collaborators. The crowd
scattered, but after a short moment of panic the solemn dedication
of the liberation of Paris proceeded to its end.

*   *   *

By August 30 our troops were crossing the Seine at many points.
Enemy losses had been tremendous: 400,000 men, half of them prisoners, 1300 tanks, 20,000 vehicles, 1500 field guns. The German
Seventh Army, and all divisions that had been sent to reinforce it, were
torn to shreds. The Allied break-out from the beachhead had been
delayed by bad weather and Hitler's mistaken resolve. But once that
battle was over, everything went with a run, and the Seine was reached
six days ahead of the planned time. There has been criticism of slowness on the British front in Normandy, and the splendid American
advances of the later stages seemed to indicate greater success on their
part than on ours. It is therefore necessary to emphasise that the
whole plan of campaign was to pivot on the British front and draw
the enemy's reserves in that direction in order to help the American
turning movement. By determination and hard fighting this was
achieved. "Without the great sacrifices made by the Anglo-Canadian
armies in the brutal, slugging battles for Caen and Falaise," wrote
General Eisenhower in his official report, "the spectacular advances
made elsewhere by the Allied forces could never have come about."

14

Italy and the Riviera Landing

L IBERATING NORMANDY was a supreme event in the European campaign of 1944, but it was only one of several concentric strokes upon Nazi Germany. In the east the Russians were flooding into Poland and the Balkans, and in the south Alexander's armies in Italy were pressing towards the river Po. Decisions had now to be taken about our next move in the Mediterranean, and it must be recorded with regret that these occasioned the first important divergence on high strategy between ourselves and our American friends.

The design for final victory in Europe had been outlined in prolonged discussion at the Teheran Conference in November 1943. Its decisions still governed our plans, and it would be well to recall them. First and foremost we had promised to carry out "Overlord." Here was the dominating task, and no one disputed that here lay our prime duty. But we still wielded powerful forces in the Mediterranean, and the question had remained, what should they do? We had resolved that they should capture Rome, whose nearby airfields were needed for bombing Southern Germany, advance up the peninsula as far as the Pisa–Rimini line, and there hold as many enemy divisions as possible. This however was not all. A third operation was also agreed upon, namely, an amphibious landing in the South of France, and it was on this project that controversy was about to descend. It was originally conceived as a feint or threat to keep German troops on the Riviera and stop them joining the battles in Normandy, but the Americans had pressed for a real attack by ten divisions, and Stalin had supported them. I accepted the change, largely to prevent undue diversions to Burma, although I contemplated other ways of exploiting success in Italy, and the plan had been given the code name "Anvil."

But there were several reservations. Many of the forces would have to come from Italy, and they had first to accomplish the arduous and important task of seizing Rome and the airfields. Until this was done little could be spared or taken from Alexander. Rome must fall before "Anvil" could start. And it must also start about the same time as "Overlord." The troops would have a long way to go before they could reach Eisenhower's armies in Normandy, and unless they landed in good time they would be too late to help and the battle of the beaches would be over. All turned on the capture of Rome. At Teheran we had confidently expected to reach it early in the spring, but this had proved impossible. The descent at Anzio to accelerate its capture had drawn eight or ten German divisions away from the vital theatre, or more than was expected to be attracted to the Riviera by "Anvil." This in effect superseded it by achieving its object. Nevertheless the Riviera project went forward as if nothing had happened.

Apart from "Anvil" hanging somewhat vaguely in the future, some of the finest divisions of the armies in Italy had rightly been assigned to the main operation of "Overlord" and had sailed for England before the end of 1943. Alexander had thus been weakened and Kesselring had been strengthened. The Germans had sent reinforcements to Italy, had parried the Anzio swoop, and had stopped us entering Rome until just before D-Day. The hard fighting had of course engulfed important enemy reserves which might otherwise have gone to France, and it certainly helped "Overlord" in its critical early stages, but none the less our advance in the Mediterranean had been gravely upset. Landing craft were another obstacle. Many of them had been sent to "Overlord." "Anvil" could not be mounted until they came back, and this in its turn depended on events in Normany. These facts had been long foreseen, and as far back as March 21 General Maitland Wilson, the Supreme Commander in the Mediterranean, reported that "Anvil" could not be launched before the end of July. Later he put it at mid-August, and declared that the best way to help "Overlord" was to abandon any attack on the Riviera and concentrate on Italy. Both he and Alexander thought their best contribution to the common end would be to press forward with all their resources into the Po valley. Thereafter, with the help of an amphibious operation against the Istrian peninsula, at the head of the Adriatic, which is dominated by and runs south from Trieste, there would be attractive prospects of advancing through the Ljubljana Gap into Austria and Hungary and striking at the heart of Germany from another direction.

When Rome fell on June 4 the problem had to be reviewed. Should we go on with "Anvil" or should we make a new plan?

General Eisenhower naturally wanted to strengthen his attack in

Northwest Europe by all available means. Strategic possibilities in Northern Italy did not attract him, but he consented to return the landing craft as soon as possible if this would lead to a speedy "Anvil." The American Chiefs of Staff agreed with Eisenhower, holding rigidly to the maxim of concentration at the decisive point, which in their eyes meant only Northwest Europe. They were supported by the President, who was mindful of the agreements made with Stalin many months before at Teheran. Yet all was changed by the delay in Italy.

Mr. Roosevelt admitted that an advance through the Ljubljana Gap might contain German troops, but it would not draw any of their divisions from France. He therefore urged that "Anvil" should be undertaken, at the expense of course of our armies in Italy, since "in my view the resources of Great Britain and the United States will not permit us to maintain two major theatres in the European war, each with decisive missions." The British Chiefs of Staff took the opposite view. Rather than land on the Riviera they preferred to send troops from Italy by sea direct to Eisenhower. With much prescience they remarked: "We think that the mounting of 'Anvil' on a scale likely to achieve success would hamstring General Alexander's remaining forces to such an extent that any further activity would be limited to something very modest."

This direct conflict of opinions, honestly held and warmly argued by either side, could only be settled, if at all, between the President and myself, and an interchange of telegrams now took place.

"The deadlock," I telegraphed on June 28, "between our Chiefs of Staff raises most serious issues. Our first wish is to help General Eisenhower in the most speedy and effective manner. But we do not think this necessarily involves the complete ruin of all our great affairs in the Mediterranean, and we take it hard that this should be demanded of us. . . . I must earnestly beg you to examine this matter in detail for yourself. . . . Please remember how you spoke to me at Teheran about Istria, and how I introduced it at the full Conference. This has sunk very deeply into my mind, although it is not by any means the immediate issue we have to decide."

Mr. Roosevelt's reply was prompt and adverse. He was resolved to carry out what he called the "grand strategy" of Teheran, namely, exploiting "Overlord" to the full, "victorious advances in Italy, and an early assault on Southern France." Political objects might be important, but military operations to achieve them must be subordinated to striking at the heart of Germany by a campaign in Europe. Stalin himself had favoured "Anvil" and had classified all other operations in the Mediterranean as of lesser importance, and Mr. Roosevelt declared he could not abandon it without consulting him. The President continued:

> My interest and hopes centre on defeating the Germans in front of Eisenhower and driving on into Germany, *rather than on limiting this action for the purpose of staging a full major effort in Italy.*[1] I am convinced we will have sufficient forces in Italy, with "Anvil" forces withdrawn, to chase Kesselring north of Pisa–Rimini and maintain heavy pressure against his army at the very least to the extent necessary to contain his present force. I cannot conceive of the Germans paying the price of ten additional divisions, estimated by General Wilson, in order to keep us out of Northern Italy.
>
> We can — and Wilson confirms this — immediately withdraw five divisions (three United States and two French) from Italy for "Anvil." *The remaining twenty-one divisions, plus numerous separate brigades, will certainly provide Alexander with adequate ground superiority.*

But it was Mr. Roosevelt's objections to a descent on the Istrian peninsula and a thrust against Vienna through the Ljubljana Gap that revealed both the rigidity of the American military plans and his own suspicion of what he called a campaign "in the Balkans." He claimed that Alexander and Smuts, who also favoured my view, "for several natural and very human reasons," were inclined to disregard two vital considerations. First, the operation infringed "the grand strategy." Secondly, it would take too long and we could probably not deploy more than six divisions. "I cannot agree," he wrote, "to the employment of United States troops against Istria *and into the Balkans,* nor can I see the French agreeing to such use of French troops. . . . For purely political considerations over here, I should never survive even a slight setback in 'Overlord' *if it were known that fairly large forces had been diverted to the Balkans.*"

No one involved in these discussions had ever thought of moving armies into the Balkans; but Istria and Trieste were strategic and political positions, which, as he saw very clearly, might exercise profound and widespread reactions, especially after the Russian advances. For the time being, however, I resigned myself, and on July 2 General Wilson was ordered to attack the South of France on August 15. Preparations began at once, but the reader should note that "Anvil" was renamed "Dragoon." This was done in case the enemy had learnt the meaning of the original code word.

*   *   *

[1] *My subsequent italics throughout.*—W.S.C.

By early August however a marked change had come over the battlefield in Normandy and great developments impended, and on the 7th I visited Eisenhower at his headquarters near Portsmouth and unfolded to him my last hope of stopping an assault on the South of France. After an agreeable luncheon we had a long and serious conversation. Eisenhower had with him Bedell Smith and Admiral Ramsay. I had brought the First Sea Lord, as the movement of shipping was the key. Briefly, what I proposed was to continue loading the "Dragoon" expedition, but when the troops were in the ships to send them through the Straits of Gibraltar and enter France at Bordeaux. The matter had been long considered by the British Chiefs of Staff, and the operation was considered feasible. I showed Eisenhower a telegram I had sent to the President, whose reply I had not yet received, and did my best to convince him. The First Sea Lord strongly supported me. Admiral Ramsay argued against any change of plan. Bedell Smith, on the contrary, declared himself strongly in favour of this sudden deflection of the attack, which would have all the surprise that sea power can bestow. Eisenhower in no way resented the views of his Chief of Staff. He always encouraged free expression of opinion in council at the summit, though of course whatever was settled would receive every loyalty in execution.

However, I was quite unable to move him, and next day I received the President's reply. "It is my considered opinion," he cabled, "that 'Dragoon' should be launched as planned at the earliest practicable date, and I have full confidence that it will be successful and of great assistance to Eisenhower in driving the Huns from France."

There was no more to be done about it. It is worth noting that we had now passed the day in July when for the first time in the war the movement of the great American armies into Europe and their growth in the Far East made their numbers in action greater than our own. Influence on Allied operations is usually increased by large reinforcements. It must also be remembered that had the British views on this strategic issue been accepted the tactical preparations might well have caused some delay, which again would have reacted on the general argument.

I now decided to go myself to Italy, where many questions could be more easily settled on the spot than by correspondence. It would be a great advantage to see the commanders and the troops from whom so much was being demanded, after so much had been taken. Alexander, though sorely weakened, was preparing his armies for a further offensive. I was anxious to meet Tito, who could easily come to Italy from the island of Vis, where we were protecting him. The Greek Prime Minister, M. Papandreou, and some of his colleagues could come from

Cairo, and plans could be made to help them back to Athens when the Germans departed. I reached Naples on the afternoon of August 11 and was installed in the palatial though somewhat dilapidated Villa Rivalta, with a glorious view of Vesuvius and the bay. Here General Wilson explained to me that all arrangements had been made for a conference next morning with Tito and Subašić, the new Yugoslav Prime Minister of King Peter's Government in London. They had already arrived in Naples, and would dine with us the next night.

On the morning of August 12 Marshal Tito came up to the villa. He wore a magnificent gold and blue uniform which was very tight under the collar and singularly unsuited to the blazing heat. The uniform had been given him by the Russians, and, as I was afterwards informed, the gold lace came from the United States. I joined him on the terrace of the villa, accompanied by Brigadier Maclean and an interpreter. I suggested that the Marshal might first like to see General Wilson's War Room, and we moved inside. The Marshal, who was attended by two ferocious-looking bodyguards, each carrying automatic pistols, wanted to bring them with him in case of treachery on our part. He was dissuaded from this with some difficulty, and proposed to bring them to guard him at dinner instead.

I led the way into a large room, where maps of the battlefronts covered the walls, and we had a long conversation. I pointed on the map to the Istrian peninsula. He was all in favour of our attacking it, and promised to help. Then and in the following days we did our best to strengthen and intensify the Yugoslav war effort and to heal the breach between him and King Peter.

On the afternoon of August 14 I flew in General Wilson's Dakota to Corsica in order to see the Riviera landing which I had tried so hard to stop, but to which I wished all success. From the British destroyer *Kimberley* we watched the long rows of boats filled with American storm troops steaming in continuously to the Bay of St. Tropez. As far as I could see or hear not a shot was fired either at the approaching flotillas or on the beaches. The battleships stopped firing, as there seemed to be nobody there. On the 16th I got back to Naples, and rested there for the night before going up to meet Alexander at the front. I had at least done the civil to "Anvil-Dragoon," and I thought it was a good thing I was near the scene to show the interest I took in it. We may here note briefly what happened.

The Seventh Army, under General Patch, had been formed to carry out the attack. Seven French and three U.S. divisions, together with a mixed American and British air-borne division, were supported by no fewer than six battleships, twenty-one cruisers, and a hundred destroyers. In the air we were overwhelmingly superior and in the midst of the Germans in Southern France over 25,000 armed men of the Resistance were ready to revolt. The assault took place early on the 15th between Cannes and Hyères. Casualties were relatively few and the Americans moved fast. On the 28th they were beyond Valence and Grenoble. The enemy made no serious attempt to stop them, except for a stiff fight at Montélimar by a Panzer division. The Allied Tactical Air Force was treating them roughly and destroying their transport. Eisenhower's pursuit from Normandy was cutting in behind them, having reached the Seine at Fontainebleau on August 20. Five days later it was well past Troyes. The surviving elements of the German Nineteenth Army, amounting to a nominal five divisions, were in full retreat, leaving 50,000 prisoners in our hands. Lyons was taken on September 3, Besançon on the 8th, and Dijon was liberated by the Resistance Movement on the 11th. On that day "Dragoon" and "Overlord" joined hands at Sombernon. In the triangle of Southwest France, trapped by these concentric thrusts, were the isolated remnants of the German First Army, over 20,000 strong, who freely gave themselves up.

To sum up the story, the original proposal at Teheran in November 1943 was for a descent in the South of France to help take the weight off "Overlord." The timing was to be either in the week before or the week after D-Day. All this was changed by what happened in the interval. The latent threat from the Mediterranean sufficed in itself to keep ten German divisions on the Riviera. Anzio alone had meant that the equivalent of four enemy divisions was lost to other fronts. When, with the help of Anzio, our whole battle line advanced, captured Rome and threatened the Gothic Line, the Germans hurried a further eight divisions to Italy. Delay in the capture of Rome and the dispatch of landing craft from the Mediterranean to help "Overlord" caused the postponement of "Anvil-Dragoon" till mid-August, or two months later than had been proposed. It therefore did not in any way affect "Overlord." When it was belatedly launched it drew no enemy down from the Normandy battle theatre. Therefore none of the reasons present in our minds at Teheran had any relation to what was done and "Dragoon" caused no diversion from the forces opposing General Eisenhower.[2] In fact, instead of helping him, he helped it by threatening the rear of the Germans retiring up the Rhone Valley. This is not to deny that the operation as carried out eventually brought important assistance to General Eisenhower by the arrival of another army on his right flank and the opening of another line of communications thither. For this a heavy price was paid. The army of Italy was deprived of its opportunity to strike a most formidable blow at the Germans, and very

[2] The first major operations in which the "Dragoon" armies took part after their junction with Eisenhower's forces was in mid-November.

possibly to reach Vienna before the Russians, with all that might have followed therefrom. But once the final decision was reached I of course gave "Anvil-Dragoon" my full support, though I had done my best to constrain or deflect it.

*   *   *

On the morning of August 17 I set out by motor to meet General Alexander. I was delighted to see him for the first time since his victory and entry into Rome. He drove me all along the old Cassino front, showing me how the battle had gone and where the main struggles had occurred. Alexander brought his chief officers to dinner, and explained to me fully his difficulties and plans. The Fifteenth Group of Armies had indeed been skinned and starved. The far-reaching projects we had cherished must now be abandoned. It was still our duty to hold the Germans in the largest numbers on our front. If this purpose was to be achieved an offensive was imperative; but the well-integrated German armies were almost as strong as ours, composed of so many different contingents and races. It was proposed to attack along the whole front early on the 26th. Our right hand would be upon the Adriatic, and our immediate objective Rimini. To the westward, under Alexander's command, lay the Fifth American Army. This had been stripped and mutilated for the sake of "Anvil," but would nevertheless advance with vigour.

On August 19 I set off to visit General Mark Clark at Leghorn. We lunched in the open air by the sea. In our friendly and confidential talks I realised how painful the tearing to pieces of this fine army had been to those who controlled it. The General seemed embittered that his army had been robbed of what he thought — and I could not disagree — was a great opportunity. Still, he would drive forward to his utmost on the British left and keep the whole front blazing. It was late and I was thoroughly tired out when I got back to the château at Siena, where Alexander came again to dine.

When one writes things on paper to decide or explain large questions affecting action there is mental stress. But all this bites much deeper when you see and feel it on the spot. Here was this splendid army, equivalent to twenty-five divisions, of which a quarter were American, reduced till it was just not strong enough to produce decisive results against the immense power of the defensive. A very little more, half of what had been taken from us, and we could have broken into the valley of the Po, with all the gleaming possibilities and prizes which lay open towards Vienna. As it was, our forces, about a million strong, could play a mere secondary part in any commanding strategic conception. They could keep the enemy on their front busy at the cost and risk of a hard offensive. They could at least do their duty. Alexander maintained his soldierly cheerfulness, but it was in a sombre mood that I went to bed. In these great matters failing to gain one's way is no escape from the responsibility for an inferior solution.

*   *   *

As Alexander's offensive could not start till the 26th I flew to Rome on the morning of the 21st. Here another set of problems and a portentous array of new personages to meet awaited me. First I had to deal with the impending Greek crisis, which had been one of the chief reasons for my Italian visit. Rumours of the German evacuation of Greece raised intense excitement and discord in M. Papandreou's Cabinet, and revealed the frail and false foundation upon which common action stood. This made it all the more necessary for me to see Papandreou and those he trusted. We met that evening. Neither his Government nor the Greek State itself had either arms or police. He asked for our help to unite Greek resistance against the Germans. At present only the wrong people had arms, and they were a minority. I told him we could make no promise and enter into no obligations about sending British forces into Greece, and that even the possibility should not be talked about in public; but I advised him to transfer his Government at once from Cairo, with its atmosphere of intrigue, to somewhere in Italy near the headquarters of the Supreme Allied Commander. This he agreed to do. As for the future, I told him we had no intention of interfering with the solemn right of the Greek people to choose between monarchy and a republic. But it must be for the Greek people as a whole, and not a handful of doctrinaires, to decide so grave an issue. Although I personally gave my loyalty to the constitutional monarchy which had taken shape in England, His Majesty's Government were quite indifferent as to which way the matter was settled for Greece provided there was a fair plebiscite. We shall in due course see what happened.

I stayed while in Rome at the Embassy, and our Ambassador, Sir Noel Charles, and his wife devoted themselves to my business and comfort. Guided by his advice, I met most of the principal figures in the debris of Italian politics produced by twenty years of dictatorship, a disastrous war, revolution, invasion, occupation, Allied control, and other evils. I had talks with, among others, Signor Bonomi and General Badoglio, also with Comrade Togliatti, who had returned to Italy at the beginning of the year after a long sojourn in Russia. The leaders of all the Italian parties were invited to meet me. None had any electoral mandate, and their party names, revived from the past, had been chosen with an eye to the future. "What is your party?" I asked one group. "We are the Christian Communists," their chief replied. I could not help saying, "It must be very inspiring to your party, having the Catacombs so handy." They did not seem to see the point, and, looking back, I am afraid their minds must have turned to the cruel mass executions which the Germans had so recently perpetrated in these ancient sepulchres. One may however be pardoned for making historical references in Rome. The Eternal City, rising on every side, majestic and apparently invulnerable, with its monuments and palaces, and with its splendour of ruins not produced by bombing, seemed to contrast markedly with the tiny and transient beings who flitted within its bounds. I also met for the first time the Crown Prince Umberto, who, as Lieutenant of the Realm, was commanding the Italian forces on our front. His powerful and engaging personality, his grasp of the whole situation, military and political, were refreshing, and gave one a more lively feeling of confidence than I had experienced in my talks with the politicians. I certainly hoped he would play his part in building up a constitutional monarchy in a free, strong, united Italy. However, this was none of my business.

Early on August 24 I returned by air to Alexander's headquarters at Siena, living in the château a few miles away, and next afternoon we flew to General Leese's battle headquarters of the Eighth Army, on the Adriatic side. Here we had tents overlooking a magnificent panorama to the northward. The Adriatic, though but twenty miles away, was hidden by the mass of Monte Maggiore. General Leese told us that the barrage to cover the advance of his troops would begin at midnight. We were well placed to watch the long line of distant gun flashes. The rapid, ceaseless thudding of the cannonade reminded me of the First World War. Artillery was certainly being used on a great scale. After an hour of this I was glad to go to bed, for Alexander had planned an early start and a long day on the front. He had also promised to take me wherever I wanted to go.

Alexander and I started together at about nine o'clock. His aide-de-camp and Tommy (Commander Thompson) came in a second car. We were thus a conveniently small party. The advance had now been in progress for six hours, and was said to be making headway. But no definite impressions could yet be formed. We first climbed by motor up a high outstanding rock pinnacle, upon the top of which a church and village were perched. The inhabitants, men and women, came out to greet us from the cellars in which they had been sheltering. It was at once plain that the place had just been bombarded. Masonry and wreckage littered the single street. "When did this stop?" Alexander asked the small crowd who gathered round us, grinning rather wryly. "About a quarter of an hour ago," they said. There was certainly a magnificent view from the ramparts of bygone centuries. The whole front of the Eighth Army offensive was visible. But apart from the smoke puffs of shells bursting seven or eight thousand yards away in a scattered fashion there was nothing to see. Presently Alexander said that we had better not stay any longer, as the enemy would naturally be firing at observation posts like this and might begin again. So we motored two or three miles to the westward, and had a picnic lunch on the broad slope of a hillside, which gave almost as good a view as the peak and was not likely to attract attention.

News was now received that our troops had pushed on a mile or two beyond the river Metauro. Here Hasdrubal's defeat had sealed the fate of Carthage, so I suggested that we should go across too. We got into our cars accordingly, and in half hour were across the river, where the road ran into undulating groves of olives, brightly patched with sunshine. Having got an officer guide from one of the battalions engaged, we pushed on through these glades till the sounds of rifle and machine-gun fire showed we were getting near to the front line. Presently warning hands brought us to a standstill. It appeared there was a mine field, and it was only safe to go where other vehicles had already gone without mishap. Alexander and his aide-de-camp now went off to reconnoitre towards a grey stone building which our troops were holding, which was said to give a good close-up view. It was evident to me that only very loose fighting was in progress. In a few minutes the aide-de-camp came back and brought me to his chief, who had found a very good place in the stone building, which was in fact an old château overlooking a rather sharp declivity. Here one certainly could see all that was possible. The Germans were firing with rifles and machine guns from thick scrub on the farther side of the valley, about five hundred yards away. Our front line was beneath us. The firing was desultory and intermittent. But this was the nearest I got to the enemy and the time I heard most bullets in the Second World War. After about half an hour we went back to our motorcars and made our way to the river, keeping very carefully to our own wheel tracks or those of other vehicles. At the river we met the supporting columns of infantry, marching up to lend weight to our thin skirmish line, and by five o'clock we were home again at General Leese's headquarters, where the news from the whole of the Army front was marked punctually on the maps. On the whole the Eighth Army had advanced since daybreak about seven thousand yards on a ten- or twelve-mile front, and the losses had not been at all heavy. This was an encouraging beginning.

*   *   *

The next morning plenty of work arrived, both by telegram and pouch. It appeared that General Eisenhower was worried by the approach of some German divisions which had been withdrawn from Italy. I was glad that our offensive, prepared under depressing conditions, had begun. I drafted a telegram to the President explaining the position as I had learned it from the generals on the spot and from my own knowledge. I wished to convey in an uncontroversial form our sense of frustration, and at the same time to indicate my hopes and ideas for the future. If only I could revive the President's interest in this sphere we might still keep alive our design of an ultimate advance to Vienna. After explaining Alexander's plan, I ended as follows:

> I have never forgotten your talks to me at Teheran about Istria, and I am sure that the arrival of a powerful army in Trieste and Istria in four or five weeks would have an effect far outside purely military values. Tito's people will be awaiting us in Istria. What the condition of Hungary will be then I cannot imagine, but we shall at any rate be in a position to take full advantage of any great new situation.

I did not send this message off till I reached Naples, whither I flew on the 28th, nor did I receive the answer till some days after I got home. Mr. Roosevelt then replied:

> I share your confidence that the Allied divisions we have in Italy are sufficient to do the task before them and that the battle commander will press the battle unrelentingly with the objective of shattering the enemy forces. . . . As to the exact employment of our forces in Italy in the future, this is a matter we can [soon] discuss. . . . With the present chaotic conditions of the Germans in Southern France, I hope that a junction of the north and south forces may be obtained at a much earlier date than was first anticipated.

We shall see that both these hopes proved vain. The army which we had landed on the Riviera at such painful cost to our operations in Italy arrived too late to help Eisenhower's first main struggle in the north, while Alexander's offensive failed, by the barest of margins, to achieve the success it deserved and we so badly needed. Italy was not to be wholly free for another eight months; the right-handed drive to Vienna was denied to us; and, except in Greece, our military power to influence the liberation of Southeastern Europe was gone.

*   *   *

The rest of the story is soon told. The Eighth Army attack prospered and augured well. It surprised the Germans, and by September 1 had penetrated the Gothic Line on a twenty-mile front. By the 18th the line had been turned at its eastern end by the Eighth Army, and pierced in the centre by the Americans.

Though at the cost of grievous casualties, great success had been achieved and the future looked hopeful. But Kesselring received further reinforcements, until his German divisions amounted to twenty-eight in all. Scraping up two divisions from quiet sectors, he started fierce counterattacks, which, added to our supply difficulties over the mountain passes, checked the Allied advance. The defence was stubborn, the ground very difficult, and it was raining hard. The climax came near Bologna between October 20 and 24, when General Mark Clark very nearly succeeded in cutting in behind the enemy facing the Eighth Army. Then, in Alexander's words, "assisted by torrential rains and winds of gale force, and the Fifth Army's exhaustion, the German line held firm." The weather was appalling. Heavy rains had swollen the numberless rivers and irrigation channels and turned the reclaimed agricultural land into the swamp it had originally been. Off the roads movement was often impossible. It was with the greatest difficulty that the troops toiled forward. Although hopes of decisive victory had faded, it remained the first duty of the armies in Italy to keep up the pressure and deter the enemy from sending help to the hard-pressed German armies on the Rhine. And so we fought forward whenever there was a spell of reasonably fine weather. But from mid-November no major offensive was possible. Small advances were made as opportunity offered, but not until the spring were the armies rewarded with the victory they had so well earned, and so nearly won, in the autumn.

15

The Russian Victories

THE READER MUST NOW hark back to the Russian struggle, which in scale far exceeded the operations with which my
account has hitherto been concerned, and formed of course the foundation upon which the British and American Armies had approached the
climax of the war. The Russians had given their enemy little time to
recover from their severe reverses of the early winter of 1943. In mid-
January 1944 they attacked on a 120-mile front from Lake Ilmen to
Leningrad and pierced the defences before the city. Farther south by
the end of February the Germans had been driven back to the shores of
Lake Peipus. Leningrad was freed once and for all, and the Russians
stood on the borders of the Baltic States. Further onslaughts to the west
of Kiev forced the Germans back towards the old Polish frontier. The
whole southern front was aflame and the German line deeply penetrated
at many points. One great pocket of surrounded Germans was left behind at Kersun, from which few escaped. Throughout March the Russians pressed their advantage all along the line and in the air. From
Gomel to the Black Sea the invaders were in full retreat, which did not
end until they had been thrust across the Dniester, back into Rumania
and Poland. Then the spring thaw brought them a short respite. In the
Crimea however operations were still possible, and in April the Russians
set about destroying the Seventeenth Germany Army and regaining
Sebastopol.

The magnitude of these victories raised issues of far-reaching importance. The Red Army now loomed over Central and Eastern Europe.
What was to happen to Poland, Hungary, Rumania, Bulgaria, and above
all, to Greece, for whom we had tried so hard and sacrificed so much?
Would Turkey come in on our side? Would Yugoslavia be engulfed
in the Russian flood? Postwar Europe seemed to be taking shape

and some political arrangement with the Soviets was becoming urgent.

On May 18 the Soviet Ambassador in London had called at the Foreign Office to discuss a general suggestion which Mr. Eden had made that the U.S.S.R. should temporarily regard Rumanian affairs as mainly their concern under war conditions while leaving Greece to us. The Russians were prepared to accept this, but wished to know if we had consulted the United States. If so they would agree. On the 31st I accordingly sent a personal telegram to Mr. Roosevelt:

> . . . I hope you may feel able to give this proposal your blessing. We do not of course wish to carve up the Balkans into spheres of influence, and in agreeing to the arrangement we should make it clear that it applied only to war conditions and did not affect the rights and responsibilities which each of the three Great Powers will have to exercise at the peace settlement and afterwards in regard to the whole of Europe. The arrangement would of course involve no change in the present collaboration between you and us in the formulation and execution of Allied policy towards these countries. We feel however that the arrangement now proposed would be a useful device for preventing any divergence of policy between ourselves and them in the Balkans.

The first reactions of the State Department were cool. Mr. Hull was nervous of any suggestion that "might appear to savour of the creation or acceptance of the idea of spheres of influence," and on June 11 the President cabled:

> . . . Briefly, we acknowledge that the military responsible Government in any given territory will inevitably make decisions required by military developments, but are convinced that the natural tendency for such decisions to extend to other than military fields would be strengthened by an agreement of the type suggested. In our opinion, this would certainly result in the persistence of differences between you and the Soviets and in the division of the Balkan region into spheres of influence despite the declared intention to limit the arrangement to military matters.
>
> We believe efforts should preferably be made to establish consultative machinery to dispel misunderstandings and restrain the tendency toward the development of exclusive spheres.

I was much concerned at this message and replied on the same day:

> . . . Action is paralysed if everybody is to consult everybody else about everything before it is taken. Events will always outstrip the changing situations in these Balkan regions. Somebody must have

the power to plan and act. A Consultative Committee would be a mere obstruction, always overridden in any case of emergency by direct interchanges between you and me, or either of us and Stalin.

See, now, what happened at Easter. We were able to cope with this mutiny of the Greek forces entirely in accordance with your own views. This was because I was able to give constant orders to the military commanders, who at the beginning advocated conciliation, and above all no use or even threat of force. Very little life was lost. The Greek situation has been immensely improved, and, if firmness is maintained, will be rescued from confusion and disaster. The Russians are ready to let us take the lead in the Greek business, which means that E.A.M.[1] and all its malice can be controlled by the national forces of Greece. . . . If in these difficulties we had had to consult other Powers and a set of triangular or quadrangular telegrams got started the only result would have been chaos or impotence.

It seems to me, considering the Russians are about to invade Rumania in great force and are going to help Rumania recapture part of Transylvania from Hungary, provided the Rumanians play, which they may, considering all that, it would be a good thing to follow the Soviet leadership, considering that neither you nor we have any troops there at all and that they will probably do what they like anyhow. . . . To sum up, I propose that we agree that the arrangements I set forth in my message of May 31 may have a trial of three months, after which it must be reviewed by the three Powers.

On June 13 the President agreed to this proposal, but added: "We must be careful to make it clear that we are not establishing any postwar spheres of influence." I shared his view, and replied the next day:

> I am deeply grateful to you for telegram. I have asked the Foreign Secretary to convey the information to Molotov and to make it clear that the reason for the three months' limit is in order that we should not prejudge the question of establishing postwar spheres of influence.

I reported the situation to the War Cabinet that afternoon, and it was agreed that, subject to the time limit of three months, the Foreign Secretary should inform the Soviet Government that we accepted this general division of responsibility. This was done on June 19. The President

---

[1] "The National Liberation Front," known by its initials in Greek as E.A.M. and predominantly under Communist control.

however was not happy about the way we had acted, and I received a pained message saying "we were disturbed that your people took this matter up with us only after it had been put up to the Russians." On June 23 accordingly I outlined to the President, in reply to his rebuke, the situation as I saw it from London:

> The Russians are the only Power that can do anything in Rumania. . . . On the other hand, the Greek burden rests almost entirely upon us, and has done so since we lost 40,000 men in a vain endeavour to help them in 1941. Similarly, you have let us play the hand in Turkey, but we have always consulted you on policy, and I think we have been agreed on the line to be followed. It would be quite easy for me, on the general principle of slithering to the Left, which is so popular in foreign policy, to let things rip, when the King of Greece would probably be forced to abdicate and E.A.M. would work a reign of terror in Greece, forcing the villagers and many other classes to form Security Battalions under German auspices to prevent utter anarchy. The only way I can prevent this is by persuading the Russians to quit boosting E.A.M. and ramming it forward with all their force. Therefore I proposed to the Russians a temporary working arrangement for the better conduct of the war. This was only a proposal, and had to be referred to you for your agreement.
>
> I have also taken action to try to bring together a union of the Tito forces with those in Serbia, and with all adhering to the Royal Yugoslav Government, which we have both recognised. You have been informed at every stage of how we are bearing this heavy burden, which at present rests mainly on us. Here again nothing would be easier than to throw the King and the Royal Yugoslav Government to the wolves and let a civil war break out in Yugoslavia, to the joy of the Germans. I am struggling to bring order out of chaos in both cases and concentrate all efforts against the common foe. I am keeping you constantly informed, and I hope to have your confidence and help within the spheres of action in which initiative is assigned to us.

Mr. Roosevelt's reply settled this argument between friends. "It appears," he cabled, "that both of us have inadvertently taken unilateral action in a direction that we both now agree to have been expedient for the time being. It is essential that we should always be in agreement in matters bearing on our Allied war effort."

"You may be sure," I replied, "I shall always be looking to our agreement in all matters before, during, and after."

The difficulties however continued on a Governmental level. Stalin, as soon as he realised the Americans had doubts, insisted on consulting them direct, and in the end we were unable to reach any final agreement about dividing responsibilities in the Balkan peninsula. Early in August the Russians dispatched from Italy by a subterfuge a mission to E.L.A.S., the military component of E.A.M., in Northern Greece. In the light of American official reluctance and of this instance of Soviet bad faith, we abandoned our efforts to reach a major understanding until I met Stalin in Moscow two months later. By then much had happened on the Eastern Front.

In Finland, Russian troops, very different in quality and armament from those who had fought there in 1940, broke through the Mannerheim Line, reopened the railway from Leningrad to Murmansk, the terminal of our Arctic convoys, and by the end of August compelled the Finns to sue for an armistice. Their main attack on the German front began on June 23. Many towns and villages had been turned into strong positions, with all-round defence, but they were successively surrounded and disposed of, while the Red armies poured through the gaps between. At the end of July they had reached the Niemen at Kovno and Grodno. Here, after an advance of two hundred and fifty miles in five weeks, they were brought to a temporary halt to replenish. The German losses had been crushing. Twenty-five divisions had ceased to exist, and an equal number were cut off in Courland.[2] On July 17 alone 57,000 German prisoners were marched through Moscow — who knows whither?

To the southward of these victories lay Rumania. Till August was far advanced the German line from Cernowitz to the Black Sea barred the way to the Ploesti oilfields and the Balkans. It had been weakened by withdrawal of troops to sustain the sagging line farther north, and under violent attacks, beginning on August 22, it rapidly disintegrated. Aided by landings on the coast, the Russians made short work of the enemy. Sixteen German divisions were lost. On August 23 a *coup d'état* in Bucharest, organised by the young King Michael and his close advisers, led to a complete reversal of the whole military position. The Rumanian armies followed their King to a man. Within three days before the arrival of the Soviet troops the German forces had been disarmed or had retired over the northern frontiers. By September 1 Bucharest had been evacuated by the Germans. The Rumanian armies disintegrated and the country was overrun. The Rumanian Government capitulated. Bulgaria after a last-minute attempt to declare war on Germany, was overwhelmed. Wheeling to the west, the Russian armies drove up the valley of the Danube and through the Transylvanian Alps

[2] Heinz Guderian, *Panzer Leader*, p. 352.

OPERATIONS
ON THE
RUSSIAN FRONT

June 1944 – January 1945

Russian Front
June 1944

Major Russian
attacks

Approximate front
February 1, 1945

Surrounded enemy
Garrisons Feb. 1945

1939 Boundaries

0   100   200   300 MILES

to Murmansk

L. Onega

F I N L A N D

L. Ladoga

Viborg

Gulf of Finland

Leningrad

Tallin

ESTONIA

L. Peipus

Pskov

U. S. S. R.

B a l t i c   S e a

Riga
LATVIA

COURLAND

Vitebsk

Smolensk

LITHUANIA

Tilsit Kovno

R. Niemen

Vilna

Minsk

Mogilev

Königsberg

Danzig

E. PRUSSIA

Grodno

Stettin

Schneidemühl

Bobruisk

Gomel

Berlin

GERMANY

Posen

R. Oder

R. Vistula

WARSAW

Pinsk

R. Pripet

P O L A N D

MARSHES

Kiev

UPPER

Breslau

SILESIA

Oppeln

Prague

Kovel

Sandomir

Jaroslav

R. San

Lemberg

Przemysl

R. Dnieper

Cracow

C Z E C H O S L O V A K I A

CARPATHIAN MTS.

Stanislav

R. Bug

Cernowitz

R. Dniester

R. Danube

Vienna

AUSTRIA

Budapest

H U N G A R Y

Jassy

R. Pruth

Odessa

L. Balaton

Transylvanian
Alps

CRIMEA

R U M A N I A

Ploesti

Y U G O S L A V I A

Belgrade

Bucharest

R. Danube

B l a c k

S e a

B U L G A R I A

to the Hungarian border, while their left flank, south of the Danube, lined up on the frontier of Yugoslavia. Here they prepared for the great westerly drive which in due time was to carry them to Vienna.

In Poland there was a tragedy which requires a more detailed account.

*   *   *

By late July the Russian armies stood before the river Vistula, and all reports indicated that in the very near future Poland would be in Russian hands. The leaders of the Polish Underground Army, which owed allegiance to the London Government, had now to decide when to raise a general insurrection against the Germans, in order to speed the liberation of their country and prevent them fighting a series of bitter defensive actions on Polish territory, and particularly in Warsaw itself. The Polish commander, General Bor-Komorowski, and his civilian adviser were authorised by the Polish Government in London to proclaim a general insurrection whenever they deemed fit. The moment indeed seemed opportune. On July 20 came the news of the plot against Hitler, followed swiftly by the Allied break-out from the Normandy beachhead. About July 22 the Poles intercepted wireless messages from the German Fourth Panzer Army ordering a general withdrawal to the west of the Vistula. The Russians crossed the river on the same day, and their patrols pushed forward in the direction of Warsaw. There seemed little doubt that a general collapse was at hand.

General Bor therefore decided to stage a major rising and liberate the city. He had about forty thousand men, with reserves of food and ammunition for seven to ten days' fighting. The sound of Russian guns across the Vistula could now be heard. The Soviet Air Force began bombing the Germans in Warsaw from recently captured airfields near the capital, of which the closest was only twenty minutes' flight away. At the same time a Communist Committee of National Liberation had been formed in Eastern Poland, and the Russians announced that liberated territory would be placed under their control. Soviet broadcasting stations had for a considerable time been urging the Polish population to drop all caution and start a general revolt against the Germans. On July 29, three days before the rising began, the Moscow radio station broadcast an appeal from the Polish Communists to the people of Warsaw, saying that the guns of liberation were now within hearing, and calling upon them as in 1939 to join battle with the Germans, this time for decisive action. "For Warsaw, which did not yield but fought on, the hour of action has already arrived." After pointing out that the German plan to set up defence points would result in the gradual destruction of the city, the broadcast ended by reminding the inhabitants that "all is lost that is not saved by active effort," and that "by direct active struggle in the streets, houses, etc., of Warsaw the moment of final liberation will be hastened and the lives of our brethren saved."

On the evening of July 31 the Underground command in Warsaw got news that Soviet tanks had broken into the German defences east of the city. The German military wireless announced, "Today the Russians started a general attack on Warsaw from the southeast." Russian troops were now at points less than ten miles away. In the capital itself the Polish Underground command ordered a general insurrection at 5 P.M. on the following day. General Bor has himself described what happened:

> At exactly five o'clock thousands of windows flashed as they were flung open. From all sides a hail of bullets struck passing Germans, riddling their buildings and their marching formations. In the twinkling of an eye the remaining civilians disappeared from the streets. From the entrances of houses our men streamed out and rushed to the attack. In fifteen minutes an entire city of a million inhabitants was engulfed in the fight. Every kind of traffic ceased. As a big communications centre where roads from north, south, east, and west converged, in the immediate rear of the German front, Warsaw ceased to exist. The battle for the city was on.

The news reached London next day, and we anxiously waited for more. The Soviet radio was silent and Russian air activity ceased. On August 4 the Germans started to attack from strongpoints which they held throughout the city and suburbs. The Polish Government in London told us of the agonising urgency of sending in supplies by air. The insurgents were now opposed by five hastily concentrated German divisions. The Hermann Goering Division had also been brought from Italy, and two more S.S. divisions arrived soon afterwards.

I accordingly telegraphed to Stalin:

> At urgent request of Polish Underground Army we are dropping, subject to weather, about sixty tons of equipment and ammunition into the southwest quarter of Warsaw, where it is said a Polish revolt against the Germans is in fierce struggle. They also say that they appeal for Russian aid, which seems to be very near. They are being attacked by one and a half German divisions. This may be of help to your operation.

The reply was prompt and grim.

> I have received your message about Warsaw.
> I think that the information which has been communicated to

848   TRIUMPH AND TRAGEDY

you by the Poles is greatly exaggerated and does not inspire confidence. One could reach that conclusion even from the fact that
the Polish emigrants have already claimed for themselves that they
all but captured Vilna with a few stray units of the Home Army,
and even announced that on the radio. But that of course does
not in any way correspond with the facts. The Home Army of
the Poles consists of a few detachments, which they incorrectly
call divisions. They have neither artillery nor aircraft nor tanks.
I cannot imagine how such detachments can capture Warsaw, for
the defence of which the Germans have produced four tank divisions, among them the Hermann Goering Division.

Meanwhile the battle went on street by street against the German
"Tiger" tanks, and by August 9 the Germans had driven a wedge right
across the city through to the Vistula, breaking up the Polish-held
districts into isolated sectors. The gallant attempts of the R.A.F., with
Polish, British, and Dominion crews, to fly to the aid of Warsaw from
Italian bases were both forlorn and inadequate. Two planes appeared
on the night of August 4, and three four nights later.

*   *   *

The Polish Prime Minister, Mikolajczyk, had been in Moscow since
July 30 trying to establish some kind of terms with the Soviet Government, which had recognised the Polish Communist Committee of
National Liberation, the Lublin Committee, as we called it, as the
future administrators of the country. These negotiations were carried
on throughout the early days of the Warsaw rising. Messages from
General Bor were reaching Mikolajczyk daily, begging for ammunition and anti-tank weapons and for help from the Red Army. Meanwhile the Russians pressed for agreement upon the postwar frontiers
of Poland and the setting up of a joint Government. A last fruitless
talk took place with Stalin on August 9.

On the night of August 16 Vyshinsky asked the United States
Ambassador in Moscow to call, and, explaining that he wished to avoid
the possibility of misunderstanding, read out the following astonishing
statement:

> The Soviet Government cannot of course object to English or
> American aircraft dropping arms in the region of Warsaw, since
> this is an American and British affair. But they decidedly object
> to American or British aircraft, after dropping arms in the region
> of Warsaw, landing on Soviet territory, since the Soviet Govern-
> ment do not wish to associate themselves either directly or in-
> directly with the adventure in Warsaw.

On the same day I received the following message couched in softer terms from Stalin:

> After the conversation with M. Mikolajczyk I gave orders that the command of the Red Army should drop arms intensively in the Warsaw sector. A parachutist liaison officer was also dropped, who, according to the report of the command, did not reach his objective as he was killed by the Germans.
>
> Further, having familiarised myself more closely with the Warsaw affair, I am convinced that the Warsaw action represents a reckless and terrible adventure which is costing the population large sacrifices. This would not have been if the Soviet command had been informed before the beginning of the Warsaw action and if the Poles had maintained contact with it.
>
> In the situation which has arisen the Soviet command has come to the conclusion that it must dissociate itself from the Warsaw adventure, as it cannot take either direct or indirect responsibility for the Warsaw action.

According to Mikolajczyk's account, the first paragraph of this telegram is quite untrue. Two officers arrived safely in Warsaw and were received by the Polish command. A Soviet colonel had also been there for some days, and sent messages to Moscow via London urging support for the insurgents.

Four days later Roosevelt and I sent Stalin the following joint appeal, which the President had drafted:

> We are thinking of world opinion if the anti-Nazis in Warsaw are in effect abandoned. We believe that all three of us should do the utmost to save as many of the patriots there as possible. We hope that you will drop immediate supplies and munitions to the patriot Poles in Warsaw, or will you agree to help our planes in doing it very quickly? We hope you will approve. The time element is of extreme importance.

This was the reply we got:

> I have received the message from you and Mr. Roosevelt about Warsaw. I wish to express my opinions.
>
> Sooner or later the truth about the group of criminals who have embarked on the Warsaw adventure in order to seize power will become known to everybody. These people have exploited the good faith of the inhabitants of Warsaw, throwing many almost un-

armed people against the German guns, tanks and aircraft. A
situation has arisen in which each new day serves, not the Poles
for the liberation of Warsaw but the Hitlerites who are inhu-
manly shooting down the inhabitants of Warsaw.

From the military point of view, the situation which has arisen,
by increasingly directing the attention of the Germans to Warsaw,
is just as unprofitable for the Red Army as for the Poles. Meanwhile the Soviet troops, which have recently encountered new
and notable efforts by the Germans to go over to the counterattack, are doing everything possible to smash these counterattacks of the Hitlerites and to go over to a new wide-scale
attack in the region of Warsaw. There can be no doubt that the
Red Army is not sparing its efforts to break the Germans round
Warsaw and to free Warsaw for the Poles. That will be the
best and most effective help for the Poles who are anti-Nazi.

*   *   *

Meanwhile the agony of Warsaw reached its height. "The German
tank forces during last night" [August 11], cabled an eyewitness, "made
determined efforts to relieve some of their strong points in the city.
This is no light task however, as on the corner of every street are
built huge barricades, mostly constructed of concrete pavement slabs
torn up from the streets especially for this purpose. In most cases
the attempts failed, so the tank crews vented their disappointment
by setting fire to several houses and shelling others from a distance.
In many cases they also set fire to the dead, who litter the streets
in many places. . . .

"When the Germans were bringing supplies by tanks to one of their
outposts they drove before them 500 women and children to prevent
the [Polish] troops from taking action against them. Many of them
were killed and wounded. The same kind of action has been reported
from many other parts of the city.

"The dead are buried in backyards and squares. The food situation
is continually deteriorating, but as yet there is no starvation. Today
[August 15] there is no water at all in the pipes. It is being drawn
from the infrequent wells and house supplies. All quarters of the
town are under shell fire, and there are many fires. The dropping
of supplies has intensified the morale. Everyone wants to fight and
will fight, but the uncertainty of a speedy conclusion is depressing. . . ."

The battle also raged literally underground. The only means of
communication between the different sectors held by the Poles lay

through the sewers. The Germans threw hand grenades and gas bombs down the manholes. Battles developed in pitch-darkness between men waist-deep in excrement, fighting hand to hand at times with knives or drowning their opponents in the slime. Above ground German artillery and fighters set alight large areas of the city.

*   *   *

I had hoped that the Americans would support us in drastic action, but Mr. Roosevelt was adverse. On September 1 I received Miko-lajczyk on his return from Moscow. I had little comfort to offer. He told me that he was prepared to propose a political settlement with the Lublin Committee, offering them fourteen seats in a combined Government. These proposals were debated under fire by the repre-sentatives of the Polish Underground in Warsaw itself. The suggestion was accepted unanimously. Most of those who took part in these de-cisions were tried a year later for "treason" before a Soviet court in Moscow.

When the Cabinet met on the night of September 4 I thought the issue so important that though I had a touch of fever I went from my bed to our underground room. We had met together on many un-pleasant affairs. I do not remember any occasion when such deep anger was shown by all our members, Tory, Labour, Liberal alike. I should have liked to say, "We are sending our aeroplanes to land in your territory, after delivering supplies to Warsaw. If you do not treat them properly all convoys will be stopped from this moment by us." But the reader of these pages in after-years must realise that everyone always has to keep in mind the fortunes of millions of men fighting in a world-wide struggle, and that terrible and even humbling submissions must at times be made to the general aim. I did not therefore propose this drastic step. It might have been effective, be-cause we were dealing with men in the Kremlin who were governed by calculation and not by emotion. They did not mean to let the spirit of Poland rise again at Warsaw. Their plans were based on the Lublin Committee. That was the only Poland they cared about. The cutting off of the convoys at this critical moment in their great advance would perhaps have bulked in their minds as much as considerations of honour, humanity, decent commonplace good faith, usually count with ordinary people. The War Cabinet in their collective capacity sent Stalin the following telegram. It was the best we thought it wise to do:

The War Cabinet wish the Soviet Government to know that public opinion in this country is deeply moved by the events in Warsaw and by the terrible sufferings of the Poles there. Whatever the rights and wrongs about the beginnings of the Warsaw rising, the people of Warsaw themselves cannot be held responsible for the decision taken. Our people cannot understand why no material help has been sent from outside to the Poles in Warsaw. The fact that such help could not be sent on account of your Government's refusal to allow United States aircraft to land on aerodromes in Russian hands is now becoming publicly known. If on top of all this the Poles in Warsaw should now be overwhelmed by the Germans, as we are told they must be within two or three days, the shock to public opinion here will be incalculable. . . .

Out of regard for Marshal Stalin and for the Soviet peoples, with whom it is our earnest desire to work in future years, the War Cabinet have asked me to make this further appeal to the Soviet Government to give whatever help may be in their power, and above all to provide facilities for United States aircraft to land on your airfields for this purpose.

On September 10, after six weeks of Polish torment, the Kremlin appeared to change their tactics. That afternoon shells from the Soviet artillery began to fall upon the eastern outskirts of Warsaw, and Soviet planes appeared again over the city. Polish Communist forces, under Soviet orders, fought their way into the fringe of the capital. From September 14 onward the Soviet Air Force dropped supplies; but few of the parachutes opened and many of the containers were smashed and useless. The following day the Russians occupied the Praga, suburb, but went no farther. They wished to have the non-Communist Poles destroyed to the full, but also to keep alive the idea that they were going to their rescue. Meanwhile, house by house, the Germans proceeded with their liquidation of Polish centres of resistance throughout the city. A fearful fate befell the population. Many were deported by the Germans. General Bor's appeals to the Soviet commander, Marshal Rokossovsky, were unanswered. Famine reigned.

My efforts to get American aid led to one isolated but large-scale operation. On September 18 a hundred and four heavy bombers flew over the capital, dropping supplies. It was too late. On the evening of October 2 Mikolajczyk came to tell me that the Polish forces in Warsaw were about to surrender to the Germans. One of the last broadcasts from the heroic city was picked up in London:

This is the stark truth. We were treated worse than Hitler's satellites, worse than Italy, Rumania, Finland. May God, who is just, pass judgment on the terrible injustice suffered by the

Polish nation, and may He punish accordingly all those who are guilty.

Your heroes are the soldiers whose only weapons against tanks, planes, and guns were their revolvers and bottles filled with petrol. Your heroes are the women who tended the wounded and carried messages under fire, who cooked in bombed and ruined cellars to feed children and adults, and who soothed and comforted the dying. Your heroes are the children who went on quietly playing among the smouldering ruins. These are the people of Warsaw.

Immortal is the nation that can muster such universal heroism. For those who have died have conquered, and those who live on will fight on, will conquer and again bear witness that Poland lives when the Poles live.

These words are indelible. The struggle in Warsaw had lasted more than sixty days. Of the 40,000 men and women of the Polish Underground Army about 15,000 fell. Out of a population of a million nearly 200,000 had been stricken. The suppression of the revolt cost the German Army 10,000 killed, 7000 missing, and 9000 wounded. The proportions attest the hand-to-hand character of the fighting.

When the Russians entered the city three months later they found little but shattered streets and the unburied dead. Such was their liberation of Poland, where they now rule. But this cannot be the end of the story.

16

Burma

THE CURTAIN MUST NOW rise on a widely different scene
in Southeast Asia. For more than eighteen months the Japanese
had been masters of a vast defensive arc covering their early conquests.
This stretched from the jungle-covered mountains of Northern and
Western Burma, where our British and Indian troops were at close
grips with them, across the sea to the Andamans and the great Dutch
dependencies of Sumatra and Java, and thence in an easterly bend
along the string of lesser islands to New Guinea.

The Americans had established a bomber force in China which was
doing good work against the enemy's sea communications between the
mainland and the Philippines. They wanted to extend this effort by
basing long-range aircraft in China to attack Japan itself. The Burma
Road was cut, and they were carrying all supplies for them and the
Chinese armies by air over the southern spurs of the Himalayas, which
they called "the Hump." This was a stupendous task. The American
wish to succour China, not only by an ever-increasing air lift but
also by land, led to heavy demands upon Britain and the Indian
Empire. They pressed as a matter of the highest urgency and importance the making of a motor road from the existing roadhead at Ledo
through five hundred miles of jungles and mountains into Chinese
territory. Only one metre-gauge, single-line railway ran through Assam
to Ledo. It was already in constant use for many other needs, including
the supply of the troops who held the frontier positions, but in order
to build the road to China the Americans wanted us to reconquer
Northern Burma first and quickly.

Certainly we favoured keeping China in the war and operating air
forces from her territory, but a sense of proportion and the study of
alternatives were needed. I disliked intensely the prospect of a large-

scale campaign in Northern Burma. One could not choose a worse place for fighting the Japanese. Making a road from Ledo to China was also an immense, laborious task, unlikely to be finished until the need for it had passed. Even if it were done in time to replenish the Chinese armies while they were still engaged it would make little difference to their fighting capacity. The need to strengthen the American air bases in China would also, in our view, diminish as Allied advances in the Pacific and from Australia gained us airfields closer to Japan. On both counts therefore we argued that the enormous expenditure of manpower and material would not be worth while. But we never succeeded in deflecting the Americans from their purpose. Their national psychology is such that the bigger the Idea the more wholeheartedly and obstinately do they throw themselves into making it a success. It is an admirable characteristic, provided the Idea is good.

We of course wanted to recapture Burma, but we did not want to have to do it by land advances from slender communications and across the most forbidding fighting country imaginable. The south of Burma, with its port of Rangoon, was far more valuable than the north. But all of it was remote from Japan. I wished, on the contrary, to contain the Japanese in Burma, and to break into or through the great arc of islands forming the outer fringe of the Dutch East Indies. Our whole British Indian Imperial front would thus advance across the Bay of Bengal into close contact with the enemy by using amphibious power at every stage. This divergence of opinion, albeit honestly held and frankly discussed, and with decisions loyally executed, continued. It is against this permanent background of geography, limited resources, and clash of policies that the story of the campaign should be read.

*   *   *

It had opened in December 1943, when General Stilwell, with two Chinese divisions which he had organised and trained in India, crossed the watershed from Ledo into the jungles below the main mountain ranges. He was opposed by the renowned Japanese 18th Division, but forged ahead steadily, and by early January had penetrated forty miles, while the roadmakers toiled behind him. In the south a British Corps began to advance down the Arakan coast of the Bay of Bengal, and at the same time, with the aid of newly arrived Spitfires, we gained a degree of air superiority which was shortly to prove invaluable.

In February our advance was suddenly halted. The Japanese also had a plan. Since November they had increased their strength in Burma from five divisions to eight, and they now proposed to invade Eastern India and raise the flag of rebellion against the British. Their first stroke was a counteroffensive in the Arakan towards the port of

Chittagong, which would draw off our reserves and our attention. Holding our 5th Division frontally on the coast, they passed the better part of a division through the jungle and round the flank of the 7th Division, which was farther inland. Within a few days it was surrounded and the enemy threatened to cut the coastal road behind the 5th Division. They fully expected both divisions to withdraw, but they had reckoned without one factor, supply by air. The 7th Division grouped themselves into perimeters, stood their ground, and fought it out. For a fortnight food, water, and ammunition were delivered to them, like manna, from above. The enemy had no such facilities; they had taken with them only ten days' supply, and the obstinacy of the 7th Division prevented more reaching them. Unable to overwhelm our forward troops, pressed from the north by a division we brought from reserve, they broke up into small parties to fight their way back through the jungle, leaving five thousand dead behind. This terminated the legend of Japanese invincibility in the jungle.

More, however, was to come. In this same February of 1944 there were sure signs that our central front at Imphal would also be attacked. We ourselves were preparing to advance to the Chindwin River, and the now famous Chindits [1] were poised for a daring stroke against the enemy supply lines and communications, notably those of the Japanese division with whom Stilwell was at close grips. Although it was clear that the Japanese would get their blow in first, it was decided that Wingate's brigades should carry on with their task. One of his brigades had already started on February 5. They marched across four hundred fifty miles of mountain and jungle and were supplied solely from the air. On March 5, sustained by an American "Air Commando" of two hundred fifty machines, the fly-in of two more Brigades of British and Gurkha troops began. After assembly at their rallying point they set out and cut the railway north of Indaw.[2] Wingate did not live long to enjoy this first success or to reap its fruits. On March 24, to my great distress, he was killed in the air. With him a bright flame was extinguished.

The main enemy blow fell, as we expected, on our central front. On March 8 three Japanese divisions attacked. General Scoones withdrew his IVth Corps, also of three divisions, to the Imphal plateau, so as to fight concentrated on ground of his own choosing. The Japanese repeated the tactics they had used with misfortune in the Arakan. They counted on capturing our stores at Imphal to feed themselves. They also intended to cut not only the road to Dimapur but also the railway there, and thus sever the supply route maintaining Stilwell's

[1] "Chindits," the familiar name for Wingate's Long-Range Penetration Force.
[2] See map at page 860.

force and the United States air lift to China. Important issues were at stake.

The key lay again in transport aircraft. Mountbatten's resources, though considerable, were not nearly enough. He sought to retain twenty United States aircraft already borrowed from the "Hump" traffic, and asked for seventy more. This was a hard requirement to make or to procure. In the anxious weeks that followed I gave him my strongest support. We halted our operations on the Arakan coast, withdrew the victorious Indian divisions, and flew them to his aid. The 5th went to Imphal, where the enemy were now pressing hard on the fringes of the plain from three sides; the 7th to Dimapur. Thither by rail also came the headquarters of General Stopford's XXXIIIrd Corps, together with a British division and another two brigades. The road through the mountains was now cut, and these new forces began to fight their way upwards.

Between them and Imphal lay the roadside township of Kohima, which commanded the pass to the Assam valley, and here, on April 4, the Japanese launched yet another furious onslaught. They used a whole division. Our garrison consisted of a battalion of the Royal West Kent, a Nepalese battalion, and a battalion of the Assam Rifles, with every man, and even convalescents from the hospital, who could bear arms. They were slowly forced back into a diminishing area, and finally on to a single hill. They had no supplies except what was dropped on them by parachutes. Attacked on every side, they held on steadfastly, supported by bombing and cannon-fire from the air, until General Stopford relieved them on the 20th. Four thousand Japanese were killed. The valiant defence of Kohima against enormous odds was a fine episode.

* * *

The climax came in May 1944. Sixty thousand British and Indian soldiers, with all their modern equipment, were confined in a circle on the Imphal plain. I could feel the stress amid all other business. All depended on the transport planes. On the principle "Nothing matters but the battle" I used my authority. On the 4th I telegraphed to Admiral Mountbatten: "Let nothing go from the battle that you need for victory. I will not accept denial of this from any quarter, and will back you to the full."

In the end his needs were largely met, but for more than another month the situation was at full strain. Our Air Force was dominant, but the monsoon was hindering the air supply, on which our success depended. All four divisions of the IVth Corps were slowly pushing outwards from their encirclement. Along the Kohima road the relieving force and the besieged were fighting their way towards each other. It was a race against time. We marked their progress with tense feelings. On June 22 I telegraphed to Mountbatten:

> The Chiefs of Staff have expressed anxiety about the situation in Imphal, particularly in respect of reserves of supplies and ammunition. You are absolutely entitled to ask for all aircraft necessary to maintain the situation, whether they come from tho "Hump" or any other source. "The Hump" must be considered the current reserve, and should be drawn upon whenever necessary. . . . If you fail to make your demands in good time, invoking me if necessary to help from here, it will be no good complaining afterwards if it is not a success. Keep your hand close on job, which seems to me both serious and critical. Every good wish.

The finale came while this message was on the way. I quote his report:

> In the third week in June the situation was critical, and it seemed possible, after all the efforts of the previous two months, that early in July the IVth Corps would finally run out of reserves. But on June 22, with a week and a half in hand, the 2nd British and 5th Indian Divisions met at a point twenty-nine miles north of Imphal and the road to the plain was open. On the same day the convoys began to roll in.

Thus ended Japan's invasion of India. Their losses had been ruinous. Over thirteen thousand dead were counted on the battlefields, and, allowing for those who died of wounds, disease, or hunger, the total amounted on a Japanese estimate to 65,000 men.

The monsoon, now at its height, had in previous years brought active operations to a standstill, and the enemy doubtless counted on a pause during which they could extricate and rebuild their shattered forces. They were given no such respite. The British Indian Fourteenth Army, under the able and forceful leadership of General Slim, took the offensive. All along the mountain tracks they found evidence of disaster — quantities of abandoned guns, transport, and equipment; thousands dead or dying. Progress, measured in miles a day, was very slow. But our men were fighting in tropical rainfall, soaked to the skin by day and night. The so-called roads were mostly fair-weather dust tracks, now churned into deep mud, through which guns and vehicles had often to be manhandled. It was not the slowness of advance but the fact that any advance was made at all that should cause surprise.

The Chindits had meanwhile been reinforced, and five of their brigades were now working northward up the railway from Indaw, preventing the passage of reinforcements and destroying dumps as they went. Despite the havoc they caused, the Japanese withdrew nothing from the Imphal front and only one battalion from Stilwell's. They brought their 53rd Division from Siam and tried, at the cost of over 5400 killed, but without success, to quell the nuisance. Stilwell continued his steady progress and captured Myitkyina on August 3, thus providing a staging post for the American air lift to China. The "Hump" traffic had no longer to make the direct and often dangerous flight from Northern Assam over the great mountains to Kunming. Work proceeded on the long road from Northern Assam, destined later to link up with the former road from Burma to China, and the strain on rearward communications was relieved by a new 750-mile oil pipeline laid from Calcutta, a greater span than the famous desert pipeline from Iraq to Haifa.

*　　　*　　　*

At this juncture I was in conference with the President at Quebec, and in spite of these successes I continued to urge that it was most undesirable that the fighting in the jungles should go on indefinitely. I desired an amphibious stroke across the Bay of Bengal on Rangoon, at the base of the Burmese land mass. If the Fourteenth Army then swept down from Central Burma we might clear the way for an assault on Sumatra. But all these projects called for men and material, and there were not enough in Southeast Asia. The only place they could come from was Europe. Landing craft would have to be taken either from the Mediterranean or from "Overlord," and the troops from Italy and elsewhere, and they would have to leave soon. It was now September. Rangoon lies forty miles up a winding estuary, complicated with backwaters and mudbanks. The monsoon starts in early May, and we should therefore have to attack by April 1945 at the latest. Was it yet safe to start weakening our effort in Europe? We carried the Americans with us on the Rangoon plan, but the sanguine hopes, which I had not shared, that Germany would collapse before the end of the year faded. It became obvious that German resistance would continue into and beyond the winter, and Mountbatten was accordingly instructed, not for the first time, that he must do what he could with what he had got.

And so we forged slowly ahead in the largest land engagement with Japan which had so far been attained. The good hygiene discipline

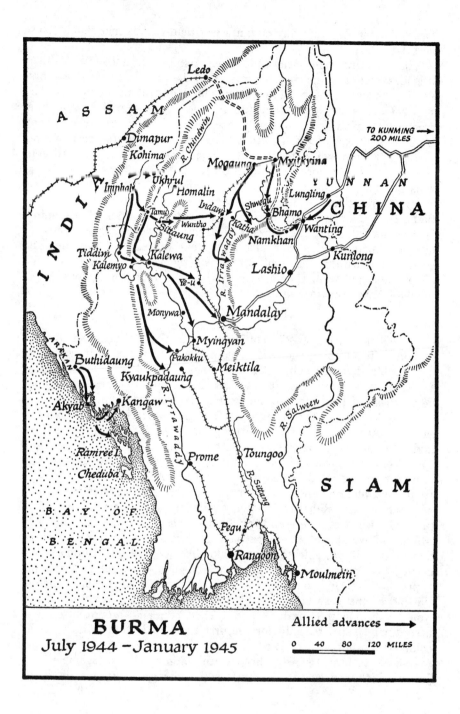

Ledo

A S S A M

Dimapur
Kohima
Ukhrul
Imphal
Tamu
Sittaung
Tiddim
Kalemyo
Kalewa

R. Chindwin

Homalin
Indaw
Wuntho

Mogaung
Myitkyina

TO KUNMING
200 MILES

Y U N N A N
Shwegu
Lungling
Katha
Bhamo
Wanting
Namkhan

C H I N A

Kunlong
Lashio

Ye-u

Monywa

Mandalay

R. Irrawaddy

Myingyan
Pakokku
Meiktila
Kyaukpadaung

Buthidaung
Kangaw
Akyab

A R A K A N

R. Irrawaddy

R. Salween

Ramree I.
Cheduba I.

Prome
Toungoo

R. Sittang

S I A M

B A Y   O F

B E N G A L

Pegu
Rangoon
Moulmein

BURMA
July 1944 – January 1945

Allied advances ⟶

0   40   80   120 MILES

now practised by all our units, the use of the new drug mepacrine, and constant spraying with D.D.T. insecticide kept the sick rate admirably low. The Japanese were not versed in these precautions and died in hundreds. The Fourteenth Army joined hands with the Chinese-American forces from the north, which now included a British division, and by early December, with two bridgeheads across the Chindwin, was poised for the main advance into the central plain of Burma.

In defiance of chronology, we may here pursue the story to its victorious conclusion. Formidable administrative problems now intervened. Far away in Southeast China, the Japanese had begun an advance on Chungking, the Generalissimo's capital, and Kunming, the delivery point of the American supply air lift. The Americans took a serious view of this situation. Their forward Air Force bases were being overrun. Chiang Kai-shek's troops gave little promise, and they now appealed for two of the Chinese divisions in North Burma, and also for more American air squadrons, in particular for three transport squadrons. These were hard tidings, but we had no choice but to accept. The loss of two good Chinese divisions was not so grave an inconvenience as parting with the transport squadrons. The Fourteenth Army was four hundred miles beyond its railhead and General Slim relied on air supply to help the tenuous road link. The squadrons had to go, and although later replaced, mostly from British sources, their absence inflicted severe delay on the campaign. In spite of all this the Fourteenth Army broke out of the hills into the plain northwest of Mandalay, and by the end of January 1945, General Sultan, who had succeeded Stilwell, reopened the land route to China.

Hard strategic decisions confronted Admiral Mountbatten when the decisive battle across the Irrawaddy began in the following month. His instructions were to liberate Burma, for which purpose he was not to expect greater resources than he already had, and then to occupy Malaya and open the Malacca Straits. Weather was dominant. The first task was to occupy the central plain of Burma and capture Rangoon before the monsoon, and the monsoon was due in early May. He could either concentrate the whole Fourteenth Army on a decisive battle in the Mandalay plain and make a swift advance to the south, or use some of his troops for an amphibious operation against Rangoon. In either case, much depended on air supply, in which United States aircraft played a big part. Aid to China still dominated American policy, and more planes might be withdrawn and his plans ruined. In face of these dangers, which were soon to become acute, Mountbatten decided on a single, fully supported operation against the main enemy body west of Mandalay, and a subsequent advance on Rangoon, which, he was advised, could be reached by April 15.

Events now moved swiftly. One of his divisions had already seized bridgeheads across the Irrawaddy about forty miles north of Mandalay, and throughout February they beat off a series of fierce counter-attacks. On February 12 the 20th Division crossed the river lower down and to the west of Mandalay. For a fortnight they had a hard fight to hold their gains, but by then they were joined by the 2nd British Division. This convinced the Japanese High Command that a decisive battle was imminent, and they sent heavy reinforcements. They did not believe that a serious flank attack was also possible, and even dispatched to Siam a division they could ill spare. This however was precisely the stroke which General Slim had prepared. On February 13 the 7th Division crossed the Irrawaddy south of Pakokku and formed a bridgehead. The enemy thought this was a mere diversion, but he was soon to be better informed. On the 21st two motorised brigades of the 17th Division and a brigade of tanks broke out from the bridgehead, and reached Meiktila on the 28th. Here was the principal administrative centre of the Japanese main front, a nodal point of their communications and the focus of several airfields. It was strongly defended, and the enemy sent two divisions posthaste to aid the garrison, but they were held at a distance until our reinforcements arrived. After a week of bitter fighting the town was in our hands, and all attempts to recapture it were repulsed. The Japanese admit losing five thousand men and as many wounded in a battle which their Commander-in-Chief has since described as "the master stroke of Allied strategy."

Far off to the northeast General Sultan was also on the move and by mid-March he had reached the road from Lashio to Mandalay. But Chiang Kai-shek now imposed a halt. He would not allow his Chinese divisions to continue. He insisted on removing them and suggested that General Slim should halt his advance when Mandalay was taken. This was precisely what Mountbatten had feared when he made his plans a month before, and in the event the Japanese were able to take two of their three divisions from this front and march them against our Fourteenth Army.

The conjoint battles of Mandalay and Meiktila raged through March. Mandalay was entered on the 9th and Mandalay Hill, 780 feet above the surrounding country, was taken in two days, but the Japanese resisted strongly and the massive walls of Fort Dufferin were impenetrable to ordinary missiles. Finally a breach was made with 2000-pound bombs, and on the 20th the enemy fled. The rest of XXXIIIrd Corps meanwhile fought on to Meiktila. They met great opposition, as the Japanese Commander-in-Chief, in spite of the in-tervention of the 17th Division behind his front, showed no signs as yet of withdrawing, and the armies were well matched. But at the end of the month the enemy gave up the struggle and began to fall back down the main road to Toungoo and Rangoon, and through the mountains to the east.

The battles, however, had lasted much longer than we expected. General Sultan was halted on the Lashio road, and there was now no prospect of the Fourteenth Army reaching Rangoon by mid-April. Indeed, it was very doubtful if they could get there before the mon-soon. Mountbatten accordingly decided to make an amphibious as-sault on the town after all. This would have to be much smaller than we had hoped, and even so it could not be launched before the first week in May. By then it might be too late.

General Slim was nevertheless determined not only to reach Ran-goon but to draw a double net down Southern Burma and trap the enemy within it. The XXXIIIrd Corps from Meiktila accordingly drove down the Irrawaddy with overlapping thrusts and reached Prome on May 2. The IVth Corps, victors at Imphal and Mandalay, advanced even more swiftly along the road and the railway to the east. An armoured column, and the mechanised brigades of the 5th and 17th Divisions, leapfrogging over each other, reached Toungoo on April 22. The next bound was to Pegu, whose capture would close the enemy's southernmost escape route from Lower Burma. Our advance troops reached it on April 29. That afternoon torrential rain fell, heralding an early monsoon. Forward air strips were out of action; tanks and vehicles could not move off the roads. The Japanese mus-tered every possible man to hold the town and the bridges over the river. On May 2 the 17th Division finally broke through, and, hop-ing to be first in Rangoon, prepared to advance the few remaining miles.

But May 2 was also the D-Day of the amphibious assault. For two days beforehand Allied heavy bombers attacked the defences which barred the entrance to Rangoon River. On May 1 a parachute battalion dropped on the defenders and the channel was opened for minesweeping. Next day ships of the 26th Division, supported by 224 Group R.A.F., reached the river-mouth. A Mosquito aircraft flew over Rangoon and saw no signs of the enemy. The crew landed at a nearby airfield, walked into the city, and were greeted by a num-ber of our prisoners of war. In the belief that an amphibious attack was no longer likely, the Japanese garrison had departed some days before to hold Pegu. That afternoon the monsoon broke in all its violence, and Rangoon fell with only a few hours to spare.

The amphibious force soon reached up to Pegu and to Prome. Many thousands of Japanese were trapped, and during the next three months greater numbers perished in attempts to escape eastward. Thus ended a long struggle in which the Fourteenth Army had fought valiantly, overcome all obstacles, and achieved the seemingly impossible.

17

The Battle of Leyte Gulf

OCEAN WAR AGAINST JAPAN had also reached its climax. From the Bay of Bengal to the Central Pacific Allied maritime power was in the ascendant. The organisation and production of the United States were in full stride, and had attained astonishing proportions. A single example may suffice to illustrate the size and success of the American effort. In the autumn of 1942 only three American aircraft carriers were afloat; a year later there were fifty; by the end of the war there were more than a hundred. This achievement had been matched by an increase in aircraft production which was no less remarkable. The advance of these great forces was animated by an aggressive strategy and an elaborate, novel, and effective tactic. The task which confronted them was formidable.

A chain of island groups, nearly two thousand miles long, stretches southward across the Pacific from Japan to the Marianas and the Carolines. Many of these islands had been fortified by the enemy and equipped with good airfields, and at the southernmost end of the chain was the Japanese naval base of Truk. Behind this shield of archipelagos lay Formosa, the Philippines, and China, and in its shelter ran the supply routes for the more advanced enemy positions. It was thus impossible to invade or bomb Japan itself. The chain must be broken first. It would take too long to conquer and subdue every fortified island, and the Americans had accordingly advanced leapfrog fashion. They seized only the more important islands and bypassed the rest; but their maritime strength was now so great and was growing so fast that they were able to establish their own lines of communication and break the enemy's, leaving the defenders of the bypassed islands immobile and powerless. Their method of assault was equally successful. First came softening attacks by planes from the aircraft carriers, then heavy and sometimes prolonged bombardment from the sea, and finally amphibious landing and the struggle ashore. When an island had been won and garrisoned, land-based planes moved in and beat off counterattacks. At the same time they helped in the next onward surge. The fleets worked in echelons. While one group waged battle another prepared for a new leap. This needed very large resources, not only for the fighting, but also for developing bases along the line of advance. The Americans took it all in their stride.

*   *   *

By June 1944 the two-pronged American thrust across the Pacific was well advanced. In the southwest General MacArthur had nearly completed his conquest of New Guinea, and in the centre Admiral Nimitz was pressing deep into the chain of fortified islands. Both were converging on the Philippines, and the struggle for this region was soon to bring about the destruction of the Japanese Fleet. It had already been much weakened and was very short of carriers, but Japan's only hope of survival lay in victory at sea. To conserve her strength for this perilous but vital hazard the main body had withdrawn from Truk and was now divided between the East Indies and her home waters; but events soon brought it to battle. At the beginning of June Admiral Spruance struck with his carriers at the Marianas, and on the 15th he landed on the fortified island of Saipan. If he captured Saipan and the adjacent islands of Tinian and Guam the enemy's defence perimeter would be broken. The threat was formidable, and the Japanese Fleet resolved to intervene. That day five of their battleships and nine carriers were sighted near the Philippines, heading east. Spruance had ample time to make his dispositions. His main purpose was to protect the landing at Saipan. This he did. He then gathered his ships, fifteen of which were carriers, and waited for the enemy to the west of the island. On June 19 Japanese carrier-borne aircraft attacked the American carrier fleet from all directions, and air fighting continued throughout the day. The Americans suffered little damage, and so shattered the Japanese air squadrons that their carriers had to withdraw.

That night Spruance searched in vain for the vanished enemy. Late in the afternoon of the 20th he found them about two hundred fifty miles away. Attacking just before sunset, the American airmen sank one carrier and damaged four others, besides a battleship and a heavy cruiser. The previous day American submarines had sunk two other large carriers. No further attack was possible, and remnants of the enemy fleet managed to escape, but its departure sealed the fate of Saipan. Though the garrison fought hard the landings continued, the build-up progressed, and by July 9 all organised resistance came to an end. The neighbouring islands of Guam and Tinian were overcome, and by the first days of August the American grip on the Marianas was complete.

The fall of Saipan was a great shock to the Japanese High Command, and led indirectly to the dismissal of General Tojo's Government. The enemy's concern was well founded. The fortress was little more than thirteen hundred miles from Tokyo. They had believed it was impregnable; now it was gone. Their southern defence regions were cut off and the American heavy bombers had gained a first-class base for attacking the very homeland of Japan. For a long time United States submarines had been sinking Japanese merchantmen along the China coast, and now the way was open for other warships to join in the onslaught. Japan's oil and raw materials would be cut off if the Americans advanced any farther. The Japanese Fleet was still powerful, but unbalanced, and so weak in destroyers, carriers, and air crews that it could no longer fight effectively without land-based planes. Fuel was scarce, and not only hampered training but made it impossible to keep the ships concentrated in one place, so that in the late summer most of the heavy vessels and cruisers lay near Singapore and the oil supplies of the Dutch East Indies, while the few surviving carriers remained in home waters, where their new air groups were completing their training.

The plight of the Japanese Army was little better. Though still strong in numbers, it sprawled over China and Southeast Asia or languished in remote islands beyond reach of support. The more sober-minded of the enemy leaders began to look for some way of ending the war; but their military machine was too strong for them. The High Command brought reinforcements from Manchuria and ordered a fight to the finish both in Formosa and the Philippines. Here and in the homeland the troops would die where they stood. The Japanese Admiralty were no less resolute. If they lost the impending battle for the islands the oil from the East Indies would be cut off. There was no purpose, they argued, in preserving ships without fuel. Steeled for sacrifice but hopeful of victory, they decided in August to send the entire Fleet into battle.

On September 15 the Americans made another advance. General MacArthur seized Morotai Island, midway between the western tip of New Guinea and the Philippines, and Admiral Halsey, who had now assumed command of the United States naval forces, captured an advanced base for his fleet in the Palau group. These simultaneous moves were of high importance. At the same time Halsey continually probed the enemy's defences with his whole force. Thus he hoped to provoke a general action at sea which would enable him to destroy the Japanese Fleet, particularly its surviving carriers. The next leap would be at the Philippines themselves, and there now occurred a dramatic change in the American plan. Till then our Allies had purposed to invade the southernmost portion of the Philippines, the island of Mindanao, and planes from Halsey's carriers had already attacked the Japanese airfields both there and in the large northern island of Luzon. They destroyed large numbers of enemy aircraft, and discovered in the clash of combat that the Japanese garrison at Leyte was unexpectedly weak. This small but now famous island, lying between the two larger but strategically less important land masses of Mindanao and Luzon, became the obvious point for the American descent. On September 13, while the Allies were still in conference at Quebec, Admiral Nimitz, at Halsey's suggestion, urged its immediate invasion. MacArthur agreed, and within two days the American Chiefs of Staff resolved to attack on October 20, two months earlier than had been planned. Such was the genesis of the Battle of Leyte Gulf.

*   *   *

The Americans opened the campaign on October 10 with raids on airfields between Japan and the Philippines. Devastating and repeated attacks on Formosa provoked the most violent resistance, and from the 12th to the 16th there followed a heavy and sustained air battle between ship-borne and land-based aircraft. The Americans inflicted grievous losses both in the air and on the ground, but suffered little themselves, and their carrier fleet withstood powerful land-based air attack. The result was decisive. The enemy's Air Force was broken before the battle for Leyte was joined. Many Japanese naval aircraft destined for the fleet carriers were improvidently sent to Formosa as reinforcements and there destroyed. Thus in the supreme naval battle which now impended the Japanese carriers were manned by little more than a hundred partially trained pilots.

To comprehend the engagements which followed a study of the accompanying maps is necessary. The two large islands of the Philippines, Luzon in the north and Mindanao in the south, are separated by a group of smaller islands, of which Leyte is the key and centre. This central group is pierced by two navigable straits, both destined to dominate this famous battle. The northerly strait is San Bernardino, and about 200 miles south of it, leading directly to Leyte, is the strait of Surigao. The Americans, as we have seen, intended to seize Leyte, and the Japanese were resolved to stop them and to destroy their fleet. The plan was simple and desperate. Four divisions under General MacArthur would land on Leyte, protected by the guns and planes of the

American fleet — so much they knew or guessed. Draw off this fleet, entice it far to the north, and engage it in a secondary battle — such was the first step. But this would be only a preliminary. As soon as the main fleet was lured away two strong columns of warships would sail through the straits, one through San Bernardino and the other through Surigao, and converge on the landings. All eyes would be on the shores of Leyte, all guns trained on the beaches, and the heavy ships and the big aircraft carriers which alone could withstand the assault would be chasing the decoy force in the far north. The plan very nearly succeeded.

On October 17 the Japanese Commander-in-Chief ordered his fleet to set sail. The decoy force, under Admiral Ozawa, the Supreme Commander, sailing direct from Japan, steered for Luzon. It was a composite force, including carriers, battleships, cruisers, and destroyers. Ozawa's task was to appear on the eastern coast of Luzon, engage the American fleet, and draw it away from the landings in Leyte Gulf. The carriers were short of both planes and pilots, but no matter. They were only bait, and bait is made to be eaten. Meanwhile the main Japanese striking forces made for the straits. The larger, or what may be termed the Centre Force, coming from Singapore, and consisting of five battleships, twelve cruisers, and fifteen destroyers, under Admiral Kurita, headed for San Bernardino to curl round Samar Island to Leyte; the smaller, or Southern Force, in two independent groups, comprising in all two battleships, four cruisers, and eight destroyers, sailed through Surigao.

On October 20 the Americans landed on Leyte. At first all went well. Resistance on shore was weak, a bridgehead was quickly formed, and General MacArthur's troops began their advance. They were supported by Admiral Kinkaid's Seventh United States Fleet, which was under MacArthur's command, and whose older battleships and small aircraft carriers were well suited to amphibious operations. Further away to the northward lay Admiral Halsey's main fleet, shielding them from attack by sea.

The crisis however was still to come. On October 23 American submarines sighted the Japanese Centre Force (Admiral Kurita) off the coast of Borneo and sank two of his heavy cruisers, one of which was Kurita's flagship, and damaged a third. Next day, October 24, planes from Admiral Halsey's carriers joined in the attack. The giant battleship *Musashi,* mounting nine 18-inch guns, was sunk, other vessels were damaged, and Kurita turned back. The reports of the American airmen were optimistic and perhaps misleading, and Halsey concluded, not without reason, that the battle was won, or at any rate this part of it. He knew that the second or Southern enemy force was approaching

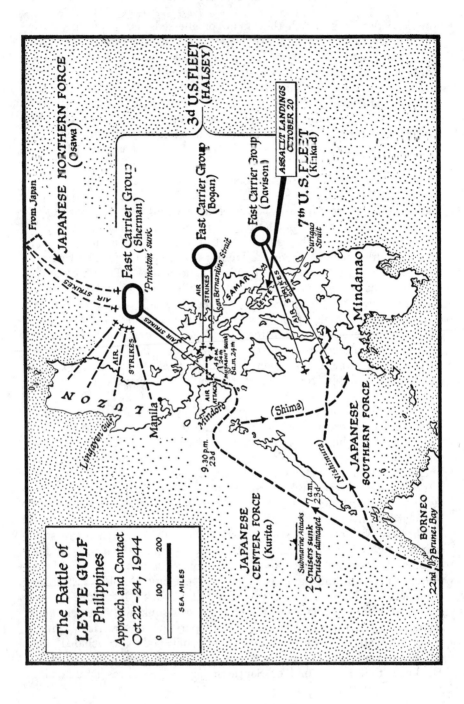

The Battle of
LEYTE GULF Philippines

Approach and Contact
Oct. 22–24, 1944

0   100   200
SEA MILES

JAPANESE NORTHERN FORCE
(Ozawa)

From Japan

AIR STRIKES

3d U.S. FLEET
(HALSEY)

Fast Carrier Group
(Sherman)
"Princeton sunk"

Fast Carrier Group
(Bogan)

Fast Carrier Group
(Davison)

ASSAULT LANDINGS
OCTOBER 20

7th U.S. FLEET
(Kinkaid)

San Bernardino Strait

SAMAR

Surigao Strait

AIR STRIKES

AIR STRIKES

AIR STRIKES

LEYTE

LUZON

AIR
STRIKES

AIR STRIKES

Manila

Lingayen Gulf

Mindoro

AIR
ATTACK

1 P.M.
24 d
"Musashi" sunk
8a.m. 24 d

9.30 p.m.
23 d

(Shima)

(Nishimura)

Mindanao

JAPANESE
SOUTHERN FORCE

JAPANESE
CENTER FORCE
(Kurita)

7 a.m.
23 d

Submarine Attacks
2 Cruisers sunk
1 Cruiser damaged

22nd   0   Brunei Bay

BORNEO

the Surigao Strait, but he judged, and rightly, that it could be repelled by Kinkaid's Seventh Fleet.

But one thing disturbed him. During the day he had been attacked by Japanese naval planes. Many of them were shot down, but the carrier *Princeton* was damaged and had later to be abandoned. The planes, he reasoned, probably came from carriers. It was most unlikely that the enemy had sailed without them, yet none had been found. The main Japanese fleet, under Kurita, had been located, and was apparently in retreat, but Kurita had no carriers, neither were there any in the Southern Force. Surely there must be a carrier force, and it was imperative to find it. He accordingly ordered a search to the north, and late in the afternoon of October 24 his flyers came upon Admiral Ozawa's decoy force, far to the northeast of Luzon and steering south. Four carriers, two battleships equipped with flying decks, three cruisers, and ten destroyers! Here, he concluded, was the source of trouble and the real target. If he could now destroy these carriers, he and his Chief of Staff, Admiral Carney, rightly considered that the power of the Japanese fleet to intervene in future operations would be broken irretrievably. This was a dominating factor in his mind, and would be of particular advantage when MacArthur came later to attack Luzon. Halsey could not know how frail was their power, nor that most of the attacks he had endured came not from carriers at all but from airfields in Luzon itself. Kurita's Centre Force was in retreat. Kinkaid could cope with the Southern Force and protect the landings at Leyte, the way was clear for a final blow, and Halsey ordered his whole fleet to steam northward and destroy Admiral Ozawa next day. Thus he fell into the trap. That same afternoon, October 24, Kurita again turned east, and sailed once more for the San Bernardino Strait. This time there was nothing to stop him.

* * *

Meanwhile the Southern Japanese Force was nearing Surigao Strait, and that night they entered it in two groups. A fierce battle followed, in which all types of vessel, from battleships to light coastal craft, were closely engaged.[1] The first group was annihilated by Kinkaid's fleet, which was concentrated at the northern exit; the second tried to break through in the darkness and confusion, but was driven back. All seemed to be going well, but the Americans had still to reckon with Admiral Kurita. While Kinkaid was fighting in the Surigao Strait and Halsey was steaming in hard pursuit of the decoy force far to the north Kurita

---

[1] Among them were two Australian warships, the cruiser *Shropshire* and the destroyer *Arunta*.

had passed unchallenged in the darkness through the Strait of San Bernardino, and in the early morning of October 25 he fell upon a group of escort carriers who were supporting General MacArthur's landings. Taken by surprise and too slow-moving to escape, they could not at once rearm their planes to repel the onslaught from the sea. For about two and a half hours the small American ships fought a valiant retreat under cover of smoke. Two of their carriers, three destroyers, and over a hundred planes were lost, one of the carriers by suicide bomber attack; but they succeeded in sinking three enemy cruisers and damaging others.[2] Help was far away. Kinkaid's heavy ships were well south of Leyte, having routed the Southern Force, and were short of ammunition and fuel. Halsey, with ten carriers and all his fast battleships, was yet more distant, and although another of his carrier groups had been detached to refuel and was now recalled it could not arrive for some hours. Victory seemed to be in Kurita's hands. There was nothing to stop him steaming into Leyte Gulf and destroying MacArthur's amphibious fleet.

But once again Kurita turned back. His reasons are obscure. Many of his ships had been bombed and scattered by Kinkaid's light escort carriers, and he now knew that the Southern Force had met with disaster. He had no information about the fortunes of the decoys in the north and was uncertain of the whereabouts of the American fleets. Intercepted signals made him think that Kinkaid and Halsey were converging on him in overwhelming strength and that MacArthur's transports had already managed to escape. Alone and unsupported, he now abandoned the desperate venture for which so much had been sacrificed and which was about to gain its prize, and, without attempting to enter Leyte Gulf, he turned about and steered once more for the San Bernardino Strait. He hoped to fight a last battle on the way with Halsey's fleet, but even this was denied him. In response to Kinkaid's repeated calls for support Halsey had indeed at last turned back with his battleships, leaving two carrier groups to continue the pursuit to the north. During the day these destroyed all four of Ozawa's carriers. But Halsey himself got back to San Bernardino too late. The fleets did not meet. Kurita escaped. Next day Halsey's and MacArthur's planes pursued the Japanese admiral and sank another cruiser and two more destroyers. This was the end of the battle. It may well be that Kurita's mind had become confused by the pressure of events. He had been under constant attack for three days, he had suffered heavy losses,

---

[2] Suicide bombers made their first appearance in the Leyte operations. The Australian cruiser *Australia*, operating with Kinkaid's fleet, had been hit by one a few days before, and had suffered casualties but no serious damage.

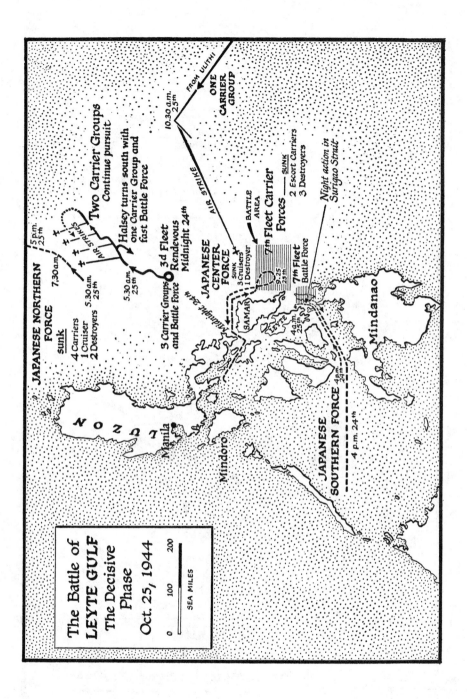

The Battle of
LEYTE GULF
The Decisive
Phase
Oct. 25, 1944

0   100   200

SEA MILES

JAPANESE NORTHERN
FORCE

sunk
4 Carriers
1 Cruiser
2 Destroyers

5 p.m.
25th

7.30 a.m.
25th

5.30 a.m.
25th

AIR STRIKE

5.30 a.m.
25th

Two Carrier Groups
Continue pursuit

Halsey turns south with
one Carrier Group and
fast Battle Force

3 Carrier Groups
and Battle Force

3d Fleet
Rendevous
Midnight 24th

FROM ULITHI

ONE
CARRIER
GROUP

10.30 a.m.
25th

AIR STRIKE

JAPANESE
CENTER
FORCE

SUNK
3 Cruisers
1 Destroyer

BATTLE
AREA

7th Fleet Carrier
Forces

SUNK
2 Escort Carriers
3 Destroyers

Night action in
Surigao Strait

9.25
25th

7th Fleet
Battle Force

(SAMAR)

LEYTE

4 a.m.
25th

Mindanao

L U Z O N

Manila

Mindoro

JAPANESE
SOUTHERN FORCE

4 p.m.
24th

4 p.m. 24th

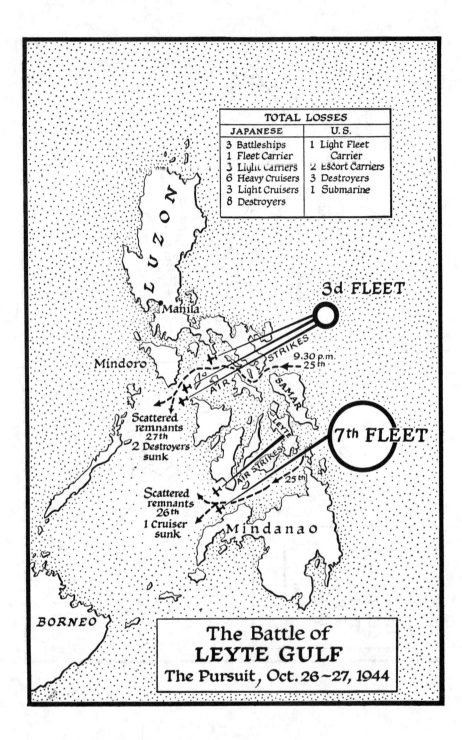

TOTAL LOSSES	
JAPANESE	**U.S.**
3 Battleships	1 Light Fleet
1 Fleet Carrier	Carrier
3 Light Carriers	2 Escort Carriers
6 Heavy Cruisers	3 Destroyers
3 Light Cruisers	1 Submarine
8 Destroyers	

LUZON

Manila

Mindoro

3d FLEET

AIR STRIKES

9.30 p.m. 25th

SAMAR

LEYTE

7th FLEET

Scattered
remnants
27th
2 Destroyers
sunk

AIR STRIKES

25th

Scattered
remnants
26th
1 Cruiser
sunk

Mindanao

BORNEO

The Battle of
LEYTE GULF
The Pursuit, Oct. 26~27, 1944

and his flagship had been sunk soon after starting from Borneo. Those who have endured a similar ordeal may judge him.

*   *   *

The Battle of Leyte Gulf was decisive. At a cost to themselves of three carriers, three destroyers, and a submarine the Americans had conquered the Japanese Fleet. The struggle had lasted from October 22 to October 27. Three battleships, four carriers, and twenty other enemy warships had been sunk, and the suicide bomber was henceforward the only effective naval weapon left to the foe. As an instrument of despair it was still deadly, but it carried no hope of victory.

Long should this victory be treasured in American history. Apart from valour, skill, and daring, it shed a light on the future more vivid and far-reaching than any we had seen. It shows a battle fought less with guns than by predominance in the air. I have told the tale fully because at the time it was almost unknown to the harassed European world. Perhaps the most important single conclusion to be derived from study of these events is the vital need for unity of command in conjoint operations of this kind in place of the concept of control by co-operation such as existed between MacArthur and Halsey at this time. The Americans learnt this lesson, and in the final operations planned against the homeland of Japan they intended that supreme command should be exercised by either Admiral Nimitz or General MacArthur, as might be advisable at any given moment.

In the following weeks the fight for the Philippines spread and grew. By the end of November nearly a quarter of a million Americans had landed in Leyte, and by mid-December Japanese resistance was broken. MacArthur pressed on with his main advance, and soon landed without opposition on Mindoro Island, little more than a hundred miles from Manila itself. On January 9, 1945, a new phase opened with the landing of four divisions in Lingayen Gulf, north of Manila, which had been the scene of the major Japanese invasion three years before. Elaborate deception measures kept the enemy guessing where the blow would fall. It came as a surprise and was only lightly opposed. As the Americans thrust towards Manila resistance stiffened, but they made two more landings on the west coast and surrounded the city. A desperate defence held out until early March, when the last survivors were killed. Sixteen thousand Japanese dead were counted in the ruins. Attacks by suicide aircraft were now inflicting considerable losses, sixteen ships being hit in a single day. The cruiser *Australia* was again unlucky, being hit five times in four days, but stayed in action. This desperate expedient however caused no check to the fleets. In mid-

January Admiral Halsey's carriers broke unmolested into the South China Sea, ranging widely along the coast and attacking airfields and shipping as far west as Saigon. At Hong Kong on January 16 widespread damage was inflicted, and great oil fires were started at Canton.

Although fighting in the islands continued for several months, command of the South China seas had already passed to the victor, and with it control of the oil and other supplies on which Japan depended.

18

The Liberation of Western Europe

ENERAL EISENHOWER, in accordance with previous and
agreed arrangements, assumed direct command of the land forces
in Northern France on September 1. These comprised the British
Twenty-first Army Group, under Field Marshal Montgomery, and the
American Twelfth Army Group, under General Omar Bradley, whose
operations Montgomery had hitherto controlled. He wielded more than
thirty-seven divisions, or over half a million fighting men, and this great
array was driving before it the remnants of the German armies in the
West, who were harassed day and night by our dominating air forces.
The enemy were still about seventeen divisions strong, but until they
could re-form and were reinforced there was little fight left in most of
them. General Speidel, Rommel's former Chief of Staff, has described
their plight:

> . . . an orderly retreat had become impossible. The Allied
> motorized armies surrounded the slow and exhausted German foot
> divisions in separate groups and smashed them up. . . . There were
> no longer any German ground forces worth mentioning in existence,
> to say nothing of air forces.[1]

Eisenhower planned to thrust northeastward in the greatest possible
strength and to the utmost limit of his supplies. The main effort was
to be made by the British Twenty-first Army Group, whose drive
along the Channel coast would not only overrun the launching sites of
the flying bomb, but also take Antwerp. Without the vast harbour of
this city no advance across the lower Rhine and into the plains of
Northern Germany was possible. The Twelfth U.S. Army Group was

[1] Hans Speidel, *Invasion 1944*, pp. 146–147.

also to pursue the enemy, its First Army keeping abreast of the British, while the remainder, bearing eastward towards Verdun and the upper Meuse, would prepare to strike towards the Saar.

Montgomery made two counterproposals, one in late August that his Army Group and the Twelfth U.S. Army Group should strike north together with a solid mass of nearly forty divisions, and the second on September 4, that only one thrust should be made, either towards the Ruhr or the Saar. Whichever was chosen the forces should be given all the resources and maintenance they needed. He urged that the rest of the front should be restrained for the benefit of the major thrust, which should be placed under one commander, himself or Bradley as the case might be. He believed it would probably reach Berlin, and considered that the Ruhr was better than the Saar.

But Eisenhower held to his plan, Germany still had reserves in the homeland, and he believed that if a relatively small force were thrust far ahead across the Rhine it would play into the enemy's hands. He thought it was better for the Twenty-first Army Group to make every effort to get a bridgehead over the Rhine, while the Twelfth advanced as far as they could against the Siegfried Line.

Strategists may long debate these issues.

The discussion caused no check in the pursuit. The number of divisions that could be sustained, and the speed and range of their advance, depended however entirely on harbours, transport, and supplies. Relatively little ammunition was being used, but food, and above all petrol, governed every movement. Cherbourg and the "Mulberry" harbour at Arromanches were the only ports we had, and these were daily being left farther behind. The front line was still sustained from Normandy, and each day about 20,000 tons of supplies had to be carried over ever-increasing distances, together with much material for mending roads and bridges and for building airfields. The Brittany ports, when captured, would be even more remote, but the Channel ports from Havre northward, and especially Antwerp, if we could capture it before it was too seriously damaged, were prizes of vital consequence.

Antwerp was thus the immediate aim of Montgomery's Army Group, which now had its first chance to show its mobility. The 11th Armoured Division captured the commander of the Seventh German Army at his breakfast in Amiens on August 31. The frontier towns so well known to the British Expeditionary Force of 1940, and, at least by name, to their predecessors a quarter of a century before — Arras, Douai, Lille, and many others — were soon reached. Brussels, hastily evacuated by the Germans, was entered by the Guards Armoured Division on September 3, and, as everywhere in Belgium, our troops had a splendid welcome and were much helped by the well-organised Resistance. Thence the Guards turned east for Louvain, and the 11th Armoured entered Antwerp on September 4, where, to our surprise and joy, they found the harbour almost intact. So swift had been the advance — over two hundred miles in under four days — that the enemy had been run off their legs and given no time for their usual and thorough demolition.

But our ships could only reach Antwerp through the winding, difficult estuary of the Scheldt, and the Germans held both banks. Hard and costly operations were needed to expel them, and the task fell principally on General Crerar's First Canadian Army.[2] Much depended on their success. By the 9th they had cleared all the Pas de Calais, with its flying-bomb launching sites. The Channel ports, Dieppe, Boulogne, Calais, Dunkirk, were either seized or invested. Havre, with a garrison over 11,000 strong, resisted fiercely, and in spite of bombardment from the sea by 15-inch guns, and more than 10,000 tons of bombs from the air, did not surrender till September 12. The Polish Armoured Division captured Ghent, only forty miles from Antwerp itself. Of course, this pace could not last. The forward leap was over and the check was now evident.

But there was still the chance of crossing the lower Rhine. Eisenhower thought this prize so valuable that he gave it priority over clearing the shores of the Scheldt estuary and opening the port of Antwerp. To renew Montgomery's effort Eisenhower gave him additional American transport and air supply. The First Airborne Army, under the American General Brereton, stood ready to strike from England and Montgomery resolved to seize a bridgehead at Arnhem. The 82nd U.S. Division was to capture the bridges at Nijmegen and Grave, while the 101st U.S. Division secured the road from Grave to Eindhoven. The XXXth Corps, led by the Guards Armoured Division, would force their way up the road to Eindhoven and thence to Arnhem along the "carpet" of air-borne troops hoping to find the bridges over the three major water obstacles already safely in their hands.

The preparations for this daring stroke, by far the greatest operation of its kind yet attempted, were complicated and urgent, because the enemy were growing stronger every day. It is remarkable that they were completed by the set date, September 17. There were not sufficient aircraft to carry the whole air-borne force simultaneously, and the movement had to be spread over three days. However, on the 17th the leading elements of the three divisions were well and truly taken to their destinations by the fine work of the Allied air forces. The 101st

---

[2] This consisted of the Ist British and the IInd Canadian Corps. The latter included the Polish Armoured Division.

U.S. Division accomplished most of their task, but a canal bridge on the road to Eindhoven was blown and they did not capture the town till the 18th. The 82nd U.S. Division also did well, but could not seize the main bridge at Nijmegen.

From Arnhem the news was scarce, but it seemed that some of our Parachute Regiment had established themselves at the north end of the bridge. The Guards Armoured Division began to advance in the afternoon up the Eindhoven road, preceded by an artillery barrage and rocket-firing planes, and protected by a corps on either flank. The road was obstinately defended, and the Guards did not reach the Americans till the afternoon of the 18th. German attacks against the narrow salient began next day and grew in strength. The 101st Division had great difficulty in keeping the road open. At times traffic had to be stopped until the enemy were beaten off. By now the news from Arnhem was bad. Our parachutists still held the northern end of the bridge, but the enemy remained in the town, and the rest of the 1st British Airborne Division, which had landed to the west, failed to break in and reinforce them.

The canal was bridged on the 18th, and early next morning the Guards had a clear run to Grave, where they found the 82nd U.S. Division. By nightfall they were close to the strongly defended Nijmegen bridge, and on the 20th there was a tremendous struggle for it. The Americans crossed the river west of the town, swung right, and seized the far end of the railway bridge. The Guards charged across the road bridge. The defenders were overwhelmed and both bridges were taken intact.

There remained the last lap to Arnhem, where bad weather had hampered the fly-in of reinforcements, food, and ammunition, and the 1st Airborne were in desperate straits. Unable to reach their bridge, the rest of the division was confined to a small perimeter on the northern bank and endured violent assaults. Every possible effort was made from the southern bank to rescue them, but the enemy were too strong. The Guards, the 43rd Division, the Polish Parachute Brigade, dropped near the road, all failed in gallant attempts at rescue. For four more days the struggle went on, in vain. On the 25th Montgomery ordered the survivors of the gallant 1st Airborne back. They had to cross the fast-flowing river at night in small craft and under close-range fire. By daybreak about 2400 men out of the original 10,000 were safely on our bank.

Heavy risks were taken in the Battle of Arnhem, but they were justified by the great prize so nearly in our grasp. Had we been more fortunate in the weather, which turned against us at critical moments and restricted our mastery in the air, it is probable that we should have succeeded. No risks daunted the brave men, including the Dutch Resistance, who fought for Arnhem, and it was not till I returned from Canada, where the glorious reports had flowed in, that I was able to understand all that had happened. General Smuts was grieved at what seemed to be a failure, and I telegraphed: "As regards Arnhem, I think you have got the position a little out of focus. The battle was a decided victory, but the leading division, asking, quite rightly, for more, was given a chop. I have not been afflicted by any feeling of disappointment over this and am glad our commanders are capable of running this kind of risk."

*   *   *

Clearing the Scheldt estuary and opening the port of Antwerp was now given first priority. During the last fortnight of September a number of preliminary actions had set the stage. Breskens "island," defended by an experienced German division, proved tough, and there was heavy fighting to cross the Leopold Canal. The hard task of capturing South Beveland was undertaken by the 2nd Canadian Division, which forced its way westward through large areas of flooding, their men often waist-deep in water. They were helped by the greater part of the 52nd Division, who were ferried across the Scheldt and landed on the southern shore. By the end of the month, after great exertions, the whole isthmus was captured. In four weeks of battle, during which the 2nd Tactical Air Force, under Air Marshal Coningham, gave them conspicuous support, they took no fewer than 12,500 German prisoners, who were anything but ready to surrender. Thus all was set for the Walcheren attack.

The island of Walcheren is shaped like a saucer and rimmed by sand dunes which stop the sea from flooding the central plain. At the western edge, near Westkapelle, is a gap in the dunes where the sea is held by a great dyke, thirty feet high and over a hundred yards wide at the base. The garrison of nearly 10,000 men was installed in strong artificial defences, and supported by about thirty batteries of artillery. Anti-tank obstacles, mines, and wire abounded, for the enemy had had four years in which to fortify the gateway to Antwerp.

Early in October the Royal Air Force struck the first blow. In a series of brilliant attacks they blew a great gap, nearly four hundred yards across, in the Westkapelle dyke. Through it poured the sea, flooding all the centre of the saucer and drowning such defences and batteries as lay within. But the most formidable emplacements and obstacles were on the saucer's rim. The attack was concentric. The main stroke was launched by three Marine Commandos. As they approached, the naval bombarding squadron opened fire. Here were

H.M.S. *Warspite* and the two 15-inch-gun monitors *Erebus* and *Roberts*, with a squadron of armed landing craft. These latter came close inshore, and, despite harsh casualties, kept up their fire until the two leading Commandos were safely ashore. The whole of the artillery of the IInd Canadian Corps, firing across the water from the Breskens shore, was brought to bear against powerful enemy guns embedded in concrete, and rocket-firing aircraft attacked the embrasures. In the gathering darkness No. 48 Commando killed or captured the defenders. Next morning it pressed on and by midday No. 47 took up the attack, and, with a weakening defence, reached the outskirts of Flushing. On November 3 they joined hands with No. 4 Commando after its stiff house-to-house fighting in the town. In a few days the whole island was in our hands, with 8000 prisoners.

Many other notable feats were performed by Commandos during the war, and though other troops and other Services played their full part in this remarkable operation the extreme gallantry of the Royal Marines stands forth. The Commando idea was once again triumphant. Minesweeping began as soon as Flushing was secure, and in the next three weeks a hundred craft were used to clear the seventy-mile channel. On November 28 the first convoy arrived, and Antwerp was opened for the British and American Armies. Flying bombs and rockets plagued the city for some time, and caused many casualties, but interfered with the furtherance of the war no more than in London.

*   *   *

On our right flank the advance beyond Paris of the Twelfth American Army Group had been conducted with all the thrustful impulse of Bradley and his ardent officers. Charleroi, Mons, and Liége fell to their grasp. In a fortnight they freed all Luxembourg and Southern Belgium, and on September 12 they closed up to the German frontier on a sixty-mile front and pierced the Siegfried Line near Aachen.

By the 16th, bridgeheads over the Moselle were won at Nancy and just south of Metz. The Sixth Army Group under General Devers, coming up from their landing in Southern France, met patrols from Patton's army west of Dijon five days before, and swinging to the east, drew level with the general advance. But here also was the end of the great pursuit. Everywhere enemy resistance was stiffening, and our supplies had been stretched to the limit. Aachen was attacked from three sides and surrendered on October 21. To the flank the Third Army were twenty miles east of the Moselle. The Seventh Army and the First French Army had drawn level and were probing towards the High Vosges and the Belfort Gap. The Americans had all but outrun their supplies in their lightning advances, and a pause was essential to build up stocks and prepare for large-scale operations in November.

*   *   *

The Strategic Air Forces played a big part in the Allied advance to the frontiers of France and Belgium. In the autumn they reverted to their primary role of bombing Germany, with oil installations and the transportation systems as specific targets. The enemy's radar screen and early-warning system had been thrust back behind his frontier, and our own navigation and bombing aids were correspondingly advanced. Our casualty rate decreased; the weight and accuracy of our attacks grew. The long-continued onslaught had forced the Germans to disperse their factories very widely. For this they now paid a heavy penalty, since they depended all the more on good communications. Urgently needed coal piled up at pitheads for lack of wagons to move it. Every day a thousand or more freight trains were halted for lack of fuel. Industry, electricity, and gas plants were beginning to close down. Oil production and reserves dropped drastically, affecting not only the mobility of the troops but also the activities and even the training of their air forces.

In August Speer had warned Hitler that the entire chemical industry was being crippled through lack of by-products from the synthetic oil plants, and the position grew worse as time went on. In November he reported that if the decline in railway traffic continued it would result in "a production catastrophe of decisive significance," and in December he paid a tribute to our "far-reaching and clever planning." [3] At long last our great bombing offensive was reaping its reward.

[3] Sir A. W. Tedder, *Air-Power in War*, pp. 118, 119.

19

October in Moscow

THE ARRANGEMENTS which I had made with President Roosevelt in the summer to divide our responsibilities for looking after particular countries affected by the movements of the armies had tided us over the three months for which our agreement ran. But as the autumn drew on, everything in Eastern Europe became more intense. I felt the need of another personal meeting with Stalin, whom I had not seen since Teheran, and with whom, in spite of the Warsaw tragedy, I felt new links since the successful opening of "Overlord." The Russian armies were now pressing heavily upon the Balkan scene, and Rumania and Bulgaria were in their power. Belgrade was soon to fall and Hitler was fighting with desperate obstinacy to keep his grip on Hungary. As the victory of the Grand Alliance became only a matter of time it was natural that Russian ambitions should grow. Communism raised its head behind the thundering Russian battlefront. Russia was the Deliverer, and Communism the gospel she brought.

I had never felt that our relations with Rumania and Bulgaria in the past called for any special sacrifices from us. But the fate of Poland and Greece struck us keenly. For Poland we had entered the war; for Greece we had made painful efforts. Both their Governments had taken refuge in London, and we considered ourselves responsible for their restoration to their own country, if that was what their peoples really wished. In the main these feelings were shared by the United States, but they were very slow in realising the upsurge of Communist influence, which slid on before, as well as followed, the onward march of the mighty armies directed from the Kremlin. I hoped to take advantage of the better relations with the Soviets to reach satisfactory solutions of these new problems opening between East and West.

Besides these grave issues which affected the whole of Central

Europe, the questions of World Organisation were also thrusting themselves upon all our minds. A lengthy conference had been held at Dumbarton Oaks, near Washington, between August and October, at which the United States, Great Britain, the U.S.S.R., and China had produced the now familiar scheme for keeping the peace of the world. The discussions had revealed many differences between the three great Allies, which will appear as this account proceeds. The Kremlin had no intention of joining an international body on which they would be outvoted by a host of small Powers, who, though they could not influence the course of the war, would certainly claim equal status in the victory. I felt sure we could only reach good decisions with Russia while we had the comradeship of a common foe as a bond. Hitler and Hitlerism were doomed; but after Hitler what?

We alighted at Moscow on the afternoon of October 9, and were received very heartily and with full ceremonial by Molotov and many high Russian personages. This time we were lodged in Moscow itself with every care and comfort. I had one small, perfectly appointed house, and Anthony Eden another nearby. We were glad to dine alone together and rest. At ten o'clock that night we held our first important meeting in the Kremlin. There were only Stalin, Molotov, Eden, and I, with Major Birse and Pavlov as interpreters. It was agreed to invite the Polish Prime Minister, M. Romer, the Foreign Minister, and M. Grabski, a grey-bearded and aged academician of much charm and quality, to Moscow at once. I telegraphed accordingly to M. Mikolajczyk that we were expecting him and his friends for discussions with the Soviet Government and ourselves, as well as with the Lublin Polish Committee. I made it clear that refusal to come to take part in the conversations would amount to a definite rejection of our advice and would relieve us from further responsibility towards the London Polish Government.

*   *   *

The moment was apt for business, so I said, "Let us settle about our affairs in the Balkans. Your armies are in Rumania and Bulgaria. We have interests, missions, and agents there. Don't let us get at cross purposes in small ways. So far as Britain and Russia are concerned, how would it do for you to have ninety per cent predominance in Rumania, for us to have ninety per cent of the say in Greece, and go fifty-fifty about Yugoslavia?" While this was being translated I wrote out on a half-sheet of paper:

Rumania

Russia 90%
The others 10%

Greece
    Great Britain 90%
        (in accord with U.S.A.)
    Russia 10%
Yugoslavia 50–50%
Hungary 50–50%
Bulgaria
    Russia 75%
    The others 25%

I pushed this across to Stalin, who had by then heard the translation. There was a slight pause. Then he took his blue pencil and made a large tick upon it, and passed it back to us. It was all settled in no more time than it takes to set down.

Of course we had long and anxiously considered our point, and were only dealing with immediate wartime arrangements. All larger questions were reserved on both sides for what we then hoped would be a peace table when the war was won.

After this there was a long silence. The pencilled paper lay in the centre of the table. At length I said, "Might it not be thought rather cynical if it seemed we had disposed of these issues, so fateful to millions of people, in such an offhand manner? Let us burn the paper." "No, you keep it," said Stalin.

"It is absolutely necessary," I reported privately to the President, "we should try to get a common mind about the Balkans, so that we may prevent civil war breaking out in several countries, when probably you and I would be in sympathy with one side and U.J. with the other. I shall keep you informed of all this, and nothing will be settled except preliminary agreements between Britain and Russia, subject to further discussion and melting down with you. On this basis I am sure you will not mind our trying to have a full meeting of minds with the Russians."

After this meeting I reflected on our relations with Russia throughout Eastern Europe, and in order to clarify my ideas drafted a letter to Stalin on the subject, enclosing a memorandum stating our interpretation of the percentages which we had accepted across the table. In the end I did not send this letter, deeming it wiser to let well alone. I print it only as an authentic account of my thought.

Moscow
October 11, 1944

I deem it profoundly important that Britain and Russia should have a common policy in the Balkans which is also acceptable to the United States. The fact that Britain and Russia have a twenty-year alliance makes it especially important for us to be in broad accord and to work together easily and trustfully and for a long time. I realise that nothing we can do here can be more than preliminary to the final decisions we shall have to take when all three of us are gathered together at the table of victory. Nevertheless I hope that we may reach understandings, and in some cases agreements, which will help us through immediate emergencies, and will afford a solid foundation for long-enduring world peace.

These percentages which I have put down are no more than a method by which in our thoughts we can see how near we are together, and then decide upon the necessary steps to bring us into full agreement. As I said, they would be considered crude, and even callous, if they were exposed to the scrutiny of the Foreign Offices and diplomats all over the world. Therefore they could not be the basis of any public document, certainly not at the present time. They might however be a good guide for the conduct of our affairs. If we manage these affairs well we shall perhaps prevent several civil wars and much bloodshed and strife in the small countries concerned. Our broad principle should be to let every country have the form of government which its people desire. We certainly do not wish to force on any Balkan State monarchic or republican institutions. We have however established certain relations of faithfulness with the Kings of Greece and Yugoslavia. They have sought our shelter from the Nazi foe, and we think that when normal tranquillity is reestablished and the enemy has been driven out the peoples of these countries should have a free and fair chance of choosing. It might even be that Commissioners of the three Great Powers should be stationed there at the time of the elections so as to see that the people have a genuine free choice. There are good precedents for this.

However, besides the institutional question there exists in all these countries the ideological issue between totalitarian forms of government and those we call free enterprise controlled by universal suffrage. We are very glad that you have declared yourselves against trying to change by force or by Communist propaganda the established systems in the various Balkan countries. Let them work out their own fortunes during the years that lie ahead. One thing however we cannot allow — Fascism or Nazism in any of their forms, which give to the toiling masses neither the securities offered by your system nor those offered by ours, but, on the contrary, lead to the build-up of tyrannies at home and aggression abroad. In principle I feel that Great Britain and Russia should feel easy about the internal government of these countries, and not worry about

them or interfere with them once conditions of tranquillity have
been restored after this terrible blood-bath which they, and indeed we, have all been through.

It is from this point of view that I have sought to adumbrate the
degrees of interest which each of us takes in these countries with
the full assent of the other, and subject to the approval of the
United States, which may go far away for a long time and then
come back again unexpectedly with gigantic strength.

In writing to you, with your experience and wisdom, I do not
need to go through a lot of arguments. Hitler has tried to exploit
the fear of an aggressive, proselytising Communism which exists
throughout Western Europe, and he is being decisively beaten to
the ground. But, as you know well, this fear exists in every country, because, whatever the merits of our different systems, no country wishes to go through the bloody revolution which will certainly
be necessary in nearly every case before so drastic a change could
be made in the life, habits and outlook of their society. At this
point, Mr. Stalin, I want to impress upon you the great desire there
is in the heart of Britain for a long, stable friendship and co-operation between our two countries, and that with the United States
we shall be able to keep the world engine on the rails.

To my colleagues at home I sent the following:

12 Oct 44

The system of percentage is not intended to prescribe the numbers sitting on commissions for the different Balkan countries, but
rather to express the interest and sentiment with which the British
and Soviet Governments approach the problems of these countries,
and so that they might reveal their minds to each other in some way
that could be comprehended. It is not intended to be more than a
guide, and of course in no way commits the United States, nor does
it attempt to set up a rigid system of spheres of interest. It may
however help the United States to see how their two principal
Allies feel about these regions when the picture is presented as a
whole.

2. Thus it is seen that quite naturally Soviet Russia has vital
interests in the countries bordering on the Black Sea, by one of
whom, Rumania, she has been most wantonly attacked with twenty-
six divisions, and with the other of whom, Bulgaria, she has ancient
ties. Great Britain feels it right to show particular respect to
Russian views about these two countries, and to the Soviet desire to
take the lead in a practical way in guiding them in the name of
the common cause.

3. Similarly, Great Britain has a long tradition of friendship with Greece, and a direct interest as a Mediterranean Power in her future. . . . Here it is understood that Great Britain will take the lead in a military sense and try to help the existing Royal Greek Government to establish itself in Athens upon as broad and united a basis as possible. Soviet Russia would be ready to concede this position and function to Great Britain in the same sort of way as Britain would recognise the intimate relationship between Russia and Rumania. This would prevent in Greece the growth of hostile factions waging civil war upon each other and involving the British and Russian Governments in vexatious arguments and conflict of policy.

4. Coming to the case of Yugoslavia, the numerical symbol 50–50 is intended to be the foundation of joint action and an agreed policy between the two Powers now closely involved, so as to favour the creation of a united Yugoslavia after all elements there have been joined together to the utmost in driving out the Nazi invaders. It is intended to prevent, for instance, armed strife between the Croats and Slovenes on the one side and powerful and numerous elements in Serbia on the other, and also to produce a joint and friendly policy towards Marshal Tito, while ensuring that weapons furnished to him are used against the common Nazi foe rather than for internal purposes. Such a policy, pursued in common by Britain and Soviet Russia, without any thought of special advantages to themselves, would be of real benefit.

5. As it is the Soviet armies which are obtaining control of Hungary, it would be natural that a major share of influence should rest with them, subject of course to agreement with Great Britain and probably the United States, who, though not actually operating in Hungary, must view it as a Central European and not a Balkan State.

6. It must be emphasised that this broad disclosure of Soviet and British feelings in the countries mentioned above is only an interim guide for the immediate wartime future, and will be surveyed by the Great Powers when they meet at the armistice or peace table to make a general settlement of Europe.

*   *   *

The Poles from London had now arrived, and at five o'clock on the evening of October 13 we assembled at the Soviet Government Hospitality House, known as Spiridonovka, to hear Mikolajczyk and his colleagues put their case. These talks were held as a preparation for a further meeting at which the British and American delegations would meet the Lublin Poles. I pressed Mikolajczyk hard to consider two

things, namely, *de facto* acceptance of the Curzon Line,[1] with interchange of population, and a friendly discussion with the Lublin Polish
Committee so that a united Poland might be established. Changes, I
said, would take place, but it would be best if unity were established
now, at this closing period of the war, and I asked the Poles to consider
the matter carefully that night. Mr. Eden and I would be at their disposal. It was essential for them to make contact with the Polish Committee and to accept the Curzon Line as a working arrangement, subject
to discussion at the Peace Conference.

At ten o'clock the same evening we met the so-called Polish National
Committee It was soon plain that the Lublin Poles were mere pawns
of Russia. They had learned and rehearsed their part so carefully that
even their masters evidently felt they were overdoing it. For instance,
M. Bierut, the leader, spoke in these terms: "We are here to demand
on behalf of Poland that Lvov shall belong to Russia. This is the will
of the Polish people." When this had been translated from Polish into
English and Russian I looked at Stalin and saw an understanding twinkle in his expressive eyes, as much as to say, "What about that for our
Soviet teaching!" The lengthy contribution of another Lublin leader,
Osóbka-Morawski, was equally depressing. Mr. Eden formed the worst
opinion of the three Lublin Poles.

The whole conference lasted over six hours, but the achievement was
small, and as the days passed only slight improvement was made with the
festering sore of Soviet-Polish affairs. The Poles from London were
willing to accept the Curzon Line "as a line of demarcation between
Russia and Poland." The Russians insisted on the words "as a basis
of frontier between Russia and Poland." Neither side would give way.
Mikolajczyk declared that he would be repudiated by his own people,
and Stalin at the end of a talk of two hours and a quarter which I had
with him alone remarked that he and Molotov were the only two of those
he worked with who were favourable to dealing "softly" with Mikolajczyk. I was sure there were strong pressures in the background, both
party and military.

Stalin was against trying to form a united Polish Government without
the frontier question being agreed. Had this been settled he would have
been quite willing that Mikolajczyk should head the new Government.
I myself thought that difficulties not less obstinate would arise in a
discussion for a merger of the Polish Government with the Lublin Poles,
whose representatives continued to make the worst possible impression
on us, and who, I told Stalin, were "only an expression of the Soviet
will." They had no doubt also the ambition to rule Poland, and were
thus a kind of Quislings. In all the circumstances the best course was for

[1] See map, page 989.

the two Polish delegations to return whence they had come. I felt very deeply the responsibility which lay on me and the Foreign Secretary in trying to frame proposals for a Russo-Polish settlement. Even forcing the Curzon Line upon Poland would excite criticism.

* * *

In other directions considerable advantages had been gained. The resolve of the Soviet Government to attack Japan on the overthrow of Hitler was obvious. This would have supreme value in shortening the whole struggle. The arrangements made about the Balkans were, I was sure, the best possible. Coupled with successful military action, they should now be effective in saving Greece, and I had no doubt that our agreement to pursue a fifty-fifty joint policy in Yugoslavia was the best solution for our difficulties in view of Tito's behaviour — having lived under our protection for three or four months, he had come secretly to Moscow to confer, without telling us where he had gone — and of the arrival of Russian and Bulgarian forces under Russian command to help his eastern flank.

There is no doubt that in our narrow circle we talked with an ease, freedom, and cordiality never before attained between our two countries. Stalin made several expressions of personal regard which I feel sure were sincere. But I became even more convinced that he was by no means alone. As I said to my colleagues at home, "Behind the horseman sits black care."

On the evening of October 17 we held our last meeting. The news had just arrived that Admiral Horthy had been arrested by the Germans as a precaution now that the whole German front in Hungary was disintegrating. I remarked that I hoped the Ljubljana Gap could be reached as fast as possible, and added that I did not think the war would be over before the spring.

20

Paris and the Ardennes

IT WAS THOUGHT FITTING that my first visit to Paris should be on Armistice Day, November 11, 1944, and this was publicly announced. There were many reports that collaborators would make attempts on my life and extreme precautions were taken. On the afternoon of November 10 I landed at Orly airfield, where de Gaulle received me with a guard of honour, and we drove together through the outskirts of Paris and into the city itself until we reached the Quai d'Orsay, where my wife and Mary and I were entertained in state. The building had long been occupied by the Germans, and I was assured I should sleep in the same bed and use the same bathroom as had Goering. Everything was mounted and serviced magnificently, and inside the palace it was difficult to believe that my last meeting there with Reynaud's Government and General Gamelin in May 1940 was anything but a bad dream. At eleven o'clock on the morning of November 11 de Gaulle conducted me in an open car across the Seine and through the Place de la Concorde, with a splendid escort of Gardes Républicains in full uniform with all their breastplates. They were several hundred strong, and provided a brilliant spectacle, on which the sun shone brightly. The whole of the famous avenue of the Champs Elysées was crowded with Parisians and lined with troops. Every window was filled with spectators and decorated with flags. We proceeded through wildly cheering multitudes to the Arc de Triomphe, where we both laid wreaths upon the tomb of the Unknown Warrior. After this ceremony was over the General and I walked together, followed by a concourse of the leading figures of French public life, for half a mile down the highway I knew so well. We then took our places on a dais, and there was a splendid march past of French and British troops. Our Guards detachment was magnificent. When this was over I laid a wreath beneath the statue of Clemenceau, who was much in my thoughts on this moving occasion.

De Gaulle entertained me at a large luncheon at the Ministry of War, and made a most flattering speech about my war services, and on the night of the 12th after dinner at the Embassy we left for Besançon. The General was anxious for me to see the attack on a considerable scale which was planned for the French Army under General de Lattre de Tassigny. All the arrangements for the journey in a luxurious special train were most carefully made, and we arrived in plenty of time for the battle. We were to go to an observation point in the mountains, but owing to bitter cold and deep snow the roads were impassable and the whole operation had to be delayed. I passed the day driving with de Gaulle, and we found plenty to talk about in a long and severe excursion, inspecting troops at intervals. The programme continued long after dark. The French soldiers seemed in the highest spirits. They marched past in great style and sang famous songs with moving enthusiasm. My personal party — my daughter Mary and my naval aide Tommy — feared that I should have another go of pneumonia, since we were out at least ten hours in terrible weather. But all went well, and in the train the dinner was pleasant and interesting. I was struck by the awe, and even apprehension, with which half a dozen high generals treated de Gaulle in spite of the fact that he had only one star on his uniform and they had lots.

During the night our train divided. De Gaulle returned to Paris, and our half went on to Reims, arriving next morning, when I went to Ike's headquarters. In the afternoon I flew back to Northolt.

*   *   *

By now the situation on the Western Front was not nearly so agreeable. There had been much preparation for the advance to the Rhine, but the November rains were the worst for many years, flooding the rivers and streams, and making quagmires through which the infantry had to struggle. In the British sector Dempsey's Second Army drove the enemy back across the Meuse. Farther south we joined hands with the Ninth U.S. Army and toiled over saturated country towards the river Roer. It would have been rash as yet to cross it, because its level was controlled by massive dams which were still in enemy hands, and by opening the sluices he could have cut off our troops on the far bank. Heavy bombers tried to burst the dams and release the water, but in spite of several direct hits no gap was made, and on December 13 the First U.S. Army had to renew their advance to capture them.

South of the Ardennes, Patton had crossed the Moselle and thrust eastward to the German frontier. Here he confronted the strongest part of the Siegfried defences. Against formidable and obstinately held fortifications his Army came to a halt. On the right of the line General Devers's Sixth Army Group forced their way through the Vosges and the Belfort Gap. The French, after a week's battle, the opening of which I had hoped to see, captured Belfort on November 22 and reached the Rhine north of Bâle. Thence they swung down river, turned the German flank in the Vosges and compelled the enemy to withdraw. Strasbourg was entered on the 23rd, and during the next few weeks the American Seventh Army cleared all Northern Alsace, wheeled up on the right of the Third Army, crossed the German frontier on a wide front, and penetrated the Siegfried Line near Wissembourg.

But these considerable successes could not mask the fact that the Western Allies had sustained a strategic reverse. Before this great movement was launched we placed on record our view that it was a mistake to attack against the whole front and that a far greater mass should have been gathered at the point of desired penetration. Montgomery's comments and predictions beforehand had in every way been borne out. "You must remember however," I cabled to Smuts, "that our armies are only about one-half the size of the American and will soon be little more than one-third. All is friendly and loyal in the military sphere in spite of the disappointment sustained. . . . But it is not so easy as it used to be for me to get things done. . . ."

On December 6 I also recounted my forebodings to the President:

> The time has come for me to place before you the serious and disappointing war situation which faces us at the close of this year. Although many fine tactical victories have been gained, . . . the fact remains that we have definitely failed to achieve the strategic object which we gave to our armies five weeks ago. We have not yet reached the Rhine in the northern part and most important sector of the front, and we shall have to continue the great battle for many weeks before we can hope to reach the Rhine and establish our bridgeheads. After that, again, we have to advance through Germany.
>
> In Italy the Germans are still keeping twenty-six divisions — equivalent to perhaps sixteen full strength or more — on our front. . . . The reason why the Fifteenth Group of Armies has not been able to inflict a decisive defeat on Kesselring is that, owing to the delay caused by the weakening of our forces for the sake of "Dragoon" [the Riviera landing in the South of France], we did not get through the Apennines till the valley of the Po had become waterlogged. Thus neither in the mountains nor on the plains have we been able to use our superiority in armour.
>
> On account of the obstinacy of the German resistance on all

fronts, we did not withdraw the five British and British-Indian divisions from Europe in order to enable Mountbatten to attack Rangoon in March, and for other reasons also this operation became impracticable. Mountbatten therefore began, as we agreed at Quebec, the general advance through Burma downstream from the north and the west, and this has made satisfactory progress. Now, owing to the advance of the Japanese in China, with its deadly threat to Kunming and perhaps Chungking, to the Generalissimo and his regime, two and possibly more Chinese divisions have to be withdrawn for the defence of China. I have little doubt that this was inevitable and right. The consequences however are serious. . . . All my ideas about a really weighty blow across the Adriatic or across the Bay of Bengal have been set back.

When we contrast these realities with the rosy expectations of our peoples, in spite of our joint efforts to damp them down, the question very definitely arises, "What are we going to do about it?" My anxiety is increased by the destruction of all hopes of an early meeting between the three of us and the indefinite postponement of another meeting of you and me with our Staffs. Our British plans are dependent on yours, our Anglo-American problems at least must be surveyed as a whole, and the telegraph and the telephone more often than not only darken counsel. Therefore I feel that if you are unable to come yourself before February I am bound to ask you whether you could not send your Chiefs of Staff over here as soon as practicable, where they would be close to your main armies and to General Eisenhower and where the whole stormy scene can be calmly and patiently studied with a view to action as closely concerned as that which signalised our campaigns of 1944.

Though sympathetic, Mr. Roosevelt did not appear to share my anxieties.

I always felt [he answered] that the occupation of Germany up to the left bank of the Rhine would be a very stiff job. Because in the old days I bicycled over most of the Rhine terrain, I have never been as optimistic as to the ease of getting across the Rhine with our joint armies as many of the commanding officers have been.

However, our agreed broad strategy is developing according to plan. You and I are now in the position of Commanders-in-Chief who have prepared their plans, issued their orders, and committed their resources to battle according to those plans and orders. For the time being, even if a little behind schedule, it seems to me the prosecution and outcome of the battles lie with our Field Commanders, in whom I have every confidence. . . .

        *   *   *

A heavy blow now impended. Within six days of sending this telegram a crisis burst upon us. The Allied decision to strike hard from Aachen in the north as well as through Alsace in the south had left our centre very weak. In the Ardennes sector a single corps, the VIIIth American, of four divisions, held a front of seventy-five miles. The risk was foreseen and deliberately accepted, but the consequences were grave and might have been graver. By a remarkable feat the enemy gathered about seventy divisions on their Western Front, of which fifteen were armoured. Many were under strength and needed rest and re-equipment, but one formation, the Sixth Panzer Army, was known to be strong and in good fettle. This potential spearhead had been carefully watched while it lay in reserve east of Aachen. When the fighting on that front died down in early December it vanished for a while from the ken of our Intelligence, and bad flying weather hindered our efforts to trace it. Eisenhower suspected that something was afoot, though its scope and violence came as a surprise.

The Germans had indeed a major plan. Rundstedt assembled two Panzer armies, the Fifth and Sixth, and the Seventh Army, a total of ten Panzer and fourteen infantry divisions. This great force, led by its armour, was intended to break through the Ardennes to the river Meuse, swing north and northwest, cut the Allied line in two, seize the port of Antwerp, and sever the lifeline of our northern armies. The stroke was planned by Hitler, who would brook no changes in it on the part of his doubting generals. The remnants of the German Air Force were assembled for a final effort, while paratroops, saboteurs, and agents in Allied uniforms were given their parts to play.

The attack began on December 16 under a heavy artillery barrage. At its northern flank the Sixth Panzer Army ran into the right of the First U.S. Army in the act of advancing towards the Roer dams. After a swaying battle the enemy were held. Farther south the Germans broke through on a narrow front, but were hindered for several critical days. The Sixth Panzer Army launched a new spearhead to strike west and then northward at the Meuse above Liége. The Fifth Panzer Army cut through the centre of the American Corps, and penetrated deeply towards the Meuse.

Although the time and weight of the attack surprised the Allied High Command its importance and purpose were quickly recognised. They resolved to strengthen the "shoulders" of the break-through, hold the Meuse crossings both east and south of Namur, and mass mobile troops

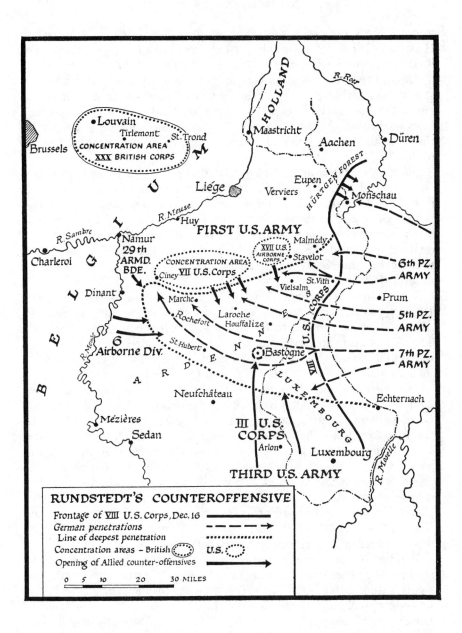

HOLLAND

R. Roer

•Louvain

Tirlemont

St. Trond

CONCENTRATION AREA

XXX BRITISH CORPS

Brussels

Maastricht

Aachen

Düren

Liége

Eupen

Verviers

HÜRTGEN FOREST

Monschau

R. Meuse

Huy

FIRST U.S. ARMY

Malmédy

R. Sambre

Namur

**29th
ARMD.
BDE.**

Charleroi

CONCENTRATION AREA

Ciney **VII** U.S. Corps

XVII U.S.
AIRBORNE
CORPS

Stavelot

**6th PZ.
ARMY**

St. Vith

Dinant

Marche•

Vielsalm

Prum

•Rochefort

Laroche
Houffalize

**5th PZ.
ARMY**

R. Meuse

**6
Airborne Div.**

St. Hubert

Bastogne

**7th PZ.
ARMY**

A R D E N N E S

LUXEMBOURG

Echternach

Neufchâteau

Mézières

Sedan

**III U.S.
CORPS**

Arlon•

Luxembourg

R. Moselle

THIRD U.S. ARMY

RUNDSTEDT'S COUNTEROFFENSIVE

Frontage of **VIII** U.S. Corps, Dec.16 ──────

German penetrations ── ── ──▶

Line of deepest penetration ••••••••••

Concentration areas – British ⟨⟩ U.S. ⟨⟩

Opening of Allied counter-offensives ━━━▶

0   5   10   20   30 MILES

to crush the salient from north and south. Eisenhower acted speedily. He stopped all Allied attacks in progress and brought up four American divisions from reserve, and six more from the south. Two air-borne divisions, one of them the 6th British, came from England. North of the salient, four divisions of the British XXXth Corps, which had just come out of the line on the river Roer, were concentrated between Liége and Louvain behind the American First and Ninth Armies. These latter threw in all their reserves to extend a defensive flank westward from Malmedy.

By covering the front of General Bradley's Twelfth Army Group the Germans had made it impossible for him to exercise effective command from his headquarters in Luxembourg over his two armies north of the bulge. General Eisenhower therefore very wisely placed Montgomery in temporary command of all Allied troops in the north, while Bradley retained the Third U.S. Army and was charged with holding and counterattacking the enemy from the south. Corresponding arrangements were made for the tactical air forces.

Three of our reinforcing divisions lined the Meuse south of Namur. Bradley concentrated a corps at Arlon and sent the American 101st Airborne Division to secure the important road junctions at Bastogne. The German armour swung north and sought to break their way northwestward, leaving their infantry to capture the town. The 101st, with some armoured units, were isolated, and for a week beat off all attacks.

The wheel of the Fifth and Sixth Panzer Armies produced bitter fighting around Marche, which lasted till December 26. By then the Germans were exhausted, although at one time they were only four miles from the Meuse and had penetrated over sixty miles. Bad weather and low ground fogs had kept our air forces out of the first week of the battle, but on December 23 flying conditions got better and they intervened with tremendous effect. Heavy bombers attacked railways and centres of movement behind the enemy lines, and tactical air forces played havoc in his forward areas, starving him of reinforcements, fuel, food, and ammunition. Strategic raids on German refineries helped to deny him petrol and slacken the advance.

Baulked of their foremost objective, the Meuse, the Panzers turned savagely on Bastogne. The 101st Division, though reinforced, was vastly outnumbered. They held the town grimly for another week, and by the end of December the German High Command must have realised, however unwillingly, that the battle was lost. A counteroffensive by Patton was steadily if slowly progressing over the snow-choked countryside. The enemy made one last bid, this time in the air. On January 1, 1945, they made a violent low-level surprise attack on all our forward airfields. Our losses, though heavy, were promptly replaced, but the Luftwaffe lost more than they could afford in their final massed attack of the Second World War.

Three days later Montgomery launched a counterstroke from the north to join Patton's advance from the south. Two American corps, with the British on their western flank, pressed down upon the enemy. Struggling through snowstorms, the two wings of the Allied attack slowly drew closer and met at Houffalize on the 16th. The Germans were forced steadily eastward and harassed continually from the air. By the end of the month they were back behind their frontiers, with nothing to show for their supreme effort except ruinous losses of material and casualties amounting to a hundred and twenty thousand men.

This was the final German offensive of the war. It caused us no little anxiety and postponed our own advance, but we benefited in the end. The Germans could not replace their losses, and our subsequent battles on the Rhine, though severe, were undoubtedly eased. Their High Command, and even Hitler, must have been disillusioned. Taken by surprise, Eisenhower and his commanders acted swiftly, but they will agree that the major credit lies elsewhere. In Montgomery's words, "The Battle of the Ardennes was won primarily by the staunch fighting qualities of the American soldier." The United States troops had indeed done almost all the fighting, and had suffered almost all the losses.

21

Christmas at Athens

T̲H̲E̲ G̲R̲E̲E̲K̲S̲ R̲I̲V̲A̲L̲ T̲H̲E̲ J̲E̲W̲S̲ in being the most politically minded race in the world. No matter how forlorn their
circumstances or how grave the peril to their country, they are always
divided into many parties, with many leaders who fight among themselves with desperate vigour. It has been well said that wherever there
are three Jews it will be found that there are two Prime Ministers and
one leader of the Opposition. The same is true of this other famous
ancient race, whose stormy and endless struggle for life stretches back
to the fountain springs of human thought. No two races have set such a
mark upon the world. Both have shown a capacity for survival, in spite
of unending perils and sufferings from external oppressors, matched only
by their own ceaseless feuds, quarrels, and convulsions. The passage of
several thousand years sees no change in their characteristics and no
diminution of their trials or their vitality. They have survived in spite
of all that the world could do against them, and all they could do against
themselves, and each of them from angles so different have left us the
inheritance of their genius and wisdom. No two cities have counted
more with mankind than Athens and Jerusalem. Their messages in
religion, philosophy, and art have been the main guiding lights of modern faith and culture. Centuries of foreign rule and indescribable, endless oppression leave them still living, active communities and forces in
the modern world, quarrelling among themselves with insatiable vivacity.
Personally I have always been on the side of both, and believed in their
invincible power to survive internal strife and the world tides threatening their extinction.

Before leaving Italy at the end of August I had asked the Chief of the
Imperial General Staff to work out the details of a British expedition
to Greece in case the Germans there collapsed.[1] We gave it the code

[1] See Chapter 14.

name "Manna," and by September our preparations were well advanced. M. Papandreou and his colleagues were brought to Italy and installed in a villa near Caserta. Here he set to work with the representatives of E.A.M. and their Nationalist rivals, E.D.E.S.[2] and, aided by Mr. Macmillan, as Minister Resident in the Mediterranean, and Mr. Leeper, our Ambassador to the Greek Government, a comprehensive agreement was signed on the 26th. It laid down that all guerrilla forces in the country should place themselves under the orders of the Greek Government, who in their turn put them under the British commander, General Scobie. The Greek guerrilla leaders declared that none of their men would take the law into their own hands. Any action in Athens would be taken only on General Scobie's direct orders. This document, known as the Caserta Agreement, governed our future action.

In October the liberation of Greece began. Commando units were sent into Southern Greece, and in the early hours of October 4 our troops occupied Patras. This was our first foothold since the tragic exit of 1941. On the 12th General Wilson learnt that the Germans were evacuating Athens, and next day British parachutists landed on the Megara airfield, about eight miles west of the capital. On the 14th the rest of the paratroopers arrived, and occupied the city on the heels of the German withdrawal. Our naval forces entered the Piraeus, bringing with them General Scobie and the main part of his force, and two days later the Greek Government arrived, together with our Ambassador.

The testing time for our arrangements had now come. At the Moscow conference I had obtained Russian abstention at a heavy price. We were pledged to support Papandreou's provisional administration, in which E.A.M. was fully represented. All parties were bound by the Caserta Agreement, and we wished to hand over authority to a stable Greek Government without loss of time. But Greece was in ruins. The Germans destroyed roads and railways as they retreated northward, and though our Air Force harassed them we could do little to interfere on land. E.L.A.S. armed bands filled the gap left by the departing invaders, and their central command made little effort to enforce the solemn promises they had given. Everywhere was want and dissension. Finances were disordered and food exhausted. Our own military resources were stretched to the limit.

At the end of the month Mr. Eden visited Athens on his way home from Moscow, and received a tumultuous welcome in memory of his efforts for Greece in 1941. With him were Lord Moyne, the Minister Resident in Cairo, and Mr. Macmillan. The whole question of relief was discussed and everything humanly possible was done. Our troops

---

[2] E.A.M., the Greek "National Liberation Front."
E.L.A.S., the Greek "People's National Army of Liberation." (Both E.A.M. and E.L.A.S. were Communist-controlled.)
E.D.E.S., the Greek "National Democratic Army."

willingly went on half-rations to increase the food supplies, and British sappers started to build emergency communications. By November 1 the Germans had evacuated Salonika and Florina. Ten days later the last of their forces had crossed the northern frontier, and apart from a few isolated island garrisons, Greece was free.

But the Athens Government had not enough troops to control the country and compel E.L.A.S. to observe the Caserta Agreement. Disorder grew and spread. A revolt by E.A.M. was imminent, and on November 15 General Scobie was directed to make counterpreparations. Athens was to be declared a military area, and authority was given to order all E.L.A.S. troops to leave it. The 4th Indian Division was sent from Italy. So also was the Greek Brigade, which became the centre of controversy between Papandreou and his E.A.M. colleagues. It was evident that the only chance of averting civil war was to disarm the guerrillas and other forces by mutual agreement and establish a new National Army and police force under the direct control of the Government in Athens.

A draft decree for the demobilisation of the guerrillas, drawn up at M. Papandreou's request by the E.A.M. Ministers themselves, was presented to the distracted Cabinet. The regular Greek Mountain Brigade and the "Sacred Squadron" of the Air Force were to remain. E.L.A.S. were to keep a brigade of their own, and E.D.E.S. were to be given a small force. But at the last moment the E.A.M. Ministers went back on their own proposals, on which they had wasted a precious week, and demanded that the Mountain Brigade should be disbanded. The Communist tactic was now in full swing. On December 1 the six Ministers associated with E.A.M. resigned, and a general strike in Athens was proclaimed for the following day. The rest of the Cabinet passed a decree dissolving the guerrillas, and the Communist Party moved its headquarters from the capital. General Scobie issued a message to the people of Greece stating that he stood firm behind the present constitutional Government "until the Greek State can be established with a legally armed force and free elections can be held." I issued a similar personal statement from London.

On Sunday, December 3, Communist supporters, engaging in a banned demonstration, collided with the police and civil war began. The next day General Scobie ordered E.L.A.S. to evacuate Athens and the Piraeus forthwith. Instead their troops and armed civilians tried to seize the capital by force.

*   *   *

At this moment I took a more direct control of the affair. On learning that the Communists had already captured almost all the police stations in Athens, murdering the bulk of their occupants not already pledged to their attack, and were within half a mile of the Government offices, I ordered General Scobie and his 5000 British troops, who ten days before had been received with rapture as deliverers by the population, to intervene and fire upon the treacherous aggressors. It is no use doing things like this by halves. The mob violence by which the Communists sought to conquer the city and present themselves to the world as the Government demanded by the Greek people could only be met by firearms. There was no time for the Cabinet to be called.

Anthony and I were together till about two o'clock, and were entirely agreed that we must open fire. Seeing how tired he was, I said to him, "If you like to go to bed, leave it to me." He did, and at about 3 A.M. I drafted the following telegram to General Scobie:

> . . . You are responsible for maintaining order in Athens and for neutralising or destroying all E.A.M.–E.L.A.S. bands approaching the city. You may make any regulations you like for the strict control of the streets or for the rounding up of any number of truculent persons. Naturally E.L.A.S. will try to put women and children in the van where shooting may occur. You must be clever about this and avoid mistakes. But do not hesitate to fire at any armed male in Athens who assails the British authority or Greek authority with which we are working. It would be well of course if your command were reinforced by the authority of some Greek Government and Papandreou is being told by Leeper to stop and help. *Do not however hesitate to act as if you were in a conquered city where a local rebellion is in progress.*[3]
>
> With regard to E.L.A.S. bands approaching from the outside, you should surely be able with your armour to give some of these a lesson which will make others unlikely to try. You may count upon my support in all reasonable and sensible action taken on this basis. *We have to hold and dominate Athens. It would be a great thing for you to succeed in this without bloodshed if possible, but also with bloodshed if necessary.*

This telegram was dispatched at 4.50 A.M. on the 5th. I must admit that it was somewhat strident in tone. I felt it so necessary to give a strong lead to the military commander that I intentionally worded it in the sharpest terms. The fact that he had such an order in his possession would not only encourage him to decisive action but gave him the certain assurance that I should be with him in any well-conceived action he might take, whatever the consequences might be. I felt grave con-

[3] My subsequent italics throughout.—W.S.C.

cern about the whole business, but I was sure that there should be no room for doubts or hedging. I had in my mind Arthur Balfour's celebrated telegram in the eighties to the British authorities in Ireland: "Don't hesitate to shoot." This was sent through the open telegraph offices. There was a furious storm about it in the House of Commons of those days, but it certainly prevented loss of life. It was one of the key steppingstones by which Balfour advanced to power and control. The setting of the scene was now entirely different. Nevertheless "Don't hesitate to shoot" hung in my mind as a prompter from those far-off days.

Now that the free world has learnt so much more than was then understood about the Communist movement in Greece and elsewhere, many readers will be astonished at the vehement attacks to which His Majesty's Government, and I in particular at its head, were subjected. The vast majority of the American press violently condemned our action, which they declared falsified the cause for which they had gone to war. The State Department, in the charge of Mr. Stettinius, issued a markedly critical pronouncement, which they were to regret, or at least reverse, in after years. In England there was much perturbation. The *Times* and the *Manchester Guardian* pronounced their censures upon what they considered our reactionary policy. Stalin however adhered strictly and faithfully to our agreement of October, and during all the long weeks of fighting the Communists in the streets of Athens not one word of reproach came from *Pravda* or *Isvestia*.

In the House of Commons there was a great stir. There was a strong current of vague opinion, and even passion, and any Government which had rested on a less solid foundation than the National Coalition might well have been shaken to pieces. But the War Cabinet stood like a rock against which all the waves and winds might beat in vain. When we recall what has happened to Poland, to Hungary, and Czechoslovakia in these later years we may be grateful to Fortune for giving us at this critical moment the calm, united strength of determined leaders of all parties. Space does not allow me to quote more than a few extracts from a speech I made on December 8.

> The charge which is made against us . . . is that we are using His Majesty's forces to disarm the friends of democracy in Greece and in other parts of Europe and to suppress those popular movements which have valorously assisted in the defeat of the enemy. . . .
>
> The question however arises, and one may be permitted to dwell on it for a moment, who are the friends of democracy, and also how is the word "democracy" to be interpreted? My idea of it is that the plain, humble, common man, just the ordinary man who keeps a

wife and family, who goes off to fight for his country when it is in trouble, goes to the poll at the appropriate time, and puts his cross on the ballot paper showing the candidate he wishes to be elected to Parliament — that he is the foundation of democracy. And it is also essential to this foundation that this man or woman should do this without fear, and without any form of intimidation or victimisation. He marks his ballot paper in strict secrecy, and then elected representatives meet and together decide what Government, or even, in times of stress, what form of government, they wish to have in their country. If that is democracy I salute it. I espouse it. I would work for it. . . . I stand upon the foundation of free elections based on universal suffrage, and that is what we consider the foundation for democracy. But I feel quite differently about a swindle democracy, a democracy which calls itself democracy because it is Left Wing. It takes all sorts to make democracy, not only Left Wing, or even Communist. I do not allow a party or a body to call themselves democrats because they are stretching farther and farther into the most extreme forms of revolution. I do not accept a party as necessarily representing democracy because it becomes more violent as it becomes less numerous.

One must have some respect for democracy and not use the word too lightly. The last thing which resembles democracy is mob law, with bands of gangsters, armed with deadly weapons, forcing their way into great cities, seizing the police stations and key points of government, endeavouring to introduce a totalitarian regime with an iron hand, and clamoring, as they can nowadays if they get the power — [*Interruption.*]

Democracy is not based on violence or terrorism, but on reason, on fair play, on freedom, on respecting the rights of other people. Democracy is no harlot to be picked up in the street by a man with a tommy gun. I trust the people, the mass of the people, in almost any country, but I like to make sure that it is the people and not a gang of bandits who think that by violence they can overturn constituted authority, in some cases ancient Parliaments, Governments, and States. . . .

Only thirty members faced us in the division lobby. Nearly three hundred voted confidence. Here again was a moment in which the House of Commons showed its enduring strength and authority.

There is no doubt that the emotional expression of American opinion and the train of thought at that time being followed by the State Department affected President Roosevelt and his immediate circle. The sentiments I had expressed in the House of Commons have now become commonplace of American doctrine and policy and command the assent of the United Nations. But in those days they had an air of novelty which was startling to those who were governed by impressions of the past and did not feel the onset of the new adverse tide in human affairs.

*   *   *

Meanwhile British troops were fighting hard in the centre of Athens, hemmed in and outnumbered. We were engaged in house-to-house combat with an enemy at least four-fifths of whom were in plain clothes. Unlike many of the Allied newspaper correspondents in Athens, our troops had no difficulty in understanding the issues involved. Papandreou and his remaining Ministers had lost all authority. Previous proposals to set up a Regency under the Archbishop Damaskinos had been rejected by the King, but on December 10 Mr. Leeper revived the idea. King George was however against it, and we were reluctant at the time to press him.

Amid these tumults Field Marshal Alexander and Mr. Harold Macmillan arrived in Athens. On December 12 the War Cabinet gave Alexander a free hand in all military measures. The 4th British Division, on passage from Italy to Egypt, was diverted, and their arrival during the latter half of the month in due course turned the scale, but in the meantime street fighting swayed to and fro on an enlarging scale. On the 15th Alexander warned me that it was most important to get a settlement quickly, and the best chance was through the Archbishop. "Otherwise," he telegraphed, "I fear if rebel resistance continues at the same intensity as at present I shall have to send further large reinforcements from the Italian front to make sure of clearing the whole of Piraeus–Athens, which is fifty square miles of houses."

A few days later I resolved to go and see for myself.

It was December 24, and we had a family and children's party for Christmas Eve. We had a Christmas tree — one sent from the President — and were all looking forward to a pleasant evening, the brighter perhaps because surrounded by dark shadows. But when I had finished reading my telegrams I felt sure I ought to fly to Athens, see the situation on the spot, and especially make the acquaintance of the Archbishop, around whom so much was turning. I therefore set the telephone working and arranged for an aeroplane to be ready at Northolt that night. I also spoilt Mr. Eden's Christmas by the proposal, which he immediately accepted, that he should come too. After having been much reproached by the family for deserting the party, I motored to meet Eden at Northolt, where the Skymaster which General Arnold had recently sent me waited, attentive and efficient. We slept soundly until about eight o'clock, when we landed at Naples to refuel. Here were several generals, and we all had breakfast together or at adjoining tables. Breakfast is not my best hour of the day, and the news we had both from the Italian front and from Athens was bleak. In an hour we were off again, and in perfect weather flew over the Peloponnese and the Straits of Corinth. Athens and the Piraeus unfolded like a map beneath us on a gigantic scale, and we gazed down upon it wondering who held what.

At about noon we landed at the Kalamaki airfield, which was guarded by about two thousand British airmen, all well armed and active. Here were Field Marshal Alexander, Mr. Leeper, and Mr. Macmillan. They came on board the plane, and we spent nearly three hours in hard discussion of the whole position, military and political. We were, I think, in complete agreement at the end, and about the immediate steps to be taken.

I and my party were to sleep on board the *Ajax*, anchored off the Piraeus, the famous light cruiser of the Plate River battle, which now seemed a long time ago. The road was reported clear, and with an escort of several armoured cars we traversed the few miles without incident. We boarded the *Ajax* before darkness fell, and I realised for the first time that it was Christmas Day. All preparations had been made by the ship's company for a jolly evening, and we certainly disturbed them as little as possible.

The sailors had a plan for a dozen of them to be dressed up in every kind of costume and disguise, as Chinese, Negroes, Red Indians, Cockneys, clowns — all to serenade the officers and warrant officers, and generally inaugurate revels suitable to the occasion. The Archbishop and his attendants arrived — an enormous tall figure in the robes and high hat of a dignitary of the Greek Church. The two parties met. The sailors thought he was part of their show of which they had not been told, and danced around him enthusiastically. The Archbishop thought this motley gang was a premeditated insult, and might well have departed to the shore but for the timely arrival of the captain, who, after some embarrassment, explained matters satisfactorily. Meanwhile I waited, wondering what had happened. But all ended happily.

He spoke with great bitterness against the atrocities of E.L.A.S. and the dark, sinister hand behind E.A.M. Listening to him, it was impossible to doubt that he greatly feared the Communist, or Trotskyite as he called it, combination in Greek affairs. He told us that he had issued an encyclical condemning E.L.A.S. for taking eight thousand hostages, middle-class people, many of them Egyptians, and shooting a few every day, and that he had said that he would report these matters to the Press of the world if the women were not released. Generally he impressed me with a good deal of confidence. He was a magnificent figure, and he immediately accepted the proposal of being chairman of the conference, which was to be held next day, to which E.L.A.S. were invited to send their representatives.

*   *   *

On the morning of the 26th, "Boxing Day," I set out for the Embassy. I remember that three or four shells from the fighting which was going on a mile away on our left raised spouts of water fairly near the *Ajax* as we were about to go ashore. Here an armoured car and military escort awaited us. We rumbled along the road to the Embassy without any trouble. I again met the Archbishop, on whom we were about to stake so much. He agreed to all that was proposed. We planned the procedure at the conference to be held in the afternoon. I was already convinced that he was the outstanding figure in the Greek turmoil. Among other things, I had learned that he had been a champion wrestler before he entered the Orthodox Church.

About six o'clock that evening, the conference opened in the Greek Foreign Office. We took our seats in a large, bleak room after darkness fell. The winter is cold in Athens. There was no heating, and a few hurricane lamps cast a dim light upon the scene. I sat on the Archbishop's right, with Mr. Eden, and Field Marshal Alexander was on his left. Mr. MacVeagh, the American Ambassador, M. Baelen, the French Minister, and the Soviet military representative had all accepted our invitation. The three Communist leaders were late. It was not their fault. There had been prolonged bickering at the outposts. After half an hour we began our work, and I was already speaking when they entered the room. They were presentable figures in British battle dress.

"It is better," I told them, "to let every effort be made to remake Greece as a factor in the victory, and to do it now. We do not intend to obstruct your deliberations. We British, and other representatives of the great united victorious Powers, will leave you Greeks to your own discussions under this most eminent and most venerable citizen, and we shall not trouble you unless you send for us again. . . . My hope is however that the conference which begins here this afternoon in Athens will restore Greece once again to her fame and power among the Allies and the peace-loving peoples of the world, will secure the Greek frontiers from any danger from the north, and will enable every Greek to make the best of himself and the best of his country before the eyes of the whole world. . . . "

Alexander added a sharp touch that Greek troops should be fighting in Italy and not against British troops in Greece.

Once we had broken the ice and got the Greeks, who had done such terrible injuries to each other, to parley round the table under the presidency of the Archbishop, and the formal speeches had been made, the British members of the conference withdrew.

Bitter and animated discussions between the Greek parties occupied all the following day. At five-thirty that evening I had a final discussion with the Archbishop. As the result of his conversations with the E.L.A.S. delegates it was agreed I should ask the King of Greece to make him Regent. He would set about forming a new Government without any Communist members. We undertook to carry on the fighting in full vigour until either E.L.A.S. accepted a truce or the Athens area was clear of them. I told him that we could not undertake any military task beyond Athens and Attica, but that we would try to keep British forces in Greece until the Greek National Army was formed.

On the following morning, December 28, Mr. Eden and I left by air. I had no chance to say good-bye to M. Papandreou before leaving. He was about to resign, and was a serious loser by the whole business. I asked our Ambassador to keep in friendly touch with him. On December 29 we arrived back in London. Mr. Eden and I sat up with the King of Greece till four-thirty in the morning, at the end of which time His Majesty agreed not to return to Greece unless summoned by a free and fair expression of the national will, and to appoint the Archbishop as Regent during the emergency. I sent the royal announcement at once to Mr. Leeper and the Archbishop replied to the King accepting his mandate as Regent. There was a new and living Greek Government. On January 3 General Plastiras, a vehement Republican, who was the leader of the Army revolt against King Constantine in 1922, became Prime Minister.

Continuous fighting in Athens during December at last drove the insurgents from the capital, and by mid-January British troops controlled all Attica. The Communists could do nothing against our men in open country, and a truce was signed on January 11.

Thus ended the six weeks' struggle for Athens, and, as it ultimately proved, for the freedom of Greece from Communist subjugation. When three million men were fighting on either side on the Western Front and vast American forces were deployed against Japan in the Pacific the spasms of Greece may seem petty, but nevertheless they stood at the nerve centre of power, law, and freedom in the Western world. It is odd, looking back on these events, now that some years have passed, to see how completely the policy for which I and my colleagues fought so stubbornly has been justified by events. Myself, I never had any doubts about it, for I saw quite plainly that Communism would be the peril civilisation would have to face after the defeat of Nazism and Fascism. It did not fall to us to end the task in Greece. I little thought however at the end of 1944 that the State Department, supported by overwhelming American opinion, would in little more than two years not only adopt and carry on the course we had opened, but would make vehement and costly exertions, even of a military character, to bring it to fruition. If Greece has escaped the fate of Czechoslovakia and survives today as one of the free nations, it is due not only to British action in 1944, but to the steadfast efforts of what was presently to become the united strength of the English-speaking world.

22

Malta and Yalta: Plans for
World Peace

A T T H E E N D O F J A N U A R Y 1945 Hitler's armies were virtually compressed within their own territory, save for a brittle
hold in Hungary and in Northern Italy, but the political situation, at
any rate in Eastern Europe, was by no means so satisfactory. A precarious tranquillity had indeed been achieved in Greece, and it seemed
that a free democratic Government, founded on universal suffrage and
secret ballot, might be established there within a reasonable time. But
Rumania and Bulgaria had passed into the grip of Soviet military occupation, Hungary and Yugoslavia lay in the shadow of the battlefield, and Poland, though liberated from the Germans, had merely
exchanged one conqueror for another. The informal and temporary
arrangement which I had made with Stalin during my October visit
to Moscow could not, and so far as I was concerned was never intended
to, govern or affect the future of these wide regions once Germany
was defeated.

The whole shape and structure of postwar Europe clamoured for
review. When the Nazis were beaten, how was Germany to be treated?
What aid could we expect from the Soviet Union in the final overthrow of Japan? And once our military aims were achieved what
measures and what organisation could the three great Allies provide
for the future peace and good governance of the world? The discussions
at Dumbarton Oaks had ended in partial disagreement. So, in a smaller
but no less vital sphere, had the negotiations between the Soviet-
sponsored "Lublin Poles" and their compatriots from London which
Mr. Eden and I had with much difficulty promoted during our visit
to the Kremlin in October 1944. An arid correspondence between the
President and Stalin, of which Mr. Roosevelt had kept me informed,
had accompanied the secession of M. Mikolajczyk from his colleagues

in London, while on January 5, contrary to the wishes of both the United States and Great Britain, the Soviets had recognised the Lublin Committee as the Provisional Government of Poland.

The President was fully convinced of the need for another meeting of "the Three," and after some urging on my part he also agreed that we should have a preliminary conference of our own at Malta. The reader will remember the anxieties which I had expressed about our operations in Northwest Europe in my telegram to the President of December 6.[1] These still weighed with me. The British and American Chiefs of Staff had great need for discussion before we met the Soviets, and on January 29, 1945, I accordingly left Northolt in the Skymaster given to me by General Arnold. My daughter Sarah and the official party, together with Mr. Martin and Mr. Rowan, my private secretaries, and Commander Thompson, travelled with me. The rest of my personal staff and some departmental officials travelled in two other planes. We arrived at Malta just before dawn on January 30, and there I learnt that one of these two aircraft had crashed near Pantelleria. Only three of the crew and two passengers survived.

On the morning of February 2 the Presidential party, on board the U.S.S. *Quincy*, steamed into Valletta harbour. It was a warm day, and under a cloudless sky I watched the scene from the deck of H.M.S. *Orion*. As the American cruiser steamed slowly past us towards her berth alongside the quay wall I could see the figure of the President seated on the bridge, and we waved to each other. With the escort of Spitfires overhead, the salutes, and the bands of the ships' companies in the harbour playing "The Star-spangled Banner," it was a splendid scene. I lunched on board the *Quincy*, and at six o'clock that evening we had our first formal meeting in the President's cabin. Here we reviewed the report of the Combined Chiefs of Staff and the military discussions which had been taking place in Malta during the previous three days. Our Staffs had done a remarkable piece of work. Their discussions had centred principally round Eisenhower's plans for carrying his forces up to and across the Rhine. There were differences of opinion on the subject, which are related in another chapter.[2] The opportunity was of course taken to review the whole span of the war, including the war against the U-boats, the future campaigns in Southeast Asia and the Pacific, and the Mediterranean situation. We reluctantly agreed to withdraw two divisions from Greece as soon as they could be spared, but I made it clear that we should not be obliged to do this until the Greek Government had built up its own military forces. Three divisions were also to be withdrawn from Italy

[1] See p. 894.
[2] Chapter 24, "Crossing the Rhine."

to reinforce Northwest Europe, but I stressed that it would be unwise to make any significant withdrawal of amphibious forces. It was very important to follow up any German surrender in Italy, and I told the President that we ought to occupy as much of Austria as possible, as it was *"undesirable that more of Western Europe than necessary should be occupied by the Russians."* [3] In all the military matters a large measure of agreement was reached, and the discussions had the useful result that the Combined Chiefs of Staff were aware of their respective points of view before engaging in talks with their Russian counterparts.

That night the exodus began. Transport planes took off at ten-minute intervals to carry some seven hundred persons, forming the British and American delegations, over fourteen hundred miles to the airfield of Saki, in the Crimea. I boarded my plane after dinner, and went to bed. After a long and cold flight we landed on the airfield, which was under deep snow. My plane was ahead of Mr. Roosevelt's, and we stood for a while awaiting him. When he was carried down the lift from the "Sacred Cow" he looked frail and ill. Together we inspected the guards of honour, the President sitting in an open car, while I walked beside him.

Presently we set off on a long drive from Saki to Yalta. Lord Moran and Mr. Martin came with me in my car. The journey took us nearly eight hours, and the road was often lined by Russian soldiers, some of them women, standing shoulder to shoulder in the village streets and on the main bridges and mountain passes, and at other points in separate detachments. As we crossed the mountains and descended towards the Black Sea we suddenly passed into warm and brilliant sunshine and a most genial climate.

*   *   *

The Soviet headquarters at Yalta were in the Yusupov Palace, and from this centre Stalin and Molotov and their generals carried on the government of Russia and the control of their immense front, now in violent action. President Roosevelt was given the more splendid Livadia Palace, close at hand, and it was here, in order to spare him physical inconvenience, that all our plenary meetings were held. This exhausted the undamaged accommodation. I and the principal members of the British delegation were assigned a very large villa about five miles away which had been built in the early nineteenth century by an English architect for a Russian Prince Vorontzov, one-time Imperial Ambassador to the Court of St. James. The rest of our delegation were put up in two rest houses about twenty minutes away, five or six people sleeping in a room, including high-ranking officers, but

---

[3] My subsequent italics.—W.S.C.

no one seemed to mind. The Germans had evacuated the neighbourhood only ten months earlier, and the surrounding buildings had been badly damaged. We were warned that the area had not been completely cleared of mines, except for the grounds of the villa, which were, as usual, heavily patrolled by Russian guards. Over a thousand men had been at work on the scene before our arrival. Windows and doors had been repaired, and furniture and stores brought down from Moscow.

The setting of our abode was impressive. Behind the villa, half Gothic and half Moorish in style, rose the mountains, covered in snow, culminating in the highest peak in the Crimea. Before us lay the dark expanse of the Black Sea, severe, but still agreeable and warm even at this time of the year. Carved white lions guarded the entrance to the house, and beyond the courtyard lay a fine park with subtropical plants and cypresses. In the dining room I recognised the two paintings hanging each side of the fireplace as copies of family portraits of the Herberts at Wilton. It appeared that Prince Vorontzov had married a daughter of the family, and had brought these pictures back with him from England. Every effort was made by our hosts to ensure our comfort, and every chance remark was noted with kindly attention. On one occasion Portal had admired a large glass tank with plants growing in it, and remarked that it contained no fish. Two days later a consignment of goldfish arrived. Another time somebody said casually that there was no lemon peel in the cocktails. The next day a lemon tree loaded with fruit was growing in the hall. All must have come by air from far away.

*   *   *

The first plenary meeting of the Conference started at a quarter past four on the afternoon of February 5. The discussion opened on the future of Germany. I had of course pondered this problem, and had thus addressed Mr. Eden a month before:

> Treatment of Germany after the war. It is much too soon for us to decide these enormous questions. Obviously, when the German organised resistance has ceased the first stage will be one of severe military control. This may well last for many months, or perhaps for a year or two, if the German underground movement is active. . . . I have been struck at every point where I have sounded opinion at the depth of the feeling that would be aroused by a policy of "putting poor Germany on her legs again." I am also well aware of the arguments about "not having a poisoned community in the heart of Europe." I do suggest that, with all the

work we have on our hands at the present moment, we should not anticipate these very grievous discussions and schisms, as they may become. We have a new Parliament to consider, whose opinions we cannot foretell.

I shall myself prefer to concentrate upon the practical issues which will occupy the next two or three years, rather than argue about the long-term relationship of Germany to Europe. . . . It is a mistake to try to write out on little pieces of paper what the vast emotions of an outraged and quivering world will be either immediately after the struggle is over or when the inevitable cold fit follows the hot. These awe-inspiring tides of feeling dominate most people's minds, and independent figures tend to become not only lonely but futile. Guidance in these mundane matters is granted to us only step by step, or at the utmost a step or two ahead. There is therefore wisdom in reserving one's decisions as long as possible and until all the facts and forces that will be potent at the moment are revealed.

So when Stalin now asked how Germany was to be dismembered, I said it was much too complicated to be settled in five or six days. It would require a very searching examination of the historical, ethnographical, and economic facts, and prolonged review by a special committee, which would go into the different proposals and advise on them. There was so much to consider. What to do with Prussia? What territory should be given to Poland and the U.S.S.R.? Who was to control the Rhine valley and the great industrial zones of the Ruhr and the Saar? A body should be set up at once to examine these matters, and we ought to have its report before reaching any final decision. Mr. Roosevelt suggested asking our Foreign Secretaries to produce a plan for studying the question within twenty-four hours and a definite plan for dismemberment within a month. Here, for a time, the matter was left.

We then arranged to meet next day and consider two topics which were to dominate our future discussions, namely, the Dumbarton Oaks scheme for world security and Poland.

*   *   *   *

As has been recorded in an earlier chapter, the conference at Dumbarton Oaks had ended without reaching complete agreement about the all-important question of voting rights in the Security Council, and space now forbids more than a reference to some of the salient points in our discussions. Stalin said he feared that, though the three Great Powers were allies today, and would none of them commit any act of aggression, in ten years or less the three leaders would disappear and a new generation would come into power which had not experienced the war and would forget what we had gone through. "All of us," he declared, "want to secure peace for at least fifty years. The greatest danger is conflict among ourselves, because if we remain united the German menace is not very important. Therefore we must now think how to secure our unity in the future, and how to guarantee that the three Great Powers (and possibly China and France) will maintain a united front. Some system must be elaborated to prevent conflict between the main Great Powers." The Russians were accused of talking too much about voting. It was true they thought it was very important, because everything would be decided by vote and they would be greatly interested in the results. Suppose, for instance, that China as a permanent member of the Security Council demanded the return of Hong Kong, or Egypt demanded the return of the Suez Canal, he assumed they would not be alone and would have friends and perhaps protectors in the Assembly or in the Council, and he feared that such disputes might break the unity of the three Great Powers.

"My colleagues in Moscow cannot forget what happened in December 1939, during the Russo-Finnish War, when the British and the French used the League of Nations against us and succeeded in isolating and expelling the Soviet Union from the League, and when they later mobilised against us and talked of a crusade against Russia. Cannot we have some guarantees that this sort of thing will not happen again?"

After much striving and explanation, we persuaded him to accept an American scheme whereby the Security Council would be virtually powerless unless the "Big Four" were unanimous. If the United States, the U.S.S.R., Great Britain, or China disagreed on any major topic, then any one of them could refuse their assent and stop the Council doing anything. Here was the Veto. Posterity may judge the results.

I myself have always held the view that the foundation of a World Instrument should be sought on a regional basis. Most of the principal regions suggest themselves — the United States, United Europe, the British Commonwealth and Empire, the Soviet Union, South America. Others are more difficult at present to define — like the Asian group or groups, or the African group — but could be developed with study. But the object would be to have many issues of fierce local controversy thrashed out in the Regional Council, which would then send three or four representatives to the Supreme Body, choosing men of the greatest eminence. This would make a Supreme Group of thirty or forty world statesmen, each responsible not only for representing their own region but for dealing with world causes, and primarily the prevention of war. What we have now is not effective for that outstanding purpose. The summoning of all nations, great and small, powerful or powerless, on even terms to the central body may be compared with the organisation of an army without any division between the High Command and the divisional and brigade commanders. All are invited to the headquarters. Babel, tempered by skilful lobbying, is all that has resulted up to the present. But we must persevere.

23

Russia and Poland: The Soviet Promise

P OLAND WAS DISCUSSED at no fewer than seven out of the eight plenary meetings of the Yalta Conference, and the British record contains an interchange on this topic of nearly eighteen thousand words between Stalin, Roosevelt, and myself. Aided by our Foreign Ministers and their subordinates, who also held tense and detailed debate at separate meetings among themselves, we finally produced a declaration [1] which represented both a promise to the world and agreement between ourselves on our future actions. The painful tale is still unfinished and the true facts are as yet imperfectly known, but what is here set down may perhaps contribute to a just appreciation of our efforts at the last but one of the wartime Conferences. The difficulties and the problems were ancient, multitudinous, and imperative. The Soviet-sponsored Lublin Government of Poland, or the "Warsaw" Government as the Russians of all names preferred to call it, viewed the London Polish Government with bitter animosity. Feeling between them had got worse, not better, since our October meeting in Moscow. Soviet troops were flooding across Poland, and the Polish Underground Army was freely charged with the murder of Russian soldiers and with sabotage and attacks on their rear areas and their lines of communication. Both access and information were denied to the Western Powers. In Italy and on the Western Front over 150,000 Poles were fighting valiantly for the final destruction of the Nazi armies. They and many others elsewhere in Europe were eagerly looking forward to the liberation of their country and a return to their homeland from voluntary and honourable exile. The large community of Poles in the United States anxiously awaited a settlement between the three Great Powers.

[1] The complete text of this declaration and a full account of the discussions at Yalta may be studied in Sir Winston's volume entitled *Triumph and Tragedy*, Book Two, Chapter 3.

The questions which we discussed may be summarised as follows:
How to form a single Provisional Government for Poland.
How and when to hold free elections.
How to settle the Polish frontiers, both in the east and the west.
How to safeguard the rear areas and lines of communication of the advancing Soviet armies.

Poland had indeed been the most urgent reason for the Yalta Conference, and was to prove the first of the great causes which led to the breakdown of the Grand Alliance. I myself was sure that a strong, free, and independent Poland was much more important than particular territorial boundaries. I wanted the Poles to be able to live freely and live their own lives in their own way. It was for this that we had gone to war against Germany in 1939. It had nearly cost us our life, not only as an Empire but as a nation, and when we met on February 6, 1945, I posed the question as follows: Could we not create a Government or governmental instrument for Poland, pending full and free elections. which could be recognised by all? Such a Government could prepare for a free vote of the Polish people on their future constitution and administration. If this could be done we should have taken one great step forward towards the future peace and prosperity of Central Europe.

In the debate which followed, Stalin claimed to understand our attitude. For the British, he said, Poland was a question of honour, but for the Russians it was a question both of honour and security; of honour because they had had many conflicts with the Poles and they wished to eliminate the causes of such conflicts; of security, because Poland was on the frontiers of Russia, and throughout history Poland had been a corridor through which Russia's enemies had passed to attack her. The Germans had done this twice during the last thirty years and they had been able to do it because Poland was weak. Russia wanted her to be strong and powerful so that she could shut this corridor of her own strength. Russia could not keep it shut from the outside. It could only be shut from the inside by Poland herself. This was a matter of life and death for the Soviet State.

As for her frontiers, Stalin went on to say that the President had suggested some modification of the Curzon Line and that Lvov and perhaps certain other districts should be given to Poland, and I had said that this would be a gesture of magnanimity. But he pointed out that the Curzon Line had not been invented by the Russians. It had been drawn up by Curzon and Clemenceau and representatives of the United States at the conference in 1918, to which Russia had not been invited. The Curzon Line had been accepted against the will of Russia on the basis of ethnographical data. Lenin had not agreed with it. The Russians had already retired from Lenin's position, and

now some people wanted Russia to take less than Curzon and Clemenceau had conceded. That would be shameful. When the Ukrainians
came to Moscow they would say that Stalin and Molotov were less
trustworthy defenders of Russia than Curzon or Clemenceau. It was
better that the war should continue a little longer, although it would
cost Russia much blood, so that Poland could be compensated at
Germany's expense. When Mikolajczyk had been in Russia during
October he had asked what frontier for Poland Russia would recognise in the west, and he had been delighted to hear that Russia thought
that the western frontier of Poland should be extended to the Neisse.
There were two rivers of that name, said Stalin, one near Breslau, and
another farther west. It was the Western Neisse he had in mind.

When we met again on February 7 I reminded my hearers that I
had always qualified the moving of the Polish frontier westward by
saying that the Poles should be free to take territory in the west, but
not more than they wished or could properly manage. It would be
a great pity to stuff the Polish goose so full of German food that it
died of indigestion. A large body of opinion in Great Britain was
shocked at the idea of moving millions of people by force. Great success had been achieved in disentangling the Greek and Turkish populations after the last war, and the two countries had enjoyed good relations ever since; but in that case under a couple of millions of people
had been moved. If Poland took East Prussia and Silesia as far as
the Oder that alone would mean moving six million Germans back
to Germany. It might be managed, subject to the moral question,
which I would have to settle with my own people.

Stalin said there were no Germans in these areas, as they had all
run away. I replied that the question was whether there was room for
them in what was left of Germany. Six or seven million Germans had
been killed and another million (Stalin suggested two millions) would
probably be killed before the end of the war. There should therefore be
room for these migrant people up to a certain point. They would be
needed to fill the vacancies. I was not afraid of the problem of transferring populations, so long as it was proportionate to what the Poles
could manage and to what could be put into Germany. But it was a
matter which required study, not as a question of principle, but of
the numbers which would have to be handled.

In these general discussions maps were not used, and the distinction between the Eastern and Western Neisse did not emerge as clearly
as it should have done. This was however soon to be made clear.[2]

*   *   *

[2] See map on page 989.

On the 8th Mr. Roosevelt agreed that the eastern boundary of Poland should be the Curzon Line, with modifications in favour of Poland in some areas of from five to eight kilometres. But he was firm and precise about the frontier in the west. Poland should certainly receive compensation at the expense of Germany, "but," he continued, *"there would appear to be little justification for extending it up to the Western Neisse."* This had always been my view, and I was to press it very hard when we met again at Potsdam five months later.

Thus at Yalta we were all united in principle about the western frontier, and the only question was where exactly the line should be drawn and how much we should say about it. The Poles should have part of East Prussia and be free to go up to the line of the Oder if they wished, but we were very doubtful about going any farther or saying anything on the question at this stage, and three days later I told the Conference that we had had a telegram from the War Cabinet which strongly deprecated any reference to a frontier as far west as the Western Neisse because the problem of moving the population was too big to manage.

We accordingly decided to insert the following in our declaration:

> The three heads of Governments consider that the eastern frontier of Poland should follow the Curzon Line, with digressions from it in some regions of five to eight kilometres in favour of Poland. They recognise that Poland must receive substantial accessions of territory in the north and west. They feel that the opinion of the new Polish Provisional Government of National Unity should be sought in due course on the extent of these accessions, and that the final delimitation of the western frontier of Poland should thereafter await the Peace Conference.

*   *   *

There remained the question of forming a Polish Government which we could all recognise and which the Polish nation would accept. Stalin began by pointing out that we could not create a Polish Government unless the Poles themselves agreed to it. Mikolajczyk and Grabski had come to Moscow during my visit there. They had met the Lublin Government, a measure of agreement had been reached, and Mikolajczyk had gone to London on the understanding that he would come back. Instead, his colleagues had turned him out of office simply because he favoured an agreement with the Lublin Government. The Polish Government in London were hostile to the very idea of the Lublin Government, and described it as a company of bandits and criminals. The Lublin Government had paid them back in their own coin, and it was now very difficult to do anything about it. "Talk to the Lublin Government if you like," he said in effect. "I will get them to meet you here or in Moscow, but they are just as democratic as de Gaulle, and they can keep the peace in Poland and stop civil war and attacks on the Red Army." The London Government could not do this. Their agents had killed Russians soldiers and had raided supply dumps to get arms. Their radio stations were operating without permission and without being registered. The agents of the Lublin Government had been helpful, and the agents of the London Government had done much evil. It was vital for the Red Army to have safe rear areas, and as a military man he would only support the Government which could guarantee to provide them.

It was now late in the evening and the President suggested adjourning till next day, but I thought it right to state that according to our information not more than one-third of the Polish people would support the Lublin Government if they were free to express their opinion. I assured Stalin that we had greatly feared a collision between the Polish Underground Army and the Lublin Government, which might lead to bitterness, bloodshed, arrests, and deportations, and that was why we had been so anxious for a joint arrangement. Attacks on the Red Army must of course be punished, but on the facts at my disposal I could not feel that the Lublin Government had a right to say that they represented the Polish nation.

The President was now anxious to end the discussion. "Poland," he remarked, "has been a source of trouble for over five hundred years." "All the more," I answered, "must we do what we can to put an end to these troubles." We then adjourned.

That night the President wrote a letter to Stalin, after consultation with and amendment by us, urging that two members of the Lublin Government and two from London or from within Poland should come to the Conference and try to agree in our presence about forming a Provisional Government which we could all recognise to hold free elections as soon as possible. But this was apparently impracticable. Molotov acclaimed the virtues of the Lublin-Warsaw Government, deplored the failings of the men from London, and said that if we tried to create a new Government the Poles themselves might never agree, so it was better to try to "enlarge" the existing one. It would only be a temporary institution, because our sole object was to hold free elections in Poland as soon as possible. How to enlarge it could best be discussed in Moscow between the American and British Ambassadors and himself. He greatly desired an agreement, and he accepted the President's proposals to invite two "non-Lublin" Poles. There was always the possibility that the Lublin Government would refuse to talk with some of them, like Mikolajczyk, but if they sent three representatives and two came from those suggested by Mr. Roosevelt conversations could start at once.

"This," I said, "is the crucial point of the Conference. The whole world is waiting for a settlement, and if we separate still recognising different Polish Governments the whole world will see that fundamental differences between us still exist. The consequences will be most lamentable, and will stamp our meeting with the seal of failure. If we brush aside the existing London Government and lend all our weight to the Lublin Government there will be a world outcry. The Poles outside of Poland will make a virtually united protest. There is under our command a Polish army of 150,000 men, who have been gathered from all who have been able to come together from outside their country. It has fought, and is still fighting, very bravely. I do not believe it will be at all reconciled to the Lublin Government, and if Great Britain transfers recognition from the Government which it has recognised since the beginning of the war they will look on it as a betrayal.

"As Marshal Stalin and M. Molotov well know," I proceeded, "I myself do not agree with the London Government's action, which has been foolish at every stage. But the formal act of transferring recognition from those whom we have hitherto recognised to this new Government would cause the gravest criticism. It would be said that His Majesty's Government have given way completely on the eastern frontier (as in fact we have) and have accepted and championed the Soviet view. It would also be said that we have broken altogether with the lawful Government of Poland, which we have recognised for these five years of war, and that we have no knowledge of what is actually going on in Poland. We cannot enter the country. We cannot see and hear what opinion is. It would be said we can only accept what the Lublin Government proclaims about the opinion of the Polish people, and we would be charged in Parliament with having altogether forsaken the cause of Poland. The debates which would follow would be most painful and embarrassing to the unity of the Allies, even supposing that we were able to agree to the proposals of my friend M. Molotov.

"I do not think," I continued, "that these proposals go nearly far enough. If we give up the Polish Government in London a new start should be made from both sides on more or less equal terms. Before His Majesty's Government ceased to recognise the London Government and transferred their recognition to another Government they would have to be satisfied that the new Government truly represented the Polish nation. I agree that this is only one point of view, as we do not fully know the facts, and all our differences will of course be removed if a free and unfettered General Election is held in Poland by ballot and with universal suffrage and free candidatures. Once this is done His Majesty's Government will salute the Government that emerges without regard to the Polish Government in London. It is the interval before the election that is causing us so much anxiety."

Molotov said that perhaps the talks in Moscow would have some useful result. The Poles would have to have their say, and it was very difficult to deal with the question without them. I agreed, but said that it was so important that the Conference should separate on a note of agreement that we must all struggle patiently to achieve it.

Stalin then took up my complaint that I had no information and no way of getting it.

"I have a certain amount," I replied.

"It doesn't agree with mine," he answered, and proceeded to make a speech, in which he assured us that the Lublin Government was really very popular, particularly Bierut and others. They had not left the country during the German occupation, but had lived all the time in Warsaw and came from the Underground movement. He did not believe they were geniuses. The London Government might well contain cleverer people, but they were not liked in Poland because they had not been seen there when the population was suffering under the Hitlerite occupation. The populace saw on the streets the members of the Provisional Government, but asked where were the London Poles. This undermined the prestige of the London Government, and was the reason why the Provisional Government, though not great men, enjoyed great popularity.

All this, he said, could not be ignored if we wanted to understand the feelings of the Polish people. I had feared the Conference would separate before we reached agreement. What then was to be done? The various Governments had different information, and drew different conclusions from it. Perhaps the first thing was to call together the Poles from the different camps and hear what they had to say. The day was near when elections could be held. Until then we must deal with the Provisional Government, as we had dealt with General de Gaulle's Government in France, which also was not elected. He did not know whether Bierut or General de Gaulle enjoyed greater authority, but it had been possible to make a treaty with General de Gaulle, so why couldn't we do the same with an enlarged Polish Government, which would be no less democratic? If we approached the matter without prejudice we should be able to find a common ground. The situation was not as tragic as I thought, and the question could be settled if too much importance was not attached to secondary matters and if we concentrated on essentials.

"How soon," asked the President, "will it be possible to hold elections?"

"Within a month," Stalin replied, "unless there is some catastrophe on the front, which is improbable."

I agreed that this would of course set our minds at rest, and we could wholeheartedly support a freely elected Government which would supersede everything else, but we must not ask for anything which would in any way hamper the military operations. These were the supreme end. If however the will of the Polish people could be ascertained in so short a time, or even within two months, the situation would be entirely different and no one could oppose it.

*   *   *

When we reassembled at four o'clock in the afternoon of February 9 Molotov produced a new formula, namely, that the Lublin Government should be *"reorganised* [as opposed to 'enlarged'] on a wider democratic basis, with the inclusion of democratic leaders from Poland itself, and also from those living abroad." He and the British and American Ambassadors should consult together in Moscow about how this would be done. Once "reorganised," the Lublin Government would be pledged to hold free elections as soon as possible, and we should then recognise whatever Government emerged.

This was a considerable advance, and I said so, but I felt it my duty to sound a general warning. This would be the last but one of our meetings.[3] There was an atmosphere of agreement, but there was also a desire to put foot in the stirrup and be off. We could not, I declared, afford to allow the settlement of these important matters to be hurried and the fruits of the Conference lost for lack of another twenty-four hours. A great prize was in view and decisions must be unhurried. These might well be among the most important days in our lives.

Mr. Roosevelt declared that the differences between us and the Russians were now largely a matter of words, but both he and I were anxious that the elections should really be fair and free. I told Stalin that we were at a great disadvantage, because we knew so little of what was going on inside Poland and yet had to take great decisions of responsibility. I knew, for instance, that there was bitter feeling among the Poles, and I had been told that the Lublin Government had openly said it would try all members of the Polish Home Army and Underground movement as traitors. Of course, I put the security of the

---

[3] Our meeting on February 11 merely approved the report on the Conference. Serious discussion ended on February 10.

Red Army first, but I begged Stalin to consider our difficulty. The British Government did not know what was going on inside Poland, except through dropping brave men by parachute and bringing members of the Underground movement out. We had no other means of knowing, and did not like getting our information in this way. Could this be remedied without hampering the movements of the Soviet troops? Could any facilities be granted to the British (and no doubt to the United States) for seeing how these Polish quarrels were being settled? Tito had said that when elections took place in Yugoslavia he would not object to Russian, British, and American observers being present to report impartially to the world that they had been carried out fairly. So far as Greece was concerned, His Majesty's Government would greatly welcome American, Russian, and British observers to make sure the elections were conducted as the people wished. The same applied to Italy — Russian, American, and British observers should be present to assure the world that everything had been done in a fair way. It was impossible, I said, to exaggerate the importance of carrying out elections fairly. For instance, would Mikolajczyk be able to go back to Poland and organise his party for the elections?

"That will have to be considered by the Ambassadors and M. Molotov when they meet the Poles," said Stalin.

I replied, "I must be able to tell the House of Commons that the elections will be free and that there will be effective guarantees that they are freely and fairly carried out."

Stalin pointed out that Mikolajczyk belonged to the Peasant Party, which, as it was not a Fascist party, could take part in the elections and put up candidates. I said that this would be still more certain if the Peasant Party were already represented in the Polish Government, and Stalin agreed that one of their representatives should be included. I added that I hoped that nothing I had said had given offence, since nothing had been further from my heart.

"We shall have to hear," he answered, "what the Poles have to say." I explained that I wanted to be able to carry the eastern frontier question through Parliament, and I thought this might be done if Parliament was satisfied that the Poles had been able to decide for themselves what they wanted.

"There are some very good people among them," he replied. "They are good fighters, and they have had some good scientists and musicians, but they are very quarrelsome."

"All I want," I answered, "is for all sides to get a fair hearing."

"The elections," said the President, "must be above criticism, like Caesar's wife. I want some kind of assurance to give to the world, and I don't want anybody to be able to question their purity. It is a matter of good politics rather than principle."

Mr. Stettinius suggested having a written pledge that the three Ambassadors in Warsaw should observe and report that the elections were really free and unfettered. "I am afraid," said Molotov, "that if we do this the Poles will feel they are not trusted. We had better discuss it with them."

I was not content with this, and resolved to raise it with Stalin later on. The opportunity presented itself next day, when Mr. Eden and I had a private conversation with him and Molotov at the Yusupov Villa. I once more explained how difficult it was for us to have no representatives in Poland who could report what was going on. The alternatives were either an Ambassador with an embassy staff or newspaper correspondents. The latter was less desirable, but I pointed out that I should be asked in Parliament about the Lublin Government and the elections and I must be able to say that I knew what was happening.

"After the new Polish Government is recognised it would be open to you to send an Ambassador to Warsaw," Stalin answered.

"Would he be free to move about the country?"

"As far as the Red Army is concerned, there will be no interference with his movements, and I promise to give the necessary instructions, but you will have to make your own arrangements with the Polish Government."

We then agreed to add the following to our declaration:

> As a consequence of the above, recognition would entail an exchange of Ambassadors, by whose reports the respective Governments would be informed about the situation in Poland.

This was the best I could get.

Sunday, February 11, was the last day of our Crimean visit. As usual at these meetings many grave issues were left unsettled. The Polish declaration laid down in general terms a policy which if carried out with loyalty and good faith might indeed have served its purpose pending the general Peace Treaty. The President was anxious to go home, and on his way to pay a visit to Egypt, where he could discuss the affairs of the Middle East with various potentates. Stalin and I lunched with him in the Czar's former billiard room at the Livadia Palace. During the meal we signed the final documents and official communiqués. All now depended upon the spirit in which they were carried out.

*   *   *

I had much looked forward to the sea voyage through the Dardanelles to Malta, but I felt it my duty to make a lightning trip to Athens and survey the Greek scene after the recent troubles. Early on February 14 accordingly we set off by car for Saki, where our aeroplane awaited us. We flew without incident to Athens, making a loop over the island of Skyros to pass over the tomb of Rupert Brooke, and were received at the airfield by the British Ambassador, Mr. Leeper, and General Scobie. Only seven weeks before, I had left the Greek capital rent by street fighting. We now drove into it in an open car, where only a thin line of kilted Greek soldiers held back a vast mob, screaming with enthusiasm, in the very streets where hundreds of men had died in the Christmas days when I had last seen the city. That evening a huge crowd of about fifty thousand people gathered in Constitution Square. The evening light was wonderful as it fell on these classic scenes. I had no time to prepare a speech. Our security services had thought it important that we should arrive with hardly any notice. I addressed them with a short harangue. That evening I dined at our shot-scarred Embassy, and in the early hours of February 15 we took off in my plane for Egypt.

Late that morning the American cruiser *Quincy* steamed into Alexandria harbour, and shortly before noon I went on board for what was to be my last talk with the President. We gathered afterwards in his cabin for an informal family luncheon. I was accompanied by Sarah and Randolph, and Mr. Roosevelt's daughter, Mrs. Boettiger, joined us, together with Harry Hopkins and Mr. Winant. The President seemed placid and frail. I felt that he had a slender contact with life. I was not to see him again. We bade affectionate farewells. That afternoon the Presidential party sailed for home. On February 19 I flew back to England. Northolt was fogbound, and our plane was diverted to Lyneham. I drove on to London by car, stopping at Reading to join my wife, who had come to meet me.

At noon on February 27 I asked the House of Commons to approve the results of the Crimea Conference. The general reaction was unqualified support for the attitude we had taken. There was however intense moral feeling about our obligations to the Poles, who had suffered so much at German hands and on whose behalf as a last resort we had gone to war. A group of about thirty Members felt so strongly on this matter that some of them spoke in opposition to the motion which I had moved. There was a sense of anguish lest we should have to face the enslavement of a heroic nation. Mr. Eden supported me. In the division on the second day we had an overwhelming majority, but twenty-five Members, most of them Conservatives, voted against the Government, and in addition eleven members of the Government abstained.

It is not permitted to those charged with dealing with events in times of war or crisis to confine themselves purely to the statement of broad general principles on which good people agree. They have to take definite decisions from day to day. They have to adopt postures which must be solidly maintained, otherwise how can any combinations for action be maintained? It is easy, after the Germans are beaten, to condemn those who did their best to hearten the Russian military effort and to keep in harmonious contact with our great Ally, who had suffered so frightfully. What would have happened if we had quarrelled with Russia while the Germans still had two or three hundred divisions on the fighting front? Our hopeful assumptions were soon to be falsified. Still, they were the only ones possible at the time.

24

Crossing the Rhine

DESPITE THEIR DEFEAT in the Ardennes,[1] the Germans decided to give battle west of the Rhine, instead of withdrawing across it to gain a breathing space, and throughout February and most of March, Field Marshal Montgomery had conducted a long and arduous struggle in the north. The defences were strong and obstinately held, the ground was sodden, and both the Rhine and the Meuse had overflowed their banks. The Germans smashed open the valves on the great dams on the Roer and the river became uncrossable until the end of February, but on March 10 eighteen German divisions were all back across the Rhine. Farther south, General Bradley cleared the whole of the eighty-mile stretch between Düsseldorf and Koblenz in a brief, swift campaign. On the 7th a stroke of fortune was boldly accepted. The 9th Armoured Division of the First U.S. Army found the railway bridge at Remagen partly destroyed but still usable. They promptly threw their advance guard across, other troops quickly followed, and soon over four divisions were on the far bank and a bridgehead several miles deep established. This was no part of Eisenhower's plan, but it proved an excellent adjunct, and the Germans had to divert considerable forces from farther north to hold the Americans in check. Patton cut off and crushed the last enemy salient around Trier. The defenders of the renowned and dreaded Siegfried Line were surrounded, and in a few days all organised resistance came to an end. As a by-product of victory the 5th U.S. Division made an unpremeditated crossing of the Rhine fifteen miles south of Mainz, which soon expanded into a deep bridgehead pointing towards Frankfurt.

Thus ended the last great German stand in the West. Six weeks of successive battles along a front of over two hundred and fifty miles had

[1] See Chapter 20.

driven the enemy across the Rhine with irreplaceable losses in men and material. The Allied Air Forces played a part of supreme importance. Constant attacks by the Tactical Air Forces aggravated the defeat and disorganisation and freed us from the dwindling Luftwaffe. Frequent patrols over the airfields containing the enemy's new jet-propelled fighters minimised a threat that had caused us anxiety. Continuing raids by our heavy bombers had reduced the German oil output to a critical point, ruined many of their airfields, and so heavily damaged their factories and transportation system as to bring them almost to a standstill.

*   *   *

I desired to be with our armies at the crossing, and Montgomery made me welcome. Taking only my secretary, Jock Colville, and "Tommy"[2] with me, I flew in the afternoon of March 23 by Dakota from Northolt to the British headquarters near Venlo. The Commander-in-Chief conducted me to the caravan in which he lived and moved. I found myself in the comfortable wagon I had used before. We dined at seven o'clock, and an hour later we repaired with strict punctuality to Montgomery's map wagon. Here were displayed all the maps kept from hour to hour by a select group of officers. The whole plan of our deployment and attack was easily comprehended. We were to force a passage over the river at ten points on a twenty-mile front from Rheinsberg to Rees. All our resources were to be used. Eighty thousand men, the advance guard of armies a million strong, were to be hurled forward. Masses of boats and pontoons lay ready. On the far side stood the Germans, entrenched and organised in all the strength of modern fire power.

Everything I had seen or studied in war, or read, made me doubt that a river could be a good barrier of defence against superior force. In Hamley's *Operations of War,* which I had pondered over ever since Sandhurst days, he argues the truth that a river running parallel to the line of advance is a much more dangerous feature than one which lies squarely athwart it, and he illustrates this theory by Napoleon's marvellous campaign of 1814. I was therefore in good hopes of the battle even before the Field Marshal explained his plans to me. Moreover, we had now the measureless advantage of mastery in the air. The episode which the Commander-in-Chief particularly wished me to see was the drop next morning of two air-borne divisions, comprising fourteen thousand men, with artillery and much other offensive equipment, behind the enemy lines. Accordingly we all went to bed before ten o'clock.

The honour of leading the attack fell to our 51st and 15th Divisions and the American 30th and 79th. Four battalions of the 51st were the

[2] Commander C. R. Thompson, R.N., my naval aide.

first to set forth, and a few minutes later they had reached the far side. Throughout the night the attacking divisions poured across, meeting little resistance at first, as the bank itself was lightly defended. At dawn bridgeheads, shallow as yet, were firmly held, and the Commandos were already at grips in Wesel.

In the morning Montgomery had arranged for me to witness from a hill-top amid rolling downland the great fly-in. It was full daylight before the subdued but intense roar and rumbling of swarms of aircraft stole upon us. After that in the course of half an hour over two thousand aircraft streamed overhead in their formations. My viewpoint had been well chosen. The light was clear enough to enable one to see where the descent on the enemy took place. The aircraft faded from sight, and then almost immediately afterwards returned towards us at a different level. The parachutists were invisible even to the best field-glasses. But now there was a double murmur and roar of reinforcements arriving and of those who had delivered their attacks returning. Soon one saw with a sense of tragedy aircraft in twos and threes coming back askew, asmoke, or even in flames. Also at this time tiny specks came floating to earth. Imagination built on a good deal of experience told a hard and painful tale. It seemed however that nineteen out of every twenty of the aircraft that had started came back in good order, having discharged their mission. This was confirmed by what we heard an hour later when we got back to headquarters.

The assault was now in progress along the whole front, and I was conducted by motor on a long tour from one point to another, and to the various corps headquarters. Things went well all that day. The four assaulting divisions were safely across and established in bridge-heads 5000 yards deep. The air-borne divisions were going strong and our air operations were most successful. The strike of the Allied Air Forces, second only to that of D-Day in Normandy, included not only the Strategic Air Forces in Britain, but also heavy bombers from Italy, who made deep penetrations into Germany.

At 8 P.M. we repaired to the map wagon, and I now had an excellent opportunity of seeing Montgomery's methods of conducting a battle on this gigantic scale. For nearly two hours a succession of young officers, of about the rank of major, presented themselves. Each had come back from a different sector of the front. They were the direct personal representatives of the Commander-in-Chief, and could go anywhere and see anything and ask any questions they liked of any commander. As in turn they made their reports and were searchingly questioned by their chief the whole story of the day's battle was unfolded. This gave Monty a complete account of what had happened by highly competent men whom he knew well and whose eyes he trusted.

It afforded an invaluable cross check to the reports from all the various headquarters and from the commanders, all of which had already been sifted and weighed by General de Guingand, his Chief of Staff, and were known to Montgomery. By this process he was able to form a more vivid, direct, and sometimes more accurate picture. The officers ran great risks, and of the seven or eight to whom I listened on this and succeeding nights two were killed in the next few weeks. I thought the system admirable, and indeed the only way in which a modern Commander-in-Chief could see as well as read what was going on in every part of the front. This process having finished, Montgomery gave a series of directions to de Guingand, which were turned into immediate action by the Staff machine. And so to bed.

*　　　*　　　*

The next day, March 25, we went to meet Eisenhower. On our way I told Montgomery how his system resembled that of Marlborough and the conduct of battles in the eighteenth century, where the Commander-in-Chief acted through his lieutenant generals. Then the Commander-in-Chief sat on his horse and directed by word of mouth a battle on a five- or six-mile front, which ended in a day and settled the fortunes of great nations, sometimes for years or generations to come. In order to make his will effective he had four or five lieutenant generals posted at different points on the front, who knew his whole mind and were concerned with the execution of his plan. These officers commanded no troops and were intended to be offshoots and expressions of the Supreme Commander. In modern times the general must sit in his office conducting a battle ranging over ten times the front and lasting often for a week or ten days. In these changed conditions Montgomery's method of personal eyewitnesses, who were naturally treated with the utmost consideration by the front-line commanders of every grade, was an interesting though partial revival of old days.

We met Eisenhower before noon. Here a number of American generals were gathered. After various interchanges we had a brief lunch, in the course of which Eisenhower said that there was a house about ten miles away on our side of the Rhine, which the Americans had sandbagged, from which a fine view of the river and of the opposite bank could be obtained. He proposed that we should visit it, and conducted us there himself. The Rhine — here about four hundred yards broad — flowed at our feet. There was a smooth, flat expanse of meadows on the enemy's side. The officers told us that the far bank was unoccupied so far as they knew, and we gazed and gaped at it for a while. With appropriate precautions we were led into the building. Then

the Supreme Commander had to depart on other business, and Montgomery and I were about to follow his example when I saw a small
launch come close by to moor. So I said to Montgomery, "Why
don't we go across and have a look at the other side?" Somewhat
to my surprise he answered, "Why not?" After he had made some inquiries we started across the river with three or four American commanders and half a dozen armed men. We landed in brilliant sunshine
and perfect peace on the German shore, and walked about for half
an hour or so unmolested.

As we came back Montgomery said to the captain of the launch,
"Can't we go down the river towards Wesel, where there is something
going on?" The captain replied that there was a chain across the river
half a mile away to prevent floating mines interfering with our operations, and several of these might be held up by it. Montgomery pressed
him hard, but was at length satisfied that the risk was too great. As we
landed he said to me, "Let's go down to the railway bridge at Wesel,
where we can see what is going on on the spot." So we got into his car,
and, accompanied by the Americans, who were delighted at the prospect, we went to the big iron-girder railway bridge, which was broken
in the middle but whose twisted ironwork offered good perches. The
Germans were replying to our fire, and their shells fell in salvos of
four about a mile away. Presently they came nearer. Then one salvo
came overhead and plunged in the water on our side of the bridge.
The shells seemed to explode on impact with the bottom, and raised
great fountains of spray about a hundred yards away. Several other
shells fell among the motorcars which were concealed not far behind
us, and it was decided we ought to depart. I clambered down and
joined my adventurous host for our two hours' drive back to his headquarters.

*   *   *

During the next few days we continued to gain ground and by the
end of the month we possessed a springboard east of the Rhine from
which to launch major operations deep into Northern Germany. In
the south, the American armies, though not opposed so strongly, had
made astonishing progress. The two bridgeheads which were the reward
of their boldness being daily reinforced and enlarged, and more
crossings were made south of Koblenz and at Worms. On March
29 the American Third Army was in Frankfurt. The Ruhr and its
325,000 defenders were encircled. Germany's Western Front had collapsed.

The question thus arose: Where should we go next? All kinds of rumors were rife about Hitler's future plans. It seemed possible that after losing Berlin and Northern Germany he might retire to the mountainous and wooded parts of Southern Germany and endeavour to prolong the fight there. The strange resistance he made at Budapest, and the retention of Kesselring's army in Italy so long, seemed in harmony with such an intention. Although nothing could be positive, the general conclusion of our Chiefs of Staff was that a prolonged German campaign, or even guerrilla, in the mountains was unlikely on any serious scale. The possibility was therefore relegated by us, as it proved rightly, to the shades. On this basis I inquired about the strategy for the advance of the Anglo-American armies as foreseen at Allied Headquarters.

> I propose [telegraphed General Eisenhower] driving eastward to join hands with Russians or to attain general line of Elbe. Subject to Russian intentions, the axis Kassel–Leipzig is the best for the drive, as it will ensure the overrunning of that important industrial area, into which German Ministries are believed to be moving; it will cut the German forces approximately in half, and it will not involve us in crossing of Elbe. It is designed to divide and destroy the major part of remaining enemy forces in West.
>
> This will be my main thrust, and until it is quite clear that concentration of all our effort on it alone will not be necessary I am prepared to direct all my forces to ensuring its success. . . .
>
> Once the success of main thrust is assured I propose to take action to clear the northern ports, which in the case of Kiel will entail forcing the Elbe. Montgomery will be responsible for these tasks, and I propose to increase his forces if that should seem necessary for the purpose.

About the same time we learned that Eisenhower had announced his policy in a direct telegram to Stalin on March 28 in which he said that after isolating the Ruhr he proposed to make his main thrust along the axis Erfurt–Leipzig–Dresden, which, by joining hands with the Russians, would cut in two the remaining German forces. A secondary advance through Regensburg to Linz, where also he expected to meet the Russians, would prevent "the consolidation of German resistance in the redoubt in Southern Germany." Stalin agreed readily. He said that the proposal "entirely coincides with the plan of the Soviet High Command." "Berlin," he added, "has lost its former strategic importance. The Soviet High Command therefore plans to allot secondary forces in the direction of Berlin." This statement was not borne out by events.

This seemed so important that on April 1 I sent a personal telegram to the President:

> . . . Obviously, laying aside every impediment and shunning every diversion, the Allied armies of the North and Centre should now march at the highest speed towards the Elbe. Hitherto the axis has been upon Berlin. General Eisenhower, on his estimate of the enemy's resistance, to which I attach the greatest importance, now wishes to shift the axis somewhat to the southward and strike through Leipzig, even perhaps as far south as Dresden. . . . I say quite frankly that Berlin remains of high strategic importance. Nothing will exert a psychological effect of despair upon all German forces of resistance equal to that of the fall of Berlin. It will be the supreme signal of defeat to the German people. On the other hand, if left to itself to maintain a siege by the Russians among its ruins, and as long as the German flag flies there, it will animate the resistance of all Germans under arms.
>
> There is moreover another aspect which it is proper for you and me to consider. The Russian armies will no doubt overrun all Austria and enter Vienna. If they also take Berlin will not their impression that they have been the overwhelming contributor to our common victory be unduly imprinted in their minds, and may this not lead them into a mood which will raise grave and formidable difficulties in the future? *I therefore consider that from a political standpoint we should march as far east into Germany as possible, and that should Berlin be in our grasp we should certainly take it.* This also appears sound on military grounds.

Actually, though I did not realise it, the President's health was now so feeble that it was General Marshall who had to deal with these grave questions, and the United States Chiefs replied in substance that Eisenhower's plan appeared to accord with agreed strategy and with his directive. He was deploying across the Rhine in the north the maximum forces which could be used. The secondary effort in the south was achieving an outstanding success, and was being exploited as much as supplies would permit. They were confident that the Supreme Commander's action would secure the ports and everything else mentioned by the British more quickly and more decisively than the plan urged by them.

The Battle of Germany, they said, was at a point where it was for the Field Commander to judge the measures which should be taken. To turn away deliberately from the exploitation of the enemy's weakness did not appear sound. The single objective should be quick and complete victory. While recognising that there were factors not of direct concern to the Supreme Commander, the United States Chiefs considered his strategic concept was sound.

Eisenhower himself assured me that he had never lost sight of the great importance of the drive to the northernmost coast, ". . . although your telegram did introduce a new idea respecting the political importance of the early attainment of particular objectives. I clearly see your point in this matter. The only difference between your suggestions and my plan is one of timing. . . . In order to assure the success of each of my planned efforts, I concentrate first in the Centre to gain the position I need. As it looks to me now, the next move thereafter should be to have Montgomery cross the Elbe, reinforced as necessary by American troops, and reach at least a line including Lübeck on the coast. If German resistance from now on should progressively and definitely crumble you can see that there would be little if any difference in time between gaining central position and crossing the Elbe. On the other hand, if resistance tends to stiffen at all I can see that it is vitally necessary that I concentrate for each effort, and do not allow myself to be dispersed by attempting to do all these projects at once.

"Quite naturally, if at any moment collapse should suddenly come about everywhere along the front we would rush forward, and Lübeck and Berlin would be included in our important targets."

"Thank you again for your most kind telegram," I answered, " . . . I am however all the more impressed with the importance of entering Berlin, which may well be open to us, by the reply from Moscow to you, which . . . says, 'Berlin has lost its former strategic importance.' This should be read in the light of what I mentioned of the political aspects. I deem it highly important that we should shake hands with the Russians as far to the east as possible. . . . Much may happen in the West before the date of Stalin's main offensive."

I thought it my duty to end this correspondence between friends, and the changes in the main plan, as I said to Roosevelt at the time, were much less than we at first supposed, but I must place on record my conviction that in Washington especially longer and wider views should have prevailed. As war waged by a coalition draws to its end political aspects have a mounting importance. It is true that American thought is at least disinterested in matters which seem to relate to territorial acquisitions, but when wolves are about the shepherd must guard his flock, even if he does not himself care for mutton. At this time the points at issue did not seem to the United States Chiefs of Staff to be of capital importance. They were of course unnoticed by and unknown to the public, and were all soon swamped, and for the time being effaced, by the flowing tide of victory. Nevertheless, as will not now be disputed, they played a dominating part in the destiny of Europe, and may well have denied us all the lasting peace for which we had fought so long and hard. We can now see the deadly hiatus which existed between the fading of President Roosevelt's strength and the growth of President Truman's grip of the vast world problem. In this melancholy void one President could not act and the other could not know. Neither the military chiefs nor the State Department received the guidance they required. The former confined themselves to their professional sphere; the latter did not comprehend the issues involved. The indispensable political direction was lacking at the moment when it was most needed. The United States stood on the scene of victory, master of world fortunes, but without a true and coherent design. Britain, though still very powerful, could not act decisively alone. I could at this stage only warn and plead. Thus this climax of apparently measureless success was to me a most unhappy time. I moved amid cheering crowds, or sat at a table adorned with congratulations and blessings from every part of the Grand Alliance, with an aching heart and a mind oppressed by forebodings.

The destruction of German military power had brought with it a fundamental change in the relations between Communist Russia and the Western democracies. They had lost their common enemy, which was almost their sole bond of union. Henceforward Russian imperialism and the Communist creed saw and set no bounds to their progress and ultimate dominion, and more than two years were to pass before they were confronted again with an equal will power. I should not tell this tale now when all is plain in glaring light if I had not known it and felt it when all was dim, and when abounding triumph only intensified the inner darkness of human affairs. Of this the reader must be the judge.

25

The Iron Curtain

As the weeks passed after Yalta it became clear that the Soviet Government was doing nothing to carry out our agreements about broadening the Polish Government to include all Polish parties and both sides. Molotov steadily refused to give an opinion about the Poles we mentioned, and not one of them was allowed to come even to a preliminary round-table discussion. He had offered to allow us to send observers to Poland, and had been disconcerted by the readiness and speed with which we had accepted, arguing, among other things, that it might affect the prestige of the Lublin Provisional Government. No progress of any kind was made in the talks at Moscow. Time was on the side of the Russians and their Polish adherents, who were fastening their grip upon the country by all kinds of severe measures, which they did not wish outside observers to see. Every day's delay was a gain to these hard forces.

On the very evening when I was speaking in the House of Commons upon the results of our labours at Yalta the first violation by the Russians both of the spirit and letter of our agreements took place in Rumania. We were all committed by the Declaration on Liberated Europe, so recently signed, to see that both free elections and democratic Governments were established in the countries occupied by Allied armies. On February 27 Vyshinsky, who had appeared in Bucharest without warning on the previous day, demanded an audience of King Michael and insisted that he should dismiss the all-party Government which had been formed after the royal *coup d'état* of August 1944 and had led to the expulsion of the Germans from Rumania. The young monarch, backed by his Foreign Minister, Visoianu, resisted these demands until the following day. Vyshinsky called again, and, brushing aside the King's request at least to be allowed to consult the leaders of the political parties, banged his fist on the table, shouted for an immediate acquiescence, and walked out of the room, slamming the door. At the same time Soviet tanks and troops deployed in the streets of the capital, and on March 6 a Soviet-nominated Administration took office.

I was deeply disturbed by this news, which was to prove a pattern of things to come, but we were hampered in our protests because Eden and I during our October visit to Moscow had recognised that Russia should have a largely predominant voice in Rumania and Bulgaria while we took the lead in Greece. Stalin had kept very strictly to this understanding during the six weeks' fighting against the Communists and E.L.A.S. in the city of Athens, in spite of the fact that all this was most disagreeable to him and those around him. Peace had now been restored, and, though many difficulties lay before us, I hoped that in a few months we should be able to hold free, unfettered elections, preferably under British, American, and Russian supervision, and that thereafter a constitution and Government would be erected on the indisputable will of the Greek people.

But in the two Black Sea Balkan countries Stalin was now pursuing the opposite course, and one which was absolutely contrary to all democratic ideas. He had subscribed on paper to the principles of Yalta, and now they were being trampled down in Rumania. But if I pressed him too much he might say, "I did not interfere with your action in Greece; why do you not give me the same latitude in Rumania?" Neither side would convince the other, and having regard to my personal relations with Stalin, I was sure it would be a mistake to embark on such an argument. I nevertheless felt we should tell him of our distress at the forceful installation of a Communist minority Government. I was particularly afraid it might lead to an indiscriminate purge of anti-Communist Rumanians, who would be accused of Fascism much on the lines of what had been happening in Bulgaria.

Meanwhile, the deadlock over Poland continued. All through March I was engaged in tense correspondence with Mr. Roosevelt, but although I had no exact information about his state of health I had the feeling that, except for occasional flashes of courage and insight, the telegrams he was sending us were not his own. The Soviet policy became daily more plain, as also did the use they were making of their unbridled and unobserved control of Poland. They asked that Poland should be represented at the forthcoming United Nations Conference at San Francisco only by the Lublin Government. When the Western Powers would not agree the Soviets refused to let Molotov attend. This threatened to make all progress at San Francisco, and even the Conference itself, impossible. Molotov persisted that the Yalta communiqué merely meant adding a few other Poles to the existing Administration of Russian puppets, and that these puppets should be consulted first. He maintained his right to veto Mikolajczyk and any other Poles we might suggest, and pretended he had insufficient information about the names we had put forward long before. It was as plain as a pikestaff that his tactics were to drag the business out while the Lublin Committee consolidated their power. Negotiations by our Ambassadors held no promise of an honest Polish settlement. They merely meant that our communications would be side-tracked and time would be wasted on finding formulae which did not decide vital points.

I was sure that the only way to stop Molotov was to send Stalin a personal message, and I therefore appealed to the President in the hope that we could address Stalin jointly on the highest level. A lengthy correspondence followed between us, but at this critical time Roosevelt's health and strength had faded. In my long telegrams I thought I was talking to my trusted friend and colleague as I had done all these years. I was no longer being fully heard by him. I did not know how ill he was, or I might have felt it cruel to press him. The President's devoted aides were anxious to keep their knowledge of his condition within the narrowest circle, and various hands drafted in combination the answers which were sent in his name. To these, as his life ebbed, Roosevelt could only give general guidance and approval. This was an heroic effort. The tendency of the State Department was naturally to avoid bringing matters to a head while the President was physically so frail and to leave the burden on the Ambassadors in Moscow. Harry Hopkins, who might have given personal help, was himself seriously ailing, and frequently absent or uninvited. These were costly weeks for all.

*   *   *

All this time a far more bitter and important interchange was taking place between the British and American Governments and the Soviets on a very different issue. The advance of the Soviet armies, Alexander's victories in Italy, the failure of their counterstroke in the Ardennes, and Eisenhower's march to the Rhine had convinced all but Hitler and his closest followers that surrender was imminent and unavoidable. The question was, surrender to whom? Germany could no longer make war on two fronts. Peace with the Soviets was evidently impossible. The rulers of Germany were too familiar with totalitarian oppression to invite its importation from the East. There remained the Allies in the West. Might it not be possible, they argued, to make a bargain with Great Britain and the United States? If a truce could be made in the West they could concentrate their troops against the Soviet advance. Hitler alone was obstinate. The Third Reich was finished and he would die with it. But several of his followers tried to make secret approaches to the English-speaking Allies. All these proposals were of course rejected. Our terms were unconditional surrender on all fronts. At the same time our commanders in the field were always fully authorised to accept purely military capitulations of the enemy forces which opposed them, and an attempt to arrange this while we were fighting on the Rhine led to a harsh exchange between the Russians and the President, whom I supported.

In February General Karl Wolff, the commander of the S.S. in Italy, had got into touch through Italian intermediaries with the American Intelligence Service in Switzerland. It was decided to examine the credentials of the persons involved, and the link was given the code name "Crossword." On March 8 General Wolff himself appeared at Zürich, and met Mr. Allen Dulles, the head of the American organisation. Wolff was bluntly told that there was no question of negotiations, and that if the matter were pursued it could only be on the basis of unconditional surrender. This information was speedily conveyed to Allied Headquarters in Italy and to the American, British, and Soviet Governments. On March 15 the British and American Chiefs of Staff at Caserta arrived in Switzerland in disguise, and four days later, on March 19, a second exploratory meeting was held with General Wolff.

I realised at once that the Soviet Government might be suspicious of a separate military surrender in the South, which would enable our armies to advance against reduced opposition to Vienna and beyond, or indeed towards the Elbe or Berlin. Moreover, as all our fronts round Germany were part of the whole Allied war the Russians would naturally be affected by anything done on any one of them. If any contacts were made with the enemy, formal or informal, they ought to be told in good time. This rule was scrupulously followed. On March 12 the British Ambassador in Moscow had informed the Soviet Government of this link with the German emissaries, and said that no contact would be made until we received the Russian reply. There was at no stage any question of concealing anything from the Russians. The Allied representatives then in Switzerland even explored ways of smuggling a Russian officer in to join them if the Soviet Government wished to send someone. This however proved impractical, and on March 13 the Russians were informed that if "Crossword" proved to be of serious import we would welcome their representatives at Alexander's headquarters. Three days later Molotov informed the British Ambassador in Moscow that the Soviet Government found the attitude of the British Government "entirely inexplicable and incomprehensible in denying facilities to the Russians to send the representative to Berne." A similar communication was passed to the American Ambassador.

On the 21st our Ambassador in Moscow was instructed to inform the Soviet Government yet again that the only object of the meetings was to make sure that the Germans had authority to negotiate a military surrender and to invite Russian delegates to Allied headquarters at Caserta. This he did. Next day Molotov handed him a written reply, which contained the following expressions:

"In Berne for two weeks, behind the backs of the Soviet Union, which is bearing the brunt of the war against Germany, negotiations have been going on between the representatives of the German military command on the one hand and representatives of the English and American commands on the other."

Sir Archibald Clark Kerr naturally explained that the Soviets had misunderstood what had occurred and that these "negotiations" were no more than an attempt to test the credentials and authority of General Wolff. Molotov's comment was blunt and insulting. "In this instance," he wrote, "the Soviet Government sees not a misunderstanding but something worse." He attacked the Americans just as bitterly.

In the face of so astonishing a charge it seemed to me that silence was better than a contest in abuse, but at the same time it was necessary to warn our military commanders in the West. I accordingly showed Molotov's insulting letter both to Montgomery and to Eisenhower, with whom I at this time was watching the crossing of the Rhine.

General Eisenhower was much upset, and seemed deeply stirred with anger at what he considered most unjust and unfounded charges about our good faith. He said that as a military commander he would accept the unconditional surrender of any body of enemy troops on his front, from a company to the entire Army, that he regarded this as a purely military matter, and that he had full authority to accept such a surrender without asking anybody's opinion. If however political matters arose he would immediately consult the Governments. He feared that if the Russians were brought into a possible surrender of Kesselring's forces what could be settled by himself in an hour might be prolonged for three or four weeks, with heavy losses to our troops. He made it clear that he would insist upon all the troops under the officer making the surrender laying down their arms and standing still until they received further orders, so that there would be no chance of their being transferred across Germany to withstand the Russians. At the same time he would advance through these surrendered troops as fast as possible to the East.

I thought myself that these matters should be left to his discretion, and that the Governments should only intervene if any political issues arose. I did not see why we should break our hearts if, owing to a mass surrender in the West, we got to the Elbe, or even farther, before Stalin.

Jock Colville reminds me that I said to him that evening, "I hardly like to consider dismembering Germany until my doubts about Russia's intentions have been cleared away."

On April 5 I received from the President the startling text of his interchanges with Stalin:

> You are absolutely right [wrote Stalin] that, in connection with the affair regarding negotiations of the Anglo-American command with the German command, somewhere in Berne or some other place, "has developed an atmosphere of fear and distrust deserving regrets."
>
> You insist that there have been no negotiations yet. It may be assumed that you have not yet been fully informed. . . . My military colleagues do not have any doubts that the negotiations have taken place, and that they have ended in an agreement with the Germans, on the basis of which the German commander on the Western Front, Marshal Kesselring, has agreed to open the front and permit the Anglo-American troops to advance to the east, and the Anglo-Americans have promised in return to ease for the Germans the peace terms.
>
> As a result of this at the present moment the Germans on the Western Front in fact have ceased the war against England and the United States. At the same time the Germans continue the war with Russia, the Ally of England and the United States. . . .

This accusation angered the President deeply. His strength did not allow him to draft his own reply. General Marshall framed the answer, with Roosevelt's approval. It certainly did not lack vigour.

> . . . With a confidence in your belief in my personal reliability, [he retorted] and in my determination to bring about together with you an unconditional surrender of the Nazis, it is astonishing that a belief seems to have reached the Soviet Government that I have entered into an agreement with the enemy without first obtaining your full agreement. Finally I would say this: it would be one of the great tragedies of history if at the very moment of the victory now within our grasp such distrust, such lack of faith, should prejudice the entire undertaking after the colossal losses of life, material, and treasure involved.
>
> *Frankly, I cannot avoid a feeling of bitter resentment toward your informers, whoever they are, for such vile misrepresentations of my actions or those of my trusted subordinates.*

I was deeply struck by this last sentence, which I print in italics. I felt that although Mr. Roosevelt did not draft the whole message he might well have added this final stroke himself. It looked like an addition or summing up, and it seemed like Roosevelt himself in anger.

I wrote at once to him and to Stalin and a few days later got the semblance of an apology from the Russian dictator. "I would minimise the general Soviet problem as much as possible," the President cabled to me on April 12, "because these problems, in one form or another, seem to arise every day, and most of them straighten out, as in the case of the Berne meeting. We must be firm, however, and our course thus far is correct."

*   *   *

President Roosevelt died suddenly that afternoon on Thursday, April 12, 1945, at Warm Springs, Georgia. He was sixty-three. While he was having his portrait painted, he suddenly collapsed, and died a few hours later without regaining consciousness. When I received these tidings early in the morning of Friday, the 13th, I felt as if I had been struck a physical blow. My relations with this shining personality had played so large a part in the long, terrible years we had worked together. Now they had come to an end, and I was overpowered by a sense of deep and irreparable loss. I went down to the House of Commons, which met at eleven o'clock, and in a few sentences proposed that we should pay our respects to the memory of our great friend by immediately adjourning. This unprecedented step on the occasion of the death of the head of a foreign State was in accordance with the unanimous wish of the Members, who filed slowly out of the chamber after a sitting which had lasted only eight minutes.

My first impulse was to fly over to the funeral, and I had already ordered an aeroplane. Lord Halifax telegraphed that both Hopkins and Stettinius were much moved by my thought of possibly coming over, and both warmly agreed with my judgment of the immense effect for good that would be produced. Mr. Truman had asked him to say how greatly he would personally value the opportunity of meeting me as early as possible. His idea was that after the funeral I might have had two or three days' talk with him.

Much pressure was however put on me not to leave the country at this most critical and difficult moment, and I yielded to the wishes of my friends. In the afterlight I regret that I did not adopt the new President's suggestion. I had never met him, and I feel that there were many points on which personal talks would have been of the greatest value, especially if they had been spread over several days and were not hurried or formalised. It seemed to me extraordinary, especially during the last few months, that Roosevelt had not made his deputy and potential successor thoroughly acquainted with the whole story and brought him into the decisions which were being taken. This proved of grave disadvantage to our affairs. There is no comparison between reading about events afterwards and living through them from hour to hour. In Mr. Eden I had a colleague who knew everything and could at any moment take over the entire direction, although I was myself in good health and full activity. But the Vice-President of the United States steps at a bound from a position where he has little information and less power into supreme authority. How could Mr. Truman know and weigh the issues at stake at this climax of the war? Everything that we have learnt about him since shows him to be a resolute and fearless man, capable of taking the greatest decisions. In these early months his position was one of extreme difficulty, and did not enable him to bring his outstanding qualities fully into action.

Mr. Truman's first political act which concerned us was to take up the Polish question from the point where it stood when Roosevelt died, only forty-eight hours earlier. He proposed a joint declaration by us both to Stalin. The document in which this was set forth must of course have been far advanced in preparation by the State Department at the moment when the new President succeeded. Nevertheless it is remarkable that he felt able so promptly to commit himself to it amid the formalities of assuming office and the funeral of his predecessor.

He admitted that Stalin's attitude was not very hopeful, but felt we should "have another go," and he accordingly proposed telling Stalin that our Ambassadors in Moscow had agreed without question to the three leaders of the Warsaw Government being invited to Moscow for consultation and assuring him we had never denied they would play a prominent part in forming the new Provisional Government of National Unity. Our Ambassadors were not demanding the right to invite an unlimited number of Poles from abroad and from within Poland. The real issue was whether the Warsaw Government could veto individual candidates for consultation, and in our opinion the Yalta agreement did not entitle them to do so.

Our joint message was sent on the 15th. Meanwhile M. Mikolajczyk confirmed that he accepted the Crimea decision on Poland, including the establishment of Poland's eastern frontier on the Curzon Line, and I so informed Stalin. As I got no answer it may be assumed that the Dictator was for the moment content. Other points were open. Mr. Eden telegraphed from Washington that he and Stettinius agreed that we should renew our demand for the entry of observers into Poland, and that we should once more press the Soviet Government to hold up their negotiations for a treaty with the Lublin Poles. But shortly after deciding this, news arrived that the treaty had been concluded.

On April 29, when it seemed evident we were getting nowhere, I put my whole case to Stalin in a lengthy telegram of which the following paragraphs may be deemed material:

It is quite true that about Poland we have reached a definite line of action with the Americans. This is because we agree naturally upon the subject, and both sincerely feel that we have been rather ill-treated . . . since the Crimea Conference. No doubt these things seem different when looked at from the opposite point of view. But we are absolutely agreed that the pledge we have given for a sovereign, free, independent Poland, with a Government fully and adequately representing all the democratic elements among Poles, is for us a matter of honour and duty. I do not think there is the slightest chance of any change in the attitude of our two Powers, and when we are agreed we are bound to say so. After all, we have joined with you, largely on my original initiative, early in 1944, in proclaiming the Polish-Russian frontier which you desired, namely, the Curzon Line, including Lvov for Russia. We think you ought to meet us with regard to the other half of the policy which you equally with us have proclaimed, namely, the sovereignity. independence, and freedom of Poland, provided it is a Poland friendly to Russia. . . .

Also, difficulties arise at the present moment because all sorts of stories are brought out of Poland which are eagerly listened to by many Members of Parliament, and which at any time may be violently raised in Parliament or the Press in spite of my deprecating such action, and on which M. Molotov will vouchsafe us no information at all in spite of repeated requests. *For instance, there is the talk of the fifteen Poles who were said to have met the Russian authorities for discussion over four weeks ago . . . and there are many other statements of deportations, etc.*[1] How can I contradict such complaints when you give me no information whatever and when neither I nor the Americans are allowed to send anyone into Poland to find out for themselves the true state of affairs? There is no part of our occupied or liberated territory into which you are not free to send delegations, and people do not see why you should have any reasons against similar visits by British delegations to foreign countries liberated by you.

There is not much comfort in looking into a future where you and the countries you dominate, plus the Communist Parties in

[1] My italics.—W.S.C.

many other States, are all drawn up on one side, and those who
rally to the English-speaking nations and their associates or Dominions are on the other. It is quite obvious that their quarrel would
tear the world to pieces and that all of us leading men on either
side who had anything to do with that would be shamed before
history. Even embarking on a long period of suspicions, of abuse
and counterabuse, and of opposing policies would be a disaster
hampering the great developments of world prosperity for the
masses which are attainable only by our trinity. I hope there is no
word or phrase in this outpouring of my heart to you which un-
wittingly gives offence. If so, let me know. But do not, I beg you,
my friend Stalin, underrate the divergences which are opening about
matters which you may think are small to us but which are symbolic of the way the English-speaking democracies look at life.

*   *   *

The incident of the missing Poles mentioned in the second paragraph
now requires to be recorded, although it carries us somewhat ahead of
the general narrative. At the beginning of March 1945 the Polish Underground were invited by the Russian Political Police to send a delegation to Moscow to discuss the formation of a united Polish Government along the lines of the Yalta agreement. This was followed by a
written guarantee of personal safety and it was understood that the
party would later be allowed if the negotiations were successful to travel
to London for talks with the Polish Government in exile. On March 27
General Leopold Okulicki, the successor of General Bor-Komorowski
in command of the Underground Army, two other leaders, and an interpreter had a meeting in the suburbs of Warsaw with a Soviet representative. They were joined the following day by eleven leaders representing
the major political parties in Poland. One other Polish leader was already in Russian hands. No one returned from the rendezvous. On
April 6 the Polish Government in exile issued a statement in London
giving the outline of this sinister episode. The most valuable representatives of the Polish Underground had disappeared without a trace
in spite of the formal Russian offer of safe-conduct. Questions were
asked in Parliament and stories have since spread of the shooting of
local Polish leaders in the areas at this time occupied by the Soviet
armies and particularly of one episode at Siedlce in Eastern Poland.
It was not until May 4 that Molotov admitted at San Francisco that
these men were being held in Russia, and an official Russian news
agency stated next day that they were awaiting trial on charges of
"diversionary tactics in the rear of the Red Army."
On May 18 Stalin publicly denied that the arrested Polish leaders had

ever been invited to Moscow and asserted that they were mere "diversionists" who would be dealt with according to "a law similar to the British Defence of the Realm Act." The Soviet Government refused to move from this position. Nothing more was heard of the victims of the trap until the case against them opened on June 18. It was conducted in the usual Communist manner. The prisoners were accused of subversion, terrorism, and espionage, and all except one admitted wholly or in part the charges against them. Thirteen were found guilty, and sentenced to terms of imprisonment ranging from four months to ten years, and three were acquitted. This was in fact the judicial liquidation of the leadership of the Polish Underground which had fought so heroically against Hitler. The rank and file had already died in the ruins of Warsaw.

*　　　*　　　*

Meanwhile, I received a most disheartening reply from Stalin to the appeal I had made to him on April 29. It was dated May 5 and read as follows:

> I am obliged to say that I cannot agree with the arguments which you advance in support of your position. . . . I am unable to share your views . . . in the passage where you suggest that the three Powers should supervise elections. Such supervision in relation to the people of an Allied State could not be regarded otherwise than as an insult to that people and a flagrant interference with its internal life. Such supervision is unnecessary in relation to the former satellite States which have subsequently declared war on Germany and joined the Allies, as has been shown by the experience of the elections which have taken place, for instance, in Finland; here elections have been held without any outside intervention and have led to constructive results. . . . Poland's peculiar position as a neighbour State of the Soviet Union . . . demands that the future Polish Government should actively strive for friendly relations between Poland and the Soviet Union, which is likewise in the interests of all other peace-loving nations. . . . The United Nations are concerned that there should be a firm and lasting friendship between the Soviet Union and Poland. Consequently we cannot be satisfied that persons should be associated with the formation of the future Polish Government who, as you express it, "are not fundamentally anti-Soviet," or that only those persons should be excluded from participation in this work who are in your opinion "extremely unfriendly towards Russia." Neither of these criteria can satisfy us. *We insist, and shall insist, that there should be*

brought into consultation on the formation of the future Polish Government only those persons who have actively shown a friendly attitude towards the Soviet Union and who are honestly and sincerely prepared to co-operate with the Soviet State.[2]

I must comment especially on [another] point of your message, in which you mention difficulties arising as a result of rumours of the arrest of fifteen Poles, of deportations and so forth.

As to this, I can inform you that the group of Poles to which you refer consists not of fifteen but of sixteen persons, and is headed by the well-known Polish general Okulicki. In view of his especially odious character the British Information Service is careful to be silent on the subject of this Polish general, who "disappeared" together with the fifteen other Poles who are said to have done likewise. But we do not propose to be silent on this subject. This party of sixteen individuals headed by General Okulicki was arrested by the military authorities on the Soviet front and is undergoing investigation in Moscow. General Okulicki's group, and especially the General himself, are accused of planning and carrying out diversionary acts in the rear of the Red Army, which resulted in the loss of over 100 fighters and officers of that Army, and are also accused of maintaining illegal wireless transmitting stations in the rear of our troops, which is contrary to law. All or some of them, according to the results of the investigation, will be handed over for trial. This is the manner in which it is necessary for the Red Army to defend its troops and its rear from diversionists and disturbers of order.

The British Information Service is disseminating rumours of the murder or shooting of Poles in Siedlce. These statements of the British Information Service are complete fabrications, and have evidently been suggested to it by [anti-Soviet] agents. . . .

It appears from your message that you are not prepared to regard the Polish Provisional Government as the foundation of the future Government of National Unity, and that you are not prepared to accord it its rightful position in that Government. I must say frankly that such an attitude excludes the possibility of an agreed solution of the Polish question.

I repeated this forbidding message to President Truman, with the following comment: "It seems to me that matters can hardly be carried further by correspondence, and that as soon as possible there should be a meeting of the three heads of Governments. *Meanwhile we should hold firmly to the existing position obtained or being obtained by our*

[2] My italics.—W.S.C.

armies in Yugoslavia, in Austria, in Czechoslovakia, on the main central
United States front, and on the British front, reaching up to Lübeck,
including Denmark [3] *. . ."* On May 4 I drew the European scene as
I saw it for Mr. Eden, who was at the San Francisco Conference, in
daily touch with Stettinius and Molotov, and soon to revisit the President
at Washington:

> I consider that the Polish deadlock can now probably only be
> resolved at a conference between the three heads of Governments in
> some unshattered town in Germany, if such can be found. This
> should take place at latest at the beginning of July. I propose to
> telegraph a suggestion to President Truman about his visit here and
> the further indispensable meeting of the three major Powers.
>
> 2. The Polish problem may be easier to settle when set in relation
> to the now numerous outstanding questions of the utmost gravity
> which require urgent settlement with the Russians. I fear terrible
> things have happened during the Russian advance through Ger-
> many to the Elbe. The proposed withdrawal of the United States
> Army to the occupational lines which were arranged with the Rus-
> sians and Americans in Quebec, and which were marked in yellow
> on the maps we studied there, would mean the tide of Russian
> domination sweeping forward 120 miles on a front of 300 or 400
> miles. This would be an event which, if it occurred, would be one
> of the most melancholy in history. After it was over and the terri-
> tory occupied by the Russians Poland would be completely en-
> gulfed and buried deep in Russian-occupied lands. What would in
> fact be the Russian frontier would run from the North Cape in
> Norway, along the Finnish-Swedish frontier, across the Baltic to a
> point just east of Lübeck, along the at present agreed line of occu-
> pation and along the frontier between Bavaria to Czechoslovakia
> to the frontiers of Austria, which is nominally to be in quadruple
> occupation, and halfway across that country to the Isonzo River,
> behind which Tito and Russia will claim everything to the east.
> Thus the territories under Russian control would include the Baltic
> provinces, all of Germany to the occupational line, all Czecho-
> slovakia, a large part of Austria, the whole of Yugoslavia, Hungary,
> Rumania, Bulgaria, until Greece in her present tottering condition
> is reached. It would include all the great capitals of Middle Eu-
> rope, including Berlin, Vienna, Budapest, Belgrade, Bucharest, and
> Sofia. The position of Turkey and Constantinople will certainly
> come immediately into discussion.
>
> 3. This constitutes an event in the history of Europe to which

[3] My italics.—W.S.C.

there has been no parallel, and which has not been faced by the Allies in their long and hazardous struggle. The Russian demands on Germany for reparations alone will be such as to enable her to prolong the occupation almost indefinitely, at any rate for many years, during which time Poland will sink with many other States into the vast zone of Russian-controlled Europe, not necessarily economically Sovietised, but police-governed.

4. It is just about time that these formidable issues were examined between the principal Powers as a whole. We have several powerful bargaining counters on our side, the use of which might make for a peaceful agreement. *First, the Allies ought not to retreat from their present positions to the occupational line until we are satisfied about Poland, and also about the temporary character of the Russian occupation of Germany, and the conditions to be established in the Russianised or Russian-controlled countries in the Danube valley, particularly Austria and Czechoslovakia, and the Balkans.*[4] Secondly, we may be able to please them about the exits from the Black Sea and the Baltic as part of a general statement. All these matters can only be settled before the United States armies in Europe are weakened. If they are not settled before the United States armies withdraw from Europe and the Western world folds up its war machines there are no prospects of a satisfactory solution and very little of preventing a third World War. It is to this early and speedy show-down and settlement with Russia that we must now turn our hopes. Meanwhile I am against weakening our claim against Russia on behalf of Poland in any way. I think it should stand where it was put in the telegrams from the President and me.

"Nothing," I added the next day, "can save us from the great catastrophe but a meeting and a show-down as early as possible at some point in Germany which is under American and British control and affords reasonable accommodation."

[4] My italics.—W.S.C.

26

The German Surrender

GLEAMING SUCCESSES marked the end of our campaigns in the Mediterranean. In December Alexander had succeeded Wilson as Supreme Commander, while Mark Clark took command of the Fifteenth Army Group. After their strenuous efforts of the autumn our armies in Italy needed a pause to reorganise and restore their offensive power.

The long, obstinate, and unexpected German resistance on all fronts had made us and the Americans very short of artillery ammunition, and our hard experiences of winter campaigning in Italy forced us to postpone a general offensive till the spring. But the Allied Air Forces, under General Eaker, and later under General Cannon, used their thirty-to-one superiority in merciless attacks on the supply lines which nourished the German armies. The most important one, from Verona to the Brenner Pass, where Hitler and Mussolini used to meet in their happier days, was blocked in many places for nearly the whole of March. Other passes were often closed for weeks at a time, and two divisions being transferred to the Russian front were delayed almost a month.

The enemy had enough ammunition and supplies, but lacked fuel. Units were generally up to strength, and their spirit was high in spite of Hitler's reverses on the Rhine and the Oder. The German High Command might have had little to fear had it not been for the dominance of our Air Forces, the fact that we had the initiative and could strike where we pleased, and their own ill-chosen defensive position, with the broad Po at their backs. They would have done better to yield Northern Italy and withdraw to the strong defences of the Adige, where they could have held us with much smaller forces, and sent troops to help their overmatched armies elsewhere, or have made a firm southern face for the National Redoubt in the Tyrol mountains, which Hitler may have had in mind as his "last ditch."

But defeat south of the Po spelt disaster. This must have been obvious to Kesselring, and was doubtless one of the reasons for the negotiations recorded in the previous chapter. Hitler was of course the stumbling block and when Vietinghoff, who succeeded Kesselring, proposed a tactical withdrawal he was thus rebuffed: "The Fuehrer expects, now as before, the utmost steadiness in the fulfilment of your present mission to defend every inch of the North Italian areas entrusted to your command."

In the evening of April 9, after a day of mass air attacks and artillery bombardment, the Eighth Army attacked. By the 14th there was good news all along the front. The Fifth Army, after a week of hard fighting, backed by the full weight of the Allied Air Forces, broke out from the mountains, crossed the main road west of Bologna, and struck north. On the 20th Vietinghoff, despite Hitler's commands, ordered a withdrawal. It was too late. The Fifth Army pressed towards the Po, with the Tactical Air Force making havoc along the roads ahead. Trapped behind them were many thousand Germans, cut off from retreat, pouring into prisoners' cages or being marched to the rear. We crossed the Po on a broad front at the heels of the enemy. All the permanent bridges had been destroyed by our air forces, and the ferries and temporary crossings were attacked with such effect that the enemy were thrown into confusion. The remnants who struggled across, leaving all their heavy equipment behind, were unable to reorganise on the far bank. The Allied armies pursued them to the Adige. Italian partisans had long harassed the enemy in the mountains and their back areas. On April 25 the signal was given for a general rising and they made widespread attacks. In many cities and towns, notably Milan and Venice, they seized control. Surrenders in Northwest Italy became wholesale. The garrison of Genoa, four thousand strong, gave themselves up to a British liaison officer and the Partisans.

There was a pause before the force of facts overcame German hesitancies, but on April 24 Wolff reappeared in Switzerland with full powers from Vietinghoff. Two plenipotentiaries were brought to Alexander's headquarters, and on April 29 they signed the instrument of unconditional surrender in the presence of high British, American, and Russian officers. On May 2 nearly a million Germans surrendered as prisoners of war, and the war in Italy ended.

Thus ended our twenty months' campaign. Our losses had been grievous, but those of the enemy, even before the final surrender, far heavier. The principal task of our armies had been to draw off and contain the greatest possible number of Germans. This had been admirably fulfilled. Except for a short period in the summer of 1944, the enemy had always outnumbered us. At the time of their crisis in August of that

year no fewer than fifty-five German divisions were deployed along the
Mediterranean fronts. Nor was this all. Our forces rounded off their
task by devouring the larger army they had been ordered to contain.
There have been few campaigns with a finer culmination.

*   *   *

For Mussolini also the end had come. Like Hitler he seems to have
kept his illusions until almost the last moment. Late in March he had
paid a final visit to his German partner, and returned to his headquarters
on Lake Garda buoyed up with the thought of the secret weapons which
could still lead to victory. But the rapid Allied advance from the Apen-
nines made these hopes vain. There was hectic talk of a last stand in
the mountainous areas of the Italo-Swiss frontier. But there was no
will to fight left in the Italian Socialist Republic.

On April 25 Mussolini decided to disband the remnants of his armed
forces and to ask the Cardinal Archbishop of Milan to arrange a meeting with the underground Military Committee of the Italian National
Liberation Movement. That afternoon talks took place in the Archbishop's palace, but with a last furious gesture of independence Mussolini walked out. In the evening, followed by a convoy of thirty vehicles,
containing most of the surviving leaders of Italian Fascism, he drove to
the prefecture at Como. He had no coherent plan, and as discussion
became useless it was each man for himself. Accompanied by a handful of supporters, he attached himself to a small German convoy heading towards the Swiss frontier. The commander of the column was not
anxious for trouble with Italian Partisans. The Duce was persuaded
to put on a German greatcoat and helmet. But the little party was
stopped by partisan patrols; Mussolini was recognised and taken into
custody. Other members, including his mistress, Signorina Petacci, were
also arrested. On Communist instructions the Duce and his mistress
were taken out in a car next day and shot. Their bodies, together with
others, were sent to Milan and strung up head downward on meat hooks
in a petrol station on the Piazzale Loreto, where a group of Italian partisans had lately been shot in public.

Such was the fate of the Italian dictator. A photograph of the final
scene was sent to me, and I was profoundly shocked. But at least
the world was spared an Italian Nuremberg.

*   *   *

In Germany the invading armies drove onward in their might and the
space between them narrowed daily. By early April Eisenhower was
across the Rhine and thrusting deep into Germany and Central Europe

against an enemy who in places resisted fiercely but was quite unable to stem our triumphant onrush. Many political and military prizes still hung in the balance. Poland was beyond our succour. So also was Vienna, where our opportunity of forestalling the Russians by an advance from Italy had been abandoned eight months earlier when Alexander's forces had been stripped for the landing in the south of France. The Russians moved on the city from east and south, and by April 13 were in full possession. But there seemed nothing to stop the Western Allies from taking Berlin. The Russians were only thirty-five miles away, but the Germans were entrenched on the Oder and much hard fighting was to take place before they could force a crossing and resume their advance. The Ninth U.S. Army, on the other hand, had moved so speedily that on April 12 they had crossed the Elbe near Magdeburg and were about sixty miles from the capital. But here they halted. Four days later the Russians started their attack and surrounded Berlin on the 25th. Stalin had told Eisenhower that his main blow against Germany would be made in "approximately the second half of May," but he was able to advance a whole month earlier. Perhaps our swift approach to the Elbe had something to do with it.

On this same 25th day of April, 1945, spearheads of the United States First Army from Leipzig met the Russians near Torgau, on the Elbe. Germany was cut in two. The German Army was disintegrating before our eyes. Over a million prisoners were taken in the first three weeks of April, but Eisenhower believed that fanatical Nazis would attempt to establish themselves in the mountains of Bavaria and Western Austria, and he swung the Third U.S. Army southward. Its left penetrated into Czechoslovakia as far as Budějovice, Pilsen, and Karlsbad. Prague was still within our reach and there was no agreement to debar him from occupying it if it were militarily feasible. On April 30 I suggested to the President that he should do so, but Mr. Truman seemed adverse. A week later I also telegraphed personally to Eisenhower, but his plan was to halt his advance generally on the west bank of the Elbe and along the 1937 boundary of Czechoslovakia. If the situation warranted he would cross it to the general line Karlsbad–Pilsen–Budějovice. The Russians agreed to this and the movement was made. But on May 4 they reacted strongly to a fresh proposal to continue the advance of the Third U.S. Army to the river Vltava, which flows through Prague. This would not have suited them at all. So the Americans "halted while the Red Army cleared the east and west banks of the Moldau river and occupied Prague."[1] The city fell on May 9, two days after the general surrender was signed at Reims.

*   *   *

[1] Eisenhower, "Report to Combined Chiefs of Staff," p. 140.

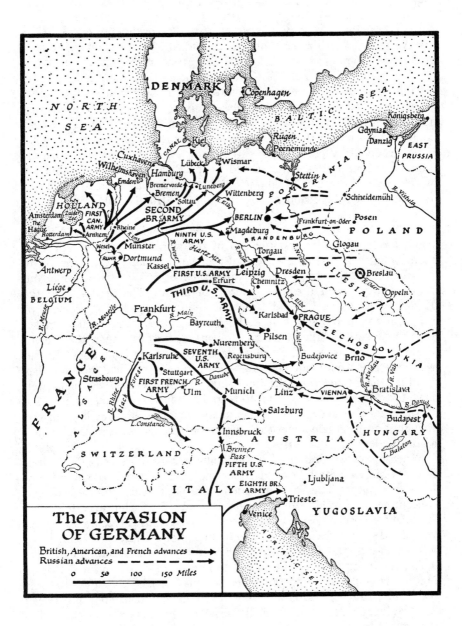

NORTH SEA

DENMARK

Copenhagen

BALTIC SEA

Königsberg

Rügen

Gdynia

Peenemünde

Danzig

EAST PRUSSIA

Cuxhaven

Kiel

CANAL

Lübeck

Wismar

Stettin

R. Vistula

Wilhelmshaven

Hamburg

Schneidemühl

Emden

Bremervorde

Bremen

Luneberg

Wittenberg

POMERANIA

HOLLAND

Soltau

R. Elbe

Amsterdam

Zuider

FIRST CAN. ARMY

SECOND BR. ARMY

BERLIN

Frankfurt-on-Oder

Posen

Zee

The Hague

Rotterdam

Rheine

R. Ems

Magdeburg

POLAND

NINTH U.S. ARMY

BRANDENBURG

Wesel

Arnhem

Münster

Glogau

Torgau

Wittenberg

Antwerp

RUHR

Dortmund

R. Weser

Hartz Mts

R. Mulde

SILESIA

Breslau

Kassel

Leipzig

Dresden

R. Oder

Liège

FIRST U.S. ARMY

Chemnitz

Oppeln

BELGIUM

Erfurt

THIRD U.S. ARMY

Frankfurt

Karlsbad

PRAGUE

R. Maas

R. Moselle

R. Main

Bayreuth

Pilsen

CZECHOSLOVAKIA

R. Vltava

Nuremberg

FRANCE

Karlsruhe

SEVENTH U.S. ARMY

Regensburg

Budejovice

Brno

Strasbourg

Stuttgart

ALSACE

FIRST FRENCH ARMY

R. Danube

Ulm

Munich

Linz

VIENNA

Bratislava

R. Rhine

Black Forest

Salzburg

R. Danube

L. Constance

Budapest

Innsbruck

AUSTRIA

HUNGARY

SWITZERLAND

Brenner Pass

L. Balaton

ITALY

FIFTH U.S. ARMY

EIGHTH BR. ARMY

Ljubljana

Venice

Trieste

YUGOSLAVIA

ADRIATIC SEA

The INVASION OF GERMANY

British, American, and French advances ⟶
Russian advances ─ ─ ─ ⟶

0   50   100   150 Miles

At this point a retrospect is necessary. The occupation of Germany by the principal Allies had long been studied. In the summer of 1943 a Cabinet Committee which I had set up under Mr. Attlee, in agreement with the Chiefs of Staff, recommended that the whole country should be occupied if Germany was to be effectively disarmed, and that our forces should be disposed in three main zones of roughly equal size, the British in the northwest, the Americans in the south and southwest, and the Russians in the eastern zone. Berlin should be a separate joint zone, occupied by each of the three major Allies. These recommendations were approved and forwarded to the European Advisory Council, which then consisted of M. Gousev, the Soviet Ambassador, Mr. Winant, the American Ambassador, and Sir William Strang of the Foreign Office.

At this time the subject seemed to be purely theoretical. No one could foresee when or how the end of the war would come. The German armies held immense areas of European Russia. A year was yet to pass before British or American troops set foot in Western Europe, and nearly two years before they entered Germany. The proposals of the European Advisory Council were not thought sufficiently pressing or practical to be brought before the War Cabinet. Like many praiseworthy efforts to make plans for the future, they lay upon the shelves while the war crashed on. In those days a common opinion about Russia was that she would not continue the war once she had regained her frontiers, and that when the time came the Western Allies might well have to try to persuade her not to relax her efforts. The question of the Russian zone of occupation in Germany therefore did not bulk in our thoughts or in Anglo-American discussions, nor was it raised by any of the leaders at Teheran.

When we met in Cairo on the way home in November 1943 the United States Chiefs of Staff brought it forward, but not on account of any Russian request. The Russian zone of Germany remained an academic conception, if anything too good to be true. I was however told that President Roosevelt wished the British and American zones to be reversed. He wanted the lines of communication of any American force in Germany to rest directly on the sea and not to run through France. This issue involved a lot of detailed technical argument and had a bearing at many points upon the plans for "Overlord." No decision was reached at Cairo, but later a considerable correspondence began between the President and myself. The British Staff thought the original plan the better, and also saw many inconveniences and complications in making the change. I had the impression that their American colleagues rather shared their view. At the Quebec Conference in September 1944 we reached a firm agreement between us.

NORTH SEA

DENMARK

BALTIC SEA

Königsberg

EAST PRUSSIA

Bremerhaven

R. Elbe

Lübeck

Hamburg

Rostock

Swinemünde

Danzig

INTERNATIONAL OCCUPATION

Stettin

R. Oder

R. Vistula

Warsaw

U.S. CONTROL

Bremen

Wittenberg

R. Weser

RUSSIAN

BERLIN

ZONE

Magdeburg

POLAND

HOLLAND

BELGIUM

Arnhem

Brunswick

Cologne

Kassel

Leipzig

Erfurt

R. Mulde

R. Elbe

Dresden

R. Neisse

Breslau

R. Oder

Cracow

BRITISH

Frankfurt

Mainz

Würzburg

Nuremberg

Karlsbad

R. Vltava

Prague

Pilsen

CZECHOSLOVAKIA

AMERICAN

Karlsruhe

Regensburg

R. Danube

Budejovice

FRANCE

R. Rhine

Stuttgart

ZONE

Munich

Linz

Salzburg

Vienna

R. Danube

Budapest

L. Constance

Innsbruck

AUSTRIA

HUNGARY

SWITZERLAND

Klagenfurt

ITALY

YUGOSLAVIA

OCCUPATION ZONES IN GERMANY
AS AGREED AT QUEBEC, September 1944

Boundaries of Zones ──────────
National Frontiers, 1937 ─··─··─··─

0   50   100   150   200 MILES

The President, evidently convinced by the military view, had a large map unfolded on his knees. One afternoon, most of the Combined Chiefs of Staff being present, he agreed verbally with me that the existing arrangement should stand subject to the United States armies having a nearby direct outlet to the sea across the British zone. Bremen and its subsidiary Bremerhaven seemed to meet the American needs, and their control over this zone was adopted. This decision is illustrated on the accompanying map. We all felt it was too early as yet to provide for a French zone in Germany, and no one as much as mentioned Russia.

At Yalta in February 1945 the Quebec plan was accepted without further consideration as the working basis for the inconclusive discussions about the future eastern frontier of Germany. This was reserved for the Peace Treaty. The Soviet armies were at this very moment swarming over the prewar frontiers, and we wished them all success. We proposed an agreement about the zones of occupation in Austria. Stalin, after some persuasion, agreed to my strong appeal that the French should be allotted part of the American and British zones and given a seat on the Allied Control Commission. It was well understood by everyone that the agreed occupational zones must not hamper the operational movements of the armies. Berlin, Prague, and Vienna could be taken by whoever got there first. We separated in the Crimea not only as Allies but as friends facing a still mighty foe with whom all our armies were struggling in fierce and ceaseless battle.

The two months that had passed since then had seen tremendous changes cutting to the very roots of thought. Hitler's Germany was doomed and he himself about to perish. The Russians were fighting in Berlin. Vienna and most of Austria was in their hands. The whole relationship of Russia with the Western Allies was in flux. Every question about the future was unsettled between us. The agreements and understandings of Yalta, such as they were had already been broken or brushed aside by the triumphant Kremlin. New perils, perhaps as terrible as those we had surmounted, loomed and glared upon the torn and harassed world.

My concern at these ominous developments was apparent even before the President's death. He himself, as we have seen, was also anxious and disturbed. His anger at Molotov's accusations over the Berne affair has been recorded. In spite of the victorious advance of Eisenhower's armies, President Truman found himself faced in the last half of April with a formidable crisis. I had for some time past tried my utmost to impress the United States Government with the vast changes which were taking place both in the military and political spheres. Our Western armies would soon be carried well beyond the boundaries of our occupation zones, as both the Western and Eastern Allied fronts approached one another, penning the Germans between them.

Telegrams which I have published elsewhere show that I never suggested going back on our word over the agreed zones provided other agreements were also respected. I became convinced however that before we halted, or still more withdrew, our troops, we ought to seek a meeting with Stalin face to face and make sure that an agreement was reached about the whole front. It would indeed be a disaster if we kept all our agreements in strict good faith while the Soviets laid their hands upon all they could get without the slightest regard for the obligations into which they had entered.

General Eisenhower had proposed that while the armies in the west and the east should advance irrespective of demarcation lines, in any area where the armies had made contact either side should be free to suggest that the other should withdraw behind the boundaries of their occupation zone. Discretion to request and to order such withdrawals would rest with Army Group commanders. Subject to the dictates of operational necessity, the retirement would then take place. I considered that this proposal was premature and that it exceeded the immediate military needs. Action was taken accordingly, and on April 18 I addressed myself to the new President. Mr. Truman was of course only newly aware at second hand of all the complications that faced us, and had to lean heavily on his advisers. The purely military view therefore received an emphasis beyond its proper proportion. I cabled him as follows:

> ... I am quite prepared to adhere to the occupational zones, but I do not wish our Allied troops or your American troops to be hustled back at any point by some crude assertion of a local Russian general. This must be provided against by an agreement between Governments so as to give Eisenhower a fair chance to settle on the spot in his own admirable way.
>
> ... The occupational zones were decided rather hastily at Quebec in September 1944, when it was not foreseen that General Eisenhower's armies would make such a mighty inroad into Germany. The zones cannot be altered except by agreement with the Russians. But the moment VE-Day [Victory in Europe Day] has occurred we should try to set up the Allied Control Commission in Berlin and should insist upon a fair distribution of the food produced in Germany between all parts of Germany. As it stands at present the Russian occupational zone has the smallest proportion of people and grows by far the largest proportion of food, the Americans have a not very satisfactory proportion of food to conquered population, and we poor British are to take over all the ruined Ruhr and large manufacturing districts, which are, like ourselves, in normal times large importers of food. ...

Mr. Eden was in Washington, and fully agreed with the views I telegraphed to him, but Mr. Truman's reply carried us little further. He proposed that the Allied troops should retire to their agreed zones in Germany and Austria as soon as the military situation allowed.

*   *   *

Hitler had meanwhile pondered where to make his last stand. As late as April 20 he still thought of leaving Berlin for the "Southern Redoubt" in the Bavarian Alps. That day he held a meeting of the principal Nazi leaders. As the German double front, east and west, was in imminent danger of being cut in twain by the spearpoint thrust of the Allies, he agreed to set up two separate commands. Admiral Doenitz was to take charge in the north both of the military and civil authorities, with the particular task of bringing back to German soil nearly two million refugees from the east. In the south General Kesselring was to command the remaining German armies. These arrangements were to take effect if Berlin fell.

Two days later, on April 22, Hitler made his final and supreme decision to stay in Berlin to the end. The capital was soon completely encircled by the Russians and the Fuehrer had lost all power to control events. It remained for him to organise his own death amid the ruins of the city. He announced to the Nazi leaders who remained with him that he would die in Berlin. Goering and Himmler had both left after the conference of the 20th, with thoughts of peace negotiations in their minds. Goering, who had gone south, assumed that Hitler had in fact abdicated by his resolve to stay in Berlin, and asked for confirmation that he should act formally as the successor to the Fuehrer. The reply was his instant dismissal from all his offices. In a remote mountain village of the Tyrol he and nearly a hundred of the more senior officers of the Luftwaffe were taken prisoner by the Americans. Retribution had come at last.

The last scenes at Hitler's headquarters have been described elsewhere in much detail. Of the personalities of his regime only Goebbels and Bormann remained with him to the end. The Russian troops were now fighting in the streets of Berlin. In the early hours of April 29 Hitler made his will. The day opened with the normal routine of work in the air-raid shelter under the Chancellery. News arrived of Mussolini's end. The timing was grimly appropriate. On the 30th Hitler lunched quietly with his suite, and at the end of the meal shook hands with those present and retired to his private room. At half past three a shot was heard, and members of his personal staff entered the room to find him lying on the sofa with a revolver by his side. He had shot himself through the mouth. Eva Braun, whom he had married secretly during these last days, lay dead beside him. She had taken poison. The bodies were burnt in the courtyard, and Hitler's funeral pyre, with the din of the Russian guns growing ever louder, made a lurid end to the Third Reich.

The leaders who were left held a final conference. Last-minute attempts were made to negotiate with the Russians, but Zhukov demanded unconditional surrender. Bormann tried to break through the Russian lines, and disappeared without trace. Goebbels poisoned his six children and then ordered an S.S. guard to shoot his wife and himself. The remaining staff of Hitler's headquarters fell into Russian hands.

That evening a telegram reached Admiral Doenitz at his headquarters in Holstein:

> In place of the former Reich-Marshal Goering the Fuehrer appoints you, Herr Grand Admiral, as his successor. Written authority is on its way. You will immediately take all such measures as the situation requires. BORMANN.

Chaos descended. Doenitz had been in touch with Himmler, who, he assumed, would be nominated as Hitler's successor if Berlin fell, and now supreme authority was suddenly thrust upon him without warning and he faced the task of organising the surrender.

For Himmler a less spectacular end was reserved. He had gone to the Eastern Front and for some months had been urged to make personal contact with the Western Allies on his own initiative in the hope of negotiating a separate surrender. He now tried to do so through Count Bernadotte, the head of the Swedish Red Cross, but we repulsed his offers. No more was heard of him till May 21, when he was arrested by a British control post at Bremervörde. He was disguised and was not recognised, but his papers made the sentries suspicious and he was taken to a camp near Second Army Headquarters. He then told the commandant who he was. He was put under armed guard, stripped, and searched for poison by a doctor. During the final stage of the examination he bit open a phial of cyanide, which he had apparently hidden in his mouth for some hours. He died almost instantly, just after eleven o'clock at night on Wednesday, May 23.

*   *   *

In the northwest the drama closed less sensationally. On May 2 news arrived of the surrender in Italy. On the same day our troops reached Lübeck, on the Baltic, making contact with the Russians and cutting off all the Germans in Denmark and Norway. On the 3rd we entered Hamburg without opposition and the garrison surrendered

unconditionally. A German delegation came to Montgomery's headquarters on Luneberg Heath. It was headed by Admiral Friedeburg,
Doenitz's emissary, who sought a surrender agreement to include German troops in the North who were facing the Russians. This was
rejected as being beyond the authority of an Army Group commander,
who could deal only with his own front. Next day, having received
fresh instructions from his superiors, Friedeburg signed the surrender
of all German forces in Northwest Germany, Holland, the Islands,
Schleswig-Holstein, and Denmark.

Friedeburg went on to Eisenhower's headquarters at Reims, where
he was joined by General Jodl on May 6. They played for time to
allow as many soldiers and refugees as possible to disentangle themselves
from the Russians and come over to the Western Allies and they tried
to surrender the Western Front separately. Eisenhower imposed a
time-limit and insisted on a general capitulation. Jodl reported to
Doenitz: "General Eisenhower insists that we sign today. If not, the
Allied fronts will be closed to persons seeking to surrender individually.
I see no alternative — chaos or signature. I ask you to confirm to me
immediately by wireless that I have full powers to sign capitulation."

The instrument of total, unconditional surrender was signed by
Lieutenant General Bedell Smith and General Jodl, with French and
Russian officers as witnesses, at 2.41 A.M., on May 7. Thereby all
hostilities ceased at midnight on May 8. The formal ratification by the
German High Command took place in Berlin, under Russian arrange-
ments, in the early hours of May 9. Air Chief Marshal Tedder signed
on behalf of Eisenhower, Marshal Zhukov for the Russians, and Field-
Marshal Keitel for Germany.

*   *   *

The immense scale of events on land and in the air has tended to
obscure the no less impressive victory at sea. The whole Anglo-American campaign in Europe depended upon the movement of convoys
across the Atlantic, and we may here carry the story of the U-boats to
its conclusion. In spite of appalling losses to themselves they continued
to attack, but with diminishing success, and the flow of shipping was
unchecked. Even after the autumn of 1944, when they were forced
to abandon their bases in the Bay of Biscay, they did not despair.
The Schnorkel-fitted boats now in service, breathing through a tube
while charging their batteries submerged, were but an introduction to
the new pattern of U-boat warfare which Doenitz had planned. He
was counting on the advent of the new type of boat, of which very
many were now being built, and the first were already under trial.
Their high submerged speed threatened us with new problems, and

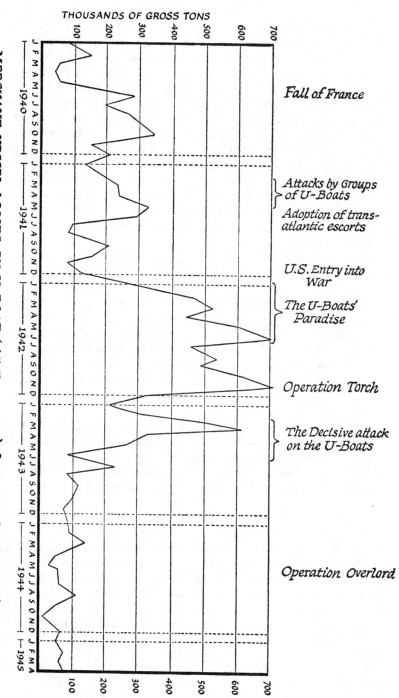

THOUSANDS OF GROSS TONS

700 600 500 400 300 200 100

Fall of France

Attacks by Groups
of U-Boats

Adoption of trans-
atlantic escorts

U.S. Entry into
War

The U-Boats'
Paradise

Operation Torch

The Decisive attack
on the U-Boats

Operation Overlord

MERCHANT VESSEL LOSSES BY U-BOAT (All Tonnages) January 1940 – April 1945 would indeed, as Doenitz predicted, have revolutionised U-boat warfare. His plans failed mainly because the special materials needed to construct these vessels became very scarce and their design had constantly to be changed. But ordinary U-boats were still being made piecemeal all over Germany and assembled in bomb-proof shelters at the ports, and in spite of the intense and continuing efforts of Allied bombers the Germans built more submarines in November 1944 than in any other month of the war. By stupendous efforts and in spite of all losses about sixty or seventy U-boats remained in action until almost the end. Their achievements were not large, but they carried the undying hope of stalemate at sea. The now revolutionary submarines never played their part in the Second World War. It had been planned to complete 350 of them during 1945, but only a few came into service before the capitulation. This weapon in Soviet hands lies among the hazards of the future.

Allied air attacks destroyed many U-boats at their berths. Nevertheless when Doenitz ordered them to surrender, no fewer than 49 were still at sea. Over a hundred more gave themselves up in harbour, and about 220 were scuttled or destroyed by their crews. Such was the persistence of Germany's effort and the fortitude of the U-boat service.

In sixty-eight months of fighting 781 German U-boats were lost. For more than half this time the enemy held the initiative. After 1942 the tables were turned; the destruction of U-boats rose and our losses fell. In the final count British and British-controlled forces destroyed 500 out of the 632 submarines known to have been sunk at sea by the Allies.

In the First World War eleven million tons of shipping were sunk, and in the second fourteen and a half million tons, by U-boats alone. If we add the loss from other causes the totals become twelve and three-quarter million and twenty-one and a half million. Of this the British bore over sixty per cent in the first war and over half in the second.

*   *   *

The unconditional surrender of our enemies was the signal for the greatest outburst of joy in the history of mankind. The Second World War had indeed been fought to the bitter end in Europe. The vanquished as well as the victors felt inexpressible relief. But for us in Britain and the British Empire, who had alone been in the struggle from the first day to the last and staked our existence on the result, there was a meaning beyond what even our most powerful and most valiant Allies could feel. Weary and worn, impoverished but undaunted and now triumphant, we had a moment that was sublime. We gave thanks to God for the noblest of all His blessings, the sense that we had done our duty.

When in these tumultuous days of rejoicing I was asked to speak to the nation I had borne the chief responsibility in our island for almost exactly five years. Yet it may well be there were few whose hearts were more heavily burdened with anxiety than mine. After reviewing the varied tale of our fortunes I struck a sombre note which may be recorded here.

"I wish," I said, "I could tell you tonight that all our toils and troubles were over. Then indeed I could end my five years' service happily, and if you thought that you had had enough of me and that I ought to be put out to grass I would take it with the best of grace. But, on the contrary, I must warn you, as I did when I began this five years' task — and no one knew then that it would last so long — that there is still a lot to do, and that you must be prepared for further efforts of mind and body and further sacrifices to great causes if you are not to fall back into the rut of inertia, the confusion of aim, and the craven fear of being great. You must not weaken in any way in your alert and vigilant frame of mind. Though holiday rejoicing is necessary to the human spirit, yet it must add to the strength and resilience with which every man and woman turns again to the work they have to do, and also to the outlook and watch they have to keep on public affairs.

"On the continent of Europe we have yet to make sure that the simple and honourable purposes for which we entered the war are not brushed aside or overlooked in the months following our success, and that the words 'freedom,' 'democracy,' and 'liberation' are not distorted from their true meaning as we have understood them. There would be little use in punishing the Hitlerites for their crimes if law and justice did not rule, and if totalitarian or police Governments were to take the place of the German invaders. We seek nothing for ourselves. But we must make sure that those causes which we fought for find recognition at the peace table in facts as well as words, and above all we must labour to ensure that the World Organisation which the United Nations are creating at San Francisco does not become an idle name, does not become a shield for the strong and a mockery for the weak. It is the victors who must search their hearts in their glowing hours, and be worthy by their nobility of the immense forces that they wield.

"We must never forget that beyond all lurks Japan, harrassed and failing. but still a people of a hundred millions, for whose warriors death has few terrors. I cannot tell you tonight how much time or what exertions will be required to compel the Japanese to make amends for their odious treachery and cruelty. We, like China, so long undaunted, have received horrible injuries from them ourselves, and we are bound by the ties of honour and fraternal loyalty to the United

States to fight this great war at the other end of the world at their side without flagging or failing. We must remember that Australia and New Zealand and Canada were and are all directly menaced by this evil Power. These Dominions come to our aid in our dark times, and we must not leave unfinished any task which concerns their safety and their future. I told you hard things at the beginning of these last five years; you did not shrink, and I should be unworthy of your confidence and generosity if I did not still cry: Forward, unflinching, unswerving, indomitable, till the whole task is done and the whole world is safe and clean."

27

The Chasm Opens

A PPREHENSION FOR THE FUTURE and many perplexi-
ties filled my mind as I moved about among the cheering
crowds of Londoners in their hour of well-won rejoicing after all they
had gone through. The Hitler peril, with its ordeals and privations,
seemed to most of them to have vanished in a blaze of glory. The
tremendous foe they had fought for more than five years had surrendered unconditionally. All that remained for the three victorious Powers
was to make a just and durable peace, guarded by a World Instument,
to bring the soldiers home to their longing loved ones, and to enter
upon a Golden Age of prosperity and progress. No more, and surely,
thought their peoples, no less.

However, there was another side to the picture. Japan was still unconquered. The atomic bomb was still unborn. The world was in
confusion. The main bond of common danger which had united the
Great Allies had vanished overnight. The Soviet menace, to my eyes,
had already replaced the Nazi foe. But no comradeship against it
existed. At home the foundations of national unity, upon which the
wartime Government had stood so firmly, were also gone. Our strength,
which had overcome so many storms, would no longer continue in the
sunshine. How then could we reach that final settlement which alone
could reward the toils and sufferings of the struggle? I could not
rid my mind of the fear that the victorious armies of democracy would
soon disperse and that the real and hardest test still lay before us.
I had seen it all before. I remembered that other joy-day nearly
thirty years before, when I had driven with my wife from the Ministry
of Munitions through similar multitudes convulsed with enthusiasm
to Downing Street to congratulate the Prime Minister. Then, as at

this time, I understood the world situation as a whole. But then at least there was no mighty army that we need fear.

*　　　*　　　*

My prime thought was a meeting of the three Great Powers and I hoped that President Truman would come through London on his way. As will be seen, very different ideas were being pressed upon the new President from influential quarters in Washington. The sort of mood and outlook which had been noticed at Yalta had been strengthened. The United States, it was argued, must be careful not to let herself be drawn into any antagonism with Soviet Russia. This, it was thought, would stimulate British ambition and would make a new gulf in Europe. The right policy, on the other hand, should be for the United States to stand between Britain and Russia as a friendly mediator, or even arbiter, trying to reduce their differences about Poland or Austria and make things settle down into a quiet and happy peace, enabling American forces to be concentrated against Japan. These pressures must have been very strong upon Truman. His natural instinct, as his historic actions have shown, may well have been different. I could not of course measure the forces at work in the brain centre of our closest Ally, though I was soon conscious of them. I could only feel the vast manifestation of Soviet and Russian imperialism rolling forward over helpless lands.

Obviously the first aim must be a conference with Stalin. Within three days of the German surrender I cabled the President that we should invite him to a conference. *"Meanwhile I earnestly hope that the American front will not recede from the now agreed tactical lines.*[1] He replied at once that he would rather have Stalin propose the meeting, and he hoped our Ambassadors would induce him to suggest it. Mr. Truman then declared that he and I ought to go to the meeting separately, so as to avoid any suspicion of "ganging up." When the Conference ended he hoped to visit England if his duties in America permitted. I did not fail to notice the difference of view which this telegram conveyed, but I accepted the procedure he proposed.

In these same days I also sent what may be called the "Iron Curtain" telegram to President Truman. Of all the public documents I have written on this issue I would rather be judged by this.

> I am profoundly concerned about the European situation. I learn that half the American Air Force in Europe has already begun to move to the Pacific theatre. The newspapers are full of the great movements of the American armies out of Europe. Our

---

[1] My subsequent italics.—W.S.C.

armies also are, under previous arrangements, likely to undergo a marked reduction. The Canadian Army will certainly leave. The French are weak and difficult to deal with. Anyone can see that in a very short space of time our armed power on the Continent will have vanished, except for moderate forces to hold down Germany.

2. Meanwhile what is to happen about Russia? I have always worked for friendship with Russia, but, like you, I feel deep anxiety because of their misrepresentation of the Yalta decisions, their attitude towards Poland, their overwhelming influence in the Balkans, excepting Greece, the difficulties they make about Vienna, the combination of Russian power and the territories under their control or occupied, coupled with the Communist technique in so many other countries, and above all their power to maintain very large armies in the field for a long time. What will be the position in a year or two, when the British and American Armies have melted and the French has not yet been formed on any major scale, when we may have a handful of divisions, mostly French, and when Russia may choose to keep two or three hundred on active service?

3. An iron curtain is drawn down upon their front. We do not know what is going on behind. There seems little doubt that the whole of the regions east of the line Lübeck–Trieste–Corfu will soon be completely in their hands. To this must be added the further enormous area conquered by the American armies between Eisenach and the Elbe, which will, I suppose, in a few weeks be occupied, when the Americans retreat, by the Russian power. All kinds of arrangements will have to be made by General Eisenhower to prevent another immense flight of the German population westward as this enormous Muscovite advance into the centre of Europe takes place. And then the curtain will descend again to a very large extent, if not entirely. Thus a broad band of many hundreds of miles of Russian-occupied territory will isolate us from Poland.

4. Meanwhile the attention of our peoples will be occupied in inflicting severities upon Germany, which is ruined and prostrate, and it would be open to the Russians in a very short time to advance if they chose to the waters of the North Sea and the Atlantic.

5. Surely it is vital now to come to an understanding with Russia, or see where we are with her, before we weaken our armies mortally or retire to the zones of occupation. This can only be done by a personal meeting. I should be most grateful for your

opinion and advice. Of course we may take the view that Russia will behave impeccably, and no doubt that offers the most convenient solution. To sum up, this issue of a settlement with Russia before our strength has gone seems to me to dwarf all others.

A week passed before I heard again from Mr. Truman on the major issues. Then on May 22 he cabled that he had asked Mr. Joseph E. Davies to come to see me before the Triple Conference, about a number of matters he preferred not to handle by cable.

Mr. Davies had been the American Ambassador in Russia before the war, and was known to be most sympathetic to the regime. He had in fact written a book on his mission to Moscow which was also produced as a film which seemed in many ways to palliate the Soviet system. I of course made immediate arrangements to receive him, and he spent the night of the 26th at Chequers. I had a very long talk with him. The crux of what he had to propose was that the President should meet Stalin first somewhere in Europe before he saw me. I was indeed astonished at this suggestion. I had not liked the President's use in his earlier message of the term "ganging up" as applied to any meeting between him and me. Britain and the United States were united by bonds of principle and by agreement upon policy in many directions, and we were both at profound difference with the Soviets on many of the greatest issues. For the President and the British Prime Minister to talk together upon this common ground, as we had so often done in Roosevelt's day, could not now deserve the disparaging expression "ganging up." On the other hand, for the President to by-pass Great Britain and meet the head of the Soviet State alone would have been, not indeed a case of "ganging up" — for that was impossible — but an attempt to reach a singlehanded understanding with Russia on the main issues upon which we and the Americans were united. I would not agree in any circumstances to what seemed to be an affront, however unintentional, to our country after its faithful service in the cause of freedom from the first day of the war. I objected to the implicit idea that the new disputes now opening with the Soviets lay between Britain and Russia. The United States was as fully concerned and committed as ourselves. I made this quite clear to Mr. Davies in our conversation, which also ranged over the whole field of Eastern and Southern European affairs, and in order that there should be no misconception I drafted and gave him a formal minute in this sense. The President received it in a kindly and understanding spirit, and I was very glad to learn that all was well and that the justice of our view was not unrecognised by our cherished friends.

About the same time as President Truman sent Mr. Davies to see me he had asked Harry Hopkins to go as his special envoy to Moscow to make another attempt to reach a working agreement on the Polish question. Although far from well, Hopkins set out gallantly for Moscow. His friendship for Russia was well known, and he received a most friendly welcome. Certainly for the first time some progress was made. Stalin agreed to invite Mikolajczyk and two of his colleagues to Moscow from London for consultation, in conformity with our interpretation of the Yalta agreement. He also agreed to invite some important non-Lublin Poles from inside Poland.

In a telegram to me the President said he felt this was a very encouraging, positive stage in the negotiations. Most of the arrested Polish leaders were apparently only charged with operating illegal radio transmitters, and Hopkins was pressing Stalin to grant them an amnesty so that the consultations could be conducted in the most favourable atmosphere possible. He asked me to urge Mikolajczyk to accept Stalin's invitation. I persuaded Mikolajczyk to go to Moscow, and in the upshot a new Polish Provisional Government was set up. At Truman's request this was recognised by both Britain and the United States on July 5.

It is difficult to see what more we could have done. For five months the Soviets had fought every inch of the road. They had gained their object by delay. During all this time the Lublin Administration, under Bierut, sustained by the might of the Russian armies, had given them a complete control of Poland, enforced by the usual deportations and liquidations. They had denied us all the access for our observers which they had promised. All the Polish parties, except their own Communist puppets, were in a hopeless minority in the new recognised Polish Provisional Government. We were as far as ever from any real and fair attempt to obtain the will of the Polish nation by free elections. There was still a hope — and it was the only hope — that the meeting of "the Three," now impending, would enable a genuine and honourable settlement to be achieved. So far only dust and ashes have been gathered, and these are all that remain to us today of Polish national freedom.

*   *   *

On June 1 President Truman told me that Marshal Stalin was agreeable to a meeting of what he called "the Three" in Berlin about July 15. I replied at once that I would gladly go to Berlin with a British delegation, but I thought that July 15, which Truman had suggested, was much too late for the urgent questions demanding attention between us, and that we should do an injury to world hopes and unity if we allowed personal or national requirements to stand in the way of an earlier meeting. "Although," I cabled, "I am in the midst of a hotly

contested election I would not consider my tasks here as comparable
to a meeting between the three of us. If June 15 is not possible why not
July 1, 2, or 3?" Mr. Truman replied that after full consideration
July 15 was the earliest for him, and that arrangements were being
made accordingly. Stalin did not wish to hasten the date. I could not
press the matter further.

The main reason why I had been anxious to hasten the date of the
meeting was of course the impending retirement of the American Army
from the line which it had gained in the fighting to the zone prescribed
in the occupation agreement. The story of the agreement about the
zones and the arguments for and against changing them are recorded
in the previous chapter. I feared that any day a decision might be
taken in Washington to yield up this enormous area — 400 miles long
and 120 at its greatest depth. It contained many millions of Germans
and Czechs. Its abandonment would place a broader gulf of territory
between us and Poland, and practically end our power to influence her
fate. The changed demeanour of Russia towards us, the constant
breaches of the understandings reached at Yalta, the dart for Denmark,
happily frustrated by Montgomery's timely action, the encroachments
in Austria, Marshal Tito's menacing pressure at Trieste, all seemed to
me and my advisers to create an entirely different situation from that
in which the zones of occupation had been prescribed two years earlier.
Surely all these issues should be considered as a whole, and *now* was the
time. Now, while the British and American armies and air forces were
still a mighty armed power, and before they melted away under
demobilisation and the heavy claims of the Japanese war — now, at
the very latest, was the time for a general settlement.

A month earlier would have been better. But it was not yet too late.
On the other hand, to give up the whole centre and heart of Germany
— nay, the centre and keystone of Europe — as an isolated act seemed
to me to be a grave and improvident decision. If it were done at all
it could only be as part of a general and lasting settlement. We should
go to Potsdam with nothing to bargain with, and all the prospects of the
future peace of Europe might well go by default. The matter however
did not rest with me. Our own retirement to the occupation frontier
was inconsiderable. The American Army was three millions to our one.
All I could do was to plead, first, for advancing the date of the meeting of "the Three," and, secondly, when that failed, to postpone the
withdrawal until we could confront all our problems as a whole, together, face to face, and on equal terms.

How stands the scene after eight years have passed? The Russian
occupation line in Europe runs from Lübeck to Linz. Czechoslovakia
has been engulfed. The Baltic states, Poland, Rumania, and Bulgaria

have been reduced to satellite states under totalitarian Communist rule. Yugoslavia has broken loose. Greece alone is saved. Our armies are gone, and it will be a long time before even sixty divisions can be once again assembled opposite Russian forces, which in armour and manpower are in overwhelming strength. This also takes no account of all that has happened in the Far East. The danger of a third World War, under conditions at the outset of grave disadvantage, casts its lurid shadow over the free nations of the world. Thus in the moment of victory was our best, and what might prove to have been our last, chance of durable world peace allowed composedly to fade away.[2] On June 4 I cabled to the President these words, which few would now dispute:

> I am sure you understand the reason why I am anxious for an earlier date, say the 3rd or 4th [of July]. I view with profound misgivings the retreat of the American Army to our line of occupation in the central sector, thus bringing Soviet power into the heart of Western Europe and the descent of an iron curtain between us and everything to the eastward. I hoped that this retreat, if it has to be made, would be accompanied by the settlement of many great things which would be the true foundation of world peace. Nothing really important has been settled yet, and you and I will have to bear great responsibility for the future. I still hope therefore that the date will be advanced.

Mr. Truman replied on June 12. He said that the tripartite agreement about the occupation of Germany, approved by President Roosevelt after "long consideration and detailed discussion" with me, made it impossible to delay the withdrawal of American troops from the Soviet Zone in order to press the settlement of other problems. The Allied Control Council could not begin to function until they left, and the military government exercised by the Allied Supreme Commander should be terminated without delay and divided between Eisenhower and Montgomery. He had been advised, he said, that it would harm our relations with the Soviet to postpone action until our meeting in July, and he accordingly proposed sending a message to Stalin.

This document suggested that we should at once instruct our armies to occupy their respective zones. He was ready to order all American troops to begin withdrawing from Germany on June 21. The military commanders should arrange for the simultaneous occupation of Berlin and for free access thereto by road, rail, and air from Frankfurt

[2] Written in 1953.—W.S.C.

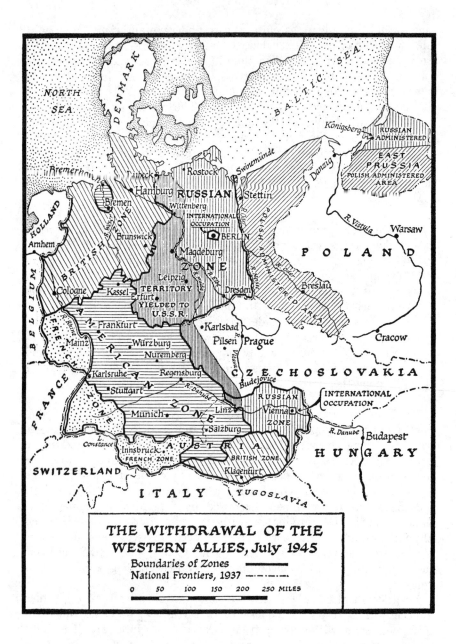

NORTH
SEA

DENMARK

BALTIC SEA

Königsberg
RUSSIAN
ADMINISTERED

EAST
PRUSSIA
POLISH ADMINISTERED
AREA

Bremerhaven
Lübeck
Rostock
Swinemünde
Hamburg
RUSSIAN
Stettin
Wittenberg
INTERNATIONAL
OCCUPATION
BERLIN

Danzig

R. Vistula
Warsaw

HOLLAND
BELGIUM
Arnhem
Bremen
BRITISH ZONE
Brunswick
Magdeburg
ZONE

P O L A N D

POLISH ADMINISTERED AREA

Cologne
AMERICAN
Kassel
Leipzig
Erfurt
TERRITORY
YIELDED TO
U.S.S.R.
Dresden
Breslau
Cracow

FRANCE
FRENCH
Frankfurt
Mainz
ZONE
Würzburg
Nuremberg
Karlsruhe
Regensburg
Stuttgart
R. Danube
ZONE
Munich
ZONE
Linz
Salzburg

Karlsbad
Pilsen
Prague
Budejovice

C Z E C H O S L O V A K I A

INTERNATIONAL
OCCUPATION

RUSSIAN
ZONE
Vienna

R. Danube
Budapest

H U N G A R Y

SWITZERLAND
L.
Constance
Innsbruck
FRENCH ZONE
A U S T R I A
Klagenfurt
BRITISH ZONE

I T A L Y
YUGOSLAVIA

THE WITHDRAWAL OF THE
WESTERN ALLIES, July 1945

Boundaries of Zones ————
National Frontiers, 1937 —·—·—·—

0   50   100   150   200   250 MILES

and Bremen for the United States forces. In Austria arrangements could be completed more quickly and satisfactorily by making the local commanders responsible for defining the zones both there and in Vienna, only referring to their Governments such matters as they were unable to resolve themselves.

This struck a knell in my breast. But I had no choice but to submit. There was nothing more that I could do. It must not be overlooked that Mr. Truman had not been concerned or consulted in the original fixing of the zones. The case as presented to him so soon after his accession to power was whether or not to depart from and in a sense repudiate the policy of the American and British Governments agreed under his illustrious predecessor. He was, I have no doubt, supported in his action by his advisers, military and civil. His responsibility at this point was limited to deciding whether circumstances had changed so fundamentally that an entirely different procedure should be adopted, with the likelihood of having to face accusations of breach of faith. Those who are only wise after the event should hold their peace.

On July 1 the United States and British Armies began their withdrawal to their allotted zones, followed by masses of refugees. Soviet Russia was established in the heart of Europe. This was a fateful milestone for mankind.

*   *   *

While all this was passing I was plunged into the turmoil of the general election, which began in earnest in the first week of June. This month was therefore hard to live through. Strenuous motor tours to the greatest cities of England and Scotland, with three or four speeches a day to enormous and, it seemed, enthusiastic crowds, and, above all, four laboriously prepared broadcasts, consumed my time and strength. All the while I felt that much we had fought for in our long struggle in Europe was slipping away and that the hopes of an early and lasting peace were receding. The days were passed amid the clamour of multitudes, and when at night, tired out, I got back to my headquarters train, where a considerable staff and all the incoming telegrams awaited me, I had to toil for many hours. The incongruity of party excitement and clatter with the sombre background which filled my mind was in itself an affront to reality and proportion. I was glad indeed when polling day at last arrived and the ballot papers were safely sealed for three weeks in their boxes.

I was resolved to have a week of sunshine to myself before the Conference. On July 7, two days after polling day, I flew to Bordeaux with Mrs. Churchill and Mary, and found myself agreeably installed at General Brutinel's villa near the Spanish frontier at Hendaye, with lovely bathing and beautiful surroundings. I spent most of the mornings in bed reading a very good account, by an excellent French writer, of the Bordeaux armistice and its tragic sequel at Oran. It was strange to revive my own memories of five years before and to learn of many things which I had not known at that time. In the afternoons I even sallied forth with my elaborate painting outfit, and found attractive subjects on the river Nive and the Bay of St. Jean de Luz. I found a gifted companion of the brush in Mrs. Nairn, the wife of the British Consul at Bordeaux, with whom I made friends at Marrakesh a year before. I dealt only with a few telegrams about the impending Conference, and strove to put party politics out of my head. And yet I must confess the mystery of the ballot boxes and their contents had an ugly trick of knocking on the door and peering in at the windows. When the palette was spread and I had a paintbrush in my hand it was easy to drive these intruders away.

The Basque people were everywhere warm in their welcome. They had endured a long spell of German occupation and were joyful to breathe freely again. I did not need to prepare myself for the Conference, for I carried so much of it in my head, and was happy to cast it off, if only for these few fleeting days. The President was at sea in the United States cruiser *Augusta,* the same ship which had carried Roosevelt to our Atlantic meeting in 1941. On the 15th I motored through the forests to the Bordeaux airfield, and my Skymaster took me to Berlin.

28

The Atomic Bomb

PRESIDENT TRUMAN arrived in Berlin the same day as I did. I was eager to meet a potentate with whom my cordial relations, in spite of differences, had already been established by correspondence. I called on him the morning after our arrival, and was impressed with his gay, precise, sparkling manner and obvious power of decision.

On July 16 both the President and I made separate tours of Berlin. The city was nothing but a chaos of ruins. No notice had of course been given of our visit and the streets had only the ordinary passers-by. In the square in front of the Chancellery there was however a considerable crowd. When I got out of the car and walked about among them, except for one old man who shook his head disapprovingly, they all began to cheer. My hate had died with their surrender and I was much moved by their demonstrations, and also by their haggard looks and threadbare clothes. Then we entered the Chancellery, and for quite a long time walked through its shattered galleries and halls. Our Russian guides then took us to Hitler's air-raid shelter. I went down to the bottom and saw the room in which he and his wife had committed suicide, and when we came up again they showed us the place where his body had been burned. We were given the best firsthand accounts available at that time of what had happened in these final scenes.

The course Hitler had taken was much more convenient for us than the one I had feared. At any time in the last few months of the war he could have flown to England and surrendered himself, saying, "Do what you will with me, but spare my misguided people." I have no doubt that he would have shared the fate of the Nuremberg criminals. The moral principles of modern civilisation seem to prescribe that the leaders of a nation defeated in war shall be put to death by the victors.

This will certainly stir them to fight to the bitter end in any future war, and no matter how many lives are needlessly sacrificed it costs them no more. It is the masses of the people who have so little to say about the starting or ending of wars who pay the additional cost. The Romans followed the opposite principle, and their conquests were due almost as much to their clemency as to their prowess.

*   *   *

On July 17 world-shaking news arrived. In the afternoon Stimson called at my abode and laid before me a sheet of paper on which was written, "Babies satisfactorily born." By his manner I saw something extraordinary had happened. "It means," he said, "that the experiment in the Mexican desert has come off. The atomic bomb is a reality." Although we had followed this dire quest with every scrap of information imparted to us, we had not been told beforehand, or at any rate I did not know, the date of the decisive trial. No responsible scientist would predict what would happen when the first full-scale atomic explosion was tried. Were these bombs useless or were they annihilating? Now we knew. The "babies" had been "satisfactorily born." No one could yet measure the immediate military consequences of the discovery, and no one has yet measured anything else about it.

Next morning a plane arrived with a full description of this tremendous event in the human story. Stimson brought me the report. I tell the tale as I recall it. The bomb, or its equivalent, had been detonated at the top of a pylon one hundred feet high. Everyone had been cleared away for ten miles round, and the scientists and their staffs crouched behind massive concrete shields and shelters at about that distance. The blast had been terrific. An enormous column of flame and smoke shot up to the fringe of the atmosphere of our poor earth. Devastation inside a one-mile circle was absolute. Here then was a speedy end to the Second World War, and perhaps to much else besides.

The President invited me to confer with him forthwith. He had with him General Marshall and Admiral Leahy. Up to this moment we had shaped our ideas towards an assault upon the homeland of Japan by terrific air bombing and by the invasion of very large armies. We had contemplated the desperate resistance of the Japanese fighting to the death with Samurai devotion, not only in pitched battles, but in every cave and dugout. I had in my mind the spectacle of Okinawa island, where many thousands of Japanese, rather than surrender, had drawn up in line and destroyed themselves by hand-grenades after their leaders had solemnly performed the rite of hara-kiri. To quell the Japanese resistance man by man and conquer the country yard by yard might well require the loss of a million American lives and half that number of British — or more if we could get them there: for we were resolved to share the agony. Now all this nightmare picture had vanished. In its place was the vision — fair and bright indeed it seemed — of the end of the whole war in one or two violent shocks. I thought immediately myself of how the Japanese people, whose courage I had always admired, might find in the apparition of this almost supernatural weapon an excuse which would save their honour and release them from their obligation of being killed to the last fighting man.

Moreover, we should not need the Russians. The end of the Japanese war no longer depended upon the pouring in of their armies for the final and perhaps protracted slaughter. We had no need to ask favours of them. The array of European problems could therefore be faced on their merits and according to the broad principles of the United Nations. We seemed suddenly to have become possessed of a merciful abridgment of the slaughter in the East and of a far happier prospect in Europe. I have no doubt that these thoughts were present in the minds of my American friends. At any rate, there never was a moment's discussion as to whether the atomic bomb should be used or not. To avert a vast, indefinite butchery, to bring the war to an end, to give peace to the world, to lay healing hands upon its tortured peoples by a manifestation of overwhelming power at the cost of a few explosions, seemed, after all our toils and perils, a miracle of deliverance.

British consent in principle to the use of the weapon had been given on July 4, before the test had taken place. The final decision now lay in the main with President Truman, who had the weapon; but I never doubted what it would be, nor have I ever doubted since that he was right. The historic fact remains, and must be judged in the aftertime, that the decision whether or not to use the atomic bomb to compel the surrender of Japan was never even an issue. There was unanimous, automatic, unquestioned agreement around our table; nor did I ever hear the slightest suggestion that we should do otherwise.

A more intricate question was what to tell Stalin. The President and I no longer felt that we needed his aid to conquer Japan. His word had been given at Teheran and Yalta that Soviet Russia would attack Japan as soon as the German Army was defeated, and in fulfilment of this a continuous movement of Russian troops to the Far East had been in progress over the Siberian Railway since the beginning of May. In our opinion they were not likely to be needed, and Stalin's bargaining power, which he had used with effect upon the Americans at Yalta, was therefore gone. Still, he had been a magnificent ally in the war against Hitler, and we both felt that he must be informed of the great New Fact which now dominated the scene, but not of any particulars. How should this news be imparted to him? Should it be in writing or by word of mouth? Should it be at a formal and special meeting, or in the course of our daily conferences, or after one of them? The conclusion which the President came to was the last of these alternatives. "I think," he said, "I had best just tell him after one of our meetings that we have an entirely novel form of bomb, something quite out of the ordinary, which we think will have decisive effects upon the Japanese will to continue the war." I agreed to this procedure.

*   *   *

Meanwhile the devastating attack on Japan had continued from the air and the sea. By the end of July the Japanese Navy had virtually ceased to exist. The homeland was in chaos and on the verge of collapse. The professional diplomats were convinced that only immediate surrender under the authority of the Emperor could save Japan from complete disintegration, but power still lay almost entirely in the hands of a military clique determined to commit the nation to mass suicide rather than accept defeat. The appalling destruction confronting them made no impression on this fanatical hierarchy, who continued to profess belief in some miracle which would turn the scale in their favour.

In several lengthy talks with the President alone, or with his advisers present, I discussed what to do. I dwelt upon the tremendous cost in American and to a smaller extent in British life if we enforced "unconditional surrender" upon the Japanese. It was for him to consider whether this might not be expressed in some other way, so that we got all the essentials for future peace and security and yet left them some show of saving their military honour and some assurance of their national existence, after they had complied with all safeguards necessary for the conqueror. The President replied bluntly that he did not think the Japanese had any military honour after Pearl Harbour. I contented myself with saying that at any rate they had something for which they were ready to face certain death in very large numbers, and this might not be so important to us as it was to them. He then became quite sympathetic, and spoke, as had Mr. Stimson, of the terrible responsibilities that rested upon him for the unlimited effusion of American blood.

Eventually it was decided to send an ultimatum calling for an immediate unconditional surrender of the armed forces of Japan. This document was published on July 26. Its terms were rejected by the military rulers of Japan, and the United States Air Force made its plans accordingly to cast one atomic bomb on Hiroshima and one on Nagasaki.

We agreed to give every chance to the inhabitants. The procedure was developed in detail. In order to minimise the loss of life eleven Japanese cities were warned by leaflets on July 27 that they would be subjected to intensive air bombardment. Next day six of them were attacked. Twelve more were warned on July 31, and four were bombed on August 1. The last warning was given on August 5. By then the Superfortresses claimed to have dropped a million and a half leaflets every day and three million copies of the ultimatum. The first atomic bomb was not cast till August 6.

On August 9 the Hiroshima bomb was followed by a second, this time on the city of Nagasaki. Next day, despite an insurrection by some military extremists, the Japanese Government agreed to accept the ultimatum, provided this did not prejudice the prerogative of the Emperor as a sovereign ruler. The Allied Fleets entered Tokyo Bay, and on the morning of September 2 the formal instrument of surrender was signed on board the United States battleship *Missouri*. Russia had declared war on August 8, only a week before the enemy's collapse. None the less she claimed her full rights as a belligerent.

It would be a mistake to suppose that the fate of Japan was settled by the atomic bomb. Her defeat was certain before the first bomb fell, and was brought about by overwhelming maritime power. This alone had made it possible to seize ocean bases from which to launch the final attack and force her metropolitan army to capitulate without striking a blow. Her shipping had been destroyed. She had entered the war with over five and a half million tons, later much augmented by captures and new construction, but her convoy system and escorts were inadequate and ill organised. Over eight and a half million tons of Japanese shipping were sunk, of which five million fell to submarines. We, an island Power, equally dependent on the sea, can read the lesson and understand our own fate had we failed to master the U-boats.

*   *   *

Frustration was the fate of this final Conference of "the Three." I shall not attempt to describe all the questions which were raised though not settled at our various meetings. I content myself with telling the tale, so far as I was then aware of it, of the atomic bomb and outlining the terrible issue of the German-Polish frontiers. These events dwell with us today.

We had agreed at Yalta that Russia should advance her western frontier into Poland as far as the Curzon Line. We had always recognised that Poland in her turn should receive substantial accessions of German territory. The question was, how much? How far into Germany should she go? There had been much disagreement. Stalin had

wanted to extend the western frontier of Poland along the river Oder
to where it joined the Western Neisse; Roosevelt, Eden, and I had
insisted it should stop at the Eastern Neisse. All three heads of Governments had publicly bound themselves at Yalta to consult the Polish
Government, and to leave it to the Peace Conference for final settlement. This was the best we had been able to do. But in July 1945 we
faced a new situation. Russia had advanced her frontier to the Curzon
Line. This meant, as Roosevelt and I had realised, that the three or
four million Poles who lived on the wrong side of the line would
have to be moved to the west. Now we were confronted with something much worse. The Soviet dominated Government of Poland had
also pressed forward, not to the Eastern Neisse, but to the Western.
Much of this territory was inhabited by Germans, and although several
millions had fled many had stayed behind. What was to be done with
them? Moving three or four million Poles was bad enough. Were we
to move more than eight million Germans as well? Even if such a
transfer could be contemplated, there was not enough food for them in
what was left of Germany. Much of Germany's grain came from the
very land which the Poles had seized, and if this was denied us the
Western Allies would be left with wrecked industrial zones and a
starved and swollen population. For the future peace of Europe here
was a wrong beside which Alsace-Lorraine and the Danzig Corridor
were trifles. One day the Germans would want their territory back,
and the Poles would not be able to stop them.

*　　　　*　　　　*

It remains for me only to mention some of the social and personal
contacts which relieved our sombre debates. Each of the three great
delegations entertained the other two. First was the United States.
When it came to my turn I proposed the toast of "The Leader of the
Opposition," adding, "whoever he may be." Mr. Attlee, whom I had
invited to the Conference in accordance with my conviction that every
head of Government in periods of crisis should have a deputy who
knows everything and can thus preserve continuity should accidents
occur, was much amused by this. So indeed were the company. The
Soviets' dinner was equally agreeable, and a very fine concert, at which
leading Russian artists performed, carried the proceedings so late that
I slipped away.

It fell to me to give the final banquet on the night of the 23rd. I
planned this on a larger scale, inviting the chief commanders as well as
the delegates. I placed the President on my right and Stalin on my left.
There were many speeches, and Stalin, without even ensuring that all
the waiters and orderlies had left the room, proposed that our next

meeting should be in Tokyo. There was no doubt that the Russian declaration of war upon Japan would come at any moment, and already their large armies were massed upon the frontier ready to overrun the much weaker Japanese front line in Manchuria. To lighten the proceedings we changed places from time to time, and the President sat opposite me. I had another very friendly talk with Stalin, who was in the best of tempers and seemed to have no inkling of the momentous information about the new bomb the President had given me. He spoke with enthusiasm about the Russian intervention against Japan, and seemed to expect a good many months of war, which Russia would wage on an ever-increasing scale, governed only by the Trans-Siberian Railway.

Then a very odd thing happened. My formidable guest got up from his seat with the bill-of-fare card in his hand and went round the table collecting the signatures of many of those who were present. I never thought to see him as an autograph-hunter! When he came back to me I wrote my name as he desired, and we both looked at each other and laughed. Stalin's eyes twinkled with mirth and good humour. I have mentioned before how the toasts at these banquets were always drunk by the Soviet representatives out of tiny glasses, and Stalin had never varied from this practice. But now I thought I would take him on a step. So I filled a small-sized claret glass with brandy for him and another for myself. I looked at him significantly. We both drained our glasses at a stroke and gazed approvingly at one another. After a pause Stalin said, "If you find it impossible to give us a fortified position in the Marmora, could we not have a base at Dedeagatch?" I contented myself with saying, "I will always support Russia in her claim to the freedom of the seas all the year round."

Next day, July 24, after our plenary meeting had ended and we all got up from the round table and stood about in twos and threes before dispersing, I saw the President go up to Stalin, and the two conversed alone with only their interpreters. I was perhaps five yards away, and I watched with the closest attention the momentous talk. I knew what the President was going to do. What was vital to measure was its effect on Stalin. I can see it all if it were yesterday. He seemed to be delighted. A new bomb! Of extraordinary power! Probably decisive on the whole Japanese war! What a bit of luck! This was my impression at the moment, and I was sure he had no idea of the significance of what he was being told. Evidently in his intense toils and stresses the atomic bomb had played no part. If he had had the slightest idea of the revolution in world affairs which was in progress his reactions would have been obvious. Nothing would have been easier than for him to say, "Thank you so much for telling me about your new bomb. I of course have no technical knowledge. May I send my expert in these nuclear sciences to see your expert tomorrow morning?" But his face remained gay and genial and the talk between these two potentates soon came to an end. As we were waiting for our cars I found myself near Truman. "How did it go?" I asked. "He never asked a question," he replied.

On the morning of the 25th the Conference met again. This was the last meeting I attended. I urged once more that Poland's western frontier could not be settled without taking into account the million and a quarter Germans who were still in the area, and the President emphasised that any peace treaty could only be ratified with the advice and consent of the Senate. We must, he said, find a solution which he could honestly recommend to the American people. I said that if the Poles were allowed to assume the position of a fifth occupying Power without arrangements being made for spreading the food produced in Germany equally over the whole German population, and without our agreeing about reparations or war booty, the Conference would have failed. This network of problems lay at the very heart of our work, and so far we had come to no agreement. The wrangle went on. Stalin said that getting coal and metal from the Ruhr was more important than food. I said they would have to be bartered against supplies from the East. How else could the miners win coal? "They have imported food from abroad before, and can do so again," was the answer. And how could they pay reparations? "There is still a good deal of fat left in Germany," was the grim reply. I refused to accept starvation in the Ruhr because the Poles held all the grain lands in the East. Britain herself was short of coal. "Then use German prisoners in the mines; that is what I am doing," said Stalin. "There are forty thousand German troops still in Norway, and you can get them from there." "We are exporting our own coal," I said, "to France, Holland, and Belgium. Why should the Poles sell coal to Sweden while Britain is denying herself for the liberated countries?" "But that is Russian coal," Stalin answered. "Our position is even more difficult than yours. We lost over five million men in the war, and we are desperately short of labour." I put my point once again. "We will send coal from the Ruhr to Poland or anywhere else providing we get in exchange food for the miners who produce it."

This seemed to make Stalin pause. He said the whole problem needed consideration. I agreed, and said I only wanted to point out the difficulties in front of us. Here, so far as I am concerned, was the end of the matter.

*   *   *

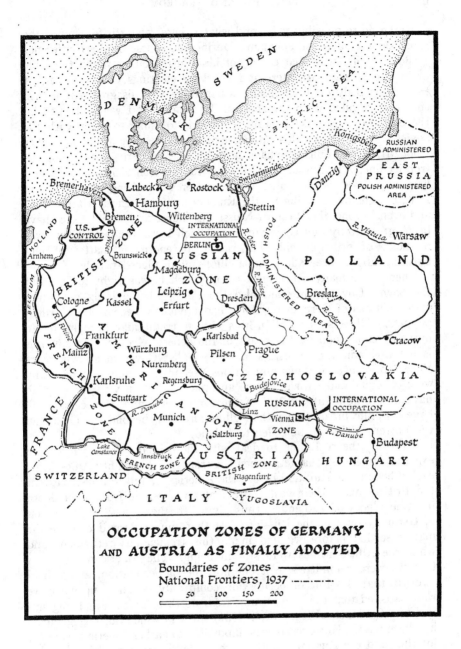

OCCUPATION ZONES OF GERMANY AND AUSTRIA AS FINALLY ADOPTED

Boundaries of Zones ————
National Frontiers, 1937 —‥—‥—

0   50   100   150   200

I take no responsibility beyond what is here set forth for any of the conclusions reached at Potsdam. During the course of the Conference I allowed differences that could not be adjusted either round the table or by the Foreign Ministers at their daily meetings to stand over. A formidable body of questions on which there was disagreement was in consequence piled upon the shelves. I intended, if I were returned by the electorate, as was generally expected, to come to grips with the Soviet Government on this catalogue of decisions. For instance, neither I nor Mr. Eden would ever have agreed to the Western Neisse being the frontier line. The line of the Oder and the Eastern Neisse had already been recognised as the Polish compensation for retiring to the Curzon Line, but the overrunning by the Russian armies of the territory up to and even beyond the Western Neisse was never and would never have been agreed to by any Government of which I was the head. Here was no point of principle only but rather an enormous matter of fact affecting about three additional millions of displaced people.

There were many other matters on which it was right to confront the Soviet Government, and also the Poles, who, gulping down immense chunks of German territory, had obviously become their ardent puppets. All this negotiation was cut in twain and brought to an untimely conclusion by the result of the general election. To say this is not to blame the Ministers of the Socialist Government, who were forced to go over without any serious preparation, and who naturally were unacquainted with the ideas and plans I had in view, namely, to have a showdown at the end of the Conference, and, if necessary, to have a public break rather than allow anything beyond the Oder and the Eastern Neisse to be ceded to Poland.

However, the real time to deal with these issues was, as has been explained in earlier chapters, when the fronts of the mighty Allies faced each other in the field, and before the Americans, and to a lesser extent the British, made their vast retirement on 400-mile front to a depth in some places of 120 miles, thus giving the heart and a great mass of Germany over to the Russians. At that time I desired to have the matter settled before we had made this tremendous retirement and while the Allied armies were still in being. The American view was that we were committed to a definite line of occupation, and I held strongly that this line of occupation could only be taken up when we were satisfied that the whole front, from north to south, was being settled in accordance with the desires and spirit in which our engagements had been made. However, it was impossible to gather American support for this, and the Russians, pushing the Poles in front of them, wended on, driving the Germans before them and depopulating large areas of Germany, whose food supplies they had seized, while chasing a multitude

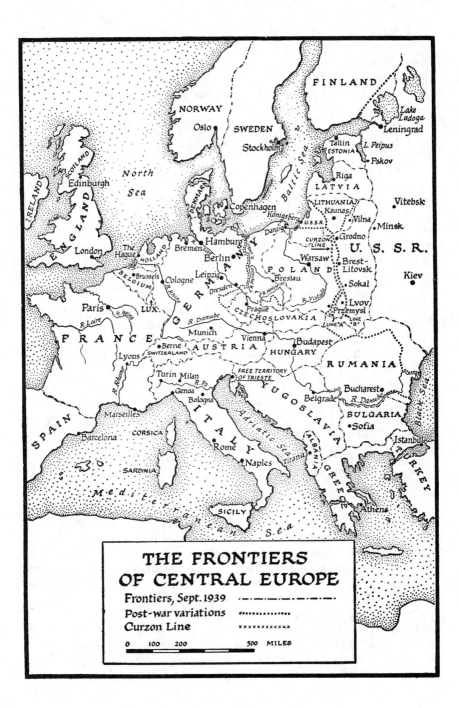

THE FRONTIERS
OF CENTRAL EUROPE

Frontiers, Sept. 1939	–·–·–·–·–·
Post-war variations	•••••••••••
Curzon Line	××××××××××

0   100   200   500   MILES

of mouths into the overcrowded British and American zones. Even at Potsdam the matter might perhaps have been recovered, but the destruction of the British National Government and my removal from the scene at the time when I still had much influence and power rendered it impossible for satisfactory solutions to be reached.

I flew home with my daughter Mary on the afternoon of July 25. My wife met me at Northolt, and we all dined quietly together.

Excellent arrangements had been made by Captain Pim and the staff of the Map Room to present a continuous tale of election results as they came in next day. The latest view of the Conservative Central Office was that we should retain a substantial majority. I had not burdened myself unduly with the subject while occupied with the grave business of the Conference. On the whole I accepted the view of the party managers, and went to bed in the belief that the British people would wish me to continue my work. My hope was that it would be possible to reconstitute the National Coalition Government in the proportions of the new House of Commons. Thus slumber. However, just before dawn I woke suddenly with a sharp stab of almost physical pain. A hitherto subconscious conviction that we were beaten broke forth and dominated my mind. All the pressure of great events, on and against which I had mentally so long maintained my "flying speed," would cease and I should fall. The power to shape the future would be denied me. The knowledge and experience I had gathered, the authority and goodwill I had gained in so many countries, would vanish. I was discontented at the prospect, and turned over at once to sleep again. I did not wake till nine o'clock, and when I went into the Map Room the first results had begun to come in. They were, as I now expected, unfavourable. By noon it was clear that the Socialists would have a majority. At luncheon my wife said to me, "It may well be a blessing in disguise." I replied, "At the moment it seems quite effectively disguised."

In ordinary circumstances I should have felt free to take a few days to wind up the affairs of the Government in the usual manner. Constitutionally I could have awaited the meeting of Parliament in a few days' time, and taken my dismissal from the House of Commons. This would have enabled me to present before resignation the unconditional surrender of Japan to the nation. The need for Britain being immediately represented with proper authority at the Conference, where all the great issues we had discussed were now to come to a head, made all delay contrary to the public interest. Moreover, the verdict of the electors had been so overwhelmingly expressed that I did not wish to remain even for an hour responsible for their affairs. At seven o'clock therefore, having asked for an audience, I drove to the Palace, tendered my resignation to the King, and advised His Majesty to send for Mr. Attlee.

I issued to the nation the following message, with which this account may close:

26 July 45

The decision of the British people has been recorded in the votes counted today. I have therefore laid down the charge which was placed upon me in darker times. I regret that I have not been permitted to finish the work against Japan. For this however all plans and preparations have been made, and the results may come much quicker than we have hitherto been entitled to expect. Immense responsibilities abroad and at home fall upon the new Government, and we must all hope that they will be successful in bearing them.

It only remains for me to express to the British people, for whom I have acted in these perilous years, my profound gratitude for the unflinching, unswerving support which they have given me during my task, and for the many expressions of kindness which they have shown towards their servant.

Epilogue

Epilogue

THE LONG TASK I set myself in writing the six volumes of the *Second World War* will now appear in an abridged form for the use of those who wish to know what happened without being cumbered with too much detail, especially military detail.

This gives me an opportunity to look back and express my views on some of the major events of the last twelve years.

When I left Potsdam on the 25th of July, 1945, I certainly expected that the election figures would leave me a reasonable majority, and it was startling to be confronted with the facts. Entirely absorbed as I had been in the prosecution of the war and the situation at its victorious close, I did not understand what had taken place in the British Isles. Otherwise, I thought and still think I could have arranged things differently. Above all, the opinion in the mass of the Army, after so many signs of goodwill, was a great surprise to me. The election results and figures were an even greater surprise to Europe and America, and indeed to the U.S.S.R. They naturally thought that the steadfastness of the British peoples, having survived the grim ordeals of 1940 and having come triumphantly through the five years' struggle, would remain unshaken, and that there would be no change of Government.

During the course of the Conference at Potsdam I had not so far sought to come to grips with Russia. Since Yalta she had behaved in an astonishing fashion. I had earnestly hoped that the Americans would not withdraw from the wide territories in Central Europe they had conquered before we met. This was the one card that the Allies held when the fighting stopped by which to arrange a level settlement. Britain sought nothing for herself, but I was sure she would view the vast advance which Russia was making in all directions as far exceeding what was fair. The Americans seemed quite unconscious of the situation, and the satellite states, as they came to be called, were occupied by Russian troops. Berlin was already in their hands, though Montgomery could have taken it had he been permitted. Vienna was Russian-held, and representatives of the Allies, even as individuals, were denied access to this key capital. As for the Balkans, Bulgaria and Rumania had already been conquered. Yugoslavia quivered under Tito, her famous patriotic leader. The Russians had occupied Prague with, as it seemed, the approval of the Americans. They held Poland, whose western boundary, it was agreed, should be moved into the heart of Europe at the expense of Germany. All these steps had in fact been taken by the Russians while their armies were still advancing. Yet the American view seemed to be that all this was a necessary part of the process of holding down Germany, and that the great national object of the United States was not to get drawn into siding too closely with Britain against Russia.

*   *   *

When the winter came along I went to the United States and remained in that country for several months. I visited the White House and the State Department. I there received an invitation to address the Westminster College in Fulton, Missouri, in March 1946. Mr. Truman had said he would himself preside. This was several months ahead, and I kept myself as fully informed as was possible. I made inquiries both at the White House and at the State Department in order to learn whether certain topics would cause embarrassment, and having been assured that I could say what I liked I devoted myself to the careful preparation of a speech. Meanwhile the dire situation with which the insatiable appetites of Russia and of international Communism were confronting us was at last beginning to make a strong impression in American circles. I showed the notes I had prepared to Mr. Byrnes, then Secretary of State, and found that he was very much in agreement with me. The President invited me to travel with him in his train the long night's journey to Fulton. We had an enjoyable game of poker. That was the only topic which I remember. However, as I was quite sure that his Secretary of State had imparted my general line to the President, and he seemed quite happy about it, I decided to go ahead. One always has to be very careful about speeches which you make in other people's countries. This is from what I said:

> A shadow has fallen upon the scenes so lately lighted by the Allied victory. Nobody knows what Soviet Russia and its Communist international organisation intends to do in the immediate future, or what are the limits, if any, to their expansive and

proselytising tendencies. I have a strong admiration and regard for the valiant Russian people and for my wartime comrade, Marshal Stalin. There is deep sympathy and goodwill in Britain — and I doubt not here also — towards the peoples of all the Russias, and a resolve to persevere through many differences and rebuffs in establishing lasting friendships. We understand the Russian need to be secure on her western frontiers by the removal of all possibility of German aggression. We welcome Russia to her rightful place among the leading nations of the world. We welcome her flag upon the seas. Above all, we welcome constant, frequent and growing contacts between the Russian people and our own people on both sides of the Atlantic. It is my duty, however, for I am sure you would wish me to state the facts as I see them to you, to place before you certain facts about the present position in Europe.

From Stettin in the Baltic to Trieste in the Adriatic, an iron curtain has descended across the Continent. Behind that line lie all the capitals of the ancient states of Central and Eastern Europe. Warsaw, Berlin, Prague, Vienna, Budapest, Belgrade, Bucharest, and Sofia, all these famous cities and the populations around them lie in what I must call the Soviet sphere, and all are subject in one form or another, not only to Soviet influence but to a very high and, in many cases, increasing measure of control from Moscow. Athens alone — Greece with its immortal glories — is free to decide its future at an election under British, American, and French observation. The Russian-dominated Polish Government has been encouraged to make enormous and wrongful inroads upon Germany, and mass expulsions of millions of Germans on a scale grievous and undreamed of are now taking place. The Communist parties, which were very small in all these Eastern States of Europe, have been raised to pre-eminence and power far beyond their numbers and are seeking everywhere to obtain totalitarian control. Police governments are prevailing in nearly every case, and so far, except in Czechoslovakia, there is no true democracy.

Turkey and Persia are both profoundly alarmed and disturbed at the claims which are being made upon them and at the pressure being exerted by the Moscow Government. An attempt is being made by the Russians in Berlin to build up a quasi-Communist party in their zone of Occupied Germany by showing special favours to groups of left-wing German leaders. At the end of the fighting last June, the American and British armies withdrew westward, in accordance with an earlier agreement, to a depth at some points of one hundred and fifty miles upon a front of nearly four hundred miles, in order to allow our Russian allies to occupy this vast expanse of territory which the Western democracies had conquered.

If now the Soviet Government tries, by separate action, to build up a pro-Communist Germany in their areas, this will cause new serious difficulties in the British and American zones, and will give the defeated Germans the power of putting themselves up to auction between the Soviets and the Western democracies. Whatever conclusions may be drawn from these facts — and facts they are — this is certainly not the Liberated Europe we fought to build up. Nor is it one which contains the essentials of permanent peace.

The audience listened with great attention, and the President and Mr. Byrnes both expressed their approval. The newspapers, however, were very varied in their comments. When the news reached Russia, it was ill received, and both Stalin and the *Pravda* responded as might be expected. The *Pravda* denounced me as "an anti-Soviet warmonger," and said I was trying to destroy the United Nations. Stalin in a newspaper interview accused me of calling for war against the Soviet Union and compared me with Hitler. Questions were also asked in the House of Commons, to which Mr. Attlee, now Prime Minister, replied that the Government was not called upon to express any opinion on a speech delivered in another country by a private individual.

I had another speech to deliver a few days later in New York, where I was the guest of the Mayor and civic authorities. All round the Waldorf Astoria Hotel, at the dinner where it was delivered, were marching pickets of Communists, and I was somewhat surprised to learn that Mr. Dean Acheson, the Under-Secretary of State, was not coming. When Mr. John Winant heard of this change of plan in Washington in the afternoon, he caught a train to New York and arrived in the middle of dinner to support me, and made a most friendly speech. I expressed myself as follows:

> When I spoke at Fulton ten days ago I felt it was necessary for someone in an unofficial position to speak in arresting terms about the present plight of the world. I do not wish to withdraw or modify a single word. I was invited to give my counsel freely in this free country and I am sure that the hope which I expressed for the increasing association of our two countries will come to pass, not because of any speech which may be made, but because of the tides that flow in human affairs and in the course of the unfolding destiny of the world. The only question which in my opinion is open is whether the necessary harmony of thought and action between the American and British peoples will be reached in a suf-

ficiently plain and clear manner and in good time to prevent a new world struggle or whether it will come about, as it has done before, only in the course of that struggle. . . .

. . . Let me declare, however, that the progress and freedom of all the people of the world, under a reign of law enforced by a world organisation, will not come to pass, nor will the age of plenty begin, without the persistent, faithful, and above all fearless exertions of the British and American systems of society.

The agitation in the newspapers and the general interest, and even excitement, continued to grow.

*		*		*

I spent the early autumn of 1946 painting in a lovely villa by the Lake of Geneva with Mont Blanc over the water in the background. When it became time to go I paid a very pleasant visit to Zurich University, and made them a speech about the tragedy of Europe and the plight to which she had been reduced, and I urged the foundation of a kind of United States of Europe, or as much of it as could be done.

I was very glad to read in the newspapers two days ago that my friend President Truman had expressed his interest and sympathy with this great design. There is no reason why a regional organisation should in any way conflict with the world organisation of the United Nations. On the contrary, I believe that the larger synthesis will only survive if it is founded upon coherent natural groupings. There is already a natural grouping in the Western Hemisphere. We British have our own Commonwealth of Nations. These do not weaken, on the contrary they strengthen, the world organisation. They are in fact its main support. And why should there not be a European group which could give a sense of enlarged patriotism and common citizenship to the distracted peoples of this turbulent and mighty continent and why should it not take its rightful place with other great groupings in shaping the destinies of men? In order that this should be accomplished there must be an act of faith in which millions of families speaking many languages must consciously take part.

We all know that the two world wars through which we have passed arose out of the vain passion of a newly united Germany to play the dominating part in the world. . . . Germany must be deprived of the power to rearm and make another aggressive war. But when all this has been done, as it will be done, as it is being

done, there must be an end to retribution. There must be what Mr.
Gladstone many years ago called "a blessed act of oblivion." We
must all turn our backs upon the horrors of the past. We must
look to the future. We cannot afford to drag forward across the
years that are to come the hatreds and revenges which have sprung
from the injuries of the past. If Europe is to be saved from infinite
misery, and indeed from final doom, there must be an act of faith
in the European family and an act of oblivion against all the crimes
and follies of the past.

   . . . I am now going to say something that will astonish you. The
first step in the re-creation of the European family must be a
partnership between France and Germany. In this way only can
France recover the moral leadership of Europe. There can be no
revival of Europe without a spiritually great France and a spiritually
great Germany. The structure of the United States of Europe, if
well and truly built, will be such as to make the material strength
of a single state less important. Small nations will count as much
as large ones and gain their honour by their contribution to the
common cause. The ancient states and principalities of Germany,
freely joined together for mutual convenience in a federal system,
might each take their individual place among the United States of
Europe. I shall not try to make a detailed programme for hundreds of millions of people who want to be happy and free, pros-
perous and safe, who wish to enjoy the four freedoms of which the
great President Roosevelt spoke, and live in accordance with the
principles embodied in the Atlantic Charter. If this is their wish,
they have only to say so, and means can certainly be found, and
machinery erected, to carry that wish into full fruition.

   But I must give you a warning. Time may be short. At present
there is a breathing space. The cannon have ceased firing. The
fighting has stopped; but the dangers have not stopped. If we are
to form the United States of Europe or whatever name or form it
may take, we must begin now.

Thus ran my thoughts in 1946. To tortured France, lately occupied
and humiliated, the spectacle of close association with her finally vanquished executioner seemed at first unthinkable. By degrees, however,
the flow of European fraternity was restored in French veins, and natural Gallic pliant good sense overcame the bitterness of the past.
   I have always held, and hold, the valiant Russian people in high
regard. But their shadow loomed disastrously over the postwar scene.
There was no visible limit to the harm they might do. Intent on victory
over the Axis Powers, Britain and America had laid no sufficient plans

for the fate and future of occupied Europe. We had gone to war in defence not only of the independence of smaller countries but to proclaim and endorse the individual rights and freedoms on which this greater morality is based. Soviet Russia had other and less disinterested aims. Her grip tightened on the territories her armies had overrun. In all the satellite states behind the Iron Curtain, coalition governments had been set up, including Communists. It was hoped that democracy in some form would be preserved. But in one country after another the Communists seized the key posts, harried and suppressed the other political parties, and drove their leaders into exile. There were trials and purges. Rumania, Hungary, and Bulgaria were soon engulfed. At Yalta and Potsdam I had fought hard for Poland, but it was in vain. In Czechoslovakia a sudden coup was carried out by the Communist Ministers, which sharply alerted world opinion. Freedom was crushed within and free intercourse with the West was forbidden. Thanks largely to Britain, Greece remained precariously independent. With British and later American aid, she fought a long civil war against the insurgent Communists. When all had been said and done, and after the long agonies and efforts of the Second World War, it seemed that half Europe had merely exchanged one despot for another.

Today, these points seem commonplace. The prolonged and not altogether unsuccessful struggle to halt the destroying tide of Russian and Russian-inspired incursion has become part of our daily lives. Indeed, as always with a good cause, it has sometimes been necessary to temper enthusiasm and to disregard opportunism. But it was not easy at the time to turn from the contemplation of a great and exhausting victory over one tyranny to the prospect of a tedious and expensive campaign against another.

The United Nations Organisation was still very young, but already it was clear that its defects might prove grave enough to vitiate the purposes for which it was created. At any rate it could not provide quickly and effectively the union and the armed forces which Free Europe and the United States needed for self-preservation. At Fulton I had suggested that the United Nations Organisation should forthwith be equipped with an international armed force. But both for the immediate future and the long term I had urged the continuation of the special Anglo-American relationship which has been one of the main themes of my political life.

Neither the sure prevention of war nor the continuous rise of world organisation will be gained without what I have called the fraternal association of the English-speaking peoples. This means a special relationship between the British Commonwealth and Empire and the United States. . . . It should carry with it a continuance of the present facilities for mutual security by the joint use of all naval and air force bases in the possession of either country all over the world. . . . The United States has already a permanent defence agreement with the Dominion of Canada . . . this principle should be extended to all British Commonwealths with full reciprocity.

The next three years were to see the unfolding of a design that approached but has not yet attained this ideal.

               *   *   *

I do not wish to claim a monopoly of credit for these conceptions. One of the advantages of being in Opposition is that one can outdistance in imagination those whose fortune it is to put plans into practical effect. The British Government, much inspired by the stouthearted and wise Mr. Ernest Bevin, took the lead in rebuilding something of the Concert of Europe, at least in what was left of Europe. Initial thoughts were mainly of the dangers of a resurrected Germany. In 1947 Britain and France signed the Treaty of Dunkirk binding each to come to the other's assistance if there was another German attack. But already the grim realities of the present were overshadowing the fears of the past. After many months of diplomatic activity the Brussels Treaty was signed in 1948. France, Great Britain, the Netherlands, the Belgians, and Luxembourg undertook to assist one another against aggression, from whatever quarter it might come. Germany was not mentioned. Moreover, the beginnings of a military organisation were set up under the chairmanship of Field Marshal Montgomery to assess the resources available for defence and to draw up a plan with what little was available. This became known as the Western Union. I endorsed these measures, but vehemently hoped that the United States, without whose aid they would be woefully incomplete, would soon be brought into the association. We were fortunate at the time to have in the American Secretary of State the far-sighted and devoted General Marshall, with whom we had worked in closest comradeship and confidence in the war years. Within the limits imposed by American congressional and public opinion, President Truman and he sought to add weight to what was being done in Europe. The efforts on both sides of the Atlantic bore fruit, and in April 1949 the North Atlantic Treaty was signed, in which for the first time in history the United States bound herself, subject always to the constitutional prerogative of Congress, to aid her allies if they were attacked. The European signatories, besides the Brussels Treaty Powers, included Norway, Denmark, Iceland, Italy and Portugal. Canada also acceded to the Treaty, thereby giving additional proof to the faith we in Britain have always entertained for her friendship and loyalty.

The work that followed was complex. It resulted in the setting up of the North Atlantic Treaty Organisation, headed by a military planning staff under General Eisenhower at Versailles. From the efforts of the Supreme Headquarters Atlantic Powers Europe, or SHAPE as it was called, there gradually grew a sober confidence that invasion from the East could be met by an effective resistance. Certainly in its early stages the Atlantic Treaty achieved more by being than by doing. It gave renewed confidence to Europe, particularly to the territories near Soviet Russia and the satellites. This was marked by a recession in the Communist parties in the threatened countries, and by a resurgence of healthy national vigour in Western Germany.

The association of Germany with the Atlantic Treaty remained in the forefront of Western plans. But it was very difficult to overcome French fears of a revived German army, and the topic was a fruitful one for the misguided as well as the mischievous. Within the space of seventy years the French had been invaded three times from across the Rhine. It was hard to forget Sedan, the blood bath at Verdun, the collapse in 1940, the long, grinding occupation of the Second World War, which had sundered so many loyalties, and in which Frenchman had fought Frenchman. In Britain I was conscious of a wide hostility to giving weapons, even under the strictest safeguards, to the new German Republic. But it was unlikely that a Soviet invasion of Western Europe could ever be repulsed without the help of the Germans. Many schemes were tried and failed. The French had taken the lead in the closer integration of Western Europe in civil matters, and they sponsored a scheme for a European army with a common uniform, into which German units would be merged without risk to their neighbours. I did not care for this idea. A sludgy amalgam of half a dozen nationalities would find it difficult to share common loyalties and the trust which is essential among comrades in battle. It was not for some years that the final simplicity of a direct German contribution through a national army to the strength of the West was achieved. Even today little has been done to put it into effect. I myself have never seen the disadvantage of making friends with your enemy when the war is over, with all that that implies in co-operation against an outside menace.

Side by side with these developments, many of them lying only in the paper sphere, the United States continued to manifest her determination to assist Europe, and thus herself. Long before the Atlantic Treaty was signed American aircraft were stationed in East Anglia in substantial numbers. Here was a most practical deterrent. Alas, the splendid structure of the Anglo-American Combined Chiefs of Staff, who had been the architects of so much of our victorious war planning,

had been dismantled at American instigation. Nothing has subsequently equalled it, and the best of the NATO arrangements are but a
poor shadow of the fraternal and closely knit organisation that formerly
existed.

The crucial test came in June 1948 when the Russians cut off Berlin
from the outside world. Their object was to incorporate the whole of
Berlin in the Communist state which they had promoted In Eastern
Germany. It seemed that Britain, France, and America must either
abandon the city or try to force supply convoys in from Western Germany, as was their legitimate right. Fortunately a solution was found
which avoided many perils. The Airlift began, and by February 1949
over a million tons of supplies had been flown in by American and
British aircraft during the preceding eight months of the blockade. This
original conception was highly successful. In due course the Russians
had to yield, and they were forced to abandon the blockade entirely.

Economic assistance to the Allies was also vital. We in Britain had
spent so much money in the war that, even if the greatest skill and
economy had been exercised, we would have been very hard pressed.
In spite of a huge American loan the position was getting ever more
serious. The rest of Europe was also suffering in varying degrees. General Marshall gave his name to a remarkable plan for economic aid and
mutual co-operation among sixteen free European countries. The benefits were offered to the Soviet bloc, but were refused. The Organisation
for European and Economic Co-operation has rendered the greatest
service to us all. But without the massive dollar aid provided by the
American administration, in spite of some hostility on the part of Congress, Europe might well have foundered into ruin and misery in which
the seeds of Communism would have grown at a deadly pace. General
Marshall's decision was on the highest level of statesmanship, and it was
a source of great pleasure, but not surprise, to me that my old friend
should have presided in America over the two great enterprises of the
Marshall Plan and the Atlantic Treaty.

*   *   *

There was another aspect to our conceptions and hopes for the
unifying and strengthening of Europe against external aggression and
internal subversion. The ideas that I had opened at Fulton were translated to a great extent into reality through governmental action and
the chain of treaties and official organisations which I have briefly
described. It was also important for more far-reaching conceptions of
the final ideal of a United Europe to find a forum in which they could
be discussed and examined. Many distinguished European statesmen
and leaders of thought held the same views, and in 1947 the European

Movement was launched to devote itself to the propagation of the theme of European unity and the examination of ways in which it could gradually be put into effect. I say gradually. There were many different opinions among those concerned, and some wanted to go faster than others. In large enterprises it is a mistake to try to settle everything at once. In matters of this kind it was not possible to plan movements as in a military operation. We were not acting in the field of force, but in the domain of opinion. I several times stressed my views on this point. It was of importance that when the inevitable lulls, delays and obstacles occurred we should not be considered to have abandoned our ultimate goal. Moreover, I did not wish to compete with governments in the executive sphere. The task was to build up moral, cultural, sentimental and social unities and affinities throughout Europe.

The European Movement gained greatly in vigour and strength and played a substantial part in governmental thinking. General Marshall referred to the concept as one of the reasons which had led him to his plan for economic aid to Europe. The culmination of the many discussions that took place came in the creation of the Council of Europe in 1949, with its seat at Strasbourg. With varying fortunes and shades of publicity much useful work was done at Strasbourg. There are those who are disappointed that the rapid creation of a federation of European States did not ensue, but there is every justification for a slow and empirical approach. Such weighty matters cannot be imposed on the people from above, however brilliant the planning. They must grow gradually from genuine and widely held convictions. Thus the Council of Europe is serving its purpose and playing an honourable part in a great enterprise.

* * *

As the stark and glaring background to all our cogitations on defence lay man's final possession of the perfected means of human destruction: the atomic weapon and its monstrous child, the hydrogen bomb. In the early days of the war Britain and the United States had agreed to pool their knowledge, and experiments in nuclear research, and the fruits of years of discovery by the English pioneer physicists were offered as a priceless contribution to the vast and most secret joint enterprise set on foot in the United States and Canada. Those who created the weapons possessed for a few years the monopoly of a power which might in less scrupulous hands have been used to dominate and enslave the entire world. They proved themselves worthy of their responsibilities, but secrets were soon disclosed to the Soviet Union which greatly helped Russian scientists in their researches. Henceforward most of the accepted theories of strategy were seen to be out of date, and a new undreamed-of balance of power was created, a balance based on the owner-ship of the means of mutual extermination.

At the end of the war I felt reasonably content that the best possible arrangement had been made in the agreement which I concluded with President Roosevelt in Quebec in 1943. Therein Britain and America affirmed that they would never use the weapon against each other, that they would not use it against third parties without each other's approval, that they would not communicate information on the subject to third parties except by mutual consent, and that they would exchange in-formation on technical developments. I do not think that one could have asked for more.

However, in 1946 a measure was passed by the American Congress which most severely curtailed any chance of the United States provid-ing us with information. Senator McMahon, who sponsored the bill, was at the time unaware of the Quebec Agreement, and he informed me in 1952 that if he had seen it there would have been no McMahon Act. The British Socialist Government certainly made some sort of a protest, but they felt unable to press it home and they did not insist on the revelation of the Quebec Agreement, at least to the McMahon Com-mittee, which would have vindicated our position and perhaps saved us many years of wearisome and expensive research and development. Thus, deprived of our share of the knowledge to which we had a most certain right, Britain had to fall back on her own resources. The So-cialist Government thereupon devoted vast sums to research, but it was not until 1952 that we were able to explode our first atomic bomb. The relative stages of research and development remain unknown, but ex-perimental explosions are not the sole criterion, and we may perhaps in some ways claim to have outdistanced even the United States. But research is one thing, production and possession another.

It was in this, then, the American possession or preponderance of nuclear weapons, that the surest foundation of our hopes for peace lay. The armies of the Western Powers were of comparative insignificance when faced with the innumerable Russian divisions that could be de-ployed from the Baltic to the Yugoslav frontier. But the certain knowl-edge that an advance on land would unleash the devouring destruction of strategic air attack was and is the most certain of deterrents.

For a time, when the United States was the sole effective possessor of nuclear weapons, there had been a chance of a general and permanent settlement with the Soviet Union. But it is not the nature of democra-cies to use their advantages in threatening or dictatorial ways. Certainly the state of opinion that prevailed in those years would not have tol-erated anything in the way of rough words to our late ally, though this might well have forestalled many unpleasant developments. Instead the

United States, with our support, chose a most reasonable and liberal attitude to the problems of controlling the use of nuclear weapons. Soviet opposition to efficient methods of supervision brought this to nothing. In former days no country could hope to build up in secret military forces vast enough to overwhelm a neighbour. Now the means of destruction of many millions can be concealed in the space of a few cubic yards.

Every aspect of military and political planning was altered by these developments. The vast bases needed to sustain the armies of the two world wars have become the most vulnerable of targets. All the workshops and stores on the Suez Canal, which had fed the Eighth Army in the desert, could vanish in a flash at the stroke of a single aircraft. Harbours, even when guarded by anti-aircraft guns and fighter planes, could become the graveyard of the fleets they had once protected. The evacuation of non-combatants from the cities was a practical proposition even in the days of highly developed bombing methods in the last war. Now, desirable though they may be, such measures are a mere palliative to the flaring ruin of nuclear attack. The whole structure of defence had to be altered to meet the new situation. Conventional forces were still needed to keep order in our possssions, and to fight what people call the small wars, but we could not afford enough of them because nuclear weapons and the means of delivering them were so expensive.

The nuclear age transformed the relations between the Great Powers. For a time I doubted whether the Kremlin accurately realised what would happen to their country in the event of war. It seemed possible that they neither knew the full effect of the atomic missiles nor how efficient were the means of delivering them. It even occurred to me that an announced but peaceful aerial demonstration over the main Soviet cities, coupled with the outlining to the Soviet leaders of some of our newest inventions, would produce in them a more friendly and sober attitude. Of course such a gesture could not have been accompanied by any formal demands, or it would have taken on the appearance of a threat and ultimatum. But Russian production of these weapons and the remarkable strides of their air force have long since removed the point of this idea. Their military and political leaders must now be well aware of what each of us could do to the other.

*   *   *

Hopes of more friendly contacts with Russia remained much in my mind, and the death of Stalin in March 1953 seemed to bring a chance. I was again Prime Minister. I regarded Stalin's death as a milestone in Russian history. His tyranny had brought fearful suffering to his own country and to much else of the world. In their fight against Hitler, the Russian peoples had built up an immense goodwill in the West, not least of all in the United States. All this had been impaired. In the dark politics of the Kremlin, none could tell who would take Stalin's place. Fourteen men and one hundred and eighty million people lost their master. The Soviet leaders must not be judged too harshly. Three times in the space of just over a century Russia has been invaded by Europe. Borodino, Tannenberg, and Stalingrad are not to be forgotten easily. Napoleon's onslaught is still remembered. Imperial and Nazi Germany were not forgiven. But security can never be achieved in isolation. Stalin tried not only to shield the Soviet Republics behind an iron curtain, military, political, and cultural. He also attempted to construct an outpost line of satellite states, deep in Central Europe, harshly controlled from Moscow, subservient to the economic needs of the Soviet Union and forbidden all contact or communion with the free world, or even with each other. No one can believe that this will last forever. Hungary has paid a terrible forfeit. But to all thinking men, certain hopeful features of the present situation must surely be clear. The doctrine of Communism is slowly being separated from the Russian military machine. Nations will continue to rebel against the Soviet Colonial Empire, not because it is Communist, but because it is alien and oppressive. An arms race, even conducted with nuclear weapons and guided missiles, will bring no security or even peace of mind to the great powers which dominate the land masses of Asia and North America, or to the countries which lie between them. I make no plea for disarmament. Disarmament is a consequence and a manifestation of free intercourse between free peoples. It is the mind which controls the weapon, and it is to the minds of the peoples of Russia and her associates that the free nations should address themselves.

But after Stalin's death it seemed that a milder climate might prevail. At all events it merited investigating, and I so expressed myself in the House of Commons on May 11, 1953. An entirely informal conference between the heads of the leading powers might succeed where repeated acrimonious exchanges at lower levels had failed. I made it plain that this could not be accompanied by any relaxation of the comradeship and preparations of the free nations, for any slackening of our defence efforts would paralyse every beneficial tendency towards peace. This is true today. What I sought was never fully accomplished. Nevertheless for a time a gentler breeze seemed to blow upon our affairs. Further opportunities will doubtless present themselves, and they must not be neglected.

*   *   *

It is not my purpose to attempt to assign blame in any quarter for the many disagreeable things that have occurred since 1945. Certainly those who were responsible in Great Britain for the direction of our affairs in the years that followed the war were beset by the most complex and malignant problems both at home and abroad. The methods by which they chose to solve them were often forced upon them by circumstances or by predetermined doctrinaire policies, and their results were not always felicitous either for Britain or the free world.

The granting of independence to the Indian subcontinent had long been in the forefront of British political thought. I had contributed a good deal to the subject in the years between the wars. Supported by seventy Conservative members, I had fought it in its early stages with all my strength. When I was at the head of the Coalition Government I was induced to modify my former views. Undoubtedly we came out of the desperate world struggle committed to Dominion status for India, including the right to secede from the Commonwealth. I thought however that the method of setting up the new Government should have given the great majority of the Indian people the power and the right to choose freely for themselves. I believed that a constitutional conference in which all the real elements of strength in India could participate would have shown us the way to produce a truly representative self-governing India which would adhere to the British Empire. The "untouchables," the Rajahs, the loyalists, of whom hundreds of millions existed, and many other different, vital, living interests, would all have had their share in the new scheme. It must be remembered that in the last year of the war we had had a revolt of the extremists in the Indian Congress party which was put down without difficulty, and with very little loss of life. The British Socialist Party took a violently factional view. They believed that the advantage lay in the granting of self-government within the shortest space of time. And they gave it without hesitation — almost identifiably — to the forces which we had vanquished so easily. Within two years of the end of the war they had achieved their purpose. On the 18th of August, 1947, Indian independence was declared. All efforts to preserve the unity of India had broken down, and Pakistan became a separate state. Four hundred million inhabitants of the subcontinent, mainly divided between Moslem and Hindu, flung themselves at one another. Two centuries of British rule in India were followed by greater bloodshed and loss of life than had ever occurred during our ameliorating tenure. In spite of the efforts of the Boundary Commission, the lines drawn between India and Pakistan were inevitably and devastatingly cruel to the areas through which the new frontiers passed. The result was a series of massacres arising out of the interchange of Moslem and Hindu population which may have run into four or five hundred thousand men, women and children. The vast majority of these were harmless people whose only fault lay in their religion.

Fortunately at the head of the larger of the two new states erected on this bloody foundation was a man of singular qualities. Nehru had languished for years in jail or other forms of confinement. He now emerged as the leader of a tiny minority of the foes of British rule, largely free alike from two of the worst faults of human nature, Hate and Fear. Gandhi, who had so long led the cause of Indian independence, was murdered by a fanatic shortly after Nehru's installation as head of the Government. Jinnah presided over the Moslem state, Pakistan. We are on easy terms with the two Republics which have come into being. Their leaders attend the meetings of the Commonwealth, and their power for good or evil in Asia and the world is undeniable. I will not attempt to prejudge the future.

In the year of Indian independence Burma was also severed from the Commonwealth. It had been the main theatre of land operations in the war in the Far East, and we had put forward a major effort to recover it from the Japanese, who had driven us out in 1942. The Nationalist elements, most of whom at some stage in the war to achieve their aims had collaborated with the Japanese invaders against the Allies, were established in the government of the country. Their control was far from full, and to this day the Burmese Government's writ runs but incompletely through its territories. They too, however, are a firmly established entity with whom our relations are friendly, and where the long and honourable tradition of British authority and its legacies of justice and order have borne fruit.

Both in India and Burma the conflict between Communism and the Free World was of relative unimportance in the immediate postwar years. Certainly Russia rejoiced at every sign of the diminution of our influence in the world and sought by all the means in her power to expedite and bedevil the birth of the new nations. She did great mischief in Indo-China and Malaya. On the whole, however, her interest was more concentrated on China, where amid confusion and slaughter a new pattern was emerging. The regime of Chiang Kai-shek, our friend and ally in the war, was gradually losing its hold. The United States attempted by every means short of armed intervention to halt the advance of Communism. But the Chinese Government carried within it the seeds of its own destruction. In spite of many years of resistance to the Japanese, the corruption and inefficiency of its sprawling system encouraged and supported the advance of the Communist armies. The process was slow, but by the end of 1949 all was over. The "People's Government," as it is called, henceforth ruled in Peking, and controlled the whole Chinese mainland. Chiang Kai-shek fled to Formosa, where his independence was secured by the American fleet and Air Force. Thus the world's most populous state passed into Communist hands and it will no doubt wield an effective force in world affairs. In this period the influence of China was mainly exerted in Korea, and in Indo-China. The wrangles over her admission to the United Nations have demonstrated one of the many weaknesses of that organisation, and China's traditional friendship with America has been suspended.

In the next year Communist attempts to harass the West, to exploit nationalist feeling in Asia and to seize upon exposed salients culminated in the peninsula of Korea. Previously their efforts had been less direct. In Indo-China the principal opponent of the French, Ho Chi Minh, had indeed been Moscow-trained, but material support for his guerrillas had not been on a large scale. In Malaya comparatively few terrorists, by murdering planters and loyal Malays and Chinese, had tied down disproportionate forces to restore order. But they, too, in general owed only their training, ideology, and moral support to the Communist States.

At Cairo, in 1943, President Roosevelt, Chiang Kai-shek and I had recorded our determination that Korea should be free and independent. At the end of the war the country had been liberated from the Japanese and occupied by American troops to the south and Russian to the north. Two separate Korean states were set up, and relations between them became increasingly strained and embittered. The 38th parallel formed an uneasy frontier, and the two states were very much like Eastern and Western Germany. Efforts by the United Nations to reunite the country had been frustrated by Soviet opposition. Tension and border incidents grew. On the 25th of June, 1950, North Korean forces invaded South Korea and advanced with great rapidity. The United Nations called on the aggressors to withdraw, and asked all member states to help. That the Soviet veto in the Security Council did not on this occasion render impotent the United Nations' intentions was due to good fortune. The faults of the system remained to be exploited again and again in later years. On this occasion the United Nations merely provided the framework in which the effective action of the United States was cast.

These bare facts encompass a momentous and historic decision by President Truman. Within the briefest interval of the news of the invasion, he had reached the conclusion that only immediate intervention by the armed forces of the United States could meet the situation. They were the nearest to the scene as well as by far the most numerous, but this was not the point. As he has said in his memoirs, "I felt certain that if South Korea was allowed to fall, Communist leaders would be emboldened to override nations closer to our own shores. If this were allowed to go unchallenged it would mean a third world war." His celerity, wisdom, and courage in this crisis make him worthy, in my estimation, to be numbered among the greatest of American Presidents. In Britain the Government endorsed and sustained the Americans, and made offer of naval units. By December, British ground forces were also in Korea. In the House of Commons on the 5th of July the Opposition supported Mr. Attlee, then Prime Minister, and I myself as its leader said that I was "fully able to associate myself with . . . his broad conclusion that the action which had been taken by the United States gives on the whole the best chance of maintaining the peace of the world." The left wing of the Socialist Party, true to their traditions, alone stood out from the courage and wisdom of what was being done.

The course of the war was difficult, bloody, and frustrating. American and Allied troops halted the Northern invaders, and the intervention of the air forces began to prove effective. General MacArthur acted with vigour and dash, and by the 14th of March, 1951, Seoul, capital of South Korea, was recaptured. Two months later the 38th parallel was crossed. Meanwhile, Chinese "volunteers" began to arrive in massive quantities. Reinforcements poured in from across the Yalu River, where the vast Chinese manpower was formed into indifferently equipped but numerically formidable armies. The American generals found it difficult to tolerate the existence of the "privileged sanctuary" beyond the Manchurian frontier. Here also lay the bases of Soviet-made jet aircraft which intervened repeatedly in the fighting. Pressure grew for permission to attack Chinese territory from the air. President Truman, however, stood firm and in a much publicised series of disagreements with General MacArthur resisted this most dangerous step. " The Reds," he has said, "were probing for weaknesses in our armour; we had to meet their thrust without getting embroiled in a world-wide war." I myself followed with some anxiety the same train of thought. On the 30th of November I pointed out to the House of Commons: "It is in Europe that the world cause will be decided. It is there that the mortal danger lies." I forbore from pressing my views too strongly lest they should be construed as criticism of United States commanders and hamper their efforts or weaken the ties that bound our fates together. British and Commonwealth forces made a small though robust contribution, but America carried almost the whole burden and paid for it with almost a hundred thousand casualties.

I will not dwell on the pendulum of military success and failure in Korea. The outcome can scarcely be thought of as satisfactory. However, South Korea remained independent and free, the aggressor suffered a costly repulse and, most important of all, the United States showed that she was not afraid to use armed force in defence of freedom, even in so remote an outpost.

Elsewhere in the continent of Asia the Western empires crumbled. Our allies the Dutch had been hustled out of the East Indies, which they had made a model of effective administration. The French endured years of frustrating and debilitating warfare in Indo-China, where casualties absorbed more officers in each year than the output of their military college at St. Cyr. Communist armies, mightily reinforced from China, gradually won control of the north of the country. In spite of heroic episodes of resistance, the French were compelled to leave this great and populous area. After long and painstaking negotiation something was saved from the wreckage of their hopes. Three states, South Viet-Nam, Laos, and Cambodia, came into existence, their independence assured, their future uncertain. North Viet-Nam, like North Korea, maintained a separate Communist Government. Partition was once more the answer in the conflict of Communist and Western interests. All these new countries were rent by factions within and overshadowed by their gigantic neighbour to the north.

The changes in Asia are immeasurable. Perhaps they were inevitable. If a note of regret is to be found in this brief account, let it not be supposed that it is in hostility to the right of Asian peoples to self-determination, or a reflection on their present standing and integrity. But the means by which the present situation was reached give pause. Was so much bloodshed necessary? Without the haste engendered by foreign pressure and the loss of influence inherent in our early defeats in the Far Eastern war, might progress to the same end have been happier, and the end itself more stable?

*   *   *

A great part of the Second World War had run its course to defend the land bridge where Africa and Asia meet, to maintain our oil supplies, and guard the Suez Canal. In the process the Middle Eastern countries, and notably Egypt, had enjoyed the advantage of protection from German and Italian invasion at no cost to themselves. There followed a further increase in the number of independent states that existed in the former domains of the Ottoman Empire. The departure of the French from Syria and the Lebanon was bitter to them but inevitable. No one can claim that we ourselves have derived any advantage there. Throughout this region the world has witnessed a surge of nationalist feeling, the consequences of which have yet to run their course. From Indonesia to Morocco the Moslem peoples are in ferment. Their assertiveness has confronted the Western Powers, and especially those with overseas responsibilities, with problems of peculiar difficulty. Amid jubilant cries for self-government and independence, it is easy to forget the many substantial benefits that have been conferred by Western rule. It is also hard to replace the orderliness which the Colonial Powers exercised over these large areas by a stable new system of sovereign states.

The most intractable of all the difficulties that faced Britain in these regions was Palestine. Ever since the Balfour Declaration of 1917 I have been a faithful supporter of the Zionist cause. I never felt that the Arab countries had had anything from us but fair play. To Britain, and Britain alone, they owed their very existence as nations. We created them; British money and British advisers set the pace of their advance; British arms protected them. We had, and I hope have, many loyal and courageous friends in the area. The late King Abdullah was a most wise ruler. His assassination removed a chance of a peaceful settlement of the Palestinian tumult. King Ibn Saud was a most staunch ally. In Iraq I followed with admiration the sagacious and brave conduct of Nuri es-Said, who most faithfully served his monarch and led his country on a path of wisdom, unaffected by threats from without or foreign-bought clamour at home. Unfortunately these men were exceptions.

As mandatory power Great Britain was confronted with the tortuous problem of combining Jewish immigration to their national home and safeguarding the rights of the Arab inhabitants. Few of us could blame the Jewish people for their violent views on the subject. A race that has suffered the virtual extermination of its national existence cannot be expected to be entirely reasonable. But the activities of terrorists, who tried to gain their ends by the assassination of British officials and soldiers, were an odious act of ingratitude that left a profound impression. There is no country in the world less fit for a conflict with terrorism than Great Britain. This is not because of weakness or cowardice; it is because of restraint and virtue, and the way of life which we have lived in our successfully defended island. Stung by the murders in Palestine, abused by the Middle Eastern countries, and even by our allies, it was not unnatural that the British Government of the day should finally wash its hands of the problem and in 1948 leave the Jews to find their own salvation. The brief war that ensued dramatically dispelled the confidence of the Arab countries who closed in for an easy kill.

The infective violence of the birth of the State of Israel has sharpened the difficulties of the Middle East ever since. I look with admiration on the work done there in building up a nation, reclaiming the desert and receiving so many unfortunates from Jewish communities all over the world. But the outlook is sombre. The position of the hundreds of thousands of Arabs driven from their homes and existing precariously in the no man's land created round Israel's frontiers is cruel and dangerous. The frontiers of Israel flicker with murder and armed raids, and the Arab countries profess irreconcilable hostility to the new State. The more far-sighted Arab leaders cannot voice counsels of moderation without being howled down and threatened with assassination. It is a black and threatening scene of unlimited violence and folly. One thing is clear. Both honour and wisdom demand that the State of Israel should be preserved, and that this brave, dynamic, and complex race should be allowed to live in peace with its neighbours. They can bring to the area an invaluable contribution of scientific knowledge, industriousness and productivity. They must be given an opportunity of doing so in the interest of the whole Middle East.

*   *   *

Before I complete this brief survey of the things that have struck me since the war, let us have a look at the United Nations. The machinery of international government may easily fail in its purpose. My idea as the end of the war approached was that the greatest minds and the greatest thoughts possessed by men should govern the world. This entailed, if all countries great and small were to be represented, that they must be graded. The spectacle presented by the United Nations is no more than a vain assertion of equality of influence and power which has no relation to the actual facts. The result is that a process of ingenious lobbying has attempted to take possession of the government of the world. I say attempted, because the vote of a country of a million or two inhabitants cannot decide or even sway the actions of powerful states. The United Nations in its present form has to cringe to dictatorships and bully the weak. Small states have no right to speak for the whole of mankind. They must accept, and they would accept, a more intimate but lower rank. The world should be ruled by the leading men of groups of countries formed geographically. The mere process of letting the groups shape themselves and not judging by their power or their numbers would tell its own tale.

I do not intend to suggest that all the efforts and sacrifices of Britain and her allies recorded in the six volumes of my War Memoirs have come to nothing and led only to a state of affairs more dangerous and gloomy than at the beginning. On the contrary, I hold strongly to the belief that we have not tried in vain. Russia is becoming a great commercial country. Her people experience every day in growing vigour those complications and palliatives of human life that will render the schemes of Karl Marx more out of date and smaller in relation to world problems than they have ever been before. The natural forces are working with greater freedom and greater opportunity to fertilise and vary the thoughts and the power of individual men and women. They are far bigger and more pliant in the vast structure of a mighty empire than could ever have been conceived by Marx in his hovel. And when war is itself fenced about with mutual extermination it seems likely that it will be increasingly postponed. Quarrels between nations, or continents, or combinations of nations there will no doubt continually be. But in the main human society will grow in many forms not comprehended by a party machine. As long therefore as the free world holds together, and especially Britain and the United States, and maintains its strength, Russia will find that Peace and Plenty have more to offer than exterminatory war. The broadening of thought is a process which acquires momentum by seeking Opportunity for all who claim it. And it may well be if wisdom and patience are practiced that Opportunity-for-All will conquer the minds and restrain the passions of mankind.

WINSTON S. CHURCHILL

Chartwell,
  Westerham,
    Kent
February 10, 1957

Index

Index

☆